30th EDITION
GUNS ILLUSTRATED®
1998

Edited by Harold A. Murtz
and the Editors of Gun Digest

DBI BOOKS
a division of Krause Publications, Inc.

STAFF

GUNS ILLUSTRATED

EDITOR
Harold A. Murtz

SENIOR STAFF EDITOR
Ray Ordorica

ASSOCIATE EDITOR
Robert S.L. Anderson

PRODUCTION MANAGER
John L. Duoba

EDITORIAL/PRODUCTION ASSOCIATE
Karen Rasmussen

ASSISTANT TO THE EDITOR
Lilo Anderson

ELECTRONIC PUBLISHING MANAGER
Nancy J. Mellem

ELECTRONIC PUBLISHING ASSOCIATE
Laura M. Mielzynski

GRAPHIC DESIGN
John L. Duoba
Bill Limbaugh

MANAGING EDITOR
Pamela J. Johnson

PUBLISHER
Charles T. Hartigan

Copyright © 1997 by Krause Publications, Inc., 700 E. State St., Iola, WI 54990. All rights reserved. Printed in the United States of America.

No part of this publication may be reproduced, stored in a retrieval system, or transmitted in any form or by any means, electronic, mechanical, photocopying, recording or otherwise, without the prior written permission of the publisher.

The views and opinions contained herein are those of the authors. The editor and publisher disclaim all responsibility for the accuracy or correctness of the authors' views.

Manuscripts, contributions and inquiries, including first class return postage, should be sent to the GUNS ILLUSTRATED Editorial Offices, 935 Lakeview Parkway, Suite 101, Vernon Hills, IL 60061. All materials received will receive reasonable care, but we will not be responsible for their safe return. Material accepted is subject to our requirements for editing and revisions. Author payment covers all rights and title to the accepted material, including photos, drawings and other illustrations. Payment is at our current rates.

CAUTION: Technical data presented here, particularly technical data on handloading and on firearms adjustment and alteration, inevitably reflects individual experience with particular equipment and components under specific circumstances the reader cannot duplicate exactly. Such data presentations therefore should be used for guidance only and with caution. Krause Publications, Inc., accepts no responsibility for results obtained using this data.

ISBN 0-87349-193-9   Library of Congress Catalog #69-11342

About Our Covers

With the number of states granting concealed carry permits these days, specialized combat guns suitable for everyday duty are becoming increasingly important. Featured on our covers are two full-size Colt Government Models (left and right) and a Lightweight Officer's Model—all three combat/carry guns built by The Robar Companies, Inc., of Phoenix, Arizona. Essentially, the same modifications have been performed on all three pistols, making them totally reliable combat weapons.

All are far from stock. The slides are finished in Roguard, a black wear- and corrosion-resistant finish that is also quite slick, while the frames are treated with NP3, a silver/gray, highly protective metal coating.

Sights on the gun at left are a Robar Combat front that has been dovetailed into the slide, and a fully-adjustable Bo-Mar rear. Note that the top of the slide on each gun has been flattened and stippled, creating an integral rib effect.

Speaking of the slide, all three guns shown received extensive work here to ensure *total* reliability. They've been deburred on the outside and polished on the inside; forward cocking serrations have been machined in; the ejection port has been opened and flared; and the breech face has been polished. In addition, the guns have heavy-duty spring kits, the feed ramps have been polished, the extractors tuned, and the barrels throated.

There are subtle differences between the two full-size guns. For instance, the one at left has a standard beavertail grip safety and checkered frontstrap, while at right is a Memory Groove beavertail, stippled frame and Robar Combat fixed sights. Both have the checkered, flat mainspring housing; match aluminum trigger; and lowered, extended thumb safety.

The middle gun, the Colt Lightweight Officer's Model, has had all of the above, in addition to a host of extras such as a special guide rod kit and bushing, Slim-Tech synthetic stocks, and other options too numerous to mention here. A superb carry gun, for sure.

Complete reliability is the goal of Robar's gunsmiths, and these guns have that and more—like good looks and accuracy. If you're in the market for a new and serious carry gun, check out Robar's work—you'll be impressed and well armed.

Photo by John Hanusin.

CONTENTS

FEATURES

Thoughts on Building a Concealed Carry Weapon
    by Robbie Barrkman 4

Gun Writers Who Made a Difference—Jack O'Connor
    by Sam Fadala 9

Knob Creek Machine Guns
    by Frank James 15

The Buckshot Trials
    by Wiley Clapp 20

Modern Gun Sleuthing
    by J.I. Galan 29

Practical Revolvers Before Colt
    by Charles W. Karwan 36

Engraving: The Good, Bad & Ugly
    by Doc O'Meara 41

Machine Pistols
    by Chuck Taylor 49

Smokeless Powder: The Early Years
    by Paul Schiffelbein, Ph.D 54

The Invalides Museum
    by Wm. Hovey Smith 62

The French SACM Modele 1935A Pistol
    by Paul Scarlata 69

The 256 Newton Sporting Springfield
    by Jim Foral 74

DEPARTMENTS

GUNDEX® 82

Handguns
    Semi-Automatics 91
    Competition 121
    Double-Action Revolvers 130
    Single-Action Revolvers 141
    Miscellaneous 148

Rifles—Centerfire
    Semi-Automatics 152
    Lever & Slide Actions 157
    Bolt Actions 162
    Single Shots 179
    Drillings, Combos, Double Rifles 184

Rifles—Rimfire
    Semi-Automatics 187
    Lever & Slide Actions 190
    Bolt Actions & Single Shots 192
    Competition Centerfires & Rimfires 197

Shotguns
    Semi-Automatics 204
    Slide Actions 209
    Over/Unders 214
    Side-by-Sides 226
    Bolt Actions & Single Shots 230
    Military & Police 234

Blackpowder Guns
    Single Shot Pistols 237
    Revolvers 239
    Muskets & Rifles 243
    Shotguns 257

Air Guns
    Handguns 259
    Long Guns 265

Scopes & Mounts 277

Scope Mounts 287

Directory of the Arms Trade 291
    Product Directory 292
    Manufacturers' Directory 309

THOUGHTS ON BUILDING A CONCEALED CARRY WEAPON

The gun hand ideally should be as close to the weapon as possible. Avoid systems that require two hands to get access to the gun.

Once you've decided to carry a gun for self-defense you'd better make it a good one. Here are some cogent thoughts on what you do and don't need.

by ROBBIE BARRKMAN

THE DECISION TO carry a concealed weapon is something the average law-abiding citizen must take very seriously. It is a responsibility that not only affects each of us personally, but can also have a major impact on our loved ones, friends and even total strangers. The ramifications of shooting another person are very possibly with us for the rest of our lives. All this is, of course, based on the presumption that we won the fight.

To win the fight we are going to have to employ a number of skills, the most important of which is the ability to hit what we are shooting at. To hit requires a weapon that we can handle comfortably and safely and—important point here—that we have it at the time our lives are threatened. Those who know me realize that I am a strong proponent of large-caliber guns, but the fact is that it is not always possible or practical to carry a large gun in a concealed manner. Okay, I said it—the 45 automatic is *not* the only handgun to consider for self-defense; I think that having *any* gun available for self-defense is better than no gun at all.

Picture this: You are alone in a poorly lit parking lot, heading for

Dovetailing the front sight is the best installation, especially if with tritium night sights. Note the forward cocking serrations and replacement bushing.

Extended safeties or ambidextrous safeties are a personal choice, but remember to keep it simple and reliable.

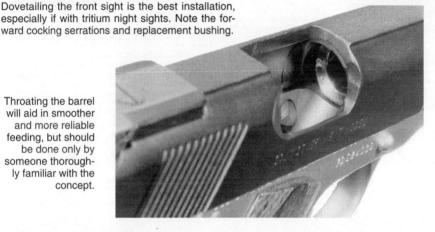

Throating the barrel will aid in smoother and more reliable feeding, but should be done only by someone thoroughly familiar with the concept.

your vehicle when you become aware of two unsavory characters closing in on you. One of them most eloquently asks for your wallet and proceeds to enlighten you as to what will happen to your genitalia should you refuse to comply. Now here's the question (remember, you only have a fraction of a second to make a decision): Would you like to have any gun at this time or no gun at all? If your reply is no gun, I would guess that you: 1) have never been in a fight; 2) have a morbid fascination with the sight of your own blood; 3) suffer from a raging case of testosterone overload. The fact is, your environment will, in most cases, dictate what type of handgun you can realistically carry. An 88 Magnum, safely tucked away at home in its fluffy, wooly, custom-made blanky, is hardly any help when you are having the poop beaten out of you in the parking lot. Enough said.

Choosing a gun for the task of saving your hide is not easy, considering all the factors involved. Again, it is important to remember that your life is on the line so don't be pressured into a system that does not fit your environment by a well-meaning but, perhaps, overzealous salesman. Start by considering the size of the gun relative to your normal mode of dress. Don't buy a gun that requires a major and expensive wardrobe change. (Lest your spouse becomes the first person you have to defend yourself against!) Buy the largest caliber gun that you can comfortably carry. I can't stress this factor enough. Remember, if it is not comfortable, you are not going to carry it—period!

Next, consider a system that is easy for you to operate. Some handgun mechanisms are more complicated than others and require more training on an on-going basis. For example, single-action semi-automatic pistols carried in the cocked and locked condition require far more handling ability than a Glock or double-action snubby revolver. In condition one (cocked and locked), any unfamiliarity with the operation of the gun, especially under stress, can have dire results. I have personally been witness to two serious injuries from mishandling a pistol in this condition. The best solution is to get as much formal training as possible with the system you intend to use from an accredited school.

Once you find a gun that meets your personal criteria, you are faced with two more questions: where to carry the gun and what to carry the gun in. In an ideal situation the gun should be as close to the shooting hand as possible. For example, if you are right handed, the gun would be on the belt on the right side centerline of your body. I should point out that this is not the only way to carry, but from a fighting perspective it's the most efficient. Crossdraw holsters, shoulder holsters, fanny packs, and other alternatives are all viable systems if it means that you have the gun with you. The role of a holster is to provide a safe and secure method for transporting and securing the gun while giving you easy and unrestricted access in an emergency. Together, the gun, the holster, your familiarity with the system and ability to use it, become the driving force behind effective self-defense.

One particular system that I do not recommend is the S.O.B. (Small of Back) carry. In my opinion, carrying a gun without a holster is nothing but a disaster waiting to happen. Just when you need the gun it will slip into the deep recesses of your butt, never to be seen again. The only reason you might survive such an encounter is that your attacker will be totally confused while you frantically dig in your crotch to try and recover your hardware.

It is important to remember that the clothes you wear need to hide the gun and holster without restricting or impeding your ability to get the gun into action with one hand. The other hand will most likely be occupied with fending off your attacker,

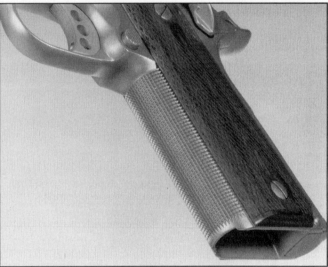

The gun should fit the shooter, meaning the middle of the trigger finger pad should rest naturally on the trigger.

Opening the ejection port will help reliable extraction and prevent dings on the brass. Also note the unique low mount Bo-Mar sight installation.

Beveling the magazine well will help magazine insertion in the dark during reload or in "eyes off" situations. Checkering or stippling the frontstrap gives better weapon retention during stress, or if you are fighting for physical possession of the gun.

parrying a blow, or just trying to get him off of you.

One of the most popular pistols for concealed carry is the Colt Lightweight Officer's Model in, what else, 45 ACP. It is *my* first choice in a defensive fighting tool. My pistol has had only those modifications that I considered necessary for this particular type of gun for this use. Each modification suits *my* shooting style, *my* environment, *my* philosophy, *my* technique and *my* wallet. The point I am trying to convey is, don't get talked into a bucket load of bolt-on garbage you don't really need. The single most important aspect of a self-defense gun is that it works every time, without fail. The more stuff you glue, screw and bolt on to the pistol the greater the chance that it will break, fall off, or worse still, poke your eye out! If ever the K.I.S.S. (Keep It Simple Stupid) principal applies, it is here when considering the modification of your concealed carry weapon.

Once you have purchased a pistol, there are a couple of additional things to take care of. Before you rush off to an appointment with Dr. Dainty and his Dreaded Dremel tool, go to the range and fire the gun. If the gun is new or was bought from a reputable dealer it will still be under warranty, and should you experience any problems with it, you will still have some recourse. Repairs or alterations to the gun will in almost all cases void any claim under the warranty agreement. The range sessions will also bring to your attention any inherent flaws that make the gun unpleasant or difficult to shoot, like hammer bite, sharp edges cutting your thumb, or a rough trigger. Fire at least 200 rounds before turning it in for work.

Before leaving for the range, be sure to code your magazines by numbering or lettering them. Most malfunctions in semi-auto pistols can be traced back to magazine or ammunition. This will give you a clue as to which magazine is causing the problem, if it is indeed a magazine at fault. If you own a Dremel tool you can engrave numbers or letters on the bottom of the magazine. Don't engrave on the sides, as this could raise burrs which could cause a tolerance problem with insertion or extraction of the magazine. There is a way to "tune" magazines but, frankly, the cost does not make sense considering the low cost of most magazines. If you have some of those dreaded high-capacity assault-type magazines, you may opt to have them tuned due to the lack of availability and their high cost. Buy at least ten magazines and *test them all*. If they are new and don't work exchange them, and if you can't exchange them, get rid of them! The magazine spring is another area to consider. Remember to unload the magazine when not using it. This will take the tension off of the spring and hopefully extend its working life. The Wolff Spring Company makes replacement spring kits, and it would be a good idea to stock up on some of

The beavertail grip safety not only prevents hammer bite, but also helps guide the hand to the grip in the drawing sequence.

Adjustable sights are great on target or general-purpose guns, but could let you down in a fight should they be knocked out of whack. Keep it simple—fixed sights are preferable.

these. For a system you stake your life on, it's wise to replace your magazine springs at least once a year if you keep the magazine loaded permanently. Always thoroughly test your magazines before trusting your life to them.

With all this behind you, two facts should be emerging. The first is that you and the gun have started to establish the beginnings of a rather close relationship. Second, you are now aching to spend big bucks on your newfound friend. Before doing anything, let me explain the three rules of building a self-protection gun. Rule 1: The gun MUST be reliable. Rule 2: The gun must be TOTALLY reliable. Rule 3: The gun must be ABSOLUTELY reliable. The easiest way to decide what you need is to divide the potential modifications into three categories: 1. Things that make the gun reliable. 2. Things that you absolutely must have. 3. Things that are nice to have.

In case I have not mentioned it, reliability is very important so let's begin here. In keeping with the theory of K.I.S.S., the following modifications should be all you need. Many of these can be done on pistols other than the Colt style. It is important to consult a qualified gunsmith to ensure that these modifications are necessary on your particular weapon system. Opening the ejection port will help ensure a clean exit of the fired case and avoid dinging of the case, a consideration if you are a reloader. Throating the barrel and polishing the feed ramp will help a smoother transition of the cartridge from the magazine to the chamber. Throating requires the alteration of the angle, or reprofiling of the throat or entry area of the chamber. The bottom edge of the barrel throat is moved forward, thereby giving the incoming cartridge a lower angle of entry into the chamber. This gives more reliability and less resistance to the slide allowing good clean lockup. Make sure you tell your gunsmith not to move the ramp too far forward. The area of the cartridge case over the ramp is not supported and too much work here can cause the case to rupture.

Tuning the extractor will help in a smooth transition of the cartridge up the breech face and under the extractor, aiding in easier closing of the slide and lessening the chance of the slide not going fully into battery. The correct extractor tension is imperative to good extraction of the fired case. Some pistols may require polishing of the breech face due to heavy machine marks, but this will vary from gun to gun. Heavy machine or tooling marks can retard the progress of the brass case up the breech face by catching or grabbing onto the softer brass as it slides up into the battery position.

Now let's talk about the must haves. The rules of marksmanship demand sights you can see and a trigger you can manage. The market is rife with sight systems to choose from. Look for sights that are strong, snag-proof and have no sharp corners or edges to catch on clothing. Sharp corners are also going to cut you, especially when practicing (Oh no, not that!) malfunction drills. For defensive work fixed sights are more than adequate and, in fact, preferable. The distance in a lethal situation is typically body contact to three to four yards. Anything further can hopefully be dealt with by applying the Nike Principle...if at all possible RUN! Sights such as the Novak Low Mounts or Robar's Combat Sights are good examples of sights that would be more than adequate. Adjustable sights are inherently more fragile and typically more bulky and serve a useful role in other applications, but not here. If it is at all possible, get the front sight dovetailed. It is stronger than any other system and allows for easy removal in the event of refinishing or repairs. If you have tritium inserts or night sights, you will really benefit by being able to remove the sight. Almost all metal finishing processes will destroy the Tritium inserts and they are not cheap to replace.

Now the trigger. Long or short? Heavy or light? Remember, this is your gun and it must fit *you*. The trigger should ideally fall on the center of the pad of the end digit of your trigger finger. Test your gun to determine if you need to change the trigger length. Take a good firm firing grip on the gun and, without moving your hand, see where the finger falls. There are all kinds of triggers on the market, each doing more and better things than everyone else's, just ask them. If the standard length works, why change it? But if your finger is longer, then by all means go to a long trigger. As for the best weight, it is important to remember the role of this pistol is *fighting*. In a lethal situation you are under stress and probably physically compromised. It is my opinion that a light trigger is an invitation to doing the Tango with Bubba in the big house. A trigger pull weight of $4^1/_2$ to 5 pounds makes a lot of sense in this case. What is more important than weight is that the trigger pull is clean and crisp. Have your gunsmith work on it if that's not the case.

The third category, nice to have, is

obviously very subjective and controlled more by the thickness of your wallet. Let me remind you that this is a fighting gun so K.I.S.S.! You may find some of the following items especially useful and even necessary in some instances. Dehorning or deburring is actually a must if you are going to carry and shoot the gun. This consists of removing all sharp corners and edges from all working surfaces. This makes the gun more user-friendly and won't tear up your clothing, hands or furniture. For my hand shape, a beavertail safety is another absolute necessity as I suffer from acute hammer bite. Pinning the grip safety or using one of the built-up safeties is an area that is a must for my hand shape. Does your grip safety engage every time you shoot the gun? If it does, leave it alone. If it does not, fix it. The shape of the grip safety also helps to locate the gun in the firing grip more readily. If you are of smaller stature or have small hands, you may want to consider a flat mainspring housing. Couple this with the new Slimtech stocks and you will find a significant difference in the feel and controllability of the gun. If the grip is on the small side, try installing some Pachmayr rubber stocks or, if you are really adventurous, you can have some custom stocks made specifically for your hand size.

Another feature some people find useful is either checkering or stippling on the frontstrap. This aids in a good grip on the gun and helps prevent slipping due to sweat, rain, etc. Stippling is gentler on the hands and is generally kinder to your clothing. If you can, try shooting a gun that has been checkered and one that has been stippled, so you can make an informed decision. Extended safeties are nice to have but the factory safety will work just fine. Ambidextrous safeties are an absolute no-no. There is no system I am aware of that won't break or come apart after prolonged use. Almost all the safeties available use some sort of tongue and groove to join the two halves and therein is the weakness in the system. This creates a dilemma for lefties. Solution: Buy three or four sets of safeties and get your gunsmith to fit them all, so as they loosen or break you can switch them out and go about your business.

One other thing to consider is a good finish for your investment. The two main types of finishes are metallic such as electroless nickel and hard chrome, and polymer such as Black T and Roguard. The metallic finishes are much harder and generally more resistant to holster wear. They are,

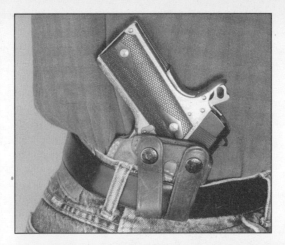

(Left) Ideally, the gun should be in a good holster that is easily accessible and will provide good retention in a fight.

(Below) Any gun is better than no gun. This small revolver packs a real surprise for a would-be attacker. Skill is more important than caliber.

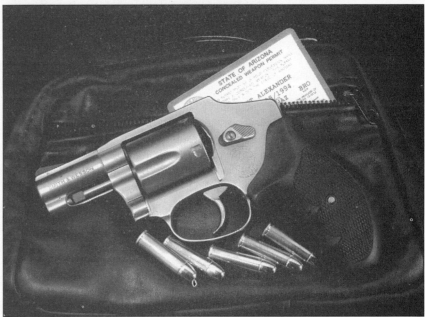

however, usually silver or gray, so if the gun needs to be black you will have to go to one of the polymer finishes. Polymers are, in actual fact, far more corrosion resistant than the metallic finishes but are more prone to abrasion in highly stressed areas. Phone calls are cheap so go ahead and call the different metal finishing companies for some input. Get all the literature and choose a finish that will fit your budget and environment. Buy a good finish up front because refinishing the gun is expensive the second time and, of course, you will be without your life insurance for the time the gun is at the shop.

A last thought. You now have a gun. You now have a CCW permit. Do yourself and your family a really big favor: Go to school and get an education. Go to a reputable school where you will not only learn about marksmanship, but how to fight with your own weapon system. The foremost school available to the public today is undoubtedly Thunder Ranch in Texas. It is run by Clint and Deb Smith who are consummate professionals in the art of teaching you how to save your life with a handgun. There are a number of other good schools around, but one is unique in that it is a traveling school that will come to you. This one is run by Louis Awerbuck and Lee Lambert and is called Yavapai Firearms Academy. Louis is without question one of the best instructors I have had the pleasure of working with. If you are unable to travel, Y.F.A. is definitely worth a call.

Fighting with a pistol is more complicated than just yanking on the trigger. It is a lifestyle change, a psychological change and a philosophical change all rolled into one. A good education will not only make you safer, it will give you the mental and physical tools to be comfortable, competent and, most importantly, to put you in control of your environment. ●

GUNWRITERS WHO MADE A DIFFERENCE

Jack O'Connor

Who were the writers who made an impact on the shooting sports? This is the first in a series that examines the famous names and personalities that shaped our guns and gear—and attitudes.

by SAM FADALA

Over the next several GUNS ILLUSTRATED annuals, we'll look at the lives of famous gunwriters. Our interest is more than biographical; we want to know how these gunwriters impacted the sport of shooting. Did they promote a certain cartridge? Did they change the way we think about shooting or hunting? What effect did they have? The series opens with Jack O'Connor, one of the most admired gunwriters of all time. Some called him a gunwriter's gunwriter. Colonel Townsend Whelen, Elmer Keith, Warren Page, Charles Askins, Paul Curtis, and others will follow.

JOHN WOOLF "JACK" O'Connor was born in Nogales, Arizona, on January 22, 1902, but he was raised in Tempe, Arizona, a boyhood paradise. Traces of the Old West lingered in the region, with tales of bad men (and women) fresh in the memories of the older residents. The horse was still a major means of transportation, and when O'Connor wrote his biographical book in 1969, he called it *Horse* and Buggy West: A Boyhood on the Last Frontier, a well-written title that deserved a wider audience. Although Jack and his parents lived in a San Francisco apartment when Jack was a little shaver, Tempe formed his early life. He loved writing. He loved hunting. But the two did not merge into a career for many years. The hunting came first.

Imagine the lonely Salt River Valley in the early 1900s. Jack got his first deer at age twelve, a mule deer buck taken with a 30-40 Krag rifle. These were good times for the nearly fatherless boy. Earlier, a personal earthquake shook Jack's young life when his mother and father announced their divorce. Jack was only five years old. His mother, the former Ida Woolf, taught at Tempe Normal School. Andy O'Connor, Jack's father, was a baseball fan, his gnarled fingers attesting to the position he played—catcher, when a mitt was little more than a scrap of leather. Andy was a well-read man, a teacher himself, and an intelligent person. But he was also a dreamer seeking to capture one big deal in life; he never made that fortune. Jack saw his father from time to time, but the relationship was distant.

He attended Arizona State Teachers College in Tempe from 1921 to 1923, also doing a one-year stint at the University of Arizona in Tucson, but he graduated from neither. He took his bachelor's degree from the University of Arkansas in 1925, following in his parents' footsteps as a teacher. In 1927, he earned a Master of Arts degree in English from the University of Missouri. In between the two degrees, he married Eleanor Bradford Barry. Eleanor was the light of his life, and Jack was hers. The couple ended up back in Arizona, where Jack taught at the then-small teachers' college in Flagstaff, now Northern Arizona University. He also taught at the University of Arizona with his master's degree.

But writing was his calling. His first novel, *Conquest* (1930), was printed, but didn't do very well. His second novel, *Boom Town*, published in 1938, sold better. Jack remained on the faculty of the University of Arizona for eleven years, freelancing his writing on the side. "I started peddling stuff in 1929," he wrote to the author in a letter dated November 8, 1971, "the year the Depression hit. Between 1930 and 1933, when the magazines ran out of old manuscripts, Dante couldn't have peddled *Inferno*." Jack hit it hard after the Depression. "I made $10,000 the first six months—about $600 the second," he wrote in another letter to the author dated August 20, 1973. By 1941, O'Connor began writing his "Getting the Range" and "Shooting" columns for *Outdoor Life* following the suicide of Captain E.C. Crossman, who worked for the magazine at the time of his death.

O'Connor's impact on the world of shooting was immediate. Jack had his own ideas. Of course, he wasn't always right, no more than great scientists are always correct. But his commonsense approach and keen powers of observation made his work potent. His books were filled with good ideas, and they were well-written. An early title, *Game in the Desert*, promoted Jack as a major writer. The excellent old work fetches several hundred dollars per copy today from collectors. O'Connor understood the

"I was resting my rifle on an anthill sticking up in the arid and barren plain that had been miserably overgrazed and eroded when Caesar was playing footsie with Cleopatra."—No wonder O'Connor was known as the gunwriter's gunwriter!

"average" shooter/hunter. That is where he made his mark. He knew this fellow (women weren't shooting much at the time) needed a middle-sized cartridge for most of his target shooting and game taking, and when the 270 Winchester came along, Jack adopted it.

Today, anyone who knows O'Connor ties the 270 to his name. Jack truly loved the cartridge because it shot a medium-weight bullet at high velocity with modest recoil and a telling effect in the field on game as large as moose. No one can say for sure, but O'Connor may have singly made the 270 Winchester as famous as it remains, ranking only behind the 30-

O'Connor hunted big game throughout five continents. While gun editor of *Outdoor Life*, he shot this greater kudu in Africa in 1956. He considered it Africa's best trophy.

Though not considered an expert with the shotgun, O'Connor fully appreciated a fine double and did well with them. Here, in about 1968, he discusses double guns with Spanish gunmaker Ignacio Ugartechea.

06 Springfield in popularity among Western hunters today. Jack summarized his commitment to the 270 in a 1951 *Shooters Bible* with a short piece called "Rifles for the Mountains." He said, "With the 130-gr. bullet in the 270 at the high muzzle velocity of 3,140 f.p.s. the Western hunter is prepared for just about anything he'll find. In spite of the light bullet the .270 will almost always kill a deer, sheep, or antelope in his tracks, and even a moose or grizzly hit with that wicked bullet won't go far." Jack admitted that the ".30/06 is also an excellent bet," but "not quite in the .270 class."

O'Connor did little with handguns. He admired the shotgun, dearly appreciating beauties like the Winchester Model 21 side-by-side, and he was a dedicated birdhunter. But his major interest was the rifle. He especially liked the big game rifle, bolt-action, scoped. We are so used to the wide acceptance of telescopic rifle sights today that it may be difficult for us to believe that O'Connor was a great promoter of the glass sight when a multitude of American riflemen didn't trust scopes any more than they would a rattler curled up in the bottom of their sleeping bags. While other gunwriters did more to promote scopes than Jack, his writing did make a difference. It was O'Connor who preached that "no one can shoot better than he can see." In his 1964 *The Rifle Book*, he wrote, "The *magnification* of the scope, which many consider all-important, is just one of the scope's many assets." He went on to point out that scopes were wonderful in low light; they put target and reticle on the same optical plane; they identified the target better for safety; and they allowed greater precision of shot placement. While not a fan of the high-power scope for big game, Jack did a lot to promote the glass sight.

His training as a teacher colored all of his writing. You couldn't read a Jack O'Connor piece without learning something, perhaps as small, yet significant, as how to shoot at big game from the proper position. While thousands of hunters thought you had to stand up and fire away, Jack told them this was wrong. Whenever possible, you sat and steadied the rifle for a far better chance of efficient bullet placement. He trained thousands of people how to use the peep sight, too, telling them not to try consciously centering the front sight in the circle, but to look *through*, not at, the aperture, letting the eye automatically do the job.

O'Connor wrote with a style all his own. He knew his audience, and he never tried to show off by writing above the average reader. His sentences were plain and accurate, but with a certain interest-increasing flair. While one writer might say "I used an anthill to steady my rifle," O'Connor wrote, "I was resting my rifle on an anthill sticking up in the arid and barren plain that had been miserably overgrazed and eroded when Caesar was playing footsie with Cleopatra." No wonder he was known as the gunwriter's gunwriter. His work deserves to be studied by anyone planning to write about firearms. He penned with a sense of humor, but didn't get carried away with it. A rifle might have enough muzzle blast to knock the ears off a brass monkey, but Jack didn't overuse such phrases, which can get pretty boring.

He kept a reader's interest as he wrote, but he was always out to inform. He was both storyteller and researcher. In a March 1957 *Outdoor Life* "Shooting" column, he taught us about "Bullets in the Breeze." You can bet at that time the average big game hunter paid very little attention to what the wind did to his projectile from muzzle to game. Jack pointed out that the target shooter was sunk without a knowledge of "reading" the wind, and the big game hunter better learn from it. In his easy-going style, he spoke of the wind in Wyoming and how "...it behooves the hunter to do a little wind doping before he touches Old Betsy off." I read that article, took it to heart, and more than once made solid hits that would have been misses had I not doped the wind, especially on antelope.

Jack O'Connor made a terrific impact on the design and manufacture of truly good big game bullets, basing his ideas on field data. He taught hunters to choose bullets carefully, especially for medium-velocity cartridges, which Jack said needed missiles that opened up quickly, even when meeting very little resistance. He was right. You remembered what he told you, and you believed him. That's because he didn't go overboard on anything, not even his beloved 270 Winchester. For example, he wrote about "The Indestructible .30/06" in 1962, saying that "for all kinds of jobs in the open and in the timber, on big animals and small, at long range and short, there isn't anything any more versatile than this perpetual best seller, the 56-year old [at the time] .30/06."

While O'Connor promoted the 270 Winchester above all other rounds, he also covered the bases. For example, he considered the 7mm Mauser a fine number. His wife certainly used it quite often. In his article, "The Tough Little 7x57," written in 1957, Jack praised it as "the liveliest corpse among centerfire rifle cartridges," going on to say that, "People have been burying it for 20 years or more, but it's still able to sit up and take nourishment. The .270 was supposed to kill it, and the death rattle was in its throat when Western Cartridge Co. dropped the 7 mm. load with the 130-gr. bullet." Jack always had an experience to draw from. In writing about the 7mm Mauser, Jack said, "let me tell a tale," recalling that a rifle he had made up for Princess Perisima of Iran recorded a fine record of one-shot harvests on many different game animals with the mild cartridge.

About the 300 Weatherby Magnum, O'Connor admitted that, "Conservative sportsmen with extensive experience back in the 20's and 30's wouldn't be caught dead with .300 Weatherbys," but went on to praise the cartridge for its ability in its own niche. And while he may not have entirely liked the idea of a 243 Winchester, he reported on it in his column entitled "Two New Shorties: .243, .358" in 1955. He even called it excellent, although being careful to remind us that the previous 250 Savage and 257 Roberts rounds were just as good when loaded to similar pressures. He also wrote about "The Big Berthas," telling American shooters about rounds they would never, as a group, fire in their lives, but would probably like to know something about, such as the 505 Gibbs and 500 Black Powder Express.

Jack was not a trained ballistician, and now and then he got into a little bit of trouble with certain notions. One concerned the 220 Swift, whose bullets, the writer surmised, were hot little molten bottles of lead due to such high velocity. Of course, the time factor made this unlikely, the bullet being in the bore for so short a period as to get hot, but not that hot. So who hasn't harbored a wrong idea? Jack had very few of them, as time has proved. His promotion of the all-around rifle, for example, was so correct that we still seek out a rifle/cartridge combination that performs many functions, no matter if we call it a mountain rifle or a medium-weight big game rifle. That was his practical side coming out again—what's best for that average sportsman under average conditions? For example, he felt that a 22-inch barrel for standard cartridges was just about right. His word was taken as gospel by many gun factories.

He also promoted the custom rifle to a great extent. If you read O'Connor, you knew about Al Biesen and Alvin Linden. You understood that the factory rifle was downright good, but the custom was something special, and you wanted one someday. I wonder how many custom rifles were sold in this country based upon a "written recommendation" by Jack O'Connor? Many, I guarantee. His leaning was, again, toward the conservative. His custom rifles were not fancy; they were classic. He had a special sense about rifle stocks, and his gunmakers paid attention to his ideas. In his title, *The Complete Book of Rifles and Shotguns*, he said this about stock design: There are "some stocks that are handsome and efficient, some that are efficient but not so handsome, and some that not only are inefficient but so fantastically ugly as to cause a miscarriage in a lady crocodile."

Jack was, for the most part, a guided hunter. He certainly began his career as a freelancer, doing hunts on his own, especially for the wiley Coues deer of Arizona, which O'Connor praised as our smartest big game animal. But he lived and worked in the golden age of gunwriting, when a prominent scribe was offered trip after trip by his magazine and the industry. Jack hunted five continents, taking big game on all of them, including lions, tigers, elephants and other exotic animals. He also hunted sheep, and his admiration for a good sheep guide clearly showed in his writing. Indirectly, if not directly, O'Connor promoted guided hunting. He spoke highly of the guides he admired, always giving them credit for locating game and working hard

Though a proponent of the 270 Winchester cartridge, Jack killed a lot of game with his beloved little 7mm Mauser. He shot this 38 1/2-inch gemsbok in Southwest Africa in 1972 using, of course, the 7x57.

At age 69, in 1971, O'Connor shot this 38-inch Stone ram in northern British Columbia. Unfortunately, the trophy just missed making the record book.

Eleanor O'Connor traveled extensively with Jack and was an accomplished hunter herself. She used the 7x57 to take this 53-inch greater kudu in Mozambique in 1962, with professional hunter Harry Manners.

Jack O'Connor retired from *Outdoor Life* magazine in 1972, but didn't fade from the shooting scene. There was no reason to stop writing, and he continued to promote the things he believed in, like handloading.

for the client. His impressive list of big game trophies earned O'Connor the coveted Weatherby Trophy, handed to him in 1957 by Governor Joe Foss of South Dakota. The plaque on the trophy honored O'Connor's Grand Slam, taking all four species of North American bighorn sheep. The trophy was also inscribed with words of praise for his books and articles, and for "bringing to the public his vast knowledge of firearms, hunting, and real conservation."

Finally, in July of 1972, O'Connor wrote a piece he called "Hail and Farewell," saying "This is my last appearance as shooting editor of *Outdoor Life*. I was 70 years old last January, and I am about five years overdue for the pasture. I have lived out my Biblical threescore and ten, and my remaining years on this earth are gravy." Actually, his goodbye was not. There had been changes at the magazine, and a bit of friction came with them. Jack was not retired for very long, however. Quite soon he had a regular column with a gun magazine. He had long before moved to Lewiston, Idaho, with his wife and family, and he was quite content in his nice home on prime acreage. There was no reason to stop writing altogether.

He continued to promote the things he believed in, such as handloading. For a time, there was a definite trend toward downgrading home-rolled fodder, and I think Jack did a lot to put down that movement. At least one company noted on its boxes of ammunition that handloading was a dangerous practice. That company recanted later, even offering reloading components to the public. They couldn't overcome what gunwriters, including O'Connor, had established—that handloading was as safe as the person doing the job. He continued to praise the standard cartridge, never turning his back on his favorite 270 Winchester. His last writings were, if anything, better than his earlier pieces, and his following of readers continued to the end.

That end came on January 20, 1978, two days before Jack's seventy-sixth birthday. He was aboard the S. Mariposa. He and his wife were returning from Hawaii when a heart attack claimed the gunwriter's life. His beloved wife soon followed, as can be the case when two people live a long and devoted life together. O'Connor was gone, but his influence remained. He certainly got the attention of one young writer he took under his wing. "Don't you know what a paragraph is?" he demanded in one note, and when the new writer was crestfallen by a rejected manuscript, Jack advised, "Faint heart never sold a magazine article." In other words, buck up and try again.

Jack O'Connor was fun to read. He was smart, studious and a good communicator. He told a story, and as you followed his words, you learned things, sometimes without knowing it until you put words into practice. Yes, he made an impact. As *Outdoor Life* magazine prepared to move its quarters in early 1997, something was discovered in an old filing cabinet. It was a previously unpublished Jack O'Connor article. Everyone in the office was excited. The editor of the magazine wrote about the event, saying that O'Connor "was a writer of great precision in his use of the language, and his knowledge of guns was such that it is still a strong influence in the market, even though he died of a heart attack in 1978, at the age of 75." The editor of the magazine may not have known that he was writing another epitath for one of the country's greatest gunwriters. But he was.

Knob Creek Machine Guns

It's a Family Affair

by FRANK JAMES

The night shoot at Knob Creek is something to behold. The downrange infusion of thousands of tracer rounds, the explosion of set charges, and the yellow bursts of exploding fuel barrels create a scenario that rivals any war movie. But this experience comes complete with the irregular illumination from parachute flares, the rush of heat against the face, and the reverberations of almost a hundred belt-fed machineguns working in unison.

WHILE THE IDEA may sound strange to many, civilian machinegun ownership is becoming a family-oriented activity, and no place is this fact more evident than at the Knob Creek Machine Gun Shoot held twice a year near Westpoint, Kentucky. Knob Creek is the mecca of civilian machinegunning, and a recent visit showed ample evidence of families and parents with children enjoying this growing pastime.

Civilian machinegun ownership is, in several respects, not for the faint of heart, nor the skimpy pocketbook. Perhaps, this is why many individuals owning machineguns are seen with their family members on the firing line at this, the granddaddy of all machinegun shoots. The whole family comes along to enjoy what represents a major purchase for any income level, because machineguns today are incredibly expensive.

A number of states permit the civilian ownership of legally registered

More and more shooters are bringing their families to this mecca of automatic fire to share the fun.

machineguns, suppressors and other National Firearms Act (NFA) devices. This federal law, enacted in 1934, makes the *unregistered* possession of certain specified weapons a felony. It provides for civilian possession, but it also requires payment of a federal tax. The federal tax is $200 for the transfer of a fully automatic weapon or a silencer. The tax is not included in the cost of these firearms, which most often cost in multiples of thousands of dollars.

The elevated cost of civilian-legal machineguns can be traced back to 1986 when the future manufacture of new machineguns for civilian sale was prohibited by a provision of the Firearms Owners Protection Act—an ironic title if ever there was one.

Now, all machineguns owned by civilians have to have been registered

At Knob Creek, you're likely to see almost every conceivable type of machinegun. Here are three Browning water-cooled belt-feds and a Vickers water-cooled with the corrugated water jacket. There is also an FN Model "D" BAR at the far end, and just in between the last Browning and the Vickers is a 1919A4 light 30 air-cooled. The liquid-cooled guns are probably asking each other how "those guys" got onto the line?

prior to May 19, 1986. The only exceptions are the guns maintained as samples by Class III, or machinegun, dealers. These guns, known as dealer samples, carry many more BATF restrictions and are unavailable to the recreational-oriented civilian non-dealer.

Knob Creek Range is privately owned by the Sumner family. Kenny Sumner is the man in charge, but it was his family who purchased this former military test range and converted it into a civilian shooting range decades ago. No one is exactly sure when someone got the bright idea for a machinegun shoot, but I first heard about it back in the early 1980s and finally went down for my first visit in April of 1986.

Back then, the whole affair was pretty informal. The shooters mounted their belt-fed machineguns atop the concrete firing benches and blazed away whenever the mood struck them. Ceasefires were called when someone wanted to go downrange to put up more dynamite or place another gasoline-can target. The whole atmosphere was pretty relaxed. Everyone thought it amazing that almost 450 people paid just to watch a hundred belt-fed machineguns work through box upon metal box of ammunition over the course of a weekend.

The decade since then has seen a number of changes. First, the guns themselves are far more valuable. Second, all the KCR (Knob Creek Range) procedures are tighter and far more controlled, and the spectators now number in the thousands, not the hundreds. (One of the 1996 shoots had an estimated crowd of 30,000 spectators over three days.) Everyone seems fascinated to watch multiple projectiles reduce old motor vehicles to burned out, bent and broken sheet metal.

Targets for a machinegun shoot usually are a problem. What do you get? And where can you get several big enough to service a line with multiples of belt-feds? The answer in this case was the local Pepsi bottler, who found an easy way to dispose of obsolete vending machines. Or maybe they were donated by a Coca-Cola vender?

The management of KCR has worked hard to improve the facility, and now it offers more than just a simple burn-'em-off opportunity to the frustrated belt-fed owner. There are a number of side events ranging from an IPSC-style combat pistol and rifle match to a sub-gun jungle lane and other competitions.

Additionally, when the main range is shut down, other firing ranges are now provided elsewhere on the property. Granted, these secondary ranges don't offer the downrange distance found on the main firing line, but most machinegunners just want a place to shoot, and they really don't care how far away the target is.

The main firing line at KCR is covered with a sheet metal roof and consists of a series of concrete benches. The firing line itself is positioned well forward of benches that are used as work tables and storage places during the firing sessions. The general spectator public is kept behind an orange snow fence at the back of the covered line, but they are still able to see all the action as shooters work their family treasures through magazine after magazine, or belt after belt, of ammunition.

Knob Creek is also host to one of the largest military weapons shows in existence. In the overall scheme of things, this is a new feature and was noticeably missing when I first ventured onto the property more than a

In a former life, this was a lawn ornament for a veterans' fraternal organization, but an enterprising soul purchased it, reactivated its dewat paper work with the BATF, and then after approval worked until the gun was safe to shoot once again. The result? A working example of a 40mm Bofors anti-aircraft gun from WWII.

The worst thing about reactivating a 40mm Bofors gun from WWII is finding the ammo, and then once you find it figuring out a way to pay for it. In any event, the Bofors was fired sans sights (the owner is still looking for them). Nothing more elaborate than a white line down the top of the barrel was used for sighting-in on old refrigerators and discarded stoves.

decade ago. Amazingly, it is really an excellent gun show. You won't find any turquoise or vintage dolls or toy trucks at this show. This is a genuine gun show with guns on each and every table, and it is so good the show itself would support the event even without the hours of continuous full-auto fire. If you're interested in military weapons or Class III devices of every conceivable type and description, you'll find what you want at this show.

The gun show is held in an open-sided pole barn just behind the firing line, and take my advice, don't wear white. Also, be sure to bring plenty of hearing protection—when the main line opens up you would be well advised to "double plug" each ear. The sound is overpowering, especially when the downrange pyrotechnics start blowing off.

About wearing white: Dust and dirt are thrown into the air whenever hundreds of machineguns open up simultaneously. The earth moves, both literally and figuratively, and eventually it filters down upon everything, including the spectators and the merchandise inside the pole barn.

But hey, machineguns are not for wimps! And neither is shooting them.

Civilians who own machineguns most frequently pay homage to a number of things military. Military uniforms, often mismatched, both old and new, foreign and domestic, abound as they are worn by participants, young and old, and are offered for sale by gun show exhibitors. Probably the most visually intriguing aspect of this interest in military artifacts is the number of former military vehicles that are seen parked behind the firing line and in the parking lot. Military re-enactors often camp in the outlying parking lots, and it's not unusual to walk through a bivouac scene from Patton's drive to the Rhine at the edge of the grass parking lot.

Not all of this interest in ex-military vehicles is without purpose, either. Many exhibitors at the military gun show offer large volumes of military surplus ammunition at significant discounts or, at best, reasonable prices when purchased in volume. (You have to remember this is a machinegun shoot and ammunition is expended in exponential amounts.) For those not actually shooting, but who purchase a case of military ammo, youngsters driving former Marine Corps motorized "mules" will deliver the heavy burden to the parked vehicle—for a small fee, of course. In fact, among the World War II-era Jeeps and light trucks, it was the motorized mules that most often caught my attention with their four-wheel steering and flat-bed platforms, as they seemed to be in constant motion throughout the weekend of my visit.

But, it is the guns that attract the shooters and spectators to Knob Creek. The lure of the machinegun, and the chance to watch a belt-fed Browning in action, or any of the World War II machineguns that grandpa talked about, is what many find so fascinating. These are the guns that many mistakenly believe are forbidden and have no legitimate purpose. The fact is, their most legitimate purpose is their use in illustrating the brutality of war and the fearsomeness of a night-time L-shaped ambush.

The night shooting sessions at Knob Creek are the most spectacular of all. Flares fall and light up the impact area as red, green and yellow tracer ammo fills the smoke-filled sky. Flame throwers are often demonstrated during these late-evening sessions, and the results are sobering. The resulting flames from burning vehicles or exploding fuel drums lend an orange cast to everything, but the colors are constantly changing. Once the flares die and the firing stops, then the darkness returns. That is, until the firing line lights are turned back on and a cease-fire is called.

The main range at Knob Creek is approximately 440 yards long. A horseshoe-shaped ridge is the impact area, but the ground on bottom is flat and level. The management of KCR has always provided a variety of large-size targets. I've seen everything from washers and dryers; kitchen stoves; automobiles of every size, shape and description; up to semi-tractors and old school buses. Targets have also included discarded fiberglass boats, and once there was even an old-fashioned deep bowel bathtub with a mannequin inside dressed to resemble Muammar Qaddafi. It didn't last very long, probably all of thirty seconds from the assembled line of multiple belt-fed machineguns.

Because of the 1986 law, machineguns now represent a major investment. Probably the least expensive one to be found on the KCR firing line would be an Ingram or MAC-10 submachine gun, any of the Sten guns made by various American manufacturers before 1986, or the 45-caliber Reising submachine gun that so many Marines cursed. Prices will vary, of course, but most current examples of these three weapons can be purchased for less than $1000—tax not included.

The top of the scale for submachine gun aficionados is the Heckler & Koch MP5. This 9mm, select-fire subgun is the acknowledged "best in class" and often carries the prices to prove it. All the civilian-legal guns, often referred to as "transferable," are really replicas of the original MP5s. To be transferable, they had to be manufactured in this country prior to 1986 or imported before 1968. Almost all of them seen today in civilian hands are conversions made from the Heckler & Koch Model 94 9mm semi-auto carbine that was formerly imported. Many of the conversations were completed as near duplicates of the original, and it's hard for even the experts to tell the difference between a genuine nontransferable MP5 and one of the better remanufactured transferable guns. Nonetheless, any MP5 legally available to civilians carries a high price. The average, as this is written, is approximately $5000 or more per gun, and you can add another $1000 to those made to duplicate the MP5SD, the silenced version of this popular and extremely successful 9mm submachine gun.

Belt-fed guns as a general rule are not quite as expensive. There seems to be a number of reasons behind this economic fact of the machinegun trade. First, it's hard for many shooters to find a range that allows full-auto belt-fed fire. The plain fact is most civilian ranges are not equipped to withstand the impact damage. Second, and this is solely my opinion, belt-feds can be a terrible pain in the butt. They're heavy and hard to maneuver, or even move from one firing point to another. Third, the cleaning procedures are involved and time consuming. That's not to say the guns aren't popular, it's just that it takes a special individual to own a belt-fed. Most shooters appreciate the convenience of the smaller machineguns because they are easier to store securely (a major concern) and easier to handle, plus it's easier to find a place to shoot them.

And when it comes to convenience, the bipod-mounted machineguns and modern assault rifles are extremely popular. They often come with lower price tags than the higher-end MP5 submachineguns, but offer increased performance.

As for the specific weapons seen at Knob Creek, my best response would be it's a living encyclopedia of military weaponry. I've seen French 75 artillery pieces that were former lawn ornaments firing cast zinc projectiles through refrigerators at 100 yards without the use of sights. I've watched a 40mm Bofors hurl both live rounds and spent cases downrange, because the Bofors ejects out of the front of the gun.

Among the machineguns, it's hard to think of a one I haven't seen at some point over the past decade on the KCR firing line. The common guns, of course, are those with an American heritage—the immortal Thompson submachine gun, the Ingram-inspired MAC-10 buzz guns, and all the Browning-designed guns like the BAR, the Browning light 30s, and the Ma-Deuce 50-caliber BMG. German guns are popular. There are always several examples of the MP-38 or MP-40 to be seen along with the belt-fed MG-42. As previously mentioned, Heckler & Koch weapons are very popular, and this includes the HK33 and G-3 clones in select-fire offerings. The Israeli Uzi and the Smith & Wesson Model 76 are always seen somewhere. Among the British guns, perhaps the two best are the Bren guns and the various water-cooled Vickers guns that can be found operating up and down the line. Modern guns like the motorized Gatling or GE Mini-Gun are fre-

(Above) The Knob Creek Machine Gun Shoot boasts one of the best military-oriented gun shows in the United States. Shown here are assault rifles from Nazi Germany, Communist China, Czechoslovakia, and the Soviet Union.

The gun shown here on the short ski-like stand is a Lahti anti-tank rifle. Semi-automatic and firing a 20mm round, these guns were imported from Finland and sold in a coffin-like box, along with a few rounds of ammunition, for around $100 back in the 1950s. The lever on the right side of the receiver is to *cock* the weapon. Note the disassembled MG-42 in the background.

quently heard with their rapid, but intensive bursts of fire. The list of guns goes on and on.

For those interested in visiting the Knob Creek Machine Gun Shoot, it takes place twice each year: once in April and again in October, on the second weekend of the month. The gun show is open on Friday 9 a.m. to 9 p.m., Saturday 9 a.m. to 10 p.m., and Sunday 9 a.m. to 6 p.m. Camping is available on a first-come, first-served basis, but there are no hookups, and the facilities are primitive. The surrounding area has a wide number of motels because of the nearby Fort Knox military base, but be sure to book your room well in advance. Daily admission charge is reasonable for adults, with children under 12 paying half-fare. Weekend admission rates are also offered.

For those who live in states that prohibit the possession of Class III weapons, machinegun rental (under supervision) is available, and this gives everyone the chance to enjoy their own full-auto fire.

Knob Creek Range is located 1 mile off highway 31W on Highway 44 in Bullitt County, Kentucky. If you go, tell 'em I sent ya. I'm sure it will be an experience you won't soon forget.

30th EDITION, 1998 **19**

THE BUCKSHOT TRIALS

Lots of shooting and hard work went into determining which of the many buckshot loads perform best. This is a long, hard look at the little ball that could save your life.

THEY DON'T CALL 'em "scatterguns" without reason. It's dated terminology from our frontier past, but it's still accurately descriptive. A scattergun is a gun that scatters its projectiles in a pattern that (hopefully) will evenly cover a given area at a given distance. Shotgun is the more common term and the modern shotgun has many uses in hunting and various types of clay target competition.

Through the use of charges of fine shot and appropriate chokes, we have developed the ability to hit small, fast moving targets with just a portion of the circular pattern. Even just a few of the shot pellets serve to bust the clay target, stop the running rabbit or drop the game bird in flight. In this context, the essence of the shotgun is to hit an area that corresponds to the location of the target. In most cases, the target is moving, so accurate placement of the pellet pattern just ahead of the target ensures a hit. In frontier America, meat was sometimes hard to come by and the shotgun was a great tool for taking small game and birds. Ergo, the shotgun was probably as deserving of the term "the gun that won the West" as was the Peacemaker Colt or Winchester '73.

Still, early on in the history of the American shotgun, some frontiersmen began to take advantage of another characteristic of the gun. In so doing, they gave even more credence to the shotgun as a West-winning survival tool. A shotgun man afield for birds or rabbits might find himself faced with predators of either two- or four-legged variety. He had but to break open the shotgun and replace the birdshot with another load. The load he dropped into the twin chambers of his trusty side-by-side was some form of buckshot. Buckshot had a long history of use in the republic, dating back to "buck-and-ball" loads of the Revolutionary War. The name comes from the special oversized shot intended for short range deer hunting in the forested portions of the frontier.

With a few notable exceptions, most shotgunners choose size #00 buckshot for hunting and defense purposes. In this article, we'll take a long look at 12-gauge buckshot loads. That means we have to define the various kinds of buckshot and loads, i.e. their relative sizes and weights, the number of pellets per shell and the velocities at which they leave typical shotguns. At the heart of the matter, we want to find out exactly which of the many types of buckshot

There are currently fifty different 12-gauge buckshot loads made by the three major ammunition makers to cover law enforcement and hunting needs.

by WILEY CLAPP

actually perform best. While some of the material we may develop might be of great value to the deer hunters of the Eastern United States who are limited to shotguns, the major thrust of the article is to get a handle on buckshot for anti-personnel use. In short, we are talking about the combat shotgun, as used by police, military personnel and concerned civilians arming for their own defense.

At the outset, please understand that the wingshooter and the competitor want their shotguns to do things almost diametrically opposed to the needs of the combat shotgunner. The former use a shotgun and fine shot to increase the probability of a hit. They only ask for a small percentage of the shot charge to strike a vital area or powder a clay. The fighting shotgunner, on the other hand, uses a shotgun for terminal effect. He wants to center a charge of buckshot on an enemy in order to stop the fight—right damned now. In this regard, remember the case of Doc Holiday striding down that dusty street in Tombstone with the Earp brothers. He carried a sawed-off 10-bore Meteor shotgun because he knew it was a fight-stopper. It worked, too—history tells us he darned near cut Frank McLowery in half with the thing.

There is a tremendous amount of short range power in a modern 12-gauge shotgun and most police agencies include one of them as permanent equipment in every patrol car. They're emergency equipment, racked up in the front seat for those desperate crisis situations where a felon must be stopped from continuing his life-threatening criminal behavior. Our military forces have used shotguns for similar duties for years. The trench shotguns of WWI came out of storage for close-quarters use in WWII and Korea. The heavy brush of Vietnam was an ideal setting for combat shotguns and even today, the military issues thousands of them to our forces deployed around the world.

And that means lots of buckshot. When you peruse the catalogs of the three major makers of shotgun ammunition (Federal, Remington, Winchester), you come away a bit dazzled by the variety. Buckshot comes in five pellet sizes and is loaded in four different types of shells. They are standard velocity $2^{3}/_{4}$-inch loads, $2^{3}/_{4}$-inch magnum loads which typically carry a heavier payload at the same velocity, then 3-inch magnums which require a longer chamber, but deliver even more shot. Law Enforcement low recoil ammo is the fourth type, a recent phenomenon that has been widely accepted in police circles, since it makes the shotgun an effective tool for smaller shooters who are troubled by heavier recoil.

The most popular shot size is #00, but other sizes are available. The smallest is #4 which is a .24-inch ball weighing about 21 grains. Next up in size is #1, .30-inch pellets weighing about 40 grains each. Then we have #0, .32-inch pellets at approximately 48 grains apiece. The legendary #00 pellet measures about .33-inch and weighs some 54 grains. Finally, there's the sleeper, #000 at .36-inch and 70 grains in weight. Since they are all fired from virtually the same shells, it follows that there are more pellets per shell when the small sizes are used—41 pellets per shell with #4 buck in a 3-inch magnum load, for example. Conversely, the 3-inch magnum #000 only delivers ten pellets.

The immutable laws of exterior ballistics apply to every pellet once it leaves the muzzle of the shotgun. Each is a separate projectile in flight and each behaves consistently with its weight. Since each pellet is a sphere and not a particularly efficient missile shape, each loses velocity quickly. But—and that's an important but—the heavier a pellet is to start, the longer it will retain velocity. Look at it this way: If you cast a bore-sized ball of lead and fired it from a Cylinder-bore gun, it would retain velocity somewhat longer than the same sized ball you cast from a lighter zinc alloy. This phenomenon has serious implications in two ways. Obviously, it will markedly impact the range you can expect from your load. But it will also affect the depth of penetration in tissue and, potentially, the lethality of the shot.

Before we get into the analysis of our buckshot shooting, we need to make another point very clear. Buckshot is serious stuff and has a lethal potential. Because it has a somewhat limited range, as do all shotguns, it can be an excellent choice for decisive use in short-range confrontations. But because it does still scatter pellets, the shooter (particularly the police officer) has a responsibility to ensure his pellets hit the mark and don't go astray. There's more than a little problem associated with shotgun marksmanship and a considerable problem with what the police administrators now call "pellet accountability." The

For testing with a Cylinder-bore gun, Winchester's Defender, a version of the Model 1200, was chosen. It has an extended magazine, matte finish and synthetic stock.

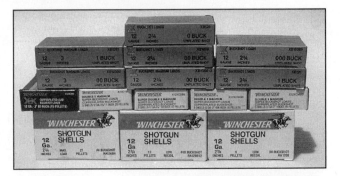

Winchester makes fifteen different buckshot loads, including the three law enforcement types on the bottom row.

marksmanship problem is mostly outside the scope of this story; the size and shape of the pattern is exactly what we are talking about. Assuming proper marksmanship, the smallest, tightest pattern delivered to the target at normal shotgun ranges is best because it will have a more decisive terminal effect and will have less chance of sending pellets astray.

You get that tight pattern effect by selecting the right load and firing it in the right gun. We are going to fire everything made in buckshot in America, but before we do, we are going to have to look at another aspect of shotguns before we can intelligently burn powder.

No one is exactly sure when it started in shotgun history, but most authorities attribute the popularizing of chokes to American shooter Fred Kimball in the late 1870s. As applied to a shotgun, choke means a slight reduction in bore diameter near the muzzle. It has a similar effect as the nozzle on a garden hose. It extends the range of the gun a little and tightens the pattern. In current use, chokes come in varying degrees of constriction. They run from none at all to increasing degrees of tightness called Improved Cylinder, Modified, Improved Modified and the tightest of them all, Full choke. There are also some special ones (Skeet, Extra Full, etc.), but they aren't really germane to our discussion.

Combat shotguns are built or mod-

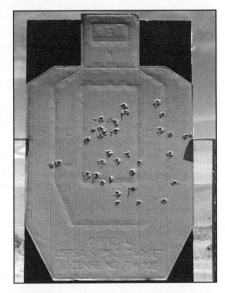

Size #000 is the answer if you're saddled with a Cylinder-bore gun. This four-shot group with Winchester's standard #000 shows all thirty-two pellets on target and 84 percent in the C-zone.

ified to be handled rapidly for close range work and they have shorter barrels than the field hunting guns. On the frontier, it was common to find shotguns with barrels as short as 12 inches used for stagecoach defense and law enforcement. Most of these were cut down sporting guns (ergo, the term "sawed-off" shotgun was born) and if there was any choke in the barrel to start with, the hacksaw took it away. I can't prove it, but I

Table 1: Buckshot Trials
Winchester Defender
18-inch Cylinder-bore Barrel

Load Number	Load Type (ins.)	Maker's Code	Pellets Size/Count	4-Shot Average Total Pellets/Total Hits/Percentage	MV (fps)	C-Zone Hits/Percentage
Remington Buckshot						
1.	2 3/4 Buckshot	SP124BK	#4/27	108/80/74	1312	54/50
2.	2 3/4 Buckshot	SP121BK	#1/16	64/49/76	1201	29/45
3.	2 3/4 Buckshot	SP120BK	#0/12	48/27/56	1188	23/47
4.	2 3/4 Buckshot	SP1200BK	#00/9	36/30/83	1312	21/58
5.	2 3/4 Buckshot	SP12000Bk	#000/8	32/29/90	1212	24/75
6.	2 3/4 Magnum	SP12SMAG1BK	#1/20	80/49/61	1081	38/47
7.	2 3/4 Magnum	SP12SMAG00BK	#00/12	48/33/68	1203	23/47
8.	2 3/4 Premier	PR12SNBK	#00/12	48/38/81	1211	20/41
9.	3 Buckshot	SP12HMAG4	#4/41	164/84/51	990	53/32
10.	3 Buckshot	SP12HMAG1BK	#1/24	96/46/47	1011	29/30
11.	3 Buckshot	SP12HMAG00	#00/15	60/41/68	1104	24/40
12.	3 Buckshot	SP12HMAG000	#000/10/40	32/80	1202	18/45
13.	2 3/4 Law Enforcement Express	SPL1200BK	#00/9	36/35/97	1279	28/87
14.	2 3/4 Red. Recoil	RR128K800	#00/8	32/32/100	1179	28/87
15.	2 3/4 Red. Recoil	RR1200BK	#00/9	36/32/88	1225	25/69
Winchester Buckshot						
1.	2 3/4 Super-X	XB121	#1/16	64/44/68	1155	32/50
2.	2 3/4 Super-X	XB120	#0/12	48/39/81	1187	30/62
3.	2 3/4 Super-X	XB1200	#00/9	36/25/69	1287	20/55
4.	2 3/4 Super-X	XB12000	#000/8	32/32/100	1211	27/84
5.	2 3/4 Super-X (M)	XB120012	#00/12	48/32/66	1186	23/47
6.	3 Super-X (M)	XB1231	#1/20	80/52/65	933	34/42
7.	3 Super-X (M)	XB12300	#00/15	60/48/80	1102	30/50
8.	2 3/4 XX Magnum	X12C1B	#1/20	80/62/77	1056	37/46
9.	2 3/4 XX Magnum	X12XCOB5	#00/12	48/37/77	1233	25/52
10.	3 XX Magnum	X12XCMB5	#4/41	164/94/57	1167	50/30
11.	3 XX Magnum	X12XC3B5	#00/15	60/43/71	1211	23/38
12.	3 XX Magnum	X123C000B	#000/10	40/38/95	1164	19/47
13.	2 3/4 Low Recoil	RA1200	#00/9	36/29/80	1161	15/41
14.	2 3/4 Low Recoil	RA120012	#00/12	48/36/75	1135	24/50
15.	2 3/4 Low Recoil	RA124BK	#4/27	108/78/72	1255	45/41
Federal Buckshot						
1.	2 3/4 Classic	F1274B	#4/27	108/75/69	1252	53/49
2.	2 3/4 Classic	F1271B	#1/16	64/41/64	1170	32/50
3.	2 3/4 Classic	F1270B	#0/12	48/30/62	1215	24/50
4.	2 3/4 Classic	F12700	#00/9	36/23/63	1263	18/50
5.	2 3/4 Classic	F127000	#000/8	32/31/96	1239	20/62
6.	2 3/4 Classic Magnum	F1304B	#4/24	96/74/77	1137	48/50
7.	2 3/4 Classic Magnum	F1301B	#1/20	80/58/72	961	45/56
8.	2 3/4 Classic Magnum	F13000	#00/12	48/36/75	1202	18/37
9.	3 Classic Magnum	F1314B	#4/41	164/119/72	993	82/50
10.	3 Classic Magnum	F13100	#00/15	60/50/83	1108	39/65
11.	3 Classic Magnum	F131000	#000/10	40/34/85	1213	19/47
12.	2 3/4 Premium	P15400	#00/9	36/32/88	1231	27/75
13.	2 3/4 Premium Magnum	P1564B	#4/34	136/108/79	1125	64/47
14.	2 3/4 Premium Magnum	P15600	#00/12	48/40/83	1138	14/29
15.	3 Premium Magnum	P1584B	#4/41	164/107/65	1097	73/44
16.	3 Premium Magnum	P1581B	#1/24	96/60/62	1048	34/35
17.	3 Premium Magnum	P15800	#00/15	60/41/68	1037	29/48
18.	3 Premium Magnum	P158000	#000/10	40/34/85	1237	22/55
19.	2 3/4 Tactical	H1324B	#4/27	108/80/74	1126	55/50
20.	2 3/4 Tactical	H13200	#00/9	36/35/97	1083	27/75

believe this is why fighting shotguns have traditionally had no choke. That's a bit surprising, since they would benefit from a bit of choke now and would have equally benefited a century ago. I make a point of this simply because the majority of combat shotguns made in modern times follow the trend and have no choke. Like it or not, we have to face the problem. In essence, choking is *critical* for a combat shotgun.

For the evaluation shooting, I obtained sample quantities of all 12-gauge buckshot loads (except the waterfowling 3 1/2-inchers) made in the United States. That's fifteen loads each from Remington and Winchester, plus twenty different loads from Federal. Shot sizes ran from #4 all the way up to #000. The most popular shot size was #00, with no less than twenty-one of the fifty total loads made with those 33-caliber pellets. There are ten #4 loads, nine #1s, seven #000s and a slim trio of #0s. While this might seem like a surprisingly wide variety, you have to recall that there are four distinctly different kinds of buckshot loads—standard, short magnum, 3-inch magnum and tactical or law enforcement loads.

Since the main idea of this shoot was to evaluate the ammunition (and, to some degree, the chokes), I came up with the following procedure. We fired all of the various loads at standard IPSC silhouette targets. This is a humanoid cardboard silhouette that's 30 inches high and 18 inches wide. That dimension includes a simulated head measuring about 6 by 6 inches.

Within the chest and extending into the head area there is a C-zone for scoring in competitive matches. The C-zone is about $11\frac{1}{2}$ inches wide by $23\frac{1}{2}$ inches high. Generally, the target is about the size of a human opponent and the rough dimensions of a man's vital area are delineated by the C-zone. With the targets placed at 25 yards, we fired a carefully aimed four-shot series with each load at one of these targets. Four shots displays the load potential better than a single round in the sense that if a load is going to throw erratic flyers, it's likely that it will do it in four rounds, but possible that it might not in one or even two. After some 600 rounds downrange, I am satisfied we had a very good representation of what all the buckshot loads will do.

To see what effect choke has on the different buckshot loads, we used three different shotguns with different choke systems. First came a current production Winchester Defender pump gun with an 18-inch barrel. This one had the Cylinder bore system of no choke at all. On the other end of the shoot came a Browning BPS pump shotgun equipped with the Browning Invector interchangeable choke tube system. For our evaluation of the buckshot loads, I screwed a Full choke tube into the Browning's 22-inch barrel. For something in the middle, I included a specimen of the familiar Remington 870. This is a somewhat customized gun and the major feature of the customizing process is the Vang-Comp system installed in the barrel. This is a combination of relieving the forcing cone and a special form of jug choking that throws patterns generally in the range of a Modified choke. The Vang-Comp also has a series of ports near the muzzle that reduce muzzle flip substantially. Hans Vang's system reduces pellet deformation, reduces recoil and muzzle rise, and reduces pattern size.

So we sat down at a sturdy bench with the three guns and fifty different loads. A Pro-Chrono chronograph was set out front about 12 feet and the target frame was out at 25 yards. That distance is generally accepted as being the maximum effective range of a combat shotgun. After conducting

The author's Remington Model 870 sports ghost-ring sights from LPA and Gunsite Training Center, and the Vang-Comp barrel system.

Remington's buckshot lineup includes three law enforcement loads in an assortment that runs to fifteen types.

this shoot, I have to believe we are underestimating a good gun and good ammo just a little, but we'll get back to that point later. Three guns times four rounds of each of fifty different loads adds up to 600 rounds fired. It went uneventfully if somewhat uncomfortably (OK, I had help) but when the test was over, I had a tall stack of cardboard targets and the basis for some interesting data.

The first step was to count pellet holes. That was a pretty tedious chore but it had to be done. Any hit on the cardboard counted, even on the very edge. For every load used, I computed the number of possible hits. For example, Winchester's load number 7 (in the accompanying table) was a 3-inch magnum #00 buckshot with fifteen pellets per shell. Four shells should deliver a total of sixty pellets in the target. In actuality, that particular load produced forty-eight, fifty-five and fifty-nine pellet hits from Cylinder, Vang-Comp and Full chokes respectively. Remember that we are concerned with pellet accountability and terminal effect, so the more centered hits the better. For that reason, we went back and counted the hits in that inner C-zone, which in the case of the same Winchester load, were thirty, forty-three and forty-nine from the same three chokes. I then calculated the *percentage* of hits in each target. Once again, the percentages for Winchester load number 7 from the three different guns ran 80, 91

A few targets looked like this—a four-shot pattern with a gaping hole in the middle. This was Remington's #0 buck from the Full choke barrel.

and 98 percent on the target with 50, 71 and 81 percent in the C-zone.

By providing the reader with the data in the accompanying tables, we have given him at least some of the means to eliminate guess work in selecting shotgun loads and the chokes through which they are to be fired.

The material is presented in tabular form with the number of the load in firing sequence, its description and manufacturer's code. Then we have a

Table 2: Buckshot Trials
Remington Model 870
Vang-Comp Barrel

Load Number	Load Type (ins.)	Maker's Code	Pellets Size/Count	4-Shot Average Total Pellets/Total Hits/Percentage	MV (fps)	C-Zone Hits/Percentage
Remington Buckshot						
1.	2¾ Buckshot	SP124BK	#4/27	108/87/80	1312	55/50
2.	2¾ Buckshot	SP121BK	#1/16	64/48/75	1199	35/54
3.	2¾ Buckshot	SP120BK	#0/12	48/40/83	1128	25/52
4.	2¾ Buckshot	SP1200BK	#00/9	36/35/97	1196	27/75
5.	2¾ Buckshot	SP12000Bk	#000/8	32/32/100	1228	31/98
6.	2¾ Magnum	SP12SMAG1BK	#1/20	80/61/76	1051	37/46
7.	2¾ Magnum	SP12SMAG00BK	#00/12	48/45/93	1238	33/68
8.	2¾ Premier	PR12SNBK	#00/12	48/48/100	1244	47/97
9.	3 Buckshot	SP12HMAG4	#4/41	164/148/90	1116	111/82
10.	3 Buckshot	SP12HMAG1BK	#1/24	96/90/93	892	79/82
11.	3 Buckshot	SP12HMAG00	#00/15	60/54/90	1171	47/78
12.	3 Buckshot	SP12HMAG000	#000/10	40/39/97	1204	38/95
13.	2¾ Law Enforcement Express	SPL1200BK	#00/9	36/36/100	1294	36/100
14.	2¾ Red. Recoil	RR128K800	#00/8	32/32/100	1159	31/96
15.	2¾ Red. Recoil	RR1200BK	#00/9	36/32/88	1082	23/63
Winchester Buckshot						
1.	2¾ Super-X	XB121	#1/16	64/61/95	1156	59/92
2.	2¾ Super-X	XB120	#0/12	48/46/95	1203	33/68
3.	2¾ Super-X	XB1200	#00/9	36/34/94	1297	24/66
4.	2¾ Super-X	XB12000	#000/8	32/31/96	1225	28/87
5.	2¾ Super-X (M)	XB120012	#00/12	48/46/95	1211	38/79
6.	3 Super-X (M)	XB1231	#1/20	80/80/100	948	62/77
7.	3 Super-X (M)	XB12300	#00/15	60/55/91	1136	43/71
8.	2¾ XX Magnum	X12C1B	#1/20	80/78/97	1038	59/73
9.	2¾ XX Magnum	X12XCOB5	#00/12	48/41/85	1252	34/70
10.	3 XX Magnum	X12XCMB5	#4/41	164/141/85	1147	95/57
11.	3 XX Magnum	X12XC3B5	#00/15	60/55/91	1114	46/76
12.	3 XX Magnum	X123C000B	#000/10	40/40/100	1180	33/82
13.	2¾ Low Recoil	RA1200	#00/9	36/36/100	1170	32/88
14.	2¾ Low Recoil	RA120012	#00/12	48/46/95	1131	42/87
15.	2¾ Low Recoil	RA124BK	#4/27	108/94/87	1110	67/62
Federal Buckshot						
1.	2¾ Classic	F1274B	#4/27	108/95/87	1275	53/49
2.	2¾ Classic	F1271B	#1/16	64/56/87	972	40/62
3.	2¾ Classic	F1270B	#0/12	48/44/91	1245	28/58
4.	2¾ Classic	F12700	#00/9	36/36/100	1329	32/88
5.	2¾ Classic	F127000	#000/8	32/30/93	1232	27/84
6.	2¾ Classic Magnum	F1304B	#4/24	96/95/98	1166	55/57
7.	2¾ Classic Magnum	F1301B	#1/20	80/67/83	914	53/66
8.	2¾ Classic Magnum	F13000	#00/12	48/44/91	1043	30/62
9.	3 Classic Magnum	F1314B	#4/41	164/101/61	945	72/43
10.	3 Classic Magnum	F13100	#00/15	60/53/88	1112	29/48
11.	3 Classic Magnum	F131000	#000/10	40/35/87	1227	25/62
12.	2¾ Premium	P15400	#00/9	36/36/100	1255	32/88
13.	2¾ Premium Magnum	P1564B	#4/34	136/127/93	1146	118/86
14.	2¾ Premium Magnum	P15600	#00/12	48/46/95	1168	43/89
15.	3 Premium Magnum	P1584B	#4/41	164/146/89	1115	115/70
16.	3 Premium Magnum	P1581B	#1/24	96/89/92	1096	56/58
17.	3 Premium Magnum	P15800	#00/15	60/60/100	1059	40/66
18.	3 Premium Magnum	P158000	#000/10	40/40/100	1241	37/92
19.	2¾ Tactical	H1324B	#4/27	108/108/100	1192	88/81
20.	2¾ Tactical	H13200	#00/9	36/36/100	1123	35/97

column showing the buckshot size and number of pellets per shell. Next is a column showing the number of pellets and hits (from four shots) on the IPSC target and the percentage achieved of possible hits. Remington's load number 1, for example, delivered eighty out of 108 #4 pellets to the target from the Cylinder bore barrel, or 74 percent performance. Our next column reports the velocity—a four-shot average—of the load as measured by a Pro-Chrono chronograph. The final column reports the number of pellets in the C-zone and the percentage of possible hits. A load and choke combination that delivers 90 percent of its pellets to the target and 80 percent to the C-zone would seem to be a pretty decent performer, and I adopted that as the standard.

While I was primarily intrigued with the selection of pellet size in general, and specific load in particular, the matter of shotgun choke is also something we need to look at closely. Probably the most glaring fact coming off the columns of figures is the performance of a Cylinder-bore barrel. It may be the traditional standard, but Cylinder bore is the worst possible choice for the shooter who wants his pattern tightly centered on a combat target in order that the pellets do their work. In the un-choked Cylinder-bore barrel, only seven of the fifty loads were 90 percent performers. Interestingly, four of them were #000s and three were #00s. There were three loads that delivered 80 percent or more of their pellets to the C-zone. The shooter who uses a Cylinder-bore barrel had better plan on a very large number of pellets going into things other than the target. Remington, the leader in the production of police shotguns, fully understands this and no longer makes the 18-inch police barrel with a Cylinder bore. All of their police shotguns have an Improved Cylinder choke.

At the opposite end of the choke spectrum we have the Full choke system of a Browning BPS. Please note that velocities from this barrel run a trifle higher than from the other two. That's because it's slightly longer at 22 inches. It is a vastly better performer than the Cylinder-bore gun, delivering excellent performance from a good range of loads. Of the fifty different loads, twenty-eight delivered 90 percent or better of their pellets to the target and twelve loads did 80 percent in the C-zone. For some unknown reason, the Full choked BPS really liked the Winchester brand of buckshot, as only one load out of Winchester's fifteen was not a 90 percent performer. Full choked fighting shotguns are not common, although many were made for the Strategic Hamlet program in Vietnam during the late '60s. If you have one, you can be sure of a greater number of hits than with the Cylinder bore. But they do have two limitations. One is the tendency of the tightly constricted barrel to deform pellets. We found a large number of out-of-round pellet holes in the targets, particularly with #000 and #00. Second, there were a few four-shot patterns that had gaping holes in their centers. With a carefully selected load, Full choke might be OK, but across the board, I would prefer to see a reduced degree of choke.

That's exactly what you get in Hans Vang's Vang-Comp system. It is a specialized form of custom shotgun, but is becoming so deservedly popular in fighting shotgun circles as to be the test gun for this shoot. The Vang-Comp is a combination of forcing cone relief, back-boring and a sort of jug choking. It babies the shot down the barrel in such a way as to avoid pellet deformation and in the process, reduce recoil. The system is literally optimized for buckshot. It works exceptionally well—thirty-five out of fifty loads put 90 percent or

To get a short-barreled shotgun that had a Full choke, the author used a Browning BPS with screw-in Invector choke tubes.

Federal buckshot comes in a wide variety of types and sizes. There are twenty different loads.

more of their pellets on the target and twenty drove 80 percent or better into the C-zone. A full dozen loads actually kept *all* of their pellets on the target. One load, Remington's Law Enforcement Express #00, fired 100 percent into the C-zone. If your goal is to shoot tight, controlled buckshot patterns from a combat shotgun, this would seem to be the gun to use. It is an outstanding performer.

Now let's look at the various loads and how they perform. The overwhelming majority of buckshot is #00 and that tends to skew the options quite a bit. But there are enough choices in #1, #4 and #000 to make them viable options. The #4 buckshot enjoys a certain vogue in police circles and both Winchester and Federal make a low recoil Tactical load in this, the smallest of buckshot sizes. The pellets are about .24-inch to .25-inch and they are used in the belief that the pattern will be more dense. True enough, but there will be more pellets that miss completely and the individual hits will be less disabling. But since the pellets are smaller and lighter, their range will be shorter and that could be a consideration in urban areas. The same is true of #1 buckshot (nine different loads of .30-inch pellets) and even #0 (three loads using .32-inch balls). But if you consider the real purpose of a fighting shotgun—to deliver a tight pattern of decisive shot into a man-sized target—the #00 and #000 loads really shine.

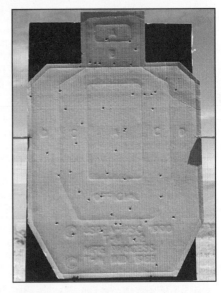

Federal's 2³⁄₄-inch #1 magnum load puts a lot of pellets (ten per shell) in the air, but when fired from a Full choke barrel almost half don't make it to the target, and only 37 percent hit the C-zone.

As a matter of fact, they are unquestionably the best ammo to use in a Cylinder-bore gun. Remember, only seven loads fired from the Cylinder barrel were 90 percent performers on the man-sized IPSC target. Four of them were #000 and three were #00. In a Full choke barrel, there were twenty-eight 90 percent loads and twenty-three were either #00 or #000. The superb Vang-Comp system gave us thirty-five out of fifty loads as 90

Table 3: Buckshot Trials
Browning BPS
22-inch Full Choke Tube

Load Number	Load Type (ins.)	Maker's Code	Pellets Size/Count	Total Pellets/Total Hits/Percentage	MV (fps)	C-Zone Hits/Percentage
Remington Buckshot						
1.	2¾ Buckshot	SP124BK	#4/27	108/72/66	1203	42/38
2.	2¾ Buckshot	SP121BK	#1/16	64/41/64	1226	19/29
3.	2¾ Buckshot	SP120BK	#0/12	48/40/83	1180	20/31
4.	2¾ Buckshot	SP1200BK	#00/9	36/33/91	1267	26/72
5.	2¾ Buckshot	SP12000Bk	#000/8	32/32/100	1271	31/96
6.	2¾ Magnum	SP12SMAG1BK	#1/20	80/35/43	1058	19/23
7.	2¾ Magnum	SP12SMAG00BK	#00/12	48/36/75	1211	21/43
8.	2¾ Premier	PR12SNBK	#00/12	48/48/100	1280	42/87
9.	3 Buckshot	SP12HMAG4	#4/41	164/162/98	1242	110/67
10.	3 Buckshot	SP12HMAG1BK	#1/24	96/73/76	1066	45/46
11.	3 Buckshot	SP12HMAG00	#00/15	60/58/96	1250	52/86
12.	3 Buckshot	SP12HMAG000	#000/10	40/38/95	1295	32/80
13.	2¾ Law Enforcement Express	SPL1200BK	#00/9	36/36/100	1327	36/100
14.	2¾ Red. Recoil	RR128K800	#00/8	32/32/100	1161	31/96
15.	2¾ Red. Recoil	RR1200BK	#00/9	36/33/91	1180	19/52
Winchester Buckshot						
1.	2¾ Super-X	XB121	#1/16	64/62/96	1261	55/85
2.	2¾ Super-X	XB120	#0/12	48/18/100	1302	38/79
3.	2¾ Super-X	XB1200	#00/9	36/35/97	1404	30/83
4.	2¾ Super-X	XB12000	#000/8	32/32/100	1332	27/84
5.	2¾ Super-X (M)	XB120012	#00/12	48/48/100	1237	38/79
6.	3 Super-X (M)	XB1231	#1/20	80/73/91	1004	58/72
7.	3 Super-X (M)	XB12300	#00/15	60/59/98	1012	49/81
8.	2¾ XX Magnum	X12C1B	#1/20	80/71/88	1095	58/72
9.	2¾ XX Magnum	X12XCOB5	#00/12	48/48/100	1309	36/75
10.	3 XX Magnum	X12XCMB5	#4/41	164/146/89	1200	84/51
11.	3 XX Magnum	X12XC3B5	#00/15	60/54/90	1190	37/61
12.	3 XX Magnum	X123C000B	#000/10	40/40/100	1187	32/80
13.	2¾ Low Recoil	RA1200	#00/9	36/36/100	1178	28/77
14.	2¾ Low Recoil	RA120012	#00/12	48/46/95	1171	38/79
15.	2¾ Low Recoil	RA124BK	#4/27	108/104/96	1251	85/78
Federal Buckshot						
1.	2¾ Classic	F1274B	#4/27	108/87/80	1384	63/58
2.	2¾ Classic	F1271B	#1/16	64/49/76	1280	27/42
3.	2¾ Classic	F1270B	#0/12	48/38/79	1205	19/39
4.	2¾ Classic	F12700	#00/9	36/32/88	1304	22/615
5.	2¾ Classic	F127000	#000/8	32/32/100	1247	21/65
6.	2¾ Classic Magnum	F1304B	#4/24	96/71/73	1220	33/34
7.	2¾ Classic Magnum	F1301B	#1/20	80/44/55	1026	30/37
8.	2¾ Classic Magnum	F13000	#00/12	48/42/87	1295	18/37
9.	3 Classic Magnum	F1314B	#4/41	164/57/34	1044	33/20
10.	3 Classic Magnum	F13100	#00/15	60/40/66	1095	13/21
11.	3 Classic Magnum	F131000	#000/10	40/36/90	1270	14/35
12.	2¾ Premium	P15400	#00/9	36/33/91	1190	27/75
13.	2¾ Premium Magnum	P1564B	#4/34	136/105/77	1197	64/47
14.	2¾ Premium Magnum	P15600	#00/12	48/46/95	1131	26/54
15.	3 Premium Magnum	P1584B	#4/41	164/121/73	995	73/44
16.	3 Premium Magnum	P1581B	#1/24	96/61/63	1185	38/39
17.	3 Premium Magnum	P15800	#00/15	60/50/83	1281	38/63
18.	3 Premium Magnum	P158000	#000/10	40/38/95	1367	30/75
19.	2¾ Tactical	H1324B	#4/27	108/92/85	1211	71/65
20.	2¾ Tactical	H13200	#00/9	36/36/100	1169	31/86

percenters and twenty-four were either #00 or #000. There were surely some bright spots for the small buckshot sizes (100 percent performance from Federal's Tactical #4s and Winchester's #0), but the definite trend was toward the bigger buckshot. So what is better, #00 or #000? I believe #000 should be the load of choice and I have several reasons for that belief.

We fired 150 four-shot groups and seventy were 90 percent patterns. Fifty-four of those groups were fired with the two larger shot sizes, thirty-seven with #00 and seventeen with #000. But there were three times as many #00 loads as #000, so #00 got more times at bat, so to speak. To continue the baseball analogy, #000 had a better batting average. In the Vang-Comp gun, six out of seven #000 loads were 90 percent and all of them were also 80 percent in the C-zone. The #00 loads placed over 90 percent of their pellets on the target eighteen times, but only half were also 80 percenters. In Full choke, all seven #000s were 90 percenters and four were C-zone 80 percenters. It was sixteen and seven for the #00s, respectively, in Full choke.

To put it another way and considering only patterning ability, #000 is better than #00 because it tends to put more pellets in a tight circle in the center. There are a lot more options in #00 and that might be a factor. The single best grouping #00 load was Remington's Law Enforcement Express, which grouped 97, 100 and 100 percent in Cylinder, Vang-Comp and Full chokes respectively, and 87, 100 and 100 percent in the C-zone. Still, the trend is for #000 to perform a little better in across-the-board pattern testing. But there are two more reasons why I feel #000 is a superior load. Each pellet is significantly larger and heavier than any other buckshot. That means more penetration in tissue and better terminal effect. It also means greater range. As an example, we fired a single #000 shot at a distant silhouette and saw every pellet hit the target, with four in the C-zone. That target was a full 65 yards away.

Several other points relative to buckshot performance came up in the course of our shooting. I can't completely prove it, but plated shot seems to perform a bit better than the plain lead type, particularly in a Full choke gun. Chokes seem to batter the larger pellets to some degree and the harder the shot, the less pellet deformation. Velocity also has an effect on patterning performance, but it also impacts felt recoil. The new low recoil ammo is a recent phenomenon and all of it

This is the result of a lot of shooting. The author (with help) fired 600 rounds over the chronograph screens at 150 targets. That's a lot of hole counting and figuring.

is loaded to reduced velocity. I would never have believed there was such a call for this kind of ammo, but people who currently train police officers are sky-high on the stuff. My friend Bob Hoelscher down in Miami has a special program going for the female officers and it's based on low-recoil loads. Whatever serves to get the pellets where they are needed is fine with me, so reduced recoil ammo has a definite place in the scheme of things.

I have long felt that #000 buckshot was the sleeper in shotgun ammunition and I am even more convinced

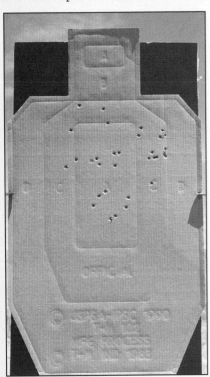

Exceptional performance was shown by the experimental XS load, which is a tactical low recoil type with eight #000 pellets. Four rounds kept all pellets in the C-zone at 25 yards.

The preferred combat load of the future? Eight #000 pellets loaded to 1100 fps for moderate recoil. The terminal ballistics would be horrendous: sort of like getting whacked with eight lightweight 38 Specials at once.

after this rather extensive exercise. A #000 hit has to be better than any other, and the ammo performs in such a way as to give more centered hits in any kind of gun, including those plain old Cylinder bores. There's plenty of reason to believe the #000 load in a good gun extends the gun's range by a good bit. Therefore, in my opinion, the ideal combat load would be hard-plated #000 in an eight-pellet low recoil load. Nobody makes that.

Or do they? I was able to put hands on a special experimental load I have chosen to call "XS" and which I can't identify as to source. It's all the things I just described: eight hard copper-plated .36-inch #000 pellets loaded to approximately 1100 fps. It's a sweetheart to shoot and we ran it through the same sequence as the other loads. In the Vang-Comp, XS was 100 percent in the C-zone and it was 96/87 in the Full choke gun. This is a tight-shooting, hard-hitting buckshot load and I hope you'll see it on the market soon. If not, I'd give serious thought to using one of the other #000 loads for whatever combat scenarios might come your way.

"Triple-ought" is the toughest buckshot of them all.

THE FORENSIC SCIENCE of firearms identification is one of those exotic, if not downright mysterious topics that seldom make their way into cocktail-party chit-chat. Like autopsy protocols, forensic "ballistics" is just not most folks' cup of tea, even among gun enthusiasts.

This state of affairs, however, might have changed drastically during the last three years if the victims in the O.J. Simpson murder case had been shot instead of butchered with a shooter's hand and other related matters, depending upon the type of firearm involved.

So what, then, is the forensic science of firearms identification, as practiced in most police crime labs across the U.S. and abroad these days?

As a fully recognized, legitimate part of the forensic sciences, firearms identification is a relatively young discipline. It developed over a period of several decades that began in the was no reliable means of demonstrating this fact. Expert firearms identification testimony from those long-ago days centered primarily on similarities as to caliber, rifling type and measurements between the rifling on the evidence bullet(s) and the rifling marks produced by the suspect gun. Although this wasn't by any means an entirely reliable way of establishing positive identification, it provided a good start in demonstrating a degree of *probability*.

Modern Gun Sleuthing

by J.I. GALAN

Forensic science has come a long way in 100 years, allowing investigators to prove—or disprove—firearms connections.

knife. Just think of the huge exposure that media coverage of the O.J. trials gave to the forensic disciplines of DNA, blood spatter and bloody shoe impressions. If a gun had been chosen instead to commit those gruesome murders, odds are that we would have been bombarded with months of testimony regarding shell-casing identification, powder patterns, wound ballistics, firearm function, trigger pull pressure, bullet trajectories, primer residue on the late 19th century. In many ways, firearms identification evolved as a result of certain events, through the efforts of a small but dedicated number of high-minded individuals determined to see truth and justice win out in criminal cases involving firearms.

The basic concept that rifled bores impart highly individual tool marks to bullets passing through them began to take shape among some criminal investigators of the late 1800s. However, at that time, there By the turn of the century, certain comparative techniques involving photomicrographs had been developed that, while more scientific in approach, were very time consuming. In addition, they still lacked the essential elements of precise magnification and three-dimensional viewing necessary for the detailed observation and comparison of the myriad microscopic *individual* or random markings transferred from the gun's bore to the surface of the bullet.

30th EDITION, 1998 **29**

A modern, comfortable and well-equipped work area is an ideal setting to conduct forensic firearms identification. A good reference library is a must.

These microscopic defects or characteristics are unique to each barrel, even consecutively rifled barrels, and, like fingerprints, can establish each barrel's individuality beyond a reasonable scientific doubt. Individual characteristics are initially created during the manufacturing and rifling process, as a result of boring, cutting or swaging the grooves, polishing, etc. The surface of the lands and grooves acquires literally thousands of random microscopic marks that are transferred to the softer copper or lead surface of each bullet fired through that barrel. Age and use can also add to the individuality of each gun to a certain degree.

Brass casings, likewise, acquire a reverse set of class and individual tool marks from such areas as the breech face, firing pin, extractor, ejector and even from the chamber walls. All of these marks are also unique to each firearm, even in guns coming consecutively off the assembly line at the factory. These are scientifically proven facts and constitute the foundation of modern firearms identification. Therefore, when a *qualified* firearms examiner testifies in court that an evidence bullet or casing was positively fired in a given gun, he is saying that no other gun in the world could have fired that evidence. This is, indeed, an ironclad expert opinion. Incidentally, the process whereby all those identifying tool marks are transferred to bullets and casings is called internal ballistics, a term seldom encountered in the popular gun literature.

The early years of this century provided ample opportunity for competent practitioners of firearms identification to prove themselves and their fledgling discipline. In 1915, the Stielow case in New York state was a landmark event in this regard. In that case, the testimony of a bogus "gun expert" was proven false through the efforts of scientific gun sleuths that included a fellow named Charles Waite, now regarded as one of the real pioneers of firearms identification. This led to the eventual exoneration of the two innocent men who had been wrongfully convicted of a murder they did not commit.

Encouraged by the results of this work, Waite associated himself with Col. Calvin Goddard, a noted firearms expert of those days; John Fisher, an expert on precision instruments; and Phillip Gravelle, a microscopist. Together, the four engaged in years of research and by 1925 had developed the modern technique for the identification of firearms through the use of the comparison microscope. In the process, they also founded the Bureau of Forensic Ballistics in New York City, offering their services to law enforcement agencies.

Two highly publicized murder cases that took place during the 1920s suddenly put firearms identification and some of its early practitioners in the center of the public eye. These were the Sacco/Vanzetti murder case in 1920 and the St. Valentine's Day

Most large crime laboratories nowadays have a comprehensive reference collection of firearms. (Left) The author inspects a World War II-vintage Japanese machinegun. (Right) In recent years, the distaff side has made sizable advances in this initially all-male profession.

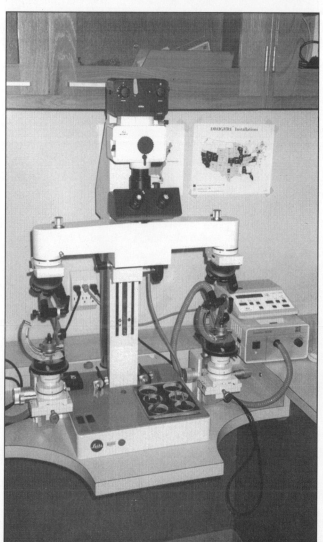

(Left) The comparison microscope is the principal tool of the firearms identification examiner.

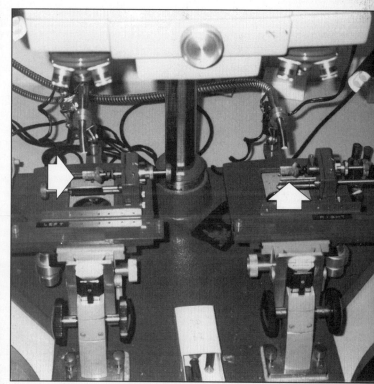

(Above) The evidence bullet is on the left spindle, while one of the test slugs fired in the lab from the suspect gun is placed on the right spindle. Both can then be viewed and compared together through the microscope.

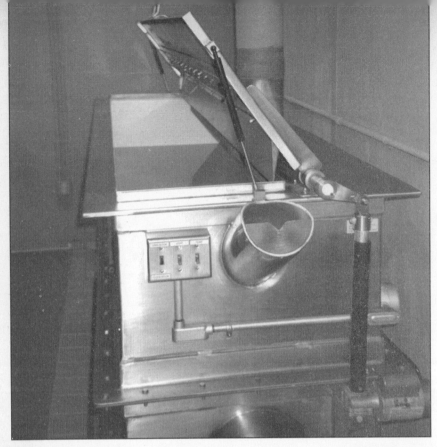

In order to obtain fired bullets with pristine rifling characteristics, guns are testfired into specially designed water tanks to prevent damage to the projectile.

Massacre in 1929. In the latter, Calvin Goddard's extensive work and detailed conclusions received worldwide attention. This led directly to the foundation of a scientific crime detection laboratory associated with the Northwestern University School of Law in Chicago. In addition, Goddard's work so impressed many foreign crime labs that the comparative method developed by him and his associates soon became the generally accepted technique, something that is still true today.

The firearms identification expert nowadays is called upon to examine and render expert opinions on a wide range of areas dealing with guns, ammunition and what is popularly and loosely called "ballistics." Besides determining whether or not bullets, casings and shotshells were fired from a specific gun, the examiner may be asked to opine on bullet trajectory (external ballistics), casing ejection patterns, gun-to-victim distance based on powder patterns on clothing and/or on the victim's skin, gun safety practices, silencers, and even wound ballistics. The latter is really the province of the Medical

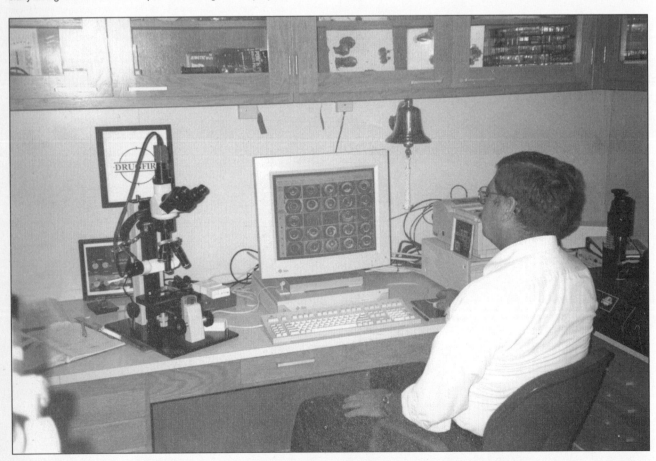

The Drugfire system is a tremendously effective new tool used by forensic firearms examiners in their task of helping to solve shooting-related crimes. The computer screen shows a selection of primers with distinctive firing pin impressions.

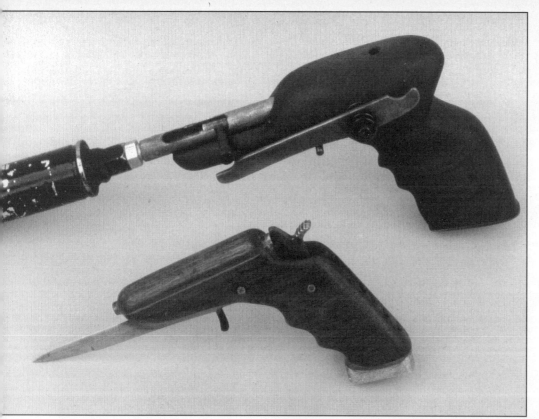

Confiscated homemade firearms (zip guns) are routinely brought to the crime lab for examination and evaluation. Extreme precautions must be taken when testfiring these devices. These examples show pretty fair workmanship and design.

(Below) The firearms examiner also must attempt to restore altered or obliterated serial numbers. There's a pretty good ratio of success.

Examiner, but there is certainly an overlapping of expertise here.

The individual who chooses this profession truly has his work cut out for him or her. This task cannot be taken lightly, either, as an identification involving, say, bullets and casings mistakenly "matched" to a given firearm may result in the wrongful conviction of an innocent person or, conversely, cause the real perpetrator to go free. In this crucial area, there's really no room for error, and the true professional is acutely aware of this. The wrong "call" in such a case will mean, in all likelihood, the end of the examiner's credibility as an expert witness in these matters.

A trainee in this field of endeavor usually undergoes a minimum of two years of on-the-job training and close scrutiny under the direct supervision of a senior examiner. Applicants, in most cases, must have at a minimum a four-year college degree, preferably in one of the physical sciences, just to be considered. During the training period, the individual will also attend as many specialized courses as possible. These include armorer courses offered by several major gun makers, such as Smith & Wesson, Ruger, Beretta and others. There are also courses in different areas of firearms identification offered by the FBI Academy in Quantico, Virginia. Attending at least a couple of pertinent FBI courses truly enhances the firearms examiner's expertise and professional credentials. In fact, training is really an ongoing process in this field, as the examiner—whether rookie or seasoned practitioner—must stay current regarding new guns and ammunition as well as the latest instrumentation and techniques.

Typically, the firearms examiner may be exposed to an incredible variety of firearms and ammunition employed in all manner of criminal activity, from first-degree murder all the way down to non-shooting incidents. This is particularly the case for examiners working for large metropolitan police agencies. In the course

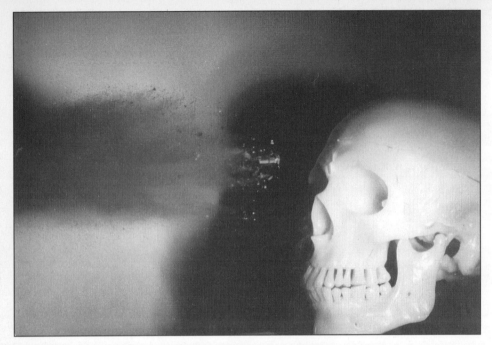

Bullets passing through glass and other objects can transfer a significant amount of foreign materials to the victim. Note the cloud of matter following this bullet.

of a single day, he or she may examine and testfire everything from pristine top-name guns to rickety relics and even homemade zip guns. In some crime labs, the examiner is also responsible for restoring serial numbers altered or removed from firearms. This is accomplished through the use of techniques involving certain acids. The results, in many cases, are truly impressive.

At this time, firearms examiners number somewhere around 700 to 800 in the U.S., with perhaps another 300 or so around the world. As can be readily seen, this is a rather limited and exclusive club. Most of these individuals belong to the prestigious Association of Firearm and Toolmark Examiners (AFTE). This organization, founded in 1969, has a weeklong annual convention that attracts a sizable number of members from across the U.S. as well as from several other countries. A wide range of technical and scientific lectures are presented during that annual meeting, with a great deal of new and useful exchange of information taking place among the attending members. In addition, the quarterly *AFTE Journal*, likewise, is always full of timely technical articles of special interest to practicing examiners.

One of the most effective crime-fighting resources made available to the forensic firearms examiner in recent years is called Drugfire. This FBI-sponsored program is, in a nutshell, a database-driven multimedia imaging system intended mostly for fired cartridge casings retrieved from shooting scenes. The monicker Drugfire apparently was chosen because of the large percentage of drug-related drive-by shootings that occur in many large cities. As a result, the vast majority of casing images put in the Drugfire database are from self-loading firearms, although other repeaters that normally eject their fired casings on the scene are also entered.

Drugfire basically takes a picture of the primer with the firing pin/breech face impressions and stores it in its computer database. Thus, within each caliber group there is a sizable open-case file instantly accessible to examiners for comparison against new entries, as well as against tests fired in guns coming into police custody. The system also allows interjurisdictional sharing of database information and images. In other words, labs within a given state in the Drugfire network are able to review the entire open-case database

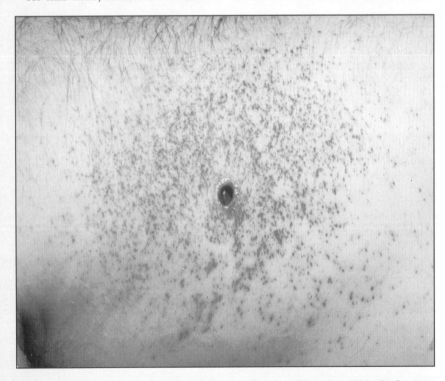

This bullet entrance wound in the side of the chest reveals a rather dense "stippling" pattern of gunshot residues that indicates: (1) the victim wore no shirt and (2) the range was probably no more than about 18 inches.

(Left) This firing pin impression on a rimfire casing reveals an abundance of individual characteristics that'll make it easy to obtain a match.

(Above) This peculiar firing pin impression is typical of Glock and S&W Sigma pistols. That can narrow the field considerably.

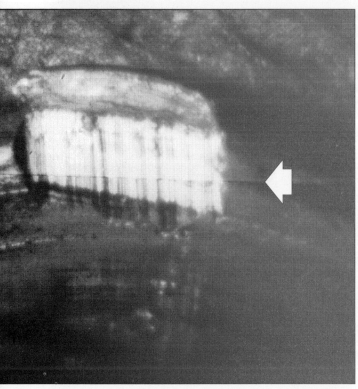

Two fired 9mm casings matched through their respective extractor marks. Arrow indicates photo dividing line.

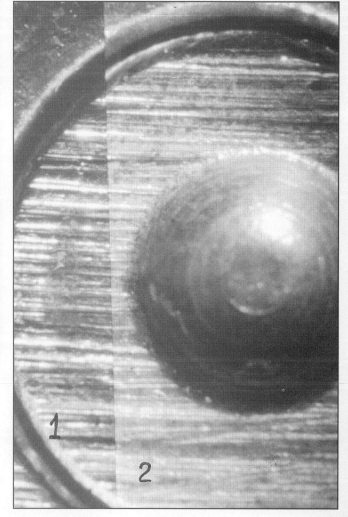

Viewed through the comparison microscope, casings 1 and 2 (the dividing line can be clearly seen) appear almost as one single object. The individual characteristics make this a textbook match.

for that state. The results are often spectacular, as shootings that are seemingly unrelated are quickly shown to be linked—and in many cases solved outright—through the use of Drugfire.

Regardless of all the computer systems and other state-of-the-art equipment available to him or her, the bottom line is that the firearms identification examiner's professionalism and impartiality must be of the highest order. His findings and opinions must be based on scientific methods, going no further than what the evidence allows, and his court testimony must always reflect those principles. •

30th EDITION, 1998 **35**

by CHARLES W. KARWAN

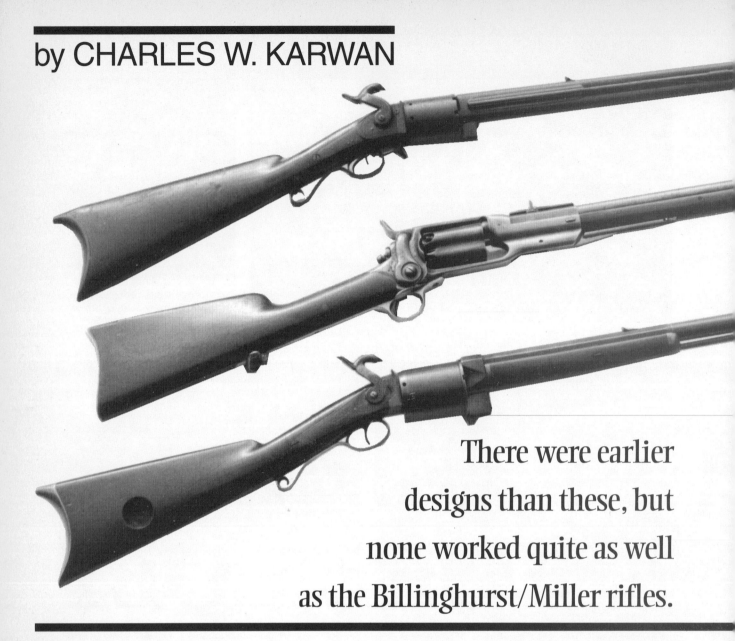

There were earlier designs than these, but none worked quite as well as the Billinghurst/Miller rifles.

IF YOU ASK most gun enthusiasts who designed the first practical revolving firearms, the answer in most cases would be Sam Colt. It is well known that there were a number of crude and sometimes pretty well-made revolvers built in the matchlock, wheellock and flintlock eras, decades or even centuries before Colt produced his guns. However, for one reason or another, usually centered on their ignition or cylinder indexing systems, they were not terribly practical. This, combined with the obvious marketing success of the Colt line, results in the perception that Colt made the first practical revolvers.

To be honest, that was also my belief until the error of my ways was pointed out by my friend Al Perry when he showed me a couple of interesting revolving rifles in his collection. Al is known as a "gun guru" in our area of Oregon, and I have to admit it is a rare visit with him that I don't learn something about one gun or another. The pieces Al showed me were extremely practical revolving rifles that pre-dated Colt's efforts by at least ten years.

Subsequent research revealed to me that there were quite a number of revolving firearms, particularly rifles, designed and made in the U.S. well before Colt came up with his. Probably the best designed, best manufactured and most successful of these pre-Colt revolving rifles was the one introduced to me by Al Perry. These were the revolving rifles produced by famed New York rifle maker William Billinghurst under the 1829 patent of James Miller.

In the mid-1820s, James Miller, of Rochester, with the help of his brother John, developed a clever revolving rifle that was decidedly practical and actually had several features superior to the Colt revolving rifles that would come later. It was officially patented in 1829, but, unfortunately, the original patent was destroyed in a Patent Office fire the same year. Fortunately for collectors and historians, a copy of the original patent surfaced in recent times so that we know its details.

The main features of the patent center on the revolving cylinder and the stout locking mechanism that indexes the cylinder in each firing position. This cylinder locking mechanism is located at the lower front of the cylinder and engages substantial notches in the end of the cylinder. So the notches fall between

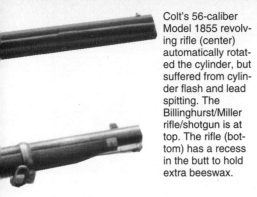

Colt's 56-caliber Model 1855 revolving rifle (center) automatically rotated the cylinder, but suffered from cylinder flash and lead spitting. The Billinghurst/Miller rifle/shotgun is at top. The rifle (bottom) has a recess in the butt to hold extra beeswax.

The Billinghurst cylinder has the seven chambers and locking notches between them. This is a simple and strong system.

PRACTICAL REVOLVERS BEFORE COLT

the chambers of the cylinder, the cylinders on these guns will invariably have an odd number of chambers, typically five, seven or nine depending on caliber. Calibers seem to vary from a small of about 34 to a large of about 45.

While Miller made some of these rifles under the name of J.&J. Miller, it fell to the shop of the very capable William Billinghurst to produce most of the Miller-patent revolving rifles. Billinghurst is best known for his manufacture of superbly accurate under-hammer target and benchrest match rifles. He operated a shop with several skilled employees in Rochester, New York, not far from that of James Miller. Miller's rifle design required a great deal of precision in its manufacture, just the type of precision that Billinghurst was known for. It is likely that the Miller brothers turned to Billinghurst for much of the work to make the early examples of the rifle, and that Billinghurst took over the production to help meet the demand for the rifles. Indeed, Billinghurst was by far the most prolific manufacturer of the Miller-patent revolving rifles. Consequently, this style of revolving rifle is commonly known among collectors as a Billinghurst type rather than by the name of its inventor Miller.

While the Miller-patent rifles were never mass-produced, they were quite successful. Demand was sufficient that they were made by at least thirteen different gunsmiths besides Miller and Billinghurst. Most of these gunsmiths were located in upstate New York, indicating a strong regional demand for the rifles. Two makers of Billinghurst/Miller rifles were former Billinghurst employees. One of these was Benjamin Bigelow, who took his skill West and set up shop in northern California. Bigelow is considered by some authorities as the second most prolific manufacturer of this type of rifle.

The other former Billinghurst employee was W.H. Smith, who worked for Billinghurst for many years and whose name is often found on the bottom of the barrels of Billinghurst rifles. Close inspection of one of Al Perry's Billinghurst revolving rifles shows Smith's name. Smith went on to set up shop for himself in New York City, where he continued to make some of the Miller-patent revolving rifles for his own customers.

It is important to realize that at

30th EDITION, 1998 **37**

(Above) Two of the rare Billinghurst revolving rifles that pre-dated Colt's efforts by as much as a decade. On top is the super-rare combination version with a shotgun barrel that serves as the axis for the cylinder.

This view clearly shows the holes in the cylinder where the priming pills were placed and sealed in with wax. Below the barrel, in front of the cylinder, is the robust latch that indexes into the large notches and locks the cylinder in position for firing.

the time the Miller rifle was introduced there were no large commercial gun companies like Remington, Colt or Winchester that mass-produced firearms for the civilian market. Virtually all of the firearms made for civilian consumption were made in small shops, and many of these were made to order. Thus, the fact that these Miller-patent revolving rifles were made in at least fifteen different shops would indicate that they were in quite high demand and popular with those who could afford them.

Undoubtedly, the main reason they were not more popular was because they were much more difficult and expensive to make than a simple muzzle-loading rifle. This period was one where there was a lot of activity with Westward expansion and exploration. It does not require much imagination to see the tremendous advantage to having a five-, seven- or nine-shot repeating rifle instead of a single shot muzzleloader, particularly if being attacked by a force that outnumbers your own.

The majority of the Billinghurst/Miller revolving rifles use a pill-lock ignition system. This was before the percussion cap had achieved dominance in the firearms of the time. Each chamber of the cylinder had a small cavity in its rear that vented to the charge in its respective chamber. First, the chambers were loaded from the front either through the muzzle or the front of the cylinder with a powder charge and a patched ball. Then a percussion-sensitive priming "pill" of fulminate was dropped into the cavity at the rear of each chamber and sealed into position with a dab of beeswax. With the front of the chambers sealed with grease, and the ignition pills sealed with wax, the shooter had a waterproof firing system. This was a huge advantage during a period when most people carried flintlock firearms that were very sensitive to rain and damp weather.

The nose of the rifle's hammer has a sort of firing pin projecting from it. On pulling the trigger and releasing the hammer to fall, this projection enters the priming hole to smash the ignition pill. This caused it to explode and in turn ignite the powder charge in the chamber. To fire the next shot, the shooter must first recock the hammer into either the half-cock or full-cock position. Then he pushes forward on the cylinder lock, manually turns the cylinder to the next firing position, and releases the lock to engage the cylinder notch. I estimate that a coordinated person could achieve a rate of fire as fast as one shot every five seconds until the rifle was empty. Once the cylinder was empty, the shooter could continue to load and fire the weapon as a single shot muzzleloader until a pause in the action would allow recharging and repriming the chambers of the cylinder.

Some of the Billinghurst/Miller rifles will be found set up to use per-

(Above) The rare Billinghurst revolving rifle/shotgun combination was a very versatile and effective weapon for its time. Functionally, it was identical to its revolving rifle brother, except for the addition of the shotgun barrel, the underhammer, and the trigger to fire it.

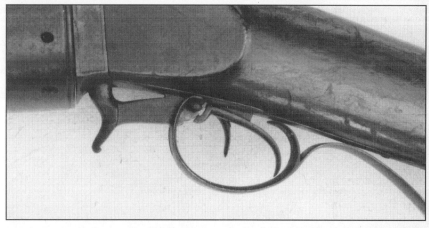

Left view of the Billinghurst combination gun. Note the extremely simple lockwork for the shotgun hammer. The trigger guard is the hammer spring, and the rear trigger engages a notch in the hammer.

cussion caps instead of the priming pills. These could be conversions of the earlier system or later production rifles originally made to take percussion caps. Having the percussion nipples and caps protruding from the periphery of the cylinder would make the rifle much more awkward to handle than the original pill-lock system.

Colt's only significant improvement to this system in his 1836 patent was to have the action of cocking the hammer also turn the cylinder to the next charge and lock it in position. This speeds up the rate of fire, but also makes the mechanism much more complicated and fragile. It also requires a larger cylinder gap to allow the cylinder to turn freely once the blackpowder fouling starts to build up. With the Billinghurst rifle, the cylinder-to-barrel gap can be adjusted down to nearly zero and still be rotated manually without difficulty. This minimizes the cylinder flash and spitting at the cylinder gap, the main objection to all the Colt revolving rifles. The cylinder of the Billinghurst rifles typically have fine grooves front and rear around the circumference of the cylinder to aid in gripping the cylinder when turning it. The Billinghurst mechanism even allows for wear compensation through the use of a tapered cylinder pin and a threaded nut.

Some of the early Billinghurst rifles have the chambers of the cylinder rifled to match that of the barrel. To accomplish this, they must have rifled the barrel and the chambers at the same time. Obviously each rifling cutter pass would have to be repeated for each chamber, which added greatly to the time and expense of making one of these rifles. Once it was realized that rifling the chambers to match the barrel was not necessary, production was simplified and cylinders could be built with blind chambers rather than being bored through then threaded and plugged.

My friend Al had another surprise for me because one of his Billinghurst rifles has two barrels. The upper one is a rifle barrel that the chambers of the cylinder lined up with, while the lower barrel was a 44-caliber shotgun barrel that also served double duty as the pivot for the cylinder of the rifle. This is basically the same system used much later by LeMat in his famous "grapeshot" revolving pistols of Civil War fame. The main difference is that LeMat used one hammer with a selector on the nose that would fire either a cylinder chamber or the shotgun barrel, depending on the position of the hammer nose, while Billinghurst used a separate hammer for the shotgun barrel.

Not surprising considering Billinghurst's considerable experience with under-hammer rifles, he used a classic under-hammer action for the shotgun barrel using the trigger guard as the hammer spring, as is typically done with that system. As a result, there are two hammers and two triggers.

The Billinghurst revolving rifle partially disassembled. Note the robust simplicity, the tapered arbor that the cylinder rotates on, and the threaded nut that allows adjustment for wear.

The top hammer and the front trigger fire the rifle, and the bottom hammer and the rear trigger fire the shotgun barrel. There are also two ramrods, one on each side of the barrels, one for the rifle and one for the shotgun.

On Al's gun, the shotgun barrel uses a percussion cap, while the rifle part uses priming pills as described earlier. This is probably because the priming pills would be difficult to use in the underside position considering the distance the shotgun chamber is inside the gun. Since LeMat's patent is dated 1856, it would appear that Billinghurst used a shotgun barrel as the axis for a revolving cylinder some twenty-five years earlier—yet few know it.

The only other gunmaker to make a Miller patent-type revolving rifle with a shotgun barrel as the cylinder pivot, beside Billinghurst, was his former employee Bigelow, who appears to have also built a few after he moved West. I can imagine that they were a big hit.

What an incredibly versatile weapon one of these double-barreled Billinghurst revolving rifle/shotgun combinations must have been. The average person of the period carried either a single shot rifle or a single shot smoothbore musket or fowler for hunting and for defense. Here in one gun was a practical repeating rifle and a practical smoothbore shotgun. One extremely important advantage was that in a fight, when both the cylinder and shotgun barrel were empty, the shooter could reload the shotgun barrel with either shot or ball quicker than any muzzle-loading rifle could be reloaded.

It is pretty clear that the Billinghurst rifle/shotgun was the "assault rifle" of its day. It was both a versatile hunting tool and an awesome weapon. Few were made mostly because they had to be extremely expensive. Needless to say, surviving specimens are extremely valuable and hard to find.

It appears that Billinghurst and other gunsmiths continued to make the Miller-patent revolving rifles nearly up to the Civil War. You have to remember that Colt's first company in Paterson, New Jersey, went out of business rather quickly and produced only a limited number of revolving rifles between 1837 and 1841. It was not until 1855 that Colt revolving rifles became available again, so anyone wanting a practical repeating rifle up until that time had to turn to gunmakers like Billinghurst and his contemporaries.

Once the Spencer and Henry cartridge repeaters were introduced in the early 1860s, the revolving rifle was obsolescent and well on the way out. The demand for the Billinghurst-type rifles disappeared.

The Billinghurst revolving rifles and combination guns were made during a very exciting phase in American history. This was the time of Western expansion, the mountain man, the Texas fight for independence, the Oregon Trail, the Mexican War, and the California Gold Rush. Considering the advantages of having a practical repeating rifle during those trying times, it does not surprise me that the survival rate of these rifles is quite small. I am sure that most just plain got used up. Even the ones that survived that period were subject to being tossed away once more practical repeaters showed up. However, I strongly suspect that the Billinghurst/Miller revolving rifles were more popular and saw more use than current histories would have us believe.

It is also important to give Miller and Billinghurst the credit they deserve. While Colt and LeMat are well-known names in the firearms field, Miller and Billinghurst pre-dated them by many years with a practical revolving rifle and a practical revolving rifle/shotgun combination. Probably the major reason they are not as well known is that their efforts were in the civilian market, while Colt's and LeMat's were in the much larger military market. Maybe this little treatise will help correct that oversight.

●

The Good, The Bad— And The Ugly

Firearms engraving is an art form unto itself, but basic competence with hammer and chisel is a must. Here are some tips on what to look for and examples of three types of work.

by DOC O'MEARA

NOT LONG AGO a friend of a friend came to my table at a gun show so I could look at a revolver that he had just had engraved. He was delighted with how his 6-inch Taurus 357 Magnum turned out and waited expectantly as I examined it. A few seconds of gazing at it under a magnifying glass and I'd seen enough. I passed it back to him with a slight smile and a noncommittal, "That's nice." He beamed and went excitedly on his way to show it to others. I simply shook my head in sorrow as he disappeared into the crowd.

I just didn't have the heart to burst his bubble. Some local "artist" had scratched some designs on it and called it engraving. It might be considered art if you truly believe that the crayon renderings of your six year old, pasted to the refrigerator door, also qualify as art. But it's like the old joke about the man who bought his first yacht and called himself its Captain. His mother asked, "By you you're a Captain, by me you're a Captain, but by a Captain are you a Captain?"

On another occasion an acquaintance brought a Model 21 Winchester shotgun to the house for me see. The scrollwork had been impeccably executed in the small, tight, English style and was beautifully accented by color case-hardening. Gold border inlays formed a frame along the sides of the receiver. On the right side there was inlaid a hunter shooting at a rising pheasant, with a pair of setters watching for the hit and fall, so they could mark and retrieve the bird. These had been nicely done. The observer could actually feel as though he were a part of the scene.

The Good

The relatively deep cuts on the cylinder of this Bisley Model Colt revolver, by Ben Shostle, could be regarded as bold by some critics and heavy-handed by others. This is a good illustration of the part played by personal taste in judging such work.

Victorian-style scrollwork with well-executed gold border inlay on this Model 1849 Pocket Model Colt revolver matches the fashion of its historical period. The skills of the engraver, in this instance, Sam Welch, of Moab, Utah, can sometimes give a worn antique a fresh, new lease on life.

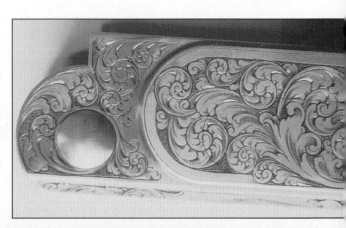

The scrollwork on this shotgun receiver was beautifully executed. Close examination shows that the continuity from the body of the receiver to the sideplate is very fluid. Even though the curves in the center of the three largest scroll patterns show minor angularity, this is still very nice work.

But on the left side, a pointer, having marked the location of a covey of quail hiding behind some brush looked expectantly over his shoulder at his master, waiting for him to move in for the flush. Here, the beauty of the gun was ruined. The quail looked more like guinea fowl and the only way the dog might have assumed the position in which he was depicted would be if his neck were broken.

One of the most striking things that one realizes when examining engraved firearms of many periods during the past century and a half, or so, is that the art is as evolutionary as the guns upon which it has been practiced. For example, works by such artists as Sanford and the Ulrich brothers, that were considered superb examples in the 1860s, would be regarded as not just amateurish, but downright cloddish by today's standards. Often, even the designs are simplistic and crude. Yet, there are a few examples from that period and the next few decades that would be accepted as masterful in any age. The works of such 19th century masters as Gustav Young and Cuno Helfrecht are excellent examples.

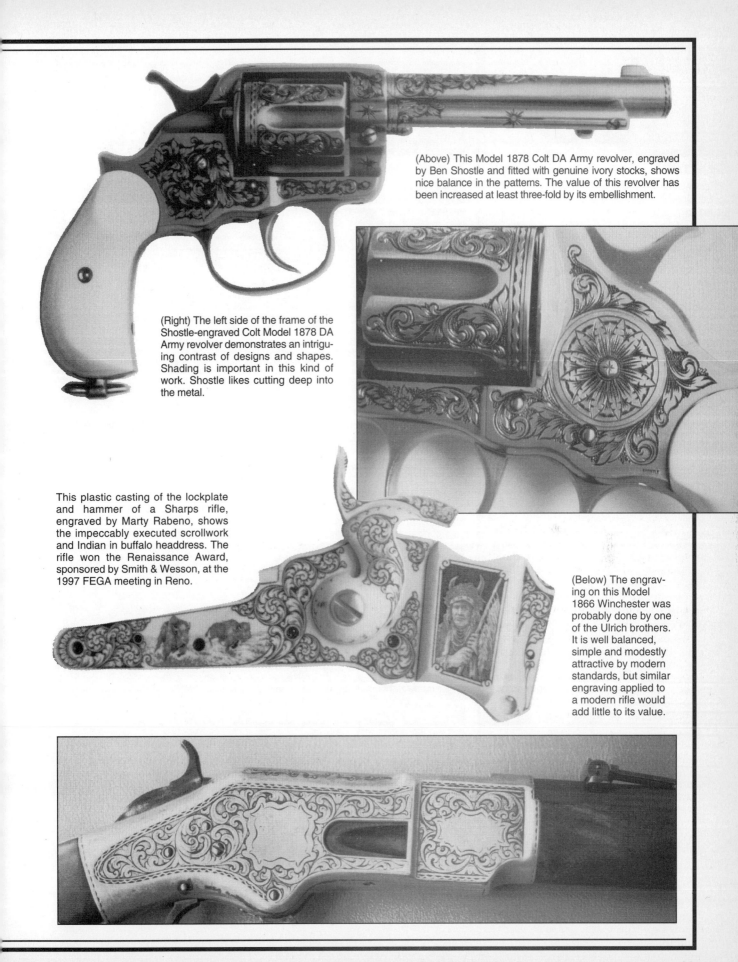

(Above) This Model 1878 Colt DA Army revolver, engraved by Ben Shostle and fitted with genuine ivory stocks, shows nice balance in the patterns. The value of this revolver has been increased at least three-fold by its embellishment.

(Right) The left side of the frame of the Shostle-engraved Colt Model 1878 DA Army revolver demonstrates an intriguing contrast of designs and shapes. Shading is important in this kind of work. Shostle likes cutting deep into the metal.

This plastic casting of the lockplate and hammer of a Sharps rifle, engraved by Marty Rabeno, shows the impeccably executed scrollwork and Indian in buffalo headdress. The rifle won the Renaissance Award, sponsored by Smith & Wesson, at the 1997 FEGA meeting in Reno.

(Below) The engraving on this Model 1866 Winchester was probably done by one of the Ulrich brothers. It is well balanced, simple and modestly attractive by modern standards, but similar engraving applied to a modern rifle would add little to its value.

The Bad

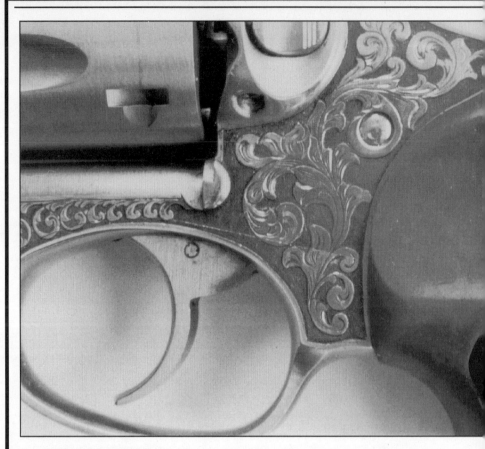

Seen in its entirety, this Ruger SP-101 isn't *too* bad. However, close scrutiny shows serious angularity to the curves of the scrollwork, and the whole thing looks stilted and amateurish. High-relief scroll such as this is especially difficult to execute with precision.

Somewhat stylized, this bighorn sheep lacks any anatomical detail. That might be all right if the borders surrounding it were not so uneven, the fleur-de-lis wasn't tilted and off center, and the ornaments at the four corners weren't so flat. This isn't high-quality work.

Frankly, I do not profess to be an authority on the subject, and I couldn't begin to practice the engraver's art myself. I have neither the digital dexterity, nor the artistic talent. But over the years I've come to know quite a number of professional engravers and have had the opportunity to examine their work and that of many others, on contemporary firearms and antiques. These talented artisans have shown me things to look for in order to make judgements as to the quality of an individual work. I'd like to share their insights with you.

First, let's consider cost. That will depend upon such factors as the amount and style of coverage to be applied to the gun. The more complex designs will, of course, require more of the artist's time and skill and, therefore, cost more than a simple scroll pattern. Light or modest coverage will cost less than full coverage. Add border inlays of silver, gold or platinum and the cost of the precious metals is added to the expense of installing them. Include figured inlays of animals, people or symbols and the amount to be spent will increase.

This is a simple pattern with excellent potential, but the scrollwork is, nevertheless, amateurish. However, coaching and training could turn the novice who did this into a real artist. Note that the curves are pretty even and there is good symmetry. There is potential here.

This is a mixed bag of varied merit. The scrollwork is nicely done, and the fighting moose aren't too bad. However, the grizzly bear at right lacks definition, and the mule deer at left has a body more in keeping with that of a caribou—or a goat. Animals should look like what they really are.

The lack of muscular definition of this "mountain lion" makes it look more like a road kill than a live animal. The accompanying scrollwork, however, is even more crude. The depth and width of the cuts vary radically, making what might otherwise be an attractive rifle magazine plate look...bad.

The scrollwork on this plate is the work of a talented student. There are many places where it is obvious that the individual stopped and started again. Vines vary considerably in their width, and curves are marred with numerous angles. However, the innate talent of the budding artist is illustrated by the balance and integrity of the design.

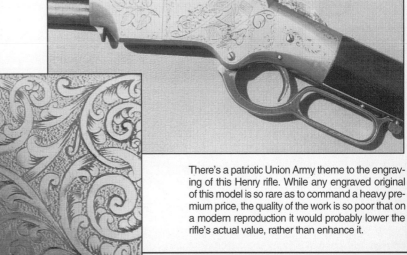

There's a patriotic Union Army theme to the engraving of this Henry rifle. While any engraved original of this model is so rare as to command a heavy premium price, the quality of the work is so poor that on a modern reproduction it would probably lower the rifle's actual value, rather than enhance it.

The Ugly

The hair of the mountain goat on the sideplate of this S&W revolver looks as though it would require a lawnmower rather than a clipping shear to cut. It's the early work of an artisan whose work is highly regarded today.

A storm god? Father Wind? Not being able to determine exactly what the simplistic central figure represents is just the beginning of this piece's problems. The scrollwork, with its elongated curves, is crude, at best. All together, it's quite bad.

Some gun makers offer embellished firearms right from the factory, and a check of the price list for their engraving services will give you a pretty fair idea of a starting point. But if you are dealing with an independent artisan, charges may vary widely, depending upon his skill and reputation. The better known engravers, whose accomplishments are well recognized and whose work is highly sought after, can justify their worth and will not work cheap. As it should be. "A workman is worthy of his hire."

Assuming that the artist whom you engage does a good job, what will your embellished firearm be worth? A good rule of thumb is to add together the cost of the gun, the cost of the engraving, plus 30 percent, to establish a fair retail price. If the work is done by a highly skilled professional with a widespread reputation, and whose work is in great demand, its value may be even greater. Market forces will dictate prices in that case.

Styles vary considerably. The depth of cuts in some examples may

It's supposed to be a wolf, but you have to look closely to be sure it's not a raccoon. Engraving cartoon-like animals onto firearms should only be done if the owner really wants it decorated with a Disney or Loony-Tunes character. Even the stippled background is poorly done.

(Right) This is a very simple scroll design that's too large to be applied to a firearm. The job shows uneven cuts, angular curves and lack of detail. It's very amateurish. The student who did this plate has a long way to go before quitting his day job.

This inlay looks more like a duck than the other bird nearby, but the wings are shaped incorrectly. The borders are crudely done and the cattail plant background is too simplistic. Overall, it's poorly done and would certainly lower the value of a gun so treated.

This is supposed to be a duck, but the wings are radically high and the shape of the body is more consistent with that of a cormorant. Birds and animals should be made to look like the creatures they are supposed to represent. Pretty poor quality here.

30th EDITION, 1998 **47**

seem heavy-handed to some critics. Others will consider the same work "bold." By the same token, the variety of patterns found on some examples may give the impression of clutter. Personal taste will dictate preference.

The basics are really quite simple. Angles should be sharp and precise, curves should be rounded and without corners or jaggedness. Cuts should be smooth and continuous, showing no indication of the places where the cut starts and stops as the tool carves through the metal. Animals must be anatomically correct in their musculature, and sized properly in relation to the background.

While one may be justifiably critical of the execution of a piece, or particular parts of it, it is important to view the gun in its entirety. Even if there are obvious faults it may be that the overall effect remains sufficiently attractive that it is worthwhile. Essentially, this amounts to looking at the proverbial forest rather than the trees. For example, a design pattern may be simple to the point of being cliché, but if it's well balanced and well executed, it may be very much worth having.

These factors are best illustrated rather than spoken of. For that reason, much of this piece will take the form of a photo essay. At the same time, the names of the engravers associated with some of the works shown here are being withheld at the request of the individuals responsible, to avoid embarrassment. Many prefer not to be connected with their early efforts, because the quality of work they now deliver is so much improved compared to that which they did as novices.

One of the most important considerations to be understood is that each artist is an individual, one whose work has its own unique characteristics and is different from all others. Just as you might prefer the paintings of Raphael over those of Da Vinci, or the sculptures of Michelangelo rather than Rodin, you may find the engraving of Marty Rabeno preferable to that of Paul Mobley or vice versa. As in any art form, preferences are often a matter of taste, rather than of the competence of the artist.

Another matter of importance is the terminology used in describing the skill level of the artisan. In Europe, engravers are usually trained in much the same fashion as was done under the guild system developed hundreds of years ago. An apprentice was taught by the journeyman and the journeyman by the master. When the journeyman reached a sufficient level of skill, in whatever trade in which he might be training, the master and his peers would examine his work and declare the individual to be, himself, a master. This system remains in place in parts of Europe to this day. An unfortunate consequence of such programs of instruction is that the judgement of one's peers is often based upon how well the student mimics the teacher's work. The end result may be an art form that stagnates. The result is a sameness of the product that, while it may be beautifully executed, can become boring.

In the United States engraving is largely a self-taught skill. There are, to be sure, a few gunsmithing schools and engraving houses that provide formal training, but few of the best known American practitioners of this art form have availed themselves of this sort of instruction. Most begin by drawing, then applying their ideas to metal. With coaching from others in the trade, and a long period of trial and error, a few develop a level of proficiency that comes to be recognized for its superb quality. Others never quite make the grade. But one of the most meaningful aspects of American engraving and the products of its better artisans is a sense of freedom of expression that is virtually unknown in Europe.

Because there is little in the way of formal training available to engravers in this country and none of the European-style schooling, the Firearms Engravers Guild of America (FEGA) has come to use a different terminology in describing its members. While many of the major firearms manufacturers continue to refer to the artisans in their employ as journeymen or masters, the Guild refers to its members in terms of their status within the organization, using the terms, Associate, defined as individuals and companies who follow the events and development of the organization; Regular, either full- or part-time engravers who are not required to meet specific levels of skill; and Professional, the status achieved by review of one's peers that for all intents and purposes equates to the term, Master.

However, FEGA studiously avoids the use of the term, Master, and frowns upon its use by its members. But in my opinion, if the organization is willing to refer to itself as a guild, it would only be reasonable to use the terms associated with such an institution. One good reason for which is that there are few members who by any standard are, indeed, masters of their craft.

Whether or not you agree with that opinion, there is one indisputable fact: Engraving is a skill that requires levels of artistic sensibility, skill and manual dexterity quite as demanding as any other graphic art. But, unlike many other art forms, gun engraving is art for the individual, the owner of the gun upon which it has been rendered, rather than for the masses.

As you look over the nearby photos, remember that we are showing engraving that represents the spectrum of work you'll likely see at gun shows, in friends' collections, and, for the older guns, in museums. There is good, bad, and some ugly work here. Our intent is to show you the differences and, perhaps, you'll be more critical when it comes time to buy an engraved gun of your own. But make no mistake; it truly is one of the great art forms.

> While one may be justifiably critical of the execution of a piece, or particular parts of it, it is important to view the gun in its entirety. Even if there are obvious faults, it may be that the overall effect remains sufficiently attractive that it is worthwhile.

MACHINE PISTOLS

Inventor's Fantasy Or Serious Fighting Weapon?

Once thought useful, the intriguing machine pistol has its place in specialized applications.

by CHUCK TAYLOR

WITHOUT QUESTION, few hand-held firearms have attained the controversial status of the machine pistol (MP). Much cussed and discussed in endless arguments between military personnel, police officers and civilian gun buffs alike, this type of gun continues to ride a wave of shooter consciousness that has remained vehement for more than 75 years.

First, what is a so-called "machine pistol?" Many, even some who claim the title of "expert," lack this critical understanding. Simply put, it is a handgun, sometimes fitted with a detachable buttstock/holster, sometimes not, that is capable of fully automatic fire via a selector switch. Because of its very definition, a machine pistol is not a submachine gun, though some less well informed writers have often categorized the two together.

Their error comes from the mistaken notion that because both weapons utilize a pistol cartridge, they're one and the same. Not so. The machine pistol is a *handgun*, while the submachine gun (SMG) is a *shoulder weapon* more resembling a carbine. Mechanically and tactically, they bear little or no relation to each other.

The mistake is compounded by a German ruse used in World War I to

Though they first appeared more than eighty years ago and have periodically resurfaced since, the machine pistol has failed to achieve any degree of proliferation as a serious weapon. Here, an NCO from the 7th Special Forces Group at Ft. Bragg, North Carolina, fires the Star Model M 45 ACP. Note the severe muzzle climb, one of the machine pistol's most serious drawbacks.

mask that nation's pioneering SMG research and development program from Allied intelligence agents. Since the MP had already existed for over a decade and had already been rejected by virtually every nation in Europe as being unsatisfactory for military needs, they were able to successfully conceal their SMG program by calling it a "maschinen pistolen."

The designator "MP" had nothing whatsoever to do with the weapon itself, which was actually a submachine gun—the Schmeisser-designed 9mm MP-18 to be precise.

If you doubt this, consider that they did *exactly* the same thing twenty years later during World War II to successfully conceal their assault rifle program, resulting in a weapon first known as the *MP*-44, then, finally, the famous 7.92mm StG-44—the world's first assault rifle!

Some say the machine pistol is worthless, that it possesses all the weaknesses of both the handgun and submachine gun, but none of their assets. Others say that they're a terrific weapon, lighter and more compact than a SMG, but more tactically versatile than a standard handgun. Its detractors cite a long list of weaknesses, only to be countered by its proponents with an equally long list of its strong points.

Who's right? Well, from my viewpoint, it depends entirely upon your perspective. As compared to a true submachine gun, the MP is without question the weaker weapon. Being a form of handgun itself, a machine

The slide-mounted selector switch of this Star Model M 45 ACP is in the up or semi-automatic position. The down position would allow the switch to contact the sear-trip protruding upward from stock panel, allowing fully automatic fire.

pistol is certainly less effective than a SMG. It has a considerably shorter sight radius than a submachine gun, magnifying sight alignment error and making it more difficult to shoot well under stress. As well, the velocities its shorter barrel produces reduce its stopping power and severely limit the expansion potential of frangible bullets.

If used without a buttstock in the fully automatic mode, it is much less controllable than a SMG, resulting in lower hit ratios and causing what many consider to be a tactically unacceptable waste of ammunition and, in so doing, an equally unacceptable level of civil liability.

If used with a buttstock attached, the MP's controllability and accuracy in both the semi- and full-auto modes are dramatically increased, although not to the point where they rival the SMG. On the other hand, a machine pistol so configured is certainly more accurate and controllable than a standard handgun. To this, detractors of the machine pistol usually respond that the buttstock normally doubles as a holster, thus making it relatively inaccessible in the field because it must be removed from the belt, a time-consuming process to say the least.

As you can see, the argument seesaws back and forth, with each side's viewpoint being valid, if considered solely from an academic perspective. So, how do we break the deadlock? Well, first let's define the missions of each weapon system. It seems to me that we really can't evaluate or compare them until this task is complete. The mission, the very purpose of the handgun is to provide its wearer with the means by which to regain control of his immediate surroundings when attacked. It *must* be capable of satisfying this requirement within extremely short time frames, and with as few shots fired as possible, preferably one per target.

Therefore, we can say that the handgun is a close-range, *reactive*, easily concealed *defensive* weapon, intended to protect its wearer from unexpected attack, and its inherent physical characteristics support this conclusion. It is the smallest, lightest, least powerful of the small arms and is the most difficult to use well, making its tactical employment quite limited.

The submachine gun, though it also utilizes a pistol cartridge, is primarily an *offensive* arm. It has a longer sight radius than a handgun, a longer barrel, which produces higher performance with the same handgun ammunition, a larger capacity magazine and a rigid buttstock to aid in rapid sight acquisition and alignment, making it quite easy to use quickly and efficiently under stress. This means that it is more tactically versatile than a handgun and can perform effectively at substantially longer ranges.

And here is where the confusion begins. The machine pistol—clearly a handgun—is capable of satisfying classic handgun mission requirements when utilized without the buttstock/holster attached and in the semi-automatic mode. However, it must also be said that this can only occur if it is presented from a traditional handgun holster worn in typical fashion and fired in the semi-automatic mode. If utilized in the fully automatic mode without a buttstock attached, it is virtually uncontrollable, wasteful of ammunition and considerably less effective than a normal handgun.

If the buttstock is attached, concealment is negatively affected. As well, attachment of the stock is time-consuming, meaning that it cannot be brought into action quickly enough to be considered equal to a normal handgun. Remember, handgun encounters take place at very close range (7-10 *feet* is the norm) and in very short time frames (2.5-3 *seconds* average). This means that rapid weapon presentation is essential, an impossible feat if a buttstock must be attached.

If we compare the machine pistol to the submachine gun, we find that

(Left) Author Taylor fires the HK VP-70M 9mm machine pistol in three-round burst. Note the high cyclic rate typical of the MP by the three spent cartridge cases just outside the weapon's ejection port. Rates as high as 1800 rpm are not uncommon with machine pistols.

(Right) A fairly recent machine pistol is the 9mm Heckler & Koch VP-70M, shown with shoulder stock/holster attached. In an effort to provide better controllability, the gun came with a selector switch (arrow) allowing either standard semi-automatic fire or three-shot bursts, rather than unrestricted fully automatic fire.

its physical characteristics—short sight radius, making it less tactically flexible, reduced ballistic performance due to a short barrel and resulting shorter effective range—limit its capacities. On the other hand, it is usually smaller, lighter and more compact than a SMG, making it easier to carry and conceal. But, fired in the fully automatic mode, with or without a buttstock attached, it is noticeably less controllable. In addition, if a larger capacity magazine is used, then its concealability and portability are reduced.

I think this is where the "neither fish nor fowl" concept espoused by those who oppose the machine pistol concept originates. Compared point by point, characteristic by characteristic, the MP does appear to lack a sufficient number of positive attributes to compare well with either the handgun or SMG.

Does this mean that it is useless, as some claim? I don't think so, but I do feel that the MP is clearly a hybrid *special-purpose* weapon. In certain situations it can certainly outperform a standard handgun, and anything that can launch aimed bullets downrange can hardly be considered to be worthless! Putting it another way, would I grab one in a fight and use it if I had to? Absolutely! Wouldn't you?

Conversely, like any other firearm, when taken out of its realm, the machine pistol compares poorly to weapons specifically intended for general-purpose employment within that realm. To me, the problem with the machine pistol is that because it is such a special-purpose weapon, for many folks it becomes too easy to simply dismiss it in preference to some other, more easily understood type of gun.

Let me give you an example. A couple of years ago, while conducting a Defensive Handgun Instructor's Course class for the firearms training officers from all of the police departments in the metropolitan Atlanta area, the representative for a major firearms manufacturing company (who also produced a machine pistol) came to the police academy range where our class was in progress. A tremendously congenial fellow and excellent businessman, he asked if he could show me and my students (themselves all highly qualified handgunners and instructors) his company's new machine pistol.

Machine pistols continued to appear long after their limited utility became apparent. This is the Soviet Stechkin APS with plastic buttstock/holster in place. A dominant handgun in the Soviet military during the '70s, it was finally abandoned and replaced by the smaller Makarov, a copy of the Walther PPK.

The most recent MP to appear is the Glock Model 18 9mm, shown here in full-auto fire. Time-lapse photography shows the rapid and pronounced muzzle rise that makes them so hard to control.

52 GUNS ILLUSTRATED

Because of ignorance some writers insist on calling them machine pistols, even though submachine guns such as this Sterling L-2A3 9mm are anything but handguns. The SMG is of completely different design and configuration than a handgun, and is a far more versatile and effective weapon than any machine pistol.

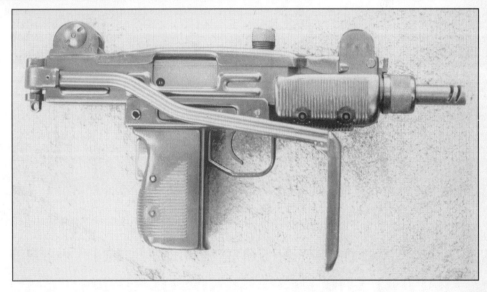

Highly compact versions of the SMG are also often erroneously thought to be machine pistols. However, such is not the case. This 9mm Mini-Uzi more closely resembles its namesake than any handgun.

Since we were on schedule and my students enthusiastically nodded their assent, I agreed to give him 30 minutes to brief them on and demonstrate the weapon, which he masterfully proceeded to do. Once this was accomplished, he allowed each of them to fire a full 32-round magazine at a 4x4-inch cardboard target placed on the berm behind the target line. As you can probably imagine, once thirty people have tried a 32-round magazine apiece, the ground was nearly ankle deep in expended cases. Yet, when examined, the target only had *one* hole in it! The area surrounding the target, however, was torn to bits for two feet in every direction.

When asked by both the factory rep and students what I thought about the weapon, I smiled and responded by quickly presenting my own handgun—coincidentally a regular semi-auto-only version of the same weapon—and firing one round, which center-punched the target and flipped it over the berm. I then reholstered, turned to face them and asked them this question: "Now that you have a perspective, what do you think?"

A stunned silence followed for about ten seconds, then a gale of laughter began. "OK," they said. "We get it!" Any form of evaluation should always have something to give it a sense of proportion and context, don't you think?

I've always said that apples and oranges don't compare very well—if you prefer apples, then oranges will always be considered second-rate. In the field of weapons and tactics, this attitude is too often encountered, which is unfortunate because it prevents objective analysis, the assimilation of useful, valid data, and subsequent advancement of the state of the art. No, the machine pistol isn't my favorite weapon; frankly, I prefer either the standard handgun or SMG. Nonetheless, in spite of its shortcomings, it is a valid weapon and one which deserves to be evaluated objectively. After all, it has been around for nearly a hundred years! If it were as useless as some would have us believe, this could not have happened.

THE ONLY EXPLOSIVE known until the 19th century was ordinary gunpowder, a mixture of potassium nitrate (saltpeter), sulfur and charcoal.[1] The sulfur and charcoal act as fuel, the potassium nitrate as the oxidizer. The composition of early gunpowder varied considerably; every powderman had his own formula. This wasn't too important at first, as the powder (called serpentine powder from the gun or "serpent" from which it was fired) was a loose mixture that burned slowly and inefficiently. In-

---

[1] The explosive properties of fulminates of gold and silver were known by 17th century alchemists, and the preparation of mercury fulminate appears in a 1690 text. These materials were not of practical use at the time and no further discoveries of other primary explosives were made until the development of modern chemistry.

reputed to make the best and strongest powder). Corning solved or partially solved several problems associated with gunpowder: keeping the mixture from separating upon storage or during transportation, preventing the mixture from absorbing moisture, and controlling the rate of burning. This improved both the reliability and power of the gunpowder.

By the end of the 18th century gunpowder was a fairly standard mixture containing 75 percent potassium nitrate, 10 percent sulfur and 15 percent charcoal. The science of chemistry was just coming of age, however, and new compounds were being synthesized in the laboratory. One notable attempt to improve gunpowder occurred in 1788, when the French chemist Claude Berthollet found that potassium chlorate could be substituted for saltpeter to produce a more powerful explosive

mill and the daughter of one of the gunpowder commissioners. While Lavoisier offered to continue research on potassium chlorate explosives, Berthollet and the other commissioners had lost their enthusiasm and the project was never pursued any further. More than anything else, this experiment helped to dissuade others from tampering with the basic formulation of gunpowder.

The gunpowder formulation may have seemed secure, but a genuine revolution was soon to occur in firearms technology. In 1836 Samuel Colt invented the revolver. The principle of rifling the inside of a gun barrel, long known, was at last being applied to increase the range and accuracy of firearms over the old smoothbore weapons. Metallic cartridges and breech-loading magazine rifles were under experiment. The Gatling gun appeared in 1865. Four

SMOKELESS POWDER

The road to a safe propellant was dangerous during the developmental years of smokeless powder, and more than a few experimenters were blown up while concocting the new blends.

by PAUL SCHIFFELBEIN, Ph.D.

gredients were ground and mixed by hand with mortar and pestle.

An important improvement was the introduction of wet mixing which made the powder granular or "corned." Each bowl of dry ingredients was moistened with a little water, or sometimes with vinegar or urine (in the middle of the 15th century the urine of a wine drinker was

mixture. Lavoisier, chemist at the Royal Arsenal, agreed to run tests on the new gunpowder at the French Royal Factory at Essone. On a Sunday morning the powder was being mixed in a stamp mill while a group who had come to witness the new powder went to breakfast. As they returned the charge exploded, killing the manager of the gunpowder

years later, the American gun maker Benjamin Hotchkiss patented a breech-loading magazine action for small caliber cannon on much the same principle. Hiram Maxim, an American who achieved his success in England, invented the first machinegun in 1885; it utilized the force of recoil from the previous round to actuate the reloading mechanism.

Cotton is first chemically purified, then run through a teasing machine to remove knots, debris and dust. This machine tears the cotton apart with rapidly spinning toothed rollers, which leaves the cotton in a fluffy state.

THE EARLY YEARS

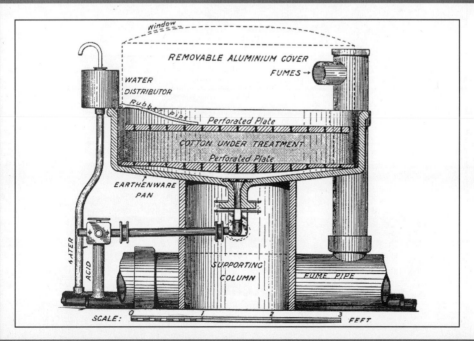

In a displacement nitrator, the mixed acids are introduced from below. Cotton is immersed in the acid then covered with a perforated aluminum plate. Water is then carefully added through a rubber pipe lying on the aluminum plate, which forms a hydraulic seal. After the nitration is complete, the acid is slowly drained while water is added from the top. After washing, the nitrocellulose is ready for pulping.

Against these changes stood the fact that gunpowder had undergone no corresponding improvement in many years. Until new powders were compounded to accommodate the new firearms, the advances made by the arms designers would be only partially realized. Any attempt to alter the gunpowder formulation by the substitution of other compounds for one of the three basic constituents did not seem to be a promising solution, as the disastrous experiment with potassium chlorate had shown. The need for an entirely new chemical compound which could serve as a propellant explosive was clear.

Nitrocellulose[2]

The first recorded attempt to create a smokeless propellant is in a treatise entitled, *Buchenmeysterei* by Abraham von Memingen (circa 1410). This master gunner mixed six parts nitric acid, two parts liquid ammonia, two parts sulfuric acid, and two parts "oleum benedictum" (a crude tar oil), and loaded a cannon with this "water." He promised a range of 3000 paces, but recommended running away from the gun as rapidly as possible after lighting the fuse!

While nitric acid was used by the alchemists as early as the 13th century for the separation of gold and silver, sound experimental work on nitrated organic substances did not begin until the early 19th century. In 1833, the French chemist Henri Braconnot studied the action of nitric acid on starch, producing a substance

[2]Nitrocellulose or cellulose nitrate is a white solid which results from the replacement of hydroxyl groups on the cellulose molecule with nitrate groups. Three types of cellulose nitrate are distinguished according to nitrogen content: pyroxylin or collodion contains 8-12 percent N, pyrocellulose has 12.6 percent N, and guncotton must have a minimum of 13.0 percent N.

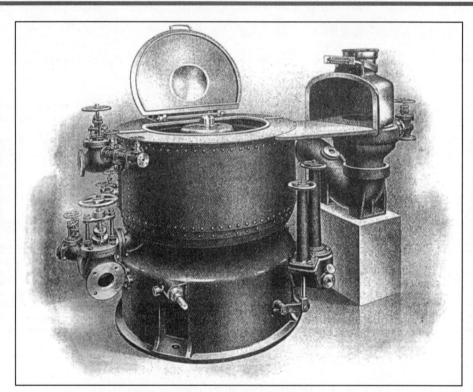

The nitration centrifuge is a cylindrical steel vessel containing a perforated stainless steel basket on a vertical rotating shaft. Acid is introduced through a pipe in the bottom of the vessel. After the cotton is added, the centrifuge is closed and run at slow speed. The rotary movement forces the acid to circulate through the basket, which quickly and uniformly nitrates the cotton. After the nitration is complete, the acid is drained and the centrifuge is run at high speed to remove the waste acid. The nitrocellulose is then transferred through a large pipe to the wash facility, after which it is ready for pulping.

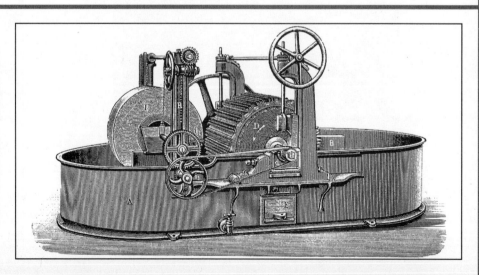

The pulping machine, or "beater," was adapted from the paper industry. It is a cast iron tub (A) divided lengthwise by a vertical partition (B). A large cutting drum (D) has teeth that mesh with fixed teeth on the bottom of the vessel. The tub is filled with nitrocellulose and water. High-speed rotation of the drum keeps the slurry moving around the tub. When the slurry passes between the teeth, the fibers are cut by the knives. Chemicals are added to neutralize freed acid. The pulp is then taken to a boiler for further washing.

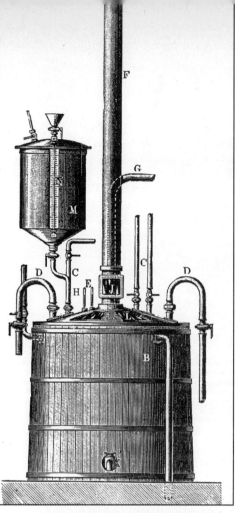

The Nobel nitrator is made of sheet lead and has an internal cooling coil and a water-cooled jacket made of wood (B). The cooling water is turned on (D) and the mixed acid (M) is measured (N) and released into the nitrator. Compressed air (C) agitates the mixture. Glycerin (G) is added to the vessel, and the reaction is watched through windows (L, J) and measured by thermometers (E). Acid vapors and mixing air are removed through a ventilator duct (F). When the reaction is complete, the contents are removed (via K) and taken to the separator.

he named xyloïdine. The property that most intrigued him was the varnish-like coating it forms when applied to paper or other fibrous surfaces, a coating which is essentially waterproof. Further studies by Theophile Pélouze in 1838 found that if paper, linen or cotton were dipped in concentrated nitric acid for two or three minutes, a surface coating is formed which is impermeable to moisture and extremely combustible. But then nothing more was heard from him or anyone else on this topic until 1846, when Christian Schönbein, a Swiss professor of chemistry at the University of Basle, announced the discovery of guncotton.

Schönbein's guncotton was created by treating absorbent cotton with a mixture of nitric and sulfuric acids[3]. Schönbein at once recognized guncotton's importance as an explosive. It appears that Schönbein, like many other inventors, often used his wife's kitchen for his experiments. One day he was distilling nitric and sulfuric acids on the stove when the flask broke. He grabbed the nearest thing—his wife's apron—to wipe up the mess. After this he washed out the apron, which appeared none the worse for the treatment, and hung it up to dry in front of the kitchen fire, congratulating himself that his wife would not be any the wiser. Suddenly there was a puff, and the apron went up in flames.

Schönbein began public demonstrations of guncotton, but he kept the method of preparation secret while he tried to sell the process to various governments. His plans for the exploitation of guncotton were complicated when Rudolph Böttger, a professor of chemistry at Frankfurt, informed him that the preparation of guncotton was no longer secret. In addition to his public demonstrations, Schönbein had also published a brief note describing the wonderful characteristics of his discovery. From this information and his own knowledge of explosives and nitration, Böttger was able to duplicate Schönbein's work. Böttger threatened to publicize the process unless Schönbein entered into an agreement with him. Reluctantly, Schönbein consented to collaborate with Böttger, and the synthesis of guncotton remained unpublished.

In August of 1846, Schönbein traveled to England in order to demonstrate the new explosive before the King. These demonstrations aroused great interest as well as skepticism. Miners at a granite quarry watched with disbelief as two holes were charged, one with the ordinary amount of gunpowder, and the other with a quarter the weight of guncotton. Ridiculing the demonstration, one of the miners offered to sit on the guncotton hole for a pint of beer! Luckily his taunt went unchallenged, as the guncotton shattered the rock into small fragments.

---

[3]Muspratt and Hofmann (1846) were the first to use a mixture of nitric and sulfuric acids for nitrating nitrobenzene to dinitrobenzene. Sulfuric acid was thought to act as a "dehydrating agent," which helped drive the nitration reaction to completion.

Schönbein patented his discovery, and in October of 1846 entered into an agreement with John Hall and Son for exclusive manufacturing rights at their powder works at Faversham.

Despite this promising turn of events, Schönbein was still not successful in exploiting guncotton. In their rush to profit from guncotton, Hall and Son began large scale production without sufficient safety precautions. On July 14, 1847, the factory at Faversham was destroyed by an explosion which killed 21 people. Hall and Son immediately repudiated their contract, and no further work was done with guncotton in England for 16 years. Other explosions, in Brunswick, Vincennes, Dorpat, and Philadelphia, followed the one at Faversham. Soon only Austria persisted in the attempt to manufacture guncotton.

In 1852, Captain von Lenk purchased the guncotton process on behalf of the Austrian army. While von Lenk did not change Schönbein's nitration procedure, he went to great lengths to purify the product. Instead of merely washing with water until neutral, von Lenk washed for three weeks, then boiled with dilute potash solution for 15 minutes, washed again for several days, impregnated the yarn with water-glass, and finally dried. The Austrian government began production of Schönbein's guncotton under von Lenk's direction, but explosions in 1862 and 1865 led it to abandon manufacture. In spite of the rigorous procedure devised by von Lenk, he was still unable to wash the guncotton completely free of acid.

In 1865, Frederick Abel, the chemist of the English War Department, started guncotton manufacture on a small scale at the Royal Gunpowder Factory at Waltham Abbey. Abel's experiments showed that by pulping guncotton, in the same way that rags are prepared in the manufacture of paper, made it much easier to wash out the impurities. The pulped guncotton was also easier to press into blocks, which tames its violence. While the pulped and compressed guncotton did burn more slowly than von Lenk's yarn, it was still unacceptable. It damaged the guns, and the accuracy of shooting was unsatisfactory. It was not until some 17 years later that a successful smokeless military powder was made. In the meantime, guncotton was used only for blasting purposes.

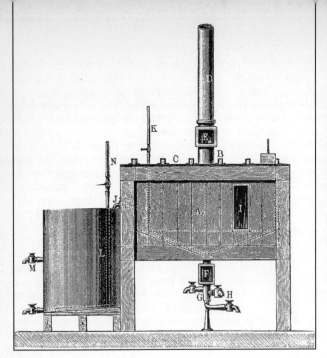

The Nobel separator is a rectangular sheet lead tank with a conical bottom. The cover is also sheet lead, reinforced with lead-coated iron rods (B), and contains several windows (C). The exhaust duct (D) also has a window (E). A compressed air line (K) can be opened in case the contents must be quickly dumped to a drowning tank (via H). When a distinct separation line is seen through the side window (I), the main yield of nitroglycerine (which is less dense than the waste acid) is released (J) into the pre-wash tank (L). The acid is then discharged through the right valve at the bottom of the separator until nitroglycerine appears in the lower window (F). The left valve is then used to remove the remaining nitroglycerine, which is added to the pre-wash tank. The pre-wash tank has two earthenware valves, an upper one (M) for removing the wash water and the lower one to run the nitroglycerine to the washing house. Air (via N) is used for agitating the mixture.

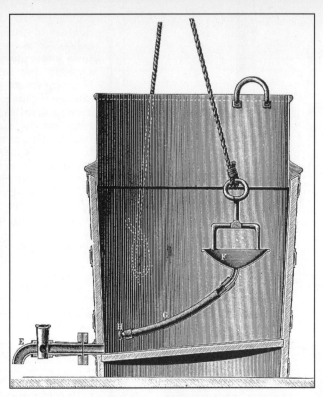

Nitroglycerine is purified in a wash tank. Wash water (which is lighter than the nitroglycerine) is skimmed from the surface using a saucer-shaped funnel (F) attached to a rubber hose (G), which leads through the side of the tank (H) to a wash-water gutter. A rope is attached to the handle of the funnel, passing over a pulley on the roof of the building, and held by a clamp. Washed nitroglycerine is drained from the bottom of the tank (E) and then filtered through salt or sponges.

Nitroglycerine[4]

Nitroglycerine was discovered in 1846 by Ascanio Sobrero, professor of Chemistry at Turin. Sobrero was a student of Pélouze in Paris, where he became familiar with his experiments on the nitration of organic substances. Sobrero found that glycerin dissolves in a cold mixture of concentrated sulfuric and nitric acids. If the solution is then poured into water, an oily liquid forms at the bottom of the container.

Sobrero warned of the migraine headaches that working with nitroglycerine caused and described an experiment of its toxicity on a dog:

> Within seven to eight minutes of ingesting approximately one teaspoon, the dog suddenly fell down. A dose of ammonia in olive oil was given in an attempt to revive the animal, but death resulted after an hour. A post mortem examination revealed that the veins in the animal's head were suffused with blood.

---

[4] Glycerin is a by-product in the manufacture of soap and candles from oil and fats.

As to its explosive properties:

> On one occasion a small quantity of an ethereal solution of nitroglycerine was certainly not more than 2 or 3 centigrams. On heating the dish over a spirit lamp a most violent explosion resulted and the dish was broken to atoms. On another occasion a drop contained in a test tube was being heated when it detonated with great violence and pieces of glass severely cut my face and hands and also injured others standing in the room some distance away.

Appalled by the deadliness and unpredictability of nitroglycerine, Sobrero warned against attempting to convert it to industrial uses.

Nitroglycerine was nearly forgotten by the time inventory Emmanuel Nobel and his son Alfred brought out their "blasting oil" in 1862. A year later they patented the mercury fulminate detonator, which provided a safe and reliable method to detonate the oil. By December 1865, nitroglycerine was being successfully used in a dozen countries, including the United States.

Nobel at first claimed that his blasting oil was a safe explosive, a misconception that led to many unnecessary accidents. One day a porter in New York City's Wyoming Hotel noticed a reddish smoke coming from the box he had been using for several months as a seat for his shoe shine operation. The box had been checked at the hotel by a German traveler, who had failed to mention that it contained 10 pounds of nitroglycerine. The porter dragged the smoking box into the street and ran. A moment later it exploded, ripping a hole 4 feet deep in the stone pavement and seriously damaging the hotel and surrounding houses.

That was just the prelude. On March 4th, 1866, two cases of nitroglycerine exploded in Sydney, Australia, demolishing a warehouse and killing many people. A month later the steamship *European*, loaded with seventy of Nobel's wood-crated cans and ammunition for South American revolutionists, blew up while unloading at Colón, Panama. Sixty died and property damage exceeded one million dollars. Only a few more days

For single-base powders, wet nitrocellulose is dehydrated by displacing the water with alcohol in a hydraulic press. The resulting blocks are broken up and incorporated with solvents in kneading machines similar to those used in bakeries. Double-base powders start with hand mixing of the nitrocellulose and nitroglycerine. The resulting paste is then incorporated with solvent. The incorporating machine has two curved blades that rotate in opposite directions, one twice as fast as the other, both going downward in the center of the trough. During operation, the machine is closed to prevent loss of solvent.

When incorporation is complete, the cover is removed, the machine is tilted up, and the blades are rotated in the opposite direction. The dough falls out and is ready for shaping.

had passed when a shipment of nitroglycerine destined for the California gold fields destroyed the Wells Fargo Express offices in San Francisco. The explosion was felt like an earthquake and killed fourteen people.

Then on September 3, 1864, there was an explosion at Nobel's Heleneborg plant which killed five workers, including Alfred's youngest brother Emil. At the same time, steamship companies began refusing to accept nitroglycerine as freight. Alfred turned his attention to the problem of stabilizing nitroglycerine. During the years following the explosion, he conducted exhaustive experiments, mixing nitroglycerine with powdered charcoal, paper pulp, brick dust, gypsum, and other substances.

In 1867, Nobel patented dynamite, a discovery that some say was accidental. Blasting oil was shipped in cans, which were packed in sawdust inside of wooden cases. The sawdust occasionally ignited through contact with acids or nitroglycerine. Nobel had recently substituted *kieselguhr*, a porous diatomaceous earth, for the packing material. One day a leaky nitroglycerine can lost most of its contents, and yet there was no sign of oil on the outside of the wooden case. It was then found that *kieselguhr* would hold three times its weight of nitroglycerine. The resulting putty-like material, "dynamite," could be kneaded and packed in cartridges and was relatively insensitive to shock.

A number of researchers including Nobel tried to use nitroglycerine as a propellant explosive, but its high temperature and pressure made it too powerful for that purpose. It seems likely that nitroglycerine would have ceased to be used as a practical explosive had not Nobel conceived of the idea of absorbing it in *kieselguhr*, thus converting it into a material which could be handled safely.

Smokeless Powder

It has already been pointed out that the early attempts of Schönbein, von Lenk and Abel to make a satisfactory smokeless powder from guncotton were unsuccessful. The primary difficulty with the use of nitrocellulose in guns was that it reacted too quickly and tended to destroy the gun barrels. Various attempts were made to regulate the rate of combustion of the nitrocellulose by adding slower-burning substances and by superficially gelatinizing or hardening the grains.

Captain Schultze of the Prussian

In some cases, the incorporated dough is rolled into sheets by passing it repeatedly through rollers resembling a papermaker's calender. The sheets are then cut into strips, which are again cut transversely to form cubes or flakes. The powder is then dried, blended and packaged.

Artillery extracted cellulose fibers from wood chips, nitrated them, and impregnated the fibers with a solution of potassium and barium nitrates. In 1865, he produced Germany's first commercial smokeless powder. The explosive was still too violent for rifles, but became very popular with shotgunners on account of the light recoil and relative absence of smoke.

In the meantime, applications of nitrocellulose in other industries were paving the way for the future propellants. Schöenbein as early as 1846 had used an ether-alcohol solution of nitrocellulose which he called "liquid glue." In 1848, J. Parker Maynard, a medical student in Boston, used a similar ether-alcohol solution that he called "Collodion" as a dressing for wounds. Scott-Archer used the same solution a few years later in photography. In 1863, a struggling young American inventor, John Hyatt, was trying to find a substance for making billiard balls. After a bottle of collodion was tipped over in the print shop where he was employed, he found that the substance when dried was like ivory. Hyatt went on to patent the use of camphor, heat and pressure to produce "Celluloid," and the plastics industry was born.

Two subsequent developments paved the way for the shaping of nitrocellulose into definite forms, which in turn led to important military and non-military uses. Daniel Spill, an Englishman, patented soluble guncotton in the shape of tubes in 1875 for use in artillery, employing a solvent which was a mixture of ether, alcohol, and nitrobenzene. During the Paris International Exposition of 1889, French Count Hilair de Chardonnet announced his successful extrusion of a pyroxylin solution. His invention, the forerunner of 20th century rayon, was first used for filaments in incandescent lamps.

These discoveries no doubt led French chemical engineer Paul Vieille to the development of a gelatinized smokeless powder in 1884. Vieille's powder was composed of 68 percent "insoluble" nitrocellulose, 30 percent "soluble" nitrocellulose, and 2 percent paraffin.[5] Additional solvents were used to gelatinize and stabilize the mixture. After incorporation, the mass was worked between heated rollers, and the rolled sheets either cut up in strips or flakes or squirted through a die into ribbons. The first powder of this type was used in the French Lebel rifle, model of 1886. The propellant was adopted by the French Army as *Poudre B* in honor of the Minister of War, General Boulanger. *Poudre B* is known as a single-based powder, since the agent used to gelatinize the nitrocellulose is not an explosive.

Nobel introduced his first gelatinous explosive in 1875. This discovery is also said to have been accidental. One day Nobel hurt his finger while working in his laboratory, and used collodion solution to cover the wound. At two o'clock the next morning he was awakened by the pain in his finger, and noticed that the collodion had formed a jelly with the nitroglycerine with which he had been experimenting. Further experiments showed that adding 7 to 10 percent of collodion directly to nitroglycerine formed a tough plastic mass. This material, termed "blasting gelatin," has two advantages compared with *kieselguhr* dynamite: it is waterproof, and the substance added is itself an explosive, which makes blasting gelatin 25 percent more powerful.

---

[5]"Soluble" nitrocelluloses such as collodion are soluble in a mixture of ether and alochol whereas guncotton is not.

It was a further development of blasting gelatin that led to Nobel's invention of a smokeless powder. Guided no doubt by the study of celluloid, Nobel found that by greatly increasing the percentage of nitrocellulose in blasting gelatin he could produce an explosive which could serve as a propellant. A mixture containing nitroglycerine and soluble nitrocellulose was kneaded between cylinders heated to 50-60°C. It was then rolled out in sheets and cut into square grains. Nobel began producing this new powder, called Ballistite, in 1889. Ballistite is known as a double-based smokeless powder because the agent used to gelatinize the nitrocellulose is an explosive.

Abel and Dewar modified Nobel's Ballistite for the British armed services in 1889. They substituted insoluble nitrocellulose (guncotton) for collodion and used acetone to gelatinize the mixture. The soft mass was extruded in a macaroni machine, and the tubular shape of its grains suggested its name: "Cordite." Adopted by the British armed services, this grain form became world-famous through its use in the Boer War and World War I. Cordite burned at a slower rate than Ballistite, developing a lower maximum temperature and gas pressure, thus reducing wear of the gun barrel. Because of this advantage, cordite, or a slight variation of the cordite formula, was soon adopted by many countries for military use.

United States Developments

The news that the military establishments of foreign powers had adopted smokeless gunpowders had a pronounced impact upon the leaders of the United States armed services. By 1889, France, Germany, and Great Britain had developed successful smokeless gunpowders and had adopted them for military use. Russia and Belgium would later adopt a modification of the French single-base powder while Italy chose a double-base variation of Ballistite. America was without a satisfactory smokeless powder of any kind, and faced a military disadvantage in any future war.

Early in 1889, the United States Navy's Bureau of Ordnance requested the Du Pont Company to send an experienced powderman on a tour of Europe to find out about the new gunpowders. Alfred I. du Pont, an assistant superintendent in the Du Pont Company's mills, sailed for France on March 20, 1889. He found the French munition industry silent concerning details of the new smokeless powder, however, but he did hear about its astonishing performance. He continued on to Belgium, for he had been told that their smokeless powder gave results comparable to that of the French. In attempting to learn Coopal and Company's formula, he used a subterfuge and infiltrated their powder works in the guise of a workman; in this manner he was able to observe portions of the process, but not enough to derive the formula or the complete method of manufacture. He then donned his "meeting clothes" and was successful in getting the Coopal officials to allow the Du Pont Company to manufacture both their brown and smokeless powders on a royalty basis.

The Belgian smokeless powder did not measure up the standard of the French product, however, and the serious task of developing U.S. smokeless powders began. How did the United States get so far behind? Most European countries had governmental control over gunpowder manufacture, and guided its direction. Various U.S. secretaries of war had urged the establishment of a national powder factory under the supervision of a joint Army and Navy board, or under that of the U.S. Ordnance Department. In 1837, President Van Buren had included such a recommendation in his annual message to Congress, but nothing came of these attempts. This lack restricted the War Department in its search for a smokeless powder, and to the testing of powders submitted by the powder companies and private individuals.

At the same time, American powder companies were reluctant to plunge into an intensive program of military powder development. War seemed remote, and for almost a century the market for industrial explosives had been increasing at a breathtaking pace. The main burden of developing a military smokeless powder in America during the 1890s therefore fell upon the Army and Navy ordnance staffs and independent inventors. The large powder makers cooperated, but their research efforts were directed more toward improving their sporting powders. Despite this slow start, by the turn of the century the United States had reached a point where it could make smokeless propellants equal to the best.

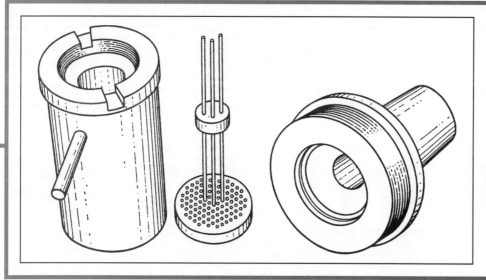

The incorporated dough can also be pressed through a water-cooled die, which is an adaptation of the machine used for making pasta. The propellant ends up as cords, strips or tubes. If desired, these can be cut into any required length. The powder is then dried, blended and packaged.

The Invalides:

IN UNIQUELY FRENCH style, The Invalides is comprised of many parts—at once it is a tomb, a church, army post, private association, museums and, in part, the home of one of the oldest firearms and artillery collections in the world. Here one may meet the Napoleons, both the canon and the men, and obtain a panoramic overview of French arms in the Musée de l'Armée.

Bronze and iron cannon representing examples of French field pieces, and trophies captured in Europe and the far edges of the former colonial empire are displayed. Cannon line both floors of the interior walls of the courtyard standing like soldiers at attention. The tubes are held muzzle up by iron staples affixed to the limestone walls. Squat mortars sit beside many entrances guarding the somber gray portals.

Some, like the bronze Napoleon smoothbore cannon by the door, are workmanlike pieces of military hardware devoid of decoration. Others are ornate, such as a baroque example from Salzburg with ornamentation cast in the shape of a couple in passionate embrace—a beautifully ironic metaphor in bronze. Many cannon have names, like Vigilant, Reliant and Defender. My wife's favorite is Hostile, a small cannon almost lost among its larger brothers.

As might be expected by the person that built much of the palace at Versailles, Louis XIV felt the need to apply appropriate decorations to his military hospital. The upper windows have carved stone moldings in the shapes of breastplates, helmets, spears and other martial motifs. The structure was begun in 1671 and completed in 1706. For Louis XIV, the decorative elements are restrained as compared to his palace at Versailles.

The Napoleons

The Invalides is crowned with a gilded dome under which is located Napoleon's tomb. There are other Napoleons there as well—Napoleon's brother Joseph, the king of Italy, and Napoleon's son, the king of Rome. The interior of the structure was modified into a resting place for Napoleon. His body was brought back from Saint Helena in September of 1840, nineteen years after his death. He was reinterred like a pharaoh in seven coffins within a red granite sarcophagus.

Almost as if too much military genius could not reside in the same family, Napoleon III was almost as inept a military leader as his uncle was successful. It was Napoleon III who embroiled France in an ill-fated attempt to capture Mexico and install Maximilian I as emperor. This venture lasted from 1864 to 1867. He tried to take advantage of the United States' involvement in the Civil War to extend his empire into the New World.

Documented within the structure are Napoleon III's campaigns in Africa, which resulted in French colonies in Algeria and sub-Saharan Africa. Moorish flintlocks and Ottoman cannon were no match for a European army equipped with percussion guns and more modern artillery.

Opposing the French were North African flintlocks that varied widely in quality, which can easily be seen in

The interior courtyard of The Invalides, with the dome of Napoleon's Tomb rising in the background. Although housed in a series of interconnected buildings, the Musée de l'Armée, the Musée de l'Ordre de la Libération, Napoleon's Tomb, a hospital, a church, and some offices of private and public functionaries are all in the same structure.

the displays here. Frequently, these guns had miquelet locks, which carried the driving spring for the hammer on the surface of the lockplate, rather than on the inside. North African guns of the period were almost invariably smoothbore and of large caliber. Guns carried by men on horse or camel were unusually long and had barrels that rapidly thinned toward the muzzle. Guns used by Muslim foot soldiers were often carbine-length and some had bell-shaped muzzles to enable them to be rapidly reloaded in the heat of combat.

French engineers have a tendency to produce innovative designs. Modern examples include the Darne shotgun and Citroen automobiles. In the time of Napoleon III, the French designed a rapid-fire gun called the *mitrailleuse*, which consisted of a number of rifle-caliber tubes within a bronze housing. Externally, the gun looked like a conventional cannon, but, functionally, it was a manually operated machinegun. A preloaded breech section was inserted into the rear of the gun and the rounds were fired in succession by turning a crank.

This case on the first-floor exhibit shows type examples of French flintlock muskets, bayonets and sabers. Included in the display are typical examples used by French infantry, cavalry and sailors.

A MULTI-FACETED ARMS MUSEUM

The famous old middle of Paris houses repository in the artifacts of interest to anyone fascinated by firearms and the military.

by WM. HOVEY SMITH

The faster the crank was turned the more rapidly the gun fired.

To enable the gun to fire more shots before reloading, more barrels could be added, but at the cost of increasing the weight and bulk of the gun. The larger mitrailleuss became so unwieldy that they were only capable of defending fixed fortications. The American Gatling gun differed in that it had several barrels rotating around an axis. The Gatling was fed by a magazine holding preloaded subchambers and later by cartridges.

Another example of period innovation was the *Chassepot*, which was a needle gun in which a long pin pierced a linen-wrapped cartridge to fire a primer held at the base of the bullet. This single shot, bolt-action breech-loader used a rubber gasket on the front of the bolt for a gas seal and was capable of being fired more rapidly than the Prussian needle gun. However, the needle gun's shorter firing needle striking a primer at the rear of the cartridge proved more reliable.

The Chassepot had several functional problems. The rubber gasket became brittle because of the effect of the sulfur in blackpowder. This caused a progressively less effective seal, and hot gasses were directed back into the user's face. The firing needle, about 2 inches long, was delicate, easily corroded and frequently

30th EDITION, 1998 **63**

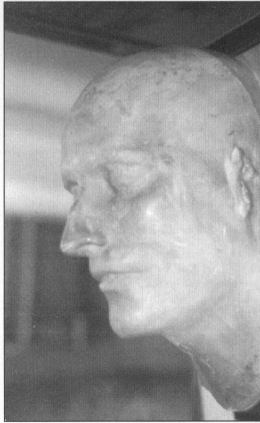

(Above) The death mask of Napoleon I. It was made on the island of St. Helena the day of Napoleon's death by an attending surgeon.

Gilded helmets, maces, flags, spears and swords complete the martial motifs on the dome above Napoleon's tomb. It is quite a sight.

broken. If a misfire occurred, the only way to remove the cartridge was to drive it from the gun with the ramrod.

On the positive side, the handling characteristics of the gun were good. It had a long loading trough and a robust cocking piece on the rear of the bolt which was easily manipulated, felt good in the hands for such a long gun, and came with a formidable sword bayonet. The author owned one for a time when they were imported into the U.S. during the late 1950s.

Napoleon III's reliance on unproven military hardware may have been a contributing factor in his capture by Prussian forces at Sedan in early September of 1870, during the Franco-Prussian War. This action essentially ended his reign, and the assembly of Bordeaux condemned him as "responsible for the ruin, invasion, and dismemberment of France."

During this time, Napoleon was seriously ill with a bladder disease, and though hardly able to move he was in command of his troops. The intense pain of the disease affected his judgement, of course, and this was probably the main reason for his capture. He lived in England thereafter and underwent surgery to cure his incapacitating disease, but died on January 9, 1873, probably as much from the surgery as the bladder problem.

Post-Napoleonic Military Developments

Seriously reexamining itself after its defeat, the French military establishment renewed its research into military technology and developed some designs that were to be vitally important in coming conflicts. Paul Vieille gelatinized guncotton in 1884. This discovery enabled a smokeless propellent to be produced. The appreciation that a smokeless-powder gun shooting a fast small-calibered bullet would have distinct military advantages progressed to the development of the 8x50mm Lebel. This rimmed, bottlenecked cartridge, adopted by the French in 1886, was the first successful smokeless powder military cartridge. Two years later, the U.S. introduced a new model of the blackpowder 45-70 single shot Springfield rifle.

Mobile and rapid-firing artillery, particularly since Napoleon, was seen as vital to any future conflict. Developments proceeded on a breechloading field artillery piece that used fixed ammunition firing an explosive high-velocity shell. From this research, the famous French 75mm field piece was designed. The 75, of course, was a mainstay of French forces and was used extensively by U.S. troops during World War I. The design was so good that the gun remained in U.S. inventory through World War II.

This was not the first time French

This double cased set (four guns) of very richly decorated flintlock pistols was owned by Napoleon. Bright with gold decorations, each of the pistols is decorated in a different style. They were difficult to photograph.

military expertise had been utilized by the United States. Although a variety of guns were used, a standard U.S. military arm during our Revolutionary War was the 69-caliber French Charleville musket. When produced in U.S. arsenals in 1795, the gun was made with metric-threaded screws. It was regular troops using smoothbore muskets that defeated General Cornwallis at Yorktown. The French Charleville was about 20-gauge, compared to the 12-gauge (approximately) British Brown Bess. The advantages of using a smaller ball were a slightly higher velocity resulting in a flatter trajectory, and the use of lighter-weight ammunition. Against human targets, there was no practical difference in being hit with a 69-caliber ball and a 75-caliber ball. In either case, the result was horrific.

The Minie ball, originally conceived by French Captain Claude Etienne Minie, was one of the most commonly used projectiles during the U.S. Civil War. The claim for this invention was disputed by British gunmaker W.W. Greener, who asserted in his book, *The Gun and Its Development*, that he developed a round-ball gun that used a tapered iron plug which was forced into a recess to expand the ball into the barrel's grooves. He was aggrieved when the British government awarded Minie 20,000 pounds in 1852 for the use of Minie projectiles with wooden expansion plugs by British forces. In 1857, Greener was

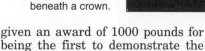

Napoleon's campaign coat and hat. Two of his personal guns are in the background. These are finely finished guns, but relatively plainly decorated. Each has on its buttstock Napoleon's monogram of an "N" beneath a crown.

given an award of 1000 pounds for being the first to demonstrate the "Minie principle" in 1836.

While the Minie and Greener projectiles used iron components to ensure the lead bullets would expand sufficiently to engage the rifling, the U.S. Springfield bullet depended on the force of gas expanding into the hollow base of the projectile to force the undersize bullet into the rifling. This effect was first recognized by another Frenchman, M. Delvigne, who patented the first elongated hollow-based projectile in 1841. The result of these developments was a rifle that could be reloaded as rapidly as a smoothbore. In trials, the Minie projectiles proved to be significantly more accurate than round balls fired from smoothbore guns in either aimed or volley fire.

Also during the U.S. Civil War, the bronze smoothbore Napoleon cannon, which was lighter in weight and more mobile than previously existing field pieces, was found to be as effective in the hands of Union and Confederate cannoneers as it had been on European battlefields. The Napoleon weighed about 500 pounds less than the field pieces it replaced and was also shorter, making it easier to manhandle on the battlefield. Some surviving Napoleons are preserved at U.S. National Military Parks, where they are used to mark the positions of the Union and Confederate forces.

The Musée de l'Armée

Cohabitating within the Invalides along with the Musée de l'Armée is the Musée de l'Order de la Libération, the Tombeau de Napoléon 1ER, a hospital, a church, a military unit, a military engineering exhibit, and a snack bar and gift shop. French museums have their own governing bodies who must ensure they are largely self-supporting. As money is available, improvements are made. Some museums have a very modern look even

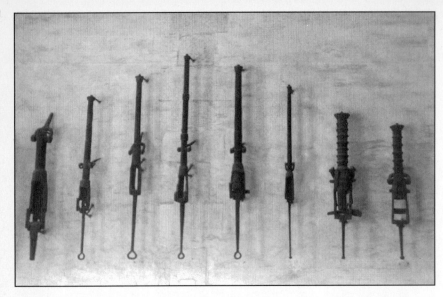

A display of eight 15th-century breech-loading iron swivel guns. These small cannon were typically used on ships. A preloaded chamber resembling a beer mug was inserted into a slot in the rear of the cannon and held fast by a wedge until the gun was fired. The barrels of these pieces were sometimes made of iron rods hooped together, in barrel fashion, by iron bands.

though housed in old buildings, while others such as the Galeries de Paléontologie and d'Anatomie Comparée appear to be unchanged since the 1880s.

The Musée de l'Armée is part historic and part modern. All exhibits are now protected under glass cases, reflecting the latest major renovation in the 1950s. Improvements are being continuously made, but the museum honestly reflects its historical evolution. Exhibits are loosely grouped by time period, although they may be somewhat scattered between the two floors of the building.

At one time, much of the complex was used for training army officer cadets. Arms collected by the French monarchs were exhibited at The Invalides as reference materials for teaching and study. During the Napoleonic era, this collection was augmented. By the end of World War I, exhibition space was almost nonexistent, but a few things were rearranged to allow a small exhibit to be presented. After World War II, there was a need to preserve the fragile relics of the French Resistance Forces, and a few rooms were allotted for that purpose on the ground floor across the courtyard from the The Musée de l'Armée.

Inside the The Musée de l'Armée, there are two floors of exhibits which seek to present French arms and soldiers from medieval times through World War I, with particular emphasis on Napoleon I and III. On the ground floor are displayed examples of French flintlock muskets covering the periods of the Monarchy, French Revolution, and Napoleonic Wars. In one transparent case, locks are displayed so they can be examined from all sides. Farther down, racks of flintlock muskets are exhibited.

At the top of the stairway in the The Musée de l'Armée, three dimensional models of towns are displayed. Every detail of the topography, city walls and individual houses are present in these exquisite models. They were constructed for training cadets on how to make assaults on fortified towns. It is worth the climb up the stairs to see them.

World Wars I and II are not forgotten. From World War I, there are a French 75, a tank, and exhibits on the first floor. Considering that most of the war was fought on French soil, the models of French soldiers and airmen with period arms seems a slight treatment, but by this time the museum had already run out of space and had to dismantle old exhibits to allow any representation of the period.

An interesting transitional piece at the museum is a bronze smoothbore cannon modified at the breech to take a screw-threaded breech section that rotates on a hinge. Equally important to the development of modern artillery was the use of heavy springs or hydraulic systems to dampen the recoil of the gun. When the recoil was

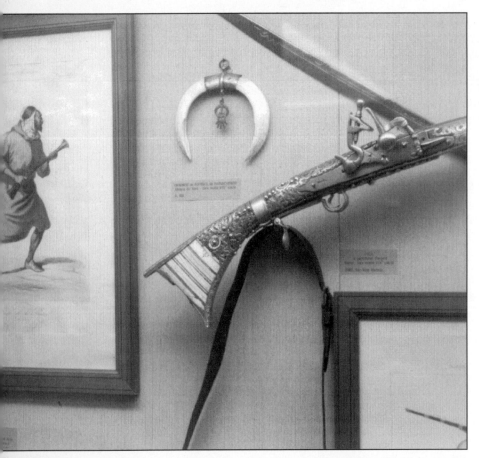

A highly decorated North African flintlock with a miquelet lock. Long smoothbores, 6 feet and more, were used by native North African forces fighting from camel or horse, whereas shorter guns such as the blunderbuss shown in the drawing were favored by foot troops.

(Above) Wheellock pistols measuring 2 feet and more were cavalry arms used by lightly armored mounted troops in the 1500s. The pistol on the left is shown with a spanner, which was used to tighten the spring that turned a hardened wheel against a piece of pyrite in the jaws of the cock. This propelled a continuous stream of sparks into the priming powder. Wheellocks were expensive to manufacture and delicate. Both factors caused them to be replaced by other systems.

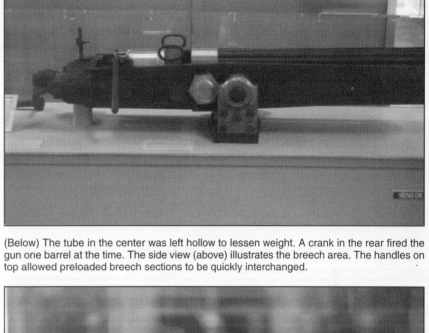

(Below) The tube in the center was left hollow to lessen weight. A crank in the rear fired the gun one barrel at the time. The side view (above) illustrates the breech area. The handles on top allowed preloaded breech sections to be quickly interchanged.

This shows to good effect the business end of a thirty-barrel, 11mm mitrailleuse.

absorbed, the gun could be automatically returned to firing position, rather than the gun crew having to pull the piece back into place and relay (re-aim) the gun after each shot.

Putting a recoil-dampened breech-loading cannon onto an armored mobile platform capable of traversing the muddy, cratered battlefields of World War I resulted in the development of the tank. French models included the Schneider, St. Chamond and the Renault. The latter was a light, two-man crawler-tracked machine, similar to the model preserved at the museum. The larger tanks mounted the 75mm gun, while the Renault used a 37mm gun. About 500 of the Renaults were produced between 1917 and 1918, and were used in at least forty battles.

Tanks are credited with hastening the end of World War I by their ability to beat down barbed wire, withstand rifle and machinegun fire, traverse narrow trenches and direct their cannon fire against resistant targets. The tanks epitomized Napoleon's concept of artillery that could be moved rapidly to take advantage of field positions by using their cannon and machineguns to focus overwhelming firepower against infantry. Countermeasures were developed to help neutralize tanks, but their initial battlefield appearances had a profound effect on modern warfare from then on.

The most striking exhibits on the first floor of the Musée de l'Armée are the life-size costumed mannequins, including horse cavalry, which are fully equipped with dress and accessories. Later exhibits were added where they could be incorporated with minimum disruption. As a consequence, a strict evolutionary series of exhibits is not present.

One wing on the second floor houses more material from Napoleon III, including a mitrailleuse, but most important are several relics of Napoleon I. These include a case containing his personal guns, some elaborately gilded such as a cased four-pistol set, probably by Nicolas-Noel Boutet, and other long arms that are finely finished but relatively

The French Charleville musket was the predecessor to the Springfield Model of 1795. This gun provided the pattern for the first U.S. musket produced by the armory at Richmond, Virginia. (Photo by the author from the N.R.A. museum, Washington, DC).

(Below) A monument in the shape of a Minie ball on the Shiloh National Battlefield site in Tennessee. Used by Union and Confederate forces alike, the French-designed Minie ball enabled rifle-like accuracy to be obtained from projectiles that could be loaded as rapidly as a smoothbore.

devoid of decoration. These are probably by Lepage, *Arquebusier de l'Empereur*. Here is also Napoleon's campaign coat and hat, his death mask, and a replica of the room on Saint Helena where he died.

The museum collections have been commonly used by scholars. W.W. Greener in his book, *The Gun and Its Development*, references several guns from the "Musée des Invalides at Paris." He also sometimes refers to the institution simply as the "Paris museum." The first edition of a predecessor work, *The Gun*, appeared in 1835, and Greener continued adding material until the ninth edition was published in 1910. In this, he depicts some elaborately decorated French wheellocks and a unique three-barreled wall piece consisting of three guns joined to a single buttstock. These guns are not currently being exhibited.

A gun described by Greener that belonged to Napoleon I was a swivel-barreled flintlock hunting carbine. Each barrel has its own frizzen and pan, and is fired by the same lock. The carbine has a carved, richly figured walnut stock featuring Napoleon's monogram and crown, as do two other guns now exhibited at the museum.

Current displays concentrate on typical arms used by the soldiers of the period rather than richly decorated "art guns." The almost sole exception to this are the personal arms of Napoleon. From the wheellock era, for example, there are some exceptionally well preserved, but unornamented, pistols displayed along with the light armor worn by the cavalry. The cavalry would charge a line of pikemen, wheel, discharge their pieces into the standing ranks and retire to reload. Unless protected by musketeers or archers, the pikemen would eventually be eliminated or their line would be broken. The long pistols—over 2 feet—were carried in saddle holsters and were most frequently issued in pairs.

Musée de l'Ordre de la Liberation

World War II is represented by a separate museum dedicated to the Partisan Resistance in France and in the former French colonies. The entrance to this museum is located across the courtyard from the Musée de l'Armée. Arranged in a group of displays are radio equipment, flags, maps, photographs and a few arms that were air-dropped into occupied France. Among these are a British 9mm STEN gun and a crude single shot 45 ACP pistol made of stamped steel parts. This gun came to be known as the Liberator pistol and was made by a division of General Motors. It was to be used until the Resistance fighter could acquire a better gun.

Visiting these exhibits, Napoleon's tomb, the Battle Gallery at Versailles, and the Louvre brings one as close to Napoleon as it is possible to get in this life. All of it is here—the excesses of the Monarchy; the terror of the Revolution; the pageantry of Napoleonic France; and Napoleon's exile, return, reexile and lonely death. However Napoleon is viewed, as a preserver of French culture and glory or as a dangerous revolutionary firebrand, his presence and ideas defined an era that still influences modern military tactics and political thought.

An admission fee of 35 Francs, about $5, is charged. The fee covers entry to the Musée de l'Armée, the Musée de l'Ordre de la Liberation, and Napoleon's tomb for two consecutive days. A gift shop and cafeteria are located on the premises. The Invalides metro and bus stations are easily accessed by the Barard or Chatillon-Montrouge metro lines, city buses, or taxis. If you are planning a visit to Paris, make it a point to see The Invalides complex.

by PAUL SCARLATA • Photos by James Walters

Bearded, axe-carrying Pioneers (combat engineers) of the 1st Cavalry Regiment of the French Foreign Legion on parade in North Africa in 1941. Each man is equipped with a Mle. 1935A in a M1937 holster. (Photo: Establissement Cinematographique et Photographique des Armes. Courtesy of Eugene Medlin.)

This profile view of the Mle. 1935A shows its elegant lines, the Browning-inspired controls, and the pivoting safety-lever at the top rear of the slide. Lanyard ring on the grip frame indicates post-WWII manufacture.

The SACM Modele 1935A Pistol

This was one of the better designs to come from the French arsenals, but was hindered by its pip-sqeak cartridge.

IT IS NO SECRET among students of military small arms that ever since the late 19th century the French soldier has been burdened with some of the oddest looking, most convoluted weapons ever to come off a deranged designer's drawing board. From the Mle. 1892 revolver with it's backwards cylinder, to the ungainly looking MAS 1936 rifle, and through the most recent contribution, the FAMAS assault rifle (nicknamed "le Clari-

Right view of the Mle. 1935A. The gun has a windage-adjustable front sight, pivoting trigger, and plastic grip panels. The rounded hammer is just visible at the rear of the slide.

The Mle. 1935A's safety-lever in the On position. The hammer can be left cocked or manually lowered.

on"—the bugle—by French soldiers because of it's odd appearance), it would seem that French weapon designers are having a good laugh at the expense of their military.

Despite having one of the largest government-run arms industries in western Europe, the French have always managed to find themselves short of small arms whenever they have become involved in a major altercation. In every war they have fought since the 1871 unpleasantness with the Prussians they have had to look outside of France for additional supplies of weapons. This was most evident in the early days of the Great War when a rapidly expanding army, and battlefield losses, quickly ate up their strategic reserves of small arms. All government facilities and domestic firms worked overtime to supply needed weapons, and contracts were placed with foreign manufacturers in the United States, Italy, and Spain.

France's standard issue sidearm, the 8mm Mle. 1892 revolver, could not be produced in sufficient quantities to meet demand, so while many older 11mm Mle. 1873 and 1874 revolvers were reissued it was only a stopgap measure. The unique conditions of the Great War, with its close-range trench fighting, made it a "handgun war" and most combatant armies issued sidearms to their troops in greater numbers then ever before or since. Officers, artillery and machine gun crews, signalers, messengers, transport drivers, military police, and every infantryman who could get his hands on one wanted, and in many cases actually needed, a pistol. In an attempt to maintain some level of standardization the French first ordered revolvers. Several gun makers in the Basque region of Spain were already producing copies of Smith & Wesson and Colt revolvers and these were easily adapted to the French 8x27R revolver cartridge. But more weapons were needed, and quickly. In response to this demand, Basque gunmakers developed a very simple semi-automatic pistol known generically as the "Eibar"-type pistols. They were produced upon a rough pattern, but the dozens of different firms involved, and the cottage-type industry used to supply parts, assured a level of interchangeability of near zero, while quality varied from acceptable to abysmal. The basic pattern was a single-action pistol, chambered for the 7.65mm Browning (32 ACP) cartridge, with a seven- or eight-round magazine. Safeties were usually simple levers above the trigger that could be rotated 90 degrees. External and internal hammer, and striker-fired designs were all bought by the desperate French. Their service record left much to be desired because of both the weak cartridge and poor reliability, but the ever-weapon-hungry French army was in no position to complain. Records show that the French purchased over 700,000 "Eibar" pistols during the war, with about 580,000 still in use or storage after the end of hostilities.

In the early 1920s the French army began trials to develop a new family of small arms to replace the worn and aging weapons left over from the war. Priority was given to the development of a new rimless cartridge and light machinegun, and after that a new infantry rifle. Handgun development was put on the slow track, especially with the financial restraints that became endemic during the Depression. In the meantime, production of the old Mle. 1892 revolver continued and stocks of Eibar-type pistols were sifted through for weapons suitable for service.

In 1927 the French announced that any new service pistol must be chambered for a 7.65mm cartridge developed by l'Etablissement d'Experience Technique de Versailles in the late 1920s. It was based on the experimental 30 Pedersen cartridge that the U.S. Army developed in 1918 for use in the semi-secret Pedersen device. In 1926 the French army purchased 50,000 rounds of 30 Pedersen ammunition from the Remington Arms Company for test purposes. Their final design, the 7.65x20 or 7.65 Longue, used an 85-grain round-nosed FMJ bullet at a velocity of 1175 fps. The new cartridge was produced at the government arsenals in Puteaux, Rennes, Tarbes, Bourges, and Toulouse, and by the private firms of la Societe Industrielle Meridionale and Gevelot. The 7.65x20 was also used in the MAS Mle. 1938 submachine gun. The adoption of that lackluster cartridge when such proven rounds as the 9x19 Parabellum and 45 ACP were available is just another example of the nationalistic desire to have something of their own winning out over common sense. Of course, this was not the exclusive domain of the French, but it appears to have persisted longer with them then with most other armies.

In 1934 the French government announced a competition to find a new semi-automatic service pistol and invited both government facilities and private firms to enter designs for consideration. From the beginning the testing commission stated as a precondition that all designs submitted must have an en-bloc firing mechanism that could be removed from the pistol as a unit when field-stripped. MAS offered a pistol, the SE MAS 1932 Type A No. 4, which would serve as the basis for the later Mle. 1935S; Fabrique National of Herstal, Belgium, provided two pistols which were prototypes of the soon to be famous M1935 High Power; a 7.65x20 Star pistol was entered by the Spanish firm; and finally, the private French firm of Societe Alsacienne de Constructions Mecaniques (SACM) entered a pistol developed by one of their engineers, Charles Gabriel Petter.

Monseiur Petter is a mysterious figure. According to French sources he was of Swiss origin and it seems apparent he was a trained engineer. Like many fellow Switzers, he had served in the French Foreign Legion during the Great War and rose to officer's rank before being wounded, some say in the Great War, while others insist it was in the 1920s in North Africa. After being discharged from the Legion he remained in France and in the early 1930s went to work for SACM, a firm specializing in steam turbines and other heavy equipment. None of my sources indicate whether he had a penchant for firearms design or if it was just another job assignment while at SACM. In 1934 he was granted several patents involving semi-automatic pistols, and it was his design that was entered in the French army trials by SACM.

The SACM/Petter pistol was based to a large degree on the Browning/Colt/FN short recoil, locked breech system in which the slide and barrel are locked together by lugs on top of the barrel mating with grooves on the inside the pistol's slide. As the slide and barrel recoil together the barrel articulates on a link and is pulled down and free of the slide, which then continues rearward, and extracts and ejects the empty cartridge case. Petter's patents covered a module that contained all the pistol's lockwork and could be removed from the pistol as a unit, greatly facilitating cleaning and repairs. He also designed the captive-type recoil spring unit with a full length guide rod for support, and the dual barrel links. Trials dragged out for so long that the French army was forced to purchase a large number of Spanish Astra 9mm M400 semi-auto pistols so as to equip their troops with a modern sidearm until a domestically-made pistol could be obtained. In 1936 the SACM design was finally adopted as the Pistolet Automatique Modele 1935A.

Unlike earlier Gallic military weapons, the Mle. 1935A looks quite

The French pistol has a loaded-chamber indicator at the top of the slide to the rear of the ejection port. It allows visual as well as tactile confirmation of a round in the chamber.

When field-stripped, the Mle. 1935A reveals its Browning ancestry. Note the captive recoil spring/guide rod unit and the removable en bloc firing mechanism with the hammer still in place.

"normal;" you might even say elegant with its slim, fine lines. The shape of the grip is especially attractive and provides a very comfortable and naturally pointing gun. It has a 4.3 inch barrel with four rifling grooves using a right-hand twist. Unlike other Browning-style pistols there are two separate articulating links on the underside of the barrel. The pistol has an overall length of 7.6 inches, a height of 4.75 inches, and it weighs 26 ounces with an empty eight-shot magazine. Controls, except for the safety, are pure Browning: a magazine release and slide stop are both located on the left behind and above, respectively, the trigger. A loaded-chamber indicator is located on the top of the slide directly behind the ejection port where it is visible to the shooter, or can be felt in the dark.

The safety consists of a small serrated lever located on the top left of the slide beneath the rear sight. The pistol is put on Safe by rotating the lever upward 90 degrees, which interposes a solid steel shaft between the hammer and firing pin. There is also a magazine disconnect safety, which prevents the pistol from being fired when the magazine is removed. The trigger is a pivoting design that moves twin trigger bars to trip the sear. Sights consist of a fixed U-notch rear and a low blade front, which is adjustable for windage. Grips were made of black plastic, and post-WWII production pistols have a prominent lanyard ring adorning the lower left corner of the grip frame. Most Mle. 1935A pistols were phospated (Parkerized) and then had a black enamel finish baked over that. Some post-1944 pistols, and arsenal rebuilds, will be found with only the dark gray phosphate finish. Materials and workmanship appear to be first class, although war-time pistols can be a bit rough. Several different patterns of leather and webbing holsters were issued to the different branches of the armed forces and gendarmerie. All had a pocket for a spare magazine and some included a cleaning rod.

SACM delivered the first pistols in late 1937. Production was fitful with several technical problems causing the assembly line to shut down for long periods of time, and by the fall of France in June 1940 only 10,700 pistols had been produced. In an attempt to acquire additional modern weapons, the French army also accepted the aforementioned MAS pistol, now designated the Mle. 1935S.

The Mle. 1935S had more angular lines and a greatly simplified locking system. Instead of the dual lugs on top of the barrel, the S model's slide and barrel were locked together by a shoulder on the top of the barrel which was raised against the forward edge of the ejection port. Stamped metal parts also replaced several that were forged in the A pistol, and as far as I can ascertain there is no interchangeability of parts between the two guns, not even the magazines. Less then 1500 MAS pistols had been delivered when the Germans occupied St. Etienne. Apparently no Mle. 1935Ss were produced during the occupation, but production of the SACM pistol continued with 23,850 units being accepted by the German Waffenamt (ordnance department) and given the Fremden Gerat designation 7.65mm Pistole 625(f). Mle. 1935As produced during the war years exhibit poorer fit and finish when compared to peacetime guns. Pistols were also supplied to the French Vichy government and security forces, while quite a number also saw service with Free French units before replaced with British and American weapons.

France continued production of both the A and S pistols after the war to supplement the large numbers of American, British, and ex-German handguns in service. SACM continued to produce the Mle. 1935A until early 1950 but Mle. 1935S production was more complicated. At first it was transferred from MAS to the private firm of Manufrance, which made about 10,000 units. After that they were made at the government arsenal at Chatellerault (MAC) and by another private concern, Societe d'Applications Generales d'Electrique et de Mecanique (SAGEM). Others were assembled from slides produced by the arsenal at Tulle (MAT) and frames made by MAC. Production numbers vary according to the source, but it appears that about 80,500 Mle. 1935As were made by 1950 and an additional 83,000 Mle. 1935Ss were produced before production ended around 1956.

Medlin and Doane quote from a training booklet printed by the military academy at St. Cyr, the French equivalent of West Point, that states officers were to be issued A-model pistols while enlisted personnel received the less elegant S models. They were both replaced in French service by the Modele 1950, which was finally chambered for a serious cartridge, the 9x19 Parabellum. But France's military commitments in Asia, the Middle East, and North Africa ensured that both M1935 pistols remained in front line service until the late 1960s. Many M1935 pistols were captured by Algerian and Viet Minh forces and used in their guerrilla war against the French, with some still turning up in the fighting against the Americans in Viet Nam. Others were supplied as military aid to former French colonies in Africa and are still in use today. French Gendarmes and some local police forces continued to use Mle. 1935 pistols well into the 1970s.

Left to right: 32 ACP, 7.65x20 French, and 9x19 Parabellum. The French round was based on the U.S. 30 Pedersen cartridge that was developed in 1918 for the Pedersen device.

DKT, Inc., and Old Western Scrounger are able to supply custom reloaded 7.65x20 ammunition. Both types use a 93-grain FMJ bullet.

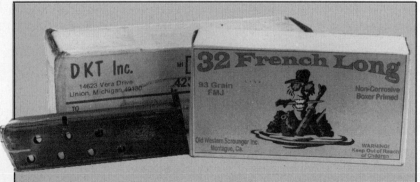

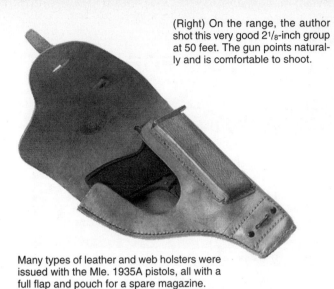

Many types of leather and web holsters were issued with the Mle. 1935A pistols, all with a full flap and pouch for a spare magazine.

(Right) On the range, the author shot this very good 2 1/8-inch group at 50 feet. The gun points naturally and is comfortable to shoot.

Stripping the Mle. 1935A

Remove the magazine and draw back the slide to verify the chamber is empty. Cock the hammer and pull the slide back about 3/16-inch and hold it with the right hand. With the forefinger of the left hand, push the slide stop lever's shaft from right to left and remove the stop, which allows you to remove the slide off the front of the frame. Push the recoil spring/guide rod unit forward slightly and lift out, and then lift the barrel out of the slide. Grasp the firing mechanism with one hand while pushing up on the back of the unit, which is fitted into the back strap. The mechanism could be a tight fit so it might require a few taps with a non-marring tool to loosen it. Reassemble in reverse order.

For this article I test-fired a Mle. 1935A pistol that I bought at a local gun show in 1995. It is in excellent mechanical condition, and has a very good bore and a dark gray phosphate finish. Marked on the left side of the frame is *Mle. 1935A*, while above the trigger is *S.A.C.M.* superimposed by the serial number. The magazine is finished in a black enamel and the baseplate is stamped *35A*. The lanyard ring indicates a post-war pistol and checking the serial number against the list in Medlin and Doane's book shows the year of manufacture to be 1948.

Naturally my first desire was to test fire my new acquisition, which is when I ran into a minor problem—ammunition. It seems that 7.65x20 ammo is virtually unobtainable on the surplus market. Fortunately I located two sources of custom reloaded ammo. Chuck Richardson of DKT, Inc., kindly sent me a supply that he makes using custom trimmed and formed Bertram brass and 93-grain FMJ bullets that were originally intended for the 7.65mm Parabellum cartridge. The Old Western Scrounger ammunition uses the same 93-grain bullet with cases made from 32 S&W Long brass that has been trimmed, had the rim removed, and an extractor groove machined in the case. Both brands of ammunition were of excellent quality and I did not experience a single misfire, failure to feed or eject in the seventy-odd rounds I fired.

I set up a target at 50 feet and fired Mle. 1935A across my pistol rest. The sights were on the narrow side and required all my attention, but while the trigger pull had a mushy take-up the let-off was surprisingly light. All my attempts put eight rounds into well-centered groups under 3 inches, with the DKT ammo providing the smallest at a nice 2 1/8 inches. Then I paced off what I guesstimated was 7 yards and set up a standard IPSC-type Milpark target. The issue holster I had, with its full flap and tight fit, was not conducive to quick draw exercises, so I held the Mle. 1935A at port arms and lifted to fire a series of rapid one-handed double taps. I'm sure this style would have met with the approval of a 1930s French training officer. Despite the minuscule sights, the pistol's elegantly curved grip provided excellent pointing characteristics which, combined with the very mild recoil of the cartridge, made shooting it a pleasure. I had no difficulty putting all of my shots on those parts of the target that counted.

In the past two years I have test-fired a number of what I diplomatically refer to as "unique" French military rifles and handguns. I found the Mle. 1935A the most "normal" looking and operating of all of these. It is a well made, elegant looking, beautiful handling, and accurate gun. But in my opinion its two glaring shortcomings—the anemic 7.65x20 cartridge and hard to operate safety—outweigh any positive qualities it may possess. I think I'll stick with my Glock 17. •

Bibliography

Ezell, Edward C. *Handguns of the World.* New York: Barnes & Noble Books, 1981.

Gander, Terry and Chamberlain, Peter. *Weapons of the Third Reich.* Garden City, NY: Doubleday and Company, Inc., 1979.

Hogg, Ian and Weeks, John. *Military Small Arms of the 20th Century.* Northbrook, IL: DBI Books, Inc., 1985.

Johnson, George B. and Lockhoven, Hans Bert. *International Armament*, Volume 1. Cologne: International Small Arms Publishers, 1965.

Josserand, Michel H. and Stevenson, Jan. A. *Pistols, Revolvers and Ammunition* (original title: *Les Pistolets, Les Revolvers et Leurs Munitions*). New York: Bonanza Books, 1972.

Medlin, Eugene and Doane, Colin. *The French 1935 Pistols.* Latham, NY: Excalibur Publications. 1995.

Medlin, Eugene and Huon, Jean. *Military Handguns of France.* Latham, NY: Excalibur Publications, 1993.

Meyer, Bernard. *Le Pistolet S.A.C.M. Modele 1935 A.* Paris: Cibles, Crepin-LeBlond Publishing, March 1995.

THE 256 NEWTON SPORTING SPRINGFIELD

by JIM FORAL

This kit-gun concept was a stepping stone to the high-pressure, high-speed rifle ammunition we shoot today.

AT THE BEGINNING of the century, rifle preferences of the mainstream big game hunter still centered on the big-bore lever actions. In a rapidly transforming America, though, the 1903 Springfield, together with the speedy 30-caliber service cartridge it fired, awakened shooters to new possibilities and stirred up new attitudes.

A small handful of independent experimenters contributed more significantly. Foremost among these was Buffalo, New York, lawyer and rifle enthusiast Chas. Newton. Between 1910 and 1914, he applied his small-bore bottleneck cartridge principle to rifle ammunition to give us the 22 Hi-Power and 250-3000 Savages. At the same time, he authored numerous magazine articles broadcasting the merits of his high-velocity rifle cartridges, all capable of an unprecedented muzzle speed of 3000 feet per second. The most notable and best promoted of these was the 256 Newton, the flatshooting wonder of the day. Newton's 256 was more powerful than the 405 Winchester, while recoiling less than a 30-30.

Newton was a fixture in that era's sporting press. This medium not only qualified him as an authority, but exposed his modern thoughts concerning rifles, allowing his ideas to be considered and accepted. He enjoyed a very sizable and loyal following.

An Idea Is Born

To capitalize on this growing and prevailing mood, Newton made a business decision to import commercial Mauser rifles chambered for his revolutionary series of high-power cartridges. He formed the Newton Arms Co. for this purpose in July of 1914. Newton and German officials at the original Mauser factory negotiated for two dozen rifles to be built around the Newton calibers. Three grades of rifles were involved. The initial shipment was ordered in May of '14 and delivered in August of that year. America's sportsmen were notified of the opportunity to buy the very first Newton rifle through display ads in the June and July issues of various outdoor magazines. The immediate reaction was intense enough that all of the original twenty-four rifles were quickly spoken for, and three more lots of Mausers were hurriedly ordered.

In the meantime, however, World War I had broken out. Germany and her entire arms manufacturing capability and output were wholly devoted

nents and manpower to relax and get the rifle into production.

A Contingency Plan

Meanwhile, persistent public inquiry motivated Newton to consider a stop-gap measure to satisfy the increasing demand for a rifle along the lines of the imported Mauser/Newton. Newton also realized the importance of keeping his now-familiar name and widely accepted high-intensity rifle/cartridge message on the minds of fickle U.S. riflemen. "How shall we avoid disappointing our friends?" was the question Newton posed to himself that summer of 1914.

He devised a plan to convert 1903 Springfield rifles into sporting rifles chambered for small-bore cartridges with his name on them. By 1914, the service rifle was familiar and accessible to nearly everyone. Blending the virtues of the two systems seemed like a saleable approach.

Newton somehow contracted with Marlin Firearms to furnish him with a supply of sporting-style stocks as well as an unspecified quantity of 256 Newton rifle barrels threaded for the

This photo of Chas. Newton was taken from his company's catalog, circa 1914.

(Below) This close up of the right side of the receiver shows the unique roller bolt lock.

to that effort. This turn of events cut off Newton's supply of rifles, effectively shutting down the importation venture. Only one shipment of Mausers was ever delivered, reportedly all in 256 Newton. Better, encouraging news was that the deluging demand for the Newton system—a strong bolt-action rifle coupled with high-velocity cartridges—became overwhelmingly evident. Alerted to this apparently underestimated market, Newton looked at manufacturing his own rifle. At first, he examined the practicality of a gun being built to his design and specifications by one of the domestic arms makers. At that time in neutral America, war orders from European governments were too abundant and too lucrative to bother with newton's trifling demands, and he decided to manufacture the rifle himself. The same wartime conditions that had tied up gunmaking plants had also dried up availability of machine tools, skilled workmen and the necessary rifle building materials. It would take some time for the unexampled demand for components receiver of the Springfield. While his rifle factory was being equipped and staffed, Newton printed up an array of circulars, flyers and catalogs marketing the Springfield conversion kits, and describing his challenges and progress in overcoming the obstacles in the way of his dream. In

The Newton series (from left): 22 Newton, 256 Newton, 30 USG (30-06), 30 Newton, 35 Newton.

(Right) Here is one of the first Newton Arms Co. advertisements touting the small-bore/high-velocity theory.

(Left) This magazine ad from *National Sportsman* magazine, September 1914, was one of the first peddling the Newton-Springfield kits.

addition, he advertised the kits in outdoor periodicals commencing in July, 1914. From a sales office at 506 Mutual Life Building in Buffalo, he shipped out sporting model stocks and 256 Newton barrels to all points of the U.S.

The 256-caliber barrels were all 24 inches long, delivered to the Newton Arms Co. office in crates containing 100 barrels each. They were threaded, chambered, blued and "ready to screw into the New Springfield action." Prospective customers were advised that the material used in the Newton barrels was the same ordnance-grade steel used by the armories to manufacture barrels for the military Springfield rifle. Barrels for the sporting Springfield were imprinted: NEWTON ARMS CO. INC. 0-0-0 BUFFALO, N.Y. 0-0-0, with the caliber designation to the right of this stamping. These Marlin-made barrels had conventional square-cut rifling and were priced at $12.50 each. The later Newton-produced segmentally rifled barrels are marked: NEWTON ARMS CO. INC. 0-0 BUFFALO N.Y. 0-0. The caliber of the rifle is underneath this marking. The words "patent pending" are found just forward of the receiver. The additional "0" for the former may have been Newton's way of identifying a barrel with one rifling style from the other. On the other hand, the difference may have distinguished contractor-supplied barrels from Newton's own.

The pistol-grip stock was fashioned from good-quality American black walnut. Like the barrels, the stocks were priced to sell at $12.50. A schnabel forend tip, very fashionable at the time, was standard. Wood finish was the common and durable varnish popular eighty years ago. Reportedly, the stock's pattern was created by Genoa, New York, master gunmaker and Newton associate Fred Adolph. It was trim, lightweight and handsomely profiled, with clean and racy lines.

The pistol grip and forend received nicely executed checkered panels. A cheekpiece that was advertised in the earliest Newton catalogs and circulars was not to be found on the stocks that were actually produced. Likewise, the advertised checkered steel buttplate was delivered as smooth. Newton's plans to have a trap in the buttplate, dependent upon a timely overseas delivery, evidently went awry. Although a pistol-grip cap was promised, it too was dropped. Instead, there is a shallow concave swale smoothing out the bottom of the grip. Stock installation was a sim-

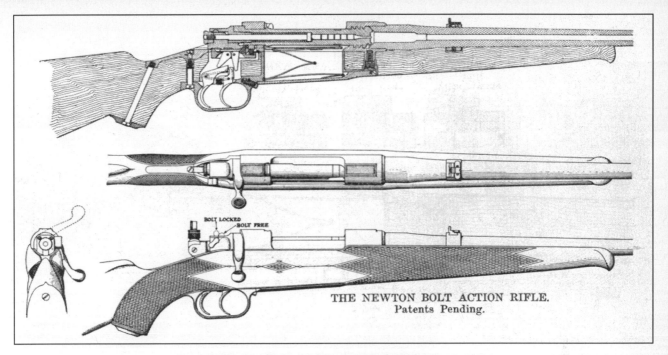

Newton Arms Co. literature of the day was the only place to see a Newton rifle, as wartime efforts slowed initial production.

ple matter. According to the catalogs, "anyone with a screwdriver can do the job."

Although no records exist, it is a safe bet the kit stocks outsold the barrels. Without doubt, more than a few riflemen passed over the offer of the 256 Newton barrel and ordered only the stock half of the package to gussy up their own Model 1903. The fad of civilian remodeling the Springfield service rifle was in full flower during this period. Men with more or less lean purses salivated over the lavish Springfield sporting stocks handmade by the ranking craftsmen just beginning to specialize in this art form. Not everyone could afford a $55 custom stock by Wundhammer or Adolph, but a fair percentage were flush enough to invest $12.50, about a week's wage for a working man, in the machine-made Newton sporting model. They bought the ready-made Newton stock, simply dropped in their arsenal Springfield barreled action, and tightened the guard screws. Aesthetically, the kit stock couldn't quite compete with the custom wood and lofty price tags of the name makers. For the cost, though, buyers of the Newton sporting stock got an affordable, stylish, semi-massproduced alternative that was far more than serviceable.

Sight selection was left to the consumer. Individual preferences in back sights varied considerably. For this reason, no rear sight was fitted to the barrel. Up front, a Springfield-style sight base was supplied, allowing the purchaser to install his favorite type of bead or blade.

Supposedly, Newton's sales facility sold and installed rear sights as an extra-cost service. The Lyman #48 was fairly new in the rifleman's marketplace at the time. It was a versatile precision sight suited to anything a hunter or target shooter might want to do with a 256 Newton. The Lyman was a favorite among the knowledgeable gun cranks of this era, and it appears that a good percentage of shooters sophisticated enough to choose a 256 Newton rifle also selected the premium receiver sight.

In 1915, a few barrels for the 256's powerful big brother, the 30 Newton cartridge, were produced and offered for sale. It is not known if these also were manufactured by Marlin or were ordinary 30-caliber Springfield barrels with the chamber enlarged to accept the larger diameter Newton shell. Folks in the market for a 30 Newton-Springfield were advised by later-edition Newton catalogs and brochures that this particular conversion required a new bolt, or the issue bolt's face had to be enlarged to accommodate the broader head of the Newton cartridge. Some modification of the weapon's extractor was also necessary.

If all a 1914-era rifleman had ever read about high-power rifle-barrel installation was from one of Newton's first catalogs, he might have gotten the impression that it was a totally uninvolved operation. Newton may have been guilty of oversimplifying the procedure. Omitted for the catalog text was any discussion about the importance of proper barrel fitting. There were no instructions or tips for removing the issue barrel, nor any mention of action wrenches, barrel vices or other specialized tools. The need for professional gunsmithing ability or assistance wasn't addressed.

Newton had a flair for salesmanship; he proposed two guns in one illusion. He advertised the enticing and perhaps misleading possibility of transforming one's Springfield into a dual-purpose rifle. For hunting, the Springfield could be converted to a 256 or 30 Newton sporter easily enough. Then, if the purchaser was prudent enough to have retained the military parts, the gun could be switched back into a 30-06 target rifle "in a few minutes" by replacing barrel and stock. A person could then take advantage of cheap government-furnished ammunition, preserving his supply of pricey Newton cartridges for when hunting season rolled around. Some would say that this versatility aspect was overstressed and oversold. The catalog's implications seem to have been that back-and-forth barrel swapping was an easily accomplished, foolproof and advisable practice. But Chas. Newton was committed to selling his barrel/stock combinations, and he was unusually accommodating to browsers of his catalogs, overcome by the high-velocity mania.

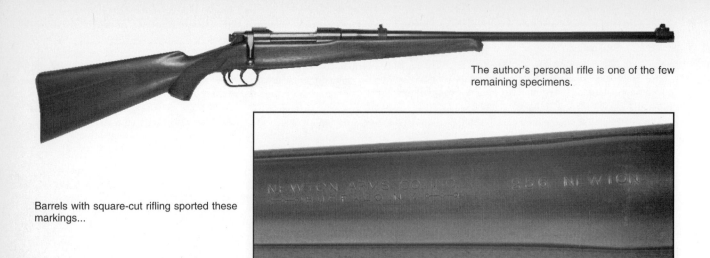

The author's personal rifle is one of the few remaining specimens.

Barrels with square-cut rifling sported these markings...

...while those with segmental rifling looked like this.

Members of an NRA-affiliated rifle club could purchase one of Uncle Sam's Springfields for $14.40 in those days. For some of the Newton-Springfield tire kickers, a new barreled action priced at $10 ("since the action is all you need") was a better value. Newton extended a personal offer to assist those interested in purchasing the Springfield, but were unsure how to go about affiliating. Used Springfields were regularly for sale at bargain prices. So Newton directed those in the market to the classified sections of the outdoor magazines as a good place to locate one. Then shooters with an action from a worn or neglected rifle could take advantage of Newton's service of refinishing Springfield actions "better than new" for $4.50.

A Personal Example

In most respects, the 256 Newton Sporting Springfield in my collection is typical of its breed. The stock conforms to the illustration in Newton's 1914-15 circulars, and the barrel is square-cut rifled in 256-caliber with the correct barrel logo. The customer-provided sights are worthy of note.

At one time, a Lyman #48 long slide receiver sight had been mounted on this rifle. The installation required the removal of some stock wood on the right side of the receiver. At some point in this Springfield's past, the Lyman was removed, and a scab of walnut expertly patched to fill the void.

However, a curious cocking piece sight still remains. This type of sight, also known as a bolt peep sight, was a familiar and effective sighting arrangement for bolt-action rifles during the early years of this century. Essentially, the bolt peep was a striker-mounted version of Lyman's popular and widely used tang sight. In terms of variety, there was a single commercial selection for the Springfield owner, and this was the Lyman #1A. These are still found on pre-1920-issue Springfield rifles converted to sporters. The Lyman sight was hinged and could be unlocked by depressing a small lever and folding it backward. The unidentified, unblued sight riding the back of this Newton-Springfield bolt follows the Lyman in basic principle. Pressing a button on the sight base allows the upright to be folded out of the way should the auxiliary rear sight be chosen.

The bolt peep sight that Chas. Newton offered as an option on his production rifle bears an extraordinarily close resemblance to the sight on this rifle. Fold-down buttons are located precisely in the same positions. The elevation adjusting systems are identical, and their locking screws happen to be in the same spots. The removable eyepiece is almost a ringer for the one later produced. Even though no maker's mark appears on this finely machined instrument, it has Newton written all over it.

If I could be allowed passage on a time machine, my first choice of destinations would be Buffalo, New York, in the fall of 1914. The mission would be to determine whose hands this rifle might have passed through.

Basically, the front sight is a narrow ivory bead suspended by dual strands of thin steel. Two sturdy inch-long ears shield this seemingly fragile arrangement. The sight is mounted on the barrel by means of a band, and these protective ears are part of an assembly fitted to the arsenal's original front sight block. The bead is centered between the ears allowing an unobstructed view of the target from all four sides. Oddly, no maker's name is stamped on this nifty piece of work. I'm not an authority on these matters, but I've never seen another like it before, nor have I bumped into it in any of the well-illustrated magazine ads or catalogs from sight manufacturers of the period.

Also present on this rifle is a unique and functional bolt lock, but a little history first. The Springfield bolt handle has the distracting habit of occasionally raising a little during hunting conditions. Pulling the trigger with an inadvertently uplifted bolt handle results in a misfire, and generally some uncharitable verbal reaction from the shooter. This is not a common or serious offense, but one that generated a measure of criticism from the most quibbling of guncranks of that day. The beloved Springfield action was under continual analysis and critique from its users since 1906, and any microscopic fault was a justifiable reason for a movement to improve the mechanism. And so when out-of-battery Springfields caused a number of shooters of that era to miss chances at game, their complaints motivated some design improvements.

This particular Newton-Springfield is equipped with someone's invention to overcome accidental raising of the bolt handle. This noiseless, spring-loaded roller device doesn't interfere with the manipulation of the bolt handle and doesn't require any conscious fiddling with to activate or disengage. With the handle lowered and the bolt locked, a small plunger fitted to the bolt lock engages a mating depression in the back of the handle. When the striker goes forward, the lock automatically disconnects, allowing follow-up shots. This gizmo prevents a coat-pocket snag or stray branch from unknowingly cracking the bolt.

While this novel and almost unknown Springfield accessory certainly dates back to the 1912-1920 era of trying to perfect that particular action, I have no idea to whom to attribute this version of the Springfield bolt lock. It is unmarked and in the white, suggesting a prototypical development.

Full Production Begins

When tooling and mechanics were in place, and material was in hand, the long-awaited production of the original Newton rifle became imminent. The focus of the entire plant was on the output of the Newton rifle and filling the substantial number of orders placed by America's performance-conscious riflemen. The Newton-Springfield conversion kit no longer fit into Newton's plans. After serving its purpose, it was abruptly phased out. Magazine ads mentioning the Newton-Springfield stocks and barrels stopped with the June 1916 issues. There wasn't the usual reference to these parts in any of the 20,000 14th Edition Newton Arms Co. catalogs issued that month. A supplement to this catalog, published shortly thereafter, promised delivery of the Newton rifle within a few months. It is here where we find the announcement that the sale of components for the 256 Newton Sporting Springfield was being discontinued.

Presumably, a few left over sporting stocks and barrels for the Springfield were in inventory when the Newton Arms Co. finally began initial production in December, 1916. Maybe this is why some early 256-caliber rifles have the non-standard square-cut rifling rather than the newly adopted segmental style. The square-cut models were cataloged as special-order options, but could be had at the regular price if the purchaser disliked the newer radical rifling type for some reason. Apparently, all one had to do was ask.

It is believed some few of the segmentally rifled barrels, together with the last of the sporting stocks, were sold to customers somewhat short of the full $40 purchase price of the trendy Newton, to be fitted to a Springfield action into hybrid Newton-Springfields. One of Newton's accountants was known to have received a factory put-together Newton-Springfield with a 30 USG (30-06) barrel with the new style segmental rifling.

Gunwriter Charles Askins Sr. assessed the Newton-Springfield for the American masses. He was disappointed that he wasn't able to file a report on the long-anticipated Newton rifle. In Newton's defense, Askins published an explanation of the conditions facing the fledgling gunmaker and urged those with rifles on order to be patient. His critique, a trifle too late to be useful, appeared in the Sept. 1916 number of *Outers Book*.

Well over a thousand shots had been fired through the barrel of Cpt. Askins' rifle in the year prior to sitting down at his typewriter. Apparently, Askins was only interested in the barrel portion of Newton's bargain and mounted the barrel in a standard-issue Springfield action and stock. His sights consisted of the favored Lyman #48 micrometer sight and the issue Springfield blade up front. Askins would have preferred to see 4 additional inches of barrel length for the increased velocity it would have afforded. He admitted to a dislike for barrels shorter than 28 inches. This petty criticism notwithstanding, Askins was convinced that the high-pressure, high-speed rifle cartridges were the wave of the future and was thoroughly impressed with Newton's small-bore high-velocity concept. He proceeded to demonstrate that the Newton system was capable of attaining a combination of flat trajectory and fine accuracy. Additionally, he provided a badly needed supply of proven loading data using available DuPont powders. There had been very little of this information prior to the publication of his *Outers Book* article.

Production figures for this model are not available, but it is presumed that the Newton-Springfield conversion kits were not sold in tremendous numbers. Chas. Askins Sr. made the same observation. "Not a great many of these arms have been issued" is what he had to say. Almost certainly, though, only a few hundred barrel/stock combinations were ever distributed.

Falling from Favor

When the original Newton rifle was released, it was well received and remained stylish for its fair share of time, but its popularity peaked and faded within a short period of years. However, the high-velocity concept Chas. Newton was instrumental in popularizing had continued to be valid and important to this day. The Newton rifle, in its various forms, as well as his proprietary series of cartridges, remained quite popular and useful through the 1920s, but new developments and other budding fads in the shooting world pushed them out of the spotlight and ultimately forced them out of fashion. Consequently, sales of the 256, 30 and 35 Newton factory ammunition became so slow that its last supplier, Western Cartridge Co., dropped all three from its product line about 1938.

Without available ammunition, the Newton-barreled Springfield's usefulness diminished. Barrels were often removed, discarded and replaced with barrels chambered for more modern cartridges. As a result, original, unbutchered specimens of the Newton/Springfield are pretty scarce items these days.

Eight decades ago, a lot of surplus Newton-made barrels and sporter stocks were attached to Springfield actions, but each ancient Springfield with a Newton-caliber barrel is not necessarily a genuine 256 Newton Sporting Springfield. To be a legitimate rifle built from a Newton kit, the gun must meet certain criteria. It must have the Marlin-made square-cut barrel, with the proper Newton Arms Co. logo placed on it. Under a limited range of circumstances, a segmentally rifled barrel could possibly be considered kosher, too. The stock, of course, must be a match for the one pictured. Any variation must be

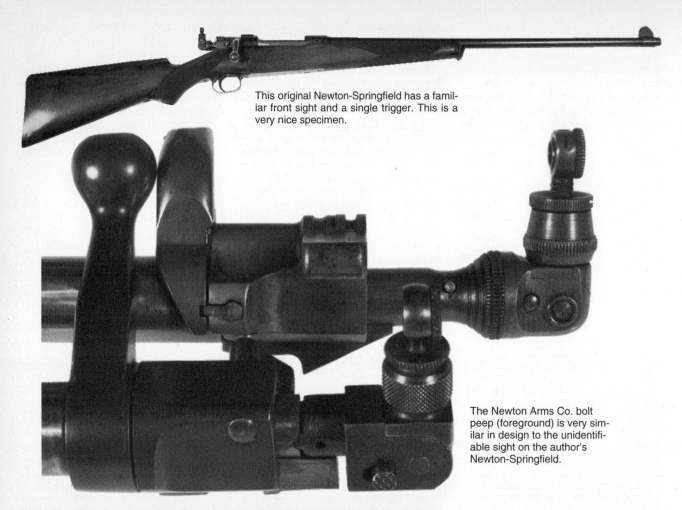

This original Newton-Springfield has a familiar front sight and a single trigger. This is a very nice specimen.

The Newton Arms Co. bolt peep (foreground) is very similar in design to the unidentifiable sight on the author's Newton-Springfield.

viewed as very suspect. Finally, the serial number of the Springfield receiver must be correct for the period of Newton's involvement in the barrel/stock sales. In other words, the receiver must have a number indicating production earlier than the 1916 discontinuance of the kits. For those made at Springfield Armory, this would be a number before 620,120. Rock Island receivers need to be lower than 234,918.

Generally speaking, it is not advisable to fire the Newton-Springfield rifles with high-pressure loads. Since the customer sometimes installed his own barrel—and there is bound to be some variance in his knowledge and skill— it may not be fitted properly. Also, the 256 Newton barrel found on an original kit rifle will certainly be screwed into one of the old brittle case-hardened "low number" Springfield receivers that we have been repeatedly warned about since 1918.

Eventually, the long awaited and longer promised Newton rifle became a reality. The consensus was that it was worth the wait. It the interim, the stop-gap Newton-Springfield had done a credible job of keeping Newton and his message in the limelight. ●

THE COMPLETE COMPACT CATALOG

GUNDEX®	.82-90
Semi-Automatic Handguns	.91-121
Competition Handguns	121-130
Double-Action Revolvers	130-141
Single-Action Revolvers	141-148
Miscellaneous Handguns	148-152
Centerfire Rifles—Semi-Automatic	152-157
Centerfire Rifles—Lever & Slide	157-161
Centerfire Rifles—Bolt Action	162-179
Centerfire Rifles—Single Shot	179-184
Drillings, Combination Guns, Double Rifles	184-187
Rimfire Rifles—Semi-Automatic	187-190
Rimfire Rifles—Lever & Slide	190-191
Rimfire Rifles—Bolt Actions & Single Shots	192-197
Competition Rifles—Centerfire & Rimfire	197-204
Shotguns—Semi-Automatic	204-209
Shotguns—Slide Actions	209-214
Shotguns—Over/Unders	214-226
Shotguns—Side By Sides	226-230
Shotguns—Bolt Actions & Single Shots	230-234
Shotguns—Military & Police	234-237
Blackpowder Single Shot Pistols	237-239
Blackpowder Revolvers	239-243
Blackpowder Muskets & Rifles	243-257
Blackpowder Shotguns	257-258
Air Guns—Handguns	259-265
Air Guns—Long Guns	265-276
Scopes & Mounts	277-286
Scope Mounts	287-290
Directory of the Arms Trade	.291
Product Directory	.292
Manufacturers' Directory	.309

GUNDEX®

A listing of all the guns in the catalog, by name and model, alphabetically and numerically.

A

A-Square Caesar Bolt-Action Rifle, 163
A-Square Hamilcar Bolt-Action Rifle, 163
A-Square Hannibal Bolt-Action Rifle, 163
A.H. Fox CE Grade Double Shotgun, 228
A.H. Fox DE Grade Double Shotgun, 228
A.H. Fox Exhibition Grade Double Shotgun, 228
A.H. Fox FE Grade Double Shotgun, 228
A.H. Fox Side-by-Side Double Shotguns, 228
A.H. Fox XE Grade Double Shotgun, 228
AA Arms AP9 Mini Pistol, 91
AA Arms AP9 Mini/5 Pistol, 91
AA Arms AP9 Pistol, 91
AA Arms AP9 Target Pistol, 91
AA Arms AR9 Semi-Auto Rifle, 152
Accu-Tek BL-9 Pistol, 91
Accu-Tek BL-380 Pistol, 91
Accu-Tek Model AT-32B Pistol, 91
Accu-Tek Model AT-32SS Pistol, 91
Accu-Tek Model AT-380B Pistol, 91
Accu-Tek Model AT-380SS Pistol, 91
Accu-Tek Model CP-9SS Pistol, 91
Accu-Tek Model CP-40SS Pistol, 91
Accu-Tek Model CP-45SS Pistol, 91
Accu-Tek Model HC-380SS Pistol, 91
Air Arms S-300 Hi-Power Air Rifle, 265
Air Arms TX-200 Air Rifle, 265
Airrow Model A-8S1P Stealth Air Gun, 265
Airrow Model A-8SRB Stealth Air Gun, 266
American Arms Bisley Single-Action Revolver, 141
American Arms Brittany Shotgun, 226
American Arms Escort Pistol, 92
American Arms Gentry Double Shotgun, 227
American Arms Mateba Auto/Revolver, 92
American Arms Regulator Single-Action Revolver, 141
American Arms Silver I Over-Under Shotgun, 214
American Arms Silver II Over-Under Shotgun, 214
American Arms Silver Sporting Over-Under Shotgun, 214
American Arms Silver Upland Lite Shotgun, 214
American Arms TS/OU 12 Shotgun, 214
American Arms TS/SS 12 Double Shotgun, 227
American Arms WS/OU 12 Shotgun, 214
American Arms WT/OU 10 Shotgun, 214
American Arms/Franchi 48/AL Auto Shotgun, 204
American Arms/Franchi 610VS Auto Shotgun, 204
American Arms/Franchi 610VSL Auto Shotgun, 204
American Arms/Franchi Alcione 2000 SX Over-Under Shotgun, 214
American Arms/Franchi Falconet 2000 Over-Under Shotgun, 215
American Arms/Franchi Sporting 2000 Over-Under Shotgun, 215
American Arms/Uberti 454 Revolver, 141
American Arms/Uberti 1860 Henry Rifle, 157
American Arms/Uberti 1866 Sporting Rifle, Carbine, 157
American Arms/Uberti 1873 Deluxe Rifle, 157
American Arms/Uberti 1873 Sporting Rifle, 157
American Arms/Uberti Yellowboy Carbine, 157
American Derringer Alaskan Surival Model Derringer, 148
American Derringer DA 38 Model, 148
American Derringer Lady Derringer, 148
American Derringer Model 4 Derringer, 148
American Derringer Model 6 Derringer, 148
American Derringer Model 7 Ultra Lightweight, 149
American Derringer Model 10 Lightweight, 149
American Derringer Model I Derringer, 148
American Derringer Texas Commemorative, 148
American Frontier 1851 Navy Concersion Single-Action Revolver, 142
American Frontier 1851 Navy Richard & Mason Conversion Revolver, 142
American Frontier 1871-1872 Open-Top Revolvers, 141
American Frontier 1871-1872 Pocket Model Revolver, 142
American Frontier Pocket Richards & Mason Navy Revolver, 141
American Frontier Remington New Army Cavalry, 142
American Frontier Remington New Model Revolver, 142
American Frontier Richards 1860 Army Single-Action Revolver, 142
AMT 22 Target Model Rifle, 197
AMT 45 ACP Hardballer Long Slide, 93
AMT 45ACP Government Model Pistol, 92
AMT 45ACP Hardballer Pistol, 92
AMT Automag II Pistol, 92
AMT Automag III Pistol, 92
AMT Automag IV Pistol, 92
AMT Back UP Double Action Only Pistol, 92
AMT Back Up II Pistol, 92
AMT Bolt-Action Rifle, 162
AMT Magnum Hunter Auto Rifle, 187
Anics A-101 Air Pistol, 259
Anics A-101 Magnum Air Pistol, 259
Anics A-201 Air Revolver, 259
Anschutz 54.18 MS REP Deluxe Silhouette Rifle, 198
Anschutz 64-MS Left Silhouette Rifle, 197
Anschutz 64-MSR Silhouette Rifle, 197
Anschutz 1416D/1516D Classic Rifles, 192
Anschutz 1416D/1516D Walnut Luxus Rifles, 192
Anschutz 1451 Super Bolt-Action Target Rifle, 198
Anschutz 1451E Sporter Target Rifle, 197
Anschutz 1451R Sporter Target Rifle, 197
Anschutz 1518D Luxus Bolt-Action Rifle, 192
Anschutz 1710D Custom Rifle, 192
Anschutz 1712D Featherweight Rifle, 192
Anschutz 1740 Monte Carlo Rifle, 162
Anschutz 1808 MSR Silhouette Rifle, 198
Anschutz 1808D-RT Super Running Target Rifle, 197
Anschutz 1827B Biathlon Rifle, 198
Anschutz 1827BT Fortner Biathlon Rifle, 198
Anschutz 1903 Match Rifle, 198
Anschutz 1907 ISU Standard Match Rifle, 198
Anschutz 1909 Target Rifle, 198
Anschutz 1911 Prone Match Rifle, 199
Anschutz 1913 Super Match Rifle, 198
Anschutz 2002 Match Air Rifle, 266
Anschutz 2002D-RT Running Target Match Air Rifle, 265
Anschutz Achiever Bolt-Action Rifle, 192
Anschutz Achiever ST Super Target Rifle, 197
Anschutz BR-50 Benchrest Rifle, 197
Anschutz Exemplar Bolt-Action Pistol, 149
Anschutz Super Match Model 2007 ISU Standard Rifle, 199
Anschutz Super Match Model 2013 Rifle, 198
Arizaga Model 31 Double Shotgun, 227
ArmaLite AR-10 (T) Rifle, 199
ArmaLite AR-10A4 Rifle, 152
ArmaLite M15A2 M4C Carbine, 153
ArmaLite M15A4 (T) Eagle Eye Rifle, 199
ArmaLite M15A4 Action Master Rifle, 199
Armoury R140 Hawken Rifle, 243
Armscor M-12Y Youth Rifle, 193
Armscor M-14P Standard Rifle, 192
Armscor M-14Y Youth Rifle, 192
Armscor M-20C Auto Carbine, 187
Armscor M-20P Standard Rifle, 187
Armscor M-30 Field Pump Shotgun, 209
Armscor M-30 Security Shotguns, 234
Armscor M-30 Special Purpose Shotguns, 234
Armscor M-30DG Special Purpose Shotgun, 234
Armscor M-30R6 Security Shotgun, 234
Armscor M-30R8 Security Shotgun, 234
Armscor M-30SAS Special Purpose Shotgun, 234
Armscor M-200DC Revolver, 130
Armscor M-1400S Classic Bolt-Action Rifle, 192
Armscor M-1400SC Super Classic Rifle, 193
Armscor M-1500S Classic Rifle, 193
Armscor M-1500SC Super Classic Rifle, 193
Armscor M-1600 Auto Rifle, 187
Armscor M-1800S Classic Bolt-Action Rifle, 163
Armscor M-1800SC Super Classic Rifle, 163
Armscor M-1911-A1P Pistol, 92
Armscor M-2000S Classic Auto Rifle, 187
Armscor M-2000SC Super Classic Rifle, 187
Armscor Model AK22 Auto Rifle, 187
Armsport 1863 Sharps Rifle, Carbine, 243
Armsport 1866 Sharps Carbine, 179
Armsport 1866 Sharps Rifle, 179
Army 1851 Percussion Revolver, 240
Army 1860 Percussion Revolver, 239
Arnold Arms African Synthetic Rifle, 162
Arnold Arms African Trophy Rifle, 162
Arnold Arms Alaskan Bush Rifle, 162
Arnold Arms Alaskan Trophy, 162
Arnold Arms Grand African Rifle, 162
Arnold Arms Grand Alaskan Rifle, 162
Arnold Arms High Country Mountain Rifle, 162
Arnold Arms Safari Rifle, 162
Arnold Arms Serengeti Synthetic Rifle, 162
Arrieta Model 557 Double Shotgun, 227
Arrieta Model 570 Double Shotgun, 227
Arrieta Model 578 Double Shotgun, 227
Arrieta Model 600 Imperial Double Shotgun, 227
Arrieta Model 600 Imperial Tiro Double Shotgun, 227
Arrieta Model 801 Double Shotgun, 227
Arrieta Model 802 Double Shotgun, 227
Arrieta Model 803 Double Shotgun, 227
Arrieta Model 871 Double Shotgun, 227
Arrieta Model 872 Double Shotgun, 227
Arrieta Model 873 Double Shotgun, 227
Arrieta Model 874 Double Shotgun, 227
Arrieta Model 875 Double Shotgun, 227
Arrieta Sidelock Double Shotguns, 227
ARS Hunting Master AR6 Air Rifle, 266
ARS/Career 707 Air Rifle, 266
ARS/Farco CO_2 Air Shotgun, 266
ARS/Farco CO_2 Stainless Steel Air Rifle, 266
ARS/Farco FP Survival Air Rifle, 266
ARS/King Hunting Master Air Rifle, 266
ARS/Magnum 6 Air Rifle, 266
ARS/QB77 Deluxe Air Rifle, 266
Astra A-70 Pistol, 93
Astra A-75 Decocker Pistol, 93
Astra A-100 Auto Pistol, 93
Autauga 32 Auto Pistol, 93
Auto-Ordnance 27 A-1 Thompson, 153
Auto-Ordnance 1911A1 45 ACP General Model Pistol, 93
Auto-Ordnance 1911A1 Competition Model Pistol, 93
Auto-Ordnance 1911A1 Competition Model Pistol, 121
Auto-Ordnance 1911A1 Pistol, 93
Auto-Ordnance 1927A1C Lightweight Thompson, 153
Auto-Ordnance Thompson M1, 153
Auto-Ordnance ZG-51 Pit Bull Pistol, 93

B

Baby Bretton Over-Under Shotgun, 215
Baby Dragoon 1848 Percussion Revolver, 240
Baby Dragoon 1849 Pocket Revolver, 240
Baby Dragoon 1849 Wells Fargo Revolver, 240
Baer 1911 Bullseye Wadcutter Pistol, 122
Baer 1911 Concept I Pistol, 94
Baer 1911 Concept II Pistol, 94
Baer 1911 Concept III Pistol, 94
Baer 1911 Concept IV Pistol, 94
Baer 1911 Concept IX Pistol, 94
Baer 1911 Concept V Pistol, 94
Baer 1911 Concept VI Pistol, 94
Baer 1911 Concept VII Pistol, 94
Baer 1911 Concept VIII Pistol, 94
Baer 1911 Concept X Pistol, 94
Baer 1911 Custom Carry Pistol, 94
Baer 1911 Premier II Pistol, 94
Baer 1911 Prowler IV Pistol, 94
Baer 1911 S.R.P. Pistol, 94
Baer 1911 Ultimate Maaaster Steel Special Pistol, 122
Baer 1911 Ultimate Master Combat Pistol, 122
Baer Lightweight 22 Pistol, 94
Baikal IJ-27M Over-Under Shotgun, 215
Baikal TOZ-34P Over-Under Shotgun, 215
Barrett Model 82A-1 Semi-Auto Rifle, 153
Barrett Model 95 Bolt-Action Rifle, 163
Beeman Bearcub Air Rifle, 267
Beeman Crow Magnum Air Rifle, 267
Beeman HW70A Air Pistol, 260
Beeman Kodiak Air Rifle, 268
Beeman Mako Air Rifle, 267
Beeman P1 Magnum Air Pistol, 259
Beeman P2 Match Air Pistol, 259
Beeman R1 Air Rifle, 268
Beeman R1 Laser MK II Air Rifle, 268
Beeman R9 Air Rifle, 268
Beeman R9 Deluxe Air Rifle, 268
Beeman R11 Air Rifle, 269
Beeman RX-1 Gas-Spring Magnum Air Rifle, 269
Beeman S1 Magnum Air Rifle, 269
Beeman Super 12 Air Rifle, 269
Beeman/Feinwerkbau 65 MKII Air Pistol, 259
Beeman/Feinwerkbau 103 Air Pistol, 259
Beeman/Feinwerkbau 300-S Mini-Match Rifle, 267
Beeman/Feinwerkbau 300-S Series Match Rifle, 267
Beeman/Feinwerkbau 603 Air Rifle, 267
Beeman/Feinwerkbau P70 Air Rifle, 267
Beeman/FWB P30 Match Air Pistol, 259
Beeman/FWS C55 CO_2 Rapid Fire Pistol, 260
Beeman/HW 97 Air Rifle, 267
Beeman/Webley Hurricane Air Pistol, 260
Beeman/Webley Nemesis Air Pistol, 260
Beeman/Webley Tempest Air Pistol, 260
Benelli Black Eagle Competition Auto Shotgun, 204
Benelli Black Eagle Shotgun, 205
Benelli Executive Series Shotguns, 205
Benelli M1 Sporting Special Auto Shotgun, 205
Benelli M1 Super 90 Camouflage Field Shotgun, 205

82 GUNS ILLUSTRATED

Benelli M1 Super 90 Field Auto Shotgun, 205
Benelli M1 Super 90 Tactical Shotgun, 235
Benelli M3 Super 90 Pump/Auto Shotgun, 235
Benelli Montefeltro Super 90 20-Gauge Shotgun, 205
Benelli Montefeltro Super 90 Shotgun, 205
Benelli MP90S Match Pistol, 122
Benelli MP95E Match Pistol, 122
Benelli Sport Shotgun, 204
Benelli Super Black Eagle Camouflage Shotgun, 205
Benelli Super Black Eagle Custom Slug Gun, 205
Benelli Super Black Eagle Limited Edition Shotgun, 205
Benelli Super Black Eagle Shotgun, 205
Benelli Super Black Eagle Shotgun, 205
Benjamin Sheridan 397C Pneumatic Carbine, 269
Benjamin Sheridan CO_2 Air Rifles, 269
Benjamin Sheridan CO_2 Pellet Pistols, 260
Benjamin Sheridan Pneumatic Air Rifle, 269
Benjamin Sheridan Pneumatic Pellet Pistols, 260
Beretta 682 Continental Course Sporting Over-Under Shotgun, 216
Beretta 682 Gold Skeet Over-Under Shotgun, 215
Beretta 682 Gold Sporting Over-Under Shotgun, 216
Beretta 682 Gold Super Trap Top Combo Shotgun, 215
Beretta 682 Gold Trap Over-Under Shotgun, 215
Beretta 686 Onyx Over-Under Shotgun, 216
Beretta 686 Onyx Sporting Over-Under Shotgun, 215
Beretta 686 Silver Essential Shotgun, 216
Beretta 686 Silver Pigeon Over-Under Shotgun, 216
Beretta 686 Silver Pigeon Skeet Shotgun, 215
Beretta 686 Silver Pigeon Sporting Shotgun, 216
Beretta 686EL Gold Perdiz Shotgun, 216
Beretta 687 Diamond Pigeon EELL Sporter Shotgun, 216
Beretta 687 EELL Diamond Pigeon Skeet Shotgun, 215
Beretta 687 EELL Diamond Pigeon Trap Shotgun, 215
Beretta 687 EELL Diamond Pigeon Trap Top Mono Shotgun, 215
Beretta 687 Silver Pigeon Sporting Combo Shotgun, 216
Beretta 687 Silver Pigeon Sporting Shotgun, 216
Beretta 687EELL Diamond Pigeon Combo Shotgun, 216
Beretta 687EELL Diamond Pigeon Shotgun, 216
Beretta 687EL Gold Pigeon Shotgun, 216
Beretta 687EL Gold Pigeon Sporting Over-Under Shotgun, 216
Beretta AL390 Gold Mallard Auto Shotgun, 206
Beretta AL390 Silver Mallard Auto Shotgun, 206
Beretta AL390 Sport Skeet Auto Shotgun, 206
Beretta AL390 Sport Sporting Auto Shotgun, 206
Beretta AL390 Sport Trap Auto Shotgun, 206
Beretta AL390 Super Skeet Auto Shotgun, 206
Beretta AL390 Super Trap Auto Shotgun, 206
Beretta AL390 Waterfowl/Turkey Auto Shotgun, 206
Beretta ASE Gold Skeet Over-Under Shotgun, 215
Beretta ASE Gold Sporting Clays Shotgun, 216
Beretta ASE Gold Trap Combo Shotgun, 215
Beretta ASE Gold Trap Over-Under Shotgun, 215
Beretta Express SS0 O/U Double Rifles, 185
Beretta Express SS06 Gold O/U Double Rifle, 185
Beretta Express SS06 O/U Double Rifle, 185
Beretta Model 21 Bobcat Pistol, 95
Beretta Model 80 Cheetah Series DA Pistols, 95
Beretta Model 84 Cheetah Pistol, 95
Beretta Model 85 Cheetah Pistol, 95
Beretta Model 86 Cheetah Pistol, 95
Beretta Model 87 Cheetah Pistol, 95
Beretta Model 89 Gold Standard Pistol, 122
Beretta Model 92D Centurion Pistol, 95
Beretta Model 92D Pistol, 95
Beretta Model 92F Stainless Pistol, 95
Beretta Model 92FS Centurion Pistol, 95
Beretta Model 92FS EL Pistol, 95
Beretta Model 92FS Pistol, 95
Beretta Model 96 Centurion Pistol, 95
Beretta Model 96 combat Pistol, 122
Beretta Model 96 Pistol, 95
Beretta Model 96 Stock Pistol, 122
Beretta Model 96D Centurion Pistol, 95
Beretta Model 96D Pistol, 95
Beretta Model 452 Sidelock Shotgun, 227
Beretta Model 455 SxS Express Rifle, 184
Beretta Model 470 Silver Hawk Shotgun, 227
Beretta Model 505 Trap, Skeet, Sporting Over-Under Shotguns, 215
Beretta Model 506 EELL Field Over-Under Shotgun, 215
Beretta Model 506 Trap, Skeet, Sporting Over-Under Shotguns, 215
Beretta Model 509 Over-Under Shotgun, 215
Beretta Model 686 Essential Over-Under Shotgun, 215
Beretta Model 950 Jet fire Pistol, 95
Beretta Model 1201FP Auto Shotgun, 235
Beretta Model 3032 Tomcat Pistol, 96
Beretta Model 8000/8040 Cougar Pistol, 96
Beretta Model 8000D/8040D Cougar Pistol, 96
Beretta Over/Under Field Shotguns, 216
Beretta Pintail Auto Shotgun, 205
Beretta Sporting Clays Shotguns, 216
Beretta Ultralight Over-Under Shotgun, 215

Bernardelli Model G9 Target Pistol, 123
Bernardelli Model USA Pistol, 96
Bernardelli P. One DA Pistol, 96
Bernardelli P. One Practical VB Pistol, 96
Bernardelli PO18 Compact DA Pistol, 96
Bernardelli PO18 DA Pistol, 96
Bersa Series 95 Pistol, 97
Bersa Thunder 22 Pistol, 97
Bersa Thunder 380 DLX Pistol, 96
Bersa Thunder 380 Pistol, 96
BF Ultimate Silhouette Pistol, 123
Bill Hanus Classic 600 Series Double Shotgun, 228
Blaser R93 Bolt-Action Rifle, 163
Blaser R93 Deluxe Rifle, 163
Blaser R93 Safari Deluxe Rifle, 163
Blaser R93 Safari Rifle, 163
Blaser R93 Safari Super Deluxe Rifle, 163
Blaser R93 Super Deluxe Rifle, 163
Bond Arms Protector Derringer, 149
Bostonian Percussion Rifle, 243
BRNO TAU-7 Match Pistol, 261
BRNO TAU-200 Air Rifle, 269
BRNO TAU-200 Junior Match Air Rifle, 269
BRNO ZKM 611 Auto Rifle, 187
Brolin L45T Compact Pistol, 97
Brolin Legend L45 Standard Pistol, 97
Brolin Legend L45C Compact Pistol, 97
Brolin Max Velocity SM100 Air Rifle, 269
Brolin P45C Compact Carry Comp Pistol, 97
Brolin Patriot P45 Comp Standard Carry-Comp Pistol, 97
Brolin Patriot P45T Comp Compact Carry-Comp Pistol, 97
Brolin Pro-Comp Competition Pistol, 123
Brolin PRO-Stock Competition Pistol, 123
Brolin TAC Series Model TAC-11 Tactical 1911 Pistol, 123
Brown Model One Single Shot Rifle, 179
Browning 40 S&W Hi-Power Mark III Pistol, 98
Browning 425 Golden Clays Shotgun, 217
Browning 425 Sporting Clays Shotgun, 217
Browning 425 WSSF Over-Under Shotgun, 217
Browning A-Bolt Hunter Shotgun, 230
Browning A-Bolt II Composite Stalker Rifle, 164
Browning A-Bolt II Eclipse M-1000 Rifle, 164
Browning A-Bolt II Eclipse Rifle, 164
Browning A-Bolt II Gold Medallion Rifle, 164
Browning A-Bolt II Hunter Rifle, 164
Browning A-Bolt II Medallion Left-Hand Rifle, 164
Browning A-Bolt II Medallion Rifle, 164
Browning A-Bolt II Micro Medallion Rifle, 164
Browning A-Bolt II Rifle, 164
Browning A-Bolt II Short Action Rifle, 164
Browning A-Bolt II Stainless Stalker Rifle, 164
Browning A-Bolt II Varmint Rifle, 164
Browning A-Bolt Shotgun, 230
Browning A-Bolt Stalker Shotgun, 230
Browning Auto-5 Light 12 and 20 Auto Shotgun, 206
Browning Auto-5 Light 12 Buck Special Auto Shotgun, 206
Browning Auto-5 Magnum 12 Auto Shotgun, 206
Browning Auto-5 Stalker Auto Shotgun, 206
Browning Auto-22 Grade VI Rifle, 187
Browning Auto-22 Rifle, 187
Browning Back Mark 22 Pistol, 98
Browning Back Mark Plus Pistol, 98
Browning BAR Mark II Lightwight Semi-Auto Rifle, 153
Browning BAR Mark II Safari Magnum Rifle, 153
Browning BAR Mark II Safari Semi-Auto Rifle, 153
Browning BDA-380 DA Pistol, 99
Browning BDM Double Mode Pistol, 98
Browning BDM Practical Pistol, 98
Browning BL-22 Lever-Action Rifle, 190
Browning BPM-D Pistol, 98
Browning BPR Pump Rifle, 157
Browning BPS Buck Special Pump Shotgun, 210
Browning BPS Game Gun Deer Special Shotgun, 210
Browning BPS Game Gun Turkey Special Shotgun, 210
Browning BPS Pigeon Grade Pump Shotgun, 210
Browning BPS Pump Shotgun Ladies and Youth, 210
Browning BPS Pump Shotgun, 210
Browning BPS Stalker Pump Shotgun, 210
Browning BRM-DAO Pistol, 98
Browning BT-100 Trap Shotgun, 231
Browning Buck Mark Bullseye Pistol, 124
Browning Buck Mark Field 5.5 Pistol, 123
Browning Buck Mark Silhouette Pistol, 123
Browning Buck Mark Target 5.5 Gold Pistol, 123
Browning Buck Mark Target 5.5 Pistol, 123
Browning Buck Mark Target 5.5 Silver Pistol, 123
Browning Buck Mark Unlimited Silhouette Pistol, 123
Browning Buck Mark Varmint Pistol, 98
Browning Capitan Hi-Power Pistol, 98
Browning Citori Grade I Hunting Over-Under Shotgun, 216
Browning Citori Grade I Lightning Over-Under Shotgun, 216

Browning Citori Grade III Lightning Over-Under Shotgun, 216
Browning Citori Grade IV Lightning Over-Under Shotgun, 216
Browning Citori Gran Lightning Over-Under Shotgun, 216
Browning Citori Over-Under Shotgun, 216
Browning Citori Special Golden Clays Shotgun, 216
Browning Citori Special Golden Clays Shotgun, 217
Browning Citori Special Skeet Over-Under Shotgun, 217
Browning Citori Special Trap Over-Under Shotgun, 216
Browning Citori Superlight Over-Under Shotgun, 216
Browning Citori Ultra Sporter Shotgun, 217
Browning Gold 10 Auto Shotgun, 207
Browning Gold 10 Stalker Auto Shotgun, 207
Browning Gold Deer Hunter Auto Shotgun, 206
Browning Gold Hunter Auto Shotgun, 206
Browning Gold Sporting Clays Auto Shotgun, 206
Browning Hi-Power 9mm Pistol, 98
Browning Hi-Power HP-Practical Pistol, 98
Browning Light Sporting 802 ES Over-Unde Shotgun, 217
Browning Lighting BLR Lever-Action Rifle, 157
Browning Lighting BLR Long Action Rifle, 157
Browning Lightning Golden Clays Shotgun, 217
Browning Lightning Sporting Clays Shotgun, 217
Browning Micro Buck Mark Pistol, 98
Browning Micro Buck Mark Plus Pistol, 98
Browning Micro Citori Lightning, 216
Browning Micro Recoilless Trap Shotgun, 231
Browning Model 1885 BPCR Rifle, 180
Browning Model 1885 High Wall Single Shot Rifle, 180
Browning Model 1885 Low Wall Rifle, 180
Browning Model 1885 Traditional Hunter Singel Shot Rifle, 180
Browning Recoilless Trap Shotgun, 231
Browning Special Sporting Clays Shotgun, 217
Browning Special Sporting Golden Clays Shotgun, 217
Bryco Model 38 Pistol, 99
Bryco Model 48 Pistol, 99
Bryco Model 58 Pistol, 99
Bryco Model 59 Pistol, 99
BSA 240 Magnum Air Pistol, 261
BSA Magnum Airsporter Crown Grade Air Rifle, 270
BSA Magnum Airsporter RB2 Air Carbine, 270
BSA Magnum Airsporter RB2 Air Rifle, 270
BSA Magnum Goldstar Air Rifle, 270
BSA Magnum Stutzen RB2 Air Rifle, 270
BSA Magnum Supersport Air Rifle, 270
BSA Magnum Superstar MKII Air Rifle, 270
BSA Magnum Superstar MKII Carbine, 270
BSA Magnum SuperTEN Air Rifle, 270
BSA Meteor MK6 Air Rifle, 270
Bushmaster M17S Bullpup Rifle, 153
Bushmaster Shorty XM-15 E2S Carbine, 153
Bushmaster XM-15 E2S Target Model Rifle, 199
Bushmaster XM-15 E2S V-Match Rifle, 199

C

C. Sharps Arms 1875 Classic Sharps Rifle, 183
C. Sharps Arms New Model 1874 Old Reliable Rifle, 183
C. Sharps Arms New Model 1875 Business Rifle, 183
C. Sharps Arms New Model 1875 Carbine, 183
C. Sharps Arms New Model 1875 Old Reliable Rifle, 183
C. Sharps Arms New Model 1875 Saddle Rifle, 183
C. Sharps Arms New Model 1875 Sporting Rifle, 183
C. Sharps Arms New Model 1875 Target & Long Range Rifle, 183
C. Sharps Arms New Model 1885 Highwall Rifle, 183
C.S. Richmond 1863 Musket, 252
Cabanas Esproncenda IV Bolt-Action Rifle, 193
Cabanas Laser Rifle, 193
Cabanas Leyre Bolt-Action Rifle, 193
Cabanas Master Bolt-Action Rifle, 193
Cabanas Mini 82 Youth Rifle, 193
Cabanas Model R83 Bolt-Action Rifle, 193
Cabanas Pony Youth Rifle, 193
Cabela's 1858 Henry Replica, 158
Cabela's 1866 Winchester Replica, 158
Cabela's 1873 Winchester Replica, 158
Cabela's Blackpowder Shotguns, 257
Cabela's Blue Ridge Rifle, 244
Cabela's Cattleman's Revolving/Carbine, 158
Cabela's Paterson Revolver, 240
Cabela's Red River Rifle, 244
Cabela's Rolling Block Muzzleloader Carbine, 244
Cabela's Rolling Block Muzzleloader Rifle, 244
Cabela's Sharps Sporting Rifle, 180
Cabela's Sharps Sporting Rifle, 244
Cabela's Sporterized Hawken Hunter Rifle, 244
Cabela's Traditional Hawken Rifle, 244
Calico Liberty 50 Carbine, 154
Calico LIberty 100 Carbine, 154
Calico M-100FS Semi-Auto Carbine, 187
Calico M-110 Pistol, 99

30th EDITION, 1998 **83**

Carbon-15 Pistol, 99
Century Centurion 14 Sporter Rifle, 165
Century Centurion 98 Sporter Rifle, 165
Century Centurion Over-Uner Shotgun, 217
Century FEG P9R Pistol, 99
Century FEG P9RK Pistol, 99
Century Gun Dist. Model 100 Single-Action Revolver, 142
Century International L1A1 Sporter Rifle, 154
Century Swedish Sporter #38 Rifle, 164
Charles Daly Diamond DL Double Shotgun, 227
Charles Daly Diamond GTX EDI Hunter Shotgun, 219
Charles Daly Diamond GTX Skeet Over-Under Shotgun, 219
Charles Daly Diamond GXT DL Hunter Shotgun, 219
Charles Daly Diamond GXT Sporting Shotgun, 219
Charles Daly Diamond GXT Trap Shotgun, 219
Charles Daly Diamond Regent DL Double Shotgun, 228
Charles Daly Diamond Regent GTX DL Hunter Shotgun, 219
Charles Daly Diamond Regent GTX EDL Hunter Shotgun, 219
Charles Daly Empire Combination Gun, 185
Charles Daly Empire DL Hunter Over-Under Shotgun, 218
Charles Daly Empire EDL Over-Under Shotgun, 218
Charles Daly Empire Hunter Double Shotgun, 227
Charles Daly Empire Skeet Over-Under Shotgun, 219
Charles Daly Empire Sporting Over-Under Shotgun, 218
Charles Daly Empire Trap Over-Under Shotgun, 218
Charles Daly Field Hunter AE Over-Under Shotgun, 218
Charles Daly Field Hunter AE-MC Shotgun, 218
Charles Daly Field Hunter Double Shotgun, 227
Charles Daly Field Hunter Over-Under Shotgun, 218
Charles Daly Field Superior Hunter AE Shotgun, 218
Charles Daly Superior Combination Gun, 185
Charles Daly Superior Hunter Double Shotgun, 227
Charles Daly Superior Skeet Over-Under Shotgun, 218
Charles Daly Superior Sporting Over-Under Shotgun, 218
Chipmunk Single Shot Rifle, 193
Churchill Turkey Auto Shotgun, 207
Cimarron 1860 Henry Replicas, 158
Cimarron 1866 Winchester Replicas, 158
Cimarron 1873 30" Express Rifle, 158
Cimarron 1873 Frontier Six Shooter Revolver, 142
Cimarron 1873 Saddle Ring Carbine, 158
Cimarron 1873 Short Rifle, 158
Cimarron 1873 Sporting Rifle, 158
Cimarron New Thunderer Revolver, 143
Cimarron rough Rider Artillery Model Single-Action Revolver, 142
Cimarron U.S. Cavalry Model Single-Action Revolver, 142
Colt 22 Automatic Pistol, 99
Colt 22 Target Pistol, 99
Colt 1847 Walker Percussion Revolver, 240
Colt 1849 Pocket Dragoon Revolver, 240
Colt 1851 Navy Percussion Revolver, 240
Colt 1860 Army Percussion Revolver, 240
Colt 1860 Cavalry Model Percussion Revolver, 240
Colt 1861 Navy Percussion Revolver, 240
Colt 1862 Pocket Police Trapper Model Revolver, 240
Colt Anaconda Revolver, 130
Colt Army Police Percussion Revolver, 239
Colt Combat Commander Pistol, 100
Colt Combat Target Model Pistol, 101
Colt DS-II Revolver, 130
Colt Game Master 50 Rifle, 244
Colt Gold Cup Trophy MK IV/Series 80 Pistol, 124
Colt Government Model 380 Pistol, 100
Colt Government Model MK IV/Series 80 Pistol, 100
Colt Government Model Pocketlite 380 Pistol, 100
Colt King Cobra Revolver, 131
Colt Lightweight Commander MK IV/Serier 80 Pistol, 100
Colt Match Target Competition HBAR II Rifle, 199
Colt Match Target Competition HBAR Rifle, 199
Colt Match Target HBAR Rifle, 199
Colt Match Target Lightweight Rifle, 154
Colt Match Target Model Rifle, 199
Colt Model 1861 Musket, 244
Colt Model 1991 A1 Commander Pistol, 100
Colt Model 1991 A1 Compact Pistol, 100
Colt Model 1991 A1 Pistol, 100
Colt Mustang 380 Pistol, 100
Colt Mustang Pocketlite Pistol, 100
Colt Officer's ACP MK IV/Series 80 Pistol, 100
Colt Pony Pistol, 100
Colt Pony Pocketlite LW Pistol, 100
Colt Python Revolver, 131
Colt Singel-Action Revolver, 143
Colt Third Model Dragoon Revolver, 241
Colt Walker 150th Anniversary Percussion Revolver, 240
Competitor Single Shot Pistol, 124

Connecticut Valley Classics Classic Field Shotgun, 218
Connecticut Valley Classics Classic Field Watefowler Shotgun, 217
Connecticut Valley Classics Classic Flyer Shotgun, 217
Connecticut Valley Classics Classic Skeet Shotgun, 218
Connecticut Valley Classics Classic Sporter Shotgun, 217
Connecticut Valley Classics Women's Classic Sporter Shotgun, 217
Cook & Brother Confederate Carbine, 244
Coonan 357 Magnum Classic Model Pistol, 101
Coonan 357 Magnum Pistol, 101
Coonan Compact Cadet 357 Magnum Pistol, 101
Cooper Arms Custom Classic Rifle, 193
Cooper Arms Model 21 Varmint Extreme Rifle, 165
Cooper Arms Model 22 Pro Varmint Extreme Rifle, 165
Cooper Arms Model 36 BR-50 Rifle, 200
Cooper Arms Model 36 Classic Sporter Rifle, 193
Cooper Arms Model 36 Featherweight Rifle, 193
Cooper Arms Model 36 Montana Trailblazer Rifle, 193
Cooper Arms Model 40 Centertire Sporter Rifle, 165
Copperhead Black Fang Pistol, 261
Copperhead Black Fire Air Rifle, 271
Copperhead Black Lightning Air Rifle, 271
Copperhead Black Serpent Air Rifle, 271
Copperhead Blsck Venom Pistol, 261
Crosman Auto Air II Pistol, 261
Crosman Model 66 Powermaster Air Rifle, 271
Crosman Model 66RT Powermaster Air Rifle, 271
Crosman Model 357 Series Air Pistol, 261
Crosman Model 664GT Powermaster Air Rifle, 271
Crosman Model 664X Powermaster Air Rifle, 271
Crosman Model 760 Pumpmaster Air Rifle, 271
Crosman Model 760SB Pumpmaster Air Rifle, 271
Crosman Model 782 Black Diamond Air Rifle, 271
Crosman Model 795 Spring Master Air Rifle, 271
Crosman Model 1008 Repeat Air Pistol, 261
Crosman Model 1008SB Repeat Air Pistol, 261
Crosman Model 1077 Repeatair Rifle, 271
Crosman Model 1077SB Silver Series Air Rifle, 271
Crosman Model 1077W Air Rifle, 271
Crosman Model 1322 Air Pistol, 262
Crosman Model 1377 Air Pistol, 262
Crosman Model 1389 Backpacker Air Rifle, 271
Crosman Model 2100 Classic Air Rifle, 271
Crosman Model 2100SB Classic Air Riflc, 271
Crosman Model 2100W Classic Air Rifle, 271
Crosman Model 2104GT Classic Air Rifle, 271
Crosman Model 2200 Magnum Air Rifle, 271
Crosman Model 2200W Magnum Air Rifle, 271
Crosman Model 6645B Powermaster Air Rifle, 271
Crossfire Shotgun/Rifle, 235
Crucelegui Hermanos Model 150 Double Shotgun, 228
Cumberland Mountain Blackpowder Rifle, 244
Cumberland Mountain Plateau Rifle, 180
CVA AccuBolt In-Line Rifle, 245
CVA AccuBolt Pro In-Line Rifle, 245
CVA Apollo Shadow Rifle, 245
CVA Bobcat Hunter Rifle, 245
CVA Bobcat Rifle, 245
CVA Brown Bear Rifle, 245
CVA Buckmaster Rifle, 245
CVA Classic Double Shotgun, 257
CVA Eclipse Rifle, 245
CVA Firebolt Bolt-Action In-Line Rifle, 245
CVA Grey Wolf Rifle, 245
CVA Hawken Pistol, 237
CVA RMEF Elk Master Rifle, 245
CVA RMEF Series Rifles, 245
CVA St. Louis Hawken Rifle, 245
CVA Staghorn Rifle, 245
CVA Staghorn Rifle, 245
CVA Trappen Percussion Shotgun, 257
CVA Varmint Rifle, 246
CZ 75 Pistol, 101
CZ 83 Double-Action Pistol, 101
CZ 85 Compact Pistol, 101
CZ 100 Pistol, 101
CZ 527 Bolt-Action Model, 165
CZ 550 Bolt-Action Rifle, 165

D

D-Max Sidewinder Revolver, 143
Daewoo DH40 Fastfire Pistol, 102
Daewoo DH380 Pistol, 102
Daewoo DP51 Fastfire Pistol, 102
Daewoo DP51C Pistol, 102
Daewoo DP51S Pistol, 102
Daewoo DP52 Pistol, 102
Daewoo DR200 Semi-Auto Rifle, 154
Daewoo DR300 Semi-Auto Rifle, 154
Daisy Model 90 Match Pistol, 262
Daisy Model 288 Air Pistol, 262
Daisy Model 850 Air Rifle, 272
Daisy Model 990 Dual-Power Air Rifle, 272

Daisy Model 1938 Red Ryder Classic BB Gun, 272
Daisy/Power Line 44 Revolver, 262
Daisy/Power Line 93 Air Pistol, 262
Daisy/Power Line 717 Pellet Pistol, 262
Daisy/Power Line 747 Air Pistol, 262
Daisy/Power Line 853 Air Rifle, 272
Daisy/Power Line 856 Pump-Up Airgun, 272
Daisy/Power Line 880 Silver Anniversary Air Rifle, 272
Daisy/Power Line 922 Air Rifle, 272
Daisy/Power Line 970/920 Air Rifle, 272
Daisy/Power Line 1000 Air Rifle, 272
Daisy/Power Line 1140 Pellet Pistol, 262
Daisy/Power Line 1170 Pellet Rifle, 272
Daisy/Power Line 1700 Air Pistol, 263
Daisy/Power Line 2002 Pellet Rifle, 273
Daisy/Power Line CO_2 1200 Pistol, 263
Daisy/Power Line CO_2 1270 Pistol, 263
Daisy/Power Line Eagle 7856 Pump-Ups Air Rifle, 273
Daisy/Power Line Match 777 Pellet Pistol, 262
Daisy/Youth Line Air Rifle, 273
Daisy/Youth Line Model 95 Air Rifle, 273
Daisy/Youth Line Model 105 Air Rifle, 273
Daisy/Youth Line Model 111 Air Rifle, 273
Dakota 22 Sporter Bolt-Action Rifle, 193
Dakota 76 Classic Bolt-Action Rifle, 166
Dakota 76 Classic Rifles, 166
Dakota 76 Safari Bolt-Action Rifle, 166
Dakota 76 Varmint Rifle, 166
Dakota 416 Rigby African Rifle, 166
Dakota Long Range Hunter Rifle, 166
Dakota Longbow Tactical E.R. Rifle, 166
Dakota Model 97 Long Range Hunter Rifle, 165
Dakota Single Shot Rifle, 180
Dan Wesson Firearms 45 Pin Gun Revolver, 139
Dan Wesson Firearms Compensated Open Hunter, 140
Dan Wesson Firearms Compensated Scoped Hunter Revolver, 140
Dan Wesson Firearms FB15 Revolver, 141
Dan Wesson Firearms FB44 Revolver, 140
Dan Wesson Firearms FB715 Revolver, 141
Dan Wesson Firearms FB744 Revolver, 140
Dan Wesson Firearms Hunter Series Revolvers, 140
Dan Wesson Firearms Model 8 Revolver, 140
Dan Wesson Firearms Model 9 Revolver, 140
Dan Wesson Firearms Model 14 Revolver, 140
Dan Wesson Firearms Model 15 Gold Series Revolver, 140
Dan Wesson Firearms Model 15 Revolver, 140
Dan Wesson Firearms Model 22 Revolver, 139
Dan Wesson Firearms Model 22 Silhouette Revolver, 139
Dan Wesson Firearms Model 32M Revolver, 140
Dan Wesson Firearms Model 40 Silhouette Revolver, 140
Dan Wesson Firearms Model 41V Revolver, 140
Dan Wesson Firearms Model 44V Revolver, 140
Dan Wesson Firearms Model 45V Revolver, 140
Dan Wesson Firearms Model 322/7322 Target Revolver, 140
Dan Wesson Firearms Model 445 Supermag Revolver, 140
Dan Wesson Firearms Model 738P Revolver, 141
Dan Wesson Firearms Open Hunter Revolver, 140
Dan Wesson Firearms Scoped Hunter, 140
Davis D-Series Derringer, 149
Davis Derringer, 149
Davis Long-Bore Derringer, 149
Davis P-32 Pistol, 102
Davis P-380 Pistol, 102
Desert Eagle Magnum Pistol, 102
Dixie 1863 Springfield Musket, 246
Dixie 1874 Sharps Blackpowder Silhouette Rifle, 180
Dixie 1874 Sharps Lightweight Hunter/Target Rifle, 180
Dixie Delux Cub Rifle, 246
Dixie English Matchlock Musket, 246
Dixie Engraved 1873 Rifle, 158
Dixie Inline Carbine, 246
Dixie Magnum Percussion Shotgun, 257
Dixie Pennsylvania Pistol, 237
Dixie Sharps New Model 1859 Military Rifle, 246
Dixie Squinnel Rifle, 246
Dixie Super Cub Rifle, 246
Dixie Tennessee Mountain Rifle, 246
Dixie U.S. Model 1816 Flintlock Musket, 246
Dixie U.S. Model 1861 Springfield Rifle, 246
Dixie Wyatt Earp Revolver, 241
Downsizer Single Shot Pistol, 149
Drulov DU-10 condor Target Pistol, 263

E

E.A.A. Big Bore Bounty Hunter Single-Action Revolver, 143
E.A.A. European Model Ladies Pistol, 103
E.A.A. European Model Pistol, 103

E.A.A. Standard Grade Revolvers, 131
E.A.A. Witness Gold Team Auto Pistol, 124
E.A.A. Witness Pistol, 102
E.A.A. Witness Silver Team Auto Pistol, 124
E.A.A./HW 660 Match Rifle, 200
E.A.A./Sabatti Model 1822 Auto Rifle, 188
E.A.A./Weihranch HW 60 Target Rifle, 200
E.M.F. 1860 Henry Rifle, 158
E.M.F. 1863 Sharps Military Carbine, 247
E.M.F. 1866 Yellowboy Lever Actions, 158
E.M.F. Model 73 Lever-Action Rifle, 159
E.M.F. Sharp Rifle, 181
Eagle Arms M15A2 Golden Eagle Rifle, 200
Eagle Arms M15A2 HBAR Rifle, 154
Eagle Arms M15A4 Carbine, 154
EMF 1875 Outlaw Revolver, 143
EMF 1890 Police Revolver, 143
EMF 1894 Bisley Single-Action Revolver, 144
EMF Hartford Express Single-Action Revolver, 144
EMF Hartford Pinkerton Single-Action Revolver, 143
EMF Hartford Pinkerton Single-Action Revolver, 144
EMF Hartford Single-Action Revolver, 143
Erma EM1 Carbine, 188
Erma ER-777 Sporting Revolver, 131
Erma ESP 85A Match Pistol, 124
Erma ESP Junior Match Pistol, 124
ERMA KGP68 Pistol, 103
Erma SR-100 Precision Rifle, 200
Euroarms 1861 Springfield Rifle, 247
Euroarms Buffalo Carbine, 247
Euroarms Volunteer Target Rifle, 247

F

FAS 601 Match Pistol, 125
FAS 603 Match Pistol, 125
FAS 607 Match Pistol, 125
FEG Mark AP22 Pistol, 103
FEG Mark II AP Pistol, 103
FEG Mark II APK Pistol, 103
FEG PJK-9HP Pistol, 103
FEG SMC-22 DA Pistol, 103
FEG SMC-380 Pistol, 103
FEG SMC-918 Pistol, 103
Fort Worth Firearms Pecos Rifle, 247
Fort Worth Firearms Rio Grande Rifle, 247
Fort Worth Firearms Sabine Rifle, 247
Fort Worth HSC Target Pistol, 125
Fort Worth HSO Target Pistol, 125
Fort Worth HSS Pistol, 103
Fort Worth HSSK Pistol, 103
Fort Worth HST Target Pistol, 125
Fort Worth HSV Target Pistol, 125
Fort Worth Matchmaster Deluxe Pistol, 125
Fort Worth Matchmaster Dovetail Pistol, 125
Fort Worth Matchmaster Standard Pistol, 125
Freedom Arms Casull Model 252 Silhouette Revolver, 125
Freedom Arms Casull Model 252 Varmint Revolver, 125
Freedom Arms Field Grade Single-Action Revolver, 144
Freedom Arms Mid Fram Revolver, 144
Freedom Arms Model 353 Field Grade Revolver, 144
Freedom Arms Model 353 Premier Grade Revolver, 144
Freedom Arms Model 353 Revolver, 144
Freedom Arms Model 353 Silhouette Revolver, 144
Freedom Arms Model 555 Revolver, 144
Freedom Arms Premier Single-Action Revolver, 144
French-Style Dueling Pistol, 237

G

Gal Compact Pistol, 104
Garbi Express Double Rifle, 185
Garbi Model 100 Double Shotgun, 228
Garbi Model 101 Double Shotgun, 228
Garbi Model 103A,B Double Shotgun, 228
Garbi Model 200 Double Shotgun, 228
Gat Air Pistol, 263
GAT Air Rifle, 273
Gaucher GN1 Silhouette Pistol, 150
Gaucher GP Silhouette Pistol, 125
Gaucher GP Silhouette Pistol, 150
Glock 17 Pistol, 104
Glock 17C Pistol, 104
Glock 17L Competition Pistol, 125
Glock 19 Pistol, 104
Glock 19C Pistol, 104
Glock 20 Pistol, 104
Glock 21 Pistol, 104
Glock 22 Pistol, 104
Glock 22C Pistol, 104
Glock 23 Pistol, 104
Glock 23C Pistol, 104
Glock 24 Competition Model Pistol, 126
Glock 24C Competition Model Pistol, 126
Glock 26 Pistol, 104
Glock 27 Pistol, 104
Glock 29 Pistol, 105
Glock 30 Pistol, 105
Golan Auto Pistol, 105
Gonic GA-87 M/L Rifle, 247
Gonic GA-93 Magnum M/L Rifle, 247
Griswold & Gunnison Percussion Revolver, 241

H

Hammerli 480 Match Air Pistol, 263
Hammerli 480K Match Air Pistol, 263
Hammerli Model 160 Free Pistol, 126
Hammerli Model 162 Free Pistol, 126
Hammerli Model 208s Pistol, 126
Hammerli Model 450 Match Air Rifle, 273
Hammerli ~Model 280 Target Pistol, 126
Harper's Ferry 1803 Flintlock Rifle, 247
Harper's Ferry 1806 Pistol, 237
Harrington & Richardson 929 Sidekick Revolver, 131
Harrington & Richardson 939 Premier Revolver, 131
Harrington & Richardson 949 Western Revolver, 131
Harrington & Richardson American Revolvers, 131
Harrington & Richardson Model 928 Ultra Slug Hunter Deluxe Shotgun, 231
Harrington & Richardson SB2-980 Ultra Slug Shotgun, 231
Harrington & Richardson Sportsman 999 Revolver, 131
Harrington & Richardson Tamer Shotgun, 232
Harrington & Richardson Topper Deluxe Model 098 Shotgun, 231
Harrington & Richardson Topper Deluxe Rifled Slug Gun, 231
Harrington & Richardson Topper Junior 098 Shotgun, 231
Harrington & Richardson Topper Junior Classic Shotgun, 231
Harrington & Richardson Topper Model 098 Shotgun, 231
Harrington & Richardson Ultra Hunter Rifle, 181
Harrington & Richardson Ultra Varmint Rifle, 181
Harris Gunwordks Antietam Sharps Rifle, 181
Harris Gunworks Classic Stainless Sporter Rifle, 167
Harris Gunworks Combo M-87 Series 50-Caliber Rifles, 201
Harris Gunworks Long Range Rifle, 201
Harris Gunworks M-86 Sniper Rifle, 200
Harris Gunworks M-87 Rifle, 201
Harris Gunworks M-87R Rifle, 201
Harris Gunworks M-89 Sniper Rifle, 201
Harris Gunworks M-92 Bullpup Rifle, 201
Harris Gunworks M-93 Rifle, 201
Harris Gunworks National Match Rifle, 200
Harris Gunworks Signature Alaskan Rifle, 167
Harris Gunworks Signature Classic Sporter Rifle, 166
Harris Gunworks Signature JR. Long Range Pistol, 126
Harris Gunworks Signature Super Varminter Rifle, 166
Harris Gunworks Signature Titanium Mountain Rifle, 166
Harris Gunworks Talon Safari Rifle, 166
Harris Gunworks Talon Sporter Rifle, 167
Hawk HP9 Combo Pump Shotgun, 210
Hawk HP9 Field Pump Shotgun, 210
Hawk HP9 Lawman Pump Shotgun, 235
Hawken Rifle, 248
Heckler & Koch Mark 23 Special Operations Pistol, 105
Heckler & Koch P7M8 Pistol, 105
Heckler & Koch PSG-1 Marksman Rifle, 201
Heckler & Koch USP 45 Match Pistol, 126
Heckler & Koch USP 45 Pistol, 105
Heckler & Koch USP Compact Pistol, 105
Heckler & Koch USP Pistol, 105
Helwan Brigadier Pistol, 105
Henry Lever-Action Rifle, 190
Heritage Rough Rider Revolver, 144
Heritage Sentry Double-Action Revolver, 132
Heritage Stealth Pistol, 106
HHF Model 101 B 12 AT-DT Over-Under Shotgun, 219
HHF Model 101 B 12 ST Over-Under Shotgun, 219
HHF Model 103 B 12 ST Over-Under Shotgun, 219
HHF Model 103 C 12 ST Over-Under Shotgun, 219
HHF Model 103 D 12 ST Over-Under Shotgun, 219
HHF Model 103 F 12 ST Over-Under Shotgun, 219
HHF Model 104 A 12 ST Over-Under Shotgun, 219
HHF Model 200 A 12 ST Side-by-Side Shotgun, 228
HHF Model 202 A 12 ST Side-by-Side Shotgun, 228
Hi-Point 9mm Carbine, 154
Hi-Point 9mm Pistol, 106
Hi-Point Firearms 9mm Compact Pistol, 106
Hi-Point Firearms 40 S&W Pistol, 106
Hi-Point Firearms 45 Caliber Pistol, 106

HJS Antigua Derringer, 150
HJS Frontier Four Derringer, 150
HJS Lone Star Derringer, 150
Howa Lighting Bolt-Action Rifle, 167

I

IAR Cowboy Shotguns, 228
IAR Model 1872 Derringer, 150
IAR Model 1873 Frontier Revolver, 144
IAR Model 1873 Six Shooter Revolver, 145
IAR Model 1888 Double Derringer, 150
IBUS M17S 223 Bullpup Rifle, 155
Intratec CAT 9 Pistol, 106
Intratec CAT 45 Pistol, 106
Intratec Double-Action Pistol, 106
Intratec Sport 22 Pistol, 106
Ithaca Deerslayer II Pump Shotgun, 211
Ithaca-Navy Hawken Rifle, 248
Ithaea Model 37 Field Pump Shotgun, 210
Ithaea Model 37 Turkeyslayer Pump Shotgun, 210

J

J.P. Henry Trade Rifle, 248
J.P. Murray 1862-1864 Cavalry Carbine, 250
Jennings J-22 Pistol, 107
Jennings J-25 Pistol, 107

K

Kahr K9 DA Pistol, 107
Kareen MK II Compact Pistol, 107
Kareen MK II Pistol, 107
Kel-Tec P-11 Pistol, 107
Kel-Tec Sub-9 Semi-Auto Rifle, 155
Kemen KM-4 Extra Gold Over-Under Shotgun, 220
Kemen KM-4 Extra Luxe-A Over-Under Shotgun, 220
Kemen KM-4 Extra Luxe-B Over-Under Shotgun, 220
Kemen KM-4 Luxe-A Over-Under Shotgun, 220
Kemen KM-4 Super Luxe-A Over-Under Shotgun, 220
Kemen Over-Under Shotguns, 220
Kentuck Flintlock Pistol, 237
Kentuck Percussion Pistol, 237
Kentuckian Rifle, Carbine, 248
Kentucky Flintlock Rifle, 248
Kentucky Percussion Rifle, 248
Kimber Classic 45 Custom Pistol, 107
Kimber Classic 45 Custom Royal Pistol, 108
Kimber Classic 45 Gold Match Pistol, 108
Kimber Classic 45 Polymer Matte Pistol, 108
Kimber Model 82C Classic Bolt-Action Rifle, 193
Kimber Model 82C Custom match Bolt-Action Rifle, 193
Kimber Model 82C HS Rifle, 194
Kimber Model 82C Stainless Classic Bolt-Action Rifle, 193
Kimber Model 82C SuperAmerica Bolt-Action Rifle, 193
Kimber Model 82C SVT Bolt-Action Rifle, 194
Kimber MOdel 84C Classic Bolt-Action Rifle, 167
Kimber Model 84C Single Shot Varmint Rifle, 167
Kimber Model 84C SuperAmerica Rifle, 167
Kimber Model 84C Varmint Stainless Rifle, 167
Kimber Model K770 Classic Rifle, 168
Kimber Model K770 Super America Bolt-Action Rifle, 168
Knight BK-92 Black Knight Rifle, 248
Knight Hawkeye Pistol, 237
Knight LK-93 Wolverine Rifle, 248
Knight MK-85 Hunter Rifle, 248
Knight MK-85 Knight Hawk Rifle, 248
Knight MK-85 Predator Rifle, 248
Knight MK-85 Rifle, 248
Knight MK-85 Stalker Rifle, 248
Knight MK-95 Magnum Elite Rifle, 249
Kodiak MK. III Double Rifle, 248
Kongsberg Classic Rifle, 168
Kongsberg Thumbhole Sporter Rifle, 168
Krico Model 260 Auto Rifle, 188
Krico Model 300 Bolt-Action Rifle, 194
Krico Model 300 Deluxe Rifle, 194
Krico Model 300 SA Rifle, 194
Krico Model 300 Stutzen Rifle, 194
Krico Model 360S2 Biathlon Rifle, 201
Krico Model 360S Biathlon Rifle, 201
Krico Model 400 Match Rifle, 201
Krico Model 600 Match Rifle, 201
Krico Model 600 Sniper Rifle, 201
Krico Model 700 Bolt-Action Rifles, 168
Krico Model 700 Deluxe Rifle, 168
Krico Model 700 Deluxe S Rifle, 168
Krico Model 700 Stutzen Rifle, 168
Krieghoff Classic Big Five Double Rifle, 185
Krieghoff Classic Double Rifle, 185
Krieghoff K-80 Combo Shotgun, 220
Krieghoff K-80 Four-Barrel Skeet Set, 220
Krieghoff K-80 International Skeet Shotgun, 220

GUNDEX

Krieghoff K-80 O/U Trap Shotgun, 220
Krieghoff K-80 Single Barrel Trap Gun, 232
Krieghoff K-80 Skeet Shotgun, 220
Krieghoff K-80 Sporting Clays Over-Under Shotgun, 220
Krieghoff K-80 Unsingle Shotgun, 220
Krieghoff KS-5 Special Shotgun, 232
Krieghoff KS-5 Trap Gun, 232

L

L.A.R. Grizzly 44 Mag MK IV Pistol, 108
L.A.R. Grizzly 50 Big Boar Rifle, 168
L.A.R. Grizzly 50 Mark V Pistol, 108
L.A.R. Grizzly Win Mag MK I Pistol, 108
Laseraim Arms Serier I Pistol, 108
Laseraim Arms Series II Pistol, 108
Laseraim Arms Series III Pistol, 108
Laurona Model 83 MG Over-Under Shotgun, 220
Laurona Model 84S Super Trap Shotgun, 220
Laurona Model 85 MS Super Piegon Shotgun, 220
Laurona Model 85 MS Super Trap Shotgun, 220
Laurona Model 85 S Super Skeet Shotgun, 220
Laurona Model 85 Super Game Shotgun, 220
Laurona Silhouette 300 Sporting Clays Shotgun, 220
Laurona Silhouette 300 Trap Over-Under Shotgun, 220
Laurona Silhouette Ultra-Magnum Shotgun, 220
Laurona Super Model Over-Under Shotguns, 220
Le Mat Percussion Revolver, 241
LePage Percussion Dueling Pistol, 238
Ljutic LM-6 Super Deluxe Over-Under Shotgun, 220
Ljutic LTX Super Deluxe, 232
Ljutic Mono Gun Single Barrel Shotgun, 232
Llama Max-I Compensator Pistols, 109
Llama Max-I Pistols, 109
Llama Max-II Compact Frame Pistol, 109
Llama Micromax 380 Pistol, 109
Llama Minimax Series Pistols, 109
Llama Minimax-II Pistol, 109
Llama Thunder 380 DLX Pistol, 109
Llama Thunder 380 Lite Pistol, 109
Llama Thunder 380 Pistol, 109
London Armory 2-Band 1858 Enfield Rifle, 249
London Armory 3-Band 1853 Enfield, 249
London Armory 1861 Enfield Musketoon, 249
Lorcin L9mm Pistol, 110
Lorcin L-22 Pistol, 110
Lorcin L-25 Pistol, 110
Lorcin L-32 Pistol, 110
Lorcin L-380 Pistol, 110
Lorcin LT-25 Pistol, 110
Lorcin Over/Under Derringer, 150
LR300 SR Light Sport Rifle, 155
Lyman Cougar In-Line Rifle, 249
Lyman Deerstalker Custom Carbine, 249
Lyman Deerstalker Rifle, 249
Lyman Great Plains Hunter Rifle, 249
Lyman Great Plains Rifle, 249
Lyman Plains Pistol, 238
Lyman Trade Rifle, 250

M

Magnum Research Lone Eagle Single Shot Pistol, 150
Magtech Model 122.2 Bolt-Action Rifle, 194
Magtech Model 122.2R Bolt-Action Rifle, 194
Magtech Model 122.2S Bolt-Action Rifle, 194
Magtech Model 122.2T Bolt-Action Rifle, 194
Magtech Model 586.2-VR Pump Shotgun, 211
Magtech MT 586P Pump Shotgun, 235
Mandall/Cabanas Pistol, 150
Manurhin 73 Revolver, 132
Manurhin MR88 Revolver, 132
Manurhin MR96 Revolver, 132
Manurhin MR 73 Sport Revolver, 132
Marksman 1010 Repeater Pistol, 263
Marksman 1710 Plainsman Air Rifle, 273
Marksman 1745 BB Repeater Air Rifle, 273
Marksman 1745S BB Repeater Air Rifle, 273
Marksman 1750 BB Biathlon Repeater Air Rifle, 273
Marksman 1790 Biathlon Trainer Air Rifle, 274
Marksman 1792 Competition Trainer Air Rifle, 273
Marksman 1795 Bolt-Action Repeater Air Rifle, 274
Marksman 2000 Repeater Pistol, 263
Marksman 2005 Laserhawk Special Edition Air Pistol, 263
Marksman 2015 Laserhawk BB Repeater Air Rifle, 274
Marlin Model 9 Camp Carbine, 155
Marlin Model 15YN Little Buckaroo Rifle, 194
Marlin Model 25MN Bolt-Action Rifle, 195
Marlin Model 25N Bolt-Action Repeater, 195
Marlin Model 30AS Lever-Action Carbine, 159
Marlin Model 39AS Golden Lever-Action Rifle, 190
Marlin Model 45 Carbine, 155
Marlin Model 50DI Bolt-Action Shotgun, 232
Marlin Model 55GDL Goose Gun Shotgun, 232

Marlin Model 60 Self-Loading Rifle, 188
Marlin Model 60SS Self-Loading Rifle, 188
Marlin Model 70PSS Stainless Rifle, 188
Marlin Model 336CS Lever-Action Carbine, 159
Marlin Model 444SS Lever-Action Sporter, 159
Marlin Model 512 Slugmaster Shotgun, 232
Marlin Model 512DL Slugmaster Shotgun, 232
Marlin Model 795 Self-Loading Rifle, 188
Marlin Model 880 Bolt-Action Rifle, 194
Marlin Model 880SQ Squirrel Rifle, 194
Marlin Model 880SS Bolt-Action Rifle, 195
Marlin Model 881 Bolt-Action Rifle, 194
Marlin Model 882 Bolt-Action Rifle, 195
Marlin Model 882L Bolt-Action Rifle, 195
Marlin Model 882SS Bolt-Action Rifle, 195
Marlin Model 882SSV Bolt-Action Rifle, 195
Marlin Model 883 Bolt-Action Rifle, 195
Marlin Model 883SS Bolt-Action Rifle, 195
Marlin Model 922 Magnum Self-Loading Rifle, 188
Marlin Model 995SS Self-Loading Rifle, 189
Marlin Model 1894 Cowboy II Lever-Action Rifle, 159
Marlin Model 1894 Cowboy Lever-Action Rifle, 159
Marlin Model 1894CS Lever-Action Carbine, 159
Marlin Model 1894S Lever-Action Carbine, 159
Marlin Model 1894S Lever-Action Carbine, 159
Marlin Model 1895SS Lever-Action Rifle, 159
Marlin Model 1897 Century Limited Rifle, 191
Marlin Model 2000L Target Rifle, 202
Marlin Model 7000 Self-Loading Rifle, 188
Marlin Model MLS-50 In-Line Rifle, 250
Marlin Model MLS-54 In-Line Rifle, 250
Marlin Model MR-7 Bolt-Action Rifle, 169
Marocchi Classic Doubles Model 92 Sporting Clays Shotgun, 221
Marocchi Conquista Skeet Over-Under Shotgun, 221
Marocchi Conquista Sporting Clays Shotguns, 221
Marocchi Conquista Trap Over-Under Shotgun, 221
Marocchi Lady Sport Over-Under Shotgun, 221
Mauser Model 94 Bolt-Action Rifle, 169
Mauser Model 96 Bolt-Action Rifle, 169
Mauser Model SR 86 Sniper Rifle, 202
Maverick Model 88 Pump Shotgun, 211
Maverick Model 95 Bolt-Action Shotgun, 233
Maximum Single Shot Pistol, 151
Merhel Boxlock Double Rifles, 185
Merhel Drillings, 185
Merhel Model 90K Drilling, 185
Merhel Model 90s Drilling, 185
Merhel Model 95K Drilling, 185
Merhel Model 95S Drilling, 185
Merhel Model 140-1 Double Rifle, 185
Merhel Model 140-1.1 Double Rifle, 185
Merhel Model 150-1 Double Rifle, 185
Merhel Model 150-1.1 Double Rifle, 185
Merhel Model 160 Side-by-Side Double Rifle, 185
Merhel Model 210E Combination Gun, 185
Merhel Model 211E Combination Gun, 185
Merhel Model 213E Combination Gun, 185
Merhel Model 313E Combination Gun, 185
Merhel Over/Under Combination Guns, 185
Merhel Over/Under Double Rifles, 186
Merkel Model 47E Side-by-Side Shotguns, 229
Merkel Model 47S Side-by-Side Shotgun, 229
Merkel Model 122 Side-by-Side Shotgun, 229
Merkel Model 147 Side-by-Side Shotgun, 229
Merkel Model 147E Side-by-Side Shotgun, 229
Merkel Model 147S Side-by-Side Shotgun, 229
Merkel Model 201E Over-Under Shotgun, 221
Merkel Model 201E Skeet, Trap Over-Under Shotgun, 221
Merkel Model 201ET Trap Shotgun, 221
Merkel Model 202E Over-Under Shotgun, 221
Merkel Model 203E Over-Under Shotgun, 221
Merkel Model 247S Side-by-Side Shotgun, 229
Merkel Model 303E Over-Under Shotgun, 221
Merkel Model 347S Side-by-Side Shotgun, 229
Merkel Model 447S Side-by-Side Shotgun, 229
Mississippi 1841 Percussion Rifle, 257
Model 1885 High Wall Rifle, 181
Morini 162E Match Air Pistol, 263
Morini Model 84E Free Pistol, 127
Mossberg Model 500 Bantam Pump Shotgun, 211
Mossberg Model 500 Camo Pump Shotgun, 211
Mossberg Model 500 Ghost-Ring Shotgun, 236
Mossberg Model 500 Mariner Pump Shotgun, 236
Mossberg Model 500 Mil-Spec Shotgun, 236
Mossberg Model 500 Persuader Cruiser Shotgun, 236
Mossberg Model 500 Persuader Security Shotguns, 236
Mossberg Model 500 Sporting Pump Shotgun, 211
Mossberg Model 500 Trophy Slugster Shotgun, 211
Mossberg Model 500 Viking Pump Shotgun, 211
Mossberg Model 590 Ghost-Ring Shotgun, 236
Mossberg Model 590 Mariner Pump Shotgun, 236
Mossberg Model 590 Shotgun, 236
Mossberg Model 695 Slugster Shotgun, 233
Mossberg Model 695 Turkey Shotgun, 233
Mossberg Model 835 American Field Pump Shotgun, 212

Mossberg Model 835 Crown Grade Ulti-Mag Pump Shotgun, 212
Mossberg Model 835 Viking Pump Shotgun, 212
Mossberg Model 9200 Bantam Auto Shotgun, 207
Mossberg Model 9200 Camo Auto Shotgun, 207
Mossberg Model 9200 Crown Grade Auto Shotgun, 207
Mossberg Model 9200 Persuader Auto Shotgun, 207
Mossberg Model 9200 Trophy Auto Shotgun, 207
Mossberg Model 9200 USST Auto Shotgun, 207
Mossberg Model 9200 Viking Auto Shotgun, 207
Mossberg Model HS410 Shotgun, 236
Mountain Eagle Compact Pistol, 110
Mountain Eagle Pistol, 110
Mountain Eagle Rifle, 169
Mountain Eagle Standard Pistol, 110
Mountain Eagle Target Pistol, 110

N

Navy 1861 Percussion Revolver, 239
Navy Arms 1777 Charleville Musket, 251
Navy Arms 1816 W.T. Wickham Musket, 251
Navy Arms 1859 Sharps Cavalry Carbine, 251
Navy Arms 1859 Sharps Infantry Rifle, 250
Navy Arms 1861 Springfield Rifle, 251
Navy Arms 1863 C.S. Richmond Rifle, 251
Navy Arms 1863 Springfield Rifle, 251
Navy Arms 1866 Yellowboy Carbine, 160
Navy Arms 1866 Yellowboy Rifle, 160
Navy Arms 1873 Single-Action Revolver, 145
Navy Arms 1873 Springfield Infantry Rifle, 181
Navy Arms 1873 U.S. Artillery Model Revolver, 145
Navy Arms 1873 U.S. Cavalry Model Revolver, 145
Navy Arms 1873 Winchester-Style Carbine, 160
Navy Arms 1873 Winchester-Style Rifle, 160
Navy Arms 1874 Sharps Cavalry Carbine, 181
Navy Arms 1874 Sharps Infantry Rifle, 181
Navy Arms 1874 Sharps Sniper Rifle, 181
Navy Arms 1875 Schofield Revolver, 145
Navy Arms 1875 U.S. Cavalry Model Revolver, 145
Navy Arms 1875 Wells Fargo Schofield Revolver, 145
Navy Arms "Pinched Frame" Single-Action Revolver, 145
Navy Arms #2 Creedmoor Rolling Block Rifle, 181
Navy Arms Berdan 1859 Sharps Rifle, 250
Navy Arms Bisley Model Single-Action Revolver, 145
Navy Arms Country Boy In-Line Rifle, 250
Navy Arms Deluxe 1858 Remington-Style Revolver, 241
Navy Arms Deputy Single-Action Revolver, 145
Navy Arms Fowler Shotgun, 258
Navy Arms Hawken Hunter Rifle, Carbine, 250
Navy Arms Henry Carbine, 159
Navy Arms Iron Frame Henry Rifle, 159
Navy Arms Kodiak MK IV Double Rifle, 186
Navy Arms Le Page Dueling Pistol, 238
Navy Arms Military Henry Rifle, 159
Navy Arms Mortimer Flintlock Rifle, 250
Navy Arms Mortimer Match Rifle, 250
Navy Arms Mortimer Shotgun, 258
Navy Arms Parker-Hale Volunteer Rifle, 251
Navy Arms Parker-Hale Whitworth Military Target Rifle, 251
Navy Arms Pennsylvania Long Rifle, 251
Navy Arms Rolling Block Buffalo Rifle, 181
Navy Arms Sharps Buffalo Rifle, 181
Navy Arms Sharps Plains Rifle, 181
Navy Arms Sharps Sporting Rifle, 181
Navy Arms Smith Carbine, 251
Navy Arms Sporting Rifle, 160
Navy Arms Steel Shot Magnum Shotgun, 258
Navy Arms T & T Shotgun, 258
Navy Arms Trapper Carbine, 159
Navy Arms Tryon Creedmoor Target Model Rifle, 256
Navy Model 1851 Percussion Revolver, 241
New Advantage Arms Derringer, 151
New England Firearms Handi-Rifle, 182
New England Firearms Lady Ultra Revolver, 132
New England Firearms Standard Pardner Shotgun, 233
New England Firearms Standard Revolvers, 132
New England Firearms Super Light Rifle, 182
New England Firearms Survivor Rifle, 182
New England Firearms Survivor Shotgun, 233
New England Firearms Tracker Slug Gun, 233
New England Firearms Turkey and Goose Gun, 233
New England Firearms Ultra Revolver, 132
New Generation Snake Charmer Shotgun, 234
New Model 1858 Army Percussion Rifle, 242
New Model 1858 Army Target Model Rifle, 242
New Model 1858 Buffalo Model Rifle, 242
New Model 1858 Hartford Rifle, 242
North American Black Widow Revolver, 146
North American Companion Percussion Revolver, 241
North American Magnum Companion Percussion Revolver, 241
North American Mini-Master Revolver, 146
North American Mini-Revolvers, 145
North American Munition Model 1996 Pistol, 110

O

Olympic Arms OA-96 AR Pistol, 110
Olympic Arms PCR Servicematch Rifle, 202
Olympic Arms PCR-1 Rifle, 202
Olympic Arms PCR-2 Rifle, 202
Olympic Arms PCR-3 Rifle, 202
Olympic Arms PCR-4 Rifle, 155
Olympic Arms PCR-5 Rifle, 155
Olympic Arms PCR-6 Rifle, 155
One Pro. 45 Pistol, 110

P

Pacific Rifle Model 1837 Zephyr Rifle, 251
Para-Ordnance P12.45ER Pistol, 110
Para-Ordnance P12.45RR Pistol, 110
Para-Ordnance P13.45RR Pistol, 110
Para-Ordnance P14.45ER Pistol, 110
Para-Ordnance P14.45RR Pistol, 110
Para-Ordnance P16.45ER Pistol, 110
Para-Ordnance P-Series Pistols, 110
Pardini GP Rapid Fire Match Pistol, 127
Pardini K50 Free Pistol, 127
Pardini K58 Match Air Pistol, 263
Pardini K60 Match Air Pistol, 263
Pardini Model HP Target Pistol, 127
Pardini Model SP Target Pistol, 127
Parker Reproductions A-1 Special Side-by-Side Shotgun, 229
Parker Reproductions D Grade Side-by-Side Shotgun, 229
Pedersoli Mang Target Pistol, 238
Peifer Model TS-93 Rifle, 252
Pennsylvania Full-Stock Rifle, 252
Perazzi Mirage Special Four-Gauge Skeet Shotgun, 222
Perazzi Mirage Special Skeet Over-Under Shotgun, 222
Perazzi Mirage Special Sporting Over-Under Shotgun, 221
Perazzi MX8 Over-Under Shotgun, 222
Perazzi MX8 Special Trap, Skeet Shotguns, 222
Perazzi MX8/20 Over-Under Shotgun, 222
Perazzi MX12 Hunting Over-Under Shotgun, 222
Perazzi MX12C Hunting Over-Under Shotgun, 222
Perazzi MX20 Hunting Over-Under Shotgun, 222
Perazzi Mx20C Hunting Over-Under Shotgun, 222
Perazzi MX28 Game Over-Under Shotgun, 222
Perazzi MX410 Game Over-Under Shotgun, 222
Perazzi Sporting Classic Over-Under Shotgun, 221
Perazzi TMX Special Single Trap Gun, 233
Phoenix Arms HP22 Pistol, 111
Phoenix Arms HP25 Pistol, 111
Phoenix Arms Model Raven Pistol, 111
Piotti King Extra Side-by-Side Shotgun, 229
Piotti King NO. 1 Side-by-Side Shotgun, 229
Piotti Lunik Side-by-Side Shotgun, 229
Piotti Piuma Side-by-Side Shotgun, 229
Piranha Autoloading Pistol, 111
Pocket Police 1862 Percussiion Revolver, 242
Prairie River Arms PRA Bullpup Rifle, 252
Prairie River Arms PRA Classic Rifle, 252
PSA-25 Auto Pistol, 111
Puotti Boss Over-Under Shotgun, 222

Q

Queen Anne Flintlock Pistol, 238

R

Record Champion Repeater Pistol, 264
Record Jumbo Deluxe Air Pistol, 264
Remington 40-XB KS Target Rifle, 202
Remington 40-XB Rangemaster Target Rifle, 202
Remington 40-XBBR KS Target Rifle, 202
Remington 40-XC KS National Match Course Rifle, 202
Remington 40-XR KS Rimfire Position Rifle, 202
Remington 90-T Super Single Shotgun, 234
Remington 396 Skeet Over-Under Shotgun, 222
Remington 396 Sporting Over-Under Shotgun, 222
Remington 541-T Bolt-Action Rifle, 195
Remington 541-T HB Bolt-Action Rifle, 195
Remington 572 BDL Fieldmaster Pump Rifle, 191
Remington Model 11-87 Premier Shotgun, 208
Remington Model 11-87 Premier Skeet Shotgun, 208
Remington Model 11-87 Premier Trap Shotgun, 208
Remington Model 11-87 SC NP Shotgun, 208
Remington Model 11-87 Special Purpose Magnum Shotgun, 208
Remington Model 11-87 Special Purpose Magnum-Turkey Shotgun, 208
Remington Model 11-87 Sporting Clays Shotgun, 207
Remington Model 11-87 SPS Cantilever Shotgun, 208
Remington Model 11-87 SPS Special Purpose Synthetic Camo Shotgun, 208
Remington Model 11-87 SPS-Deer Shotgun, 208
Remington Model 11-87 SPS-T Camo Auto Shotgun, 208
Remington Model 11-96 Euro Lightweight Auto Shotgun, 209
Remington Model 110 Special Field Shotgun, 209
Remington Model 522 Viper Autoloading Rifle, 189
Remington Model 552 BDL Speedmaste Rifle, 189
Remington Model 597 Auto Rifle, 189
Remington Model 597 LSS Auto Rifle, 189
Remington Model 597 Magnum Auto Rifle, 189
Remington Model 700 ADL Bolt-Action Rifle, 169
Remington Model 700 ADL Synthetic Rifle, 170
Remington Model 700 APR Rifle, 171
Remington Model 700 AWR Rifle, 170
Remington Model 700 BDL Bolt-Action Rifle, 169
Remington Model 700 BDL DM Rifle, 170
Remington Model 700 BDL Left-Hand Rifle, 171
Remington Model 700 BDL SS DM Rifle, 170
Remington Model 700 BDL SS DM-B Rifle, 171
Remington Model 700 BDL SS Rifle, 170
Remington Model 700 Classic Rifle, 171
Remington Model 700 Custom KS Mountan Rifle, 170
Remington Model 700 LSS Rifle, 171
Remington Model 700 ML, MLS Rifles, 252
Remington Model 700 MTN DM Rifle, 171
Remington Model 700 Safari Custom KS Rifle, 170
Remington Model 700 Safari Rifle, 170
Remington Model 700 Sendero Rifle, 170
Remington Model 700 Sendero SF Rifle, 170
Remington Model 700 VLS Rifle, 170
Remington Model 700 VS Rifle, 170
Remington Model 700 VS SF Rifle, 170
Remington Model 870 Express HD Shotgun, 212
Remington Model 870 Express Pump Shotgun, 212
Remington Model 870 Express Rifle-Sighted Deer Gun, 212
Remington Model 870 Express Small Gauge Shotgun, 212
Remington Model 870 Express Synthetic Shotgun, 212
Remington Model 870 Express Turkey Shotgun, 212
Remington Model 870 Express Youth Gun, 212
Remington Model 870 High Grade Shotguns, 213
Remington Model 870 Marine Magnum Shotgun, 212
Remington Model 870 Special Purpose Synthetic Camo Shotgun, 213
Remington Model 870 SPS Cantilever Shotgun, 213
Remington Model 870 SPS-Deer Shotgun, 213
Remington Model 870 SPS-T Camo Pump Shotgun, 213
Remington Model 870 SPS-T Special Purpose Magnum Shotgun, 213
Remington Model 870 TC Trap Gun, 212
Remington Model 870 Wingmaster Pump Shotgun, 212
Remington Model 1100 LT-20 Auto Shotgun, 208
Remington Model 1100 LT-20 Skeet Auto Shotgun, 208
Remington Model 1100 LT-20 Youth Auto Shotgun, 208
Remington Model 1100 Sporting 28 Shotgun, 209
Remington Model 1100 Synthetic Auto Shotgun, 208
Remington Model 1100V Synthetic FR CL Shotgun, 209
Remington Model 7400 Auto Rifle, 155
Remington Model 7600 Slide-Action Carbine, 160
Remington Model 7600 Slide-Action Rifle, 160
Remington Model Seven Bolt-Action Rifle, 171
Remington Model Seven Custom MS Rifle, 171
Remington Model Seven SS Rifle, 171
Remington Model Seven Youth Rifle, 171
Remington Model SP-10 Magnum Auto Shotgun, 209
Remington Model SP-10 Magnum-Camo Auto Shotgun, 209
Remington No. 1 Rolling Block Creedmoor Rifle, 182
Remington Peerless Over-Under Shotgun, 222
Republic Patriot Pistol, 111
Rizzini Artemis Over-Under Shotgun, 223
Rizzini Aurum Over-Under Shotgun, 223
Rizzini Express 90L Double Rifle, 186
Rizzini S782 EMEL Over-Under Shotgun, 223
Rizzini S790 EMEL Over-Under Shotgun, 223
Rizzini S790 Sporting EL Over-Under Shotgun, 223
Rizzini S792 EMEL Over-Under Shotgun, 223
Rizzini Sidelock Side-by-Side Shotgun, 229
Rocky Mountain Arms Patriot Pistol, 111
Rogers & Spencer Percussion Revolver, 242
Rogers & Spencer Target Percussion Revolver, 242
Rossi Cyclops Revolver, 133
Rossi Lady Rossi Revolver, 132
Rossi Model 62 SA Pump Rifle, 191
Rossi Model 62 SAC Carbine, 191
Rossi Model 68 Revolver, 133
Rossi Model 88 Revolver, 133
Rossi Model 92 Saddle-Ring Carbine, 160
Rossi Model 92 Short Carbine, 160
Rossi Model 515 Revolver, 133
Rossi Model 518 Revolver, 133
Rossi Model 720 Revolver, 133
Rossi Model 851 Revolver, 133
Rossi Model 877 Revolver, 133
Rossi Model 971 Comp Gun Revolver, 133
Rossi Model 971 Revolver, 133
Rossi Model 971 VRC Revolver, 133
Rottweil Paragon Over-Under Shotgun, 223
RPM XL Single Shot Pistol, 151
Ruger 10/22 Autoloading Carbine, 189
Ruger 10/22 Deluxe Sporter, 189
Ruger 10/22 International Carbine, 189
Ruger 10/22T Target Rifle, 189
Ruger 22/45 Mark II Pistol, 113
Ruger 77/22 Rimfire Bolt-Action Rifle, 195
Ruger 77/22R Bolt-Action Rifle, 195
Ruger 77/22RM Bolt-Action Rifle, 195
Ruger 77/22RS Bolt-Action Rifle, 195
Ruger 77/22RSM Bolt-Action Rifle, 195
Ruger 77/22VBZ Bolt-Action Rifle, 195
Ruger 77/22VMB Bolt-Action Rifle, 195
Ruger Bisley Single-Action Revolver, 146
Ruger Bisley Small Frame Revolver, 146
Ruger Blackhawk Revolver, 146
Ruger Cyclops Revolver, 133
Ruger GP-100 Revolvers, 134
Ruger GP-141 Revolver, 134
Ruger GP-160 Revolver, 134
Ruger GP-161 Revolver, 134
Ruger GP-331 Revolver, 134
Ruger GPF-340 Revolver, 134
Ruger GPF-341 Revolver, 134
Ruger K10/22RP All-Weather Rifle, 189
Ruger K77/22 Varmint Rifle, 195
Ruger K77/22RMP Bolt-Action Rifle, 195
Ruger K77/22RP Bolt-Action Rifle, 195
Ruger K77/22RSMP Bolt-Action Rifle, 195
Ruger K77/22RSP Bolt-Action Rifle, 195
Ruger KGP-141 Revolver, 134
Ruger KGP-160 Revolver, 134
Ruger KGPF-330 Revolver, 134
Ruger KGPF-331 Revolver, 134
Ruger KGPF-340 Revolver, 134
Ruger KGPF-341 Revolver, 134
Ruger KMK4 Semi-Auto Pistol, 112
Ruger KMK6 Semi-Auto Pistol, 112
Ruger KMK-10 Bull Barrel Pistol, 127
Ruger KMK-512 Bull Barrel Pistol, 127
Ruger KMK-678 Target Model Semi-Auto Pistol, 127
Ruger KMK-678G Government Target Model Pistol, 127
Ruger KMK-678GC Competition Model 22 Pistol, 128
Ruger KP4 Semi-Auto Pistol, 113
Ruger KP89 Semi-Auto Pistol, 111
Ruger KP89D Decocker Semi-Auto Pistol, 112
Ruger KP90 Safety Model Semi-Auto Pistol, 112
Ruger KP94 Semi-Auto Pistol, 112
Ruger KP94D Semi-Auto Pistol, 112
Ruger KP94DAO Semi-Auto Pistol, 112
Ruger KP95 Semi-Auto Pistol, 112
Ruger KP512 Semi-Auto Pistol, 113
Ruger KSP-221 Revolver, 134
Ruger KSP-240 Revolver, 134
Ruger KSP-241 Revolver, 134
Ruger KSP-321 Revolver, 134
Ruger KSP-331 Revolver, 134
Ruger KSP-821 Revolver, 134
Ruger KSP-831 Revolver, 134
Ruger KSP-921 Revolver, 134
Ruger KSP-931 Revolver, 134
Ruger KSP-3231 Revolver, 134
Ruger KSRH-7 Revolver, 134
Ruger KSRH-9 Revolver, 134
Ruger M77 Mark II All-Weather Stainless Rifle, 172
Ruger M77 Mark II Express Rifle, 172
Ruger M77 Mark II Magnum Rifle, 172
Ruger M77 Mark II Rifle, 171
Ruger M77/22 Hornet Bolt-Action Rifle, 172
Ruger M77RL Ultra Light Rifle, 171
Ruger M77RSI International Carbine, 171
Ruger M77VT Target Rifle, 172
Ruger Mark II Bull Barrel Pistol, 127
Ruger Mark II Government Target Model Pistol, 127
Ruger Mark II Standard Semi-Auto Pistol, 112
Ruger Mark II Target Model Semi-Auto Pistol, 127
Ruger Mini Thirty Rifle, 156
Ruger Mini-14/5 Autoloading Rifle, 156
Ruger MK4 Semi-Auto Pistol, 112
Ruger MK6 Semi-Auto Pistol, 112
Ruger MK-4B Compact Pistol, 113
Ruger MK-10 Bull Barrel Pistol, 127
Ruger MK-512 Bull Barrel Pistol, 127
Ruger MK-678 Target Model Semi-Auto Pistol, 127
Ruger MK-678G Government Target Model Pistol, 127
Ruger Model 68 Revolver, 133
Ruger Model 88 Revolver, 133
Ruger Model 96/22 Lever-Action Rifle, 191
Ruger Model 96/44 Lever-Action Rifle, 160
Ruger Model 515 Revolver, 133
Ruger Model 518 Revolver, 133
Ruger Model 720 Revolver, 133
Ruger Model 851 Revolver, 133
Ruger Model 877 Revolver, 133

Ruger Model 971 Comp Gun Revolver, 133
Ruger Model 971 Revolver, 133
Ruger Model 971 VRC Revolver, 133
Ruger New Super Bearcat Single-Action Revolver, 146
Ruger No. 1 RSI International Rifle, 183
Ruger No. 1A Light Sporter Rifle, 182
Ruger No. 1B Single Shot Rifle, 182
Ruger No. 1H Tropical Rifle, 182
Ruger No. 1S Medium Sporter Rifle, 183
Ruger No. 1V Special Varminter Rifle, 182
Ruger Old Army Percussion Rifle, 242
Ruger P4 Semi-Auto Pistol, 113
Ruger P89 Double-Action-Only Semi-Auto Pistol, 111
Ruger P89 Semi-Auto Pistol, 111
Ruger P89D Decocker Semi-Auto Pistol, 112
Ruger P90 Decocker Semi-Auto Pistol, 112
Ruger P90 Safety Model Semi-Auto Pistol, 112
Ruger P93 Compact Semi-Auto Pistol, 112
Ruger P94L Semi-Auto Pistol, 112
Ruger P95 Semi-Auto Pistol, 112
Ruger P95D Semi-Auto Pistol, 112
Ruger P512 SEmi-Auto Pistol, 113
Ruger PC4 Carbine, 156
Ruger PC9 Carbine, 156
Ruger PGP-161 Revolver, 134
Ruger Red Label English Field Over-Under Shotgun, 223
Ruger Red Label Over-Under Shotgun, 223
Ruger Redhawk Revolver, 134
Ruger SP101 Double-Action Only Revolver, 134
Ruger SP101 Revolvers, 134
Ruger Sporting Clays Over-Under Shotgun, 223
Ruger Stainless Government Competition Model 22 Pistol, 128
Ruger Super Blackhawk Revolver, 146
Ruger Super Redhawk Revolver, 134
Ruger Super Single-Six Convertible Revolver, 146
Ruger Vaquero Single-Action Revolver, 146
Ruger Woodside Over-Under Shotgun, 223
Ruger Woodside Sporting Clays Shotgun, 223
Ruger ~P95DAO Semi-Auto Pistol, 112
Ruptor Bolt-Action Rifle, 169
RWS Model 75S TO1 Match Air Rifle, 274
RWS Model CA 101 Air Rifle, 274
RWS TX200 Magnum Air Rifle, 274
RWS TX200SR Magnum Air Rifle, 274
RWS/Diana Model 5G Air Pistol, 264
RWS/Diana Model 6G Air Pistol, 264
RWS/Diana Model 6M Match Air Pistol, 264
RWS/Diana Model 24 Air Rifle, 274
RWS/Diana Model 24C Air Rifle, 274
RWS/Diana Model 34 Air Rifle, 274
RWS/Diana Model 34BC Air Rifle, 274
RWS/Diana Model 34N Air Rifle, 274
RWS/Diana Model 36 Air Rifle, 275
RWS/Diana Model 36 Carbine, 275
RWS/Diana Model 45 Air Rifle, 275
RWS/Diana Model 48 Air Rifle, 275
RWS/Diana Model 48B Air Rifle, 275
RWS/Diana Model 52 Air Rifle, 275
RWS/Diana Model 52 Deluxe Air Rifle, 275
RWS/Diana Model 54 Air King Air Rifle, 275
RWS/Diana Model 100 Match Air Rifle, 275

S

SA-85 Semi-Auto Rifle, 156
Safari Arms BIg Deuce Pistol, 128
Safari Arms Cohort Pistol, 113
Safari Arms Enforcer Pistol, 113
Safari Arms Engorcer Carrycomp I Pistol, 113
Safari Arms GI Safari Pistol, 113
Safari Arms Griffon Pistol, 113
Safari Arms Reliable 4-Star Pistol, 113
Safari Arms Reliable Pistol, 113
Safari Arms Renegade Pistol, 113
Sako 75 Deluxe Rifle, 172
Sako 75 Hunter Bolt-Action Rifle, 172
Sako 75 Stainless Synthetic Rifle, 172
Sako Finnfire Bolt-Action Rifle, 196
Sako Hunter Rifle, 173
Sako Lightweight Deluxe Rifle, 173
Sako Long-Range Hunting Rifle, 173
Sako Super Deluxe Sporter Rifle, 173
Sako TRG-21 Bolt-Action Rifle, 203
Sako TRG-41 Bolt-Action Rifle, 203
Sako TRG-S Bolt-Action Rifle, 173
Sako Varmint Heavy Barrel Rifle, 173
Sarsilmaz Cobra Pump Shotgun, 213
Sarsilmaz Over-Under Shotgun, 224
Sauer 90 Bolt-Action Rifle, 173
Sauer 202 Bolt-Action Rifle, 173
Sauer Drilling, 186
Savage 24F Predator O/U Combination Gun, 186
Savage 24F-12 Predator O/U Combination Gun, 186
Savage 24F-20 Predator O/U Combination Gun, 186
Savage Mark I-G Bolt-Action Rifle, 196
Savage Mark I-GY Bolt-Action Rifle, 196
Savage Mark II-FSS Stainless Rifle, 196
Savage Mark II-G Bolt-Action Rifle, 196
Savage Mark II-GL Bolt-Action Rifle, 196
Savage Mark II-GLY Bolt-Action Rifle, 196
Savage Mark II-GXP Package Gun, 196
Savage Mark II-GY Bolt-Action Rifle, 196
Savage Mark II-LV Heavy Barrel Rifle, 196
Savage Model 64F Auto Rifle, 190
Savage Model 64G Auto Rifle, 190
Savage Model 64GxP Package Gun, 190
Savage Model 93FSS Magnum Rifle, 196
Savage Model 93G Magnum Bolt-Action Rifle, 196
Savage Model 99C Lever-Action Rifle, 160
Savage Model 110 FM Sierra Light Weight Rifle, 173
Savage Model 110FP Tactical Rifle, 174
Savage Model 110GCXP3 Package Gun, 174
Savage Model 110GXP3 Package Gun, 174
Savage Model 111 Classic Hunter Rifles, 174
Savage Model 111F Classic Hunter Rifle, 174
Savage Model 111FAK Classic Hunter Rifle, 174
Savage Model 111FC Classic Hunter Rifle, 174
Savage Model 111FCXP3 Package Gun, 174
Savage Model 111FNS Classic Hunter Rifle, 174
Savage Model 111FXP3 Package Gun, 174
Savage Model 111G Classic Hunter Rifle, 174
Savage Model 111GC Classic Hunter Rifle, 174
Savage Model 111GNS Classic Hunter Rifle, 174
Savage Model 112BT Competition Grade Rifle, 203
Savage Model 112BT-S Competition Grade Rifle, 203
Savage Model 114C Classic Rifle, 174
Savage Model 114CE Classic European Rifle, 174
Savage Model 116 Weather Warrior Rifles, 175
Savage Model 116FCS Weather Warrior Rifle, 175
Savage Model 116FCSAK Weather Warrior Rifle, 175
Savage Model 116FSAK Weather Warrior Rifle, 175
Savage Model 116FSK Weather Warrior Rifle, 175
Savage Model 116FSS Weather Warrior Rifle, 175
Savage Model 116SE Safari Express Rifle, 174
Savage Model 116US Ultra Stainless Rifle, 174
Savage Model 210F Master Shot Slug Gun, 234
Savage Model 210FT Master Shot Slug Gun, 234
Savage Model 900B Biathlon Rifle, 203
Savage Model 900S Silhouette Rifle, 203
Savage Model 900TR Target Rifle, 203
Second Model Brown Bess Carbine, 252
Second Model Brown Bess Musket, 252
Seecamp LWS 32 DA Pistol, 113
Sharps 1874 Rifle, 184
Sheriff Model 1851 Percussion Rifle, 242
Shiloh Sharps 1874 Business Rifle, 184
Shiloh Sharps 1874 Hartford Model Rifle, 183
Shiloh Sharps 1874 Long Range Express Rifle, 183
Shiloh Sharps 1874 Montana Roughrider Rifle, 183
Shiloh Sharps 1874 Saddle Rifle, 184
Shiloh Sharps 1874 Sporting Rifle No. 1, 183
Shiloh Sharps 1874 Sporting Rifle No. 3, 183
SIG P210 Service Pistol, 114
SIG Saver P220 Service Pistol, 114
SIG Saver P225 Compact Pistol, 114
SIG Saver P226 Service Pistol, 114
SIG Saver P228 Compact Pistol, 114
SIG Saver P229 DA Pistol, 114
SIG Saver P232 Personal Size Pistol, 114
SIG Saver P239 Pistol, 114
Sigarms SA3 Over-Under Shotgun, 224
Sigarms SA5 Over-Under Shotgun, 224
Silma Model 70 Over-Under Shotgun, 224
SKB Model 385 Side-by-Side Shotgun, 229
SKB Model 485 Side-by-Side Shotgun, 229
SKB Model 505 Shotguns, 224
SKB Model 585 Over-Under Shotgun, 224
SKB Model 585 Sporting Clays Over-Under Shotgun, 224
SKB Model 785 Field Over-Under Shotgun, 224
SKB Model 785 Over-Under Shotgun, 224
SKB Model 785 Skeet Over-under Shotgun, 224
SKB Model 785 Sporting Clays Shotgun, 224
SKB Model 785 Trap Over-Under Shotgun, 224
Slavia Model 631 Air Rifle, 276
Smith & Wesson Model 10 Revolver, 134
Smith & Wesson Model 13 H.B. M&P Revolvers, 134
Smith & Wesson Model 14 Full Lug Revolvers, 135
Smith & Wesson Model 15 combat Masterpiece Revolver, 135
Smith & Wesson Model 17 K-22 Masterpiece Revolver, 135
Smith & Wesson Model 19 Combat Magnum Revolver, 135
Smith & Wesson Model 22A Sport Pistol, 115
Smith & Wesson Model 22A Target Pistol, 128
Smith & Wesson Model 22S Sport Pistol, 115
Smith & Wesson Model 22S Target Pistol, 128
Smith & Wesson Model 29 Revolver, 135
Smith & Wesson Model 36 Chiefs Special Revolver, 135
Smith & Wesson Model 36LS LadySmith Revolver, 136
Smith & Wesson Model 37 Airweight Revolver, 135
Smith & Wesson Model 38 Bodyguard Revolver, 136
Smith & Wesson Model 41 Target Pistol, 128
Smith & Wesson Model 60 3" Full Lug Revolver, 135
Smith & Wesson Model 60 Chiefs Special Stainless Revolver, 135
Smith & Wesson Model 60LS LadySmith Revolver, 136
Smith & Wesson Model 63 Kit Gun Revolver, 136
Smith & Wesson Model 64 Stainless M&P Revolver, 136
Smith & Wesson Model 65 Revolver, 134
Smith & Wesson Model 65LS LadySmith Revolver, 136
Smith & Wesson Model 66 Stainless Combat Magnum Revolver, 136
Smith & Wesson Model 67 Combat Masterpiece Revolver, 136
Smith & Wesson Model 317 Airlite Revolver, 136
Smith & Wesson Model 410 DA Pistol, 25
Smith & Wesson Model 442 Centennial Airweight Revolver, 137
Smith & Wesson Model 457 DA Pistol, 115
Smith & Wesson Model 586 Distinguiehed Combat Magnum Revolvers, 137
Smith & Wesson Model 625 Revolver, 137
Smith & Wesson Model 629 Classic DX Revolver, 135
Smith & Wesson Model 629 Classic Powerport Revolver, 135
Smith & Wesson Model 629 Classic Revolver, 135
Smith & Wesson Model 629 Revolver, 135
Smith & Wesson Model 637 Airweight Revolvers, 136
Smith & Wesson Model 638 Airweight Bodyguard Revolver, 136
Smith & Wesson Model 640 Centennial Revolver, 137
Smith & Wesson Model 642 Airweight Revolver, 137
Smith & Wesson Model 642LS LadySmith Revolver, 137
Smith & Wesson Model 649 Bodyguard Revolver, 137
Smith & Wesson Model 651 Revolver, 137
Smith & Wesson Model 657 Revolver, 138
Smith & Wesson Model 686 Distinguished Combat Magnum Revolver, 137
Smith & Wesson Model 696 Revolver, 138
Smith & Wesson Model 908 DA Pistol, 115
Smith & Wesson Model 910 DA Pistol, 115
Smith & Wesson Model 2213 Sportsman Pistol, 115
Smith & Wesson Model 2214 Sportsman Pistol, 115
Smith & Wesson Model 3913 DA Pistol, 115
Smith & Wesson Model 3913 LadySmith Pistol, 115
Smith & Wesson Model 3953 DA Pistol, 115
Smith & Wesson Model 4006 DA Pistol, 116
Smith & Wesson Model 4013 TSW Pistol, 116
Smith & Wesson Model 4046 DA Pistol, 116
Smith & Wesson Model 4056 TSW Pistol, 116
Smith & Wesson Model 4500 Series Pistol, 116
Smith & Wesson Model 4506 Pistol, 116
Smith & Wesson Model 4516 Pistol, 116
Smith & Wesson Model 4566 Pistol, 116
Smith & Wesson Model 4586 Pistol, 116
Smith & Wesson Model 5900 Series Pistol, 116
Smith & Wesson Model 5903 Pistol, 116
Smith & Wesson Model 5904 Pistol, 116
Smith & Wesson Model 5904 Pistol, 116
Smith & Wesson Model 5946 Pistol, 116
Smith & Wesson Model 6904 DA Pistol, 116
Smith & Wesson Model 6946 DA Pistol, 116
Smith & Wesson Sigma Series Pistol, 116
Smith & Wesson Sigma SW380 Pistol, 116
Smith & Wesson SW9C Sigma Pistol, 116
Smith & Wesson SW9M Sigma Pistol, 116
Smith & Wesson SW9V Sigma Pistol, 116
Smith & Wesson SW40C Sigma Pistol, 116
Smith & Wesson SW40F Sigma Pistol, 116
Smith & Wesson SW40V Sigma Pistol, 116
Snake Charmer II Shotgun, 234
Sphinx AT-380 Pistol, 117
Sphinx AT-2000C Competitor Pistol, 128
Sphinx AT-2000CS Competitor Pistol, 128
Sphinx AT-2000GM Grand Master Pistol, 128
Sphinx AT-2000GMS Grand Master Pistol, 128
Sphinx AT-2000H DA Pistol, 117
Sphinx AT-2000P DA Pistol, 117
Sphinx AT-2000PS DA Pistol, 117
Sphinx AT-2000S DA Pistol, 117
Spiller & Burr Percussion Rifle, 242
Springfield, Inc. 1911A1 Bullseye Wadcutter Pistol, 128
Springfield, Inc. 1911A1 Champion Pistol, 118
Springfield, Inc. 1911A1 Custom Carry Gun, 117
Springfield, Inc. 1911A1 Defender Pistol, 117
Springfield, Inc. 1911A1 Factory Comp Pistol, 117
Springfield, Inc. 1911A1 High Capacity Pistol, 118
Springfield, Inc. 1911A1 Lightweight Pistol, 117
Springfield, Inc. 1911A1 Mil-Spec Pistol, 117
Springfield, Inc. 1911A1 N.M. Hardball Pistol, 129
Springfield, Inc. 1911A1 Pistol, 117
Springfield, Inc. 1911A1 Standard Pistol, 117
Springfield, Inc. 1911A1 Trophy Match Pistol, 129
Springfield, Inc. Basic Competition Pistol, 128
Springfield, Inc. Distinguished Pistol, 129
Springfield, Inc. Expert Pistol, 128
Springfield, Inc. M1A National Match Rifle, 156
Springfield, Inc. M1A Rifle, 156

Springfield, Inc. M1A Super Match Rifle, 156
Springfield, Inc. M1A Super Match Rifle, 203
Springfield, Inc. M1A-A1 Bush Rifle, 156
Springfield, Inc. M1A-A1 Collector Rifle, 156
Springfield, Inc. M1A-A1 Rifle, 156
Springfield, Inc. M1A-A1 Scout Rifle, 156
Springfield, Inc. M1A/M-21 Tactical Rifle, 203
Springfield, Inc. M6 Scout Rifle/Shotgun, 186
Springfield, Inc. NRA PPC Pistol, 117
Springfield, Inc. SAR-8 Sporter Rifle, 156
Springfield, Inc. SAR-4800 Rifle, 156
Springfield, Inc. V10 Ultra Compact Pistol, 118
Star Firestar Model 45 Pistol, 118
Star Firestar Pistol, 118
Star Firestar Plus Pistol, 118
Star Ultrastar DA Pistol, 118
Steyr CO_2 Match LP1C Air Pistol, 264
Steyr LG1P Air Rifle, 276
Steyr LP5C Match Pistol, 264
Steyr LP5CP Match Pistol, 264
Steyr SBS Forester Rifle, 175
Steyr SBS Pro-Hunter Rifle, 175
Steyr SSG Bolt-Action Rifles, 175
Steyr SSG-P1 Bolt-Action Rifle, 175
Steyr SSG-PII Bolt-Action Rifle, 175
Steyr SSG-PIII Bolt-Action Rifle, 175
Steyr SSG-PIV Bolt-Action Rifle, 175
Steyr-Mannlicher SBS Rifle, 175
STI Eagle 5.1 Pistol, 129
Stoeger American Eagle Luger, 118
Stoeger Pro Series 95 Bull Barrel Pistol, 129
Stoeger Pro Series 95 Fluted Barrel Pistol, 129
Stoeger Pro Series 95 Vent Rib Pistol, 129
Stoeger/IGA Coach Gun Shotgun, 230
Stoeger/IGA Condor I Over-Under Shotgun, 225
Stoeger/IGA Condor II Over-Under Shotgun, 225
Stoeger/IGA Condor Supreme Over-Under Shotgun, 225
Stoeger/IGA Condor Waterfowl Over-Under Shotgun, 225
Stoeger/IGA Deluxe Coach Gun Shotgun, 230
Stoeger/IGA Deluxe Hunter Clays Over-Under Shotgun, 225
Stoeger/IGA Deluxe Uplander Shotgun, 230
Stoeger/IGA English Stock Double Shotgun, 230
Stoeger/IGA Turkey Double Shotgun, 230
Stoeger/IGA Turkey Model Over-Under Shotgun, 225
Stoeger/IGA Uplander Side-by-Side, 230
Stoeger/IGA Youth Double Shotgun, 230
Stone Mountain Silver Eagle Rifle, 252
Stone Mountain Silver Wolf Rifle, 252
Stoner SR-25 Lightweight Match Rifle, 203
Stoner SR-25 Match Rifle, 203
Stoner SR-35 Carbine, 156
Stoner SR-50 Long Range Precision Rifle, 157
Sundance BOA Pistol, 119
Sundance Lady Laser 25 Pistol, 119
Sundance Laser 25 Pistol, 119
Sundance MOdel A-25 Pistol, 119
Sundance Point Blank O/U Derringer, 151
Survival Arms AR-7 Explorer Rifle, 190

T

Tactical Response TR-870 Border Patrol Shotgun, 236
Tactical Response TR-870 Compact Model Shotgun, 236
Tactical Response TR-870 Entry Model Shotgun, 236
Tactical Response TR-870 Expert Model Shotgun, 236
Tactical Response TR-870 FBI Model Shotgun, 236
Tactical Response TR-870 K-9 Model Shotgun, 236
Tactical Response TR-870 Louis Awerbuck Shotgun, 236
Tactical Response TR-870 Military Model Shotgun, 236
Tactical Response TR-870 Patrol Model Shotgun, 236
Tactical Response TR-870 Practical Turkey Shotgun, 236
Tactical Response TR-870 Professional Model Shotgun, 236
Tactical Response TR-870 Standard Model Shotgun, 236
Tactical Response TR-870 SWAT Model Shotgun, 236
Tactical Response TR-870 Urban Sniper Shotgun, 236
Tanner 50 Meter Free Rifle, 204
Tanner 300 Meter Free Rifle, 204
Tanner Standard UIT Rifle, 204
Tar-Hunt RSG-12 Matchless Model Rifled Slug Gun, 234
Tar-Hunt RSG-12 Peerless Model Rifled Slug Gun, 234
Tar-Hunt RSG-12 Professional Rifled Slug Gun, 234
Tar-Hunt RSG-20 Mountaineer Rifled Slug Gun, 234
Taurus Model 44 Revolver, 138
Taurus Model 82 Heavy Barrel Revolver, 138
Taurus Model 83 Revolver, 138
Taurus Model 85 Revolver, 138
Taurus Model 85CH Revolver, 138
Taurus Model 94 Revolver, 138

Taurus Model 94 UL Revolver, 138
Taurus Model 96 Revolver, 138
Taurus Model 445 Revolver, 138
Taurus Model 445CH Revolver, 138
Taurus Model 454 Revolver, 139
Taurus Model 605 Revolver, 139
Taurus Model 605CH Revolver, 139
Taurus Model 606 Revolver, 139
Taurus Model 606CH Revolver, 139
Taurus Model 607 Revolver, 139
Taurus Model 608 Revolver, 139
Taurus Model 669 Revolver, 139
Taurus Model 689 Revolver, 139
Taurus Model 941 Revolver, 139
Taurus Model 941 UL Revolver, 139
Taurus Model PT 22 Pistol, 119
Taurus Model PT 25 Pistol, 119
Taurus Model PT 92AF Pistol, 119
Taurus Model PT 99AF Pistol, 119
Taurus Model PT-111 Pistol, 119
Taurus Model PT-911 Pistol, 119
Taurus Model PT-938 Pistol, 119
Taurus Model PT-940 Pistol, 119
Taurus Model PT-945 Pistol, 120
Texas Longhorn "The Jezebel" Pistol, 151
Texas Longhorn Arms Cased Set Revolvers, 147
Texas Longhorn Arms Grover's Improved No. Five Revolver, 147
Texas Longhorn Arms Right-Hand Single-Action Revolver, 147
Texas Longhorn Arms Sesquicentennial Model Revolver, 147
Texas Longhorn Arms Texas Border Special Revolver, 147
Texas Longhorn Arms West Texas Flat Top Target Revolver, 147
Texas Paterson 1836 Rifle, 243
The Judge Single Shot Pistol, 151
Thompson/Center Big Boar Rifle, 253
Thompson/Center Contender Carbine Youth Model, 184
Thompson/Center Contender Carbine, 184
Thompson/Center Contender Hunter Package, 152
Thompson/Center Contender Pistol, 152
Thompson/Center Encore Pistol, 151
Thompson/Center Encore Rifle, 184
Thompson/Center Fire Hawk Rifle, 253
Thompson/Center Grey Hawk Rifle, 253
Thompson/Center Hawken Rifle, 253
Thompson/Center Hawken Silver Elite Rifle, 253
Thompson/Center New Englander Rifle, 253
Thompson/Center New Englander Shotgun, 258
Thompson/Center Pennsylvania Hunter Carbine, 253
Thompson/Center Pennsylvania Hunter Rifle, 253
Thompson/Center Pennsylvania Match Rifle, 253
Thompson/Center Renegade Rifle, 253
Thompson/Center Scout Carbine, 254
Thompson/Center Scout Pistol, 238
Thompson/Center Scout Rifle, 254
Thompson/Center Stainless Contender Carbine, 184
Thompson/Center Stainless Contender, 152
Thompson/Center Stainless Super 14 Contender, 152
Thompson/Center Stainless Super 16 Contender, 152
Thompson/Center Super 14 Contender Pistol, 129
Thompson/Center Super 16 Contender Pistol, 129
Thompson/Center System 1 In-Line Rifle, 154
Thompson/Center ThunderHawk Carbine, 254
Thompson/Center ThunderHawk Shadow, 254
Tikka Continental Long Range Hunting Rifle, 176
Tikka Continental Varmint Rifle, 176
Tikka Model 512S Combination Gun, 186
Tikka Model 512S Double Rifle, 186
Tikka Model 512S Field Grade Over-Under Shotgun, 225
Tikka Whitetail Hunter Bolt-Action Rifle, 175
Tikka Whitetail Hunter Stainless Synthetic Rifle, 176
Tikka Whitetail Hunter Synthetic Rifle, 175
Traditions 1853 Three-Band Enfield, 256
Traditions 1861 U.S. Springfield Rifle, 256
Traditions Buckhunter Pro In-Line Pistol, 238
Traditions Buckhunter Pro In-Line Rifle, 254
Traditions Buckhunter Pro Shotgun, 258
Traditions Buckskinner Carbine, 254
Traditions Buckskinner Pistol, 238
Traditions Deerhunter Composite Rifle, 255
Traditions Deerhunter Rifle, 255
Traditions Deerhunter Scout Rifle, 255
Traditions Hawken Woodsman Rifle, 255
Traditions In-Line Buckhunter Series Rifles, 254
Traditions Kentucky Pistol, 239
Traditions Kentucky Rifle, 255
Traditions Lightning Bolt-Action Muzzleloader, 255
Traditions Panther Rifle, 255
Traditions Pennsylvania Rifle, 255
Traditions Pioneer Pistol, 239
Traditions Shenandoah Rifle, 255
Traditions Tennessee Rifle, 256
Traditions Trapper Pistol, 239
Traditions William Parker Pistol, 239

Tristar Model 300 Over-Under Shotgun, 225
Tristar Model 311 Double Shotgun, 230
Tristar Model 330 Over-Under Shotgun, 225
Tristar Model 333 Over-Under Shotgun, 225
Tristar Model 333L Over-Under Shotgun, 225
Tristar Model 333SC Over-Under Shotgun, 225
Tristar Model 333SCL Over-Under Shotgun, 225
Tristar Model 1887 Lever-Action Shotgun, 213
Tristar Pee-Wee 22 Bolt-Action Rifle, 196
Tryon Trailblazer Rifle, 256

U

U.S. Paten Fire-Arms 1862 Pocket Navy Rifle, 243
U.S. Patent Fire-Arms Bird Head Model Revolver, 148
U.S. Patent Fire-Arms Bisley Model Revolver, 148
U.S. Patent Fire-Arms Flattop Target Revolver, 148
U.S. Patent Fire-Arms Nettleton Cavalry Revolver, 148
U.S. Patent Fire-Arms Single Action Army Revolver, 148
Uberti 1st Model Dragoon Percussion Rifle, 243
Uberti 2nd Model Dragoon Percussion Rifle, 243
Uberti 3rd Model Dragoon Percussion Rifle, 243
Uberti 1861 Navy Percussion Revolver, 240
Uberti 1862 Pocket Navy Percussion Rifle, 243
Uberti 1873 Buckhorn Single-Action Revolver, 147
Uberti 1873 Cattleman Single-Action Revolver, 147
Uberti 1875 SA Army Outlaw Revolver, 147
Uberti 1890 Army Outlaw Revolver, 147
Uberti Rolling Block Baby Carbine, 184
Uberti Rolling Block Target Pistol, 152
UFA Grand Teton Rifle, 256
UFA Teton Blackstone Rifle, 256
UFA Teton Rifle, 256
Ultra Light Arms Model 20 REB Hunter's Pistol, 152
Ultra Light Arms Model 20 RF Bolt-Action Rifle, 197
Ultra Light Arms Model 20 Rifle, 176
Ultra Light Arms Model 24 Rifle, 176
Ultra Light Arms Model 28 Rifle, 176
Ultra Light Arms Model 40 Rifle, 176
Unique D.E.S. 32U Target Pistol, 129
Unique D.E.S. 69U Target Pistol, 130
Unique Model 96U Target Pistol, 129
Uzi Eagle Auto Pistol, 120
Uzi Eagle Compact Pistol, 120
Uzi Eagle Polymer Compact Pistlo, 120

V

Voere VEC-91 Lighting Bolt-Action Rifle, 176
Voere VEC-91BR Caseless Rifle, 176
Voere VEC-91HB Varmint Special Rifle, 176
Voere VEC-91SS Rifle, 176
Voere VEC-95CG Single Shot Pistol, 152
Voere VEC-RG Repeater Pistol, 152

W

Walker 1847 Percussion Rifle, 243
Walther CP88 competition Pellet Pistol, 265
Walther CP88 Pellet Pistol, 265
Walther CPM-1 CO_2 Match Air Pistol, 265
Walther GSP Match Pistol, 130
Walther Model TPH Pistol, 121
Walther P88 Compact Pistol, 121
Walther P99 DA Pistol, 120
Walther PP DA Pistol, 120
Walther PPK American Pistol, 120
Walther PPK/S American Pistol, 120
Weatherby Accumark Rifle, 176
Weatherby Athena Grade IV Over-Under Shotgun, 226
Weatherby Athena Grade V Classic Field Shotgun, 226
Weatherby Mark V Deluxe Bolt-Action Rifle, 177
Weatherby Mark V Euromark Rifle, 177
Weatherby Mark V Eurosport Rifle, 177
Weatherby Mark V Fluted Stainless Rifle, 177
Weatherby Mark V Fluted Synthetic Rifle, 177
Weatherby Mark V Lazermark Rifle, 177
Weatherby Mark V Light-Weight Sporter Bolt-Action Rifle, 177
Weatherby Mark V Lightweight Carbine, 178
Weatherby Mark V Lightweight Stainless Rifle, 178
Weatherby Mark V Lightweight Synthetic Rifle, 178
Weatherby Mark V SLS Stainless Laminate Sporter Rifle, 177
Weatherby Mark V Sporter Rifle, 177
Weatherby Mark V Stainless Rifle, 177
Weatherby Mark V Synthetic Rifle, 177
Weatherby Orion Grade I Over-Under Shotgun, 226
Weatherby Orion Grade II Classic Sporting Shotgun, 226
Weatherby Orion Grade II Sporting Clays Shotgun, 226
Weatherby Orion Grade III Over-Under Shotgun, 226
Weatherby Orion II Sporting Clays Over-Under Shotgun, 226
Weatherby Orion II, III Classic Field Shotguns, 226
Weatherby Orion III English Field Over-Under Shotgun, 226
Webley Patriot Air Rifle, 276
Webley Tracker Air Rifle, 276

Wesson & Harrington Buffalo Classic Rifle, 184
Whiscombe JW50 MKII Air Rifles, 276
Whiscombe JW60 MKII Air Rifles, 276
Whiscombe JW70 MKII Air Rifles, 276
Whiscombe JW75 MKII Air Rifles, 276
Whiscombe JW80 MKII Air Rifles, 276
Whiscombe JW Series Air Rifles, 276
White Muzzleloading systems "Tominator" Shotgun, 258
White Muzzleloading Systems Bison Rifle, 257
White Muzzleloading Systems Javelina Pistol, 239
White Muzzleloading Systems Super 91 Rifle, 256
White Muzzleloading Systems Super Safari Rifle, 256
White Muzzleloading Systems White Lighting Rifle, 257
White Muzzleloading Systems White Thunder Shotgun, 258
White Muzzleloading Systems Whitetail Rifle, 257
Wichita Classic Rifle, 178
Wichita Classic Silhouette Pistol, 130
Wichita Silhouette Pistol, 130
Wichita Varmint Rifle, 178
Wilderness Explorer Mult-Caliber Carbine, 178
Wildey Automatic Pistol, 121
Wilkinson Linda Pistol, 121
Wilkinson Sherry Pistol, 121
Wilkinson Terry Carbine, 157
Winchester 8-Shot Pistol Grip Pump Security Shotgun, 237
Winchester Model 52B Bolt-Action Rifle, 197

Winchester Model 63 Auto Rifle, 190
Winchester Model 70 Classic Featherweight All Terrain, 178
Winchester Model 70 Classic Featherweight Rifle, 178
Winchester Model 70 Classic Featherweight Stainless, 178
Winchester Model 70 Classic Laredo Rifle, 179
Winchester Model 70 Classic Sporter Stainless, 178
Winchester Model 70 Classic Sporter, 178
Winchester Model 70 Classic Stainless Rifle, 179
Winchester Model 70 Classic Super Express Magnum, 179
Winchester Model 70 Classic Super Grade Rifle, 179
Winchester Model 70 Synthetic Heavy Varmint Rifle, 179
Winchester Model 94 Big Bore Side Eject Lever-Action Rifle, 161
Winchester Model 94 Legacy Lever-Action Rifle, 161
Winchester Model 94 Ranger Side Eject Lever-Action Rifle, 161
Winchester Model 94 Trails End Lever-Action Rifle, 161
Winchester Model 94 Trapper Side Eject Lever-Action Rifle, 161
Winchester Model 94 Walnut Side Eject Lever-Action Rifle, 161
Winchester Model 94 Wrangler Side Eject Rifle, 161
Winchester Model 1300 Advantage Camo Deer Gun, 213
Winchester Model 1300 Black Shadow Deer Gun, 213

Winchester Model 1300 Black Shadow Field Gun, 213
Winchester Model 1300 Black Shadow Turkey Gun, 213
Winchester Model 1300 Defender Pump Gun, 236
Winchester Model 1300 Ladies/Youth Pump Shotgun, 214
Winchester Model 1300 Lady Defneder Pump Gun, 237
Winchester Model 1300 Ranger Pump Combo Deer Gun, 214
Winchester Model 1300 Ranger Pump Gun Combo, 214
Winchester Model 1300 Ranger Pump Gun, 214
Winchester Model 1300 Realtree Turkey Field Gun, 213
Winchester Model 1300 Stainless Marine Pump Gun, 237
Winchester Model 1300 Walnut Pump Shotgun, 213
Winchester Model 1892 Lever-Action Rifle, 161
Winchester Model 1895 High Grade Rifle, 161
Winchester Model 1895 Lever-Action Rifle, 161
Winchester Model 9422 25th Anniversary Edition Rifle, 191
Winchester Model 9422 Lever-Action Rifle, 191
Winchester Model 9422 Magnum Lever-Action Rifle, 191
Winchester Model 9422 Trapper Lever-Action Rifle, 191
Winchester Ranger Rifle, 179

Z

Zouave Percussion Rifle, 421

HANDGUNS—AUTOLOADERS, SERVICE & SPORT

Includes models suitable for several forms of competition and other sporting purposes.

AA ARMS AP9 PISTOL
Caliber: 9mm Para., 10-shot magazine.
Barrel: 5".
Weight: 3.5 lbs. **Length:** 12" overall.
Stocks: Checkered black synthetic.
Sights: Post front adjustable for elevation, rear adjustable for windage
Features: Ventilated barrel shroud; blue or electroless nickel finish. Made in U.S. by AA Arms.
Price: Blue $299.00
Price: Nickel $312.00
Price: AP9 Target (11" barrel) $399.00

ACCU-TEK BL-9 AUTO PISTOL
Caliber: 9mm Para., 5-shot magazine.
Barrel: 3".
Weight: 22 oz. **Length:** 5.6" overall.
Stocks: Black composition.
Sights: Fixed.
Features: Double action only; black finish. Introduced 1997. Made in U.S. by Accu-Tek.
Price: $199.00

Accu-Tek BL-380
Same as the BL-9 except chambered for 380 ACP. Introduced 1997. Made in U.S. by Accu-Tek.
Price: $199.00

ACCU-TEK MODEL CP-9SS AUTO PISTOL
Caliber: 9mm Para., 8-shot magazine.
Barrel: 3.2".
Weight: 28 oz. **Length:** 6.25" overall.
Stocks: Black checkered nylon.
Sights: Blade front, rear adjustable for windage; three-dot system.
Features: Stainless steel construction. Double action only. Firing pin block with no external safeties. Lifetime warranty. Introduced 1992. Made in U.S. by Accu-Tek.
Price: Satin stainless $265.00

Accu-Tek CP-45SS Auto Pistol
Same as the Model AT-9SS except chambered for 45 ACP, 6-shot magazine. Introduced 1995. Made in U.S. by Accu-Tek.
Price: Stainless steel $265.00

ACCU-TEK MODEL AT-380SS AUTO PISTOL
Caliber: 380 ACP, 5-shot magazine.
Barrel: 2.75".
Weight: 20 oz. **Length:** 5.6" overall.
Stocks: Grooved black composition.
Sights: Blade front, rear adjustable for windage.
Features: Stainless steel frame and slide. External hammer; manual thumb safety; firing pin block, trigger disconnect. Lifetime warranty. Introduced 1991. Made in U.S. by Accu-Tek.
Price: Satin stainless $182.00
Price: Black finish over steel (AT-380B) $187.00

Accu-Tek Model AT-32SS Auto Pistol
Same as the AT-380SS except chambered for 32 ACP. Introduced 1991.
Price: Satin stainless $176.00
Price: Black finish over steel (AT-32B) $181.00

ACCU-TEK MODEL HC-380SS AUTO PISTOL
Caliber: 380 ACP, 10-shot magazine.
Barrel: 2.75".
Weight: 28 oz. **Length:** 6" overall.
Stocks: Checkered black composition.
Sights: Blade front, rear adjustable for windage.
Features: External hammer; manual thumb safety with firing pin and trigger disconnect; bottom magazine release. Stainless finish. Introduced 1993. Made in U.S. by Accu-Tek.
Price: Satin stainless $230.00
Price: Black finish over stainless $235.00

AA Arms AP9 Mini, AP9 Mini/5 Pistol
Similar to AP9 except scaled-down dimensions with 3" or 5" barrel.
Price: 3" barrel, blue $239.00
Price: 3" barrel, electroless nickel $259.00
Price: Mini/5, 5" barrel, blue $259.00
Price: Mini/5, 5" barrel, electroless nickel $279.00

Accu-Tek BL-9

Accu-Tek CP-9SS

Accu-Tek CP-40SS Auto Pistol
Same as the Model AT-9 except chambered for 40 S&W, 7-shot magazine. Introduced 1992.
Price: Stainless $265.00

Accu-Tek AT 380SS

Accu-Tek HC-380SS

CAUTION: PRICES SHOWN ARE SUPPLIED BY THE MANUFACTURER OR IMPORTER. CHECK YOUR LOCAL GUNSHOP.

HANDGUNS—AUTOLOADERS, SERVICE & SPORT

American Arms Escort

ARMSCOR M-1911-A1P AUTOLOADING PISTOL
Caliber: 45 ACP, 7- or 10-shot magazine.
Barrel: 5".
Weight: 38 oz. **Length:** 8 3/4" overall.
Stocks: Checkered.
Sights: Blade front, rear drift adjustable for windage; three-dot system.
Features: Skeletonized combat hammer and trigger; beavertail grip safety; extended slide release; oversize thumb safety; Parkerized finish. Introduced 1996. Imported from the Philippines by K.B.I., Inc.
Price: $479.00

AMT AUTOMAG II AUTO PISTOL
Caliber: 22 WMR, 9-shot magazine (7-shot with 3 3/8" barrel).
Barrel: 3 3/8", 4 1/2", 6".
Weight: About 23 oz. **Length:** 9 3/8" overall.
Stocks: Grooved carbon fiber.
Sights: Blade front, adjustable rear.
Features: Made of stainless steel. Gas-assisted action. Exposed hammer. Slide flats have brushed finish, rest is sandblast. Squared trigger guard. Introduced 1986. From AMT.
Price: $405.95

AMT AUTOMAG III PISTOL
Caliber: 30 Carbine, 8-shot magazine.
Barrel: 6 3/8".
Weight: 43 oz. **Length:** 10 1/2" overall.
Stocks: Carbon fiber.
Sights: Blade front, adjustable rear.
Features: Stainless steel construction. Hammer-drop safety. Slide flats have brushed finish, rest is sandblasted. Introduced 1989. From AMT.
Price: $469.79

AMT AUTOMAG IV PISTOL
Caliber: 45 Winchester Magnum, 6-shot magazine.
Barrel: 6.5".
Weight: 46 oz. **Length:** 10.5" overall.
Stocks: Carbon fiber.
Sights: Blade front, adjustable rear.
Features: Made of stainless steel with brushed finish. Introduced 1990. Made in U.S. by AMT.
Price: $699.99

AMT BACK UP II AUTO PISTOL
Caliber: 380 ACP, 5-shot magazine.
Barrel: 2 1/2".
Weight: 18 oz. **Length:** 5" overall.
Stocks: Carbon fiber.
Sights: Fixed, open, recessed.
Features: Concealed hammer, blowback operation; manual and grip safeties. All stainless steel construction. Smallest domestically-produced pistol in 380. From AMT.
Price: $309.99

AMT Back Up Double Action Only Pistol
Similar to the standard Back Up except has double-action-only mechanism, enlarged trigger guard, slide is rounded ar rear. Has 5-shot magazine. Introduced 1992. From AMT.
Price: $329.99
Price: 9mm Para., 38 Super, 40 S&W, 45 ACP, 357 SIG, 400 Cor-Bon .$449.99

AMERICAN ARMS ESCORT AUTO PISTOL
Caliber: 380 ACP, 7-shot magazine.
Barrel: 3 3/8".
Weight: 19 oz. **Length:** 6 1/8" overall.
Stocks: Soft polymer.
Sights: Blade front, rear adjustable for windage.
Features: Double-action-only trigger; stainless steel construction; chamber loaded indicator. Introduced 1995. From American Arms, Inc.
Price: $349.00

AMERICAN ARMS MATEBA AUTO/REVOLVER
Caliber: 357 Mag., 6-shot.
Barrel: 4".
Weight: 2.75 lbs. **Length:** 8.77" overall.
Stocks: Smooth walnut.
Sights: Blade on ramp front, adjustable rear.
Features: Double or single action. Cylinder and slide recoil together upon firing. All-steel construction with polished blue finish. Introduced 1997. Imported from Italy by American Arms, Inc.
Price: $1,295.00

Armscor M-1911-A1P

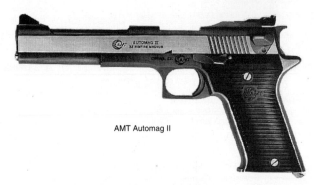

AMT Automag II

AMT 45 ACP Backup

AMT 45 ACP HARDBALLER
Caliber: 45 ACP.
Barrel: 5".
Weight: 39 oz. **Length:** 8 1/2" overall.
Stocks: Wrap-around rubber.
Sights: Adjustable.
Features: Extended combat safety, serrated matte slide rib, loaded chamber indicator, long grip safety, beveled magazine well, adjustable target trigger. All stainless steel. From AMT.
Price: $549.95
Price: Government model (as above except no rib, fixed sights) $489.99

HANDGUNS—AUTOLOADERS, SERVICE & SPORT

AMT 45 ACP HARDBALLER LONG SLIDE
Caliber: 45 ACP.
Barrel: 7". **Length:** 10½" overall.
Stocks: Wrap-around rubber.
Sights: Fully adjustable rear sight.
Features: Slide and barrel are 2" longer than the standard 45, giving less recoil, added velocity, longer sight radius. Has extended combat safety, serrated matte rib, loaded chamber indicator, wide adjustable trigger. From AMT.
Price: $595.99

ASTRA A-70 AUTO PISTOL
Caliber: 9mm Para., 8-shot; 40 S&W, 7-shot magazine.
Barrel: 3.5".
Weight: 29.3 oz. **Length:** 6.5" overall.
Stocks: Checkered black plastic.
Sights: Blade front, rear adjustable for windage.
Features: All steel frame and slide. Checkered grip straps and trigger guard. Nickel or blue finish. Introduced 1992. Imported from Spain by European American Armory.
Price: Blue, 9mm Para. $360.00
Price: Blue, 40 S&W $360.00
Price: Nickel, 9mm Para. $385.00
Price: Nickel, 40 S&W $385.00
Price: Stainless steel, 9mm $450.00
Price: Stainless steel, 40 S&W $450.00

ASTRA A-100 AUTO PISTOL
Caliber: 9mm Para., 10-shot; 40 S&W, 10-shot; 45 ACP, 9-shot magazine.
Barrel: 3.9".
Weight: 29 oz. **Length:** 7.1" overall.
Stocks: Checkered black plastic.
Sights: Blade front, interchangeable rear blades for elevation, screw adjustable for windage.
Features: Double action. Decocking lever permits lowering hammer onto locked firing pin. Automatic firing pin block. Side button magazine release. Introduced 1993. Imported from Spain by European American Armory.
Price: Blue, 9mm, 40 S&W, 45 ACP $450.00
Price: As above, nickel $475.00

> Consult our Directory pages for the location of firms mentioned.

AUTO-ORDNANCE 1911A1 AUTOMATIC PISTOL
Caliber: 9mm Para., 38 Super, 9-shot; 10mm, 45 ACP, 7-shot magazine.
Barrel: 5".
Weight: 39 oz. **Length:** 8½" overall.
Stocks: Checkered plastic with medallion.
Sights: Blade front, rear adjustable for windage.
Features: Same specs as 1911A1 military guns—parts interchangeable. Frame and slide blued; each radius has non-glare finish. Made in U.S. by Auto-Ordnance Corp.
Price: 45 ACP, blue $397.50
Price: 45 ACP, Parkerized $389.95
Price: 45 ACP, satin nickel $425.95
Price: 9mm, 38 Super $435.00
Price: 10mm (has three-dot combat sights, rubber wrap-around grips) ..$435.00
Price: 45 ACP General Model (Commander style) $465.00
Price: Duo Tone (nickel frame, blue slide, three-dot sight system, textured black wrap-around grips) $435.00

Auto-Ordnance 1911A1 Competition Model
Similar to the standard Model 1911A1 except has barrel compensator. Commander hammer, flat mainspring housing, three-dot sight system, low-profile magazine funnel, Hi-Ride beavertail grip safety, full-length recoil spring guide system, black-textured rubber, wrap-around grips, and extended slide stop, safety and magazine catch. In 45 or 38 Super. Introduced 1994. Made in U.S. by Auto-Ordnance Corp.
Price: $635.00

Auto-Ordnance ZG-51 Pit Bull Auto
Same as the 1911A1 except has 3½" barrel, weighs 36 oz. and has an over-all length of 7¼". Available in 45 ACP only; 7-shot magazine. Introduced 1989.
Price: $455.00

Astra A-75

Astra A-75 Decocker Auto Pistol
Same as the A-70 except has decocker system, double or single action, different trigger, contoured pebble-grain grips. Introduced 1993. Imported from Spain by European American Armory.
Price: Blue, 9mm or 40 S&W $415.00
Price: Nickel, 9mm or 40 S&W $440.00
Price: Blue, 45 ACP $445.00
Price: Nickel, 45 ACP $460.00
Price: Stainless steel, 9mm, 40 S&W $495.00
Price: Featherweight (23.5 oz.), 9mm, blue $440.00

Astra A-100

Auto-Ordnance 1911A1

AUTAUGA 32 AUTO PISTOL
Caliber: 32 ACP, 6-shot magazine.
Barrel: 2".
Weight: 11.3 oz. **Length:** 4.3" overall.
Stocks: Black polymer.
Sights: Fixed.
Features: Double-action-only mechansim. Stainless steel construction. Uses Winchester Silver Tip ammunition. Introduced 1996. From Autauga Arms.
Price: ....NA

HANDGUNS—AUTOLOADERS, SERVICE & SPORT

BAER 1911 CONCEPT I AUTO PISTOL
Caliber: 45 ACP, 7-shot magazine.
Barrel: 5".
Weight: 37 oz. **Length:** 8.5" overall.
Stocks: Checkered rosewood.
Sights: Baer dovetail front, Bo-Mar deluxe low-mount rear with hidden leaf.
Features: Baer forged steel frame, slide and barrel with Baer stainless bushing; slide fitted to frame; double serrated slide; Baer beavertail grip safety, checkered slide stop, tuned extractor, extended ejector, deluxe hammer and sear, match disconnector; lowered and flared ejection port; fitted recoil link; polished feed ramp, throated barrel; Baer fitted speed trigger, flat serrated mainspring housing. Blue finish. Made in U.S. by Les Baer Custom, Inc.
Price: $1,390.00
Price: Concept II (with Baer adjustable rear sight) $1,390.00

Baer 1911 Concept IX Auto Pistol
Same as the Commanche Concept VII except has Baer lightweight forged aluminum frame, blued steel slide, Baer adjustable rear sight. Chambered for 45 ACP, 7-shot magazine. Made in U.S. by Les Baer Custom, Inc.
Price: $1,598.00
Price: Concept X (as above with stainless slide) $1,598.00

Baer Premier II

Baer 1911 Prowler III Auto Pistol
Same as the Premier II except also has tapered cone stub weight and reverse recoil plug. Made in U.S. by Les Baer Custom, Inc.
Price: Standard size, blued $1,795.00

Baer Custom Carry

BAER LIGHTWEIGHT 22
Caliber: 22 LR.
Barrel: 5".
Weight: 25 oz. **Length:** 8.5" overall.
Stocks: Checkered walnut.
Sights: Blade front.
Features: Aluminum frame and slide. Baer beavertail grip safety with pad, checkered slide stop, deluxe hammer and sear, match disconnector, flat serrated mainspring housing, Baer fitted speed trigger, tuned extractor. Has total reliability tuning package, action job. Baer Ultra Coat finish. Introduced 1996. Made in U.S. by Les Baer Custom, Inc.
Price: Government model size, fixed sights $1,498.00
Price: Government model, Bo-Mar sights $1,498.00
Price: Commanche size, fixed sights $1,428.00

Baer 1911 Concept III Auto Pistol
Same as the Concept I except has forged stainless frame with blued steel slide, Bo-Mar rear sight, 30 lpi checkering on front strap. Made in U.S. by Les Baer Custom, Inc.
Price: $1,520.00
Price: Concept IV (with Baer adjustable rear sight) $1,499.00
Price: Concept V (all stainless, Bo-Mar sight, checkered front strap) ..$1,558.00
Price: Concept VI (stainless, Baer adjustable sight, checkered front strap) $1,558.00

Baer 1911 Concept VII Auto Pistol
Same as the Concept I except reduced Commanche size with 4.25" barrel, weighs 27.5 oz., 7.75" overall. Blue finish, checkered front strap. Made in U.S. by Les Baer Custom, Inc.
Price: $1,495.00
Price: Concept VIII (stainless frame and slide, Baer adjustable rear sight) $1,547.00

BAER 1911 PREMIER II AUTO PISTOL
Caliber: 45 ACP, 7- or 10-shot magazine.
Barrel: 5".
Weight: 37 oz. **Length:** 8.5" overall.
Stocks: Checkered rosewood, double diamond pattern.
Sights: Baer dovetailed front, low-mount Bo-Mar rear with hidden leaf.
Features: Baer NM forged steel frame and barrel with stainless bushing; slide fitted to frame; double serrated slide; lowered, flared ejection port; tuned, polished extractor; Baer extended ejector, checkered slide stop, aluminum speed trigger with 4-lb. pull, deluxe Commander hammer and sear, beavertail grip safety with pad, beveled magazine well, extended ambidextrous safety; flat mainspring housing; polished feed ramp and throated barrel; 30 lpi checkered front strap. Made in U.S. by Les Baer Custom, Inc.
Price: Blued $1,428.00
Price: Stainless $1,558.00
Price: 6" model, blued $1,690.00

Manufacturers' addresses in the
Directory of the Arms Trade
page 309, this issue

BAER 1911 S.R.P. PISTOL
Caliber: 45 ACP.
Barrel: 5".
Weight: 37 oz. **Length:** 8.5" overall.
Stocks: Checkered walnut.
Sights: Trijicon night sights.
Features: Similar to the F.B.I. contract gun except uses Baer forged steel frame. Has Baer match barrel with supported chamber, Wolff springs, complete tactical action job. All parts Mag-na-fluxed; deburred for tactical carry. Has Baer Ultra Coat finish. Tuned for reliability. Contact Baer for complete details. Introduced 1996. Made in U.S. by Les Baer Custom, Inc.
Price: Government or Commanche length $2,495.00

BAER 1911 CUSTOM CARRY AUTO PISTOL
Caliber: 45 ACP, 7- or 10-shot magazine.
Barrel: 5".
Weight: 37 oz. **Length:** 8.5" overall.
Stocks: Checkered walnut.
Sights: Baer improved ramp-style dovetailed front, Novak low-mount rear.
Features: Baer forged NM frame, slide and barrel with stainless bushing; fitted slide to frame; double serrated slide (full-size only); Baer speed trigger with 4-lb. pull; Baer deluxe hammer and sear, tactical-style extended ambidextrous safety, beveled magazine well; polished feed ramp and throated barrel; tuned extractor; Baer extended ejector, checkered slide stop; lowered and flared ejection port, full-length recoil guide rod; recoil buff. Made in U.S. by Les Baer Custom, Inc.
Price: Standard size, blued $1,620.00
Price: Standard size, stainless $1,690.00
Price: Commanche size, blued $1,490.00
Price: Commanche size, stainless $1,690.00
Price: Commanche size, aluminum frame, blued slide $1,890.00
Price: Commanche size, aluminum frame, stainless slide $1,995.00

HANDGUNS—AUTOLOADERS, SERVICE & SPORT

BERETTA MODEL 92FS PISTOL
Caliber: 9mm Para., 10-shot magazine.
Barrel: 4.9".
Weight: 34 oz. **Length:** 8.5" overall.
Stocks: Checkered black plastic; wood optional at extra cost.
Sights: Blade front, rear adjustable for windage. Tritium night sights available.
Features: Double action. Extractor acts as chamber loaded indicator, squared trigger guard, grooved front- and backstraps, inertia firing pin. Matte or blued finish. Introduced 1977. Made in U.S. and imported from Italy by Beretta U.S.A.
Price: With plastic grips $626.00
Price: With wood grips $647.00
Price: Tritium night sights, add $90.00
Price: Deluxe, gold or silver plated $5,429.00
Price: EL model, blued $790.00

Beretta Model 92FS

Beretta Model 92F Stainless Pistol
Same as the Model 92FS except has stainless steel barrel and slide, and frame of aluminum-zirconium alloy. Has three-dot sight system. Introduced 1992.
Price: $757.00
Price: For tritium sights, add $90.00

Beretta Model 92D Pistol
Same as the Model 92FS except double-action-only and has bobbed hammer, no external safety. Introduced 1992.
Price: With plastic grips, three-dot sights $586.00
Price: As above with tritium sights $676.00

Beretta Models 92FS/96 Centurion Pistols
Identical to the Model 92FS and 96F except uses shorter slide and barrel (4.3"). Tritium or three-dot sight systems. Plastic or wood grips. Available in 9mm or 40 S&W. Also available in D Models (double-action-only). Introduced 1992.
Price: Model 92FS Centurion, three-dot sights, plastic grips $626.00
Price: Model 92FS Centurion, wood grips $647.00
Price: Model 96 Centurion, three-dot sights, plastic grips $626.00
Price: Model 92D Centurion $586.00
Price: Model 96D Centurion $586.00
Price: For tritium sights, add $90.00

Beretta Model 96 Pistol
Same as the Model 92F except chambered for 40 S&W. Ambidextrous safety mechanism with passive firing pin catch, slide safety/decocking lever, trigger bar disconnect. Has 10-shot magazine. Available with tritium or three-dot sights. Introduced 1992.
Price: Model 96, plastic grips $626.00
Price: Model 96D, double-action-only, three-dot sights $586.00
Price: For tritium sights, add $90.00

Beretta 96D

BERETTA MODEL 80 CHEETAH SERIES DA PISTOLS
Caliber: 380 ACP, 10-shot magazine (M84); 8-shot (M85); 22 LR, 7-shot (M87).
Barrel: 3.82".
Weight: About 23 oz. (M84/85); 20.8 oz. (M87). **Length:** 6.8" overall.
Stocks: Glossy black plastic (wood optional at extra cost).
Sights: Fixed front, drift-adjustable rear.
Features: Double action, quick takedown, convenient magazine release. Introduced 1977. Imported from Italy by Beretta U.S.A.
Price: Model 84 Cheetah, plastic grips $529.00
Price: Model 84 Cheetah, wood grips $557.00
Price: Model 84 Cheetah, wood grips, nickel finish $600.00
Price: Model 85 Cheetah, plastic grips, 8-shot $500.00
Price: Model 85 Cheetah, wood grips, 8-shot $530.00
Price: Model 85 Cheetah, wood grips, nickel, 8-shot $559.00
Price: Model 87 Cheetah wood, 22 LR, 7-shot $529.00

Beretta Model 86 Cheetah
Similar to the 380-caliber Model 85 except has tip-up barrel for first-round loading. Barrel length is 4.4", overall length of 7.33". Has 8-shot magazine, walnut grips. Introduced 1989.
Price: $530.00

BERETTA MODEL 950 JETFIRE AUTO PISTOL
Caliber: 25 ACP, 8-shot.
Barrel: 2.4".
Weight: 9.9 oz. **Length:** 4.7" overall.
Stocks: Checkered black plastic or walnut.
Sights: Fixed.
Features: Single action, thumb safety; tip-up barrel for direct loading/unloading, cleaning. From Beretta U.S.A.
Price: Jetfire plastic, blue $216.00
Price: Jetfire plastic, nickel $250.00
Price: Jetfire wood, engraved $296.00
Price: Jetfire plastic, matte finish $187.00

Beretta Model 21 Bobcat Pistol
Similar to the Model 950 BS. Chambered for 22 LR or 25 ACP. Both double action. Has 2.4" barrel, 4.9" overall length; 7-round magazine on 22 cal.; 8 rounds in 25 ACP, 9.9 oz., available in nickel, matte, engraved or blue finish. Plastic or walnut grips. Introduced in 1985.
Price: Bobcat, 22-cal., blue $259.00
Price: Bobcat, nickel, 22-cal. $269.00
Price: Bobcat, 25-cal., blue $259.00
Price: Bobcat, nickel, 25-cal. $269.00
Price: Bobcat wood, engraved, 22 or 25 ....$309.00
Price: Bobcat plastic matte, 22 or 25 ....$209.00

Beretta 950 Jetfire

CAUTION: PRICES SHOWN ARE SUPPLIED BY THE MANUFACTURER OR IMPORTER. CHECK YOUR LOCAL GUNSHOP.

HANDGUNS—AUTOLOADERS, SERVICE & SPORT

BERETTA MODEL 8000/8040 COUGAR PISTOL
Caliber: 9mm Para., 10-shot, 40 S&W, 10-shot magazine.
Barrel: 3.6".
Weight: 33.5 oz. **Length:** 7" overall.
Stocks: Checkered plastic.
Sights: Blade front, rear drift adjustable for windage.
Features: Slide-mounted safety; rotating barrel; exposed hammer. Matte black Bruniton finish. Announced 1994. Imported from Italy by Beretta U.S.A.
Price: $699.00
Price: D models $663.00

Beretta M8000/8040 Cougar

BERETTA MODEL 3032 TOMCAT PISTOL
Caliber: 32 ACP, 7-shot magazine.
Barrel: 2.45".
Weight: 14.5 oz. **Length:** 5" overall.
Stocks: Checkered black plastic.
Sights: Blade front, drift-adjustable rear.
Features: Double action with exposed hammer; tip-up barrel for direct loading/unloading; thumb safety; polished or matte blue finish. Imported from Italy by Beretta U.S.A. Introduced 1996.
Price: Blue $330.00
Price: Matte $269.00

BERNARDELLI PO18 DA AUTO PISTOL
Caliber: 9mm Para., 10-shot magazine.
Barrel: 4.8".
Weight: 34.2 oz. **Length:** 8.23" overall.
Stocks: Checkered plastic; walnut optional.
Sights: Blade front, rear adjustable for windage and elevation; low profile, three-dot system.
Features: Manual thumb half-cock, magazine and auto-locking firing pin safeties. Thumb safety decocks hammer. Reversible magazine release. Imported from Italy by Armsport.
Price: Black $725.00
Price: Chrome $780.00

Bernadelli PO18

Bernardelli PO18 Compact DA Auto Pistol
Similar to the PO18 except has 4" barrel, 7.44" overall length, 10-shot magazine. Weighs 31.7 oz. Imported from Italy by Armsport.
Price: Black $725.00
Price: Chrome $780.00

BERNARDELLI MODEL USA AUTO PISTOL
Caliber: 22 LR, 10-shot, 380 ACP, 7-shot magazine.
Barrel: 3.5".
Weight: 26.5 oz. **Length:** 6.5" overall.
Stocks: Checkered plastic with thumbrest.
Sights: Ramp front, white outline rear adjustable for windage and elevation.
Features: Hammer-block slide safety; loaded chamber indicator; dual recoil buffer springs; serrated trigger; inertia-type firing pin. Imported from Italy by Armsport.
Price: Black, either caliber $499.00
Price: Chrome, either caliber $570.00

Bernadelli USA

BERNARDELLI P. ONE DA AUTO PISTOL
Caliber: 9mm Para., 16-shot, 40 S&W, 10-shot magazine.
Barrel: 4.8".
Weight: 34 oz. **Length:** 8.35" overall.
Stocks: Checkered black plastic.
Sights: Blade front, rear adjustable for windage and elevation; three dot system.
Features: Forged steel frame and slide; full-length slide rails; reversible magazine release; thumb safety/decocker; squared trigger guard. Introduced 1994. Imported from Italy by Armsport.
Price: 9mm Para., blue/black $690.00
Price: 9mm Para., chrome $750.00
Price: 40 S&W, blue/black $690.00
Price: 40 S&W, chrome $750.00

Bernardelli P. One Practical VB Pistol
Similar to the P. One except chambered for 9x21mm, two- or four-port compensator, straight trigger, micro-adjustable rear sight. Introduced 1994. Imported from Italy by Armsport.
Price: Blue/black, two-port compensator $1,600.00
Price: As above, four-port compensator $1,670.00
Price: Chrome, two-port compensator $1,675.00
Price: As above, four-port compensator $1,735.00
Price: Customized VB, four-plus-two-port compensator ...$2,375.00
Price: As above, chrome $2,450.00

Bersa Thunder 380

BERSA THUNDER 380 AUTO PISTOLS
Caliber: 380 ACP, 7-shot (Thunder 380 Lite), 9-shot magazine (Thunder 380 DLX).
Barrel: 3.5".
Weight: 25.75 oz. **Length:** 6.6" overall.
Stocks: Black polymer.
Sights: Blade front, notch rear adjustable for windage; three-dot system.
Features: Double action; firing pin and magazine safeties. Available in blue or nickel. Introduced 1995. Distributed by Eagle Imports, Inc.
Price: Thunder 380, 7-shot, deep blue finish ...$258.95
Price: As above, satin nickel $274.95
Price: Thunder 380 DLX, 9-shot, matte blue $274.95

HANDGUNS—AUTOLOADERS, SERVICE & SPORT

BERSA SERIES 95 AUTO PISTOL
Caliber: 380 ACP, 7-shot magazine.
Barrel: 3.5".
Weight: 22 oz. **Length:** 6.6" overall.
Stocks: Wrap-around textured rubber.
Sights: Blade front, rear adjustable for windage; three-dot system.
Features: Double action; firing pin and magazine safeties; combat-style trigger guard. Matte blue or satin nickel. Introduced 1992. Distributed by Eagle Imports, Inc.
Price: Matte blue $231.95
Price: Satin nickel $253.95

Bersa Series 95

BERSA THUNDER 22 AUTO PISTOL
Caliber: 22 LR, 10-shot magazine.
Barrel: 3.5".
Weight: 24.2 oz. **Length:** 6.6" overall.
Stocks: Black polymer.
Sights: Blade front, notch rear adjustable for windage; three-dot system.
Features: Double action; firing pin and magazine safeties. Available in blue or nickel. Introduced 1995. Distributed by Eagle Imports, Inc.
Price: Blue $258.95
Price: Nickel $274.95

BROLIN LEGEND L45 STANDARD PISTOL
Caliber: 45 ACP, 7-shot magazine.
Barrel: 5".
Weight: 35.9 oz. **Length:** 8.5" overall.
Stocks: Checkered walnut.
Sights: Millett High Visibility front, white outline fixed rear.
Features: Throated match barrel; polished feed ramp; lowered and flared ejection port; beveled magazine well; flat top slide; flat mainspring housing; lightened aluminum match trigger; slotted Commander hammer; matte blue finish. Introduced 1996. Made in U.S. by Brolin Arms.
Price: $459.00

Brolin Legend L45

Brolin Legend L45C Compact Pistol
Similar to the L45 Standard pistol except has 4" barrel with conical lock up; overall length 7.5"; weighs 32 oz. Matte blue finish. Introduced 1996. Made in U.S. by Brolin Arms.
Price: $489.00

Brolin L45T Compact Auto Pistol
Same as the L45 Legend except uses compact slide on the standard-size frame. Has 4" barrel, weighs 33.9 oz., and is 7.5" overall. Introduced 1996. Made in U.S. by Brolin Industries.
Price: $489.00

Brolin Patriot P45

Brolin L45T Compact

Brolin P45C Compact Carry Comp
Similar to the P45 Standard Carry Comp except has 3.25" barrel with integral milled compensator; overall length 7.5"; weighs 33 oz. Introduced 1996. Made in U.S. by Brolin Industries.
Price: Matte blue, Millett sights $689.00
Price: Two-tone finish, Millett sights $709.00
Price: Matte blue, Novak Combat sights $759.00
Price: Two-tone finish, Novak Combat sights $769.00

BROLIN PATRIOT P45 COMP STANDARD CARRY-COMP
Caliber: 45 ACP, 7-shot magazine.
Barrel: 4".
Weight: 37 oz. **Length:** 8.5" overall.
Stocks: Checkered wood.
Sights: Millett high visibility front, fixed white outline rear; or Novak sights.
Features: One-piece milled compensator with throated match barrel utilizing the conical lock-up system; polished feed ramp; lowered and flared ejection port; beveled magazine well; flat top slide; flat mainspring housing; front strap high relief cut, adjustable aluminum match trigger; custom beavertail grip safety; slotted commander hammer. Introduced 1996. Made in the U.S. by Brolin Industries.
Price: Matte blue finish with Millett sights $649.00
Price: Two-tone with Millett sights $669.00
Price: Matte blue finish with Novak Combat sights $719.00
Price: Two-tone with Noval Combat sights $729.00

Brolin Patriot P45T Comp Compact Carry-Comp
Similar to the Patriot P45 Comp Standard Carry-Comp. Has compact slide with standard frame; one-piece milled compensator with throated match barrel utilizing the integral lock-up system; one-piece guide rod; polished feed ramp; lowered and flared ejection port; beveled magazine well; flat-top slide; flat mainspring housing; front strap high relief cut; adjustable aluminum match trigger; custom beavertal safety; slotted commander hammer. Made in U.S. by Brolin Industries.
Price: Matte blue, Millett sights $699.00
Price: Two-tone finish, Millett sights $709.00
Price: Matte blue, Novak Combat sights $759.00
Price: Two-tone finish, Novak Combat sights $769.00

CAUTION: PRICES SHOWN ARE SUPPLIED BY THE MANUFACTURER OR IMPORTER. CHECK YOUR LOCAL GUNSHOP.

HANDGUNS—AUTOLOADERS, SERVICE & SPORT

Browning Capitan Hi-Power

Browning Capitan Hi-Power Pistol
Similar to the standard Hi-Power except has adjustable tangent rear sight authentic to the early-production model. Also has Commander-style hammer. Checkered walnut grips, polished blue finish. Reintroduced 1993. Imported from Belgium by Browning.
Price: 9mm only $692.95

Browning Hi-Power HP-Practical Pistol
Similar to the standard Hi-Power except has silver-chromed frame with blued slide, wrap-around Pachmayr rubber grips, round-style serrated hammer and removable front sight, fixed rear (drift-adjustable for windage). Available in 9mm Para. or 40 S&W. Introduced 1991.
Price: $629.75
Price: With fully adjustable rear sight $681.95

Browning BDM Silver Chrome

Browning Micro Buck Mark Standard

Browning Buck Mark Varmint

BROWNING HI-POWER 9mm AUTOMATIC PISTOL
Caliber: 9mm Para., 40 S&W, 10-shot magazine.
Barrel: 4$^{21}/_{32}$".
Weight: 32 oz. **Length:** 7$^{3}/_{4}$" overall.
Stocks: Walnut, hand checkered, or black Polyamide.
Sights: $^{1}/_{8}$" blade front; rear screw-adjustable for windage and elevation. Also available with fixed rear (drift-adjustable for windage).
Features: External hammer with half-cock and thumb safeties. A blow on the hammer cannot discharge a cartridge; cannot be fired with magazine removed. Fixed rear sight model available. Imported from Belgium by Browning.
Price: Fixed sight model, walnut grips $584.95
Price: 9mm with rear sight adj. for w. and e., walnut grips $635.95
Price: Mark III, standard matte black finish, fixed sight, moulded grips, ambidextrous safety $550.75
Price: Silver chrome, adjustable sight, Pachmayr grips $650.95

Browning 40 S&W Hi-Power Mark III Pistol
Similar to the standard Hi-Power except chambered for 40 S&W, 10-shot magazine, weighs 35 oz., and has 4$^{3}/_{4}$" barrel. Comes with matte blue finish, low profile front sight blade, drift-adjustable rear sight, ambidextrous safety, moulded polyamide grips with thumb rest. Introduced 1993. Imported from Belgium by Browning.
Price: Mark III $550.95

BROWNING BDM DOUBLE MODE AUTO PISTOL
Caliber: 9mm Para., 10-shot magazine.
Barrel: 4$^{3}/_{4}$".
Weight: 31 oz. **Length:** 7$^{3}/_{4}$" overall.
Stocks: Moulded black composition; checkered, with thumbrest on both sides.
Sights: Low profile removable blade front, rear screw adjustable for windage.
Features: Mode selector allows switching from DA pistol to "revolver" mode via a switch on the slide. Decocking lever/safety on the frame. Two redundant, passive, internal safety systems. All steel frame; matte black finish. Introduced 1991. Made in the U.S. From Browning.
Price: BDM Standard $550.95
Price: BDM Practical $570.95
Price: BDM Silver Chrome $570.95

Browning BPM-D Auto Pistol
Similar to the BDM Double Mode except the Browning Pistol Mode-Decocker is a conventional double action with an ambidextruous decocker/slide release lever; reversible magazine release; three-dot sight system. All dimensions are the same as the BDM. Introduced 1997. Made in the U.S. From Browning.
Price: $550.95

Browning BRM-DAO Auto Pistol
Similar to the BDM Double Mode except the Browning Revolver Mode—Double Action Only is a full-time double action. No decocking lever because the hammer always returns to the decocking position. All dimensions are the same as the BDM. Introduced 1997. Made in the U.S. From Browning.
Price: $550.95

BROWNING BUCK MARK 22 PISTOL
Caliber: 22 LR, 10-shot magazine.
Barrel: 5$^{1}/_{2}$".
Weight: 32 oz. **Length:** 9$^{1}/_{2}$" overall.
Stocks: Black moulded composite with skip-line checkering.
Sights: Ramp front, Browning Pro Target rear adjustable for windage and elevation.
Features: All steel, matte blue finish or nickel, gold-colored trigger. Buck Mark Plus has laminated wood grips. Made in U.S. Introduced 1985. From Browning.
Price: Buck Mark, blue $256.95
Price: Buck Mark, nickel finish with contoured rubber stocks $301.95
Price: Buck Mark Plus $313.95

Browning Micro Buck Mark
Same as the standard Buck Mark and Buck Mark Plus except has 4" barrel. Available in blue or nickel. Has 16-click Pro Target rear sight. Introduced 1992.
Price: Blue $256.95
Price: Nickel $301.95
Price: Buck Mark Micro Plus $313.95
Price: Buck Mark Micro Plus Nickel $342.95

Browning Buck Mark Varmint
Same as the Buck Mark except has 9$^{7}/_{8}$" heavy barrel with .900" diameter and full-length scope base (no open sights); walnut grips with optional forend, or finger-groove walnut. Overall length is 14", weighs 48 oz. Introduced 1987.
Price: $390.95

HANDGUNS—AUTOLOADERS, SERVICE & SPORT

Browning BDA 380

BRYCO MODEL 38 AUTO PISTOLS
Caliber: 22 LR, 32 ACP, 380 ACP, 6-shot magazine.
Barrel: 2.8".
Weight: 15 oz. **Length:** 5.3" overall.
Stocks: Polished resin-impregnated wood.
Sights: Fixed.
Features: Safety locks sear and slide. Choice of satin nickel, bright chrome or black Teflon finishes. Introduced 1988. From Jennings Firearms.
Price: 22 LR, 32 ACP, about $109.95
Price: 380 ACP, about $129.95

BRYCO MODEL 59 AUTO PISTOL
Caliber: 9mm Para., 10-shot magazine.
Barrel: 4".
Weight: 33 oz. **Length:** 6.5" overall.
Stocks: Black composition.
Sights: Blade front, fixed rear.
Features: Striker-fired action; manual thumb safety; polished blue finish. Comes with two magazines. Introduced 1994. From Jennings Firearms.
Price: About $169.00
Price: Model 58 (5.5" overall length, 30 oz.) $169.00

CALICO M-110 AUTO PISTOL
Caliber: 22 LR. 100-shot magazine.
Barrel: 6".
Weight: 3.7 lbs. (loaded). **Length:** 17.9" overall.
Stocks: Moulded composition.
Sights: Adjustable post front, notch rear.
Features: Aluminum alloy frame; flash suppressor; pistol grip compartment; ambidextrous safety. Uses same helical-feed magazine as M-100 Carbine. Introduced 1986. Made in U.S. From Calico.
Price: $432.00

CENTURY FEG P9R PISTOL
Caliber: 9mm Para., 10-shot magazine.
Barrel: 4.6".
Weight: 35 oz. **Length:** 8" overall.
Stocks: Checkered walnut.
Sights: Blade front, rear drift adjustable for windage.
Features: Double action with hammer-drop safety. Polished blue finish. Comes with spare magazine. Imported from Hungary by Century International Arms.
Price: About $263.00

COLT 22 AUTOMATIC PISTOL
Caliber: 22 LR, 10-shot magazine.
Barrel: 4.5".
Weight: 33 oz. **Length:** 8.62" overall.
Stocks: Textured black polymer.
Sights: Blade front, rear drift adjustable for windage.
Features: Stainless steel construction; ventilated barrel rib; single action mechanism; cocked striker indicator; push-button safety. Introduced 1994. Made in U.S. by Colt's Mfg. Co.
Price: $248.00

Colt 22 Target Pistol
Similar to the Colt 22 pistol except has 6" bull barrel, full-length sighting rib with lightening cuts and mounting rail for optical sights; fully adjustable rear sight; removable sights; two-point factory adjusted trigger travel. Stainless steel frame. Introduced 1995. Made in U.S. by Colt's Mfg. Co.
Price: $377.00

BROWNING BDA-380 DA AUTO PISTOL
Caliber: 380 ACP, 10-shot magazine.
Barrel: 3 3/16".
Weight: 32 oz. **Length:** 6 3/4" overall.
Stocks: Smooth walnut with inset Browning medallion.
Sights: Blade front, rear drift-adjustable for windage.
Features: Combination safety and decocking lever will automatically lower a cocked hammer to half-cock and can be operated by right- or left-hand shooters. Inertia firing pin. Introduced 1978. Imported from Italy by Browning.
Price: Blue $563.95
Price: Nickel $606.95

BRYCO MODEL 48 AUTO PISTOLS
Caliber: 22 LR, 32 ACP, 380 ACP, 6-shot magazine.
Barrel: 4".
Weight: 19 oz. **Length:** 6.7" overall.
Stocks: Polished resin-impregnated wood.
Sights: Fixed.
Features: Safety locks sear and slide. Choice of satin nickel, bright chrome or black Teflon finishes. Announced 1988. From Jennings Firearms.
Price: 22 LR, 32 ACP, about $139.00
Price: 380 ACP, about $139.00

Calico M-110

CARBON-15 PISTOL
Caliber: 223, 10-shot magazine.
Barrel: 7.25".
Weight: 46 oz. **Length:** 20" overall.
Stock: Checkered composite.
Sights: Ghost ring.
Features: Semi-automatic, gas-operated, rotating bolt action. Carbon fiber upper and lower receiver; chromemoly bolt carrier; fluted stainless match barrel; mil. spec. optics mounting base; uses AR-15-type magazines. Introduced 1992. From Professional Ordnance, Inc.
Price: $1,607.87

> Consult our Directory pages for the location of firms mentioned.

Century FEG P9RK Auto Pistol
Similar to the P9R except has 4.12" barrel, 7.5" overall length and weighs 33.6 oz. Checkered walnut grips, fixed sights, 10-shot magazine. Introduced 1994. Imported from Hungary by Century International Arms, Inc.
Price: About $290.00

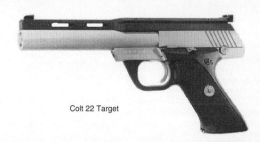

Colt 22 Target

HANDGUNS—AUTOLOADERS, SERVICE & SPORT

COLT PONY AUTOMATIC PISTOL
Caliber: 380 ACP.
Barrel: 2 3/4".
Weight: 19 oz. **Length:** 5 1/2" overall.
Stocks: Black composition.
Sights: Ramp front, fixed rear.
Features: Stainless steel construction. Double-action-only mechanism; recoil-reducing locked breech. Introduced 1997. Made in U.S. by Colt's Mfg. Co.
Price: $493.00
Price: Pocketlite LW, aluminum and stainless steel, weighs 13 oz. $493.00

COLT COMBAT COMMANDER AUTO PISTOL
Caliber: 45 ACP, 8-shot.
Barrel: 4 1/4".
Weight: 36 oz. **Length:** 7 3/4" overall.
Stocks: Checkered rubber composite.
Sights: Fixed, glare-proofed blade front, square notch rear; three-dot system.
Features: Long trigger; arched housing; grip and thumb safeties.
Price: 45, stainless $789.00

Colt Lightweight Commander MK IV/Series 80
Same as Commander except high strength aluminum alloy frame, checkered rubber composite stocks, weighs 27 1/2 oz. 45 ACP only.
Price: Blue $735.00

COLT MODEL 1991 A1 AUTO PISTOL
Caliber: 45 ACP, 7-shot magazine.
Barrel: 5".
Weight: 38 oz. **Length:** 8.5" overall.
Stocks: Checkered black composition.
Sights: Ramped blade front, fixed square notch rear, high profile.
Features: Parkerized finish. Continuation of serial number range used on original G.I. 1911 A1 guns. Comes with one magazine and moulded carrying case. Introduced 1991.
Price: $556.00
Price: Stainless $610.00

Colt Model 1991 A1 Compact Auto Pistol
Similar to the Model 1991 A1 except has 3 1/2" barrel. Overall length is 7", and gun is 3/8" shorter in height. Comes with one 6-shot magazine, moulded case. Introduced 1993.
Price: $556.00

Colt Model 1991 A1 Commander Auto Pistol
Similar to the Model 1991 A1 except has 4 1/4" barrel. Parkerized finish. 7-shot magazine. Comes in moulded case. Introduced 1993.
Price: $556.00

COLT GOVERNMENT MODEL 380
Caliber: 380 ACP, 7-shot magazine.
Barrel: 3 1/4".
Weight: 21 3/4 oz. **Length:** 6" overall.
Stocks: Checkered composition.
Sights: Ramp front, square notch rear, fixed.
Features: Scaled-down version of the 1911 A1 Colt G.M. Has thumb and internal firing pin safeties. Introduced 1983.
Price: Blue $474.00
Price: Stainless $508.00
Price: Pocketlite 380, stainless $508.00

Colt Mustang 380, Mustang Pocketlite
Similar to the standard 380 Government Model. Mustang has steel frame (18.5 oz.), Pocketlite has aluminum alloy (12.5 oz.). Both are 1/2" shorter than 380 G.M., have 2 3/4" barrel. Introduced 1987.
Price: Mustang 380, stainless $500.00
Price: Mustang Pocketlite STS/N $500.00

COLT OFFICER'S ACP MK IV/SERIES 80
Caliber: 45 ACP, 6-shot magazine.
Barrel: 3 1/2".
Weight: 34 oz. (steel frame); 24 oz. (alloy frame). **Length:** 7 1/4" overall.
Stocks: Checkered rubber composite.
Sights: Ramp blade front with white dot, square notch rear with two white dots.
Features: Trigger safety lock (thumb safety), grip safety, firing pin safety; long trigger; flat mainspring housing. Also available with lightweight alloy frame and in stainless steel. Introduced 1985.
Price: L.W., blue finish $735.00
Price: Stainless $813.00

Colt Pony

Colt Government Model

COLT GOVERNMENT MODEL MK IV/SERIES 80
Caliber: 9x23 Win., 38 Super, 9-shot; 45 ACP, 8-shot magazine.
Barrel: 5".
Weight: 38 oz. **Length:** 8 1/2" overall.
Stocks: Black composite.
Sights: Ramp front, fixed square notch rear; three-dot system.
Features: Grip and thumb safeties and internal firing pin safety, long trigger.
Price: 9x23 Win., blue $735.00
Price: 45 ACP, stainless $813.00
Price: 38 Super, blue $735.00

Colt 1991 A1 Compact

Colt Mustang 380

HANDGUNS—AUTOLOADERS, SERVICE & SPORT

COLT COMBAT TARGET MODEL
Caliber: 45 ACP, 7-shot magazine.
Barrel: 5".
Weight: 39 oz. **Length:** 8½" overall.
Stocks: Black composition.
Sights: Patridge-style front, Colt Accro adjustable rear.
Features: Steel target trigger with cut-out; flat-top slide; flared and lowered ejection port; beveled magazine well. Introduced 1996. Made in U.S. by Colt's Mfg. Co.
Price: Matte blue $768.00
Price: Matte stainless $820.00

Coonan 357 Magnum

COONAN 357 MAGNUM PISTOL
Caliber: 357 Mag., 7-shot magazine.
Barrel: 5".
Weight: 42 oz. **Length:** 8.3" overall.
Stocks: Smooth walnut.
Sights: Interchangeable ramp front, rear adjustable for windage.
Features: Stainless steel construction. Unique barrel hood improves accuracy and reliability. Linkless barrel. Many parts interchange with Colt autos. Has grip, hammer, half-cock safeties, extended slide latch. Made in U.S. by Coonan Arms, Inc.
Price: 5" barrel $735.00
Price: 6" barrel $768.00
Price: With 6" compensated barrel $1,014.00
Price: Classic model (Teflon black two-tone finish, 8-shot magazine, fully adjustable rear sight, integral compensated barrel) $1,400.00

Coonan Compact Cadet 357 Magnum Pistol
Similar to the 357 Magnum full-size gun except has 3.9" barrel, shorter frame, 6-shot magazine. Weight is 39 oz., overall length 7.8". Linkless bull barrel, full-length recoil spring guide rod, extended slide latch. Introduced 1993. Made in U.S. by Coonan Arms, Inc.
Price: $850.00

CZ 75 AUTO PISTOL
Caliber: 9mm Para., 40 S&W, 10-shot magazine.
Barrel: 4.7".
Weight: 34.3 oz. **Length:** 8.1" overall.
Stocks: High impact checkered plastic.
Sights: Square post front, rear adjustable for windage; three-dot system.
Features: Single action/double action design; choice of black polymer, matte or high-polish blue finishes. All-steel frame. Imported from the Czech Republic by CZ-USA.
Price: .... NA

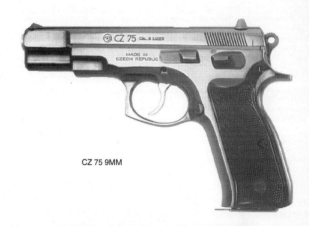

CZ 75 9MM

CZ 85 Auto Pistol
Same gun as the CZ 75 except has ambidextrous slide release and safety-levers; non-glare, ribbed slide top; squared, serrated trigger guard; trigger stop to prevent overtravel. Introduced 1986. Imported from the Czech Republic by CZ-USA.
Price: .... NA

CZ 75 Compact Auto Pistol
Similar to the CZ 75 except has 10-shot magazine, 3.9" barrel and weighs 32 oz. Has removable fromt sight, non-glare ribbed slide top. Trigger guard is squared and serrated; combat hammer. Introduced 1993. Imported from the Czech Republic by CZ-USA.
Price: .... NA

Manufacturers' addresses in the
Directory of the Arms Trade
page 309, this issue

CZ 83 DOUBLE-ACTION PISTOL
Caliber: 380 ACP, 10-shot magazine.
Barrel: 3.8".
Weight: 26.2 oz. **Length:** 6.8" overall.
Stocks: High impact checkered plastic.
Sights: Removable square post front, rear adjustable for windage; three-dot system.
Features: Single action/double action; ambidextrous magazine release and safety. Blue finish; non-glare ribbed slide top. Imported from the Czech Republic by CZ-USA.
Price: .... NA

CZ 83 380

CZ 100 AUTO PISTOL
Caliber: 9mm Para., 40 S&W, 10-shot magazine.
Barrel: 3.7".
Weight: 24 oz. **Length:** 6.9" overall.
Stocks: Grooved polymer.
Sights: Blade front with dot, white outline rear drift adjustable for windage.
Features: Double action only with firing pin block; polymer frame, steel slide; has laser sight mount. Introduced 1996. Imported from the Czech Republic by CZ-USA.
Price: .... NA

CAUTION: PRICES SHOWN ARE SUPPLIED BY THE MANUFACTURER OR IMPORTER. CHECK YOUR LOCAL GUNSHOP.

HANDGUNS—AUTOLOADERS, SERVICE & SPORT

DAEWOO DP51 FASTFIRE AUTO PISTOL
Caliber: 9mm Para., 40 S&W, 10-shot magazine.
Barrel: 4.1".
Weight: 28.2 oz. **Length:** 7.5" overall.
Stocks: Checkered composition.
Sights: $1/8$" blade front, square notch rear drift adjustable for windage. Three dot system.
Features: Patented Fastfire mechanism. Ambidextrous manual safety and magazine catch, automatic firing pin block. Alloy frame, squared trigger guard. Matte black finish. Introduced 1991. Imported from South Korea by Nationwide Sports Dist.
Price: DP51 ....$400.00
Price: DH40 (40 S&W) $450.00

Daewoo DP51C, DP51S Auto Pistols
Same as the DP51 except DP51C has 3.6" barrel, $1/4$" shorter grip frame, flat mainspring housing, and is 2 oz. lighter. Model DP51S has 3.6" barrel, same grip as standard DP51, weighs 27 oz. Introduced 1995. Imported from South Korea by Nationwide Sports Dist.
Price: DP51C $445.00
Price: DP51S $420.00

DAVIS P-32 AUTO PISTOL
Caliber: 32 ACP, 6-shot magazine.
Barrel: 2.8".
Weight: 22 oz. **Length:** 5.4" overall.
Stocks: Laminated wood.
Sights: Fixed.
Features: Choice of black Teflon or chrome finish. Announced 1986. Made in U.S. by Davis Industries.
Price: $87.50

DAVIS P-380 AUTO PISTOL
Caliber: 32 ACP, 6-shot, 380 ACP, 5-shot magazine.
Barrel: 2.8".
Weight: 22 oz. **Length:** 5.4" overall.
Stocks: Black composition.
Sights: Fixed.
Features: Choice of chrome or black Teflon finish. Introduced 1991. Made in U.S. by Davis Industries.
Price: $98.00

Desert Eagle Magnum

E.A.A. WITNESS DA AUTO PISTOL
Caliber: 9mm Para., 10-shot magazine; 38 Super, 40 S&W, 10-shot magazine; 45 ACP, 10-shot magazine.
Barrel: 4.50".
Weight: 35.33 oz. **Length:** 8.10" overall.
Stocks: Checkered rubber.
Sights: Undercut blade front, open rear adjustable for windage.
Features: Double-action trigger system; round trigger guard; frame-mounted safety. Introduced 1991. Imported from Italy by European American Armory.
Price: 9mm, blue $399.00
Price: 9mm, satin chrome $425.00
Price: 9mm Compact, blue, 10-shot $399.00
Price: As above, chrome $425.00
Price: 40 S&W, blue $425.00
Price: As above, chrome $450.00
Price: 40 S&W Compact, 8-shot, blue $425.00
Price: As above, chrome $450.00
Price: 45 ACP, blue $525.00
Price: As above, chrome $550.00

Daewoo DP51 Fastfire

DAEWOO DP52, DH380 AUTO PISTOLS
Caliber: 22 LR, 10-shot magazine.
Barrel: 3.8".
Weight: 23 oz. **Length:** 6.7" overall.
Stocks: Checkered black composition with thumbrest.
Sights: $1/8$" blade front, rear drift adjustable for windage; three-dot system.
Features: All-steel construction with polished blue finish. Dual safety system with hammer block. Introduced 1994. Imported from South Korea by Nationwide Sports Distributors.
Price: $380.00
Price: DH380 (as above except 380 ACP, 7-shot magazine) $410.00

Davis P-32

DESERT EAGLE MAGNUM PISTOL
Caliber: 357 Mag., 9-shot; 44 Mag., 8-shot; 50 Magnum, 7-shot.
Barrel: 6", 10", interchangeable.
Weight: 357 Mag.—62 oz.; 44 Mag.—69 oz.; 50 Mag.—72 oz.
Length: $10 1/4$" overall (6" bbl.).
Stocks: Hogue rubber.
Sights: Blade on ramp front, combat-style rear. Adjustable available.
Features: Rotating three-lug bolt; ambidextrous safety; combat-style trigger guard; adjustable trigger optional. Military epoxy finish. Satin, bright nickel, hard chrome, polished and blued finishes available. Made in U.S. From Magnum Research, Inc.
Price: 357, 6" bbl., standard pistol $979.00
Price: 44 Mag., 6", standard pistol $999.00
Price: 50 Magnum, 6" bbl., standard pistol $1,049.00

E.A.A. Witness

Price: 45 ACP Compact, 8-shot, blue $525.00
Price: As above, chrome $550.00
Price: 9mm/40 S&W Combo, blue, compact or full size $595.00
Price: 9mm or 40 S&W Carry Comp, blue $550.00

HANDGUNS—AUTOLOADERS, SERVICE & SPORT

E.A.A. EUROPEAN MODEL AUTO PISTOLS
Caliber: 32 ACP or 380 ACP, 7-shot magazine.
Barrel: 3.88".
Weight: 26 oz. **Length:** 7 3/8" overall.
Stocks: European hardwood.
Sights: Fixed blade front, rear drift-adjustable for windage.
Features: Chrome or blue finish; magazine, thumb and firing pin safeties; external hammer; safety-lever takedown. Imported from Italy by European American Armory.
Price: Blue $160.00
Price: Chrome $175.00
Price: Ladies Model $225.00

FEG MARK II AP PISTOL
Caliber: 380 ACP, 7-shot magazine.
Barrel: 3.9".
Weight: 27 oz. **Length:** 6.9" overall.
Stocks: Checkered black composition.
Sights: Blade front, rear adjustable for windage.
Features: Double action. All-steel construction. Polished blue finish. Comes with two magazines. Introduced 1997. Imported from Hungary by Interarms.
Price: $269.00
Price: Mark II APK, as above with 3.4" barrel, 6.4" overall length, weighs 25 oz. $269.00

FEG MARK II AP22 PISTOL
Caliber: 22 LR, 8-shot magazine.
Barrel: 3.4".
Weight: 23 oz. **Length:** 6.3" overall.
Stocks: Checkered black composition.
Sights: Blade front, rear adjustable for windage.
Features: Double action. All-steel construction. Polished blue finish. Introduced 1997. Imported from Hungary by Interarms.
Price: $269.00

FEG PJK-9HP AUTO PISTOL
Caliber: 9mm Para., 10-shot magazine.
Barrel: 4.75".
Weight: 32 oz. **Length:** 8" overall.
Stocks: Hand-checkered walnut.
Sights: Blade front, rear adjustable for windage; three dot system.
Features: Single action; polished blue or hard chrome finish; rounded combat-style serrated hammer. Comes with two magazines and cleaning rod. Imported from Hungary by K.B.I., Inc.
Price: Blue $349.00
Price: Hard chrome $429.00

FEG SMC-22 DA AUTO PISTOL
Caliber: 22 LR, 8-shot magazine.
Barrel: 3.5".
Weight: 18.5 oz. **Length:** 6.12" overall.
Stocks: Checkered composition with thumbrest.
Sights: Blade front, rear adjustable for windage.
Features: Patterned after the PPK pistol. Alloy frame, steel slide; blue finish. Comes with two magazines, cleaning rod. Introduced 1994. Imported from Hungary by K.B.I., Inc.
Price: $279.00

FEG SMC-380 AUTO PISTOL
Caliber: 380 ACP, 6-shot magazine.
Barrel: 3.5".
Weight: 18.5 oz. **Length:** 6.1" overall.
Stocks: Checkered composition with thumbrest.
Sights: Blade front, rear adjustable for windage.
Features: Patterned after the PPK pistol. Alloy frame, steel slide; double action. Blue finish. Comes with two magazines, cleaning rod. Imported from Hungary by K.B.I., Inc.
Price: $279.00

FEG SMC-918 Auto Pistol
Same as the SMC-380 except chambered for 9x18 Makarov. Alloy frame, steel slide, blue finish. Comes with two magazines, cleaning rod. Introduced 1995. Imported from Hungary by K.B.I., Inc.
Price: $279.00

ERMA KGP68 AUTO PISTOL
Caliber: 32 ACP, 6-shot, 380 ACP, 5-shot.
Barrel: 4".
Weight: 22 1/2 oz. **Length:** 7 3/8" overall.
Stocks: Checkered plastic.
Sights: Fixed.
Features: Toggle action similar to original "Luger" pistol. Action stays open after last shot. Has magazine and sear disconnect safety systems. Imported from Germany by Mandall Shooting Supplies.
Price: $499.95

FEG Mark II AP

FEG Mark II AP22

FEG PJK-9HP

FORT WORTH HSS
Caliber: 22 LR, 10-shot magazine.
Barrel: 5 1/2" bull.
Weight: 45 oz. **Length:** 10.25" overall.
Stocks: Checkered walnut.
Sights: Ramp front, slide-mounted square notch rear adjustable for windage and elevation.
Features: Stainless steel construction. Military grip. Slide lock; smooth grip straps; push-button takedown; drilled and tapped for barrel weights. Introduced 1995. Made in U.S. by Fort Worth Firearms.
Price: $379.95

FORT WORTH HSSK
Caliber: 22 LR, 10-shot magazine.
Barrel: 4 1/2" or 6 3/4".
Weight: 39 oz. (4 1/2" barrel). **Length:** 9" overall (4 1/2" barrel).
Stocks: Checkered black plastic.
Sights: Blade front, side-mounted rear adjustable for windage.
Features: Stainless steel construction, military grip; standard trigger; push-button barrel takedown. Introduced 1995. Made in U.S. by Fort Worth Firearms.
Price: $312.95

CAUTION: PRICES SHOWN ARE SUPPLIED BY THE MANUFACTURER OR IMPORTER. CHECK YOUR LOCAL GUNSHOP.

HANDGUNS—AUTOLOADERS, SERVICE & SPORT

GAL COMPACT AUTO PISTOL
Caliber: 45 ACP, 8-shot magazine.
Barrel: 4.25″.
Weight: 36 oz. **Length:** 7.75″ overall.
Stocks: Rubberized wrap-around.
Sights: Low profile, fixed, three-dot system.
Features: Forged steel frame and slide; competition trigger, hammer, slide stop magazine release, beavertail grip safety; front and rear slide grooves; two-tone finish. Introduced 1996. Imported from Israel by J.O. Arms, Inc.
Price: $480.00

Gal Compact

GLOCK 17 AUTO PISTOL
Caliber: 9mm Para., 10-shot magazine.
Barrel: 4.49″.
Weight: 21.9 oz. (without magazine). **Length:** 7.28″ overall.
Stocks: Black polymer.
Sights: Dot on front blade, white outline rear adjustable for windage.
Features: Polymer frame, steel slide; double-action trigger with "Safe Action" system; mechanical firing pin safety, drop safety; simple takedown without tools; locked breech, recoil operated action. Adopted by Austrian armed forces 1983. NATO approved 1984. Imported from Austria by Glock, Inc.
Price: Fixed sight, with extra magazine, magazine loader, cleaning kit ..$616.00
Price: Adjustable sight $644.00
Price: Model 17L (6″ barrel) $800.00
Price: Model 17C, ported barrel $646.00

Glock 19

Glock 19 Auto Pistol
Similar to the Glock 17 except has a 4″ barrel, giving an overall length of 6.85″ and weight of 20.99 oz. Magazine capacity is 10 rounds. Fixed or adjustable rear sight. Introduced 1988.
Price: Fixed sight $616.00
Price: Adjustable sight $644.00
Price: Model 19C, ported barrel $646.00

Glock 20 10mm Auto Pistol
Similar to the Glock Model 17 except chambered for 10mm Automatic cartridge. Barrel length is 4.60″, overall length is 7.59″, and weight is 26.3 oz. (without magazine). Magazine capacity is 10 rounds. Fixed or adjustable rear sight. Comes with an extra magazine, magazine loader, cleaning rod and brush. Introduced 1990. Imported from Austria by Glock, Inc.
Price: Fixed sight $668.00
Price: Adjustable sight $697.00

Glock 21 Auto Pistol
Similar to the Glock 17 except chambered for 45 ACP, 10-shot magazine. Overall length is 7.59″, weight is 25.2 oz. (without magazine). Fixed or adjustable rear sight. Introduced 1991.
Price: Fixed sight $668.00
Price: Adjustable sight $697.00

Glock 22 Auto Pistol
Similar to the Glock 17 except chambered for 40 S&W, 10-shot magazine. Overall length is 7.28″, weight is 22.3 oz. (without magazine). Fixed or adjustable rear sight. Introduced 1990.
Price: Fixed sight $616.00
Price: Adjustable sight $844.00
Price: Model 22C, ported barrel $646.00

Glock 23 Auto Pistol
Similar to the Glock 19 except chambered for 40 S&W, 10-shot magazine. Overall length is 6.85″, weight is 20.6 oz. (without magazine). Fixed or adjustable rear sight. Introduced 1990.
Price: Fixed sight $616.00
Price: Adjustable sight $644.00
Price: Model 23C, ported barrel $646.00

Consult our Directory pages for the location of firms mentioned.

Glock 21

Glock 27 40 S&W

GLOCK 26, 27 AUTO PISTOLS
Caliber: 9mm Para. (M26), 10-shot magazine; 40 S&W (M27), 9-shot magazine.
Barrel: 3.47″.
Weight: 21.75 oz. **Length:** 6.3″ overall.
Stocks: Integral. Stippled polymer.
Sights: Dot on front blade, fixed or fully adjustable white outline rear.
Features: Subcompact size. Polymer frame, steel slide; double-action trigger with "Safe Action" system, three safeties. Matte black Tenifer finish. Hammer-forged barrel. Imported from Austria by Glock, Inc. Introduced 1996.
Price: Fixed sight $616.00
Price: Adjustable sight $644.00

HANDGUNS—AUTOLOADERS, SERVICE & SPORT

GLOCK 29, 30 AUTO PISTOLS
Caliber: 10mm (M29), 45 ACP (M30), 10-shot magazine.
Barrel: 3.78".
Weight: 24 oz. **Length:** 6.7" overall.
Stocks: Integral. Stippled polymer.
Sights: Dot on front, fixed or fully adjustable white outline rear.
Features: Compact size. Polymer frame, steel slide; double recoil spring reduces recoil; Safe Action system with three safeties; Tenifer finish. Two magazines supplied. Introduced 1997. Imported from Austria by Glock, Inc.
Price: Fixed sight $668.00
Price: Adjustable sight $697.00

Golan Auto

GOLAN AUTO PISTOL
Caliber: 9mm Para., 40 S&W, 10-shot magazine.
Barrel: 3.9".
Weight: 34 oz. **Length:** 7" overall.
Stocks: Textured composition.
Sights: Fixed.
Features: Fully ambidextrous double/single action; forged steel slide, alloy frame; matte blue finish. Introduced 1994. Imported from Israel by J.O. Arms, Inc.
Price: $649.95

HECKLER & KOCH USP AUTO PISTOL
Caliber: 9mm Para., 10-shot magazine, 40 S&W, 10-shot magazine.
Barrel: 4.25".
Weight: 28 oz. (USP40). **Length:** 6.9" overall.
Stocks: Non-slip stippled black polymer.
Sights: Blade front, rear adjustable for windage.
Features: New HK design with polymer frame, modified Browning action with recoil reduction system, single control lever. Special "hostile environment" finish on all metal parts. Available in SA/DA, DAO, left- and right-hand versions. Introduced 1993. Imported from Germany by Heckler & Koch, Inc.
Price: Right-hand $636.00
Price: Left-hand $656.00
Price: Stainless steel, right-hand $681.00
Price: Stainless steel, left-hand $701.00

Heckler & Koch USP 45

Heckler & Koch USP Compact Auto Pistol
Similar to the USP except has 3.58" barrel, measures 6.81" overall, and weighs 1.60 lbs. (9mm). Available in 9mm Para. or 40 S&W with 10-shot magazine. Introduced 1996. Imported from Germany by Heckler & Koch, Inc.
Price: Blue $665.00
Price: Blue with control lever on right $685.00
Price: Stainless steel $710.00
Price: Stainless steel with control lever on right ...$730.00

Heckler & Koch USP 45 Auto Pistol
Similar to the 9mm and 40 S&W USP except chambered for 45 ACP, 10-shot magazine. Has 4.13" barrel, overall length of 7.87" and weighs 30.4 oz. Has adjustable three-dot sight system. Available in SA/DA, DAO, left- and right-hand versions. Introduced 1995. Imported from Germany by Heckler & Koch, Inc.
Price: Right-hand $696.00
Price: Left-hand $716.00
Price: Stainless steel right-hand $741.00
Price: Stainless steel left-hand $761.00

Heckler & Koch USP Compact

HECKLER & KOCH MARK 23 SPECIAL OPERATIONS PISTOL
Caliber: 45 ACP, 10-shot magazine.
Barrel: 5.87".
Weight: 43 oz. **Length:** 9.65" overall.
Stocks: Integral with frame; black polymer.
Sights: Blade front, rear drift adjustable for windage; three-dot.
Features: Polymer frame; double action; exposed hammer; short recoil, modified Browning action. Civilian version of the SOCOM pistol. Introduced 1996. Imported from Germany by Heckler & Koch, Inc.
Price: $1,995.00

Heckler & Koch Mark 23

HECKLER & KOCH P7M8 AUTO PISTOL
Caliber: 9mm Para., 8-shot magazine.
Barrel: 4.13".
Weight: 29 oz. **Length:** 6.73" overall.
Stocks: Stippled black plastic.
Sights: Blade front, adjustable rear; three dot system.
Features: Unique "squeeze cocker" in frontstrap cocks the action. Gas-retarded action. Squared combat-type trigger guard. Blue finish. Compact size. Imported from Germany by Heckler & Koch, Inc.
Price: P7M8, blued $1,187.00

HELWAN "BRIGADIER" AUTO PISTOL
Caliber: 9mm Para., 8-shot magazine.
Barrel: 4.5".
Weight: 32 oz. **Length:** 8" overall.
Stocks: Grooved plastic.
Sights: Blade front, rear adjustable for windage.
Features: Polished blue finish. Single-action design. Cross-bolt safety. Imported by Century International Arms.
Price: $209.00

HANDGUNS—AUTOLOADERS, SERVICE & SPORT

HERITAGE STEALTH AUTO PISTOL
Caliber: 9mm Para., 40 S&W, 10-shot magazine.
Barrel: 3.9".
Weight: 20.2 oz. **Length:** 6.3" overall.
Stocks: Black polymer; integral.
Sights: Blade front, rear drift adjustable for windage.
Features: Gas retarded blowback action; polymer frame, 17-4 stainless slide; frame mounted ambidextrous trigger safety, magazine safety. Introduced 1996. Made in U.S. by Heritage Mfg., Inc.
Price: $299.95

Heritage Stealth

HI-POINT FIREARMS 40 S&W AUTO
Caliber: 40 S&W, 8-shot magazine.
Barrel: 4.5".
Weight: 39 oz. **Length:** 7.72" overall.
Stocks: Checkered acetal resin.
Sights: Fixed; low profile.
Features: Internal drop-safe mechansim; all aluminum frame. Introduced 1991. From MKS Supply, Inc.
Price: Matte black $148.95

HI-POINT FIREARMS 45 CALIBER PISTOL
Caliber: 45 ACP, 7-shot magazine.
Barrel: 4.5".
Weight: 39 oz. **Length:** 7.95" overall.
Stocks: Checkered acetal resin.
Sights: Fixed; low profile.
Features: Internal drop-safe mechanism; all aluminum frame. Introduced 1991. From MKS Supply, Inc.
Price: Matte black $148.95

Hi-Point 40 S&W

HI-POINT FIREARMS 9MM AUTO PISTOL
Caliber: 9mm Para., 9-shot magazine.
Barrel: 4.5".
Weight: 39 oz. **Length:** 7.72" overall.
Stocks: Textured acetal plastic.
Sights: Fixed, low profile.
Features: Single-action design. Scratch-resistant, non-glare blue finish. Introduced 1990. From MKS Supply, Inc.
Price: Matte black $139.95

HI-POINT FIREARMS MODEL 9MM COMPACT PISTOL
Caliber: 380 ACP, 9mm Para., 8-shot magazine.
Barrel: 3.5".
Weight: 29 oz. **Length:** 6.7" overall.
Stocks: Textured acetal plastic.
Sights: Combat-style fixed three-dot system; low profile.
Features: Single-action design; frame-mounted magazine release. Scratch-resistant matte finish. Introduced 1993. From MKS Supply, Inc.
Price: $124.95
Price: With polymer frame (29 oz.), non-slip grips $132.95
Price: 380 ACP $89.95

INTRATEC DOUBLE-ACTION AUTO PISTOLS
Caliber: 22 LR, 10-shot; 25 ACP, 8-shot magazine.
Barrel: 2 1/2".
Weight: 14 oz. **Length:** 5" overall.
Stocks: Wraparound composition in gray, black or driftwood color.
Sights: Fixed.
Features: Double-action only trigger mechanism. Choice of black, satin or TEC-KOTE finish. Announced 1991. Made in U.S. by Intratec.
Price: 22 or 25, black finish $112.00
Price: 22 or 25, satin or TEC-KOTE finish $117.00

INTRATEC SPORT 22 AUTO PISTOL
Caliber: 22 LR, 10-shot magazine.
Barrel: 4".
Weight: 28 oz. **Length:** 11 3/16" overall.
Stocks: Moulded composition.
Sights: Protected post front, adjustable for windage, rear adjustable elevation.
Features: Ambidextrous cocking knobs and safety. Matte black finish. Accepts any 10/22-type magazine. Introduced 1988. Made in U.S. by Intratec.
Price: $130.00

INTRATEC CAT 9 AUTO PISTOL
Caliber: 380 ACP, 9mm Para., 7-shot magazine.
Barrel: 3".
Weight: 21 oz. **Length:** 5.5" overall.
Stocks: Textured black polymer.
Sights: Fixed channel.
Features: Black polymer frame. Introduced 1993. Made in U.S. by Intratec.
Price: About $235.00

Intratec Cat 9

INTRATEC CAT 45
Caliber: 40 S&W, 45 ACP; 6-shot magazine.
Barrel: 3.25".
Weight: 19 oz. **Length:** 6.35" overall.
Stocks: Moulded composition.
Sights: Fixed, channel.
Features: Black polymer frame. Introduced 1996. Made in U.S. by Intratec.
Price: $255.00

HANDGUNS—AUTOLOADERS, SERVICE & SPORT

Jennings J-25

JENNINGS J-22, J-25 AUTO PISTOLS
Caliber: 22 LR, 25 ACP, 6-shot magazine.
Barrel: 2 1/2".
Weight: 13 oz. (J-22). **Length:** 4 15/16" overall (J-22).
Stocks: Walnut on chrome or nickel models; grooved black Cycolac or resin-impregnated wood on Teflon model.
Sights: Fixed.
Features: Choice of bright chrome, satin nickel or black Teflon finish. Introduced 1981. From Jennings Firearms.
Price: J-22, about $79.95
Price: J-25, about $79.95

KAHR K9 DA AUTO PISTOL
Caliber: 9mm Para., 7-shot, 40 S&W, 6-shot magazine.
Barrel: 3.5".
Weight: 25 oz. **Length:** 6" overall.
Stocks: Wrap-around textured soft polymer.
Sights: Blade front, rear drift adjustable for windage; bar-dot combat style.
Features: Trigger-cocking double-action mechanism with passive firing pin block. Made of 4140 ordnance steel with matte black finish. Contact maker for complete price list. Introduced 1994. Made in U.S. by Kahr Arms.
Price: 9mm $538.00
Price: Matte black, night sights 9mm $624.00
Price: Matte nickel finish 9mm $612.00
Price: Matte nickel, night sights 9mm $699.00
Price: Matte nickel, 9mm, 4" ported barrel $650.00
Price: Matte stainless steel, 9mm $588.00
Price: 40 S&W, matte black $552.00
Price: 40 S&W, matte black, night sights $638.00
Price: 40 S&W, matte stainless $602.00
Price: Lady K9, 9mm, matte black $545.00

Kahr K9 Nickel

Kareen Mk II Compact

KAREEN MK II AUTO PISTOL
Caliber: 9mm Para., 10-shot magazine.
Barrel: 4.75".
Weight: 34 oz. **Length:** 7.85" overall.
Stocks: Textured composition.
Sights: Blade front, rear adjustable for windage.
Features: Single-action mechanism; ambidextrous external hammer safety; magazine safety; combat trigger guard. Two-tone finish. Introduced 1985. Imported from Israel by J.O. Arms & Ammunition.
Price: $425.00
Price: Kareen Mk II Compact 9mm (3.75" barrel, 30 oz., 6.75" overall length) $495.00

KEL-TEC P-11 AUTO PISTOL
Caliber: 9mm Para., 10-shot magazine.
Barrel: 3.1".
Weight: 14 oz. **Length:** 5.6" overall.
Stocks: Checkered black polymer.
Sights: Blade front, rear adjustable for windage.
Features: Ordnance steel slide, aluminum frame. Double-action-only trigger mechanism. Introduced 1995. Made in U.S. by Kel-Tec CNC Industries, Inc.
Price: Blue $309.00
Price: Stainless $407.00
Price: Parkerized $350.00

Kel-Tec P-11

Kimber Classic 45 Custom

KIMBER CLASSIC 45 CUSTOM AUTO PISTOL
Caliber: 45 ACP, 7- or 8-shot magazine.
Barrel: 5".
Weight: 38 oz. **Length:** 8.5" overall.
Stocks: Black synthetic, walnut (custom).
Sights: McCormick dovetailed front, low combat rear.
Features: Uses Kimber forged frame and slide, match-grade barrel, extended combat thumb safety, high beavertail grip safety, skeletonized lightweight composite trigger, skeletonized Commander-type hammer, and elongated Commander ejector. Bead-blasted black oxide finish; flat mainspring housing; lowered and flared ejection port; front and rear slide serrations; relief cut under trigger guard; Wolff spring set; beveled magazine well. Introduced 1995. Made in U.S. by Kimber of America, Inc.
Price: Custom $615.00
Price: Custom Stainless $690.00

HANDGUNS—AUTOLOADERS, SERVICE & SPORT

Kimber Classic 45 Polymer Matte Auto Pistol
Similar to the Classic 45 Custom except has black polymer frame with stainless steel insert, hooked trigger guard, checkered front strap. Weighs 34.4 oz., overall length 8.75". Introduced 1997. Made in U.S. by Kimber of America.
Price: With matte blue slide $815.00
Price: With stainless steel slide $890.00

Kimber Classic 45 Custom Royal Auto Pistol
Same as the Custom model except has hand-checkered diamond-pattern rosewood grips, polished blue finish, and comes with two magazines. Introduced 1995. Made in U.S. by Kimber of America, Inc.
Price: $739.00

Kimber Classic 45 Polymer

Kimber Classic 45 Gold Match Auto Pistol
Same as the Custom Royal except also has Kimber low-mount adjustable rear sight, fancy rosewood grips, hand fit slide, barrel, frame and bushing. Comes with one 10-shot and one 8-shot magazine and factory proof target. Introduced 1995. Made in U.S. by Kimber of America, Inc.
Price: $935.00

L.A.R. GRIZZLY WIN MAG MK I PISTOL
Caliber: 357 Mag., 357/45, 10mm, 44 Mag., 45 Win. Mag., 45 ACP, 7-shot magazine.
Barrel: 5.4", 6.5".
Weight: 51 oz. **Length:** 10 1/2" overall.
Stocks: Checkered rubber, non-slip combat-type.
Sights: Ramped blade front, fully adjustable rear.
Features: Uses basic Browning/Colt 1911A1 design; interchangeable calibers; beveled magazine well; combat-type flat, checkered rubber mainspring housing; lowered and back-chamfered ejection port; polished feed ramp; throated barrel; solid barrel bushings. Available in satin hard chrome, matte blue, Parkerized finishes. Introduced 1983. From L.A.R. Mfg., Inc.
Price: 45 Win. Mag. $1,000.00
Price: 357 Mag. $1,014.00
Price: Conversion units (357 Mag.) $248.00
Price: As above, 45 ACP, 10mm, 45 Win. Mag., 357/45 Win. Mag. $233.00

Kimber Classic 45 Gold Match

L.A.R. Grizzly 50 Mark V Pistol
Similar to the Grizzly Win Mag Mark I except chambered for 50 Action Express with 6-shot magazine. Weight, empty, is 56 oz., overall length 10 5/8". Choice of 5.4" or 6.5" barrel. Has same features as Mark I, IV pistols. Introduced 1993. From L.A.R. Mfg., Inc.
Price: $1,152.00

L.A.R. Grizzly 44 Mag MK IV
Similar to the Win Mag Mk I except chambered for 44 Magnum, has beavertail grip safety. Matte blue finish only. Has 5.4" or 6.5" barrel. Introduced 1991. From L.A.R. Mfg., Inc.
Price: $1,014.00

L.A.R. Girzzly MK I

Laseraim Series II

Laseraim Arms Series III Auto Pistol
Similar to the Series II except has 5" barrel only, with dual-port compensator; weighs 43 oz.; overall length is 7 5/8". Choice of fixed or adjustable rear sight. Introduced 1994. Made in U.S. by Laseraim Technologies, Inc.
Price: Fixed sight $533.95
Price: Adjustable sight $559.95
Price: Fixed sight Dream Team Laseraim laser sight $629.95

LASERAIM ARMS SERIES I AUTO PISTOL
Caliber: 10mm Auto, 8-shot, 45 ACP, 7-shot magazine.
Barrel: 6", with compensator.
Weight: 46 oz. **Length:** 9.75" overall.
Stocks: Pebble-grained black composite.
Sights: Blade front, fully adjustable rear.
Features: Single action; barrel compensator; stainless steel construction; ambidextrous safety-levers; extended slide release; matte black Teflon finish; integral mount for laser sight. Introduced 1993. Made in U.S. by Laseraim Technologies, Inc.
Price: Standard, fixed sight $552.95
Price: Standard, Compact (4 3/8" barrel), fixed sight $552.95
Price: Adjustable sight $579.95
Price: Standard, fixed sight, Auto Illusion red dot sight system $649.95
Price: Standard, fixed sight, Laseraim Laser with Hotdot $694.95

Laseraim Arms Series II Auto Pistol
Similar to the Series I except without compensator, has matte stainless finish. Standard Series II has 5" barrel, weighs 43 oz., Compact has 3 3/8" barrel, weighs 37 oz. Blade front sight, rear adjustable for windage or fixed. Introduced 1993. Made in U.S. by Laseraim Technologies, Inc.
Price: Standard or Compact (3 3/8" barrel), fixed sight $399.95
Price: Adjustable sight, 5" only $429.95
Price: Standard, fixed sight, Auto Illusion red dot sight $499.95
Price: Standard, fixed sight, Laseraim Laser $499.95

HANDGUNS—AUTOLOADERS, SERVICE & SPORT

Llama Max-I

Llama Max-1 Compensator

Llama Max-II Compact

Llama Micromax 380

Llama Minimax

LLAMA MAX-I AUTO PISTOLS
Caliber: 9mm Para., 9-shot, 45 ACP, 7-shot.
Barrel: 4 1/4" (Compact); 5 1/8" (Government).
Weight: 34 oz. (Compact); 36 oz. (Government). **Length:** 7 3/8" overall (Compact).
Stocks: Black rubber.
Sights: Blade front, rear adjustable for windage; three-dot system.
Features: Single-action trigger; skeletonized combat-style hammer; steel frame; extended manual and grip safeties. Introduced 1995. Imported from Spain by Import Sports, Inc.
Price: 9mm, 9-shot, Government model $349.95
Price: As above, Compact model $349.95
Price: 45 ACP, 7-shot, Government model $358.95
Price: As above, finish $374.95
Price: As above, Compact model $382.95

LLAMA MAX-I COMPENSATOR
Caliber: 45 ACP, 7-, 10-shot magazine.
Barrel: 4 7/8" (without compensator, 6 1/3" with).
Weight: 42 oz. (7-shot). **Length:** 9 7/8" overall.
Stocks: Checkered rubber.
Sights: Dovetail blade front, fully adjustable rear.
Features: Extended beavertail grip safety, skeletonized combat hammer, extended slide release. Introduced 1996. Imported from Spain by Import Sports, Inc.
Price: 7-shot $491.95
Price: 10-shot $516.95

LLAMA MAX-II COMPACT FRAME AUTO PISTOL
Caliber: 45 ACP, 10-shot.
Barrel: 4 1/4".
Weight: 39 oz.
Stocks: Black rubber.
Sights: Blade front, rear adjustable for windage; three-dot system.
Features: Scaled-down version of the Large Frame gun. Locked breech mechanism; manual and grip safeties. Introduced 1995. Imported from Spain by Import Sports, Inc.
Price: Matte finish $408.95
Price: Satin chrome $424.95

LLAMA THUNDER 380 AUTO PISTOL
Caliber: 380 ACP.
Barrel: 3 11/16".
Weight: 23 oz. **Length:** 6 1/2" overall.
Stocks: Checkered polymer, thumbrest.
Sights: Fixed front, adjustable notch rear.
Features: Ventilated rib, manual and grip safeties. Imported from Spain by Import Sports, Inc.
Price: Blue $258.95
Price: Thunder 380 Lite, satin nickel $274.55
Price: Thunder 380 DLX, blue finish $274.95

LLAMA MICROMAX 380 AUTO PISTOL
Caliber: 380 ACP, 7-shot magazine.
Barrel: 3 11/16".
Weight: 23 oz. **Length:** 6 1/2" overall.
Stocks: Checkered high impact polymer.
Sights: 3-dot combat.
Features: Single action design. Mini custom extended slide release; mini custom extended beavertail grip safety; combat-style hammer. Introduced 1997. Imported from Spain by Import Sports, Inc.
Price: Matte blue $258.95
Price: Satin chrome $291.95

LLAMA MINIMAX SERIES
Caliber: 9mm Para., 40 S&W, 45 ACP, 6-shot magazine.
Barrel: 3 1/2".
Weight: 35 oz. **Length:** 7 1/3" overall.
Stocks: Checkered rubber.
Sights: Three-dot combat.
Features: Single action, skeletonized combat-style hammer, extended slide release, cone-style barrel, flared ejection port. Introduced 1996. Imported from Spain by Import Sports, Inc.
Price: Blue $374.95
Price: Duo-Tone finish (45 only) $391.95
Price: Satin chrome $419.95

Llama Minimax-II Auto Pistol
Same as the Minimax except in 45 ACP only, with 10-shot staggered magazine. Introduced 1997. Imported from Spain by Import Sports, Inc.
Price: Matte blue $398.95
Price: Satin chrome $431.95

CAUTION: PRICES SHOWN ARE SUPPLIED BY THE MANUFACTURER OR IMPORTER. CHECK YOUR LOCAL GUNSHOP.

HANDGUNS—AUTOLOADERS, SERVICE & SPORT

LORCIN L-22 AUTO PISTOL
Caliber: 22 LR, 9-shot magazine.
Barrel: 2.5".
Weight: 16 oz. **Length:** 5.25" overall.
Stocks: Black combat, or pink or pearl.
Sights: Fixed three-dot system.
Features: Available in chrome or black Teflon finish. Introduced 1989. From Lorcin Engineering.
Price: About $89.00

LORCIN L9MM AUTO PISTOL
Caliber: 9mm Para., 10-shot magazine.
Barrel: 4.5".
Weight: 31 oz. **Length:** 7.5" overall.
Stocks: Grooved black composition.
Sights: Fixed; three-dot system.
Features: Matte black finish; hooked trigger guard; grip safety. Introduced 1994. Made in U.S. by Lorcin Engineering.
Price: $159.00

LORCIN L-25, LT-25 AUTO PISTOLS
Caliber: 25 ACP, 7-shot magazine.
Barrel: 2.4".
Weight: 14.5 oz. **Length:** 4.8" overall.
Stocks: Smooth composition.
Sights: Fixed.
Features: Available in choice of finishes: chrome, black Teflon or camouflage. Introduced 1989. From Lorcin Engineering.
Price: L-25 ....$69.00
Price: LT-25 ....$79.00

LORCIN L-32, L-380 AUTO PISTOLS
Caliber: 32 ACP, 380 ACP, 7-shot magazine.
Barrel: 3.5".
Weight: 27 oz. **Length:** 6.6" overall.
Stocks: Grooved composition.
Sights: Fixed.
Features: Black Teflon or chrome finish with black grips. Introduced 1992. From Lorcin Engineering.
Price: L-32 32 ACP $89.00
Price: L-380 380 ACP $100.00

Lorcin L9MM

MOUNTAIN EAGLE AUTO PISTOL
Caliber: 22 LR, 10-shot magazine.
Barrel: 4.5", 6.5", 8".
Weight: 21 oz., 23 oz. **Length:** 10.6" overall (with 6.5" barrel).
Stocks: One-piece impact-resistant polymer in "conventional contour"; checkered panels.
Sights: Serrated ramp front with interchangeable blades, rear adjustable for windage and elevation; interchangeable blades.
Features: Injection moulded grip frame, alloy receiver; hybrid composite barrel replicates shape of the Desert Eagle pistol. Flat, smooth trigger. Introduced 1992. From Magnum Research.
Price: Mountain Eagle Compact $199.00
Price: Mountain Eagle Standard $239.00
Price: Mountain Eagle Target Edition (8" barrel) $279.00

ONE PRO.45 AUTO PISTOL
Caliber: 45 ACP or 400 Cor-bon, 10-shot magazine.
Barrel: 3.75"
Weight: 31.1 oz. **Length:** 7.04" overall.
Stocks: Textured composition.
Sights: Blade front, drift-adjustable rear; three-dot system.
Features: All-steel construction; decocking lever and automatic firing pin lock; DA or DAO operation. Introduced 1997. Imported from Switzerland by Magnum Research, Inc.
Price: $649.00
Price: Conversion kit, 45 ACP/400, 400/45 ACP $249.00

Mountain Eagle Target

Para-Ordnance P16.40

NORTH AMERICAN MUNITIONS MODEL 1996
Caliber: 9mm Para., 9-shot magazine.
Barrel: 4.5".
Weight: 40 oz. **Length:** 8.38" overall.
Stocks: Black polycarbonate.
Sights: Blade front, adjustable rear; three-dot system.
Features: Gas-delayed blowback system; no external safeties; fixed 10-groove barrel. Introduced 1996. Made in U.S. From Intercontinental Munitions Distributors, Ltd.
Price: $275.00

OLYMPIC ARMS OA-96 AR PISTOL
Caliber: 223.
Barrel: 6", 4140 chromemoly steel.
Weight: 5 lbs. **Length:** 15 3/4" overall.
Stocks: A2 stowaway pistol grip; no buttstock or receiver tube.
Sights: Flat-top upper receiver, cut-down front sight base.
Features: AR-15-type receivers with special bolt carrier; short aluminum hand guard; Vortex flash hider. Introduced 1996. Made in U.S. by Olympic Arms, Inc.
Price: $940.00

PARA-ORDNANCE P-SERIES AUTO PISTOLS
Caliber: 40 S&W, 45 ACP, 10-shot magazine.
Barrel: 3", 3 1/2", 4 1/4", 5".
Weight: From 24 oz. (alloy frame). **Length:** 8.5" overall.
Stocks: Textured composition.
Sights: Blade front, rear adjustable for windage. High visibility three-dot system.
Features: Available with alloy, steel or stainless steel frame with black finish (silver or stainless gun). Steel and stainless steel frame guns weigh 40 oz. (P14.45), 36 oz. (P13.45), 34 oz. (P12.45). Grooved match trigger, rounded combat-style hammer. Beveled magazine well. Manual thumb, grip and firing pin lock safeties. Solid barrel bushing. Contact maker for full details. Introduced 1990. Made in Canada by Para-Ordnance.
Price: P14.45ER (steel frame) $750.00
Price: P14.45RR (alloy frame) $705.00
Price: P12.45RR (3 1/2" bbl., 24 oz., alloy) $705.00
Price: P13.45RR (4 1/4" barrel, 28 oz., alloy) ...$705.00
Price: P12.45ER (steel frame) $750.00
Price: P16.40ER (steel frame) $750.00

HANDGUNS—AUTOLOADERS, SERVICE & SPORT

PHOENIX ARMS HP22, HP25 AUTO PISTOLS
Caliber: 22 LR, 10-shot (HP22), 25 ACP, 10-shot (HP25).
Barrel: 3".
Weight: 20 oz. **Length:** 5 1/2" overall.
Stocks: Checkered composition.
Sights: Blade front, adjustable rear.
Features: Single action, exposed hammer; manual hold-open; button magazine release. Available in satin nickel, polished blue finish. Introduced 1993. Made in U.S. by Phoenix Arms.
Price: $99.00

Phoenix Arms HP22

PHOENIX ARMS MODEL RAVEN AUTO PISTOL
Caliber: 25 ACP, 6-shot magazine.
Barrel: 2 7/16".
Weight: 15 oz. **Length:** 4 3/4" overall.
Stocks: Ivory-colored or black slotted plastic.
Sights: Ramped front, fixed rear.
Features: Available in blue, nickel or chrome finish. Made in U.S. Available from Phoenix Arms.
Price: $79.00

PIRANHA AUTOLOADING PISTOL
Caliber: 9mm Para., 9mm Largo, 30 Luger, 10-shot magazine.
Barrel: 4", 6", 8", 10", 16".
Weight: About 2.7 lbs. **Length:** 9" overall with 4" barrel.
Stocks: Smooth walnut.
Sights: Blade front, rear adjustable for windage.
Features: Nearly recoilless action; stainless steel construction; fires from closed bolt; change caliber by changing barrel. Introduced 1996. Made in U.S. by Recoilless Technologies, Inc.
Price: $600.00

Piranha Pistol

PSA-25 AUTO PISTOL
Caliber: 25 ACP, 6-shot magazine.
Barrel: 2 1/8".
Weight: 9.5 oz. **Length:** 4 1/8" overall.
Stocks: Checkered black plastic.
Sights: Fixed.
Features: All steel construction with polished finish. Introduced 1984. Made in the U.S. by PSP.
Price: Black oxide $262.00
Price: Bead-blasted chrome over nickel and copper $298.00
Price: Featherweight $335.00
Price: Renaissance, engraved $975.00
Price: Imperiale, gold inlaid, engraved, gold accents $1,550.00

PSA-25 Pistol

REPUBLIC PATRIOT PISTOL
Caliber: 45 ACP, 6-shot magazine.
Barrel: 3".
Weight: 20 oz. **Length:** 6" overall.
Stocks: Checkered.
Sights: Blade front, drift-adjustable rear.
Features: Black polymer frame, stainless steel slide; double-action-only trigger system; squared trigger guard. Introduced 1997. Made in U.S. by Republic Arms, Inc.
Price: About $300.00

ROCKY MOUNTAIN ARMS PATRIOT PISTOL
Caliber: 223, 10-shot magazine.
Barrel: 7", with muzzle brake.
Weight: 5 lbs. **Length:** 20.5" overall.
Stocks: Black composition.
Sights: None furnished.
Features: Milled upper receiver with enhanced Weaver base; milled lower receiver from billet plate; machined aluminum National Match handguard. Finished in DuPont Teflon-S matte black or NATO green. Comes with black nylon case, one magazine. Introduced 1993. From Rocky Mountain Arms, Inc.
Price: With A-2 handle top ....$2,500.00 to $2,800.00
Price: Flat top model ....$3,000.00 to $3,500.00

Ruger KP89D

RUGER P89 SEMI-AUTO PISTOL
Caliber: 9mm Para., 10-shot magazine.
Barrel: 4.50".
Weight: 32 oz. **Length:** 7.84" overall.
Stocks: Grooved black Xenoy composition.
Sights: Square post front, square notch rear adjustable for windage, both with white dot inserts.
Features: Double action with ambidextrous slide-mounted safety-levers. Slide is 4140 chrome-moly steel or 400-series stainless steel, frame is a lightweight aluminum alloy. Ambidextrous magazine release. Blue or stainless steel. Introduced 1986; stainless introduced 1990.
Price: P89, blue, with extra magazine and magazine loading tool, plastic case with lock $410.00
Price: KP89, stainless, with extra magazine and magazine loading tool, plastic case with lock $452.00

Ruger P89 Double-Action-Only Semi-Auto Pistol
Same as the KP89 except operates only in the double-action mode. Has a spurless hammer, gripping grooves on each side of the rear of the slide; no external safety or decocking lever. An internal safety prevents forward movement of the firing pin unless the trigger is pulled. Available in 9mm Para., stainless steel only. Introduced 1991.
Price: With lockable case, extra magazine, magazine loading tool $452.00

HANDGUNS—AUTOLOADERS, SERVICE & SPORT

Ruger P89D Decocker Semi-Auto Pistol
Similar to the standard P89 except has ambidextrous decocking levers in place of the regular slide-mounted safety. The decocking levers move the firing pin inside the slide where the hammer can not reach it, while simultaneously blocking the firing pin from forward movement—allows shooter to decock a cocked pistol without manipulating the trigger. Conventional thumb decocking procedures are therefore unnecessary. Blue or stainless steel. Introduced 1990.
Price: P89D, blue with extra magazine and loader, plastic case with lock **$410.00**
Price: KP89D, stainless, with extra magazine, plastic case with lock **$452.00**

Ruger KP94 Semi-Auto Pistol
Sized midway between the full-size P-Series and the compact P93. Has 4.25" barrel, 7.5" overall length and weighs about 33 oz. KP94 is manual safety model; KP94DAO is double-action-only (both 9mm Para., 10-shot magazine); KP94D is decocker-only in 40-caliber with 10-shot magazine. Slide gripping grooves roll over top of slide. KP94 has ambidextrous safety-levers; KP94DAO has no external safety, full-cock hammer position or decocking lever; KP94D has ambidextrous decocking levers. Matte finish stainless slide, barrel, alloy frame. Introduced 1994. Made in U.S. by Sturm, Ruger & Co.
Price: KP94 (9mm), KP944 (40-caliber) $520.00
Price: KP94DAO (9mm), KP944DAO (40-caliber) $520.00
Price: KP94D (9mm), KP9440 (40-caliber) $520.00

RUGER P93 COMPACT SEMI-AUTO PISTOL
Caliber: 9mm Para., 10-shot magazine.
Barrel: 3.9".
Weight: 31 oz. **Length:** 7.3" overall.
Stocks: Grooved black Xenoy composition.
Sights: Square post front, square notch rear adjustable for windage.
Features: Front of slide is crowned with a convex curve; slide has seven finger grooves; trigger guard bow is higher for a better grip; 400-series stainless slide, lightweight alloy frame. Decocker-only or DAO-only. Introduced 1993. Made in U.S. by Sturm, Ruger & Co.
Price: $520.00

RUGER P95 SEMI-AUTO PISTOL
Caliber: 9mm Para., 10-shot magazine.
Barrel: 3.9".
Weight: 27 oz. **Length:** 7.3" overall.
Stocks: Grooved; integral with frame.
Sights: Blade front, rear drift adjustable for windage; three-dot system.
Features: Moulded polymer grip frame, stainless steel or chrome-moly slide. Suitable for +P+ ammunition. Decocker or DAO. Introduced 1996. Made in U.S. by Sturm, Ruger & Co. Comes with lockable plastic case, spare magazine, loading tool.
Price: P95 DAO double-action-only $351.00
Price: P95D decocker only $351.00
Price: KP95 stainless steel $369.00

RUGER P90 SAFETY MODEL SEMI-AUTO PISTOL
Caliber: 45 ACP, 7-shot magazine.
Barrel: 4.50".
Weight: 33.5 oz. **Length:** 7.87" overall.
Stocks: Grooved black Xenoy composition.
Sights: Square post front, square notch rear adjustable for windage, both with white dot inserts.
Features: Double action with ambidextrous slide-mounted safety-levers which move the firing pin inside the slide where the hammer can not reach it, while simultaneously blocking the firing pin from forward movement. Stainless steel only. Introduced 1991.
Price: KP90 with extra magazine, loader, plastic case with lock $488.65

Ruger P90 Decocker Semi-Auto Pistol
Similar to the P90 except has a manual decocking system. The ambidextrous decocking levers move the firing pin inside the slide where the hammer can not reach it, while simultaneously blocking the firing pin from forward movement—allows shooter to decock a cocked pistol without manipulating the trigger. Available only in stainless steel. Overall length 7.87", weighs 34 oz. Introduced 1991.
Price: P90D with lockable case, extra magazine, and magazine loading tool $488.65

Ruger P94L Semi-Auto Pistol
Same as the KP94 except mounts a laser sight in a housing cast integrally with the frame. Allen-head screws control windage and elevation adjustments. Announced 1994. Made in U.S. by Sturm, Ruger & Co.
Price: For law enforcement only . . . . NA

Ruger P93 DAO

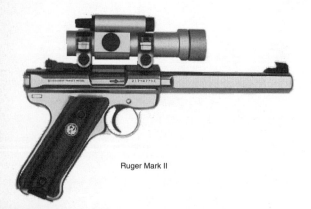

Ruger Mark II

RUGER MARK II STANDARD SEMI-AUTO PISTOL
Caliber: 22 LR, 10-shot magazine.
Barrel: 4 3/4" or 6".
Weight: 25 oz. (4 3/4" bbl.). **Length:** 8 5/16" (4 3/4" bbl.).
Stocks: Checkered plastic.
Sights: Fixed, wide blade front, square notch rear adjustable for windage.
Features: Updated design of the original Standard Auto. Has new bolt hold-open latch. 10-shot magazine, magazine catch, safety, trigger and new receiver contours. Introduced 1982.
Price: Blued (MK 4, MK 6) $252.00
Price: In stainless steel (KMK 4, KMK 6) $330.25

Manufacturers' addresses in the
Directory of the Arms Trade
page 309, this issue

Ruger KP95D

HANDGUNS—AUTOLOADERS, SERVICE & SPORT

Ruger 22/45 Mark II Pistol
Similar to the other 22 Mark II autos except has grip frame of Zytel that matches the angle and magazine latch of the Model 1911 45 ACP pistol. Available in 4", 4 3/4" standard and 5 1/2" bull barrel. Introduced 1992.
Price: P4, 4", adjustable sights $237.50
Price: KP4 (4 3/4" barrel) $280.00
Price: KP512 (5 1/2" bull barrel) $330.00
Price: P512 (5 1/2" bull barrel, all blue) $237.50

Ruger P4

Ruger MK-4B Compact

Safari Arms Enforcer

Ruger MK-4B Compact Pistol
Similar to the Mark II Standard pistol except has 4" bull barrel, Patridge-type front sight, fully adjustable rear, and smooth laminated hardwood thumbrest stocks. Weighs 38 oz., overall length of 8 3/16". Comes with extra magazine, plastic case, lock. Introduced 1996. Made in U.S. by Sturm, Ruger & Co.
Price: $336.50

SAFARI ARMS ENFORCER PISTOL
Caliber: 45 ACP, 6-shot magazine.
Barrel: 3.8", stainless.
Weight: 36 oz. **Length:** 7.3" overall.
Stocks: Smooth walnut with etched black widow spider logo.
Sights: Ramped blade front, LPA adjustable rear.
Features: Extended safety, extended slide release; Commander-style hammer; beavertail grip safety; throated, polished, tuned. Parkerized matte black or satin stainless steel finishes. Made in U.S. by Safari Arms.
Price: $750.00

Safari Arms Enforcer Carrycomp I Pistol
Similar to the Enforcer except has Wil Schueman-designed hybrid compensator system. Introduced 1993. Made in U.S. by Safari Arms, Inc.
Price: 5" barrel $1,160.00

SAFARI ARMS GI SAFARI PISTOL
Caliber: 45 ACP, 7-shot magazine.
Barrel: 5", 416 stainless.
Weight: 39.9 oz. **Length:** 8.5" overall.
Stocks: Checkered walnut.
Sights: G.I.-style blade front, drift-adjustable rear.
Features: Beavertail grip safety; extended thumb safety and slide release; Commander-style hammer. Parkerized finish. Reintroduced 1996.
Price: $595.00

SAFARI ARMS GRIFFON PISTOL
Caliber: 45 ACP, 10-shot magazine.
Barrel: 5", 416 stainless steel.
Weight: 40.5 oz. **Length:** 8.5" overall.
Stocks: Smooth walnut.
Sights: Ramped blade front, LPA adjustable rear.
Features: 10+1 1911 enhanced 45. Beavertail grip safety; long aluminum trigger; full-length recoil spring guide; Commander-style hammer. Throated, polished and tuned. Grip size comparable to standard 1911. Satin stainless steel finish. Introduced 1996. Made in U.S. by Olympic Arms, Inc.
Price: $920.00

SAFARI ARMS COHORT PISTOL
Caliber: 45 ACP, 7-shot magazine.
Barrel: 3.8", 416 stainless.
Weight: 37 oz. **Length:** 8.5" overall.
Stocks: Smooth walnut with laser-etched black widow logo.
Sights: Ramped blade front, LPA adjustable rear.
Features: Combines the Enforcer model, slide and MatchMaster frame. Beavertail grip safety; extended thumb safety and slide release; Commander-style hammer. Throated, polished and tuned. Satin stainless finish. Introduced 1996. Made in U.S. by Safari Arms, Inc.
Price: $790.00

SAFARI ARMS RELIABLE PISTOL
Caliber: 45 ACP, 7-shot magazine.
Barrel: 5", 416 stainless steel.
Weight: 39 oz. **Length:** 8.5" overall.
Stocks: Checkered walnut.
Sights: Ramped blade front, LPA adjustable rear.
Features: Beavertail grip safety; long aluminum trigger; full-length recoil spring guide; Commander-style hammer. Throated, polished and tuned. Satin stainless steel finish. Introduced 1996. Made in U.S. by Safari Arms, Inc.
Price: $825.00

Safari Arms Reliable 4-Star Pistol
Similar to the Reliable except has 4.5" barrel, 7.5" overall length, and weighs 35.7 oz. Introduced 1996. Made in U.S. by Safari Arms, Inc.
Price: $885.00

SAFARI ARMS RENEGADE
Caliber: 45 ACP, 7-shot magazine.
Barrel: 5", 416 stainless steel.
Weight: 39 oz. **Length:** 8.5" overall.
Stocks: Checkered walnut.
Sights: Ramped blade, LPA adjustable rear.
Features: True left-hand pistol. Beavertail grip safety; long aluminum trigger; full-length recoil spring guide; satin stainless finish. Throated, polished and tuned. Introduced 1996. Made in U.S. by Safari Arms, Inc.
Price: $1,085.00

SEECAMP LWS 32 STAINLESS DA AUTO
Caliber: 32 ACP Win. Silvertip, 6-shot magazine.
Barrel: 2", integral with frame.
Weight: 10.5 oz. **Length:** 4 1/8" overall.
Stocks: Glass-filled nylon.
Sights: Smooth, no-snag, contoured slide and barrel top.
Features: Aircraft quality 17-4 PH stainless steel. Inertia-operated firing pin. Hammer fired double-action-only. Hammer automatically follows slide down to safety rest position after each shot—no manual safety needed. Magazine safety disconnector. Polished stainless. Introduced 1985. From L.W. Seecamp.
Price: $425.00

HANDGUNS—AUTOLOADERS, SERVICE & SPORT

SIG SAUER P220 SERVICE AUTO PISTOL
Caliber: 38 Super, 45 ACP, (7- or 8-shot magazine).
Barrel: 4 3/8".
Weight: 28 1/4 oz. (9mm). **Length:** 7 3/4" overall.
Stocks: Checkered black plastic.
Sights: Blade front, drift adjustable rear for windage. Optional Siglite nightsights.
Features: Double action. Decocking lever permits lowering hammer onto locked firing pin. Squared combat-type trigger guard. Slide stays open after last shot. Imported from Germany by SIGARMS, Inc.
Price: Blue SA/DA or DAO $750.00
Price: Blue, Siglite night sights $845.00
Price: K-Kote or nickel slide $795.00
Price: K-Kote or nickel slide with Siglite night sights $885.00

SIG Sauer P220

SIG Sauer P226 Service Pistol
Similar to the P220 pistol except has 4.4" barrel, and weighs 26 1/2 oz. 357 SIG or 9mm. Imported from Germany by SIGARMS, Inc.
Price: Blue SA/DA or DAO $750.00
Price: With Siglite night sights $845.00
Price: Blue, SA/DA or DAO 357 SIG $795.00
Price: With Siglite night sights $885.00
Price: K-Kote finish, 9mm only or nickel slide $795.00
Price: K-Kote or nickel slide, Siglite night sights $885.00
Price: Nickel slide, 357 SIG $830.00
Price: Nickel slide, Siglite night sights $825.00

SIG Sauer P228 Compact Pistol
Similar to the P226 except has 3.86" barrel, with 7.08" overall length and 3.35" height. Chambered for 9mm Para. only, 10-shot magazine. Weight is 29.1 oz. with empty magazine. Introduced 1989. Imported from Germany by SIGARMS, Inc.
Price: Blue SA/DA or DAO $750.00
Price: Blue, with Siglite night sights $845.00
Price: K-Kote or nickel slide $875.00
Price: K-Kote, Siglite night sights $885.00
Price: K-Kote, double-action-only $795.00
Price: K-Kote, double-action-only, Siglite night sights $885.00

SIG Sauer P225

SIG SAUER P225 COMPACT PISTOL
Caliber: 9mm Para., 8-shot magazine.
Barrel: 3.8".
Weight: 26 oz. **Length:** 7 3/32" overall.
Stocks: Checkered black plastic.
Sights: Blade front, rear adjustable for windage. Optional Siglite night sights.
Features: Double action. Decocking lever permits lowering hammer onto locked firing pin. Square combat-type trigger guard. Imported from Germany by SIGARMS, Inc.
Price: Blue, SA/DA or DAO $725.00
Price: With Siglite night sights $830.00
Price: K-Kote or nickel slide $770.00
Price: K-Kote or nickel slide with Siglite night sights $860.00

SIG Sauer P229 DA Auto Pistol
Similar to the P228 except chambered for 40 S&W, 357 SIG. Has 3.86" barrel, 7.08" overall length and 3.35" height. Weight is 30.5 oz. Introduced 1991. Frame made in Germany, stainless steel slide assembly made in U.S.; pistol assembled in U.S. From SIGARMS, Inc.
Price: Blue, SA/DA or DAO $795.00
Price: With nickel slide $830.00
Price: Blue, Siglite night sights $885.00
Price: Nickel slide, Siglite night sights $925.00

SIG SAUER P232 PERSONAL SIZE PISTOL
Caliber: 380 ACP, 7-shot.
Barrel: 3 3/4".
Weight: 16 oz. **Length:** 6 1/2" overall.
Stocks: Checkered black composite.
Sights: Blade front, rear adjustable for windage.
Features: Double action/single action or DAO. Blowback operation, stationary barrel. Introduced 1997. Imported from Germany by SIGARMS, Inc.
Price: Blue, SA/DA or DAO $485.00
Price: In stainless steel $525.00
Price: With stainless steel slide, blue frame $505.00

SIG Sauer P228

SIG P210-6 SERVICE PISTOL
Caliber: 9mm Para., 8-shot magazine.
Barrel: 4 3/4".
Weight: 32 oz. **Length:** 8 1/2" overall.
Stocks: Checkered walnut.
Sights: Blade front, notch rear drift adjustable for windage.
Features: Mechanically locked, short-recoil operation; single action only; target trigger with adjustable stop; magazine safety; all-steel construction with matte blue finish. Optional 22 LR conversion kit consists of barrel, slide, recoil spring and magazine. Imported from Switzerland by SIGARMS, Inc.
Price: $2,300.00
Price: With 22LR conversion kit $2,900.00

SIG SAUER P239 PISTOL
Caliber: 9mm Para., 8-shot, 357 SIG 7-shot magazine.
Barrel: 3.6".
Weight: 25.2 oz. **Length:** 6.6" overall.
Stocks: Checkered black composite.
Sights: Blade front, rear adjustable for windage. Optional Siglite night sights.
Features: SA/DA or DAO; blackened stainless steel slide, aluminum alloy frame. Introduced 1996. Made in U.S. by SIGARMS, Inc.
Price: SA/DA or DAO $595.00
Price: SA/DA or DAO with Siglite night sights $685.00

HANDGUNS—AUTOLOADERS, SERVICE & SPORT

SMITH & WESSON MODEL 22A SPORT PISTOL
Caliber: 22 LR, 10-shot magazine.
Barrel: 4", 5 1/2", 7".
Weight: 29 oz. **Length:** 8" overall.
Stocks: Two-piece polymer.
Sights: Patridge front, fully adjustable rear.
Features: Comes with a sight bridge with Weaver-style integral optics mount; alloy frame; .312" serrated trigger; stainless steel slide and barrel with matte blue finish. Introduced 1997. Made in U.S. by Smith & Wesson.
Price: 4" $214.00
Price: 5 1/2" $237.00
Price: 7" $270.00

SMITH & WESSON MODEL 410 DA AUTO PISTOL
Caliber: 40 S&W, 10-shot magazine.
Barrel: 4".
Weight: 28.5 oz. **Length:** 7.5 oz.
Stocks: One-piece Xenoy, wrap-around with straight backstrap.
Sights: Post front, fixed rear; three-dot system.
Features: Aluminum alloy frame; blued carbon steel slide; traditional double action with left-side slide-mountd decocking lever. Introduced 1996. Made in U.S. by Smith & Wesson.
Price: $490.00

SMITH & WESSON MODEL 457 DA AUTO PISTOL
Caliber: 45 ACP, 7-shot magazine.
Barrel: 3 3/4".
Weight: 29 oz. **Length:** 7 1/4" overall.
Stocks: One-piece Xenoy, wrap-around with straight backstrap.
Sights: Post front, fixed rear, three-dot system.
Features: Aluminum alloy frame, matte blue carbon steel slide; bobbed hammer; smooth trigger. Introduced 1996. Made in U.S. by Smith & Wesson.
Price: $490.00

SMITH & WESSON MODEL 908 AUTO PISTOL
Caliber: 9mm Para., 8-shot magazine.
Barrel: 3 1/2".
Weight: 26 oz. **Length:** 6 13/16".
Stocks: One-piece Xenoy, wrap-around with straight backstrap.
Sights: Post front, fixed rear, three-dot system.
Features: Aluminum alloy frame, matte blue carbon steel slide; bobbed hammer; smooth trigger. Introduced 1996. Made in U.S. by Smith & Wesson.
Price: $443.00

SMITH & WESSON MODEL 910 DA AUTO PISTOL
Caliber: 9mm Para., 10-shot magazine.
Barrel: 4".
Weight: 28 oz. **Length:** 7 3/8" overall.
Stocks: One-piece Xenoy, wrap-around with straight backstrap.
Sights: Post front with white dot, fixed two-dot rear.
Features: Alloy frame, blue carbon steel slide. Slide-mounted decocking lever. Introduced 1995.
Price: Model 910 .... $443.00

SMITH & WESSON MODEL 2213, 2214 SPORTSMAN AUTOS
Caliber: 22 LR, 8-shot magazine.
Barrel: 3".
Weight: 18 oz. **Length:** 6 1/8" overall.
Stocks: Checkered black polymer.
Sights: Patridge front, fixed rear; three-dot system.
Features: Internal hammer; serrated trigger; single action. Model 2213 is stainless with alloy frame, Model 2214 is blued carbon steel with alloy frame. Introduced 1990. Made in U.S. by Smith & Wesson.
Price: Model 2213 .... $319.00
Price: Model 2214 .... $273.00

SMITH & WESSON MODEL 3913 DOUBLE ACTION
Caliber: 9mm Para., 8-shot magazine.
Barrel: 3 1/2".
Weight: 26 oz. **Length:** 6 13/16" overall.
Stocks: One-piece Delrin wrap-around, textured surface.
Sights: Post front with white dot, Novak LoMount Carry with two dots, adjustable for windage.
Features: Aluminum alloy frame, stainless slide (M3913) or blue steel slide (M3914). Bobbed hammer with no half-cock notch; smooth .304" trigger with rounded edges. Straight backstrap. Extra magazine included. Introduced 1989.
Price: $633.00

Smith & Wesson Model 22S Sport Pistol
Similar to the Model 22A Sport except with stainless steel frame. Available only with 5 1/2" or 7" barrel. Introduced 1997. Made in U.S. by Smith & Wesson.
Price: 5 1/2" $293.00
Price: 7 1/2" $324.00

Smith & Wesson Model 457

Smith & Wesson Model 910

Smith & Wesson 3913 LadySmith

Smith & Wesson Model 3913 LadySmith Auto
Similar to the standard Model 3913 except has frame that is upswept at the front, rounded trigger guard. Comes in frosted stainless steel with matching gray grips. Grips are ergonomically correct for a woman's hand. Novak LoMount Carry rear sight adjustable for windage, smooth edges for snag resistance. Extra magazine included. Introduced 1990.
Price: $651.00

Smith & Wesson Model 3953 DA Pistol
Same as the Model 3913 except double-action-only. Model 3953 has stainless slide with alloy frame. Overall length 7"; weighs 25.5 oz. Extra magazine included. Introduced 1990.
Price: $633.00

HANDGUNS—AUTOLOADERS, SERVICE & SPORT

SMITH & WESSON MODEL 4006 DA AUTO
Caliber: 40 S&W, 10-shot magazine.
Barrel: 4".
Weight: 38.5 oz. **Length:** 7 7/8" overall.
Stocks: Xenoy wrap-around with checkered panels.
Sights: Replaceable post front with white dot, Novak LoMount Carry fixed rear with two white dots, or micro. click adjustable rear with two white dots.
Features: Stainless steel construction with non-reflective finish. Straight backstrap. Extra magazine included. Introduced 1990.
Price: With adjustable sights $788.00
Price: With fixed sight $758.00
Price: With fixed night sights $870.00

Smith & Wesson Model 4013 TSW

Smith & Wesson Model 4046 DA Pistol
Similar to the Model 4006 except is double-action-only. Has a semi-bobbed hammer, smooth trigger, 4" barrel; Novak LoMount Carry rear sight, post front with white dot. Overall length is 7 1/2", weighs 28 oz. Extra magazine included. Introduced 1991.
Price: $758.00
Price: With fixed night sights $870.00

SMITH & WESSON MODEL 4013, 4056 TSW Autos
Caliber: 40 S&W, 9-shot magazine.
Barrel: 3 1/2".
Weight: 26.4 oz. **Length:** 6 7/8" overall.
Stocks: Checkered black polymer.
Sights: Novak three-dot system.
Features: Traditional double-action system; stainless slide, alloy frame; fixed barrel bushing; ambidextrous decocker; reversible magazine catch. Introduced 1997. Made in U.S. by Smith & Wesson.
Price: Model 4013 ....$788.00
Price: Model 4056, double-action-only $815.00

Smith & Wesson Model 4506

SMITH & WESSON MODEL 4500 SERIES AUTOS
Caliber: 45 ACP, 7-shot (M4516), 8-shot magazine for M4506, 4566/4586.
Barrel: 3 3/4" (M4516), 5" (M4506).
Weight: 41 oz. (4506). **Length:** 7 1/8" overall (4516).
Stocks: Xenoy one-piece wrap-around, arched or straight backstrap on M4506, straight only on M4516.
Sights: Post front with white dot, adjustable or fixed Novak LoMount Carry on M4506.
Features: M4506 has serrated hammer spur. Extra magazine included. Contact Smith & Wesson for complete data. Introduced 1989.
Price: Model 4506, fixed sight $787.00
Price: Model 4506, adjustable sight $819.00
Price: Model 4516, fixed sight $787.00
Price: Model 4566 (stainless, 4 1/4", traditional DA, ambidextrous safety, fixed sight) $787.00
Price: Model 4586 (stainless, 4 1/4", DA only) $787.00

Smith & Wesson Sigma SW40V

SMITH & WESSON MODEL 5900 SERIES AUTO PISTOLS
Caliber: 9mm Para., 10-shot magazine.
Barrel: 4".
Weight: 28 1/2 to 37 1/2 oz. (fixed sight); 38 oz. (adjustable sight). **Length:** 7 1/2" overall.
Stocks: Xenoy wrap-around with curved backstrap.
Sights: Post front with white dot, fixed or fully adjustable with two white dots.
Features: All stainless, stainless and alloy or carbon steel and alloy construction. Smooth .304" trigger, .260" serrated hammer. Extra magazine included. Introduced 1989.
Price: Model 5903 (stainless, alloy frame, traditional DA, fixed sight, ambidextrous safety) $701.00
Price: Model 5904 (blue, alloy frame, traditional DA, adjustable sight, ambidextrous safety) $653.00
Price: Model 5906 (stainless, traditional DA, adjustable sight, ambidextrous safety) $754.00
Price: As above, fixed sight $719.00
Price: With fixed night sights $831.00
Price: Model 5946 (as above, stainless frame and slide) $719.00

Smith & Wesson Model 6904/6906 Double-Action Autos
Similar to the Models 5904/5906 except with 3 1/2" barrel, 10-shot magazine, fixed rear sight, .260" bobbed hammer. Extra magazine included. Introduced 1989.
Price: Model 6904, blue $625.00
Price: Model 6906, stainless $688.00
Price: Model 6906 with fixed night sights $801.00
Price: Model 6946 (stainless, DA only, fixed sights) $688.00

SMITH & WESSON SIGMA SERIES PISTOLS
Caliber: 9mm Para., 40 S&W, 10-shot magazine.
Barrel: 3 1/4", 4", 4 1/2".
Weight: 26 oz. **Length:** 7.4" overall.
Stocks: Integral.
Sights: White dot front, fixed rear; three-dot system. Tritium night sights available.
Features: Ergonomic polymer frame; low barrel centerline; internal striker firing system; corrosion-resistant slide; Teflon-filled, electroless-nickel coated magazine. Introduced 1994. Made in U.S. by Smith & Wesson.
Price: SW9M, 9mm, 3 1/4" barrel, black polymer, fixed sights $356.00
Price: SW9C, 9mm, 4" barrel, black polymer, fixed night sights $531.00
Price: SW9V, 9mm, 4" barrel, black or gray polymer, stainless finish, fixed sights $372.00
Price: SW40C, 40 S&W, 4" barrel, black polymer, fixed sights $531.00
Price: SW40V, 40 S&W, 4" barrel, gray or black polymer, fixed sights ..$372.00
Price: SW40F, 40 S&W, 4 1/2" barrel, black polymer, fixed night sights ..$531.00

SMITH & WESSON SIGMA SW380 AUTO
Caliber: 380 ACP, 6-shot magazine.
Barrel: 3".
Weight: 14 oz. **Length:** 5.8" overall.
Stocks: Integral.
Sights: Fixed groove in the slide.
Features: Polymer frame; double-action-only trigger mechanism; grooved/serrated front and rear straps; two passive safeties. Introduced 1995. Made in U.S. by Smith & Wesson.
Price: $308.00

HANDGUNS—AUTOLOADERS, SERVICE & SPORT

SPHINX AT-380 AUTO PISTOL
Caliber: 380 ACP, 10-shot magazine.
Barrel: 3.27".
Weight: 25 oz. Length: 6.03" overall.
Stocks: Checkered plastic.
Sights: Fixed.
Features: Double-action-only mechanism, Chamber loaded indicator; ambidextrous magazine release and slide latch. Introduced 1993. Imported from Switzerland by Sphinx USA, Inc.
Price: Two-tone $493.95
Price: Black finish $513.95
Price: Nickel/Palladium finish $564.95

Sphinx AT-380M

SPHINX AT-2000S DOUBLE-ACTION PISTOL
Caliber: 9mm Para., 9x21mm, 40 S&W, 10-shot magazine.
Barrel: 4.53".
Weight: 36.3 oz. Length: 8.03" overall.
Stocks: Checkered neoprene.
Sights: Fixed, three-dot system.
Features: Double-action mechanism changeable to double-action-only. Stainless frame, blued slide. Ambidextrous safety, magazine release, slide latch. Introduced 1993. Imported from Switzerland by Sphinx USA, Inc.
Price: 9mm, two-tone $1,090.00
Price: 40 S&W, two-tone $1,120.00

Sphinx AT-2000P, AT-2000PS Auto Pistols
Same as the AT-2000S except AT-2000P has shortened frame, 3.74" barrel, 7.25" overall length, and weighs 34 oz. Model AT-2000PS has full-size frame. Both have stainless frame with blued slide. Introduced 1993. Imported from Switzerland by Sphinx USA, Inc.
Price: 9mm, two-tone $940.00
Price: 40 S&W, two-tone $980.00

Sphinx AT-2000P

Sphinx AT-2000H Auto Pistol
Similar to the AT-2000P except has shorter slide with 3.54" barrel, shorter frame, 10-shot magazine, with 7" overall length. Weight is 32.2 oz. Stainless frame with blued slide. Introduced 1993. Imported from Switzerland by Sphinx USA, Inc.
Price: 9mm, two-tone $940.00
Price: 40 S&W, two-tone $980.00

SPRINGFIELD, INC. 1911A1 AUTO PISTOL
Caliber: 9mm Para., 9-shot; 38 Super, 9-shot; 45 ACP, 8-shot.
Barrel: 5".
Weight: 35.6 oz. Length: 8$\frac{5}{8}$" overall.
Stocks: Checkered plastic or walnut.
Sights: Fixed three-dot system.
Features: Beveled magazine well; lowered and flared ejection port. All forged parts, including frame, barrel, slide. All new production. Introduced 1990. From Springfield, Inc.
Price: Mil-Spec 45 ACP, Parkerized $519.00
Price: Standard, 45 ACP, blued $549.00
Price: Basic, 45 ACP, stainless $589.00
Price: Lightweight (28.6 oz., matte finish) $549.00
Price: Standard, 9mm, 38 Super, blued $549.00
Price: Standard, 9mm, stainless steel $599.00

Springfield Standard

Consult our Directory pages for the location of firms mentioned.

Springfield, Inc. N.R.A. PPC Pistol
Specifically designed to comply with NRA rules for PPC competition. Has custom slide-to-frame fit; polished feed ramp; throated barrel; total internal honing; tuned extractor; recoil buffer system; fully checkered walnut grips; two fitted magazines; factory test target; custom carrying case. Introduced 1995. From Springfield, Inc.
Price: $1,649.00

Springfield, Inc. 1911A1 Custom Carry Gun
Similar to the standard 1911A1 except has fixed three-dot low profile sights, Videki speed trigger, match barrel and bushing; extended thumb safety, beavertail grip safety; beveled, polished magazine well, polished feed ramp and throated barrel; match Commander hammer and sear, tuned extractor; lowered and flared ejection port; recoil buffer system, full-length spring guide rod; walnut grips. Comes with two magazines with slam pads, plastic carrying case. Available in all popular calibers. Introduced 1992. From Springfield, Inc.
Price: $1,299.00

Springfield, Inc. 1911A1 Factory Comp
Similar to the standard 1911A1 except comes with bushing-type dual-port compensator, adjustable rear sight, extended thumb safety, Videki speed trigger, and beveled magazine well. Checkered walnut grips standard. Available in 38 Super or 45 ACP, blue only. Introduced 1992.
Price: 38 Super $979.00
Price: 45 ACP $947.00

Springfield, Inc. 1911A1 Defender Pistol
Similar to the 1911A1 Champion except has tapered cone dual-port compensator system, rubberized grips. Has reverse recoil plug, full-length recoil spring guide, serrated frontstrap, extended thumb safety, skeletonized hammer with modified grip safety to match and a Videki speed trigger. Bi-Tone finish. Introduced 1991.
Price: 45 ACP $993.00

HANDGUNS—AUTOLOADERS, SERVICE & SPORT

Springfield Champion

Springfield, Inc. 1911A1 High Capacity Pistol
Similar to the Standard 1911A1 except available in 45 ACP and 9mm with 10-shot magazine (45 ACP). Has Commander-style hammer, walnut grips, ambidextrous thumb safety, beveled magazine well, plastic carrying case. Introduced 1993. From Springfield, Inc.
Price: Mil-Spec 45 ACP $659.00
Price: 9mm $679.00
Price: 45 ACP Factory Comp $964.00
Price: 45 ACP Comp Lightweight, matte finish $840.00
Price: 45 ACP Compact, blued $609.00
Price: As above, stainless steel $648.00

Springfield, Inc. V10 Ultra Compact Pistol
Similar to the 1911A1 Compact except has shorter slide, 3.5" barrel, recoil reducing compensator built into the barrel and slide. Beavertail grip safety, beveled magazine well, "hi-viz" combat sights, Videcki speed trigger, flared ejection port, stainless steel frame, blued slide, match grade barrel, walnut grips. Introduced 1996. From Springfield, Inc.
Price: V10 45 ACP $675.00
Price: Ultra Compact (no compensator), 45 ACP $629.00

Star Firestar

Springfield, Inc. 1911A1 Champion Pistol
Similar to the standard 1911A1 except slide is 4.025". Has low-profile three-dot sight system. Comes with skeletonized hammer and walnut stocks. Available in 45 ACP only; blue or stainless. Introduced 1989.
Price: Blue $569.00
Price: Stainless $579.00
Price: Mil-Spec $519.00
Price: Champion Comp (single-port compensator) $869.00

STAR FIRESTAR AUTO PISTOL
Caliber: 9mm Para., 7-shot; 40 S&W, 6-shot.
Barrel: 3.39".
Weight: 30.35 oz. Length: 6.5" overall.
Stocks: Checkered rubber.
Sights: Blade front, fully adjustable rear; three-dot system.
Features: Low-profile, combat-style sights; ambidextrous safety. Available in blue or weather-resistant Starvel finish. Introduced 1990. Imported from Spain by Interarms.
Price: Blue, 9mm $430.00
Price: Starvel finish 9mm $450.00
Price: Blue, 40 S&W $445.00
Price: Starvel finish, 40 S&W $465.00

Star Firestar Model 45 Auto Pistol
Similar to the standard Firestar except chambered for 45 ACP with 6-shot magazine. Has 3.6" barrel, weighs 35 oz., 6.85" overall length. Reverse-taper Acculine barrel. Introduced 1992. Imported from Spain by Interarms.
Price: Blue $470.00
Price: Starvel finish $490.00

Star Firestar Plus Auto Pistol
Same as the standard Firestar except has 10-shot magazine. Introduced 1994. Imported from Spain by Interarms.
Price: Blue 9mm $460.00
Price: Starvel, 9mm $485.00

Star Firestar Plus

STAR ULTRASTAR DOUBLE-ACTION PISTOL
Caliber: 9mm Para., 9-shot magazine; 40 S&W, 8-shot.
Barrel: 3.57".
Weight: 26 oz. Length: 7" overall.
Stocks: Checkered black polymer.
Sights: Blade front, rear adjustable for windage; three-dot system.
Features: Polymer frame with inside steel slide rails; ambidextrous two-position safety (Safe and Decock). Introduced 1994. Imported from Spain by Interarms.
Price: $359.00

Stoeger American Eagle Luger

Star Ultrastar

STOEGER AMERICAN EAGLE LUGER
Caliber: 9mm Para., 7-shot magazine.
Barrel: 4", 6".
Weight: 32 oz. Length: 9.6" overall.
Stocks: Checkered walnut.
Sights: Blade front, fixed rear.
Features: Recreation of the American Eagle Luger pistol in stainless steel. Chamber loaded indicator. Introduced 1994. From Stoeger Industries.
Price: 4", or 6" Navy Model $699.00
Price: With matte black finish $789.00

HANDGUNS—AUTOLOADERS, SERVICE & SPORT

SUNDANCE MODEL A-25 AUTO PISTOL
Caliber: 25 ACP, 7-shot magazine.
Barrel: 2.5".
Weight: 16 oz. **Length:** 4 7/8" overall.
Stocks: Grooved black ABS or simulated smooth pearl; optional pink.
Sights: Fixed.
Features: Manual rotary safety; button magazine release. Bright chrome or black Teflon finish. Introduced 1989. Made in U.S. by Sundance Industries, Inc.
Price: $79.00

SUNDANCE BOA AUTO PISTOL
Caliber: 25 ACP, 7-shot magazine.
Barrel: 2 1/2".
Weight: 16 oz. **Length:** 4 7/8".
Stocks: Grooved ABS or smooth simulated pearl; optional pink.
Sights: Fixed.
Features: Patented grip safety, manual rotary safety; button magazine release; lifetime warranty. Bright chrome or black Teflon finish. Introduced 1991. Made in the U.S. by Sundance Industries, Inc.
Price: $95.00

SUNDANCE LASER 25 PISTOL
Caliber: 25 ACP, 7-shot magazine.
Barrel: 2 1/2".
Weight: 18 oz. **Length:** 4 7/8" overall.
Stocks: Grooved black ABS.
Sights: Class IIIa laser, 670 NM, 5mW, and fixed open.
Features: Factory installed and sighted laser sight activated by squeezing the grip safety; manual rotary safety; button magazine release. Bright chrome or black finish. Introduced 1995. Made in U.S. by Sundance Industries, Inc.
Price: With laser $219.95
Price: Lady Laser (as above except different name, bright chrome only) .$219.95

Sundance Laser 25

TAURUS MODEL PT 22/PT 25 AUTO PISTOLS
Caliber: 22 LR, 9-shot (PT 22); 25 ACP, 8-shot (PT 25).
Barrel: 2.75".
Weight: 12.3 oz. **Length:** 5.25" overall.
Stocks: Smooth Brazilian hardwood.
Sights: Blade front, fixed rear.
Features: Double action. Tip-up barrel for loading, cleaning. Blue or stainless. Introduced 1992. Made in U.S. by Taurus International.
Price: 22 LR or 25 ACP $162.00
Price: Nickel $171.00

Taurus PT 25

TAURUS MODEL PT 92AF AUTO PISTOL
Caliber: 9mm Para., 10-shot magazine.
Barrel: 4.92".
Weight: 34 oz. **Length:** 8.54" overall.
Stocks: Brazilian hardwood.
Sights: Fixed notch rear. Three-dot sight system.
Features: Double action, exposed hammer, chamber loaded indicator, ambidextrous safety, inertia firing pin. Imported by Taurus International.
Price: Blue $449.00
Price: Blue, Deluxe Shooter's Pak (extra magazine, case) $477.00
Price: Stainless steel $493.00
Price: Stainless, Deluxe Shooter's Pak (extra magazine, case) $522.00

Taurus Model PT 99AF Auto Pistol
Similar to the PT-92 except has fully adjustable rear sight, smooth Brazilian walnut stocks and is available in stainless steel or polished blue. Introduced 1983.
Price: Blue $471.00
Price: Blue, Deluxe Shooter's Pak (extra magazine, case) $500.00
Price: Stainless steel $518.00
Price: Stainless, Deluxe Shooter's Pak (extra magazine, case) $546.00

Taurus PT 92

TAURUS MODEL PT-111 AUTO PISTOL
Caliber: 9mm Para., 10-shot magazine.
Barrel: 3.30".
Weight: 19 oz. **Length:** 6.0" overall.
Stocks: Polymer.
Sights: Fixed. Low profile, three-dot combat.
Features: Double action only. Firing pin lock; chamber loaded indicator. Introduced 1997. Imported by Taurus International.
Price: Blue $319.00
Price: Stainless $329.00

TAURUS MODEL PT-940 AUTO PISTOL
Caliber: 40 S&W, 10-shot magazine.
Barrel: 3.35".
Weight: 28.2 oz. **Length:** 7.05" overall.
Stocks: Santoprene II.
Sights: Drift-adjustable front and rear; three-dot combat.
Features: Double action, exposed hammer; manual ambidextrous hammer-drop; inertia firing pin; chamber loaded indicator. Introduced 1996. Imported by Taurus International.
Price: Blue $437.00
Price: Stainless steel $452.00

TAURUS MODEL PT-938 AUTO PISTOL
Caliber: 380 ACP, 10-shot magazine.
Barrel: 3.72".
Weight: 27 oz. **Length:** 6.75" overall.
Stocks: Santoprene II.
Sights: Fixed. Low profile, three-dot combat.
Features: Double action only. Chamber loaded indicator; firing pin block; ambidextrous hammer drop. Introduced 1997. Imported by Taurus International.
Price: Blue $397.00
Price: Stainless $412.00

TAURUS MODEL PT-911 AUTO PISTOL
Caliber: 9mm Para., 10-shot magazine.
Barrel: 3.85".
Weight: 28.2 oz. **Length:** 7.05" overall.
Stocks: Santoprene II.
Sights: Fixed. Low profile, three-dot combat.
Features: Double action, exposed hammer; ambidextrous hammer drop; chamber loaded indicator. Introduced 1997. Imported by Taurus International.
Price: Blue $390.00
Price: Stainless $406.00

HANDGUNS—AUTOLOADERS, SERVICE & SPORT

TAURUS MODEL PT-945 AUTO PISTOL
Caliber: 45 ACP, 8-shot magazine.
Barrel: 4.25".
Weight: 29.5 oz. **Length:** 7.48" overall.
Stocks: Santoprene II.
Sights: Drift-adjustable front and rear; three-dot system.
Features: Double-action mechanism. Has manual ambidextrous hammer drop safety, intercept notch, firing pin block, chamber loaded indicator, last-shot hold-open. Introduced 1995. Imported by Taurus International.
Price: Blue $453.00
Price: Stainless $469.00
Price: Blue, Deluxe Shooter's Pak $476.00
Price: Stainless, Deluxe Shooter's Pak $490.00

Taurus PT 945

UZI EAGLE AUTO PISTOL
Caliber: 9mm Para., 40 S&W, 45 ACP, 10-shot magazine.
Barrel: 4.4".
Weight: 35 oz. **Length:** 8.1" overall.
Stocks: Textured, high-impact polymer.
Sights: Three-dot tritium night sights.
Features: Double-action mechanism with decocker; polygonal rifling; matte blue/black finish. Introduced 1997. Imported from Israel by Uzi America, Inc.
Price: Full-size, 9mm or 40 S&W $535.00
Price: Short Slide, 3.7" barrel, 7.5" overall, 9mm, 40 S&W $535.00
Price: Short Slide 45 ACP $566.00
Price: Compact, 3.5" barrel, 7.2" overall, DA or DAO 9mm or 40 S&W ..$535.00
Price: Polymer Compact, polymer frame, DA or DAO 9mm or 40 S&W ..$535.00

> *Manufacturers' addresses in the*
> **Directory of the Arms Trade**
> *page 309, this issue*

Walther P99

WALTHER P99 AUTO PISTOL
Caliber: 9mm Para., 10-shot magazine.
Barrel: 4".
Weight: 25 oz. **Length:** 7" overall.
Stocks: Textured polymer.
Sights: Blade front (comes with three interchangeable blades for elevation adjustment), micrometer rear adjustable for windage.
Features: Double-action mechanism with trigger safety, decock safety, internal striker safety; chamber loaded indicator; ambidextrous magazine release levers; polymer frame with interchangeable backstrap inserts. Comes with two magazines. Introduced 1997. Imported from Germany by Interarms.
Price: $799.00

Walther PP

WALTHER PP AUTO PISTOL
Caliber: 32 ACP, 380 ACP, 7-shot magazine.
Barrel: 3.86".
Weight: 23 1/2 oz. **Length:** 6.7" overall.
Stocks: Checkered plastic.
Sights: Fixed, white markings.
Features: Double action; manual safety blocks firing pin and drops hammer; chamber loaded indicator on 32 and 380; extra finger rest magazine provided. Imported from Germany by Interarms.
Price: 32 ....$999.00
Price: 380 ....$999.00
Price: Engraved models ....On Request

Walther PPK/S American

Walther PPK/S American Auto Pistol
Similar to Walther PP except made entirely in the United States. Has 3.27" barrel with 6.1" length overall. Introduced 1980.
Price: 380 ACP only, blue $540.00
Price: As above, stainless $540.00

Walther PPK American Auto Pistol
Similar to Walther PPK/S except weighs 21 oz., has 6-shot capacity. Made in the U.S. Introduced 1986.
Price: Stainless, 380 ACP only $540.00
Price: Blue, 380 ACP only $540.00

Walther PPK

HANDGUNS—AUTOLOADERS, SERVICE & SPORT

WALTHER MODEL TPH AUTO PISTOL
Caliber: 22 LR, 25 ACP, 6-shot magazine.
Barrel: 2 1/4".
Weight: 14 oz. **Length:** 5 3/8" overall.
Stocks: Checkered black composition.
Sights: Blade front, rear drift-adjustable for windage.
Features: Made of stainless steel. Scaled-down version of the Walther PP/PPK series. Made in U.S. Introduced 1987. From Interarms.
Price: Blue or stainless steel, 22 or 25 ....$440.00

Walther TPH

WALTHER P88 COMPACT PISTOL
Caliber: 9mm Para., 10-shot magazine.
Barrel: 3.93".
Weight: 28 oz. **Length:** NA.
Stocks: Checkered black polymer.
Sights: Blade front, drift adjustable rear.
Features: Double action with ambidextrous decocking lever and magazine release; alloy frame; loaded chamber indicator; matte blue finish. Imported from Germany by Interarms.
Price: $900.00

WILKINSON SHERRY AUTO PISTOL
Caliber: 22 LR, 8-shot magazine.
Barrel: 2 1/8".
Weight: 9 1/4 oz. **Length:** 4 3/8" overall.
Stocks: Checkered black plastic.
Sights: Fixed, groove.
Features: Cross-bolt safety locks the sear into the hammer. Available in all blue finish or blue slide and trigger with gold frame. Introduced 1985.
Price: $195.00

Walther P88 Compact

Wilkinson Sherry

WILDEY AUTOMATIC PISTOL
Caliber: 10mm Wildey Mag., 11mm Wildey Mag., 30 Wildey Mag., 357 Peterbuilt, 45 Win. Mag., 475 Wildey Mag., 7-shot magazine.
Barrel: 5", 6", 7", 8", 10", 12", 14" (45 Win. Mag.); 8", 10", 12", 14" (all other cals.). Interchangeable.
Weight: 64 oz. (5" barrel). **Length:** 11" overall (7" barrel).
Stocks: Hardwood.
Sights: Ramp front (interchangeable blades optional), fully adjustable rear. Scope base available.
Features: Gas-operated action. Made of stainless steel. Has three-lug rotary bolt. Double or single action. Polished and matte finish. Made in U.S. by Wildey, Inc.
Price: ....$1,175.00 to $1,495.00

WILKINSON LINDA AUTO PISTOL
Caliber: 9mm Para.
Barrel: 8 5/16".
Weight: 4 lbs., 13 oz. **Length:** 12 1/4" overall.
Stocks: Checkered black plastic pistol grip, walnut forend.
Sights: Protected blade front, aperture rear.
Features: Fires from closed bolt. Semi-auto only. Straight blowback action. Cross-bolt safety. Removable barrel. From Wilkinson Arms.
Price: $533.33

HANDGUNS—COMPETITION HANDGUNS

Includes models suitable for several forms of competition and other sporting purposes.

Manufacturers' addresses in the
Directory of the Arms Trade
page 309, this issue

Auto-Ordnance Competition Model

AUTO-ORDNANCE 1911A1 COMPETITION MODEL
Caliber: 45 ACP.
Barrel: 5".
Weight: 42 oz. **Length:** 10" overall.
Stocks: Black textured rubber wrap-around.
Sights: Blade front, rear adjustable for windage; three-dot system.
Features: Machined compensator, combat Commander hammer; flat mainspring housing; low profile magazine funnel; metal form magazine bumper; high-ride beavertail grip safety; full-length recoil spring guide system; extended slide stop, safety and magazine catch; Videcki adjustable speed trigger; extended combat ejector. Introduced 1994. Made in U.S. by Auto-Ordnance Corp.
Price: $635.50

CAUTION: PRICES SHOWN ARE SUPPLIED BY THE MANUFACTURER OR IMPORTER. CHECK YOUR LOCAL GUNSHOP.

HANDGUNS—COMPETITION HANDGUNS

Baer 1911 Ultimate Master

Baer 1911 Bullseye Wadcutter

BAER 1911 ULTIMATE MASTER COMBAT PISTOL
Caliber: 9x23, 38 Super, 400 Cor-Bon, 45 ACP (others available), 10-shot magazine.
Barrel: 5", 6"; Baer NM.
Weight: 37 oz. **Length:** 8.5" overall.
Stocks: Checkered rosewood.
Sights: Baer dovetail front, low-mount Bo-Mar rear with hidden leaf.
Features: Full-house competition gun. Baer forged NM blued steel frame and double serrated slide; Baer triple port, tapered cone compensator; fitted slide to frame; lowered, flared ejection port; Baer reverse recoil plug; full-length guide rod; recoil buff; beveled magazine well; Baer Commander hammer, sear; Baer extended ambidextrous safety, extended ejector, checkered slide stop, beavertail grip safety with pad, extended magazine release button; Baer speed trigger. Made in U.S. by Les Baer Custom, Inc.
Price: Compensated, open sights $2,295.00
Price: 6" Model 400 Cor-Bon $2,390.00
Price: Compensated, with Baer optics mount $3,195.00

Baer 1911 Ultimate Master Steel Special Pistol
Similar to the Ultimate Master except chambered for 38 Super with supported chamber (other calibers available), lighter slide, bushing-type compensator; two-piece guide rod. Designed for maximum 150 power factor. Comes without sights—scope and mount only. Hard chrome finish. Made in U.S. by Les Baer Custom, Inc.
Price: $2,940.00

BAER 1911 NATIONAL MATCH HARDBALL PISTOL
Caliber: 45 ACP, 7-shot magazine.
Barrel: 5".
Weight: 37 oz. **Length:** 8.5" overall.
Stocks: Checkered walnut.
Sights: Baer dovetail front with undercut post, low-mount Bo-Mar rear with hidden leaf.
Features: Baer NM forged steel frame, double serrated slide and barrel with stainless bushing; slide fitted to frame; Baer match trigger with 4-lb. pull; polished feed ramp, throated barrel; checkered front strap, arched mainspring housing; Baer beveled magazine well; lowered, flared ejection port; tuned extractor; Baer extended ejector, checkered slide stop; recoil buff. Made in U.S. by Les Baer Custom, Inc.
Price: $1,235.00

Beretta Model 89

Beretta Model 96 Combat

Baer 1911 Bullseye Wadcutter Pistol
Similar to the National Match Hardball except designed for wadcutter loads only. Has polished feed ramp and barrel throat; Bo-Mar rib on slide; full-length recoil rod; Baer speed trigger with 3 1/2-lb. pull; Baer deluxe hammer and sear; Baer beavertail grip safety with pad; flat mainspring housing checkered 20 lpi. Blue finish; checkered walnut grips. Made in U.S. by Les Baer Custom, Inc.
Price: $1,547.00
Price: With 6" barrel $1,777.00

BENELLI MP95E MATCH PISTOL
Caliber: 22 LR, 9-shot magazine, or 32 S&W WC, 5-shot magazine.
Barrel: 4.33".
Weight: 38.8 oz. **Length:** 11.81" overall.
Stocks: Checkered walnut match type; anatomically shaped.
Sights: Match type. Blade front, click-adjustable rear for windage and elevation.
Features: Removable, trigger assembly. Special internal weight box on sub-frame below barrel. Cut for scope rails. Introduced 1993. Imported from Italy by European American Armory.
Price: Blue $550.00
Price: Chrome $599.00
Price: MP90S (competition version of MP95E), 22 LR $1,295.00
Price: As above, 32 S&W $1,495.00

BERETTA MODEL 89 GOLD STANDARD PISTOL
Caliber: 22 LR, 8-shot magazine.
Barrel: 6".
Weight: 41 oz. **Length:** 9.5" overall.
Stocks: Target-type walnut with thumbrest.
Sights: Interchangeable blade front, fully adjustable rear.
Features: Single action target pistol. Matte black, Bruniton finish. Imported from Italy by Beretta U.S.A.
Price: $750.00

BERETTA MODEL 96 COMBAT PISTOL
Caliber: 40 S&W, 10-shot magazine.
Barrel: 4.9" (5.9" with weight).
Weight: 34.4 oz. **Length:** 8.5" overall.
Stocks: Checkered black plastic.
Sights: Blade front, fully adjustable target rear.
Features: Uses heavier Brigadier slide with front and rear serrations; extended frame-mounted safety; extended, reversible magazine release; single-action-only with competition-tuned trigger with extra-short let-off and over-travel adjustment. Comes with tool kit. Introduced 1997. Imported from Italy by Beretta U.S.A.
Price: $1,593.00

Beretta Model 96 Stock Pistol
Similar to the Model 96 Combat except is single/double action, with half-cock notch. Has front and rear slide serrations, rubber magazine bumper, replaceable accurizing barrel bushing, ultra-thin fine-checkered grips (aluminum optional), checkered front and back straps, radiused back strap, fitted case. Weighs 35 oz., 8.5" overall. Introduced 1997. Imported from Italy by Beretta U.S.A.
Price: $1,371.00

HANDGUNS—COMPETITION HANDGUNS

BF ULTIMATE SILHOUETTE SINGLE SHOT PISTOL
Caliber: Most popular chamberings from 17 through 50-cal.
Barrel: 10.75".
Weight: 52 oz. **Length:** NA.
Stocks: Custom Herrett finger-groove grip and forend.
Sights: Undercut Patridge front, 1/2-MOA match-quality fully adjustable RPM Iron Sight rear; barrel or receiver mounting. Drilled and tapped for scope mounting.
Features: Rigid barrel/receiver; falling block action with short lock time; automatic ejection; air-gauged match barrels; matte black oxide finish standard, electroless nickel optional. Barrel has 11-degree recessed target crown. Introduced 1988. Made in U.S. by E.A. Brown Mfg.
Price: $1,014.00

BF Single Shot

BERNARDELLI MODEL 69 TARGET PISTOL
Caliber: 22 LR, 10-shot magazine.
Barrel: 5.9".
Weight: 38 oz. **Length:** 9" overall.
Stocks: Wrap-around, hand-checkered walnut with thumbrest.
Sights: Fully adjustable and interchangeable target type.
Features: Conforms to U.I.T. regulations. Has 7.1" sight radius, .27" wide grooved trigger. Manual thumb safety and magazine safety. Introduced 1987. Imported from Italy by Mitchell Arms.
Price: $612.00

BROLIN TAC SERIES MODEL TAC-11 TACTICAL 1911
Caliber: 45 ACP, 8-shot magazine.
Barrel: 5"; conical match design.
Weight: 35.1 oz. **Length:** 8.5" overall.
Stocks: Contoured black rubber.
Sights: Ramp front, Novak Low Profile Combat Sight or Novak tritium.
Features: Throated conical barrel; polished feed ramp; lowered and flared ejection port; beveled magazine well; flat-top slide; flat mainspring housing; front strap high relief cut; lightened aluminum match trigger; slotted commander hammer; custom beavertail grip safety; Brolin "Iron Claw" extractor. Introduced 1997. Made in the U.S. by Brolin Industries.
Price: Matte blue $649.00
Price: Matte blue with Novak tritium night sights $649.00

BROLIN PRO-STOCK COMPETITION PISTOL
Caliber: 45 ACP, 8-shot magazine.
Barrel: 5".
Weight: 37 oz. **Length:** 8.5" overall.
Stocks: Checkered with Brolin logo.
Sights: High visibility front, Bo-Mar fully adjustable rear.
Features: Throated match barrel; full-length recoil spring guide; polished feed ramp; lowered and flared ejection port; beveled magazine well; flat-top slide; flat mainspring housing; high relief front strap cut; adjustable match trigger; slotted Commander hammer; beavertail grip safety; ambidextrous thumb safety; front slide serrations. Introduced 1996. Made in U.S. by Brolin Industries.
Price: Blue $779.00
Price: Two-tone $799.00

Brolin Pro-Comp Competition Pistol
Similar to the Pro-Stock model except has integral milled DPC Comp on the heavy match barrel; barrel length 4", overall length 8.5"; weighs 37 oz.; 8-shot magazine. Introduced 1996. Made in U.S. by Brolin Industries.
Price: Blue $919.00
Price: Two-tone $929.00

BROWNING BUCK MARK SILHOUETTE
Caliber: 22 LR, 10-shot magazine.
Barrel: 9 7/8".
Weight: 53 oz. **Length:** 14" overall.
Stocks: Smooth walnut stocks and forend, or finger-groove walnut.
Sights: Post-type hooded front adjustable for blade width and height; Pro Target rear fully adjustable for windage and elevation.
Features: Heavy barrel with .900" diameter; 12 1/2" sight radius. Special sighting plane forms scope base. Introduced 1987. Made in U.S. From Browning.
Price: $434.95

Browning Buck Mark Unlimited Silhouette
Same as the Buck Mark Silhouette except has 14" heavy barrel. Conforms to IHMSA 15" maximum sight radius rule. Introduced 1991.
Price: $535.95

Brolin TAC-11

Brolin Pro-Stock

Browning Buck Mark Target 5.5
Same as the Buck Mark Silhouette except has a 5 1/2" barrel with .900" diameter. Has hooded sights mounted on a scope base that accepts an optical or reflex sight. Rear sight is a Browning fully adjustable Pro Target, front sight is an adjustable post that customizes to different widths, and can be adjusted for height. Contoured walnut grips with thumbrest, or finger-groove walnut. Matte blue finish. Overall length is 9 5/8", weighs 35 1/2 oz. Has 10-shot magazine. Introduced 1990. From Browning.
Price: $411.95
Price: Target 5.5 Gold (as above with gold anodized frame and top rib) .$462.95
Price: Target 5.5 Nickel (as above with nickel frame and top rib) $462.95

Browning Buck Mark Field 5.5
Same as the Target 5.5 except has hoodless ramp-style front sight and low profile rear sight. Matte blue finish, contoured or finger-groove walnut stocks. Introduced 1991.
Price: $411.95

Browning Buck Mark Target 5.5

CAUTION: PRICES SHOWN ARE SUPPLIED BY THE MANUFACTURER OR IMPORTER. CHECK YOUR LOCAL GUNSHOP.

HANDGUNS—COMPETITION HANDGUNS

Browning Buck Mark Bullseye

Browning Buck Mark Bullseye
Similar to the Buck Mark Silhouette except has 7¼" heavy barrel with three flutes per side; trigger is adjustable from 2½ to 5 lbs.; specially designed rosewood target or three-finger-groove stocks with competition-style heel rest, or with contoured rubber grip. Overall length is 11⁵⁄₁₆", weighs 36 oz. Introduced 1996. Made in U.S. From Browning.
Price: With ambidextrous moulded composite stocks $376.95
Price: With rosewood stocks, or wrap-around finger groove $484.95

Colt Gold Cup Trophy

COLT GOLD CUP TROPHY MK IV/SERIES 80
Caliber: 45 ACP, 8-shot magazine.
Barrel: 5", with new design bushing.
Weight: 39 oz. **Length:** 8½".
Stocks: Checkered rubber composite with silver-plated medallion.
Sights: Patridge-style front, Colt-Elliason rear adjustable for windage and elevation, sight radius 6¾".
Features: Arched or flat housing; wide, grooved trigger with adjustable stop; ribbed-top slide, hand fitted, with improved ejection port.
Price: Blue $1,050.00
Price: Stainless $1,116.00

Competitor Single Shot

COMPETITOR SINGLE SHOT PISTOL
Caliber: 22 LR through 50 Action Express, including belted magnums.
Barrel: 14" standard; 10.5" silhouette; 16" optional.
Weight: About 59 oz. (14" bbl.). **Length:** 15.12" overall.
Stocks: Ambidextrous; synthetic (standard) or laminated or natural wood.
Sights: Ramp front, adjustable rear.
Features: Rotary canon-type action cocks on opening; cammed ejector; interchangeable barrels, ejectors. Adjustable single stage trigger, sliding thumb safety and trigger safety. Matte blue finish. Introduced 1988. From Competitor Corp., Inc.
Price: 14", standard calibers, synthetic grip $399.95
Price: Extra barrels, from $149.95

E.A.A. WITNESS GOLD TEAM AUTO
Caliber: 9mm Para., 9x21, 38 Super, 40 S&W, 45 ACP.
Barrel: 5.1".
Weight: 41.6 oz. **Length:** 9.6" overall.
Stocks: Checkered walnut, competition style.
Sights: Square post front, fully adjustable rear.
Features: Triple-chamber cone compensator; competition SA trigger; extended safety and magazine release; competition hammer; beveled magazine well; beavertail grip. Hand-fitted major components. Hard chrome finish. Match-grade barrel. From E.A.A. Custom Shop. Introduced 1992. From European American Armory.
Price: $2,195.00

E.A.A. Witness Gold Team

E.A.A. Witness Silver Team Auto
Similar to the Witness Gold Team except has double-chamber compensator, oval magazine release, black rubber grips, double-dip blue finish. Comes with Super Sight and drilled and tapped for scope mount. Built for the intermediate competition shooter. Introduced 1992. From European American Armory Custom Shop.
Price: 9mm Para., 9x21, 38 Super, 40 S&W, 45 ACP $975.00

> Consult our Directory pages for the location of firms mentioned.

Erma ESP Junior Match

ERMA ESP 85A MATCH PISTOL
Caliber: 22 LR, 6-shot; 32 S&W, 6-shot magazine.
Barrel: 6".
Weight: 39 oz. **Length:** 10" overall.
Stocks: Match-type of stippled walnut; adjustable.
Sights: Interchangeable blade front, micrometer adjustable rear with interchangeable leaf.
Features: Five-way adjustable trigger; exposed hammer and separate firing pin block allow unlimited dry firing practice. Blue or matte chrome; right- or left-hand. Introduced 1989. Imported from Germany by Precision Sales International.
Price: 22 LR $1,895.00
Price: 22 LR, left-hand $1,935.00
Price: 22 LR, matte chrome $2,110.00
Price: 32 S&W $2,110.00
Price: 32 S&W, left-hand $2,155.00
Price: 32 S&W, matte chrome $2,325.00
Price: 32 S&W, matte chrome, left-hand $2,375.00

Erma ESP Junior Match Pistol
Similar to the ESP 85A Match except chambered only for 22 LR, blue finish only. Stippled non-adjustable walnut match grips (adjustable grips optional). Introduced 1995. Imported from Germany by Precision Sales International.
Price: $1,460.00

HANDGUNS—COMPETITION HANDGUNS

FAS 607 MATCH PISTOL
Caliber: 22 LR, 5-shot.
Barrel: 5.6".
Weight: 37 oz. **Length:** 11" overall.
Stocks: Walnut wrap-around; sizes small, medium, large or adjustable.
Sights: Match. Blade front, open notch rear fully adjustable for windage and elevation. Sight radius is 8.66".
Features: Line of sight is only $^{11}/_{32}$" above centerline of bore; magazine is inserted from top; adjustable and removable trigger mechanism; single lever takedown. Full 5-year warranty. Imported from Italy by Nygord Precision Products.
Price: $1,175.00
Price: Model 603 (32 S&W) $1,175.00

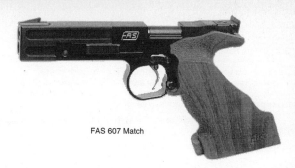

FAS 607 Match

FAS 601 Match Pistol
Similar to Model 607 except has different match stocks with adjustable palm shelf, 22 Short only for rapid fire shooting; weighs 40 oz., 5.6" bbl.; has gas ports through top of barrel and slide to reduce recoil; slightly different trigger and sear mechanisms. Imported from Italy by Nygord Precision Products.
Price: $1,250.00

FORT WORTH HSV TARGET PISTOL
Caliber: 22 LR, 10-shot magazine.
Barrel: 4 1/2" or 5 1/2"; push-button takedown.
Weight: 46 oz. **Length:** 9.5" overall.
Stocks: Checkered hardwood with thumbrest.
Sights: Undercut ramp front, micro-click rear adjustable for windage and elevation. Also available with scope mount, rings, no sights.
Features: Stainless steel construction. Full-length vent rib. Gold-plated trigger, slide lock, safety-lever and magazine release; stippled front grip and backstrap; polished slide; adjustable trigger and sear. Comes with barrel weight. Introduced 1995. From Fort Worth Firearms.
Price: $472.95
Price: With Weaver rib $537.95
Price: With 8" barrel, Weaver rib, custom grips, sights $616.95
Price: As above, 10" barrel $629.95

FORT WORTH HST TARGET PISTOL
Caliber: 22 LR, 10-shot magazine.
Barrel: 5 1/2" bull or 7 1/4" fluted.
Weight: 44 oz. **Length:** 9.5" overall.
Stocks: Checkered hardwood with thumbrest.
Sights: Undercut ramp front, frame-mounted micro-click rear adjustable for windage and elevation; drilled and tapped for scope mounting.
Features: Gold-plated trigger, slide lock, safety-lever and magazine release; stippled front grip and backstrap; adjustable trigger and sear. Barrel weights optional. Introduced 1995. From Fort Worth Firearms.
Price: 5 1/2" or 7 1/4" right-hand $410.95
Price: 5 1/2" left-hand $451.95

Fort Worth HSC Target Pistol
Same as the HST model except has nickel-plated trigger, slide lock, safety-lever, magazine release, and has slightly heavier trigger pull. Has stippled front grip and backstrap, checkered walnut thumbrest grips, adjustable trigger and sear. Matte finish. Drilled and tapped for scope mount and barrel weight. Introduced 1995. From Fort Worth Firearms.
Price: $388.95

FORT WORTH HSO AUTO PISTOL
Caliber: 22 Short, 10-shot magazine.
Barrel: 6 3/4" round tapered, with stabilizer and built-in muzzle brake.
Weight: 40 oz. **Length:** 11 1/4" overall.
Stocks: Checkered walnut with thumbrest.
Sights: Undercut ramp front, frame-mounted click adjustable square notch rear. Drilled and tapped for scope mount.
Features: Integral stabilizer with two removable weights. Trigger adjustable for pull and length of travel; stippled front and backstraps; push-button barrel takedown. Introduced 1995. Made in U.S. by Fort Worth Firearms.
Price: $599.95

GLOCK 17L COMPETITION AUTO
Caliber: 9mm Para., 10-shot magazine.
Barrel: 6.02".
Weight: 23.3 oz. **Length:** 8.85" overall.
Stocks: Black polymer.
Sights: Blade front with white dot, fixed or adjustable rear.
Features: Polymer frame, steel slide; double-action trigger with "Safe Action" system; mechanical firing pin safety, drop safety; simple takedown without tools; locked breech, recoil operated action. Introduced 1989. Imported from Austria by Glock, Inc.
Price: Fixed sight $800.00
Price: Adjustable sight $828.00

FORT WORTH MATCHMASTER STANDARD PISTOL
Caliber: 22 LR, 10-shot magazine.
Barrel: 3 7/8", 4 1/2", 5 1/2", 7 1/2", 10".
Weight: NA. **Length:** NA.
Stocks: Checkered walnut.
Sights: Ramp front, slide-mounted adjustable rear.
Features: Stainless steel construction. Double extractors; trigger finger magazine release button and standard button; beveled magazine well; grip angle equivalent to M1911; low-profile frame. Introduced 1997. From Fort Worth Firearms.
Price: $388.95

Fort Worth Matchmaster Deluxe Pistol
Same as the Matchmaster Standard except comes with Weaver-style rib mount and integral adjustable rear sight system. Introduced 1997. From Fort Worth Firearms.
Price: $537.95

Fort Worth Matchmaster Dovetail Pistol
Same as the Matchmaster Standard except has a dovetail-style mount and integral rear sight system. Available with 3 7/8", 4 1/2" or 5 1/2" barrel only. Introduced 1997. From Fort Worth Firearms.
Price: $472.95

FREEDOM ARMS CASULL MODEL 252 SILHOUETTE
Caliber: 22 LR, 5-shot cylinder.
Barrel: 9.95".
Weight: 63 oz. **Length:** NA
Stocks: Black micarta, western style.
Sights: 1/8" Patridge front, Iron Sight Gun Works silhouette rear, click adjustable for windage and elevation.
Features: Stainless steel. Built on the 454 Casull frame. Two-point firing pin, lightened hammer for fast lock time. Trigger pull is 3 to 5 lbs. with pre-set overtravel screw. Introduced 1991. From Freedom Arms.
Price: Silhouette Class $1,509.00
Price: Extra fitted 22 WMR cylinder $264.00

Freedom Arms 252 Varmint

Freedom Arms Model 252 Varmint
Similar to the Silhouette Class revolver except has 7.5" barrel, weighs 59 oz., has black and green laminated hardwood grips, and comes with brass bead front sight, express shallow V rear sight with windage and elevation adjustments. Introduced 1991. From Freedom Arms.
Price: Varmint Class $1,454.00
Price: Extra fitted 22 WMR cylinder $264.00

GAUCHER GP SILHOUETTE PISTOL
Caliber: 22 LR, single shot.
Barrel: 10".
Weight: 42.3 oz. **Length:** 15.5" overall.
Stocks: Stained hardwood.
Sights: Hooded post on ramp front, open rear adjustable for windage and elevation.
Features: Matte chrome barrel, blued bolt and sights. Other barrel lengths available on special order. Introduced 1991. Imported by Mandall Shooting Supplies.
Price: $425.00

HANDGUNS—COMPETITION HANDGUNS

GLOCK 24 COMPETITION MODEL PISTOL
Caliber: 40 S&W, 10-shot magazine.
Barrel: 6.02".
Weight: 29.5 oz. **Length:** 8.85" overall.
Stocks: Black polymer.
Sights: Blade front with dot, white outline rear adjustable for windage.
Features: Long-slide competition model available as compensated or non-compensated gun. Factory-installed competition trigger; drop-free magazine. Introduced 1994. Imported from Austria by Glock, Inc.
Price: Fixed sight $800.00
Price: Adjustable sight $828.00
Price: Model 24C, ported barrel, fixed sight $830.00
Price: Model 24C, ported barrel, adjustable sight $858.00

Glock 24 Competition

HAMMERLI MODEL 160/162 FREE PISTOLS
Caliber: 22 LR, single shot.
Barrel: 11.30".
Weight: 46.94 oz. **Length:** 17.52" overall.
Stocks: Walnut; full match style with adjustable palm shelf. Stippled surfaces.
Sights: Changeable blade front, open, fully adjustable match rear.
Features: Model 160 has mechanical set trigger; Model 162 has electronic trigger; both fully adjustable with provisions for dry firing. Introduced 1993. Imported from Switzerland by Sigarms, Inc.
Price: Model 160 ....$2,085.00
Price: Model 162 ....$2,295.00

Hammerli Model 160

HAMMERLI MODEL 208s PISTOL
Caliber: 22 LR, 8-shot magazine.
Barrel: 5.9".
Weight: 37.5 oz. **Length:** 10" overall.
Stocks: Walnut, target-type with thumbrest.
Sights: Blade front, open fully adjustable rear.
Features: Adjustable trigger, including length; interchangeable rear sight elements. Imported from Switzerland by Sigarms, Inc.
Price: $1,925.00

HAMMERLI MODEL 280 TARGET PISTOL
Caliber: 22 LR, 6-shot; 32 S&W Long WC, 5-shot.
Barrel: 4.5".
Weight: 39.1 oz. (32). **Length:** 11.8" overall.
Stocks: Walnut match-type with stippling, adjustable palm shelf.
Sights: Match sights, micrometer adjustable; interchangeable elements.
Features: Has carbon-reinforced synthetic frame and bolt/barrel housing. Trigger is adjustable for pull weight, take-up weight, let-off, and length, and is interchangeable. Interchangeable metal or carbon fiber counterweights. Sight radius of 8.8". Comes with barrel weights, spare magazine, loading tool, cleaning rods. Introduced 1990. Imported from Sigarms, Inc.
Price: 22-cal. $1,565.00
Price: 32-cal. $1,765.00
Price: Set with two calibers, one frame $2,595.00

Hammerli Model 280

HARRIS GUNWORKS SIGNATURE JR. LONG RANGE PISTOL
Caliber: Any suitable caliber.
Barrel: To customer specs.
Weight: 5 lbs.
Stock: Gunworks fiberglass.
Sights: None furnished; comes with scope rings.
Features: Right- or left-hand benchrest action of titanium or stainless steel; single shot or repeater. Comes with bipod. Introduced 1992. Made in U.S. by Harris Gunworks, Inc.
Price: $2,600.00

Hammerli Model 208s

Manufacturers' addresses in the
Directory of the Arms Trade
page 309, this issue

Heckler & Koch USP45 Match

HECKLER & KOCH USP45 MATCH PISTOL
Caliber: 45 ACP, 10-shot magazine.
Barrel: 6.02".
Weight: 2.38 lbs. **Length:** 9.45"overall.
Stocks: Textured polymer.
Sights: High profile target front, fully adjustable target rear.
Features: Adjustable trigger stop. Polymer frame, blue or stainless steel slide. Introduced 1997. Imported from Germany by Heckler & Koch, Inc.
Price: Blue $1,329.00
Price: Stainless $1,399.00

HANDGUNS—COMPETITION HANDGUNS

MORINI MODEL 84E FREE PISTOL
Caliber: 22 LR, single shot.
Barrel: 11.4".
Weight: 43.7 oz. **Length:** 19.4" overall.
Stocks: Adjustable match type with stippled surfaces.
Sights: Interchangeable blade front, match-type fully adjustable rear.
Features: Fully adjustable electronic trigger. Introduced 1995. Imported from Switzerland by Nygord Precision Products.
Price: $1,495.00

PARDINI MODEL SP, HP TARGET PISTOLS
Caliber: 22 LR, 32 S&W, 5-shot magazine.
Barrel: 4.7".
Weight: 38.9 oz. **Length:** 11.6" overall.
Stocks: Adjustable; stippled walnut; match type.
Sights: Interchangeable blade front, interchangeable, fully adjustable rear.
Features: Fully adjustable match trigger. Introduced 1995. Imported from Italy by Nygord Precision Products.
Price: Model SP (22 LR) $995.00
Price: Model HP (32 S&W) $1,095.00

PARDINI GP RAPID FIRE MATCH PISTOL
Caliber: 22 Short, 5-shot magazine.
Barrel: 4.6".
Weight: 43.3 oz. **Length:** 11.6" overall.
Stocks: Wrap-around stippled walnut.
Sights: Interchangeable post front, fully adjustable match rear.
Features: Model GP Schuman has extended rear sight for longer sight radius. Introduced 1995. Imported from Italy by Nygord Precision Products.
Price: Model GP $995.00
Price: Model GP Schuman $1,450.00

PARDINI K50 FREE PISTOL
Caliber: 22 LR, single shot.
Barrel: 9.8".
Weight: 34.6 oz. **Length:** 18.7" overall.
Stocks: Wrap-around walnut; adjustable match type.
Sights: Interchangeable post front, fully adjustable match open rear.
Features: Removable, adjustable match trigger. Barrel weights mount above the barrel. Introduced 1995. Imported from Italy by Nygord Precision Products.
Price: $1,050.00

RUGER MARK II TARGET MODEL SEMI-AUTO
Caliber: 22 LR, 10-shot magazine.
Barrel: 6 7/8".
Weight: 42 oz. **Length:** 11 1/8" overall.
Stocks: Checkered hard plastic.
Sights: .125" blade front, micro-click rear, adjustable for windage and elevation. Sight radius 9 3/8".
Features: Introduced 1982.
Price: Blued (MK-678) $310.50
Price: Stainless (KMK-678) $389.00

Ruger Mark II Government Target Model
Same gun as the Mark II Target Model except has 6 7/8" barrel, higher sights and is roll marked "Government Target Model" on the right side of the receiver below the rear sight. Identical in all aspects to the military model used for training U.S. armed forces except for markings. Comes with factory test target. Introduced 1987.
Price: Blued (MK-678G) $356.50
Price: Stainless (KMK-678G) $427.25

Ruger Mark II Target

Ruger Government Target

Ruger Mark II Bull Barrel
Same gun as the Target Model except has 5 1/2" or 10" heavy barrel (10" meets all IHMSA regulations). Weight with 5 1/2" barrel is 42 oz., with 10" barrel, 51 oz.
Price: Blued (MK-512) $310.50
Price: Blued (MK-10) $294.50
Price: Stainless (KMK-10) $373.00
Price: Stainless (KMK-512) $389.00

Ruger Stainless Government Competition Model 22 Pistol
Similar to the Mark II Government Target Model stainless pistol except has 6 7/8" slab-sided barrel; the receiver top is drilled and tapped for a Ruger scope base adaptor of blued, chrome moly steel; comes with Ruger 1" stainless scope rings with integral bases for mounting a variety of optical sights; has checkered laminated grip panels with right-hand thumbrest. Has blued open sights with 9 1/4" radius. Overall length is 11 1/8", weight 45 oz. Introduced 1991.
Price: KMK-678GC $441.00

Ruger Mark II Bull Barrel

> Consult our Directory pages for the location of firms mentioned.

SAFARI ARMS BIG DEUCE PISTOL
Caliber: 45 ACP, 7-shot magazine.
Barrel: 6", 416 stainless steel.
Weight: 40.3 oz. **Length:** 9.5" overall.
Stocks: Smooth walnut.
Sights: Ramped blade front, LPA adjustable rear.
Features: Beavertail grip safety; extended thumb safety and slide release; Commander-style hammer. Throated, polished and tuned. Parkerized matte black slide with satin stainless steel frame. Introduced 1995. Made in U.S. by Safari Arms, Inc.
Price: $854.00

Safari Arms Big Deuce

HANDGUNS—COMPETITION HANDGUNS

SMITH & WESSON MODEL 41 TARGET
Caliber: 22 LR, 10-shot clip.
Barrel: 5½", 7".
Weight: 44 oz. (5½" barrel). **Length:** 9" overall (5½" barrel).
Stocks: Checkered walnut with modified thumbrest, usable with either hand.
Sights: ⅛" Patridge on ramp base; micro-click rear adjustable for windage and elevation.
Features: ⅜" wide, grooved trigger; adjustable trigger stop.
Price: S&W Bright Blue, either barrel $768.00

SMITH & WESSON MODEL 22A TARGET PISTOL
Caliber: 22 LR, 10-shot magazine.
Barrel: 5½" bull.
Weight: 38.5 oz. **Length:** 9½" overall.
Stocks: Dymondwood with ambidextrous thumbrests and flared bottom.
Sights: Patridge front, fully adjustable rear.
Features: Sight bridge with Weaver-style integral optics mount; alloy frame, stainless barrel and slide; matte blue finish. Introduced 1997. Made in U.S. by Smith & Wesson.
Price: $300.00

Smith & Wesson Model 22S Target Pistol
Similar to the Model 22A except has stainless steel frame. Introduced 1997. Made in U.S. by Smith & Wesson.
Price: $357.00

SPHINX AT-2000C, CS COMPETITOR PISTOL
Caliber: 9mm Para., 9x21mm, 40 S&W, 10-shot.
Barrel: 5.31".
Weight: 40.56 oz. **Length:** 9.84" overall.
Stocks: Checkered neoprene.
Sights: Fully adjustable Bo-Mar or Tasco Pro-Point dot sight in Sphinx mount.
Features: Extended magazine release. Competition slide with dual-port compensated barrel. Two-tone finish only. Introduced 1993. Imported from Switzerland by Sphinx U.S.A., Inc.
Price: With Bo-Mar sights $1,902.00
Price: With Tasco Pro-Point and mount (AT-2000CS) $2,189.00

Sphinx AT-2000GM Grand Master Pistol
Similar to the AT-2000C except has single-action-only trigger mechanism, squared trigger guard, extended beavertail grip, safety and magazine release; notched competition slide for easier cocking. Two-tone finish only. Has dual-port compensated barrel. Available with fully adjustable Bo-Mar sights or Tasco Pro-Point and Sphinx mount. Introduced 1993. Imported from Switzerland by Sphinx U.S.A., Inc.
Price: With Bo-Mar sights (AT-2000GMS) $2,894.00
Price: With Tasco Pro-Point and mount (AT-2000GM) $2,972.00

Smith & Wesson Model 41

Sphinx AT-2000C Competitor

SPRINGFIELD, INC. 1911A1 BULLSEYE WADCUTTER PISTOL
Caliber: 45 ACP.
Barrel: 5".
Weight: 45 oz. **Length:** 8.59" overall (5" barrel).
Stocks: Checkered walnut.
Sights: Bo-Mar rib with undercut blade front, fully adjustable rear.
Features: Built for wadcutter loads only. Has full-length recoil spring guide rod, fitted Videki speed trigger with 3.5-lb. pull; match Commander hammer and sear; beavertail grip safety; lowered and flared ejection port; tuned extractor; fitted slide to frame; recoil buffer system; beveled and polished magazine well; checkered front strap and steel mainspring housing (flat housing standard); polished and throated National Match barrel and bushing. Comes with two magazines with slam pads, plastic carrying case, test target. Introduced 1992. From Springfield, Inc.
Price: $1,499.00

Springfield, Inc. Expert Pistol
Similar to the Competition Pistol except has triple-chamber tapered cone compensator on match barrel with dovetailed front sight; lowered and flared ejection port; fully tuned for reliability; fitted slide to frame; extended ambidextrous thumb safety, extended magazine release button; beavertail grip safety; Pachmayr wrap-around grips. Comes with two magazines, plastic carrying case. Introduced 1992. From Springfield, Inc.
Price: 45 ACP, Duotone finish $1,724.00
Price: Expert Ltd. $1,624.00

Springfield 1911A1 Trophy Match

> Consult our Directory pages for the location of firms mentioned.

Springfield, Inc. 1911A1 Trophy Match Pistol
Similar to the 1911A1 except factory accurized, Videki speed trigger, skeletonized hammer; has 4- to 5½-lb. trigger pull, click adjustable rear sight, match-grade barrel and bushing. Comes with checkered walnut grips. Introduced 1994. From Springfield, Inc.
Price: Blue $989.00
Price: Stainless steel $1,029.00

Springfield, Inc. Basic Competition Pistol
Has low-mounted Bo-Mar adjustable rear sight, undercut blade front; match throated barrel and bushing; polished feed ramp; lowered and flared ejection port; fitted Videki speed trigger with tuned 3.5-lb. pull; fitted slide to frame; recoil buffer system; checkered walnut grips; serrated, arched mainspring housing. Comes with two magazines with slam pads, plastic carrying case. Introduced 1992. From Springfield, Inc.
Price: 45 ACP, blue, 5" only $1,295.00

HANDGUNS—COMPETITION HANDGUNS

Springfield, Inc. 1911A1 N.M. Hardball Pistol
Has Bo-Mar adjustable rear sight with undercut front blade; fitted match Videki trigger with 4-lb. pull; fitted slide to frame; throated National Match barrel and bushing, polished feed ramp; recoil buffer system; tuned extractor; Herrett walnut grips. Comes with two magazines, plastic carrying case, test target. Introduced 1992. From Springfield, Inc.
Price: 45 ACP, blue $1,336.00

STI EAGLE 5.1 PISTOL
Caliber: 9mm Para., 38 Super, 40 S&W, 45 ACP, 10-shot magazine.
Barrel: 5", bull.
Weight: 34 oz. **Length:** 8.62" overall.
Stocks: Checkered polymer.
Sights: Bo-Mar blade front, Bo-Mar fully adjustable rear.
Features: Modular frame design; adjustable match trigger; skeletonized hammer; extended grip safety with locator pad; match-grade fit of all parts. Many options available. Introduced 1994. Made in U.S. by STI International.
Price: $1,792.00

STOEGER PRO SERIES 95 VENT RIB
Caliber: 22 LR, 10-shot magazine.
Barrel: 5 1/2".
Weight: 48 oz. **Length:** 9 5/8" overall.
Stocks: Pachmayr wrap-around checkered rubber.
Sights: Blade front, fully adjustable micro-click rear mounted on rib.
Features: Stainless steel construction; full-length ventilated rib; gold-plated trigger, slide lock, safety-lever and magazine release; adjustable trigger; interchangeable barrels. Introduced 1996. From Stoeger Ind.
Price: $595.00

Stoeger Pro Series 95 Bull Barrel
Similar to the Vent Rib model except has 5 1/2" bull barrel, rear sight mounted on slide bridge. Introduced 1996. From Stoeger Ind.
Price: $495.00

Unique D.E.S. 69U

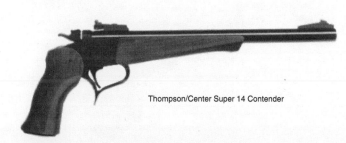
Thompson/Center Super 14 Contender

THOMPSON/CENTER SUPER 14 CONTENDER
Caliber: 22 LR, 222 Rem., 223 Rem., 7mm TCU, 7-30 Waters, 30-30 Win., 35 Rem., 357 Rem. Maximum, 44 Mag., 10mm Auto, 445 Super Mag., single shot.
Barrel: 14".
Weight: 45 oz. **Length:** 17 1/4" overall.
Stocks: T/C "Competitor Grip" (walnut and rubber).
Sights: Fully adjustable target-type.
Features: Break-open action with auto safety. Interchangeable barrels for both rimfire and centerfire calibers. Introduced 1978.
Price: Blued $473.80
Price: Stainless steel $504.70
Price: Extra barrels, blued $224.00
Price: Extra barrels, stainless steel $239.50

Springfield, Inc. Distinguished Pistol
Has all the features of the 1911A1 Expert except is full-house pistol with deluxe Bo-Mar low-mounted adjustable rear sight; full-length recoil spring guide rod and recoil spring retainer; checkered frontstrap; S&A magazine well; walnut grips. Hard chrome finish. Comes with two magazines with slam pads, plastic carrying case. From Springfield, Inc.
Price: 45 ACP $2,445.00
Price: Distinguished Limited $2,345.00

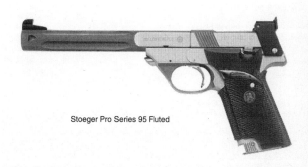
Stoeger Pro Series 95 Fluted

Stoeger Pro Series 95 Fluted Barrel
Similar to the Vent Rib model except has 7 1/2" heavy fluted barrel, rear sight mounted on slide bridge. Overall length 11 1/4", weighs 50 oz. Introduced 1996. From Stoeger Ind.
Price: $525.00

UNIQUE D.E.S. 32U TARGET PISTOL
Caliber: 32 S&W Long wadcutter.
Barrel: 5.9".
Weight: 40.2 oz.
Stocks: Anatomically shaped, adjustable stippled French walnut.
Sights: Blade front, micrometer click rear.
Features: Trigger adjustable for weight and position; dry firing mechanism; slide stop catch. Optional sleeve weights. Introduced 1990. Imported from France by Nygord Precision Products.
Price: Right-hand, about $1,350.00
Price: Left-hand, about $1,380.00

UNIQUE D.E.S. 69U TARGET PISTOL
Caliber: 22 LR, 5-shot magazine.
Barrel: 5.91".
Weight: 35.3 oz. **Length:** 10.5" overall.
Stocks: French walnut target-style with thumbrest and adjustable shelf; hand-checkered panels.
Sights: Ramp front, micro. adjustable rear mounted on frame; 8.66" sight radius.
Features: Meets U.I.T. standards. Comes with 260-gram barrel weight; 100, 150, 350-gram weights available. Fully adjustable match trigger; dry-firing safety device. Imported from France by Nygord Precision Products.
Price: Right-hand, about $1,250.00
Price: Left-hand, about $1,290.00

> Consult our Directory pages for the location of firms mentioned.

UNIQUE MODEL 96U TARGET PISTOL
Caliber: 22 LR, 5- or 6-shot magazine.
Barrel: 5.9".
Weight: 40.2 oz. **Length:** 11.2" overall.
Stocks: French walnut. Target style with thumbrest and adjustable shelf.
Sights: Blade front, micrometer rear mounted on frame.
Features: Designed for Sport Pistol and Standard U.I.T. shooting. External hammer; fully adjustable and movable trigger; dry-firing device. Introduced 1997. Imported from France by Nygord Precision Products.
Price: $1,350.00

HANDGUNS—COMPETITION HANDGUNS

Thompson/Center Super 16 Contender
Same as the T/C Super 14 Contender except has 16¼" barrel. Rear sight can be mounted at mid-barrel position (10¾" radius) or moved to the rear (using scope mount position) for 14¾" radius. Overall length is 20¼". Comes with T/C Competitor Grip of walnut and rubber. Available in 22 LR, 22 WMR, 223 Rem., 7-30 Waters, 30-30 Win., 35 Rem., 44 Mag., 45-70 Gov't. Also available with 16" vent rib barrel with internal choke, caliber 45 Colt/410 shotshell.
Price: Blue $478.90
Price: Stainless steel $509.90
Price: 45-70 Gov't., blue $484.10
Price: As above, stainless steel $530.50
Price: Super 16 Vent Rib, blued $509.90
Price: As above, stainless steel $540.80
Price: Extra 16" barrel, blued $229.20
Price: As above, stainless steel $244.50
Price: Extra 45-70 barrel, blued $234.30
Price: As above, stainless steel $265.50
Price: Extra Super 16 vent rib barrel, blue $260.10
Price: As above, stainless steel $265.50

WALTHER GSP MATCH PISTOL
Caliber: 22 LR, 32 S&W Long (GSP-C), 5-shot magazine.
Barrel: 4.22".
Weight: 44.8 oz. (22 LR), 49.4 oz. (32). **Length:** 11.8" overall.
Stocks: Walnut.
Sights: Post front, match rear adjustable for windage and elevation.
Features: Available with either 2.2-lb. (1000 gm) or 3-lb. (1360 gm) trigger. Spare magazine, barrel weight, tools supplied. Imported from Germany by Nygord Precision Products.
Price: GSP, with case $1,495.00
Price: GSP-C, with case $1,595.00

WICHITA CLASSIC SILHOUETTE PISTOL
Caliber: All standard calibers with maximum overall length of 2.800".
Barrel: 11¼".
Weight: 3 lbs., 15 oz.
Stocks: AAA American walnut with oil finish, checkered grip.
Sights: Hooded post front, open adjustable rear.
Features: Three locking lug bolt, three gas ports; completely adjustable Wichita trigger. Introduced 1981. From Wichita Arms.
Price: $3,450.00

Manufacturers' addresses in the
Directory of the Arms Trade
page 309, this issue

WICHITA SILHOUETTE PISTOL
Caliber: 308 Win. F.L., 7mm IHMSA, 7mm-308.
Barrel: 14¹⁵⁄₁₆".
Weight: 4½ lbs. **Length:** 21³⁄₈" overall.
Stock: American walnut with oil finish. Glass bedded.
Sights: Wichita Multi-Range sight system.
Features: Comes with left-hand action with right-hand grip. Round receiver and barrel. Fluted bolt, flat bolt handle. Wichita adjustable trigger. Introduced 1979. From Wichita Arms.
Price: Center grip stock $1,650.00
Price: As above except with Rear Position Stock and target-type Lightpull trigger $1,640.00

HANDGUNS—DOUBLE-ACTION REVOLVERS, SERVICE & SPORT

Includes models suitable for hunting and competitive courses for fire, both police and international.

Armscor M-200DC

Colt DS-II

ARMSCOR M-200DC REVOLVER
Caliber: 38 Spec., 6-shot cylinder.
Barrel: 2½", 4".
Weight: 22 oz. (2½" barrel). **Length:** 7³⁄₈" overall (2½" barrel).
Stocks: Checkered rubber.
Sights: Blade front, fixed notch rear.
Features: All-steel construction; floating firing pin, transfer bar ignition; shrouded ejector rod; blue finish. Reintroduced 1996. Imported from the Philippines by K.B.I., Inc.
Price: $199.99

COLT DS-II REVOLVER
Caliber: 38 Spec., 6-shot.
Barrel: 2".
Weight: 21 oz. **Length:** NA
Stocks: Combat-style rubber.
Sights: Ramp front, fixed notch rear.
Features: Stainless steel construction. Smooth combat trigger. Introduced 1997. Made in U.S. by Colt's Mfg. Co.
Price: $408.00

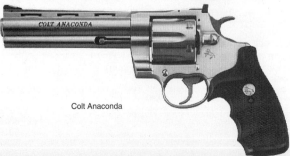
Colt Anaconda

COLT ANACONDA REVOLVER
Caliber: 44 Rem. Magnum, 45 Colt, 6-shot.
Barrel: 4", 6", 8".
Weight: 53 oz. (6" barrel). **Length:** 11⁵⁄₈" overall.
Stocks: TP combat style with finger grooves.
Sights: Red insert front, adjustable white outline rear.
Features: Stainless steel; full-length ejector rod housing; ventilated barrel rib; off-set bolt notches in cylinder; wide spur hammer. Introduced 1990.
Price: $612.00
Price: 45 Colt, 6", 8" barrel only $612.00
Price: With complete Realtree camouflage coverage $740.00
Price: As above with scope and mount $999.00

HANDGUNS—DOUBLE-ACTION REVOLVERS, SERVICE & SPORT

COLT KING COBRA REVOLVER
Caliber: 357 Magnum, 6-shot.
Barrel: 4", 6".
Weight: 42 oz. (4" bbl.). **Length:** 9" overall (4" bbl.).
Stocks: TP combat style.
Sights: Red insert ramp front, adjustable white outline rear.
Features: Full-length contoured ejector rod housing, barrel rib. Introduced 1986.
Price: Stainless $455.00

E.A.A. STANDARD GRADE REVOLVERS
Caliber: 22 LR, 22 LR/22 WMR, 8-shot; 38 Spec., 6-shot; 357 magnum, 6-shot.
Barrel: 4", 6" (22 rimfire); 2", 4" (38 Spec.).
Weight: 38 oz. (22 rimfire, 4"). **Length:** 8.8" overall (4" bbl.).
Stocks: Rubber with finger grooves.
Sights: Blade front, fixed or adjustable on rimfires; fixed only on 32, 38.
Features: Swing-out cylinder; hammer block safety; blue finish. Introduced 1991. Imported from Germany by European American Armory.
Price: 38 Special 2" $180.00
Price: 38 Special, 4" $199.00
Price: 357 Magnum $199.00
Price: 22 LR, 6" $199.00
Price: 22 LR/22 WMR combo, 4" $200.00
Price: As above, 6" $200.00

ERMA ER-777 SPORTING REVOLVER
Caliber: 357 Mag., 6-shot.
Barrel: 5 1/2".
Weight: 43.3 oz. **Length:** 9 1/2" overall (4" barrel).
Stocks: Stippled walnut service-type.
Sights: Interchangeable blade front, micro-adjustable rear for windage and elevation.
Features: Polished blue finish. Adjustable trigger. Imported from Germany by Precision Sales Int'l. Introduced 1988.
Price: $1,019.00

HARRINGTON & RICHARDSON 929 SIDEKICK
Caliber: 22 LR, 9-shot cylinder.
Barrel: 4" heavy.
Weight: 30 oz. **Length:** NA.
Stocks: Cinnamon-color laminated wood.
Sights: Blade front, notch rear.
Features: Double action; swing-out cylinder, traditional loading gate; blued frame and barrel. Comes with lockable storage case, Uncle Mike's Sidekick holster. Introduced 1996. Made in U.S. by H&R 1871, Inc.
Price: $159.95
Price: NTA Trapper Edition, special rollmark, gray laminate grips $174.95

HARRINGTON & RICHARDSON 939 PREMIER REVOLVER
Caliber: 22 LR, 9-shot cylinder.
Barrel: 6" heavy.
Weight: 36 oz. **Length:** NA.
Stocks: Walnut-finished hardwood.
Sights: Blade front, fully adjustable rear.
Features: Swing-out cylinder with plunger-type ejection; solid barrel rib; high-polish blue finish; double-action mechanism; Western-style grip. Introduced 1995. Made in U.S. by H&R 1871, Inc.
Price: $189.95

HARRINGTON & RICHARDSON 949 WESTERN REVOLVER
Caliber: 22 LR, 9-shot cylinder.
Barrel: 5 1/2", 7 1/2".
Weight: 36 oz. **Length:** NA.
Stocks: Walnut-stained hardwood.
Sights: Blade front, adjustable rear.
Features: Color case-hardened frame and backstrap, traditional loading gate and ejector rod. Introduced 1994. Made in U.S. by H&R 1871, Inc.
Price: About $189.95

HARRINGTON & RICHARDSON SPORTSMAN 999 REVOLVER
Caliber: 22 Short, Long, Long Rifle, 9-shot.
Barrel: 4", 6".
Weight: 30 oz. (4" barrel). **Length:** 8.5" overall.
Stocks: Walnut-finished hardwood.
Sights: Blade front adjustable for elevation, rear adjustable for windage.
Features: Top-break loading; polished blue finish; automatic shell ejection. Reintroduced 1992. From H&R 1871, Inc.
Price: $279.95

COLT PYTHON REVOLVER
Caliber: 357 Magnum (handles all 38 Spec.), 6-shot.
Barrel: 4", 6" or 8", with ventilated rib.
Weight: 38 oz. (4" bbl.). **Length:** 9 1/4" (4" bbl.).
Stocks: Hogue Monogrip (4"), TP combat style (6", 8").
Sights: 1/8" ramp front, adjustable notch rear.
Features: Ventilated rib; grooved, crisp trigger; swing-out cylinder; target hammer.
Price: Royal blue, 4", 6", 8" $815.00
Price: Stainless, 4", 6", 8" $904.00
Price: Bright stainless, 4", 6", 8" $935.00

E.A.A. Standard Grade

Harrington & Richardson 939

Harrington & Richardson Sportsman 999

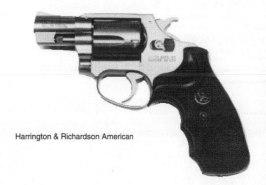

Harrington & Richardson American

HARRINGTON & RICHARDSON AMERICAN REVOLVERS
Caliber: 38 Spec., 5-shot.
Barrel: 2", 3".
Weight: 24 oz. **Length:** 7 1/8" overall.
Stocks: Pachmayr rubber.
Sights: Ramp front, fixed notch rear.
Features: Available in blue or nickel. Introduced 1996. Made in U.S. by Amtec 2000.
Price: .... NA

HANDGUNS—DOUBLE-ACTION REVOLVERS, SERVICE & SPORT

Heritage Sentry

MANURHIN MR 73 REVOLVER
Caliber: 32 S&W, 38 Spec., 357 Mag.
Barrel: 3", 4", 5 1/4", 5 3/4", 6".
Weight: 38 oz. (6" barrel). **Length:** 11" overall (6" barrel).
Stocks: Checkered hardwood.
Sights: Blade front, fully adjustable rear.
Features: Polished bright blue finish; hammer-forged barrel. Imported from France by Sphinx U.S.A., Inc.
Price: Police model, 3" or 4" barrel $1,885.00
Price: Sport model, 5 1/4" or 6" barrel, undercut blade front sight $1,885.00
Price: 38 Spec. Match, 5 3/4" barrel, single action only $1,975.00
Price: 32 S&W Match, 6" barrel, single action only $1,975.00

MANURHIN MR 88 REVOLVER
Caliber: 357 Magnum, 6-shot.
Barrel: 4", 5 1/4", 6".
Weight: 33.5 oz. **Length:** 8.1" overall.
Stocks: Checkered wood.
Sights: Blade front, fully adjustable rear.
Features: Stainless steel construction; hammer-forged barrel. Imported from France by Sphinx U.S.A., Inc.
Price: $877.95

NEW ENGLAND FIREARMS LADY ULTRA REVOLVER
Caliber: 32 H&R Mag., 5-shot.
Barrel: 3".
Weight: 31 oz. **Length:** 7.25" overall.
Stocks: Walnut-finished hardwood with NEF medallion.
Sights: Blade front, fully adjustable rear.
Features: Swing-out cylinder; polished blue finish. Comes with lockable storage case. Introduced 1992. From New England Firearms.
Price: $174.95

NEW ENGLAND FIREARMS ULTRA REVOLVER
Caliber: 22 LR, 9-shot; 22 WMR, 6-shot.
Barrel: 4", 6".
Weight: 36 oz. **Length:** 10 5/8" overall (6" barrel).
Stocks: Walnut-finished hardwood with NEF medallion.
Sights: Blade front, fully adjustable rear.
Features: Blue finish. Bull-style barrel with recessed muzzle, high "Lustre" blue/black finish. Introduced 1989. From New England Firearms.
Price: $174.95
Price: Ultra Mag 22 WMR $174.95

NEW ENGLAND FIREARMS STANDARD REVOLVERS
Caliber: 22 LR, 9-shot; 32 H&R Mag., 5-shot.
Barrel: 2 1/2", 4".
Weight: 26 oz. (22 LR, 2 1/2"). **Length:** 8 1/2" overall (4" bbl.).
Stocks: Walnut-finished American hardwood with NEF medallion.
Sights: Fixed.
Features: Choice of blue or nickel finish. Introduced 1988. From New England Firearms.
Price: 22 LR, 32 H&R Mag., blue $134.95
Price: 22 LR, 2 1/2", 4", nickel, 32 H&R Mag. 2 1/2" nickel $144.95

REXIO PUCARA M224 REVOLVER
Caliber: 22 LR, 9-shot cylinder.
Barrel: 4".
Weight: 33 oz. **Length:** 9" overall.
Stocks: Checkered hardwood.
Sights: Blade front, square notch rear adjustable for windage.
Features: Alloy frame; hammer block safety; polished blue finish. Introduced 1997. Imported from Argentina by Century International Arms.
Price: $169.00
Price: M226, 6" barrel $169.00

HERITAGE SENTRY DOUBLE-ACTION REVOLVERS
Caliber: 38 Spec., 6-shot.
Barrel: 2".
Weight: 23 oz. **Length:** 6 1/4" overall (2" barrel).
Stocks: Checkered plastic.
Sights: Ramp front, fixed rear.
Features: Pull-pin-type ejection; serrated hammer and trigger. Polished blue or nickel finish. Introduced 1993. Made in U.S. by Heritage Mfg., Inc.
Price: .... $129.95 to $139.95

MANURHIN MR 73 SPORT REVOLVER
Caliber: 357 Magnum, 6-shot cylinder.
Barrel: 6".
Weight: 37 oz. **Length:** 11.1" overall.
Stocks: Checkered walnut.
Sights: Blade front, fully adjustable rear.
Features: Double action with adjustable trigger. High-polish blue finish, straw-colored hammer and trigger. Comes with extra sight. Introduced 1984. Imported from France by Century International Arms.
Price: About $1,500.00

MANURHIN MR 96 REVOLVER
Caliber: 357 Magnum, 6-shot.
Barrel: 3", 4", 5 1/4", 6".
Weight: 38.4 oz. **Length:** 8.8" overall.
Stocks: Checkered rubber.
Sights: Blade front, fully adjustable rear.
Features: Polished blue finish; removable sideplate holds action parts; separate barrel and shroud. Introduced 1996. Imported from France by Sphinx U.S.A., Inc.
Price: $857.95

MEDUSA MODEL 47 REVOLVER
Caliber: Most 9mm, 38 and 357 caliber cartridges; 6-shot cylinder.
Barrel: 2 1/2", 3", 4", 5", 6"; fluted.
Weight: 39 oz. **Length:** 10" overall (4" barrel).
Stocks: Gripper-style rubber.
Sights: Changeable front blades, fully adjustable rear.
Features: Patented extractor allows gun to chamber, fire and extract over 25 different cartridges in the .355- to .357 range, without half-moon clips. Steel frame and cylinder; match quality barrel. Matte blue finish. Introduced 1996. Made in U.S. by Phillips & Rogers, Inc.
Price: $899.00

New England Lady Ultra

REXIO PUCARA M324 REVOLVER
Caliber: 32 S&W Long, 7-shot cylinder.
Barrel: 4".
Weight: 31 oz. **Length:** 9" overall.
Stocks: Checkered hardwood.
Sights: Blade front, square notch rear adjustable for windage.
Features: Alloy frame; polished blue finish; hammer block safety. Introduced 1997. Imported from Argentina by Century International Arms.
Price: $169.00
Price: M326, 6" barrel $169.00

REXIO PUCARA M384 REVOLVER
Caliber: 38 Spec., 6-shot.
Barrel: 4".
Weight: 30 oz. **Length:** 9" overall.
Stocks: Checkered hardwood.
Sights: Blade front, rear adjustable for windage.
Features: Alloy frame. Polished blue finish. Imported from Argentina by Century International Arms.
Price: $169.00
Price: M386, 6" barrel $169.00

HANDGUNS—DOUBLE-ACTION REVOLVERS, SERVICE & SPORT

ROSSI CYCLOPS REVOLVER
Caliber: 357 Magnum, 6-shot.
Barrel: 6", 8".
Weight: 44 oz. (6" barrel). **Length:** 11 3/4" overall (6" barrel).
Stocks: Checkered rubber.
Sights: Blade front, fully adjustable rear.
Features: Extra-heavy barrel with four recessed compensator ports on each side of the muzzle. Stainless steel construction. Comes with scope mount and rings. Polished finish. Introduced 1997. Imported from Brazil by Interarms.
Price: $429.00

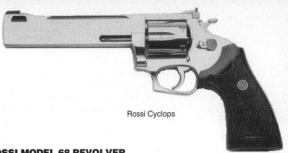
Rossi Cyclops

ROSSI LADY ROSSI REVOLVER
Caliber: 38 Spec., 5-shot.
Barrel: 2", 3".
Weight: 21 oz. **Length:** 6.5" overall (2" barrel).
Stocks: Smooth rosewood.
Sights: Fixed.
Features: High-polish stainless steel with "Lady Rossi" engraved on frame. Comes with velvet carry bag. Introduced 1995. Imported from Brazil by Interarms.
Price: $285.00

ROSSI MODEL 68 REVOLVER
Caliber: 38 Spec.
Barrel: 2", 3".
Weight: 22 oz.
Stocks: Checkered wood and rubber.
Sights: Ramp front, low profile adjustable rear.
Features: All-steel frame, thumb latch operated swing-out cylinder. Introduced 1978. Imported from Brazil by Interarms.
Price: 38, blue, 3", wood or rubber grips $225.00
Price: M68/2 (2" barrel), wood or rubber grips $225.00
Price: 3", nickel $225.00

ROSSI MODEL 88 STAINLESS REVOLVER
Caliber: 38 Spec., 5-shot.
Barrel: 2", 3".
Weight: 22 oz. **Length:** 7.5" overall.
Stocks: Checkered wood, service-style, and rubber.
Sights: Ramp front, square notch rear drift adjustable for windage.
Features: All metal parts except springs are of 440 stainless steel; matte finish; small frame for concealability. Introduced 1983. Imported from Brazil by Interarms.
Price: 3" barrel, wood or rubber grips $255.00
Price: 2" barrel, wood or rubber grips $255.00

Rossi Model 88

Rossi Model 518

ROSSI MODEL 515, 518 REVOLVERS
Caliber: 22 LR (Model 518), 22 WMR (Model 515), 6-shot.
Barrel: 4".
Weight: 30 oz. **Length:** 9" overall.
Stocks: Checkered wood and finger-groove wrap-around rubber.
Sights: Blade front with red insert, rear adjustable for windage and elevation.
Features: Small frame; stainless steel construction; solid integral barrel rib. Introduced 1994. Imported from Brazil by Interarms.
Price: Model 518, 22 LR $255.00
Price: Model 515, 22 WMR $270.00

ROSSI MODEL 720 REVOLVER
Caliber: 44 Spec., 5-shot.
Barrel: 3".
Weight: 27.5 oz. **Length:** 8" overall.
Stocks: Checkered rubber, combat style.
Sights: Red insert front on ramp, fully adjustable rear.
Features: All stainless steel construction; solid barrel rib; full ejector rod shroud. Introduced 1992. Imported from Brazil by Interarms.
Price: $290.00
Price: Model 720C, spurless hammer, DA only $290.00

ROSSI MODEL 851 REVOLVER
Caliber: 38 Spec., 6-shot.
Barrel: 3" or 4".
Weight: 27.5 oz. (3" bbl.). **Length:** 8" overall (3" bbl.).
Stocks: Checkered Brazilian hardwood.
Sights: Blade front with red insert, rear adjustable for windage.
Features: Medium-size frame; stainless steel construction; ventilated barrel rib. Introduced 1991. Imported from Brazil by Interarms.
Price: $255.00

ROSSI MODEL 877 REVOLVER
Caliber: 357 Mag., 6-shot cylinder.
Barrel: 2".
Weight: 26 oz. **Length:** NA.
Stocks: Stippled synthetic.
Sights: Blade front, fixed groove rear.
Features: Stainless steel construction; fully enclosed ejector rod. Introduced 1996. Imported from Brazil by Interarms.
Price: $290.00
Price: Model 677 (same as Model 877 except in matte blue steel) $260.00

ROSSI MODEL 971 REVOLVER
Caliber: 357 Mag., 6-shot.
Barrel: 2 1/2", 4", 6", heavy.
Weight: 36 oz. **Length:** 9" overall.
Stocks: Checkered Brazilian hardwood. Stainless models have checkered, contoured rubber.
Sights: Blade front, fully adjustable rear.
Features: Full-length ejector rod shroud; matted sight rib; target-type trigger, wide checkered hammer spur. Introduced 1988. Imported from Brazil by Interarms.
Price: 4", stainless $290.00
Price: 6", stainless $290.00
Price: 4", blue $255.00
Price: 2 1/2", stainless $290.00

Rossi Model 971 Comp Gun
Same as the Model 971 stainless except has 3 1/4" barrel with integral compensator. Overall length is 9", weighs 32 oz. Has red insert front sight, fully adjustable rear. Checkered, contoured rubber grips. Introduced 1993. Imported from Brazil by Interarms.
Price: $290.00

Rossi Model 971 VRC Revolver
Similar to the Model 971 except has Rossi's 8-port Vented Rib Compensator; checkered finger-groove rubber grips; stainless steel construction. Available with 2.5", 4", 6" barrel; weighs 30 oz. with 2 5/8" barrel. Introduced 1996. Imported from Brazil by Interarms.
Price: $340.00

HANDGUNS—DOUBLE-ACTION REVOLVERS, SERVICE & SPORT

RUGER GP-100 REVOLVERS
Caliber: 38 Spec., 357 Mag., 6-shot.
Barrel: 3", 3" heavy, 4", 4" heavy, 6", 6" heavy.
Weight: 3" barrel—35 oz., 3" heavy barrel—36 oz., 4" barrel—37 oz., 4" heavy barrel—38 oz.
Sights: Fixed; adjustable on 4" heavy, 6", 6" heavy barrels.
Stocks: Ruger Santoprene Cushioned Grip with Goncalo Alves inserts.
Features: Uses action and frame incorporating improvements and features of both the Security-Six and Redhawk revolvers. Full length and short ejector shroud. Satin blue and stainless steel. Available in high-gloss stainless steel finish. Introduced 1988.
Price: GP-141 (357, 4" heavy, adj. sights, blue) $440.00
Price: GP-160 (357, 6", adj. sights, blue) $440.00
Price: GP-161 (357, 6" heavy, adj. sights, blue) $440.00
Price: GPF-331 (357, 3" heavy), GPF-831 (38 Spec.) $423.00
Price: GPF-340 (357, 4"), GPF-840 (38 Spec.) $423.00
Price: GPF-341 (357, 4" heavy), GPF-841 (38 Spec.) $423.00
Price: KGP-141 (357, 4" heavy, adj. sights, stainless) $474.00
Price: KGP-160 (357, 6", adj. sights, stainless) $474.00
Price: KGP-161 (357, 6" heavy, adj. sights, stainless) $474.00
Price: KGPF-330 (357, 3", stainless), KGPF-830 (38 Spec.) $457.00
Price: KGPF-331 (357, 3" heavy, stainless), KGPF-831 (38 Spec.) $457.00
Price: KGPF-340 (357, 4", stainless), KGPF-840 (38 Spec.) $457.00
Price: KGPF-341 (357, 4" heavy, stainless), KGPF-841 (38 Spec.) . . . $457.00

Ruger GP-100

Ruger SP101 DAO

RUGER SP101 REVOLVERS
Caliber: 22 LR, 32 H&R Mag., 6-shot, 9mm Para., 38 Spec. +P, 357 Mag., 5-shot.
Barrel: 2 1/4", 3 1/16", 4".
Weight: 2 1/4"—25 oz.; 3 1/16"—27 oz.
Sights: Adjustable on 22, 32, fixed on others.
Stocks: Ruger Santoprene Cushioned Grip with Xenoy inserts.
Features: Incorporates improvements and features found in the GP-100 revolvers into a compact, small frame, double-action revolver. Full-length ejector shroud. Stainless steel only. Available with high-polish finish. Introduced 1988.
Price: KSP-821 (2 1/2", 38 Spec.) $443.00
Price: KSP-831 (3 1/16", 38 Spec.) $443.00
Price: KSP-221 (2 1/4", 22 LR) $443.00
Price: KSP-240 (4", 22 LR) $443.00
Price: KSP-241 (4" heavy bbl., 22 LR) $443.00
Price: KSP-3231 (3 1/16", 32 H&R) $443.00
Price: KSP-921 (2 1/4", 9mm Para.) $443.00
Price: KSP-931 (3 1/16", 9mm Para.) $443.00
Price: KSP-321 (2 1/4", 357 Mag.) $443.00
Price: KSP-331 (3 1/16", 357 Mag.) $443.00
Price: GKSP321X (2 1/4", 357 Mag.), high-gloss stainless $443.00
Price: GKSP331X (3 1/16", 357 Mag.), high-gloss stainless $443.00
Price: GKSP321XL (2 1/4", 357 MAg., spurless hammer, DAO) high-gloss stainless $443.00

Ruger SP101 Double-Action-Only Revolver
Similar to the standard SP101 except is double-action-only with no single-action sear notch. Has spurless hammer for snag-free handling, floating firing pin and Ruger's patented transfer bar safety system. Available with 2 1/4" barrel in 38 Special +P and 357 Magnum only. Weighs 25 1/2 oz., overall length 7.06". Natural brushed satin or high-polish stainless steel. Introduced 1993.
Price: KSP821L (38 Spec.), KSP321XL (357 Mag.) $443.00

RUGER REDHAWK
Caliber: 44 Rem. Mag., 6-shot.
Barrel: 5 1/2", 7 1/2".
Weight: About 54 oz. (7 1/2" bbl.). **Length:** 13" overall (7 1/2" barrel).
Stocks: Square butt Goncalo Alves.
Sights: Interchangeable Patridge-type front, rear adjustable for windage and elevation.
Features: Stainless steel, brushed satin finish, or blued ordnance steel. Has a 9 1/2" sight radius. Introduced 1979.
Price: Blued, 44 Mag., 5 1/2", 7 1/2" $490.00
Price: Blued, 44 Mag., 7 1/2", with scope mount, rings $527.00
Price: Stainless, 44 Mag., 5 1/2", 7 1/2" $547.00
Price: Stainless, 44 Mag., 7 1/2", with scope mount, rings $589.00

Ruger Super Redhawk Revolver
Similar to the standard Redhawk except has a heavy extended frame with the Ruger Integral Scope Mounting System on the wide topstrap. The wide hammer spur has been lowered for better scope clearance. Incorporates the mechanical design features and improvements of the GP-100. Choice of 7 1/2" or 9 1/2" barrel, both with ramp front sight base with Redhawk-style Interchangeable Insert sight blades, adjustable rear sight. Comes with Ruger "Cushioned Grip" panels of Santoprene with Goncalo Alves wood panels. Satin or high-polished stainless steel. Introduced 1987.
Price: KSRH-7 (7 1/2") $574.00
Price: KSRH-9 (9 1/2") $589.00

Ruger Redhawk

SMITH & WESSON MODEL 13 H.B. M&P
Caliber: 357 Mag. and 38 Spec., 6-shot.
Barrel: 4".
Weight: 34 oz. **Length:** 9 5/16" overall (4" bbl.).
Stocks: Uncle Mike's Combat soft rubber; wood optional.
Sights: 1/8" serrated ramp front, fixed square notch rear.
Features: Heavy barrel, K-frame, square butt (4"), round butt (3").
Price: Blue $400.00

Smith & Wesson Model 65 Revolver
Smiliar to the Model 13 except made of stainless steel. Has Uncle Mike's Combat grips, smooth combat trigger, fixed notch rear sight. Made in U.S. by Smith & Wesson.
Price: 3" or 4" $435.00

SMITH & WESSON MODEL 10 M&P HB REVOLVER
Caliber: 38 Spec., 6-shot.
Barrel: 4".
Weight: 33.5 oz. **Length:** 9 5/16" overall.
Stocks: Uncle Mike's Combat soft rubber; square butt. Wood optional.
Sights: Fixed; ramp front, square notch rear.
Price: Blue $397.00

HANDGUNS—DOUBLE-ACTION REVOLVERS, SERVICE & SPORT

Smith & Wesson Model 14

Smith & Wesson Model 19

Smith & Wesson Model 629 PowerPort

Smith & Wesson Model 37

SMITH & WESSON MODEL 14 FULL LUG REVOLVER
Caliber: 38 Spec., 6-shot.
Barrel: 6", full lug.
Weight: 47 oz. **Length:** 11 1/8" overall.
Stocks: Hogue soft rubber; wood optional.
Sights: Pinned Patridge front, adjustable micrometer click rear.
Features: Has .500" target hammer, .312" smooth combat trigger. Polished blue finish. Reintroduced 1991. Limited production.
Price: $473.00

SMITH & WESSON MODEL 15 COMBAT MASTERPIECE
Caliber: 38 Spec., 6-shot.
Barrel: 4".
Weight: 32 oz. **Length:** 9 5/16" (4" bbl.).
Stocks: Uncle Mike's Combat soft rubber; wood optional.
Sights: Front, Baughman Quick Draw on ramp, micro-click rear adjustable for windage and elevation.
Price: Blued $426.00

> Consult our Directory pages for the location of firms mentioned.

SMITH & WESSON MODEL 17 K-22 MASTERPIECE
Caliber: 22 LR, 10-shot cylinder.
Barrel: 6".
Weight: 42 oz. **Length:** 11 1/8" overall.
Stocks: Hogue rubber.
Sights: Pinned Patridge front, fully adjustable rear.
Features: Polished blue finish; smooth combat trigger; semi-target hammer. The 10-slot version of this model introduced 1996.
Price: $498.00

SMITH & WESSON MODEL 19 COMBAT MAGNUM
Caliber: 357 Mag. and 38 Spec., 6-shot.
Barrel: 2 1/2", 4".
Weight: 36 oz. **Length:** 9 9/16" (4" bbl.).
Stocks: Uncle Mike's Combat soft rubber; wood optional.
Sights: Serrated ramp front 2 1/2" or 4" bbl., red ramp on 4", 6" bbl., micro-click rear adjustable for windage and elevation.
Price: 2 1/2" $423.00
Price: 4" $433.00

SMITH & WESSON MODEL 29, 629 REVOLVERS
Caliber: 44 Magnum, 6-shot.
Barrel: 6", 8 3/8" (Model 29); 4", 6", 8 3/8" (Model 629).
Weight: 47 oz. (6" bbl.). **Length:** 11 3/8" overall (6" bbl.).
Stocks: Soft rubber; wood optional.
Sights: 1/8" red ramp front, micro-click rear, adjustable for windage and elevation.
Price: S&W Bright Blue, 6" $564.00
Price: S&W Bright Blue, 8 3/8" $575.00
Price: Model 629 (stainless steel), 4" $597.00
Price: Model 629, 6" $602.00
Price: Model 629, 8 3/8" barrel $616.00

Smith & Wesson Model 629 Classic Revolver
Similar to the standard Model 629 except has full-lug 5", 6 1/2" or 8 3/8" barrel; chamfered front of cylinder; interchangable red ramp front sight with adjustable white outline rear; Hogue grips with S&W monogram; the frame is drilled and tapped for scope mounting. Factory accurizing and endurance packages. Overall length with 5" barrel is 10 1/2"; weighs 51 oz. Introduced 1990.
Price: Model 629 Classic (stainless), 5", 6 1/2" $640.00
Price: As above, 8 3/8" $661.00

Smith & Wesson Model 629 Classic DX Revolver
Similar to the Model 629 Classic except offered only with 6 1/2" or 8 3/8" full-lug barrel; comes with five front sights: 50-yard red ramp; 50-yard black Patridge; 100-yard black Patridge with gold bead; 50-yard black ramp; and 50-yard black Patridge with white dot. Comes with Hogue combat-style round butt grip. Introduced 1991.
Price: Model 629 Classic DX, 6 1/2" $825.00
Price: As above, 8 3/8" $852.00

Smith & Wesson Model 629 Classic PowerPort Revolver
Similar to the Model 629 Classic with 6 1/2" full-lug barrel except has PowerPort compensator. Introduced 1996. Made in U.S. by Smith & Wesson.
Price: 6 1/2" barrel only $640.00

SMITH & WESSON MODEL 36, 37 CHIEFS SPECIAL & AIRWEIGHT
Caliber: 38 Spec., 5-shot.
Barrel: 2".
Weight: 19 1/2 oz. (2" bbl.); 13 1/2 oz. (Airweight). **Length:** 6 1/2" (2" bbl. and round butt).
Stocks: Round butt soft rubber; wood optional.
Sights: Fixed, serrated ramp front, square notch rear.
Price: Blue, standard Model 36 ....$384.00
Price: Blue, Airweight Model 37, 2" only $419.00

Smith & Wesson Model 60 Chiefs Special Stainless
Same as Model 36 except 357 Magnum or 38 Special (only). All stainless construction, 2" bbl. and round butt only.
Price: $438.00

Smith & Wesson Model 60 3" Full-Lug Revolver
Similar to the Model 60 Chief's Special except has 3" full-lug barrel, adjustable micrometer click black blade rear sight; rubber Uncle Mike's Custom Grade Boot Grip. Overall length 7 1/2"; weighs 24 1/2 oz. Introduced 1991.
Price: $466.00

HANDGUNS—DOUBLE-ACTION REVOLVERS, SERVICE & SPORT

Smith & Wesson Model 60LS

Smith & Wesson Model 638

Smith & Wesson Model 637 Airweight Revolver
Similar to the Model 37 Airweight except has alloy frame, stainless steel barrel, cylinder and yoke; Uncle Mike's Boot Grip. Weighs 15 oz. Introduced 1996. Made in U.S. by Smith & Wesson.
Price: $435.00

Smith & Wesson Model 36LS, 60LS LadySmith
Similar to the standard Model 36. Available with 2" barrel, 38 Special. Comes with smooth, contoured rosewood grips with the S&W monogram. Has a speed-loader cutout. Comes in a fitted carry/storage case. Introduced 1989.
Price: Model 36LS $415.00
Price: Model 60LS, as above except in stainless, 357 Magnum $469.00

SMITH & WESSON MODEL 38 BODYGUARD
Caliber: 38 Spec., 5-shot.
Barrel: 2".
Weight: 14 1/2 oz. **Length:** 6 5/16" overall.
Stocks: Soft rubber; wood optional.
Sights: Fixed serrated ramp front, square notch rear.
Features: Alloy frame; internal hammer.
Price: Blue $452.00

Smith & Wesson Model 638 Airweight Bodyguard Revolver
Similar to the Model 38 except has alloy frame, stainless steel cylinder and barrel; shrouded hammer. Weighs 14 oz. Has Uncle Mike's Boot Grip. Introduced 1997. Made in U.S. by Smith & Wesson.
Price: ....NA

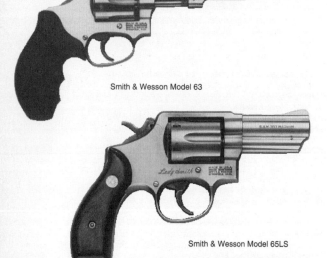

Smith & Wesson Model 63

Smith & Wesson Model 65LS

SMITH & WESSON MODEL 63 KIT GUN
Caliber: 22 LR, 6-shot.
Barrel: 2", 4".
Weight: 24 oz. (4" bbl.). **Length:** 8 3/8" (4" bbl. and round butt).
Stocks: Round butt soft rubber; wood optional.
Sights: Red ramp front, micro-click rear adjustable for windage and elevation.
Features: Stainless steel construction.
Price: 2" $466.00
Price: 4" $470.00

SMITH & WESSON MODEL 64 STAINLESS M&P
Caliber: 38 Spec., 6-shot.
Barrel: 2", 3", 4".
Weight: 34 oz. **Length:** 9 5/16" overall.
Stocks: Soft rubber; wood optional.
Sights: Fixed, 1/8" serrated ramp front, square notch rear.
Features: Satin finished stainless steel, square butt.
Price: 2" $422.00
Price: 3", 4" $430.00

SMITH & WESSON MODEL 65LS LADYSMITH
Caliber: 357 Magnum, 6-shot.
Barrel: 3".
Weight: 31 oz. **Length:** 7.94" overall.
Stocks: Rosewood, round butt.
Sights: Serrated ramp front, fixed notch rear.
Features: Stainless steel with frosted finish. Smooth combat trigger, service hammer, shrouded ejector rod. Comes with soft case. Introduced 1992.
Price: $469.00

Smith & Wesson Model 317

SMITH & WESSON MODEL 66 STAINLESS COMBAT MAGNUM
Caliber: 357 Mag. and 38 Spec., 6-shot.
Barrel: 2 1/2", 4", 6".
Weight: 36 oz. (4" barrel). **Length:** 9 9/16" overall.
Stocks: Soft rubber; wood optional.
Sights: Red ramp front, micro-click rear adjustable for windage and elevation.
Features: Satin finish stainless steel.
Price: 2 1/2" $474.00
Price: 4", 6" $480.00
Price: 6", 6" $479.00

SMITH & WESSON MODEL 317 AIRLITE REVOLVER
Caliber: 22 LR, 8-shot.
Barrel: 1 7/8".
Weight: 9.9 oz. **Length:** 6 3/16" overall.
Stocks: Dymondwood Boot or Uncle Mike's Boot.
Sights: Serrated ramp front, fixed notch rear.
Features: Aluminum alloy, carbon and stainless steels, and titanium construction. Short spur hammer, smooth combat trigger. Clear Cote finish. Introduced 1997. Made in U.S. by Smith & Wesson.
Price: With Uncle Mike's Boot grip $441.00
Price: With Dymondwood Boot grip $475.00

SMITH & WESSON MODEL 67 COMBAT MASTERPIECE
Caliber: 38 Special, 6-shot.
Barrel: 4".
Weight: 32 oz. **Length:** 9 5/16" overall.
Stocks: Soft rubber; wood optional.
Sights: Red ramp front, micro-click rear adjustable for windage and elevation.
Features: Stainless steel with satin finish. Smooth combat trigger, semi-target hammer. Introduced 1994.
Price: $475.00

HANDGUNS—DOUBLE-ACTION REVOLVERS, SERVICE & SPORT

SMITH & WESSON MODEL 586, 686 DISTINGUISHED COMBAT MAGNUMS
Caliber: 357 Magnum.
Barrel: 4", 6", full shroud.
Weight: 46 oz. (6"), 41 oz. (4").
Stocks: Soft rubber; wood optional.
Sights: Baughman red ramp front, four-position click-adjustable front, S&W micrometer click rear. Drilled and tapped for scope mount.
Features: Uses L-frame, but takes all K-frame grips. Full-length ejector rod shroud. Smooth combat-type trigger, semi-target type hammer. Trigger stop on 6" models. Also available in stainless as Model 686. Introduced 1981.
Price: Model 586, blue, 4", from $469.00
Price: Model 586, blue, 6" $474.00
Price: Model 686, 6", ported barrel $537.00
Price: Model 686, $8^{3}/_{8}$" $524.00
Price: Model 686, $2^{1}/_{2}$" $489.00

Smith & Wesson Model 686 Magnum Plus Revolver
Similar to the Model 686 except has 7-shot cylinder, $2^{1}/_{2}$", 4" or 6" barrel. Weighs $34^{1}/_{2}$ oz., overall length $7^{1}/_{2}$" ($2^{1}/_{2}$" barrel). Hogue rubber grips. Introduced 1996. Made in U.S. by Smith & Wesson.
Price: $2^{1}/_{2}$" barrel $508.00
Price: 4" barrel $516.00
Price: 6" barrel $524.00

SMITH & WESSON MODEL 625 REVOLVER
Caliber: 45 ACP, 6-shot.
Barrel: 5".
Weight: 46 oz. **Length:** 11.375" overall.
Stocks: Soft rubber; wood optional.
Sights: Patridge front on ramp, S&W micrometer click rear adjustable for windage and elevation.
Features: Stainless steel construction with .400" semi-target hammer, .312" smooth combat trigger; full lug barrel. Introduced 1989.
Price: $607.00

SMITH & WESSON MODEL 640 CENTENNIAL
Caliber: 357 Mag., 38 Spec., 5-shot.
Barrel: $2^{1}/_{8}$".
Weight: 25 oz. **Length:** $6^{3}/_{4}$" overall.
Stocks: Uncle Mike's Boot Grip.
Sights: Serrated ramp front, fixed notch rear.
Features: Stainless steel version of the original Model 40 but without the grip safety. Fully concealed hammer, snag-proof smooth edges. Introduced 1995 in 357 Magnum.
Price: $477.00
Price: Model 940 (9mm Para.) $482.00

SMITH & WESSON MODEL 617 FULL LUG REVOLVER
Caliber: 22 LR, 10-shot.
Barrel: 4", 6", $8^{3}/_{8}$".
Weight: 42 oz. (4" barrel). **Length:** NA.
Stocks: Soft rubber; wood optional.
Sights: Patridge front, adjustable rear. Drilled and tapped for scope mount.
Features: Stainless steel with satin finish; 4" has .312" smooth trigger, .375" semi-target hammer; 6" has either .312" combat or .400" serrated trigger, .375" semi-target or .500" target hammer; $8^{3}/_{8}$" with .400" serrated trigger, .500" target hammer. Introduced 1990.
Price: 4" $468.00
Price: 6", target hammer, target trigger $498.00
Price: $8^{3}/_{8}$" $510.00

Manufacturers' addresses in the
Directory of the Arms Trade
page 309, this issue

Smith & Wesson Model 625

Smith & Wesson Model 642LS LadySmith Revolver
Same as the Model 642 except has smooth combat wood grips, and comes with case; aluminum alloy frame, stainless cylinder, barrel and yoke; frosted matte finish. Weighs 15.8 oz. Introduced 1996. Made in U.S. by Smith & Wesson.
Price: $480.00

Smith & Wesson Model 442 Centennial Airweight
Similar to the Model 640 Centennial except has alloy frame giving weight of 15.8 oz. Chambered for 38 Special, 2" carbon steel barrel; carbon steel cylinder; concealed hammer; Uncle Mike's Custom Grade Santoprene grips. Fixed square notch rear sight, serrated ramp front. Introduced 1993.
Price: Blue $435.00

Smith & Wesson Model 642 Airweight Revolver
Similar to the Model 442 Centennial Airweight except has stainless steel barrel, cylinder and yoke with matte finish; Uncle Mike's Boot Grip; weights 15.8 oz. Introduced 1996. Made in U.S. by Smith & Wesson.
Price: $450.00

Smith & Wesson Model 442

SMITH & WESSON MODEL 649 BODYGUARD REVOLVER
Caliber: 357 Mag., 5-shot.
Barrel: $2^{1}/_{8}$".
Weight: 20 oz. **Length:** $6^{5}/_{16}$" overall.
Stocks: Uncle Mike's Combat.
Sights: Black pinned ramp front, fixed notch rear.
Features: Stainless steel construction; shrouded hammer; smooth combat trigger. Made in U.S. by Smith & Wesson.
Price: $477.00

Smith & Wesson Model 651

SMITH & WESSON MODEL 651 REVOLVER
Caliber: 22 WMR, 6-shot cylinder.
Barrel: 4".
Weight: $24^{1}/_{2}$ oz. **Length:** $8^{11}/_{16}$" overall.
Stocks: Soft rubber; wood optional.
Sights: Red ramp front, adjustable micrometer click rear.
Features: Stainless steel construction with semi-target hammer, smooth combat trigger. Reintroduced 1991. Limited production.
Price: $468.00

CAUTION: PRICES SHOWN ARE SUPPLIED BY THE MANUFACTURER OR IMPORTER. CHECK YOUR LOCAL GUNSHOP.

HANDGUNS—DOUBLE-ACTION REVOLVERS, SERVICE & SPORT

Smith & Wesson Model 696

SMITH & WESSON MODEL 657 REVOLVER
Caliber: 41 Mag., 6-shot.
Barrel: 6".
Weight: 48 oz. **Length:** $11^{3}/_{8}$" overall.
Stocks: Soft rubber; wood optional.
Sights: Pinned $1/_{8}$" red ramp front, micro-click rear adjustable for windage and elevation.
Features: Stainless steel construction.
Price: $537.00

SMITH & WESSON MODEL 696 REVOLVER
Caliber: 44 Spec., 5-shot.
Barrel: 3".
Weight: 35.5 oz. **Length:** $8^{1}/_{4}$" overall.
Stocks: Uncle Mike's Combat.
Sights: Red ramp front, click adjustable white outline rear.
Features: Stainless steel construction; round butt frame; satin finish. Introduced 1997. Made in U.S. by Smith & Wesson.
Price: $499.00

Taurus Model 82

TAURUS MODEL 44 REVOLVER
Caliber: 44 Mag., 6-shot.
Barrel: 3", 4", $6^{1}/_{2}$", $8^{3}/_{8}$".
Weight: $44^{3}/_{4}$ oz. (4" barrel). **Length:** NA.
Stocks: Soft black rubber.
Sights: Serrated ramp front, micro-click rear adjustable for windage and elevation.
Features: Heavy solid rib on 4", vent rib on $6^{1}/_{2}$", $8^{3}/_{8}$". Compensated barrel. Blued model has color case-hardened hammer and trigger. Introduced 1994. Imported by Taurus International.
Price: Blue, 3", 4" $447.00
Price: Blue, 3", concealed hammer $447.00
Price: Blue, $6^{1}/_{2}$", $8^{3}/_{8}$" $465.00
Price: Stainless, 3", 4" $508.00
Price: Stainless, 3", concealed hammer $508.00
Price: Stainless, $6^{1}/_{2}$", $8^{3}/_{8}$" $529.00

TAURUS MODEL 82 HEAVY BARREL REVOLVER
Caliber: 38 Spec., 6-shot.
Barrel: 3" or 4", heavy.
Weight: 34 oz. (4" bbl.). **Length:** $9^{1}/_{4}$" overall (4" bbl.).
Stocks: Soft black rubber.
Sights: Serrated ramp front, square notch rear.
Features: Imported by Taurus International.
Price: Blue $264.00
Price: Stainless $313.00

Taurus Model 85CH

TAURUS MODEL 83 REVOLVER
Caliber: 38 Spec., 6-shot.
Barrel: 4" only, heavy.
Weight: 34 oz.
Stocks: Soft black rubber.
Sights: Ramp front, micro-click rear adjustable for windage and elevation.
Features: Blue or nickel finish. Introduced 1977. Imported by Taurus International.
Price: Blue $278.00
Price: Stainless $324.00

TAURUS MODEL 85 REVOLVER
Caliber: 38 Spec., 5-shot.
Barrel: 2", 3".
Weight: 21 oz.
Stocks: Black rubber, boot grip.
Sights: Ramp front, square notch rear.
Features: Blue finish or stainless steel. Introduced 1980. Imported by Taurus International.
Price: Blue, 2", 3" $239.00
Price: Stainless steel $287.00
Price: Stainless steel $287.00
Price: Blue, 2", ported barrel $258.00
Price: Stainless, 2", ported barrel $305.00
Price: Stainless, Ultra-Lite (17 oz.), 2", ported barrel $305.00
Price: Blue, Ultra-Lite (17 oz.), 2", ported barrel $274.00
Price: Stainless, Ultra-Lite (17 oz.), 2", ported barrel $304.00

Taurus Model 85CH Revolver
Same as the Model 85 except has 2" barrel only and concealed hammer. Soft rubber boot grip. Introduced 1991. Imported by Taurus International.
Price: Blue $239.00
Price: Stainless $287.00
Price: Blue, ported barrel $258.00

TAURUS MODEL 94 REVOLVER
Caliber: 22 LR, 9-shot cylinder.
Barrel: 2", 3", 4", 5".
Weight: 25 oz.
Stocks: Soft black rubber.
Sights: Serrated ramp front, click-adjustable rear for windage and elevation.
Features: Floating firing pin, color case-hardened hammer and trigger. Introduced 1989. Imported by Taurus International.
Price: Blue $308.00
Price: Stainless $356.00
Price: Model 94 UL, 2", fixed sight, weighs 14 oz. ....NA

TAURUS MODEL 96 REVOLVER
Caliber: 22 LR, 6-shot.
Barrel: 6".
Weight: 34 oz. **Length:** NA.
Stocks: Soft black rubber.
Sights: Patridge-type front, micrometer click rear adjustable for windage and elevation.
Features: Heavy solid barrel rib; target hammer; adjustable target trigger. Blue only. Imported by Taurus International.
Price: $376.00

TAURUS MODEL 445, 445CH REVOLVERS
Caliber: 44 Special, 5-shot.
Barrel: 2".
Weight: 28.25 oz. **Length:** $6^{3}/_{4}$" overall.
Stocks: Santoprene I.
Sights: Serrated ramp front, notch rear.
Features: Blue or stainless steel. Standard or concealed hammer. Introduced 1997. Imported by Taurus International.
Price: Blue $270.00
Price: Stainless $319.00

HANDGUNS—DOUBLE-ACTION REVOLVERS, SERVICE & SPORT

TAURUS MODEL 454 REVOLVER
Caliber: 454 Casull, 5-shot.
Barrel: 6½", 8⅜".
Weight: 53 oz. (6½" barrel). Length: 12" overall (6½" barrel).
Stocks: Santoprene I or walnut.
Sights: Patridge front, micrometer click adjustable rear.
Features: Ventilated rib; integral compensating system. Introduced 1997. Imported by Taurus International.
Price: Blue $699.00
Price: Stainless $767.00

TAURUS MODEL 606, 606CH REVOLVER
Caliber: 357 Magnum, 6-shot.
Barrel: 2".
Weight: 29 oz. Length: 6¾" overall.
Stocks: Santoprene I, boot type.
Sights: Serrated ramp front, notch rear.
Features: Heavy, solid barrel rib, ejector shroud. Available with porting, concealed hammer. Introduced 1997. Imported by Taurus International.
Price: Blue, regular or concealed hammer $270.00
Price: Stainless, regular or concealed hammer $319.00
Price: Blue, ported $289.00
Price: Stainless, ported $339.00

TAURUS MODEL 607 REVOLVER
Caliber: 357 Mag., 8-shot.
Barrel: 3", 4", 6½", 8⅜.
Weight: 44 oz. Length: NA.
Stocks: Santoprene I with finger grooves.
Sights: Serrated ramp front, fully adjustable rear.
Features: Ventilated rib with built-in compensator on 6½" barrel. Available in blue or stainless. Introduced 1995. Imported by Taurus international.
Price: Blue, 3", 4" $447.00
Price: Blue, 6½", 8⅜" $465.00
Price: Stainless, 4" $508.00
Price: Stainless, 6½", 8⅜" $529.00
Price: Stainless, 3", fixed sights, ported, round butt wood stocks $554.00

TAURUS MODEL 669 REVOLVER
Caliber: 357 Mag., 6-shot.
Barrel: 4", 6".
Weight: 37 oz., (4" bbl.).
Stocks: Black rubber.
Sights: Serrated ramp front, micro-click rear adjustable for windage and elevation.
Features: Wide target-type hammer, floating firing pin, full-length barrel shroud. Introduced 1988. Imported by Taurus International.
Price: Blue, 4", 6" $344.00
Price: Stainless, 4", 6" $421.00

Taurus Model 689 Revolver
Same as the Model 669 except has full-length ventilated barrel rib. Available in blue or stainless steel. Introduced 1990. From Taurus International.
Price: Blue, 4" or 6" $358.00
Price: Stainless, 4" or 6" $435.00

DAN WESSON FIREARMS 45 PIN GUN
Caliber: 45 ACP, 6-shot.
Barrel: 5" with 1:14" twist; Taylor two-stage forcing cone; compensated shroud.
Weight: 54 oz. Length: 12.5" overall.
Stocks: Finger-groove Hogue Monogrip.
Sights: Pin front, fully adjustable rear. Has 8.375" sight radius.
Features: Based on 44 Magnum frame. Polished blue or brushed stainless steel. Uses half-moon clips with 45 ACP, or 45 Auto Rim ammunition. Reintroduced 1997. Made in U.S. by Dan Wesson Firearms.
Price: Blue, regular vent $654.00
Price: Blue, vent heavy $663.00
Price: Stainless, regular vent $713.00
Price: Stainless, vent heavy $762.00

DAN WESSON FIREARMS MODEL 22 SILHOUETTE REVOLVER
Caliber: 22 LR, 6-shot.
Barrel: 10", regular vent or vent heavy.
Weight: 53 oz.
Stocks: Combat style.
Sights: Partridge-style front, .080" narrow notch rear.
Features: Single action only. Available in blue or stainless. Reintroduced 1997. Made in U.S. by Dan Wesson Firearms.
Price: Blue, regular vent $474.00
Price: Blue, vent heavy $492.00
Price: Stainless, regular vent $504.00
Price: Stainless, vent heavy $532.00

TAURUS MODEL 605 REVOLVER
Caliber: 357 Mag., 5-shot.
Barrel: 2¼", 3".
Weight: 24.5 oz. Length: NA.
Stocks: Finger-groove Santoprene I.
Sights: Serrated ramp front, fixed notch rear.
Features: Heavy, solid rib barrel; floating firing pin. Blue or stainless. Introduced 1995. Imported by Taurus International.
Price: Blue $262.00
Price: Stainless $312.00
Price: Model 605CH (concealed hammer) 2¼", blue $262.00
Price: Model 605CH, stainless, 2¼" $312.00
Price: Blue, 2¼", ported barrel $281.00
Price: Stainless, 2¼", ported barrel $331.00
Price: Blue, 2¼", ported barrel, concealed hammer $281.00
Price: Stainless, 2¼", ported barrel, concealed hammer $331.00

Taurus Model 607

Taurus Model 608 Revolver
Same as the Model 607 except has 8-shot cylinder integral compensator. Introduced 1996. Imported by Taurus International.
Price: Blue, 4" $447.00
Price: Blue, 6½" $465.00
Price: Stainless, 4" $508.00
Price: Stainless, 6½" $529.00
Price: Stainless, 3", fixed sight, ported, round butt, wood stocks $554.00

Taurus Model 669

TAURUS MODEL 941 REVOLVER
Caliber: 22 WMR, 8-shot.
Barrel: 2", 3", 4".
Weight: 27.5 oz. (4" barrel). Length: NA.
Stocks: Soft black rubber.
Sights: Serrated ramp front, rear adjustable for windage and elevation.
Features: Solid rib heavy barrel with full-length ejector rod shroud. Blue or stainless steel. Introduced 1992. Imported by Taurus International.
Price: Blue $331.00
Price: Stainless $384.00
Price: Model 941 UL, 2", fixed sight, weighs 14 oz. .... NA

DAN WESSON FIREARMS MODEL 22 REVOLVER
Caliber: 22 LR, 22 WMR, 6-shot.
Barrel: 2½", 4", 6", 8"; interchangeable.
Weight: 36 oz. (2½"), 44 oz. (6"). Length: 9¼" overall (4" barrel).
Stocks: Checkered; undercover, service or over-size target.
Sights: ⅛" serrated, interchangeable front, white outline rear adjustable for windage and elevation.
Features: Built on the same frame as the Wesson 357; smooth, wide trigger with over-travel adjustment, wide spur hammer, with short double-action travel. Available in Brite blue or stainless steel. Reintroduced 1997. Contact Dan Wesson Firearms for complete price list.
Price: 2½" bbl., blue $357.00
Price: As above, stainless $400.00
Price: With 4", vent. rib, blue $392.00
Price: As above, stainless $432.00
Price: Blue Pistol Pac, 22 LR $653.00

CAUTION: PRICES SHOWN ARE SUPPLIED BY THE MANUFACTURER OR IMPORTER. CHECK YOUR LOCAL GUNSHOP.

HANDGUNS—DOUBLE-ACTION REVOLVERS, SERVICE & SPORT

DAN WESSON FIREARMS MODEL 322/7322 TARGET REVOLVER
Caliber: 32-20, 6-shot.
Barrel: 2.5", 4", 6", 8", standard vent, vent heavy.
Weight: 43 oz. (6" VH). **Length:** 11.25" overall.
Stocks: Checkered walnut.
Sights: Red ramp interchangeable front, fully adjustable rear.
Features: Bright blue or stainless. Reintroduced 1997. Made in U.S. by Dan Wesson Firearms.
Price: 6", blue $377.00
Price: 6", stainless $419.00
Price: 8", vent, blue $429.00
Price: 8", stainless $472.00
Price: 6", vent heavy, blue $437.00
Price: 6", vent heavy, stainless $480.00
Price: 8", vent heavy, blue $449.00
Price: 8", vent heavy, stainless $501.00

DAN WESSON FIREARMS MODEL 40 SILHOUETTE
Caliber: 357 Maximum, 6-shot.
Barrel: 4", 6", 8", 10".
Weight: 64 oz. (8" bbl.). **Length:** 14.3" overall (8" bbl.).
Stocks: Smooth walnut, target-style.
Sights: 1/8" serrated front, fully adjustable rear.
Features: Meets criteria for IHMSA competition with 8" slotted barrel. Blue or stainless steel. Made in U.S. by Dan Wesson Firearms.
Price: Blue, 4" $502.00
Price: Blue, 6" $544.00
Price: Blue, 8" $567.00
Price: Blue, 10" $597.00
Price: Stainless, 4" $567.00
Price: Stainless, 6" $610.00
Price: Stainless, 8" slotted $645.00
Price: Stainless, 10" $671.00

Dan Wesson Firearms FB44

Dan Wesson Firearms Model 15 Gold Series
Similar to the Model 15 except has smoother action to reduce DA pull to 8-10 lbs.; comes with either 6" or 8" vent heavy slotted barrel shroud with bright blue barrel. Shroud is stamped "Gold Series" with the Wesson signature engraved and gold filled. Hammer and trigger are polished bright; rosewood grips. New sights with orange dot Patridge front, white triangle on rear blade. Reintroduced 1997. Made in U.S. by Dan Wesson Firearms.
Price: 6" ....NA
Price: 8" ....NA

DAN WESSON FIREARMS MODEL 8 & MODEL 14
Caliber: 38 Special (Model 8); 357 (Model 14), both 6-shot.
Barrel: 2 1/2", 4", 6"; interchangeable.
Weight: 30 oz. (2 1/2"). **Length:** 9 1/4" overall (4" bbl.).
Stocks: Checkered, interchangeable.
Sights: 1/8" serrated front, fixed rear.
Features: Interchangeable barrels and grips; smooth, wide trigger; wide hammer spur with short double-action travel. Available in stainless or Brite blue. Reintroduced 1997. Contact Dan Wesson Firearms for complete price list.
Price: Model 8-2, 2 1/2", blue $274.00
Price: As above except in stainless $319.00

Dan Wesson Firearms Model 9, 15 & 32M Revolvers
Same as Models 8 and 14 except they have adjustable sight. Model 9 chambered for 38 Special, Model 15 for 357 Magnum. Model 32M is chambered for 32 H&R Mag. Same specs and prices as for Model 15 guns. Available in blue or stainless. Reintroduced 1997. Contact Dan Wesson Firearms for complete price list.
Price: Model 9-2 or 15-2, 2 1/2", blue $346.00
Price: As above except in stainless $376.00

Dan Wesson Firearms Model 40

Dan Wesson Firearms Model 445 Supermag Revolver
Similar size and weight as the Model 40 revolvers. Chambered for the 445 Supermag cartridge, a longer version of the 44 Magnum. Barrel lengths of 4", 6", 8", 10". Contact maker for complete price list. Reintroduced 1997. Made in the U.S. by Dan Wesson Firearms.
Price: 4", vent heavy, blue $542.00
Price: As above, stainless $621.00
Price: 8", vent heavy, blue $597.00
Price: As above, stainless $665.00
Price: 10", vent heavy, blue $619.00
Price: As above, stainless $687.00
Price: 8", vent slotted, blue $577.00
Price: As above, stainless $636.00
Price: 10", vent slotted, blue $601.00
Price: As above, stainless $661.00

DAN WESSON FIREARMS FB44, FB744 REVOLVERS
Caliber: 44 MAgnum, 6-shot.
Barrel: 4", 5", 6", 8".
Weight: 50 oz. (4" barrel). **Length:** 9 3/4" overall (4" barrel).
Stocks: Hogue finger-groove rubber.
Sights: Interchangeable blade front, fully adjustable rear.
Features: Fixed, non-vented heavy barrel shrouds, but other features same as other Wesson revolvers. Brushed stainless or polished blue finish. Reintroduced 1997. Made in the U.S. by Dan Wesson Firearms.
Price: FB44-4 (4", blue) $447.00
Price: As above, stainless (FB744-4) $493.00
Price: FB44-5 (5", blue) $450.00
Price: As above, stainless (FB744-5) $496.00
Price: FB44-6 (6", blue) $454.00
Price: As above, stainless (FB744-6) $500.00
Price: FB44-8 (8", blue) $462.00
Price: As above, stainless (FB744-8) $508.00

DAN WESSON FIREARMS MODEL 41V, 44V, 45V REVOLVERS
Caliber: 41 Mag., 44 Mag., 45 Colt, 6-shot.
Barrel: 4", 6", 8", 10"; interchangeable.
Weight: 48 oz. (4"). **Length:** 12" overall (6" bbl.)
Stocks: Smooth.
Sights: 1/8" serrated front, white outline rear adjustable for windage and elevation.
Features: Available in blue or stainless steel. Smooth, wide trigger with adjustable over-travel; wide hammer spur. Available in Pistol Pac set also. Reintroduced 1997. Contact Dan Wesson Firearms for complete price list.
Price: 41 Mag., 4", vent $447.00
Price: As above except in stainless $524.00
Price: 44 Mag., 4", blue $447.00
Price: As above except in stainless $524.00
Price: 45 Colt, 4", vent $447.00
Price: As above except in stainless $524.00

DAN WESSON FIREARMS HUNTER SERIES REVOLVERS
Caliber: 357 Supermag, 41 Mag., 44 Mag., 445 Supermag, 6-shot.
Barrel: 6", 7 1/2", depending upon model.
Weight: About 64 oz. **Length:** 14" overall.
Stocks: Hogue finger-groove rubber, wood presentation.
Sights: Blade front, dovetailed Iron Sight Gunworks rear.
Features: Fixed barrel revolvers. Barrels have 1:18.75" twist, Alan Taylor two-stage forcing cone; non-fluted cylinder; bright blue or satin stainless. Reintroduced 1997. Made in U.S. by Dan Wesson Firearms.
Price: Open Hunter (open sights, 7 1/2" barrel), blue $805.00
Price: As above, stainless $849.00
Price: Compensted Open Hunter (6" compenstaed barrel, 7" shroud), blue $837.00
Price: As above, stainless $881.00
Price: Scoped Hunter (7 1/2" barrel, no sights, comes with scope rings on shroud), blue $838.00
Price: As above, stainless $881.00
Price: Compensated Scoped Hunter (6" barrel, 7" shroud, scope rings on shroud), blue $871.00
Price: As above, stainless $914.00

HANDGUNS—DOUBLE-ACTION REVOLVERS, SERVICE & SPORT

DAN WESSON FIREARMS MODEL 738P REVOLVER
Caliber: 38 Special +P, 5-shot.
Barrel: 2".
Weight: 24.6 oz. **Length:** 6.5" overall.
Stocks: Pauferro wood or rubber.
Sights: Blade front, fixed notch rear.
Features: Designed for +P ammunition. Stainless steel construction. Reintroduced 1997. Made in U.S. by Dan Wesson Firearms.
Price: $340.00

Dan Wesson Firearms Model 738P

DAN WESSON FIREARMS FB15, FB715 REVOLVERS
Caliber: 357 Magnum, 6-shot.
Barrel: 2½", 4" (Service models), 3", 4", 5", 6" (target models).
Weight: 40 oz. (4" barrel). **Length:** 9¾" overall (4" barrel).
Stocks: Service style or Hogue rubber.
Sights: Blade front, adjustable rear (Target); fixed rear on Service.
Features: Fixed barrel, but other features same as other Wesson revolvers. Service models in brushed stainless, satin blue, Traget in brushed stainless or polished blue. Reintroduced 1997. Made in U.S. by Dan Wesson Firearms.
Price: FB14-2 (Service, 2½", blue) $289.00
Price: As above, 4" $296.00
Price: FB714-2 (Service, 2½", stainless) $313.00
Price: As above, 4" $319.00
Price: FB15-3 (Target, 3", blue) $322.00
Price: As above, 5" $331.00
Price: FB715 (Target, 4", stainless) $354.00
Price: As above, 6" $370.00

Dan Wesson Firearms FB715

HANDGUNS—SINGLE-ACTION REVOLVERS
Both classic six-shooters and modern adaptations for hunting and sport.

AMERICAN ARMS REGULATOR SINGLE-ACTIONS
Caliber: 357 Mag. 44-40, 45 Colt.
Barrel: 4¾", 5½", 7½".
Weight: 32 oz. (4¾" barrel). **Length:** 8 1/6" overall (4¾" barrel).
Stocks: Smooth walnut.
Sights: Blade front, groove rear.
Features: Blued barrel and cylinder, brass trigger guard and backstrap. Introduced 1992. Imported from Italy by American Arms, Inc.
Price: Regulator, single cylinder $365.00
Price: Regulator, dual cylinder (44-40/44 Spec. or 45 Colt/45 ACP) $435.00
Price: Regulator DLX (all steel) $425.00

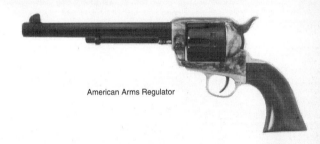
American Arms Regulator

American Arms Bisley Single-Action Revolver
Similar to the Regulator except has Bisley-style grip with steel backstrap and trigger guard, Bisley-style hammer. Color case-hardened steel frame. Hammer block safety. Available with 4¾", 5½", 7½" barrel, 45, Colt only. Introduced 1997. Imported from Italy by American Arms, Inc.
Price: $475.00

AMERICAN FRONTIER POCKET RICHARDS & MASON NAVY
Caliber: 32, 5-shot cylinder.
Barrel: 4¾", 5½".
Weight: NA. **Length:** NA.
Stocks: Varnished walnut.
Sights: Blade front, fixed rear.
Features: Shoots metallic-cartridge ammunition. Non-rebated cylinder; high-polish blue, silver-plated brass backstrap and trigger guard; ejector assembly; color case-hardened hammer and trigger. Introduced 1996. Imported from Italy by American Frontier Firearms Mfg.
Price: From $495.00

Consult our Directory pages for the location of firms mentioned.

AMERICAN ARMS/UBERTI 454 REVOLVER
Caliber: 454.
Barrel: 6", 7½".
Weight: NA. **Length:** NA.
Stocks: Smooth hardwood.
Sights: Blade front, fully adjustable rear.
Features: Porter barrels to reduce recoil; solid raised barrel rib on 6" model only; satin nickel finish; hammer block safety; wide serrated trigger; hooked trigger guard. Introduced 1997. Imported from Italy by American Arms, Inc.
Price: $869.00

AMERICAN FRONTIER 1871-1872 OPEN-TOP REVOLVERS
Caliber: 38, 44.
Barrel: 4¾", 5½", 7½", 8" round.
Weight: NA. **Length:** NA.
Stocks: Varnished walnut.
Sights: Blade front, fixed rear.
Features: Reproduction of the early cartridge conversions from percussion. Made for metallic cartridges. High polish blued steel, silver-plated brass backstrap and trigger guard, color case-hardened hammer; straight non-rebated cylinder with naval engagement engraving; stamped with original patent dates. Does not have conversion breechplate. Introduced 1996. Imported from Italy by American Frontier Firearms Mfg.
Price: $795.00
Price: Tiffany model with Tiffany grips, silver and gold finish with engraving $995.00

American Frontier 1871-1872 Open-Top

CAUTION: PRICES SHOWN ARE SUPPLIED BY THE MANUFACTURER OR IMPORTER. CHECK YOUR LOCAL GUNSHOP.

HANDGUNS—SINGLE-ACTION REVOLVERS

AMERICAN FRONTIER REMINGTON NEW ARMY CAVALRY
Caliber: 38, 44, 45.
Barrel: 5 1/2", 7 1/2", 8".
Weight: NA. **Length:** NA.
Stocks: Varnished walnut
Sights: Blade front, fixed rear.
Features: High polish blue finish; color case-hardened hammer. Has ejector assembly, loading gate. Government inspector's cartouche on left grip, sub-inspector's initials on various parts. Introduced 1997. Imported from Italy by American Frontier Firearms Mfg.
Price: $795.00
Price: Artillery model (5 1/2" barrel only) $795.00

AMERICAN FRONTIER 1871-1872 POCKET MODEL REVOLVER
Caliber: 32, 5-shot cylinder.
Barrel: 4 3/4", 5 1/2" round.
Weight: NA. **Length:** NA.
Stocks: Varnished walnut or Tiffany.
Sights: Blade front, fixed rear.
Features: Based on the 1862 Police percussion revolver converted to metallic cartridge. High polish blue finish with silver-plated brass backstrap and trigger guard, color case-hardened hammer. Introduced 1996. Imported from Italy by American Frontier Firearms Mfg.
Price: From $350.00

AMERICAN FRONTIER REMINGTON NEW MODEL REVOLVER
Caliber: 38, 44.
Barrel: 5 1/2", 7 1/2"
Weight: NA. **Length:** NA.
Stocks: Varnished walnut.
Sights: Blade front, fixed rear.
Features: Replica of the factory conversions by Remington between 1863 and 1875. High polish blue or silver finish with color case-hardened hammer; has original loading lever and no gate or ejector assembly. Introduced 1996. Imported from Italy by American Frontier Firearms Mfg.
Price: $695.00

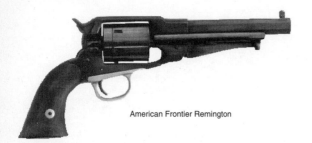
American Frontier Remington

AMERICAN FRONTIER 1851 NAVY CONVERSION
Caliber: 38, 44.
Barrel: 4 3/4", 5 1/2", 7 1/2", octagon.
Weight: NA. **Length:** NA.
Stocks: Varnished walnut, Navy size.
Sights: Blade front, fixed rear.
Features: Shoots metallic cartridge ammunition. Non-rebated cylinder; blued steel backstrap and trigger guard; color case-hardened hammer, trigger, ramrod, plunger; no ejector rod assembly. Introduced 1996. Imported from Italy by American Frontier Firearms Mfg.
Price: $695.00

American Frontier 1851 Navy Richards & Mason Conversion
Similar to the 1851 Navy Conversion except has Mason ejector assembly. Introduced 1996. Imported from Italy by American Frontier Firearms Mfg.
Price: $695.00

AMERICAN FRONTIER RICHARDS 1860 ARMY
Caliber: 38, 44.
Barrel: 4 3/4", 5 1/2", 7 1/2", round.
Weight: NA. **Length:** NA.
Stocks: Varnished walnut, Army size.
Sights: Blade front, fixed rear.
Features: Shoots metallic cartridge ammunition. Rebated cylinder; available with or without ejector assembly; high-polish blue including backstrap; silver-plated trigger guard; color case-hardened hammer and trigger. Introduced 1996. Imported from Italy by American Frontier Firearms Mfg.
Price: $695.00

American Frontier 1851 Mason

Century Model 100

CENTURY GUN DIST. MODEL 100 SINGLE-ACTION
Caliber: 30-30, 375 Win., 444 Marlin, 45-70, 50-70.
Barrel: 6 1/2" (standard), 8", 10".
Weight: 6 lbs. (loaded). **Length:** 15" overall (8" bbl.).
Stocks: Smooth walnut.
Sights: Ramp front, Millett adjustable square notch rear.
Features: Highly polished high tensile strength manganese bronze frame, blue cylinder and barrel; coil spring trigger mechanism. Calibers other than 45-70 start at $2,000.00. Contact maker for full price information. Introduced 1975. Made in U.S. From Century Gun Dist., Inc.
Price: 6 1/2" barrel, 45-70 ....$1,250.00

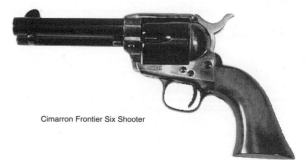
Cimarron Frontier Six Shooter

CIMARRON U.S. CAVALRY MODEL SINGLE-ACTION
Caliber: 45 Colt
Barrel: 7 1/2".
Weight: 42 oz. **Length:** 13 1/2" overall.
Stocks: Walnut.
Sights: Fixed.
Features: Has "A.P. Casey" markings; "U.S." plus patent dates on frame, serial number on backstrap, trigger guard, frame and cylinder, "APC" cartouche on left grip; color case-hardened frame and hammer, rest charcoal blue. Exact copy of the original. Imported by Cimarron Arms.
Price: $499.00

Cimarron Rough Rider Artillery Model Single-Action
Similar to the U.S. Cavalry model except has 5 1/2" barrel, weighs 39 oz., and is 11 1/2" overall. U.S. markings and cartouche, case-hardened frame and hammer; 45 Colt only.
Price: $499.00

CIMARRON 1873 FRONTIER SIX SHOOTER
Caliber: 38 WCF, 357 Mag., 44 WCF, 44 Spec., 45 Colt.
Barrel: 4 3/4", 5 1/2", 7 1/2".
Weight: 39 oz. **Length:** 10" overall (4" barrel).
Stocks: Walnut.
Sights: Blade front, fixed or adjustable rear.
Features: Uses "old model" blackpowder frame with "Bullseye" ejector or New Model frame. Imported by Cimarron Arms.
Price: 4 3/4" barrel $469.00
Price: 5 1/2" barrel $469.00
Price: 7 1/2" barrel $469.00

HANDGUNS—SINGLE-ACTION REVOLVERS

CIMARRON NEW THUNDERER REVOLVER
Caliber: 357 Mag., 44 WCF, 44 Spec., 45 Colt, 6-shot.
Barrel: 3½", 4¾", with ejector.
Weight: 38 oz. (3½" barrel). **Length:** NA.
Stocks: Hand-checkered walnut.
Sights: Blade front, notch rear.
Features: Thunderer grip; color case-hardened frame with balance blued, or nickel finish. Introduced 1993. Imported by Cimarron Arms.
Price: Color case-hardened **$489.00**
Price: Nickeled **$639.00**

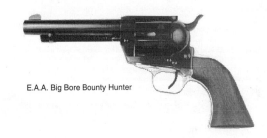
Colt Single Action Army

COLT SINGLE ACTION-ARMY REVOLVER
Caliber: 44-40, 45 Colt, 6-shot.
Barrel: 4¾", 5½", 7½".
Weight: 40 oz. (4¾" barrel). **Length:** 10¼" overall (4¾" barrel).
Stocks: Black Eagle composite.
Sights: Blade front, notch rear.
Features: Available in full nickel finish with nickel grip medallions, or Royal Blue with color case-hardened frame, gold grip medallions. Reintroduced 1992.
Price: **$1,213.00**

D-MAX SIDEWINDER REVOLVER
Caliber: 45 Colt/410 shotshell, 6-shot.
Barrel: 6.5", 7.5".
Weight: 57 oz. (6.5"). **Length:** 14.1" (6.5" barrel).
Stocks: Hogue black rubber with finger grooves.
Sights: Blade on ramp front, fully adjustable rear.
Features: Stainless steel construction. Has removable choke for firing shotshells. Grooved, wide-spur hammer; transfer bar ignition; satin stainless finish. Introduced 1992. Made in U.S. by D-Max, Inc.
Price: **$750.00**

D-Max Sidewinder

E.A.A. Big Bore Bounty Hunter

E.A.A. BIG BORE BOUNTY HUNTER SA REVOLVERS
Caliber: 357 Mag., 44 Mag., 45 Colt, 6-shot.
Barrel: 4½", 7½".
Weight: 2.5 lbs. **Length:** 11" overall (4⅝" barrel).
Stocks: Smooth walnut.
Sights: Blade front, grooved topstrap rear.
Features: Transfer bar safety; three position hammer; hammer forged barrel. Introduced 1992. Imported by European American Armory.
Price: Blue **$299.00**
Price: Color case-hardened frame **$310.00**

EMF 1875 OUTLAW REVOLVER
Caliber: 357 Mag., 44-40, 45 Colt.
Barrel: 7½".
Weight: 46 oz. **Length:** 13½" overall.
Stocks: Smooth walnut.
Sights: Blade front, fixed groove rear.
Features: Authentic copy of 1875 Remington with firing pin in hammer; color case-hardened frame, blue cylinder, barrel, steel backstrap and brass trigger guard. Also available in nickel, factory engraved. Imported by E.M.F.
Price: All calibers **$465.00**
Price: Nickel **$550.00**
Price: Engraved **$600.00**
Price: Engraved Nickel **$710.00**

EMF 1890 Police Revolver
Similar to the 1875 Outlaw except has 5½" barrel, weighs 40 oz., with 12½" overall length. Has lanyard ring in butt. No web under barrel. Calibers 357, 44-40, 45 Colt. Imported by E.M.F.
Price: All calibers **$470.00**
Price: Nickel **$560.00**
Price: Engraved **$620.00**
Price: Engraved nickel **$725.00**

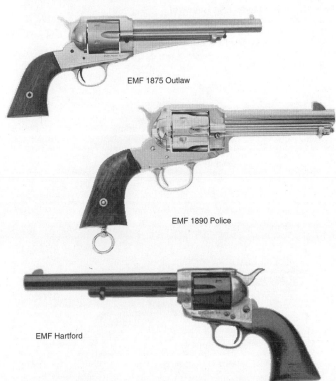
EMF 1875 Outlaw
EMF 1890 Police
EMF Hartford

> Consult our Directory pages for the location of firms mentioned.

EMF HARTFORD SINGLE-ACTION REVOLVERS
Caliber: 22 LR, 357 Mag., 32-20, 38-40, 44-40, 44 Spec., 45 Colt.
Barrel: 4¾", 5½", 7½".
Weight: 45 oz. **Length:** 13" overall (7½" barrel).
Stocks: Smooth walnut.
Sights: Blade front, fixed rear.
Features: Identical to the original Colts with inspector cartouche on left grip, original patent dates and U.S. markings. All major parts serial numbered using original Colt-style lettering, numbering. Bullseye ejector head and color case-hardening on frame and hammer. Introduced 1990. From E.M.F.
Price: **$600.00**
Price: Cavalry or Artillery **$655.00**
Price: Nickel plated **$725.00**
Price: Engraved, nickel plated **$840.00**

HANDGUNS—SINGLE-ACTION REVOLVERS

EMF 1894 Bisley Revolver
Similar to the Hartford single-action revolver except has special grip frame and trigger guard, wide spur hammer; available in 45 Colt only, 5½" or 7½" barrel. Introduced 1995. Imported by E.M.F.
Price: Blue $680.00
Price: Nickel $805.00

EMF Hartford Express Single-Action Revolver
Same as the regular Hartford model except uses grip of the Colt Lightning revolver. Barrel lengths of 4", 4¾", 5½". Introduced 1997. Imported by E.M.F.
Price: $475.00

EMF Hartford Pinkerton Single-Action Revolver
Same as the regular Hartford except has 4" barrel with ejector tube and birds' head grip. Calibers 32-20, 38-40, 44-40, 44 Special, 45 Colt. Introduced 1997. Imported by E.M.F.
Price: $475.00

FREEDOM ARMS MID FRAME REVOLVER
Caliber: 357 Mag., 6-shot cylinder.
Barrel: 5½", 7½".
Weight: 40 oz.(5½"barrel). **Length:** 10¾"overall(5½"barrel).
Stocks: Wine wood.
Sights: Blade on ramp front, fixed or fully adjustable rear.
Features: Made of stainless steel; polished cylinder, matte frame. Introduced 1997. Made in U.S. by Freedom Arms.
Price: ....NA

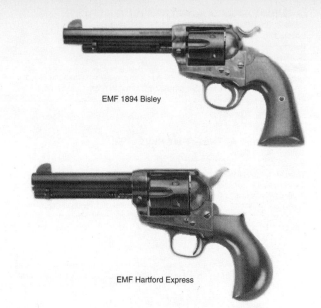

EMF 1894 Bisley

EMF Hartford Express

Freedom Arms Premier

Freedom Arms Model 353 Revolver
Similar to the Premier 454 Casull except chambered for 357 Magnum with 5-shot cylinder; 4¾", 6", 7½" or 9" barrel. Weighs 59 oz. with 7½" barrel. Field grade model has adjustable sights, matte finish, Pachmayr grips, 7½" or 10" barrel; Silhouette has 9" barrel, Patridge front sight, Iron Sight Gun Works Silhouette adjustable rear, Pachmayr grips, trigger over-travel adjustment screw. All stainless steel. Introduced 1992.
Price: Field Grade $1,269.00
Price: Premier Grade (brushed finish, impregnated hardwood grips, Premier Grade sights) $1,673.00
Price: Silhouette (9", 357 Mag., 10", 44 Mag.) $1,376.85

HERITAGE ROUGH RIDER REVOLVER
Caliber: 22 LR, 22 LR/22 WMR combo, 6-shot.
Barrel: 2¾", 3½", 4¾", 6½", 9".
Weight: 31 to 38 oz. **Length:** NA
Stocks: Exotic hardwood.
Sights: Blade front, fixed rear.
Features: Hammer block safety. High polish blue or nickel finish. Introduced 1993. Made in U.S. by Heritage Mfg., Inc.
Price: ....$109.95 to $169.95
Price: 2¾", 3½", 4¾" birdshead grip ....$129.95 to $149.95

IAR MODEL 1873 FRONTIER REVOLVER
Caliber: 22 RL, 22 LR/22 WMR.
Barrel: 4¾".
Weight: 45 oz. **Length:** 10½"overall.
Stocks: One-piece walnut with inspector's cartouche.
Sights: Blade front, notch rear.
Features: Color case-hardened frame, blued barrel, black nickel-plated brass trigger guard and backstrap. Bright nickel and engraved versions available. Introduced 1997. Imported from Italy by IAR, Inc.
Price: $400.00
Price: Nickel-plated $485.00
Price: 22 LR/22WMR combo $425.00

FREEDOM ARMS PREMIER SINGLE-ACTION REVOLVER
Caliber: 44 Mag., 454 Casull with 45 Colt, 45 ACP, 45 Win. Mag. optional cylinders, 5-shot.
Barrel: 4¾", 6", 7½", 10".
Weight: 50 oz. **Length:** 14" overall (7½" bbl.).
Stocks: Impregnated hardwood.
Sights: Blade front, notch or adjustable rear.
Features: All stainless steel construction; sliding bar safety system. Lifetime warranty. Made in U.S. by Freedom Arms, Inc.
Price: Field Grade (matte finish, Pachmayr grips), adjustable sights, 4¾", 6", 7½", 10" $1,328.00
Price: Field Grade, fixed sights, 4¾", 6", 7½", 10" $1,218.00
Price: Field Grade, 44 Rem. Mag., adjustable sights, all lengths $1,269.00
Price: Premier Grade 454 (brush finish, impregnated hardwood grips) adjustable sights, 4¾", 6", 7½", 10" $1,724.00
Price: Premier Grade, fixed sights, all barrel lengths $1,620.00
Price: Premier Grade, 44 Rem. Mag., adjustable sights, all lengths ...$1,673.00
Price: Fitted 45 ACP, 45 Colt or 45 Win. Mag cylinder, add $343.00

Freedom Arms Model 555 Revolver
Same as the 454 Casull except chambered for the 50 A.E. (Action Express) cartridge. Offered in Premier and Field Grades with adjustable sights, 4¾", 6", 7½" or 10" barrel. Introduced 1994. Made in U.S. by Freedom Arms, Inc.
Price: Premier Grade $1,724.00
Price: Field Grade $1,328.00

Heritage Rough Rider

IAR Model 1873 Frontier

HANDGUNS—SINGLE-ACTION REVOLVERS

IAR MODEL 1873 SIX SHOOTER
Caliber: 22 LR/22 WMR combo
Barrel: 5 1/2".
Weight: 36 1/2 oz. **Length:** 11 3/8" overall.
Stocks: One-piece walnut.
Sights: Blade front, notch rear.
Features: A 3/4-scale reproduction. Color case-hardened frame, blued barrel. All-Steel construction. Made by Uberti. Imported from Italy by IAR, Inc.
Price: $400.00

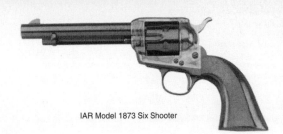

IAR Model 1873 Six Shooter

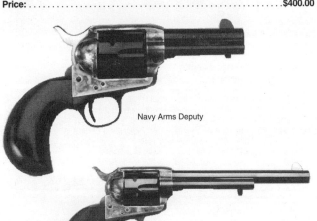

Navy Arms Deputy

NAVY ARMS DEPUTY SINGLE-ACTION REVOLVER
Caliber: 44-40 or 45 Colt, 6-shot cylinder
Barrel: 3 1/2", 4", 4 3/4".
Weight: 33 oz. **Length:** 8 1/2" overall (3 1/2" barrel).
Stocks: Smooth walnut
Sights: Blade front, notch rear.
Features: Replica of the Colt Thunderer. Polished blue finish with color case-hardened frame. Introduced 1997. Imported by Navy Arms.
Price: $405.00

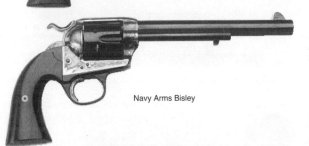

Navy Arms Pinched Frame

NAVY ARMS "PINCHED FRAME" SINGLE-ACTION REVOLVER
Caliber: 45 Colt.
Barrel: 7 1/2".
Weight: 37 oz. **Length:** 13" overall.
Stocks: Smooth walnut
Sights: German silver blade, notch rear.
Features: Replica of Colt's original Peacemaker. Color case-hardened frame, hammer, rest blued. Introduced 1997. Imported by Navy Arms.
Price: $405.00

Navy Arms Bisley

NAVY ARMS BISLEY MODEL SINGLE-ACTION REVOLVER
Caliber: 44-40 or 45 Colt, 6-shot cylinder.
Barrel: 4 3/4", 5 1/2", 7 1/2".
Weight: 40 oz. **Length:** 12 1/2" overall (7 1/2" barrel).
Stocks: Smooth walnut.
Sights: Blade front, notch rear.
Features: Replica of Colt's Bisley Model. Polished blue finish, color case-hardened frame. Introduced 1997. Imported by Navy Arms.
Price: $445.00

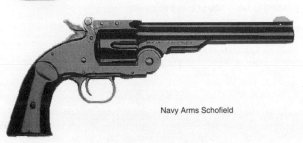

Navy Arms 1873

NAVY ARMS 1873 SINGLE-ACTION REVOLVER
Caliber: 44-40, 45 Colt, 6-shot cylinder.
Barrel: 3", 4 3/4", 5 1/2", 7 1/2".
Weight: 36 oz. **Length:** 10 3/4" overall (5 1/2" barrel).
Stocks: Smooth walnut.
Sights: Blade front, groove in topstrap rear.
Features: Blue with color case-hardened frame, or nickel. Introduced 1991. Imported by Navy Arms.
Price: Blue $390.00
Price: Nickel $455.00
Price: 1873 U.S. Cavalry Model (7 1/2", 45 Colt, arsenal markings) $480.00
Price: 1895 U.S. Artillery Model (as above, 5 1/2" barrel) $480.00

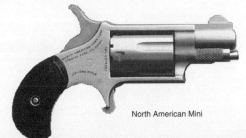

Navy Arms Schofield

NAVY ARMS 1875 SCHOFIELD REVOLVER
Caliber: 44-40, 44 S&W Spec., 45 Colt, 6-shot cylinder.
Barrel: 5", 7".
Weight: 39 oz. **Length:** 10 3/4" overall (5" barrel).
Stocks: Smooth walnut.
Sights: Blade front, notch rear.
Features: Replica of Smith & Wesson Model 3 Schofield. Single-action, top-break with automatic ejection. Polished blue finish. Introduced 1994. Imported by Navy Arms.
Price: Wells Fargo (5" barrel, Wells Fargo markings) $795.00
Price: U.S. Cavalry model (7" barrel, military markings) $795.00

North American Mini

NORTH AMERICAN MINI-REVOLVERS
Caliber: 22 Short, 22 LR, 22 WMR, 5-shot.
Barrel: 1 1/8", 1 5/8".
Weight: 4 to 6.6 oz. **Length:** 3 5/8" to 6 1/8" overall.
Stocks: Laminated wood.
Sights: Blade front, notch fixed rear.
Features: All stainless steel construction. Polished satin and matte finish. Engraved models available. From North American Arms.
Price: 22 Short, 22 LR, 1 1/8" bbl. $157.00
Price: 22 LR, 1 5/8" bbl. $157.00
Price: 22 WMR, 1 5/8" bbl. $178.00
Price: 22 WMR, 1 1/8" or 1 5/8" bbl. with extra 22 LR cylinder $210.00

HANDGUNS—SINGLE-ACTION REVOLVERS

NORTH AMERICAN MINI-MASTER
Caliber: 22 LR, 22 WMR, 5-shot cylinder.
Barrel: 4".
Weight: 10.7 oz. **Length:** 7.75" overall.
Stocks: Checkered hard black rubber.
Sights: Blade front, white outline rear adjustable for elevation, or fixed.
Features: Heavy vent barrel; full-size grips. Non-fluted cylinder. Introduced 1989.
Price: Adjustable sight, 22 WMR or 22 LR $279.00
Price: As above with extra WMR/LR cylinder $317.00
Price: Fixed sight, 22 WMR or 22 LR $264.00
Price: As above with extra WMR/LR cylinder $302.00

North American Black Widow Revolver
Similar to the Mini-Master except has 2" heavy vent barrel. Built on the 22 WMR frame. Non-fluted cylinder, black rubber grips. Available with either Millett Low Profile fixed sights or Millett sight adjustable for elevation only. Overall length 5 7/8", weighs 8.8 oz. From North American Arms.
Price: Adjustable sight, 22 LR or 22 WMR $249.00
Price: As above with extra WMR/LR cylinder $285.00
Price: Fixed sight, 22 LR or 22 WMR $235.00
Price: As above with extra WMR/LR cylinder $270.00

RUGER BLACKHAWK REVOLVER
Caliber: 357 Mag./38 Spec., 45 Colt, 6-shot.
Barrel: 4 5/8" or 6 1/2", either caliber; 7 1/2" (45 Colt only).
Weight: 42 oz. (6 1/2" bbl.). **Length:** 12 1/4" overall (6 1/2" bbl.).
Stocks: American walnut.
Sights: 1/8" ramp front, micro-click rear adjustable for windage and elevation.
Features: Ruger transfer bar safety system, independent firing pin, hardened chrome-moly steel frame, music wire springs throughout.
Price: Blue, 357 Mag. (4 5/8", 6 1/2"), BN34, BN36 ....$360.00
Price: Blue, 357/9mm Convertible (4 5/8", 6 1/2"), BN34X, BN36X ...$360.00
Price: Blue, 45 Colt (4 5/8", 6 1/2"), BN42, BN45 ....$360.00
Price: Stainless, 357 Mag. (4 5/8", 6 1/2"), KBN34, KBN36 .....$443.00

Ruger Blackhawk

Ruger Bisley Single-Action Revolver
Similar to standard Blackhawk except the hammer is lower with a smoothly curved, deeply checkered wide spur. The trigger is strongly curved with a wide smooth surface. Longer grip frame has a hand-filling shape. Adjustable rear sight, ramp-style front. Has an unfluted cylinder and roll engraving, adjustable sights. Chambered for 357, 44 Mags. and 45 Colt; 7 1/2" barrel; overall length of 13". Introduced 1985.
Price: $450.00

RUGER SUPER BLACKHAWK
Caliber: 44 Mag., 6-shot. Also fires 44 Spec.
Barrel: 4 5/8", 5 1/2", 7 1/2", 10 1/2".
Weight: 48 oz. (7 1/2" bbl.), 51 oz. (10 1/2" bbl.). **Length:** 13 3/8" overall (7 1/2" bbl.).
Stocks: American walnut.
Sights: 1/8" ramp front, micro-click rear adjustable for windage and elevation.
Features: Ruger transfer bar safety system, non-fluted cylinder, steel grip and cylinder frame, square back trigger guard, wide serrated trigger and wide spur hammer.
Price: Blue (S45N, S47N, S411N) $413.00
Price: Stainless (KS45N, KS47N, KS411N) $450.00

Ruger Bisley Single-Action

Ruger New Super Bearcat

RUGER NEW SUPER BEARCAT SINGLE-ACTION
Caliber: 22 LR, 6-shot.
Barrel: 4".
Weight: 23 oz. **Length:** 8 7/8" overall.
Stocks: Smooth rosewood with Ruger medallion.
Sights: Blade front, fixed notch rear.
Features: Reintroduction of the Ruger Super Bearcat with slightly lengthened frame, Ruger patented transfer bar safety system. Available in blue only. Introduced 1993. From Sturm, Ruger & Co.
Price: SBC4, blue $320.00

Ruger Vaquero

RUGER VAQUERO SINGLE-ACTION REVOLVER
Caliber: 357 Mag., 44-40, 44 Mag., 45 Colt, 6-shot.
Barrel: 4 5/8", 5 1/2", 7 1/2".
Weight: 41 oz. **Length:** 13 3/8" overall (7 1/2" barrel).
Stocks: Smooth rosewood with Ruger medallion.
Sights: Blade front, fixed notch rear.
Features: Uses Ruger's patented transfer bar safety system and loading gate interlock with classic styling. Blued model has color case-hardened finish on the frame, the rest polished and blued. Stainless model has high-gloss polish. Introduced 1993. From Sturm, Ruger & Co.
Price: 357 Mag. BNV34 (4 5/8"), BNV35 (5 1/2") $434.00
Price: 357 Mag. KBNV34 (4 5/8"), KBNV35 (5 1/2") stainless ...$434.00
Price: BNV44 (4 5/8"), BNV445 (5 1/2"), BNV45 (7 1/2"), blue ..$434.00
Price: KBNV44 (4 5/8"), KBNV455 (5 1/2"), KBNV45 (7 1/2"), stainless ...$434.00
Price: 44 Mag. BNV475IE (engraved cylinder, simulated ivory grips), blue, 5 1/2" $583.00
Price: 44 Mag. KBNV475IE (engraved cylinder, simulated ivory grips), stainless, 5 1/2" $583.00
Price: 45 Colt BNV455IE (engraved cylinder, simulated ivory grips), blue, 5 1/2" $583.00
Price: 45 Colt KBNV455IE (engraved cylinder, simulated ivory grips), stainless, 5 1/2" $583.00

RUGER SUPER SINGLE-SIX CONVERTIBLE
Caliber: 22 LR, 6-shot; 22 WMR in extra cylinder.
Barrel: 4 5/8", 5 1/2", 6 1/2", or 9 1/2" (6-groove).
Weight: 34 1/2 oz. (6 1/2" bbl.). **Length:** 11 13/16" overall (6 1/2" bbl.).
Stocks: Smooth American walnut.
Sights: Improved Patridge front on ramp, fully adjustable rear protected by integral frame ribs; or fixed sight.
Features: Ruger transfer bar safety system, gate-controlled loading, hardened chrome-moly steel frame, wide trigger, music wire springs throughout, independent firing pin.
Price: 4 5/8", 5 1/2", 6 1/2", 9 1/2" barrel, blue, fixed or adjustable sight (5 1/2", 6 1/2") $313.00
Price: 5 1/2", 6 1/2" bbl. only, high-gloss stainless steel, fixed or adjustable sight $393.00

Ruger Bisley Small Frame Revolver
Similar to the Single-Six except frame is styled after the classic Bisley "flat-top." Most mechanical parts are unchanged. Hammer is lower and smoothly curved with a deeply checkered spur. Trigger is strongly curved with a wide smooth surface. Longer grip frame designed with a hand-filling shape, and the trigger guard is a large oval. Adjustable dovetail rear sight; front sight base accepts interchangeable square blades of various heights and styles. Has an unfluted cylinder and roll engraving. Weighs about 41 oz. Chambered for 22 LR, 6 1/2" barrel only. Introduced 1985.
Price: $380.00

Ruger Bisley-Vaquero Single-Action Revolver
Similar to the Vaquero except has Bisley-style hammer, grip and trigger, and is available in 44 Magnum and 45 Colt only, with 5 1/2" barrel. Has smooth rosewood grips with Ruger medallion. Introduced 1997. From Sturm, Ruger & Co.
Price: $434.00

HANDGUNS—SINGLE-ACTION REVOLVERS

TEXAS LONGHORN ARMS GROVER'S IMPROVED NO. FIVE
Caliber: 44 Mag., 6-shot.
Barrel: 5½".
Weight: 44 oz. **Length:** 11½" overall.
Stocks: Smooth walnut.
Sights: Square blade front on ramp, fully adjustable rear.
Features: Music wire coil spring action with double locking bolt; polished blue finish. Handmade in limited 1,200-gun production. Grip contour, straps, over-sized base pin, lever latch and lockwork identical copies of Elmer Keith design. Lifetime warranty to original owner. Introduced 1988.
Price: $1,195.00

TEXAS LONGHORN ARMS RIGHT-HAND SINGLE-ACTION
Caliber: All centerfire pistol calibers.
Barrel: 4¾".
Weight: 40 oz. **Length:** 10¼" overall.
Stocks: One-piece fancy walnut.
Sights: Blade front, grooved topstrap rear.
Features: Loading gate and ejector housing on left side of gun. Cylinder rotates to the left. All steel construction; color case-hardened frame; high polish blue; music wire coil springs. Lifetime guarantee to original owner. Introduced 1984. From Texas Longhorn Arms.
Price: South Texas Army Limited Edition—handmade, only 1,000 to be produced; "One of One Thousand" engraved on barrel $1,595.00

Texas Longhorn Arms Texas Border Special
Similar to the South Texas Army Limited Edition except has 4" barrel, bird's-head style grip. Same special features. Introduced 1984.
Price: $1,595.00

Texas Longhorn Arms West Texas Flat Top Target
Similar to the South Texas Army Limited Edition except choice of barrel length from 7½" through 15"; flat-top style frame; 1/8" contoured ramp front sight, old model steel micro-click rear adjustable for windage and elevation. Same special features. Introduced 1984.
Price: $1,595.00

Texas Longhorn Arms Cased Set
Set contains one each of the Texas Longhorn Right-Hand Single-Actions, all in the same caliber, same serial numbers (100, 200, 300, 400, 500, 600, 700, 800, 900). Ten sets to be made. All other specs same as Limited Edition guns. Introduced 1984.
Price: $5,750.00
Price: With ¾-coverage "C-style" engraving $7,650.00

Texas Longhorn Grover's No. Five

Texas Longhorn Arms Sesquicentennial Model Revolver
Similar to the South Texas Army Model except has ¾-coverage Nimschke-style engraving, antique golden nickel plate finish, one-piece elephant ivory grips. Comes with handmade solid walnut presentation case, factory letter to owner. Limited edition of 150 units. Introduced 1986.
Price: $2,500.00

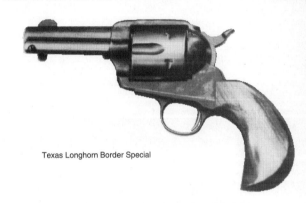
Texas Longhorn Border Special

Manufacturers' addresses in the
Directory of the Arms Trade
page 309, this issue

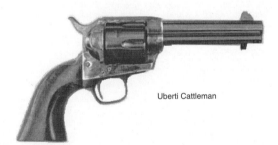
Uberti Cattleman

UBERTI 1875 SA ARMY OUTLAW REVOLVER
Caliber: 357 Mag., 44-40, 45 Colt, 45 Colt/45 ACP convertible, 6-shot.
Barrel: 5½", 7½".
Weight: 44 oz. **Length:** 13¾" overall.
Stocks: Smooth walnut.
Sights: Blade front, notch rear.
Features: Replica of the 1875 Remington S.A. Army revolver. Brass trigger guard, color case-hardened frame, rest blued. Imported by Uberti U.S.A.
Price: $435.00
Price: 45 Colt/45 ACP convertible $475.00

UBERTI 1890 ARMY OUTLAW REVOLVER
Caliber: 357 Mag., 44-40, 45 Colt, 45 Colt/45 ACP convertible, 6-shot.
Barrel: 5½", 7½".
Weight: 37 oz. **Length:** 12½" overall.
Stocks: American walnut.
Sights: Blade front, groove rear.
Features: Replica of the 1890 Remington single-action. Brass trigger guard, rest is blued. Imported by Uberti U.S.A.
Price: $435.00
Price: 45 Colt/45 ACP convertible $475.00

UBERTI 1873 CATTLEMAN SINGLE-ACTIONS
Caliber: 22 LR/22 WMR, 38 Spec., 357 Mag., 44 Spec., 44-40, 45 Colt/45 ACP, 6-shot.
Barrel: 4¾", 5½", 7½"; 44-40, 45 Colt also with 3", 3½", 4".
Weight: 38 oz. (5½" bbl.). **Length:** 10¾" overall (5½" bbl.).
Stocks: One-piece smooth walnut.
Sights: Blade front, groove rear; fully adjustable rear available.
Features: Steel or brass backstrap, trigger guard; color case-hardened frame, blued barrel, cylinder. Imported from Italy by Uberti U.S.A.
Price: Steel backstrap, trigger guard, fixed sights $435.00
Price: Brass backstrap, trigger guard, fixed sights $365.00
Price: Bisley model $435.00

Uberti 1873 Buckhorn Single-Action
A slightly larger version of the Cattleman revolver. Available in 44 Magnum or 44 Magnum/44-40 convertible, otherwise has same specs.
Price: Steel backstrap, trigger guard, fixed sights $410.00
Price: Convertible (two cylinders) $475.00

Uberti 1875 Army

CAUTION: PRICES SHOWN ARE SUPPLIED BY THE MANUFACTURER OR IMPORTER. CHECK YOUR LOCAL GUNSHOP.

HANDGUNS—SINGLE-ACTION REVOLVERS

U.S. PATENT FIRE-ARMS BISLEY MODEL REVOLVER
Caliber: 4 Colt, 6-shot cylinder.
Barrel: 4 3/4", 5 1/2", 7 1/2", 10".
Weight: 38 oz. (5 1/2" barrel). **Length:** NA.
Stocks: Smooth walnut.
Sights: Blade front, notch rear.
Features: Available in all-blue, blue with color case-hardening, or full nickel plate finish. Made in Italy; available from United States Patent Fire-Arms Mfg. Co.
Price: 4 3/4", blue $652.00
Price: 5 1/2", blue/case-colors $750.50
Price: 7 1/2", blue/case-colors $756.00
Price: 10", nickel $862.50

U.S. Patent Fire-Arms Flattop Target Revolver
Similar to the Single Action Army except 4 3/4", 5 1/2" or 7 1/2" barrel, two-piece hard rubber stocks, flat top frame, adjustable rear sight. Made in Italy; available from United States Patent Fire-Arms Mfg. Co.
Price: 4 3/4", blue, polished hammer $690.00
Price: 4 3/4", blue, case-colored hammer $813.00
Price: 5 1/2", blue, case-colored hammer $816.00
Price: 5 1/2", nickel-plated $765.00
Price: 7 1/2", blue, polished hammer $717.00
Price: 7 1/2", blue, case-colored hammer $822.00

U.S. PATENT FIRE-ARMS SINGLE ACTION ARMY REVOLVER
Caliber: 22 LR, 22 WMR, 357 Mag., 44 Russian, 38-40, 44-40, 45 Colt, 6-shot cylinder.
Barrel: 3", 4", 4 3/4", 5 1/2", 7 1/2", 10".
Weight: 37 oz. **Length:** NA.
Stocks: Smooth walnut.
Sights: Blade front, notch rear.
Features: Recreation of original guns; 3" and 4" have no ejector. Available with all-blue, blue with color case-hardening, or full nickel-plate finish. Made in Italy; available from United States Patent Fire-Arms Mfg. Co.
Price: 3" blue $600.00
Price: 4 3/4", blue/cased-colors $732.00
Price: 7 1/2", blue/case-colors $739.00
Price: 10", nickel $847.50

U.S. Patent Fire-Arms Nettleton Cavalry Revolver
Similar to the single Action Army, except in 45 Colt only, with 7 1/2" barrel, color case-hardened/blue finish, and has old-style hand numbering, exact cartouche branding and correct inspector hand-stamp markings. Made in Italy, available from United States Patent Fire-Arms Mfg. Co.
Price: $950.00
Price: Artillery Model, 5 1/2" barrel $950.00

U.S. Patent Fire-Arms Bird Head Model Revolver
Similar to the Single Action Army except has bird's-head grip and comes with 3 1/2", 4" or 4 1/2" barrel. Made in Italy; available from United States Patent Fire-Arms Mfg. Co.
Price: 3 1/2", blue $635.50
Price: 4", blue with color case-hardening $735.00
Price: 4 1/2", nickel-plated $795.50

HANDGUNS—MISCELLANEOUS
Specially adapted single-shot and multi-barrel arms.

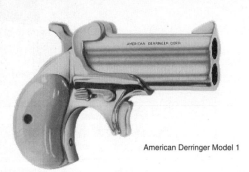
American Derringer Model 1

American Derringer Lady Derringer
Same as the Model 1 except has tuned action, is fitted with scrimshawed synthetic ivory grips; chambered for 32 H&R Mag. and 38 Spec.; 357 Mag., 45 Colt. Deluxe Grade is highly polished; Deluxe Engraved is engraved in a pattern similar to that used on 1880s derringers. All come in a French fitted jewelry box. Introduced 1991.
Price: 32 H&R Mag. $285.00
Price: 357 Mag. $305.00
Price: 38 Spec. $205.00
Price: 45 Colt $350.00

American Derringer Texas Commemorative
A Model 1 Derringer with solid brass frame, stainless steel barrel and rosewood grips. Available in 38 Spec., 44-40 Win., or 45 Colt. Introduced 1987.
Price: 38 Spec. $285.00
Price: 44-40 or 45 Colt $350.00

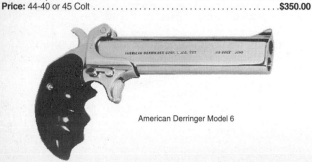
American Derringer Model 6

AMERICAN DERRINGER MODEL 1
Caliber: 22 LR, 22 WMR, 30 Carbine, 30 Luger, 30-30 Win., 32 H&R Mag., 32-20, 380 ACP, 38 Super, 38 Spec., 38 Spec. shotshell, 38 Spec. +P, 9mm Para., 357 Mag., 357 Mag./45/410, 357 Maximum, 10mm, 40 S&W, 41 Mag., 38-40, 44-40 Win., 44 Spec., 44 Mag., 45 Colt, 45 Win. Mag., 45 ACP, 45 Colt/410, 45-70 single shot.
Barrel: 3".
Weight: 15 1/2 oz. (38 Spec.). **Length:** 4.82" overall.
Stocks: Rosewood, Zebra wood.
Sights: Blade front.
Features: Made of stainless steel with high-polish or satin finish. Two-shot capacity. Manual hammer block safety. Introduced 1980. Available in almost any pistol caliber. Contact the factory for complete list of available calibers and prices. From American Derringer Corp.
Price: 22 LR $250.00
Price: 38 Spec. $250.00
Price: 357 Maximum $270.00
Price: 357 Mag. $262.00
Price: 9mm, 380, ...$250.00
Price: 40 S&W $262.00
Price: 44 Spec., ...$325.00
Price: 44-40 Win., 45 Colt $325.00
Price: 30-30, 41, 44 Mags., 45 Win. Mag. ....$380.00 to $390.00
Price: 45-70, single shot $317.00
Price: 45 Colt, 410, 2 1/2" $325.00
Price: 45 ACP, 10mm Auto $262.00

American Derringer Model 4
Similar to the Model 1 except has 4.1" barrel, overall length of 6", and weighs 16 1/2 oz.; chambered for 357 Mag., 357 Maximum, 45-70, 3" 410-bore shotshells or 45 Colt or 44 Mag. Made of stainless steel. Manual hammer block safety. Introduced 1985.
Price: 3" 410/45 Colt $362.00
Price: 3" 410/45 Colt or 45-70 (Alaskan Survival model) $393.00
Price: 44 Mag. with oversize grips $427.00
Price: Alaskan Survival model (45-70 upper, 410 or 45 Colt lower) $393.00

American Derringer Model 6
Similar to the Model 1 except has 6" barrel chambered for 3" 410 shotshells or 22 WMR, 357 Mag., 45 ACP, 45 Colt; rosewood stocks; 8.2" o.a.l. and weighs 21 oz. Shoots either round for each barrel. Manual hammer block safety. Introduced 1986.
Price: 22 WMR $305.00
Price: 357 Mag. $305.00
Price: 45 Colt/410 ....$36800
Price: 45 ACP $350.00

148 GUNS ILLUSTRATED

HANDGUNS—MISCELLANEOUS

American Derringer Model 7 Ultra Lightweight
Similar to Model 1 except made of high strength aircraft aluminum. Weighs 7½ oz., 4.82″ o.a.l., rosewood stocks. Available in 22 LR, 22 WMR, 32 H&R Mag., 380 ACP, 38 Spec., 44 Spec. Introduced 1986.
Price: 22 LR, WMR $245.00
Price: 38 Spec. $245.00
Price: 380 ACP $245.00
Price: 32 H&R Mag. $245.00
Price: 44 Spec. $505.00

American Derringer Model 10 Lightweight
Similar to the Model 1 except frame is of aluminum, giving weight of 10 oz. Stainless barrels. Available in 38 Spec., 45 Colt or 45 ACP only. Matte gray finish. Introduced 1989.
Price: 45 Colt $325.00
Price: 45 ACP $262.00
Price: 38 Spec. $245.00

ANSCHUTZ EXEMPLAR BOLT-ACTION PISTOL
Caliber: 22 LR, 5-shot.
Barrel: 10″.
Weight: 3½ lbs. **Length:** 17″ overall.
Stock: European walnut with stippled grip and forend.
Sights: Hooded front on ramp, open notch rear adjustable for windage and elevation.
Features: Uses Match 64 action with left-hand bolt; Anschutz #5091 two-stage trigger set at 9.85 oz. Receiver grooved for scope mounting; open sights easily removed. The 22 Hornet version uses Match 54 action with left-hand bolt, Anschutz #5099 two-stage trigger set at 19.6 oz. Introduced 1987. Imported from Germany by AcuSport Corp.
Price: 22 LR $579.95
Price: 22 LR, left-hand $469.95
Price: 22 LR single shot $486.95

BOND ARMS PROTECTOR DERRINGER
Caliber: 9mm Para, 38 Spec./357 Mag., 44 Spec./44 Mag., 45 Colt/410 shotshell.
Barrel: 3″.
Weight: 21 oz. **Length:** 5″ overall.
Stocks: Laminated black wood or rosewood.
Sights: Blade front, fixed rear.
Features: Interchangeable barrels; retracting firing pins; rebounding firing pins; cross-bolt safety; removable trigger guard; automatic extractor for rimmed calibers. Stainless steel construction with blasted/polished and ground combination finish. Introduced 1997. Made in U.S. by Bond Arms, Inc.
Price: .... NA

DAVIS DERRINGERS
Caliber: 22 LR, 22 WMR, 25 ACP, 32 ACP.
Barrel: 2.4″.
Weight: 9.5 oz. **Length:** 4″ overall.
Stocks: Laminated wood.
Sights: Blade front, fixed notch rear.
Features: Choice of black Teflon or chrome finish; spur trigger. Introduced 1986. Made in U.S. by Davis Industries.
Price: $75.00

Davis D-38 Derringer

DOWNSIZER SINGLE SHOT PISTOL
Caliber: 9mm Para, 357 Magnum, 40 S&W, 45 ACP.
Barrel: 2.10″
Weight: 11 oz. **Length:** 3.25″ overall.
Stocks: Black composite.
Sights: None.
Features: Single shot, tip-up barrel without extractor. Double action only. Stainless steel construction. Measures .900″ thick. Introduced 1997. From Downsizer Corp.
Price: $299.00

AMERICAN DERRINGER DA 38 MODEL
Caliber: 22 LR, 9mm Para., 38 Spec., 357 Mag., 40 S&W.
Barrel: 3″.
Weight: 14.5 oz. **Length:** 4.8″ overall.
Stocks: Rosewood, walnut or other hardwoods.
Sights: Fixed.
Features: Double-action only; two-shots. Manual safety. Made of satin-finished stainless steel and aluminum. Introduced 1989. From American Derringer Corp.
Price: 22 LR, 38 Spec. $305.00
Price: 9mm Para. $330.00
Price: 357 Mag., 40 S&W $355.00

Anschutz Exemplar

Bond Arms Protector

DAVIS LONG-BORE DERRINGERS
Caliber: 22 WMR, 32 H&R Mag., 38 Spec., 9mm Para.
Barrel: 3.5″.
Weight: 16 oz. **Length:** 5.4″ overall.
Stocks: Textured black synthetic.
Sights: Fixed.
Features: Chrome or black teflon finish. Larger than Davis D-Series models. Introduced 1995. Made in U.S. by Davis Industries.
Price: $104.00
Price: Big-Bore models (same calibers, ¾″ shorter barrels) $98.00

DAVIS D-SERIES DERRINGERS
Caliber: 22 WMR, 32 H&R, 38 Spec..
Barrel: 2.75″.
Weight: 11.5 oz. **Length:** 4.65″ overall.
Stocks: Textured black synthetic.
Sights: Blade front, fixed notch rear.
Features: Alloy frame, steel-lined barrels, steel breech block. Plunger-type safety with integral hammer block. Chrome or black Teflon finish. Introduced 1992. Made in U.S. by Davis Industries.
Price: $98.00

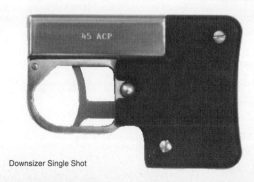

Downsizer Single Shot

HANDGUNS—MISCELLANEOUS

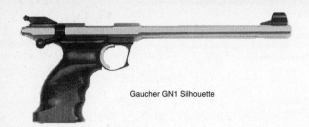

Gaucher GN1 Silhouette

GAUCHER GN1 SILHOUETTE PISTOL
Caliber: 22 LR, single shot.
Barrel: 10″.
Weight: 2.4 lbs. **Length:** 15.5″ overall.
Stocks: European hardwood.
Sights: Blade front, open adjustable rear.
Features: Bolt action, adjustable trigger. Introduced 1990. Imported from France by Mandall Shooting Supplies.
Price: About $525.00
Price: Model GP Silhouette $425.00

HJS FRONTIER FOUR DERRINGER
Caliber: 22 LR.
Barrel: 2″.
Weight: 5 1/2 oz. **Length:** 3 15/16″ overall.
Stocks: Brown plastic.
Sights: None.
Features: Four barrels fire with rotating firing pin. Stainless steel construction. Introduced 1993. Made in U.S. by HJS Arms, Inc.
Price: $165.00

HJS LONE STAR DERRINGER
Caliber: 380 ACP.
Barrel: 2″.
Weight: 6 oz. **Length:** 3 15/16″ overall.
Stocks: Brown plastic.
Sights: Groove.
Features: Stainless steel construction. Beryllium copper firing pin. Button-rifled barrel. Introduced 1993. Made in U.S. by HJS Arms, Inc.
Price: $185.00

IAR MODEL 1888 DOUBLE DERRINGER
Caliber: 38 Special.
Barrel: 2 3/4″.
Weight: 16 oz. **Length:** NA.
Stocks: Smooth walnut.
Sights: Blade front, notch rear.
Features: All steel construction. Blue barrel, color case-hardened frame. Uses original designs and tooling for the Uberti New Maverick Derringer. Introduced 1997. Made in U.S. by IAR, Inc.
Price: $225.00

HJS Frontier Four

HJS Antigua Derringer
Same as the Frontier Four except blued barrel, brass frame, brass pivot pins. Brown plastic grips. Introduced 1994. Made in U.S. by HJS Arms, Inc.
Price: $180.00

IAR MODEL 1872 DERRINGER
Caliber: 22 Short.
Barrel: 2 3/8″.
Weight: 7 oz. **Length:** 5 1/8″ overall.
Stocks: Smooth walnut.
Sights: Blade front, notch rear.
Features: Gold or nickel frame with blue barrel. Reintroduced 1996 using original Colt designs and tooling for the Colt model 4 Derringer. Made in U.S. by IAR, Inc.
Price: $85.00
Price: Single cased gun $115.00
Price: Double cased set $189.00

> Consult our Directory pages for the location of firms mentioned.

IAR Model 1888 Derringer

MAGNUM RESEARCH LONE EAGLE SINGLE SHOT PISTOL
Caliber: 22 Hornet, 223, 22-250, 243, 7mm BR, 7mm-08, 30-30, 7.62x39, 308, 30-06, 357 Max., 35 Rem., 358 Win., 44 Mag., 444 Marlin.
Barrel: 14″, interchangable.
Weight: 4lbs., 3 oz. to 4 lbs., 7 oz. **Length:** 15″ overall.
Stocks: Ambidextrous.
Sights: None furnished; drilled and tapped for scope mounting and open sights. Open sights optional.
Features: Cannon-type rotating breech with spring-activated ejector. Ordnance steel with matte blue finish. Cross-bolt safety. External cocking lever on left side of gun. Muzzle break optional. Introduced 1991. Available from Magnum Research, Inc.
Price: Complete pistol, black $408.00
Price: Barreled action only, black $289.00
Price: Complete pistol, chrome $438.00
Price: Barreled action, chrome $319.00
Price: Scope base $14.00
Price: Adjustable open sights $35.00

MANDALL/CABANAS PISTOL
Caliber: 177, pellet or round ball; single shot.
Barrel: 9″.
Weight: 51 oz. **Length:** 19″ overall.
Stock: Smooth wood with thumbrest.
Sights: Blade front on ramp, open adjustable rear.
Features: Fires round ball or pellets with 22 blank cartridge. Automatic safety; muzzlebrake. Imported from Mexico by Mandall Shooting Supplies.
Price: $139.95

LORCIN OVER/UNDER DERRINGER
Caliber: 38 Spec./357 Mag., 45 ACP.
Barrel: 3.5″.
Weight: NA. **Length:** 6.5″ overall.
Stocks: Black composition.
Sights: Blade front, fixed rear.
Features: Stainless steel construction. Rebounding hammer. Introduced 1996. Made in U.S. by Lorcin Engineering.
Price: $129.00

Magnum Research Lone Eagle

HANDGUNS—MISCELLANEOUS

Maximum Single Shot

New Advantage Derringer

MAXIMUM SINGLE SHOT PISTOL
Caliber: 22 LR, 22 Hornet, 22 BR, 22 PPC, 223 Rem., 22-250, 6mm BR, 6mm PPC, 243, 250 Savage, 6.5mm-35M, 270 MAX, 270 Win., 7mm TCU, 7mm BR, 7mm-35, 7mm INT-R, 7mm-08, 7mm Rocket, 7mm Super Mag., 30 Herrett, 30 Carbine, 30-30, 308 Win., 30x39, 32-20, 350 Rem. Mag., 357 Mag., 357 Maximum, 358 Win., 44 Mag., 454 Casull.
Barrel: 8 3/4", 10 1/2", 14".
Weight: 61 oz. (10 1/2" bbl.); 78 oz. (14" bbl.). **Length:** 15", 18 1/2" overall (with 10 1/2" and 14" bbl., respectively).
Stocks: Smooth walnut stocks and forend. Also available with 17° finger groove grip.
Sights: Ramp front, fully adjustable open rear.
Features: Falling block action; drilled and tapped for M.O.A. scope mounts; integral grip frame/receiver; adjustable trigger; Douglas barrel (interchangeable). Introduced 1983. Made in U.S. by M.O.A. Corp.
Price: Stainless receiver, blue barrel $653.00
Price: Stainless receiver, stainless barrel $711.00
Price: Extra blued barrel $164.00
Price: Extra stainless barrel $222.00
Price: Scope mount $52.00

NEW ADVANTAGE ARMS DERRINGER
Caliber: 22 LR, 22 WMR, 4-shot.
Barrel: 2 1/2".
Weight: 15 oz. **Length:** 4 1/2" overall.
Stocks: Smooth walnut.
Sights: Fixed.
Features: Double-action mechanism, four barrels, revolving hammer with four firing pins. Rebounding hammer. Blue or stainless. Reintroduced 1989. From New Advantage Arms Corp.
Price: 22 LR, 22 WMR, blue, about $249.99
Price: As above, stainless, about $249.99

RPM XL SINGLE SHOT PISTOL
Caliber: 22 LR through 45-70.
Barrel: 8", 10 3/4", 12", 14".
Weight: About 60 oz. **Length:** NA.
Stocks: Smooth Goncalo Alves with thumb and heel rests.
Sights: Hooded front with interchangeable post, or Patridge; ISGW rear adjustable for windage and elevation.
Features: Barrel drilled and tapped for scope mount. Visible cocking indicator. Spring-loaded barrel lock, positive hammer-block safety. Trigger adjustable for weight of pull and over-travel. Contact maker for complete price list. Made in U.S. by RPM.
Price: Hunter model (stainless frame, 5/16" underlug, latch lever and positive extractor) $1,295.00
Price: Extra barrel, 8" through 10 3/4" $287.50
Price: Muzzle brake $100.00

SUNDANCE POINT BLANK O/U DERRINGER
Caliber: 22 LR, 2-shot.
Barrel: 3".
Weight: 8 oz. **Length:** 4.6" overall.
Stocks: Grooved composition.
Sights: Blade front, fixed notch rear.
Features: Double-action trigger, push-bar safety, automatic chamber selection. Fully enclosed hammer. Matte black finish. Introduced 1994. Made in U.S. by Sundance Industries.
Price: $99.00

TEXAS LONGHORN "THE JEZEBEL" PISTOL
Caliber: 22 Short, Long, Long Rifle, single shot.
Barrel: 6".
Weight: 15 oz. **Length:** 8" overall.
Stocks: One-piece fancy walnut grip (right- or left-hand), walnut forend.
Sights: Bead front, fixed rear.
Features: Handmade gun. Top-break action; all stainless steel; automatic hammer block safety; music wire coil springs. Barrel is half-round, half-octagon. Announced 1986. From Texas Longhorn Arms.
Price: About $250.00

RPM XL Pistol

Sundance Point Blank

THE JUDGE SINGLE SHOT PISTOL
Caliber: 22 Hornet, 22 K-Hornet, 218 Bee, 7-30 Waters, 30-30.
Barrel: 10" or 16.2".
Weight: NA. **Length:** NA.
Stocks: Walnut.
Sights: Bead on ramp front, open adjustable rear.
Features: Break-open design; made of 17-4 stainless steel. Also available as a kit. Introduced 1995. Made in U.S. by Cumberland Mountain Arms.
Price: .... NA

THOMPSON/CENTER ENCORE PISTOL
Caliber: 22-250, 223, 7mmBR, 7mm-08, 243, 7.62x39, 308, 270, 30-06, 44 Mag., 444 Marlin single shot.
Barrel: 10", 15", tapered round.
Weight: NA. **Length:** 19" overall with 10" barrel.
Stocks: American walnut with finger grooves, walnut forend.
Sights: Blade on ramp front, adjustable rear, or none.
Features: Interchangeable barrels; action opens by squeezing the trigger guard; drilled and tapped for scope mounting; blue finish. Announced 1996. Made in U.S. by Thompson/Center Arms.
Price: $495.00
Price: Extra 10" barrels $215.00

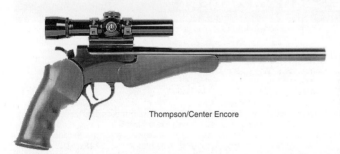

Thompson/Center Encore

HANDGUNS—MISCELLANEOUS

T/C Contender

THOMPSON/CENTER CONTENDER
Caliber: 7mm TCU, 30-30 Win., 22 LR, 22 WMR, 22 Hornet, 223 Rem., 270 Ren, 7-30 Waters, 32-20 Win., 357 Mag., 357 Rem. Max., 44 Mag., 10mm Auto, 445 Super Mag., 45/410, single shot.
Barrel: 10″, tapered octagon, bull barrel and vent. rib.
Weight: 43 oz. (10″ bbl.). **Length:** 13 1/4″ (10″ bbl.).
Stocks: T/C "Competitor Grip." Right or left hand.
Sights: Under-cut blade ramp front, rear adjustable for windage and elevation.
Features: Break-open action with automatic safety. Single-action only. Interchangeable bbls., both caliber (rim & centerfire), and length. Drilled and tapped for scope. Engraved frame. See T/C catalog for exact barrel/caliber availability.
Price: Blued (rimfire cals.) $463.50
Price: Blued (centerfire cals.) $463.50
Price: Extra bbls. (standard octagon) $213.70
Price: 45/410, internal choke bbl. $218.90

Thompson/Center Stainless Super 14, Super 16 Contender
Same as the standard Super 14 and Super 16 except they are made of stainless steel with blued sights. Both models have black Rynite forend and fingergroove, ambidextrous grip with a built-in rubber recoil cushion that has a sealed-in air pocket. Receiver has a different cougar etching. Available in 22 LR, 22 LR Match, 22 Hornet, 223 Rem., 30-30 Win., 35 Rem. (Super 14), 45-70 (Super 16 only), 45 Colt/410. Introduced 1993.
Price: 14″ bull barrel $504.70
Price: 16 1/4″ bull barrel $509.90
Price: 45 Colt/410, 14″ $535.60
Price: 45 Colt/410, 16″ $540.80

Thompson/Center Stainless Contender
Same as the standard Contender except made of stainless steel with blued sights, black Rynite forend and ambidextrous finger-groove grip with a built-in rubber recoil cushion that has a sealed-in air pocket. Receiver has a different cougar etching. Available with 10″ bull barrel in 22 LR, 22 LR Match, 22 Hornet, 223 Rem., 30-30 Win., 357 Mag., 44 Mag., 45 Colt/410. Introduced 1993.
Price: $494.40
Price: 45 Colt/410 .... $499.60
Price: With 22 LR match chamber $504.70

Thompson/Center Contender Hunter Package
Package contains the Contender pistol in 223, 7-30 Waters, 30-30, 375 Win., 357 Rem. Maximum, 35 Rem., 44 Mag. or 45-70 with 14″ barrel with T/C's Muzzle Tamer, a 2.5x Recoil Proof Long Eye Relief scope with lighted reticle, q.d. sling swivels with a nylon carrying sling. Comes with a suede leather case with foam padding and fleece lining. Introduced 1990. From Thompson/Center Arms.
Price: Blued $798.00
Price: Stainless $829.00

UBERTI ROLLING BLOCK TARGET PISTOL
Caliber: 22 LR, 22 WMR, 22 Hornet, 357 Mag., 45 Colt, single shot.
Barrel: 9 7/8″, half-round, half-octagon.
Weight: 44 oz. **Length:** 14″ overall.
Stocks: Walnut grip and forend.
Sights: Blade front, fully adjustable rear.
Features: Replica of the 1871 rolling block target pistol. Brass trigger guard, color case-hardened frame, blue barrel. Imported by Uberti U.S.A.
Price: $410.00

ULTRA LIGHT ARMS MODEL 20 REB HUNTER'S PISTOL
Caliber: 22-250 thru 308 Win. standard. Most silhouette calibers and others on request. 5-shot magazine.
Barrel: 14″, Douglas No. 3.
Weight: 4 lbs.
Stock: Composite Kevlar, graphite reinforced. Du Pont Imron paint in green, brown, black and camo.
Sights: None furnished. Scope mount included.
Features: Timney adjustable trigger; two-position, three-function safety; benchrest quality action; matte or bright stock and metal finish; right- or left-hand action. Shipped in hard case. Introduced 1987. From Ultra Light Arms.
Price: $1,600.00

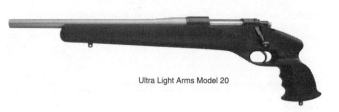
Ultra Light Arms Model 20

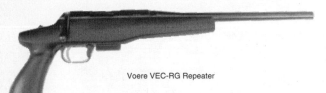
Voere VEC-RG Repeater

VOERE VEC-95CG SINGLE SHOT PISTOL
Caliber: 5.56mm, 6mm UCC caseless, single shot.
Barrel: 12″, 14″.
Weight: 3 lbs. **Length:** NA.
Stock: Black synthetic; center grip.
Sights: None furnished.
Features: Fires caseless ammunition via electronic ignition; two batteries in the grip last about 500 shots. Bolt action has two forward locking lugs. Tang safety. Drilled and tapped for scope mounting. Introduced 1995. Imported from Austria by JagerSport, Ltd.
Price: $1,495.00

Voere VEC-RG Repeater pistol
Similar to the VEC-95CG except has rear grip stock and detachable 5-shot magazine. Available with 12″ or 14″ barrel. Introduced 1995. Imported from Austria by JagerSport, Ltd.
Price: $1,495.00

CENTERFIRE RIFLES—AUTOLOADERS

Both classic arms and recent designs in American-style repeaters for sport and field shooting.

AA ARMS AR9 SEMI-AUTOMATIC RIFLE
Caliber: 9mm Para., 10-shot magazine.
Barrel: 16″.
Weight: 6 lbs. **Length:** 31″ overall.
Stock: Folding metal skeleton.
Sights: Post front adjustable for elevation, open rear for windage.
Features: Ventilated barrel shroud. Blue or electroless nickel finish. Made in U.S. by AA Arms, Inc.
Price: Blue $695.00

ARMALITE AR-10A4 RIFLE.
Caliber: 308 Win., 10-shot magazine.
Barrel: 20″ chrome-lined, 1:12″ twist.
Weight: 9.6 lbs. **Length:** 41″ overall
Stock: Green or black composition.
Sights: Detachable handle, front sight, or scope mount available; comes with international style flattop receiver with Picatinny rail.
Features: Proprietary recoil check. Forged upper receiver with case deflector. Receivers are hard-coat anodized. Introduced 1995. Made in U.S. by ArmaLite, Inc.
Price: $1,325.00

CENTERFIRE RIFLES—AUTOLOADERS

AMALITE M15A2 M4C CARBINE
Caliber: 223, 7-shot magazine.
Barrel: 16" heavy chrome lined; 1:9" twist.
Weight: 7 lbs. **Length:** 35$^{11}/_{16}$" overall.
Stock: Black retractable.
Sights: Standard A2.
Features: Upper and lower receivers have push-type pivot pin; hard coat anodized; A2-style forward assist; M16A2-type raised fence around magazine release button. Made in U.S. by ArmaLite, Inc.
Price: ....NA

AUTO-ORDNANCE 27 A-1 THOMPSON
Caliber: 45 ACP, 30-shot magazine.
Barrel: 16".
Weight: 11$^1/_2$ lbs. **Length:** About 42" overall (Deluxe).
Stock: Walnut stock and vertical forend.
Sights: Blade front, open rear adjustable for windage.
Features: Recreation of Thompson Model 1927. Semi-auto only. Deluxe model has finned barrel, adjustable rear sight and compensator; Standard model has plain barrel and military sight. From Auto-Ordnance Corp.
Price: Deluxe $795.00
Price: 1927A1C Lightweight model $767.00

Thompson M1

Auto-Ordnance Thompson M1
Similar to the Model 27 A-1 except is in the M-1 configuration with side cocking knob, horizontal forend, smooth unfinned barrel, sling swivels on butt and forend. Matte black finish. Introduced 1985.
Price: $772.50

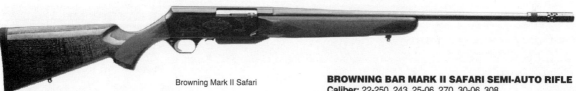
Barrett Model 82A-1

BARRETT MODEL 82A-1 SEMI-AUTOMATIC RIFLE
Caliber: 50 BMG, 10-shot detachable box magazine.
Barrel: 29".
Weight: 28.5 lbs. **Length:** 57" overall.
Stock: Composition with Sorbothane recoil pad.
Sights: Scope optional.
Features: Semi-automatic, recoil operated with recoiling barrel. Three-lug locking bolt; muzzlebrake. Self-leveling bipod. Fires same 50-cal. ammunition as the M2HB machinegun. Introduced 1985. From Barrett Firearms.
Price: From $6,800.00

Browning Mark II Safari

BROWNING BAR MARK II SAFARI SEMI-AUTO RIFLE
Caliber: 22-250, 243, 25-06, 270, 30-06, 308.
Barrel: 22" round tapered.
Weight: 7$^3/_8$ lbs. **Length:** 43" overall.
Stock: French walnut pistol grip stock and forend, hand checkered.
Sights: Gold bead on hooded ramp front, click adjustable rear, or no sights.
Features: Has new bolt release lever; removable trigger assembly with larger trigger guard; redesigned gas and buffer systems. Detachable 4-round box magazine. Scroll-engraved receiver is tapped for scope mounting. BOSS barrel vibration modulator and muzzlebrake system available only on models without sights. Mark II Safari introduced 1993. Imported from Belgium by Browning.
Price: Safari, with sights $729.95
Price: Safari, no sights $713.95
Price: Safari, no sights, BOSS $785.25

Browning BAR Mark II Safari Magnum Rifle
Same as the standard caliber model, except weighs 8$^3/_8$ lbs., 45" overall, 24" bbl., 3-round mag. Cals. 7mm Mag., 300 Win. Mag., 338 Win. Mag. BOSS barrel vibration modulator and muzzlebrake system available only on models without sights. Introduced 1993.
Price: Safari, with sights $781.95
Price: Safari, no sights $765.95
Price: Safari, no sights, BOSS $837.25

Browning BAR MARK II Lightweight Semi-Auto
Similar to the Mark II Safari except has lighter alloy receiver and 20" barrel. Available in 243, 308, 270 and 30-06, weighs 7 lbs., 2 oz.; overall length 41". Has dovetailed, gold bead front sight on hooded ramp, open rear click adjustable for windage and elevation. BOSS system not available. Introduced 1997. Imported from Belgium by Browning.
Price: $730.00

Consult our Directory pages for the location of firms mentioned.

BUSHMASTER M17S BULLPUP RIFLE
Caliber: 223, 10-shot magazine.
Barrel: 21.5", heavy; 1:9" twist.
Weight: 8.2 lbs. **Length:** 30" overall.
Stock: Fiberglass-filled nylon.
Sights: Has 25-meter open emergency sights; designed for optics mounted to rail on carrying handle for Weaver-type rings.
Features: Gas-operated, short-stroke piston system; ambidextrous magazine release. Introduced 1993. Made in U.S. by Bushmaster Firearms, Inc./Quality Parts Co.
Price: $575.00

BUSHMASTER SHORTY XM-15 E2S CARBINE
Caliber: 223, 30-shot magazine.
Barrel: 16", heavy; 1:9" twist.
Weight: 7.3 lbs. **Length:** 34.5" overall.
Stock: Fixed black composition.
Sights: Adjustable post front, adjustable aperture rear.
Features: Patterned after Colt M-16A2. Chrome-lined barrel with manganese phosphate finish. "Shorty" handguards. Has E-2 lower receiver with push-pin. Made in U.S. by Bushmaster Firearms Inc./Quality Parts Co.
Price: $730.00
Price: XM-15 E-2S Dissipator ("Dissipator") full-length handguard) $740.00

CENTERFIRE RIFLES—AUTOLOADERS

CALICO LIBERTY 50, 100 CARBINES
Caliber: 9mm Para.
Barrel: 16.1".
Weight: 7 lbs. **Length:** 34.5" overall.
Stock: Glass-filled, impact resistant polymer.
Sights: Adjustable front post, fixed notch and aperture flip rear.
Features: Helical feed magazine; ambidextrous, rotating sear/striker block safety; static cocking handle; retarded blowback action; aluminum alloy receiver. Introduced 1995. Made in U.S. by Calico.
Price: Liberty 50 ....$648.00
Price: Liberty 100 ....$684.00

Calico Liberty 50

Century L1A1 Sporter

CENTURY INTERNATIONAL L1A1 SPORTER RIFLE
Caliber: 308 Win.
Barrel: 20.75".
Weight: 9 lbs., 13 oz. **Length:** 41.125" overall.
Stock: Bell & Carlson thumbhole sporter.
Sights: Protected post front, adjustable aperture rear.
Features: Matte blue finish; rubber butt pad. From Century International Arms.
Price: About $595.00

Colt Match Target Lightweight

COLT MATCH TARGET LIGHTWEIGHT RIFLE
Caliber: 9mm Para., 223 Rem., 5-shot magazine.
Barrel: 16".
Weight: 6.7 lbs. (223); 7.1 lbs. (9mm Para.). **Length:** 34.5" overall.
Stock: Composition stock, grip, forend.
Sights: Post front, rear adjustable for windage and elevation.
Features: 5-round detachable box magazine, flash suppressor, sling swivels. Forward bolt assist included. Introduced 1991.
Price: $987.00

Daewoo DR200

DAEWOO DR200, DR300 AUTOLOADING RIFLES
Caliber: 223 Rem., 7.62x39mm, 6-shot magazine.
Barrel: 18.3".
Weight: 9 lbs. **Length:** 39.2" overall.
Stock: Synthetic thumbhole style with rubber buttpad.
Sights: Post front in ring, aperture rear adjustable for windage and elevation.
Features: Forged aluminum receiver; bolt, bolt carrier, firing pin, piston and recoil spring contained in one assembly. Rotating bolt locking. Uses all AR-15 magazines. Introduced 1995. Imported from Korea by Kimber of America, Inc.
Price: DR200, 223 Rem. $535.00
Price: DR300, 7.62x39mm $750.00

EAGLE ARMS M15A2 HBAR
Caliber: 223, 30-shot magazine.
Barrel: 20" chrome moly premium heavy; 1:9" twist.
Weight: 8 lbs. **Length:** 39$^{5}/_{8}$" overall.
Stock: Black or green composition.
Sights: Standard A2.
Features: Pre-ban rifle with full front sight housing with bayonet lug; threaded barrel; flash suppressor; push-type pivot pin; hard coat anodized A2-style forward assist; fence-type magazine release. Made in U.S. by Armalite, Inc.
Price: $1,100.00

EAGLE ARMS M15A4 CARBINE
Caliber: 223, 30-shot magazine.
Barrel: 16" heavy chrome moly; 1:9" twist.
Weight: 6.2 lbs. **Length:** 30$^{5}/_{8}$" overall.
Stock: Retractable; black composition.
Sights: One-piece international-style flattop receiver with Picatinny rail; comes with detachable carry handle with N.M. sights.
Features: Pre-ban rifle with full front sight housing with bayonet lug, threaded barrel with 4.33" fixed flash suppressor; hard coat anodized; A1-style forward assist; M16A2-type raised fence around magazine release. Made in U.S. by ArmaLite, Inc.
Price: $1,160.00

Hi-Point 9mm Carbine

HI-POINT 9mm CARBINE
Caliber: 9mm Para., 10-shot magazine.
Barrel: 16$^{1}/_{2}$".
Weight: NA. **Length:** 31$^{1}/_{2}$" overall.
Stock: Black polymer.
Sights: Protected post front, aperture rear. Integral scope mount.
Features: Grip-mounted magazine release. Parkerized or chrome finish. Sling swivels. Introduced 1996. Made in U.S. by MKS Supply, Inc.
Price: $169.00

CENTERFIRE RIFLES—AUTOLOADERS

IBUS M17S 223 BULLPUP RIFLE
Caliber: 223, 10-shot magazine.
Barrel: 21.5".
Weight: 8.2 lbs. **Length:** 30" overall.
Stock: Zytel glass-filled nylon.
Sights: None furnished. Comes with scope mount for Weaver-type rings.
Features: Gas-operated, short-stroke piston system. Ambidextrous magazine release. Introduced 1993. Made in U.S. by Bushmaster Firearms Inc./Quality Parts Co.
Price: $975.00

Kel-Tec Sub-9

KEL-TEC SUB-9 AUTO RIFLE
Caliber: 9mm Para.
Barrel: 16.1".
Weight: 4.6 lbs. **Length:** 30" overall (extended), 15.9" (closed).
Stock: Metal tube; grooved rubber butt pad.
Sights: Hooded post front, flip-up rear. Interchangeable grip assemblies allow use of most double-column high capacity pistol magazines.
Features: Barrel folds back over the butt for transport and storage. Introduced 1997. Made in U.S. by Kel-Tec CNC Industries, Inc.
Price: $700.00

LR 300 SR LIGHT SPORT RIFLE
Caliber: 223.
Barrel: 16 1/4"; 1:9" twist.
Weight: 7.2 lbs. **Length:** 36" overall (extended stock), 26 1/4" (stock folded).
Stock: Folding, tubular steel, with thumbhold-type grip.
Sights: Trijicon post front, Trijicon rear.
Features: Uses AR-15 type upper and lower receivers; flat-top receiver with weaver base. Accepts all AR-15/M-16 magazines. Introduced 1996. made in U.S. from Z-M weapons.
Price: $2,550.00

Marlin Model 45

MARLIN MODEL 9 CAMP CARBINE
Caliber: 9mm Para., 12-shot magazine.
Barrel: 16 1/2", Micro-Groove® rifling.
Weight: 6 3/4 lbs. **Length:** 35 1/2" overall.
Stock: Press-checkered walnut-finished Maine birch; rubber buttpad; Mar-Shield™ finish; swivel studs.
Sights: Ramp front with orange post, cutaway Wide-Scan™ hood, adjustable open rear.
Features: Manual bolt hold-open; Garand-type safety, magazine safety; loaded chamber indicator; receiver drilled, tapped for scope mounting. Introduced 1985.
Price: $438.00

Marlin Model 45 Carbine
Similar to the Model 9 except chambered for 45 ACP, 7-shot magazine. Introduced 1986.
Price: $438.00

Olympic PCR-5

OLYMPIC ARMS PCR-5, PCR-6 RIFLES
Caliber: 9mm Para., 40 S&W, 45 ACP, 223, 7.62x39mm (PCR-6), 10-shot magazine.
Barrel: 16".
Weight: 7 lbs. **Length:** 34.75" overall.
Stock: A2 stowaway grip, trapdoor buttstock.
Sights: Post front, A1 rear adjustable for windage.
Features: Based on the CAR-15. No bayonet lug. Button-cut rifling. Introduced 1994. Made in U.S. by Olympic Arms, Inc.
Price: 9mm Para., 40 S&W, 45 ACP $830.00
Price: 223 Rem. $785.00
Price: 7.62x39mm (PCR-6) $845.00

OLYMPIC ARMS PCR-4 RIFLE
Caliber: 223, 10-shot magazine.
Barrel: 20".
Weight: 8 lbs., 5 oz. **Length:** 38.25" overall.
Stock: A2 stowaway grip, trapdoor buttstock.
Sights: Post front, A1 rear adjustable for windage.
Features: Based on the AR-15 rifle. Barrel is button rifled with 1:9" twist. No bayonet lug. Introduced 1994. Made in U.S. by Olympic Arms, Inc.
Price: $820.00

Remington Model 7400p

REMINGTON MODEL 7400 AUTO RIFLE
Caliber: 243 Win., 270 Win., 280 Rem., 308 Win., 30-06, 4-shot magazine.
Barrel: 22" round tapered.
Weight: 7 1/2 lbs. **Length:** 42" overall.
Stock: Walnut, deluxe cut checkered pistol grip and forend. Satin or high-gloss finish.
Sights: Gold bead front sight on ramp; step rear sight with windage adjustable.
Features: Redesigned and improved version of the Model 742. Positive cross-bolt safety. Receiver tapped for scope mount. Comes with green Remington hard case. Introduced 1981.
Price: About $573.00
Price: Carbine (18 1/2" bbl., 30-06 only) $573.00

CAUTION: PRICES SHOWN ARE SUPPLIED BY THE MANUFACTURER OR IMPORTER. CHECK YOUR LOCAL GUNSHOP.

CENTERFIRE RIFLES—AUTOLOADERS

Ruger PC4 Carbine

RUGER PC4, PC9 CARBINES
Caliber: 9mm Para., 40 S&W, 10-shot magazine.
Barrel: 16.25".
Weight: 6 lbs., 4oz. **Length:** 34.75" overall.
Stock: Black DuPont (Zytel) with checkered grip and forend.
Sights: Blade front, open adjustable rear; integral Ruger scope mounts.
Features: Delayed blowback action; manual push-button cross bolt safety and internal firing pin block safety automatic slide lock. Introduced 1997. Made in U.S. by Sturm, Ruger & Co.
Price: PC4, 40 S&W $550.00
Price: PC9, 9mm Para $550.00

Ruger Mini-14/5

Ruger Mini Thirty Rifle
Similar to the Mini-14 Ranch Rifle except modified to chamber the 7.62x39 Russian service round. Weight is about 7 lbs., 3 oz. Has 6-groove barrel with 1:10" twist, Ruger Integral Scope Mount bases and folding peep rear sight. Detachable 5-shot staggered box magazine. Blued finish. Introduced 1987.
Price: Blue $556.00
Price: Stainless $609.00

RUGER MINI-14/5 AUTOLOADING RIFLE
Caliber: 223 Rem., 5-shot detachable box magazine.
Barrel: 18 1/2". Rifling twist 1:9".
Weight: 6.4 lbs. **Length:** 37 1/4" overall.
Stock: American hardwood, steel reinforced.
Sights: Ramp front, fully adjustable rear.
Features: Fixed piston gas-operated, positive primary extraction. New buffer system, redesigned ejector system. Ruger S100RH scope rings included. 20-, 30- shot magazine available to police departments and government agencies only.
Price: Mini-14/5R, Ranch Rifle, blued, scope rings $556.00
Price: K-Mini-14/5R, Ranch Rifle, stainless, scope rings $609.00
Price: Mini-14/5, blued, no scope rings $516.00
Price: K-Mini-14/5, stainless, no scope rings $569.00

Springfield M1A

SPRINGFIELD, INC. SAR-4800 RIFLE
Caliber: 5.56, 7.62 NATO (308 Win.), 20-shot magazine.
Barrel: 21".
Weight: 9.5 lbs. **Length:** 43.3" overall.
Stock: Fiberglass forend, composite thumbhole butt.
Sights: Protected post front, adjustable peep rear.
Features: New production. Reintroduced 1995. From Springfield, Inc.
Price: $1,249.00

SPRINGFIELD, INC. M1A RIFLE
Caliber: 7.62mm NATO (308), 5-, 10- or 20-shot box magazine.
Barrel: 25 1/16" with flash suppressor, 22" without suppressor.
Weight: 8 3/4 lbs. **Length:** 44 1/4" overall.
Stock: American walnut with walnut-colored heat-resistant fiberglass handguard. Matching walnut handguard available. Also available with fiberglass stock.
Sights: Military, square blade front, full click-adjustable aperture rear.
Features: Commercial equivalent of the U.S. M-14 service rifle with no provision for automatic firing. From Springfield, Inc.
Price: M1A-A1, Scout Rifle, black fiberglass stock $1,459.00
Price: Standard M1A rifle, about $1,381.00
Price: National Match, about $1,729.00
Price: Super Match (heavy premium barrel), about $2,050.00
Price: M1A-A1 Bush Rifle, walnut stock, about $1,359.00
Price: M1A-A1 Collector, G.I. stock $1,307.00

Springfield SAR-8

SPRINGFIELD, INC. SAR-8 SPORTER RIFLE
Caliber: 308 Win., 20-shot magazine.
Barrel: 18".
Weight: 8.7 lbs. **Length:** 40.3" overall.
Stock: Black composition, thumbhole buttstock.
Sights: Protected post front, rotary-style adjustable rear.
Features: Delayed roller-lock action; fluted chamber; matte black finish. Reintroduced 1995. From Springfield, Inc.
Price: $1,204.00

SA-85M SEMI-AUTO RIFLE
Caliber: 7.62x39mm, 6-shot magazine.
Barrel: 16.3".
Weight: 7.6 lbs. **Length:** 34.7" overall.
Stock: European hardwood; thumbhole design.
Sights: Post front, lpen adjustable rear.
Features: BATF-approved version of the Kalashnikov rifle. Gas operated. Black phosphate finish. Comes with one magazine, cleaning rod, cleaning/tool kit. Introduced 1995. Imported from Hungary by K.B.I., Inc.
Price: $399.00

STONER SR-25 CARBINE
Caliber: 7.62 NATO, 10-shot steel magazine.
Barrel: 16" free-floating.
Weight: 7 3/4 lbs. **Length:** 35.75" overall.
Stock: Black synthetic.
Sights: Integral Weaver-style rail. Scope rings, iron sights optional.
Features: Shortened, non-slip handguard; removable carrying handle. Matte black finish. Introduced 1995. Made in U.S. by Knight's Mfg. Co.
Price: $2,995.00

CENTERFIRE RIFLES—AUTOLOADERS

STONER SR-50 LONG RANGE PRECISION RIFLE
Caliber: 50 BMG, 10-shot magazine.
Barrel: 35.5".
Weight: 31.5 lbs. **Length:** 58.37" overall.
Stock: Tubular steel.
Sights: Scope mount.
Features: Gas-operated semi-automatic action; two-stage target-type trigger; M-16-type safety lever; easily removable barrel. Introduced 1996. Made in U.S. by Knight's Mfg. Co.
Price: ....NA

WILKINSON TERRY CARBINE
Caliber: 9mm Para., 31-shot magazine.
Barrel: 16 3/16".
Weight: 6 lbs., 3 oz. **Length:** 30" overall.
Stock: Maple stock and forend.
Sights: Protected post front, aperture rear.
Features: Semi-automatic blowback action fires from a closed breech. Bolt-type safety and magazine catch. Ejection port has automatic trap door. Receiver equipped with dovetail for scope mounting. Made in U.S. From Wilkinson Arms.
Price: $636.29

CENTERFIRE RIFLES—LEVER & SLIDE

Both classic arms and recent designs in American-style repeaters for sport and field shooting.

American Arms/Uberti 1866 Sporting

AMERICAN ARMS/UBERTI 1873 SPORTING RIFLE
Caliber: 44-40, 45 Colt.
Barrel: 24 1/4", octagonal.
Weight: 8.1 lbs. **Length:** 43 1/4" overall.
Stock: Walnut.
Sights: Blade front adjustable for windage, open rear adjustable for elevation.
Features: Color case-hardened frame, blued barrel, hammer, lever, buttplate, brass elevator. Imported from Italy by American Arms, Inc.
Price: $984.00

AMERICAN ARMS/UBERTI 1860 HENRY RIFLE
Caliber: 44-40, 45 Colt.
Barrel: 24 1/4", half-octagon.
Weight: 9.2 lbs. **Length:** 43 3/4" overall.
Stock: American walnut.

AMERICAN ARMS/UBERTI 1866 SPORTING RIFLE, CARBINE
Caliber: 22 LR, 22 WMR, 38 Spec., 44-40, 45 Colt.
Barrel: 24 1/4", octagonal.
Weight: 8.1 lbs. **Length:** 43 1/4" overall.
Stock: Walnut.
Sights: Blade front adjustable for windage, rear adjustable for elevation.
Features: Frame, buttplate, forend cap of polished brass, balance charcoal blued. Imported by American Arms, Inc.
Price: $829.00
Price: Yellowboy Carbine (19" round bbl.) $797.00

American Arms/Uberti 1873 Deluxe
Similar to the 1873 Sporting Rifle except has checkered pistol grip stock of select walnut. Imported from Italy by American Arms, Inc.
Price: $1,299.00

Sights: Blade front, rear adjustable for elevation.
Features: Frame, elevator, magazine follower, buttplate are brass, balance blue (also available in polished steel). Imported by American Arms, Inc.
Price: $996.00

Browning BPR

BROWNING BPR PUMP RIFLE
Caliber: 243, 308 (short action); 270, 30-06, 7mm Rem. Mag., 300 Win. Mag., 4-shot magazine (3 for magnums).
Barrel: 22"; 24" for magnum calibers.
Weight: 7 lbs., 3 oz. **Length:** 43" overall (22" barrel).
Stock: Select walnut with full pistol grip, high gloss finish.
Sights: Gold bead on hooded ramp front, open click adjustable rear.
Features: Slide-action mechanism cams forend down away from the barrel. Seven-lug rotary bolt; cross-bolt safety behind trigger; removable magazine; alloy receiver. Introduced 1997. Imported from Belgium by Browning.
Price: Standard calibers $689.95
Price: Magnum calibers $741.95

Consult our Directory pages for the location of firms mentioned.

Browning Lightning BLR

Browning Lightning BLR Long Action
Similar to the standard Lightning BLR except has long action to accept 30-06, 270, 7mm Rem. Mag. and 300 Win. Mag. Barrel lengths are 22" for 30-06 and 270, 24" for 7mm Rem. Mag. and 300 Win. Mag. Has six-lug rotary bolt; bolt and receiver are full-length fluted. Fold-down hammer at half-cock. Weighs about 7 lbs., overall length 42 7/8" (22" barrel). Introduced 1996.
Price: $608.95

BROWNING LIGHTNING BLR LEVER-ACTION RIFLE
Caliber: 223, 22-250, 243, 7mm-08, 308 Win., 4-shot detachable magazine.
Barrel: 20" round tapered.
Weight: 6 lbs., 8 oz. **Length:** 39 1/2" overall.
Stock: Walnut. Checkered grip and forend, high-gloss finish.
Sights: Gold bead on ramp front; low profile square notch adjustable rear.
Features: Wide, grooved trigger; half-cock hammer safety; fold-down hammer. Receiver tapped for scope mount. Recoil pad installed. Introduced 1996. Imported from Japan by Browning.
Price: $576.95

CENTERFIRE RIFLES—LEVER & SLIDE

CABELA'S CATTLEMAN'S CARBINE
Caliber: 44-40, 6-shot.
Barrel: 18".
Weight: 4 lbs. **Length:** 34" overall.
Stock: European walnut.
Sights: Blade front, notch rear.
Features: Revolving carbine. Color case-hardened frame, rest blued. Introduced 1994. Imported by Cabela's.
Price: $299.95

CABELA'S 1866 WINCHESTER REPLICA
Caliber: 44-40, 13-shot.
Barrel: 24 1/4".
Weight: 9 lbs. **Length:** 43" overall.
Stock: European walnut.
Sights: Bead front, open adjustable rear.
Features: Solid brass receiver, buttplate, forend cap. Octagonal barrel. Faithful to the original Winchester `66 rifle. Introduced 1994. Imported by Cabela's.
Price: $499.95

CIMARRON 1866 WINCHESTER REPLICAS
Caliber: 22 LR, 22 WMR, 38 Spec., 44 WCF.
Barrel: 24 1/4" (rifle), 19" (carbine).
Weight: 9 lbs. **Length:** 43" overall (rifle).
Stock: European walnut.
Sights: Bead front, open adjustable rear.
Features: Solid brass receiver, buttplate, forend cap. Octagonal barrel. Faithful to the original Winchester `66 rifle. Introduced 1991. Imported by Cimarron Arms.
Price: Rifle $839.00
Price: Carbine $829.00

CABELA'S 1858 HENRY REPLICA
Caliber: 44-40, 13-shot magazine.
Barrel: 24 1/4".
Weight: 9.5 lbs. **Length:** 43" overall.
Stock: European walnut.
Sights: Bead front, open adjustable rear.
Features: Brass receiver and buttplate. Uses original Henry loading system. Faithful to the original rifle. Introduced 1994. Imported by Mitchell Arms, Inc.
Price: $649.95

CABELA'S 1873 WINCHESTER REPLICA
Caliber: 44-40, 45 Colt, 13-shot.
Barrel: 24 1/4", 30".
Weight: 8.5 lbs. **Length:** 43 1/4" overall.
Stock: European walnut.
Sights: Bead front, open adjustable rear; globe front, tang rear.
Features: Color case-hardened steel receiver. Faithful to the original Model 1873 rifle. Introduced 1994. Imported by Cabela's.
Price: With tang sight, globe front $639.95
Price: Sporting model, 30" barrel, 44-40, 45 Colt $599.95
Price: With half-round/half-octagon barrel, half magazine $639.95

CIMARRON 1860 HENRY REPLICA
Caliber: 44 WCF, 13-shot magazine.
Barrel: 24 1/4" (rifle), 22" (carbine).
Weight: 9 1/2 lbs. **Length:** 43" overall (rifle).
Stock: European walnut.
Sights: Bead front, open adjustable rear.
Features: Brass receiver and buttplate. Uses original Henry loading system. Faithful to the original rifle. Introduced 1991. Imported by Cimarron Arms.
Price: $899.95

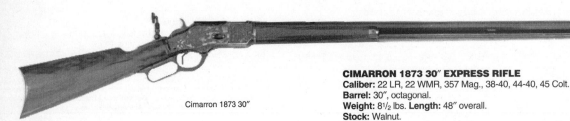

Cimarron 1873 30"

Cimarron 1873 Sporting Rifle
Similar to the 1873 Express except has 24" barrel with half-magazine.
Price: $949.00
Price: 1873 Saddle Ring Carbine, 19" barrel $949.00

CIMARRON 1873 SHORT RIFLE
Caliber: 22 LR, 22 WMR, 357 Mag., 44-40, 45 Colt.
Barrel: 20" tapered octagon.
Weight: 7.5 lbs. **Length:** 39" overall.
Stock: Walnut.

CIMARRON 1873 30" EXPRESS RIFLE
Caliber: 22 LR, 22 WMR, 357 Mag., 38-40, 44-40, 45 Colt.
Barrel: 30", octagonal.
Weight: 8 1/2 lbs. **Length:** 48" overall.
Stock: Walnut.
Sights: Blade front, semi-buckhorn ramp rear. Tang sight optional.
Features: Color case-hardened frame; choice of modern blue-black or charcoal blue for other parts. Barrel marked "Kings Improvement." From Cimarron Arms.
Price: $999.00

Sights: Bead front, adjustable semi-buckhorn rear.
Features: Has half "button" magazine. Original-type markings, including caliber, on barrel and elevator and "Kings" patent. From Cimarron Arms.
Price: $949.00

Dixie 1873

DIXIE ENGRAVED 1873 RIFLE
Caliber: 44-40, 11-shot magazine.
Barrel: 20", round.
Weight: 7 3/4 lbs. **Length:** 39" overall.
Stock: Walnut.
Sights: Blade front, adjustable rear.
Features: Engraved and case-hardened frame. Duplicate of Winchester 1873. Made in Italy. From Dixie Gun Works.
Price: $1,295.00
Price: Plain, blued carbine $850.00

E.M.F. 1866 YELLOWBOY LEVER ACTIONS
Caliber: 38 Spec., 44-40.
Barrel: 19" (carbine), 24" (rifle).
Weight: 9 lbs. **Length:** 43" overall (rifle).
Stock: European walnut.
Sights: Bead front, open adjustable rear.
Features: Solid brass frame, blued barrel, lever, hammer, buttplate. Imported from Italy by E.M.F.
Price: Rifle $848.00
Price: Carbine $825.00

E.M.F. 1860 HENRY RIFLE
Caliber: 44-40 or 44 rimfire.
Barrel: 24.25".
Weight: About 9 lbs. **Length:** About 43.75" overall.
Stock: Oil-stained American walnut.
Sights: Blade front, rear adjustable for elevation.
Features: Reproduction of the original Henry rifle with brass frame and buttplate, rest blued. From E.M.F.
Price: Standard $1,100.00

CENTERFIRE RIFLES—LEVER & SLIDE

Marlin Model 336CS

E.M.F. MODEL 73 LEVER-ACTION RIFLE
Caliber: 357 Mag., 44-40, 45 Colt.
Barrel: 24".
Weight: 8 lbs. **Length:** 43 1/4" overall.
Stock: European walnut.
Sights: Bead front, rear adjustable for windage and elevation.
Features: Color case-hardened frame (blue on carbine). Imported by E.M.F.
Price: Rifle $1,050.00
Price: Carbine, 19" barrel $1,020.00

MARLIN MODEL 1894S LEVER-ACTION CARBINE
Caliber: 44 Spec./44 Mag., 10-shot tubular magazine.
Barrel: 20" Micro-Groove®.
Weight: 6 lbs. **Length:** 37 1/2" overall.
Stock: Checkered American black walnut, straight grip and forend. Mar-Shield® finish. Rubber rifle buttpad; swivel studs.
Sights: Wide-Scan™ hooded ramp front, semi-buckhorn folding rear adjustable for windage and elevation.
Features: Hammer-block safety. Receiver tapped for scope mount, offset hammer spur, solid top receiver sand blasted to prevent glare.
Price: $477.00

MARLIN MODEL 336CS LEVER-ACTION CARBINE
Caliber: 30-30 or 35 Rem., 6-shot tubular magazine.
Barrel: 20" Micro-Groove®.
Weight: 7 lbs. **Length:** 38 1/2" overall.
Stock: Checkered American black walnut, capped pistol grip with white line spacers. Mar-Shield® finish; rubber buttpad; swivel studs.
Sights: Ramp front with Wide-Scan™ hood, semi-buckhorn folding rear adjustable for windage and elevation.
Features: Hammer-block safety. Receiver tapped for scope mount, offset hammer spur; top of receiver sand blasted to prevent glare.
Price: $459.00

Marlin Model 30AS Lever-Action Carbine
Same as the Marlin 336CS except has press-checkered, walnut-finished Maine birch pistol grip stock, 30-30 only, 6-shot. Hammer-block safety. Adjustable rear sight, brass bead front.
Price: $392.00

Marlin Model 1894CS Carbine
Similar to the standard Model 1894S except chambered for 38 Spec./357 Mag. with full-length 9-shot magazine, 18 1/2" barrel, hammer-block safety, brass bead front sight. Introduced 1983.
Price: $477.00

Marlin Model 1894 Cowboy II

MARLIN MODEL 444SS LEVER-ACTION SPORTER
Caliber: 444 Marlin, 5-shot tubular magazine.
Barrel: 22" Micro-Groove®.
Weight: 7 1/2 lbs. **Length:** 40 1/2" overall.
Stock: Checkered American black walnut, capped pistol grip with white line spacers, rubber rifle buttpad. Mar-Shield® finish; swivel studs.
Sights: Hooded ramp front, folding semi-buckhorn rear adjustable for windage and elevation.
Features: Hammer-block safety. Receiver tapped for scope mount; offset hammer spur.
Price: $543.00

Navy Arms Henry Trapper
Similar to the Military Henry Rifle except has 16 1/2" barrel, weighs 7 1/2 lbs. Brass frame and buttplate, rest blued. Introduced 1991. Imported from Italy by Navy Arms.
Price: $875.00

MARLIN MODEL 1894 COWBOY, COWBOY II
Caliber: 357 Mag., 44 Mag., 44-40, 45 Colt, 10-shot magazine.
Barrel: 24" tapered octagon, deep cut rifling.
Weight: 7 1/2 lbs. **Length:** 41 1/2" overall.
Stock: Straight grip American black walnut with cut checkering, hard rubber buttplate, Mar-Shield® finish.
Sights: Marble carbine front, adjustable Marble semi-buckhorn rear.
Features: Squared finger lever; straight grip stock; blued steel forend tip. Designed for Cowboy Shooting events. Introduced 1996. Made in U.S. by Marlin.
Price: Cowboy I, 45 Colt $691.00
Price: Cowboy II, 357 Mag., 44 Mag., 44-40 ....$691.00

MARLIN MODEL 1895SS LEVER-ACTION RIFLE
Caliber: 45-70, 4-shot tubular magazine.
Barrel: 22" round.
Weight: 7 1/2 lbs. **Length:** 40 1/2" overall.
Stock: Checkered American black walnut, full pistol grip. Mar-Shield® finish; rubber buttpad; quick detachable swivel studs.
Sights: Bead front with Wide-Scan™ hood, semi-buckhorn folding rear adjustable for windage and elevation.
Features: Hammer-block safety. Solid receiver tapped for scope mounts or receiver sights; offset hammer spur.
Price: $543.00

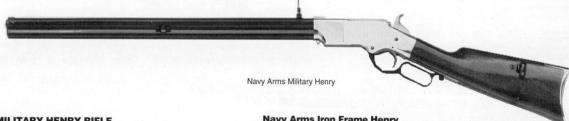
Navy Arms Military Henry

NAVY ARMS MILITARY HENRY RIFLE
Caliber: 44-40, 12-shot magazine.
Barrel: 24 1/4".
Weight: 9 lbs., 4 oz.
Stock: European walnut.
Sights: Blade front, adjustable ladder-type rear.
Features: Brass frame, buttplate, rest blued. Recreation of the model used by cavalry units in the Civil War. Has full-length magazine tube, sling swivels; no forend. Imported from Italy by Navy Arms.
Price: $895.00

Navy Arms Iron Frame Henry
Similar to the Military Henry Rifle except receiver is blued or color case-hardened steel. Imported by Navy Arms.
Price: $945.00

Navy Arms Henry Carbine
Similar to the Military Henry rifle except has 22" barrel, weighs 8 lbs., 12 oz., is 41" overall; no sling swivels. Caliber 44-40. Introduced 1992. Imported from Italy by Navy Arms.
Price: $875.00

CENTERFIRE RIFLES—LEVER & SLIDE

NAVY ARMS 1866 YELLOWBOY RIFLE
Caliber: 44-40, 12-shot magazine.
Barrel: 24", full octagon.
Weight: 8½ lbs. **Length:** 42½" overall.
Stock: European walnut.
Sights: Blade front, adjustable ladder-type rear.
Features: Brass frame, forend tip, buttplate, blued barrel, lever, hammer. Introduced 1991. Imported from Italy by Navy Arms.
Price: $680.00
Price: Carbine, 19" barrel $670.00

Navy Arms 1873 Winchester Style

NAVY ARMS 1873 WINCHESTER-STYLE RIFLE
Caliber: 44-40, 45 Colt, 12-shot magazine.
Barrel: 24".
Weight: 8¼ lbs. **Length:** 43" overall.
Stock: European walnut.
Sights: Blade front, buckhorn rear.
Features: Color case-hardened frame, rest blued. Full-octagon barrel. Imported by Navy Arms.
Price: $820.00
Price: Carbine, 19" barrel $800.00

Navy Arms 1873 Sporting Rifle
Similar to the 1873 Winchester-Style rifle except has checkered pistol grip stock, 30" octagonal barrel (24" available). Introduced 1992. Imported by Navy Arms.
Price: 30" barrel $960.00
Price: 24" barrel $930.00

Remington 7600 Rifle

REMINGTON MODEL 7600 SLIDE ACTION
Caliber: 243, 270, 280, 30-06, 308.
Barrel: 22" round tapered.
Weight: 7½ lbs. **Length:** 42" overall.
Stock: Cut-checkered walnut pistol grip and forend, Monte Carlo with full cheekpiece. Satin or high-gloss finish.
Sights: Gold bead front sight on matted ramp, open step adjustable sporting rear.
Features: Redesigned and improved version of the Model 760. Detachable 4-shot clip. Cross-bolt safety. Receiver tapped for scope mount. Also available in high grade versions. Comes with green Remington hard case. Introduced 1981.
Price: About $540.00
Price: Carbine (18½" bbl., 30-06 only) $540.00

Rossi Model 92 Large Loop

ROSSI MODEL 92 SADDLE-RING CARBINE
Caliber: 38 Spec./357 Mag., 44 Spec./44-40, 44 Mag., 45 Colt, 10-shot magazine.
Barrel: 20".
Weight: 5¾ lbs. **Length:** 37" overall.
Stock: Walnut.
Sights: Blade front, buckhorn rear.
Features: Recreation of the famous lever-action carbine. Handles 38 and 357 interchangeably. Introduced 1978. Imported by Interarms.
Price: $360.00
Price: Stainless steel, 38/357 only $415.00
Price: With 24" half-octagon barrel, 13-shot magazine, brass blade front sight, 45 Colt only, blue only. $429.00

Rossi Model 92 Short Carbine
Similar to the standard M92 except has 16" barrel, overall length of 33½", weighs 5½ lbs. Introduced 1986.
Price: $360.00
Price: Model 92 Large Loop (has oversize cocking lever loop and saddle rings, 44 Mag., 45 Colt only) $360.00

Ruger Model 96/44

RUGER MODEL 96/44 LEVER-ACTION RIFLE
Caliber: 44 Mag., 4-shot rotary magazine.
Barrel: 18½".
Weight: 5⅞ lbs. **Length:** 37 5/16" overall.
Stock: American hardwood.
Sights: Gold bead front, folding leaf rear.
Features: Manual cross-bolt safety, visible cocking indicator; short-throw lever action; integral scope mount; blued finish. Introduced 1996. Made In U.S. by Sturm, Ruger & Co.
Price: $419.00

Savage Model 99C

SAVAGE MODEL 99C LEVER-ACTION RIFLE
Caliber: 243, 308, 4-shot detachable box magazine.
Barrel: 22".
Weight: 7¾ lbs. **Length:** 45½" overall.
Stock: American walnut; Monte Carlo comb; cut-checkered grip and forend.
Sights: Bead on blade front, open fully adjsutable rear. Drilled and tapped for scope mounts.
Features: Polished blue finish; solid red buttpad; swivel studs. From Savage Arms. Reintroduced 1996.
Price: $665.00

CENTERFIRE RIFLES—LEVER & SLIDE

WINCHESTER MODEL 1892 LEVER-ACTION RIFLE
Caliber: 45 Colt, 11-shot magazine.
Barrel: 24", round.
Weight: 6 1/4 lbs. **Length:** 41 1/4" overall.
Stockp: Smooth walnut.
Sights: Bead front, ramp-adjustable buckhorn-style rear.
Features: Recreation of the Model 1892. Tang-mounted manual hammer stop; blued crescent butt plate; full magazine tube; straight-grip stock. Reintroduced 1997. From U.S. Repeating Arms Co.
Price: Grade I $722.00
Price: High Grade, select walnut, engraved receiver with gold highlights (1,000 made in 1997 only) $1,285.00

WINCHESTER MODEL 94 BIG BORE SIDE EJECT
Caliber: 307 Win., 356 Win., 6-shot magazine.
Barrel: 20".
Weight: 7 lbs. **Length:** 38 5/8" overall.
Stock: American walnut. Satin finish.
Sights: Hooded ramp front, semi-buckhorn rear adjustable for windage and elevation.
Features: All external metal parts have Winchester's deep blue finish. Rifling twist 1:12". Rubber recoil pad fitted to buttstock. Introduced 1983. From U.S. Repeating Arms Co., Inc.
Price: $404.00

Winchester 94 Side Eject

WINCHESTER MODEL 94 WALNUT SIDE EJECT LEVER-ACTION RIFLE
Caliber: 30-30 Win., 6-shot tubular magazine.
Barrel: 20".
Weight: 6 1/2 lbs. **Length:** 37 3/4" overall.
Stock: Straight grip walnut stock and forend.
Sights: Hooded blade front, semi-buckhorn rear. Drilled and tapped for scope mount. Post front sight on Trapper model.
Features: Solid frame, forged steel receiver; side ejection, exposed rebounding hammer with automatic trigger-activated transfer bar. Introduced 1984.
Price: Checkered walnut $393.00
Price: No checkering, walnut $363.00

Winchester Model 94 Ranger Side Eject Lever-Action Rifle
Same as Model 94 Side Eject except has 6-shot magazine, American hardwood stock and forend, post front sight. Introduced 1985.
Price: $320.00
Price: With 4x32 Bushnell scope, mounts $376.00

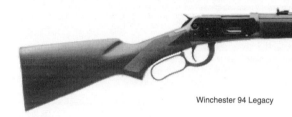
Winchester 94 Legacy

Winchester Model 94 Legacy
Similar to the Model 94 Side Eject except has half pistol grip walnut stock, checkered grip and forend. Chambered for 30-30, 357 Mag., 44 Mag., 45 Colt; 20" or 24" barrel. Introduced 1995. Made in U.S. by U.S. Repeating Arms Co., Inc.
Price: With 20" barrel $393.00
Price: With 24" barrel $407.00

Winchester Model 94 Trails End
Similar to the Model 94 Walnut except chambered only for 357 Mag., 44 Mag., 45 Colt; 11-shot magazine. Available with standard or large lever loop. Introduced 1997. From U.S. Repeating Arms Co.
Price: With standard lever loop $398.00
Price: With large lever loop $420.00

Winchester Model 94 Trapper Side Eject
Same as the Model 94 except has 16" barrel, 5-shot magazine in 30-30, 9-shot in 357 Mag., 44 Magnum/44 Special, 45 Colt. Has stainless steel claw extractor, saddle ring, hammer spur extension, walnut wood.
Price: 30-30 .... $363.00
Price: 357 Mag., 44 Mag./44 Spec., 45 Colt $384.00

Winchester Model 94 Wrangler

Winchester Model 94 Wrangler Side Eject
Same as the Model 94 except has 16" barrel and large loop lever for large and/or gloved hands. Has 9-shot capacity (5-shot for 30-30), stainless steel claw extractor. Available in 30-30, 44 Magnum/44 Special. Specially inscribed with "1894-1994" on the receiver. Reintroduced 1992.
Price: 30-30 .... $384.00
Price: 44 Magnum/44 Special $404.00

Winchester Model 1895

Winchester Model 1895 High Grade Rifle
Same as the Grade I except has silvered receiver with extensive engraving: right side shows two scenes portraying large big horn sheep; left side has bull elk and cow elk. Gold borders accent the scenes. Magazine and cocking lever also engraved. Has classic Winchester H-style checkering pattern on fancy grade American walnut. Only 4000 rifles made. Introduced 1995. From U.S. Repeating Arms Co., Inc.
Price: $1,360.00

WINCHESTER MODEL 1895 LEVER-ACTION RIFLE
Caliber: 30-06, 270, 4-shot magazine.
Barrel: 24", round.
Weight: 8 lbs. **Length:** 42" overall.
Stock: American walnut.
Sights: Gold bead front, buckhorn rear adjustable for elevation.
Features: Recreation of the original Model 1895. Polished blue finish with Nimschke-style scroll engraving on receiver. Scalloped receiver, two-piece cocking lever, schnabel forend, straight-grip stock. Introduced 1995. Only 4000 rifles made in 30-06 only. From U.S. Repeating Arms Co., Inc.
Price: Grade I, 30-06 .... $853.00
Price: Grade I, 270 .... $909.00

CENTERFIRE RIFLES—BOLT ACTION

Includes models for a wide variety of sporting and competitive purposes and uses.

AMT BOLT-ACTION RIFLE
Caliber: Single shot—22 Hornet, 222, 223, 22-250, 243 Win., 243 A, 22 PPC, 6mm PPC, 6.5x08, 7mm-08, 308; repeater—223, 22-250, 243 Win., 243 A, 6mm PPC, 25-06, 6.5x08, 270, 7mm-08, 308, 30-06, 7mm Rem. Mag; 300 Win. Mag., 338 Win. Mag., 375 H&H, 416 Rem; 458 Win. Mag., 416 Rigby, 7.62x39, 7x57.
Barrel: Up to 28", #3 contour.
Weight: About 8 1/2 lbs.
Stock: Classic composite on Standard grade; McMillan or H-S Precision on Deluxe.
Sights: None furnished; drilled and tapped for scope mounting.
Features: Single shot uses cone breach action with post-64-type extractor, pre-64-type three-position safety; repeater has Mauser-type extractor and magazine, pre-64 three-position safety; plunger-type ejector; short, medium, long action, right- or left-handed. Introduced 1996. Made in U.S. by AMT. Deluxe has Mauser controlled feed action with plunger ejector, claw-type extractor; Standard uses push-feed post-64 Winchester-type action.
Price: Single shot $1,499.00

Arnold Arms High Country Mountain Rifle
Simliar to the Alaskan Bush rifle except chambered for 257 to 338 Magnum; choice of AA English walnut or synthetic stock; scope mount only. Introduced 1996. Made in U.S. by Arnold Arms Co.
Price: Chrome-moly steel, synthetic stock $2,995.00
Price: Stainless steel, synthetic stock $3,170.00
Price: Chrome-moly steel, walnut stock $4,489.00
Price: Stainless steel, walnut stock $4,839.00

Arnold Arms Grand Alaskan Rifle
Similar to the Alaskan Bush rifle except has AAA fancy select or exhibition-grade English walnut; barrel band swivel; comes with iron sights and scope mount; 24" to 26" barrel; 300 Magnum to 458 Win. Mag. Introduced 1996. Made in U.S. by Arnold Arms Co.
Price: Chrome-moly steel, from $6,550.00
Price: Stainless steel, from $6,710.00

Arnold Arms Serengeti Synthetic Rifle
Similar to the Safari except has Fibergrain synthetic stock in classic or Monte Carlo style; traditional checkering pattern or stipple finish; polished or matte blue or bead-blast stainless finish; chambered for 243 to 300 Magnum. Introduced 1996. Made in U.S. by Arnold Arms Co.
Price: Chrome-moly steel $2,995.00
Price: Stainless steel $3,170.00

Arnold Arms Grand African Rifle
Similar to the Safari rifle except has Exhibition Grade stock; polished blue chrome-moly steel or bead-blasted or teflon-coated stainless; barrel band; scope mount, express sights; calibers 338 Magnum to 458 Win. Mag.; 24" to 26" barrel. Introduced 1996. Made in U.S. by Arnold Arms Co.
Price: Chrome-moly steel $7,630.00
Price: Stainless steel $7,780.00

ARNOLD ARMS ALASKAN BUSH RIFLE
Caliber: 223 to 338 Magnum.
Barrel: 22" to 26".
Weight: NA. **Length:** NA.
Stock: Synthetic; black, woodland or arctic camouflage.
Sights: Optional; drilled and tapped for scope mounting.
Features: Uses the Apollo action with controlled round feed or push feed; chrome-moly steel or stainless; one-piece bolt, handle, knob; cone head bolt and breech; three-position safety; fully adjustable trigger. Introduced 1996. Made in U.S. by Arnold Arms Co.
Price: Chrome-moly steel $2,995.00
Price: Stainless steel $3,145.00

Arnold Arms Alaskan Trophy Rifle
Similar to the Alaskan Bush rifle except chambered for 300 Magnums to 458 Win. Mag.; 24" to 26" barrel; Fibergrain or black synthetic stock, or AA English walnut; comes with barrel band on 375 H&H and larger; scope mount; iron sights. Introduced 1996. Made in U.S. by Arnold Arms Co.
Price: Chrome-moly steel $3,525.00
Price: Stainless steel $3,990.00
Price: Chrome-moly steel, walnut stock $5,140.00
Price: Stainless steel, walnut stock $5,299.00

ARNOLD ARMS SAFARI RIFLE
Caliber: 243 to 458 Win. Mag.
Barrel: 22" to 26".
Weight: NA. **Length:** NA.
Stock: Grade A and AA Fancy English walnut.
Sights: Optional; drilled and tapped for scope mounting.
Features: Uses the Apollo action with controlled or push round feed; one-piece bolt, handle, knob; cone head bolt and breech; three-position safety; fully adjustable trigger; chrome-moly steel in matte blue, polished, or bead blasted stainless. Introduced 1996. Made in U.S. by Arnold Arms Co.
Price: Grade A walnut, chrome-moly $4,435.00
Price: Grade A walnut, stainless steel $4,695.00
Price: Grade AA walnut, chrome-moly steel $4,690.00
Price: Grade AA walnut, stainless steel $4,840.00

Arnold Arms African Trophy Rifle
Similar to the Safari rifle except has AAA Extra Fancy English walnut stock with wrap-around checkering; matte blue chrome-moly or polished or bead blasted stainless steel; scope mount standard or optional Express sights. Introduced 1996. Made in U.S. by Arnold Arms Co.
Price: Blued chrome-moly steel $6,098.00
Price: Stainless steel $6,255.00

Arnold Arms African Synthetic Rifle
Similar to the Safari except has Fibergrain synthetic stock with or without cheek-piece and traditional checkering pattern, or stipple finish; standard iron sights or Express folding leaf optional; chambered for 338 Magnum to 458 Win. Mag.; 24" to 26" barrel. Introduced 1996. Made in U.S. by Arnold Arms Co.
Price: Chrome-moly steel $2,995.00
Price: Stainless steel $3,170.00

Anschutz 1740 Monte Carlo

ANSCHUTZ 1740 MONTE CARLO RIFLE
Caliber: 22 Hornet, 5-shot clip; 222 Rem., 3-shot clip.
Barrel: 24".
Weight: 6 1/2 lbs. **Length:** 43.25" overall.
Stock: Select European walnut.
Sights: Hooded ramp front, folding leaf rear; drilled and tapped for scope mounting.
Features: Uses match 54 action. Adjustable single stage trigger. Stock has roll-over Monte Carlo cheekpiece, slim forend with Schnabel tip, Wundhammer palm swell on grip, rosewood grip cap with white diamond insert. Skip-line checkering on grip and forend. Introduced 1997. Imported from Germany by AcuSport.
Price: $1,294.95
Price: Model 1730 Monte Carlo, as above except in 22 Hornet . . . $1,294.95

Anschutz 1733D Rifle
Similar to the 1740 Monte Carlo except has full-length, walnut, Mannlicher-style stock with skip-line checkering, rosewood schnabel tip, and is chambered for 22 Hornet. Weighs 6.4 lbs., overall length 39", barrel length 19.7". Imported from Germany by AcuSport.
Price: $2,089.95

CENTERFIRE RIFLES—BOLT ACTION

ARMSCOR M-1800S CLASSIC BOLT-ACTION RIFLE
Caliber: 22 Hornet, 5-shot magazine.
Barrel: 22.6".
Weight: 6.6 lbs. **Length:** 41.25" overall.
Stock: Walnut-finished hardwood with Monte Carlo comb and checkpiece.
Sights: Ramped blade front, fully adjustable open rear.
Features: Receiver dovetailed for tip-off scope mount. Introduced 1996. Imported from the Philippines by K.B.I., Inc.
Price: $358.00

Armscor M-1800SC Super Classic Rifle
Similar to the M-1800S except has oil-finished American walnut stock with 18 lpi hand checkering; black hardwood grip cap and forend tip; highly polished barreled action; jewelled bolt; recoil pad; swivel studs. Imported from the Philippines by K.B.I., Inc.
Price: $486.00

ANSCHUTZ 1743D BOLT-ACTION RIFLE
Caliber: 222 Rem., 3-shot magazine.
Barrel: 19.7".
Weight: 6.4 lbs. **Length:** 39" overall.
Stock: European walnut.
Sights: Hooded blade front, folding leaf rear.
Features: Receiver grooved for scope mounting; single stage trigger; claw extractor; sliging safety; sling swivels. Imported from Germany by AcuSport Corp.
Price: $1,469.95

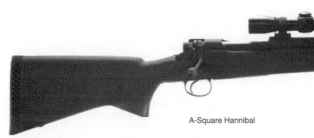
A-Square Hannibal

A-SQUARE HANNIBAL BOLT-ACTION RIFLE
Caliber: 7mm Rem. Mag., 7mm STW, 30-06, 300 Win. Mag., 300 H&H, 300 Wea. Mag., 8mm Rem. Mag., 338 Win. Mag., 340 Wea. Mag., 338 A-Square Mag., 9.3x62, 9.3x64, 375 H&H, 375 Wea. Mag., 375 JRS, 375 A-Square Mag., 378 Wea. Mag., 416 Taylor, 416 Rem. Mag., 416 Hoffman, 416 Rigby, 416 Wea. Mag., 404 Jeffery, 425 Express, 458 Win. Mag., 458 Lott, 450 Ackley, 460 Short A-Square Mag., 460 Wea. Mag., 470 Capstick, 495 A-Square Mag., 500 A-Square Mag.
Barrel: 20" to 26" (no-cost customer option).
Weight: 9 to 11 3/4 lbs.
Stock: Claro walnut with hand-rubbed oil finish; classic style with A-Square Coil-Chek® features for reduced recoil; flush detachable swivels. Customer choice of length of pull. Available with synthetic stock.
Sights: Choice of three-leaf express, forward or normal-mount scope, or combination (at extra cost).
Features: Matte non-reflective blue, double cross-bolts, steel and fiberglass reinforcement of wood from tang to forend tip; Mauser-style claw extractor; expanded magazine capacity; two-position safety; three-way target trigger. Right-hand only. Introduced 1983. Made in U.S. by A-Square Co., Inc.
Price: Walnut stock $2,995.00
Price: Synthetic stock $3,345.00

A-SQUARE CAESAR BOLT-ACTION RIFLE
Caliber: 7mm Rem. Mag., 7mm STW, 30-06, 300 Win. Mag., 300 H&H, 300 Wea. Mag., 8mm Rem. Mag., 338 Win. Mag., 340 Wea. Mag., 338 A-Square, 9.3x62, 9.3x64, 375 Wea. Mag., 375 H&H, 375 JRS, 375 A-Square, 416 Hoffman, 416 Rem. Mag., 416 Taylor, 404 Jeffery, 425 Express, 458 Win. Mag., 458 Lott, 450 Ackley, 460 Short A-Square, 470 Capstick, 495 A-Square.
Barrel: 20" to 26" (no-cost customer option).
Weight: 8 1/2 to 11 lbs.
Stock: Claro walnut with hand-rubbed oil finish; classic style with A-Square Coil-Chek® features for reduced recoil; flush detachable swivels. Customer choice of length of pull.
Sights: Choice of three-leaf express, forward or normal-mount scope, or combination (at extra cost).
Features: Matte non-reflective blue, double cross-bolts, steel and fiberglass reinforcement of wood from tang to forend tip; three-position positive safety; three-way adjustable trigger; expanded magazine capacity. Right- or left-hand. Introduced 1984. Made in U.S. by A-Square Co., Inc.
Price: Walnut stock $2,995.00
Price: Synthetic stock $3,345.00

A-Square Hamilcar Bolt-Action Rifle
Similar to the A-Square Hannibal rifle except chambered for 25-06, 6.5x55, 270 Win., 7x57, 280 Rem., 30-06, 338-06, 9.3x62, 257 Wea. Mag., 264 Win. Mag., 270 Wea. Mag., 7mm Rem. Mag., 7mm Wea. Mag., 7mm STW, 300 Win. Mag., 300 Wea. Mag. Weighs 8-8 1/2 lbs. Introduced 1994. From A-Square Co., Inc.
Price: $2,995.00

Barrett Model 95

BARRETT MODEL 95 BOLT-ACTION RIFLE
Caliber: 50 BMG, 5-shot magazine.
Barrel: 29".
Weight: 22 lbs. **Length:** 45" overall.
Stock: Sorbothane recoil pad.
Sights: Scope optional.
Features: Updated version of the Model 90. Bolt-action, bullpup design. Disassembles without tools; extendable bipod legs; match-grade barrel; high efficiency muzzlebrake. Introduced 1995. From Barrett Firearms Mfg., Inc.
Price: From $4,700.00

Blaser R93

BLASER R93 BOLT-ACTION RIFLE
Caliber: 222, 243, 6.5x55, 270, 7x57, 308, 30-06, 7mm Rem. Mag., 300 Win. Mag., 300 Wea. Mag., 338 Win. Mag., 375 H&H, 416 Rem. Mag., 3-shot magazine.
Barrel: 22" (standard calibers), 24" (magnum calibers).
Weight: 6.5 to 7.5 lbs. **Length:** 40" overall (22" barrel).
Stock: Two-piece European walnut.
Sights: Blade front on ramp, open rear, or no sights.
Features: Straight-pull bolt action with thumb-activated safety slide/cocking mechanism. Interchangeable barrels and bolt heads. Introduced 1994. Imported from Germany by Autumn Sales, Inc.
Price: Standard $2,800.00
Price: Deluxe (better wood, engraving) $3,100.00
Price: Super Deluxe (best wood, gold animal inlays) $3,500.00
Price: Safari, standard grade, 375 H&H, 416 Rem. Mag. $3,300.00
Price: Safari Deluxe $3,600.00
Price: Safari Super Deluxe $4,000.00

CENTERFIRE RIFLES—BOLT ACTION

Browning A-Bolt II Medallion

Browning A-Bolt II Medallion Left-Hand
Same as the Medallion model A-Bolt except has left-hand action and is available in 25-06, 270, 280, 30-06, 7mm Rem. Mag., 300 Win. Mag., 338 Win. Mag., 375 H&H. Introduced 1987.
Price: $734.95
Price: With BOSS $832.95
Price: 375 H&H, with sights $846.95

Browning A-Bolt II Stainless Stalker
Similar to the Hunter model A-Bolt except receiver and barrel are made of stainless steel; the rest of the exposed metal surfaces are finished with a durable matte silver-gray. Graphite-fiberglass composite textured stock. No sights are furnished. Available in 223, 22-250, 243, 308, 7mm-08, 270, 30-06, 7mm Rem. Mag., 375 H&H. Introduced 1987.
Price: $708.25
Price: With BOSS $769.75
Price: Left-hand, no sights $730.75
Price: With BOSS $792.25
Price: 375 H&H, with sights $806.35
Price: 375 H&H, left-hand, with sights $893.05

BROWNING A-BOLT II RIFLE
Caliber: 25-06, 270, 30-06, 280, 7mm Rem. Mag., 300 Win. Mag., 338 Win. Mag., 375 H&H Mag.
Barrel: 22" medium sporter weight with recessed muzzle; 26" on mag. cals.
Weight: 6$1/2$ to 7$1/2$ lbs. **Length:** 44$3/4$" overall (magnum and standard); 41$3/4$" (short action).
Stock: Classic style American walnut; recoil pad standard on magnum calibers.
Features: Short-throw (60°) fluted bolt, three locking lugs, plunger-type ejector; adjustable trigger is grooved and gold-plated. Hinged floorplate, detachable box magazine (4 rounds std. cals., 3 for magnums). Slide tang safety. Medallion has glossy stock finish, rosewood grip and forend caps, high polish blue. BOSS barrel vibration modulator and muzzlebrake system not available in 375 H&H. Introduced 1985. Imported from Japan by Browning.
Price: Medallion, no sights $623.25
Price: Hunter, no sights $545.35
Price: Hunter, with sights $613.75
Price: Medallion, 375 H&H Mag., with sights $737.05
Price: For BOSS add $61.70

Browning A-Bolt II Composite Stalker
Similar to the A-Bolt II Hunter except has black graphite-fiberglass stock with textured finish. Matte blue finish on all exposed metal surfaces. Available in 223, 22-250, 243, 7mm-08, 308, 30-06, 270, 280, 25-06, 7mm Rem. Mag., 300 Win. Mag., 338 Win. Mag. BOSS barrel vibration modulator and muzzlebrake system offered in all calibers. Introduced 1994.
Price: No sights $562.45
Price: No sights, BOSS $623.95

Browning A-Bolt II Eclipse

Browning A-Bolt II Gold Medallion
Similar to the standard A-Bolt except has select walnut stock with brass spacers between rubber recoil pad and between the rosewood grip cap and forend tip; gold-filled barrel inscription; palm-swell pistol grip, Monte Carlo comb, 22 lpi checkering with double borders; engraved receiver flats. In 270, 30-06, 7mm Rem. Mag. only. Introduced 1988.
Price: $854.95
Price: For BOSS, add $61.60

Browning A-Bolt II Varmint Rifle
Same as the A-Bolt II Hunter except has heavy varmint/target barrel, laminated wood stock with special dimensions, flat forend and palm swell grip. Chambered only for 223, 22-250, 308. Comes with BOSS barrel vibration modulator and muzzlebrake system. Introduced 1994.
Price: With BOSS, gloss or matte finish $819.25

Browning A-Bolt II Eclipse
Similar to the A-Bolt II except has gray/black laminated, thumbhole stock, BOSS barrel vibration modulator and muzzlebrake. Available in long and short action with standard weight barrel, or short-action Varmint with heavy barrel. Introduced 1996. Imported from Japan by Browning.
Price: Standard barrel $895.75
Price: Varmint $922.75

Browning A-Bolt II Short Action
Similar to the standard A-Bolt except has short action for 223, 22-250, 243, 257 Roberts, 7mm-08, 284 Win., 308 chamberings. Available in Hunter or Medallion grades. Weighs 6$1/2$ lbs. Other specs essentially the same. BOSS barrel vibration modulator and muzzlebrake system optional. Introduced 1985.
Price: Medallion, no sights $636.95
Price: Hunter, no sights $545.35
Price: Hunter, with sights $613.75
Price: Composite, no sights $562.45
Price: For BOSS, add $61.70

Browning A-Bolt II Micro Medallion
Similar to the standard A-Bolt except is a scaled-down version. Comes with 20" barrel, shortened length of pull (13$5/16$"); three-shot magazine capacity; weighs 6 lbs., 1 oz. Available in 22 Hornet, 243, 308, 7mm-08, 257 Roberts, 223, 22-250. BOSS feature not available for this model. Introduced 1988.
Price: No sights $636.25

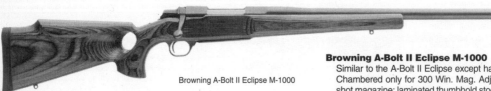

Browning A-Bolt II Eclipse M-1000

Browning A-Bolt II Eclipse M-1000
Similar to the A-Bolt II Eclipse except has long action and heavy target barrel. Chambered only for 300 Win. Mag. Adjustable trigger, bench-style forend, 3-shot magazine; laminated thumbhold stock; BOSS system standard. Introduced 1997. Imported for Japan by Browning.
Price: $923.00

CENTURY SWEDISH SPORTER #38
Caliber: 6.5x55 Swede, 5-shot magazine.
Barrel: 24".
Weight: NA. **Length:** 44.1" overall.
Stock: Walnut-finished European hardwood with checkered pistol grip and forend; Monte Carlo comb.
Sights: Blade front, adjustable rear.
Features: Uses M38 Swedish Mauser action; comes with Holden Ironsighter see-through scope mount. Introduced 1987. From Century International Arms.
Price: About $249.00

CENTERFIRE RIFLES—BOLT ACTION

Century Centurion 14

CENTURY CENTURION 98 SPORTER RIFLE
Caliber: 270 Win., 5-shot magazine.
Barrel: 22"
Weight: 7 lbs., 13 oz. **Length:** 44" overall.
Stock: Black synthetic.
Sights: None furnished; comes with scope base.
Features: Low-swing safety; polished blue finish. From Century International Arms.
Price: $279.00

COOPER ARMS MODEL 40 CENTERFIRE SPORTER
Caliber: 22 Hornet, 22 K-Hornet, 5-shot magazine.
Barrel: 23".
Weight: 7 lbs. **Length:** 42½" overall.
Stock: AAA Claro walnut with 22 lpi borderless wrap-around ribbon checkering, oil finish, steel grip cap, Pachmayr pad.
Sights: None furnished.
Features: Action has three mid-bolt locking lugs, 45-degree bolt rotation; fully adjustable trigger; swivel studs. Pachmayr butt pad. Introduced 1994. Made in U.S. by Cooper Arms.
Price: Classic $1,825.00
Price: Custom Classic (AAA Claro walnut, Monte Carlo beaded cheekpiece, oil finish) $2,025.00

CENTURY CENTURION 14 SPORTER
Caliber: 7mm Rem. Mag., 5-shot magazine.
Barrel: 24".
Weight: NA. **Length:** 43.3" overall.
Stock: Walnut-finished European hardwood. Checkered pistol grip and forend. Monte Carlo comb.
Sights: None furnished.
Features: Uses modified Pattern 14 Enfield action. Drilled and tapped; scope base mounted. Blue finish. From Century International Arms.
Price: About $275.00

COOPER ARMS MODEL 21 VARMINT EXTREME RIFLE
Caliber: 17 Rem., 17 Mach IV, 221 Fireball, 222, 222 Rem. Mag., 223, 22 PPC, single shot.
Barrel: 23.75"; stainless steel, with competition step crown; free-floated.
Weight: NA. **Length:** NA.
Stock: AAA Claro walnut with flared oval forend, ambidextrous palm swell, 22 lpi checkering, oil finish, Pachmayr buttpad.
Sights: None furnished; drilled and tapped for scope mounting.
Features: Action has three mid-bolt locking lugs; adjustable trigger; glass bedded; swivel studs. Introduced 1994. Made in U.S. by Cooper Arms.
Price: $1,675.00
Price: Benchrest with Jewell trigger $2,140.00
Price: Classic model $1,675.00
Price: Custom Classic $1,960.00

Cooper Model 22 PV

COOPER ARMS MODEL 22 PRO VARMINT EXTREME
Caliber: 22-250, 220 Swift, 243, 25-06, 6mm PPC, 308, single shot.
Barrel: 26"; stainless steel match grade, straight taper; free-floated.
Weight: NA. **Length:** NA.
Stock: AAA Claro walnut, oil finish, 22 lpi wrap-around borderless ribbon checkering, beaded cheekpiece, steel grip cap, flared varminter forend, Pachmayr pad.
Sights: None furnished; drilled and tapped for scope mounting.
Features: Uses a three front locking lug system. Available with sterling silver inlaid medallion, skeleton grip cap, and French walnut. Introduced 1995. Made in U.S. by Cooper Arms.
Price: $1,785.00
Price: Benchrest model with Jewell trigger $2,140.00
Price: Black Jack model (McMillan synthetic stock) $1,575.00

CZ 550 BOLT-ACTION RIFLE
Caliber: 243, 308 (4-shot detachable magazine), 308, 270, 30-06, 7mm Rem. Mag., 300 Win. Mag. (5-shot internal magazine).
Barrel: 23.6".
Weight: 7.2 lbs. **Length:** 44.7" overall.
Stock: Walnut with high comb; checkered grip and forend.
Sights: None furnished; drilled and tapped for Remington 700-style bases.
Features: Polished blue finish. Introduced 1995. Imported from the Czech Republic by CZ-USA.
Price: ....NA

CZ 527

Dakota Model 97

CZ 527 BOLT-ACTION RIFLE
Caliber: 22 Hornet, 222 Rem., 223 Rem., detachable 5-shot magazine.
Barrel: 23½"; standard or heavy barrel.
Weight: 6 lbs., 1 oz. **Length:** 42½" overall.
Stock: European walnut with Monte Carlo.
Sights: Hooded front, open adjustable rear.
Features: Improved mini-Mauser action with non-rotating claw extractor; grooved receiver. Imported from the Czech Republic by CZ-USA.
Price: ....NA

DAKOTA MODEL 97 LONG RANGE HUNTER RIFLE
Caliber: 25-06, 257 Roberts, 270 Win., 280, 7mm Rem. Mag., 7mm Dakota Mag., 30-06, 300 Win. Mag., 300 Dakota Mag., 338 Win. Mag., 330 Dakota Mag., 375 H&H, 375 Dakota Mag.
Barrel: 24", 26"; medium-weight match quality.
Weight: 7.7 lbs. **Length:** 45" to 47" overall.
Stock: Composite, black H-S Precision with one-piece bedding block system.
Sights: None supplied. Drilled and tapped for scope mounts.
Features: Dakota controlled round feeding; three-position striker blocking safety; Mauser-style extractor; hand-polished chamber. Introduced 1997. Made in U.S. by Dakota Arms.
Price: $1,595.00

CENTERFIRE RIFLES—BOLT ACTION

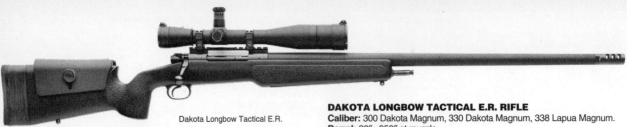
Dakota Longbow Tactical E.R.

DAKOTA 76 CLASSIC BOLT-ACTION RIFLE
Caliber: 257 Roberts, 270, 280, 30-06, 7mm Rem. Mag., 338 Win. Mag., 300 Win. Mag., 375 H&H, 458 Win. Mag.
Barrel: 23".
Weight: 7½ lbs. **Length:** 43½" overall.
Stock: Medium fancy grade walnut in classic style. Checkered pistol grip and forend; solid buttpad.
Sights: None furnished; drilled and tapped for scope mounts.
Features: Has many features of the original Model 70 Winchester. One-piece rail trigger guard assembly; steel grip cap. Model 70-style trigger. Many options available. Left-hand rifle available at same price. Introduced 1988. From Dakota Arms, Inc.
Price: $2,995.00

Dakota 76 Classic Rifles
A scaled-down version of the standard Model 76. Standard chamberings are 22-250, 243, 6mm Rem., 250-3000, 7mm-08, 308, others on special order. Short Classic Grade has 21" barrel; Alpine Grade is lighter (6½ lbs.), has a blind magazine and slimmer stock. Introduced 1989.
Price: Short Classic $2,995.00

Dakota 76 Varmint Rifle
Similar to the Dakota 76 except is a single shot with heavy barrel contour and special stock dimensions for varmint shooting. Chambered for 17 Rem., 22 BR, 222 Rem., 22-250, 220 Swift, 223, 6mm BR, 6mm PPC. Introduced 1994. Made in U.S. by Dakota Arms, Inc.
Price: $2,500.00

DAKOTA LONGBOW TACTICAL E.R. RIFLE
Caliber: 300 Dakota Magnum, 330 Dakota Magnum, 338 Lapua Magnum.
Barrel: 28", .950" at muzzle.
Weight: 13.7 lbs. **Length:** 50" to 52" overall.
Stock: Ambidextrous McMillan A-2 fiberglass, black or olive green color; adjustable cheekpiece and buttplate.
Sights: None furnished. comes with Picatinny one-piece optical rail.
Features: Uses the Dakota 76 actiion with controlled-round feed; three-position firing pin block safety; claw extractor; Model 70-style trigger. comes with bipod, case tool kit. Introduced 1997. Made in U.S. by Dakota Arms, Inc.
Price: $4,250.00

DAKOTA LONG RANGE HUNTER RIFLE
Caliber: 25-06, 257 Roberts, 270 Win., 280 Rem., 7mm Rem. Mag., 7mm Dakota Mag., 30-06, 300 Win. Mag., 300 Dakota Mag., 338 Win. Mag., 330 Dakota Mag., 375 H&H Mag., 375 Dakota Mag.
Barrel: 24", 26", match-quality; free-floating.
Weight: 7.7 lbs. **Length:** 45" to 47" overall.
Stock: H-S Precision black synthetic, with one-piece bedding block system.
Sights: None furnished. Drilled and tapped for scope mounting.
Features: Cylindrical machined receiver. controlled round feed; Mauser-style extractor; three-position striker blocking safety; fully adjustable match trigger. Introduced 1997. Made in U.S. by Dakota Arms, Inc.
Price: $1,595.00

> Consult our Directory pages for the location of firms mentioned.

Dakota 416 Rigby

Dakota 416 Rigby African
Similar to the 76 Safari except chambered for 404 Jeffery, 416 Rigby, 416 Dakota, 450 Dakota, 4-round magazine, select wood, two stock cross-bolts. Has 24" barrel, weight of 9-10 lbs. Ramp front sight, standing leaf rear. Introduced 1989.
Price: $4,495.00

HARRIS GUNWORKS TALON SAFARI RIFLE
Caliber: 300 Win. Mag., 300 Wea. Mag., 300 Phoenix, 338 Win. Mag., 30/378, 338 Lapua, 300 H&H, 340 Wea. Mag., 375 H&H, 404 Jeffery, 416 Rem. Mag., 458 Win. Mag. (Safari Magnum); 378 Wea. Mag., 416 Rigby, 416 Wea. Mag., 460 Wea. Mag. (Safari Super Magnum).
Barrel: 24".
Weight: About 9-10 lbs. **Length:** 43" overall.
Stock: Gunworks fiberglass Safari.
Sights: Barrel band front ramp, multi-leaf express rear.
Features: Uses Harris Gunworks Safari action. Has quick detachable 1" scope mounts, positive locking steel floorplate, barrel band sling swivel. Match-grade barrel. Matte black finish standard. Introduced 1989. From Harris Gunworks, Inc.
Price: Talon Safari Magnum $3,900.00
Price: Talon Safari Super Magnum $4,200.00

Harris Gunworks Signature Titanium Mountain Rifle
Similar to the Classic Sporter except action made of titanium alloy, barrel of chrome-moly steel. Stock is of graphite reinforced fiberglass. Weight is 5½ lbs. Chambered for 270, 280 Rem., 30-06, 7mm Rem. Mag., 300 Win. Mag. Fiberglass stock optional. Introduced 1989.
Price: $3,200.00

DAKOTA 76 SAFARI BOLT-ACTION RIFLE
Caliber: 270 Win., 7x57, 280, 30-06, 7mm Dakota, 7mm Rem. Mag., 300 Dakota, 300 Win. Mag., 330 Dakota, 338 Win. Mag., 375 Dakota, 458 Win. Mag., 300 H&H, 375 H&H, 416 Rem.
Barrel: 23".
Weight: 8½ lbs. **Length:** 43½" overall.
Stock: XXX fancy walnut with ebony forend tip; point-pattern with wrap-around forend checkering.
Sights: Ramp front, standing leaf rear.
Features: Has many features of the original Model 70 Winchester. Barrel band front swivel, inletted rear. Cheekpiece with shadow line. Steel grip cap. Introduced 1988. From Dakota Arms, Inc.
Price: Wood stock $3,995.00

HARRIS GUNWORKS SIGNATURE CLASSIC SPORTER
Caliber: 22-250, 243, 6mm Rem., 7mm-08, 284, 308 (short action); 25-06, 270, 280 Rem., 30-06, 7mm Rem. Mag., 300 Win. Mag., 300 Wea. (long action); 338 Win. Mag., 340 Wea., 375 H&H (magnum action).
Barrel: 22", 24", 26".
Weight: 7 lbs. (short action).
Stock: Fiberglass in green, beige, brown or black. Recoil pad and 1" swivels installed. Length of pull up to 14¼".
Sights: None furnished. Comes with 1" rings and bases.
Features: Uses right- or left-hand action with matte black finish. Trigger pull set at 3 lbs. Four-round magazine for standard calibers; three for magnums. Aluminum floorplate. Wood stock optional. Introduced 1987. From Harris Gunworks, Inc.
Price: $2,600.00

Harris Gunworks Signature Super Varminter
Similar to the Classic Sporter except has heavy contoured barrel, adjustable trigger, field bipod and special hand-bedded fiberglass stock. Chambered for 223, 22-250, 220 Swift, 243, 6mm Rem., 25-06, 7mm-08, 7mm BR, 308, 350 Rem. Mag. Comes with 1" rings and bases. Introduced 1989.
Price: $2,600.00

CENTERFIRE RIFLES—BOLT ACTION

Harris Gunworks Alaskan

Harris Gunworks Signature Alaskan
Similar to the Classic Sporter except has match-grade barrel with single leaf rear sight, barrel band front, 1" detachable rings and mounts, steel floorplate, electroless nickel finish. Has wood Monte Carlo stock with cheekpiece, palm-swell grip, solid buttpad. Chambered for 270, 280 Rem., 30-06, 7mm Rem. Mag., 300 Win. Mag., 300 Wea., 358 Win., 340 Wea., 375 H&H. Introduced 1989.
Price: $3,700.00

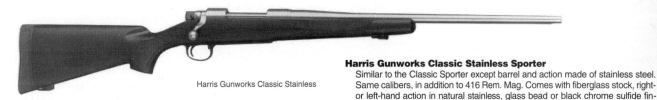
Harris Gunworks Classic Stainless

Harris Gunworks Classic Stainless Sporter
Similar to the Classic Sporter except barrel and action made of stainless steel. Same calibers, in addition to 416 Rem. Mag. Comes with fiberglass stock, right- or left-hand action in natural stainless, glass bead or black chrome sulfide finishes. Introduced 1990. From Harris Gunworks, Inc.
Price: $2,600.00

HARRIS GUNWORKS TALON SPORTER RIFLE
Caliber: 22-250, 243, 6mm Rem., 6mm BR, 7mm BR, 7mm-08, 25-06, 270, 280 Rem., 284, 308, 30-06, 350 Rem. Mag. (Long Action); 7mm Rem. Mag., 7mm STW, 300 Win. Mag., 300 Wea. Mag., 300 H&H, 338 Win. Mag., 340 Wea. Mag., 375 H&H, 416 Rem. Mag.
Barrel: 24" (standard).
Weight: About 7 1/2 lbs. **Length:** NA.

Stock: Choice of walnut or fiberglass.
Sights: None furnished; comes with rings and bases. Open sights optional.
Features: Uses pre-'64 Model 70-type action with cone breech, controlled feed, claw extractor and three-position safety. Barrel and action are of stainless steel; chrome-moly optional. Introduced 1991. From Harris Gunworks, Inc.
Price: $2,800.00

Howa Lightning

HOWA LIGHTNING BOLT-ACTION RIFLE
Caliber: 223, 22-250, 243, 270, 308, 30-06, 7mm Rem. Mag., 300 Win. Mag., 338 Win. Mag.
Barrel: 22", 24" magnum calibers.

Weight: 7 1/2 lbs. **Length:** 42" overall (22" barrel).
Stock: Black Bell & Carlson Carbelite composite with Monte Carlo comb; checkered grip and forend.
Sights: None furnished. Drilled and tapped for scope mounting.
Features: Sliding thumb safety; hinged floorplate; polished blue/black finish. Introduced 1993. From Interarms.
Price: Standard calibers $425.00
Price: Magnum calibers $425.00

Kimber Model 84C Varmint

Kimber Model 84C Varmint Stainless
Similar to the Model 84C Classic except chambered for 223 Rem-only; 24" medium-heavy, match-grade stainless steel, fluted barrel; with recessed target crown; weighs 7 1/2 lbs.; measures 42 1/2" overall. The Claro walnut stock has 18 lpi hand-checkering. Introduced 1997. Made in U.S. by Kimber of America.
Price: $1,215.00

KIMBER MODEL 84C CLASSIC BOLT-ACTION RIFLE
Caliber: 222, 223, 5-shot magazine.
Barrel: 22" match-grade sporter weight.
Weight: 6 3/4 lbs. **Length:** 40 1/2" overall.
Stock: Select A Claro walnut.
Sights: None furnished; drilled and tapped for Warne, Leupold or Millett scope mounts.
Features: Controlled round feed with Mauser-style extractor; pillar-bedded action; free-floating barrel; fully adjustable trigger; steel floorplate and trigger guard. Reintroduced 1996. Made in U.S. by Kimber of America, Inc.
Price: $1,145.00

Kimber 84C Single Shot

Kimber Model 84C SuperAmerica
Similar to the Model 84C Classic except has AAA Claro walnut stock with beaded checkpiece, ebony forend tip, wrap-around 22 lpi checkering, and black rubber butt pad. Chambered for 17 Rem., 222 Rem., 223 Rem. Reintroduced 1996. Made in U.S. by Kimber of America, Inc.
Price: $1,595.00

Kimber Model 84C Single Shot Varmint
Similar to the Model 84C except is a single shot chambered only for 17 Rem. and 223 Rem.; 25" fluted match-grade stainless barrel with target crown; and has varmint-profile stock with wide forend. Introduced 1996. Made in U.S. by Kimber of America, Inc.
Price: 17 Rem. $1,075.00
Price: 223 Rem. $999.00

CENTERFIRE RIFLES—BOLT ACTION

Kimber Model K770

Kimber Model K770 SuperAmerica Bolt-Action Rifle
Similar to the K770 Custom except has AAA Fancy Claro walnut stock with beaded checkpiece, ebony forend tip, and wrap-around hand-cut 22 lpi checkering. Introduced 1996.
Price: 270, 30-06 .... $1,260.00
Price: 7mm Rem., 300 Win. Mag., 338 Win. Mag. $1,275.00

KONGSBERG CLASSIC RIFLE
Caliber: 22-250, 243, 6.5x55, 270 Win., 30-06, 308 Win., 4-shot magazine; 7mm Rem. Mag., 300 Win. Mag., 338 Win. Mag., 3-shot magazine.
Barrel: 23" in standard calibers, 26" for magnums.
Weight: About 7 1/2 lbs. **Length:** 44" overall (23" barrel).
Stock: Oil-finished European walnut with straight fluted comb; 18 lpi checkering; rubber buttpad.
Sights: Hooded blade front, open adjustable rear. Receiver dovetailed for Weaver-type scope mount, and drilled and tapped.
Features: Rotary magazine; adjustable trigger; three-position safety; 60° bolt throw; claw extractor. Introduced 1993. Imported from Norway by Kongsberg America L.L.C.
Price: Right-hand, standard calibers $995.00
Price: Right-hand, magnum calibers $1,109.00
Price: Left-hand, standard calibers $1,133.00
Price: Left-hand, magnum calibers $1,245.00

KIMBER MODEL K770 CLASSIC RIFLE
Caliber: 270 Win., 30-06, 7mm Rem. Mag., 300 Win. Mag., 338 Win. Mag.
Barrel: 24" match grade, sporter weight.
Weight: About 7 1/2 lbs. **Length:** 43" overall.
Stock: Classic-style select Claro walnut, hand-cut panel checkering, solid rubber recoil pad, blued steel grip cap.
Sights: None furnished; drilled and tapped for Warne, Leupold or Millett scope mounts.
Features: Bolt locks into barrel breach; 60° bolt throw; pillar bedding; free-floated barrel; hinged floorplate. Introduced 1996. Made in U.S. by Kimber of America, Inc.
Price: 270, 30-06 .... $745.00
Price: 7mm Rem. Mag., 300 Win. Mag., 338 Win. Mag. $760.00

> *Manufacturers' addresses in the*
> **Directory of the Arms Trade**
> *page 309, this issue*

Kongsberg Thumbhole Sporter

KONGSBERG THUMBHOLE SPORTER RIFLE
Caliber: 22-250, 308 Win., 4-shot magazine.
Barrel: 23" heavy barrel (.750" muzzle).
Weight: About 8 1/2 lbs. **Length:** NA.
Stock: Oil-finished American walnut with stippled thumbhole grip, wide stippled forend, cheekpiece fully adjustable for height.
Sights: None furnished. Receiver dovetailed for scope mounting, and is drilled and tapped.
Features: Large bolt knob; rotary magazine; adjustable trigger; three-position safety; 60° bolt throw; claw extractor. Introduced 1993. Imported from Norway by Kongsberg America L.L.C.
Price: Right-hand $1,580.00
Price: Left-hand $1,718.00

KRICO MODEL 700 BOLT-ACTION RIFLES
Caliber: 17 Rem., 222, 222 Rem. Mag., 223, 5.6x50 Mag., 243, 308, 5.6x57 RWS, 22-250, 6.5x55, 6.5x57, 7x57, 270, 7x64, 30-06, 9.3x62, 6.5x68, 7mm Rem. Mag., 300 Win. Mag., 8x68S, 7.5 Swiss, 9.3x64, 6x62 Freres.
Barrel: 23.6" (std. cals.); 25.5" (mag. cals.).
Weight: 7 lbs. **Length:** 43.3" overall (23.6" bbl.).
Stock: European walnut, Bavarian cheekpiece.
Sights: Blade on ramp front, open adjustable rear.
Features: Removable box magazine; sliding safety. Drilled and tapped for scope mounting. Imported from Germany by Mandall Shooting Supplies.
Price: Model 700 .... $995.00
Price: Model 700 Deluxe S $1,495.00
Price: Model 700 Deluxe $1,025.00
Price: Model 700 Stutzen (full stock) $1,249.00

L.A.R. Grizzly 50

L.A.R. GRIZZLY 50 BIG BOAR RIFLE
Caliber: 50 BMG, single shot.
Barrel: 36".
Weight: 28.4 lbs. **Length:** 45.5" overall.
Stock: Integral. Ventilated rubber recoil pad.
Sights: None furnished; scope mount.
Features: Bolt-action bullpup design; thumb safety. All-steel construction. Introduced 1994. Made in U.S. by L.A.R. Mfg., Inc.
Price: $2,570.00

CENTERFIRE RIFLES—BOLT ACTION

Marlin Model MR-7

MARLIN MODEL MR-7 BOLT-ACTION RIFLE
Caliber: 25-06, 270, 30-06, 4-shot detachable box magazine.
Barrel: 22"; six-groove rifling.
Weight: 7 1/2 lbs. **Length:** 43" overall.
Stock: American black walnut with cut-checkered grip and forend, rubber buttpad, Mar-Shield® finish.
Sights: Bead on ramp front, adjustable rear, or no sights. (25-06 only available without sights.)
Features: Three-position safety; shrouded striker; red cocking indicator; adjustable 3-6 lb. trigger; quick-detachable swivel studs. Introduced 1996. Made in U.S. by Marlin.
Price: 30-06, 270, with sights $638.00
Price: 25-06, 270, no sights $598.00

Mauser Model 96

MAUSER MODEL 96 BOLT-ACTION RIFLE
Caliber: 25-06, 270, 7x64, 308, 30-06, 7mm Rem. Mag., 5-shot magazine.
Barrel: 22" standard, 24" magnum calibers.
Weight: 6 1/4 lbs. **Length:** 42" overall.
Stock: Checkered walnut.
Sights: None furnished. Drilled and tapped for Remington 700 mounts.
Features: "Slide-bolt" straight pull action with 16 locking lugs; three-position tang safety; quick-detachable 1" swivels. Introduced 1996. Imported from Germany by GSI, Inc.
Price: $699.00

MAUSER MODEL 94 BOLT-ACTION RIFLE
Caliber: 243, 7x64, 308, 270, 30-06, 9.3x62, 7mm Rem. Mag., 300 Win. Mag., 8x68S. detachable, 4-shot magazine.
Barrel: 22" standard, 24 magnum calibers.
Weight: 7 1/4 lbs. **Length:** 42 1/2" overall.
Stock: Select European walnut; hog back comb, Bavarian cheekpiece.
Sights: Blade on ramp front, open rear adjustable for windage.
Features: Interchangeable barrel system; 60-degree bolt throw; 6 locking lugs; aluminum alloy receiver; aluminum bedding black in stock. Introduced 1997. Imported from Germany by GSI, Inc.
Price: $1,850.00
Price: Interchangeable barrels $799.00

Mountain Eagle Varmint

MOUNTAIN EAGLE RIFLE
Caliber: 222 Rem., 223 Rem. (Varmint); 270, 280, 30-06 (long action); 7mm Rem. Mag., 300 Win. Mag., 338 Win. Mag., 300 Wea. Mag., 340 Wea. Mag., 375 H&H, 416 Rem. Mag. (magnum action).
Barrel: 24", 26" (Varmint); match-grade; fluted stainless on Varmint. Free floating.
Weight: 7 lbs., 13 oz. **Length:** 44" overall (24" barrel).
Stock: Kevlar-graphite with aluminum bedding block, high comb, recoil pad, swivel studs; made by H-S Precision.
Sights: None furnished; accepts any Remington 700-type base.
Features: Special Sako action with one-piece forged bolt, hinged steel floorplate, lengthened receiver ring; adjustable trigger. Cut-rifled benchrest barrel. Introduced 1996. From Magnum Research, Inc.
Price: Right-hand $1,499.00
Price: Left-hand $1,549.00
Price: Varmint Edition $1,629.00
Price: 375 H&H, 416 Rem., add $300.00

Raptor Bolt-Action

RAPTOR BOLT-ACTION RIFLE
Caliber: 270, 30-06, 243, 25-06, 308; 4-shot magazine.
Barrel: 22".
Weight: 7 lbs. **Length:** 42.5" overall.
Stock: Black synthetic, fiberglass reinforced; checkered grip and forend; vented recoil pad; Monte Carlo cheekpiece.
Sights: None furnished; drilled and tapped for scope mounts.
Features: Rust-resistant "Taloncote" treated barreled action; pillar bedded; stainless bolt with three locking lugs; adjustable trigger. Announced 1997. Made in U.S. by Raptor Arms Co., Inc.
Price: $249.00

REMINGTON MODEL 700 ADL BOLT-ACTION RIFLE
Caliber: 243, 270, 308, 30-06 and 7mm Rem. Mag.
Barrel: 22" or 24" round tapered.
Weight: 7 lbs. **Length:** 41 1/2" to 43 1/2" overall.
Stock: Walnut. Satin-finished pistol grip stock with fine-line cut checkering, Monte Carlo.
Sights: Gold bead ramp front; removable, step-adjustable rear with windage screw.
Features: Side safety, receiver tapped for scope mounts.
Price: From $485.00

Remington Model 700 BDL Bolt-Action Rifle
Same as the 700 ADL except chambered for 222, 223 (short action, 24" barrel), 22-250, 25-06, 6mm Rem. (short action, 22" barrel), 243, 270, 7mm-08, 280, 300 Savage, 30-06, 308; skip-line checkering; black forend tip and grip cap with white line spacers. Matted receiver top, fine-line engraving, quick-release floorplate. Hooded ramp front sight; quick detachable swivels.
Price: About $583.00
Also available in 17 Rem., 7mm Rem. Mag., 300 Win. Mag. (long action, 24" barrel), 338 Win. Mag., 35 Whelen (long action, 22" barrel). Overall length 44 1/2", weight about 7 1/2 lbs.
Price: About $609.00

CENTERFIRE RIFLES—BOLT ACTION

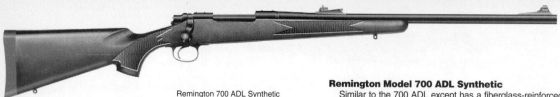
Remington 700 ADL Synthetic

Remington Model 700 VLS Varmint Laminated Stock
Similar to the 700 BDL except has 26" heavy barrel without sights, brown laminated stock with forend tip, grip cap, rubber buttpad. Available in 222 Rem., 223 Rem., 22-250, 7mm-08, 243, 308. Polished blue finish. Introduced 1995.
Price: $625.00

Remington Model 700 Custom KS Mountain Rifle
Similar to the 700 BDL except custom finished with Kevlar reinforced resin synthetic stock. Available in both left- and right-hand versions. Chambered for 270 Win., 280 Rem., 30-06, 7mm Rem. Mag., 300 Win. Mag., 300 Wea. Mag., 35 Whelen, 338 Win. Mag., 8mm Rem. Mag., 375 H&H, all with 24" barrel only. Weighs 6 lbs., 6 oz. Introduced 1986.
Price: From $1,100.00

Remington Model 700 BDL DM Rifle
Same as the 700 BDL except has detachable box magazine (4-shot, standard calibers, 3-shot for magnums). Has glossy stock finish, fine-line engraving, open sights, recoil pad, sling swivels. Right-hand action calibers: 6mm, 243, 25-06, 270, 280, 7mm-08, 30-06, 308, 7mm Rem. Mag., 300 Win. Mag., 338 Win. Mag.; left-hand calibers: 270, 30-06, 7mm Rem. Mag., 300 Win. Mag. Introduced 1995.
Price: From $636.00

Remington Model 700 ADL Synthetic
Similar to the 700 ADL except has a fiberglass-reinforced synthetic stock with straight comb, raised cheekpiece, positive checkering, and black rubber buttpad. Metal has matte finish. Available in 243, 270, 308, 30-06 with 22" barrel, 7mm Rem. Mag. with 24" barrel. Introduced 1996.
Price: From $425.00

Remington Model 700 Safari
Similar to the 700 BDL except custom finished and tuned. In 8mm Rem. Mag., 375 H&H, 416 Rem. Mag. or 458 Win. Mag. calibers only with heavy barrel. Hand checkered, oil-finished stock in classic or Monte Carlo style with recoil pad installed. Delivery time is about 5 months.
Price: About $1,093.00
Price: Classic stock, left-hand $1,160.00
Price: Safari Custom KS (Kevlar stock), right-hand $1,433.00
Price: As above, left-hand $1,326.00

Remington Model 700 AWR Alaskan Wilderness Rifle
Similar to the Model 700 BDL except has stainless barreled action with satin blue finish; special 24" Custom Shop barrel profile; matte gray stock of fiberglass and graphite, reinforced with DuPont Kevlar, straight comb with raised cheekpiece, magnum-grade black rubber recoil pad. Chambered for 7mm Rem. Mag., 300 Win. Mag., 300 Wea. Mag., 338 Win. Mag., 375 H&H. Introduced 1994.
Price: $1,345.00

Remington 700 Varmint Synthetic

Remington Model 700 VS Varmint Synthetic Rifle
Similar to the 700 BDL Varmint Laminated except has composite stock reinforced with DuPont Kevlar, fiberglass and graphite. Has aluminum bedding block that runs the full length of the receiver. Free-floating 26" barrel. Metal has black matte finish; stock has textured black and gray finish and swivel studs. Available in 220 Swift, 223, 22-250, 243, 308. Introduced 1992.
Price: $705.00

Remington Model 700 VS SF Rifle
Similar to the Model 700 Varmint Synthetic except has satin-finish stainless barreled action with 26" fluted barrel, spherical concave muzzle crown. Chambered for 223, 220 Swift, 22-250, 308. Introduced 1994.
Price: $852.00

Remington Model 700 Sendero Rifle
Similar to the Model 700 Varmint Synthetic except has long action for magnum calibers. Has 26" heavy varmint barrel with spherical concave crown. Chambered for 25-06, 270, 7mm Rem. Mag., 300 Win. Mag. Introduced 1994.
Price: From $705.00

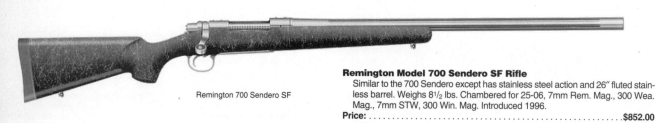
Remington 700 Sendero SF

Remington Model 700 Sendero SF Rifle
Similar to the 700 Sendero except has stainless steel action and 26" fluted stainless barrel. Weighs 8 1/2 lbs. Chambered for 25-06, 7mm Rem. Mag., 300 Wea. Mag., 7mm STW, 300 Win. Mag. Introduced 1996.
Price: $852.00

Remington 700 BDL LSS

Remington Model 700 BDL SS DM Rifle
Same as the 700 BDL SS except has detachable box magazine. Barrel, receiver and bolt made of #416 stainless steel; black synthetic stock, fine-line engraving. Available in 243, 25-06, 260 Rem., 270, 280, 7mm-08, 308, 30-06, 7mm Rem. Mag., 300 Win. Mag., 300 Wea. Mag., 338 Win. Mag. Introduced 1995.
Price: From $702.00

Remington Model 700 BDL SS Rifle
Similar to the 700 BDL rifle except has hinged floorplate, 24" standard weight barrel in all calibers; magnum calibers have magnum-contour barrel. No sights supplied, but comes drilled and tapped. Has corrosion-resistant follower and fire control, stainless BDL-style barreled action with fine matte finish. Synthetic stock has straight comb and cheekpiece, textured finish, positive checkering, plated swivel studs. Calibers—270, 30-06; magnums—7mm Rem. Mag., 300 Win. Mag., 338 Win. Mag., 375 H&H. Weighs 6 3/4-7 lbs. Introduced 1993.
Price: From $641.00

CENTERFIRE RIFLES—BOLT ACTION

Remington 700 BDL SS DM-B

Remington Model 700 BDL Left Hand
Same as 700 BDL except mirror-image left-hand action, stock. Available in 270, 30-06, 7mm Rem. Mag.
Price: About $609.00
Price: 7mm Rem. Mag. $636.00

Remington Model 700 LSS Rifle
Similar to the 700 BDL except has stainless steel barreled action, gray laminated wood stock with Monte Carlo comb and cheekpiece. No sights furnished. Available in 7mm Rem. Mag. and 300 Win. Mag. Introduced 1996.
Price: $714.00

REMINGTON MODEL 700 CLASSIC RIFLE
Caliber: 280 Rem.
Barrel: 24".
Weight: About 7 3/4 lbs. **Length:** 44 1/2" overall.
Stock: American walnut, 20 lpi checkering on pistol grip and forend. Classic styling. Satin finish.
Sights: None furnished. Receiver drilled and tapped for scope mounting.
Features: A "classic" version of the M700 ADL with straight comb stock. Fitted with rubber recoil pad. Sling swivel studs installed. Hinged floorplate. Limited production in 1997 only.
Price: $583.00

Remington Model 700 BDL SS DM-B
Same as the 700 BDL SS DM except has muzzlebrake, fine-line engraving. Available only in 7mm Rem. Mag., 7mm STW, 300 Win. Mag., 300 Wea. Mag., 338 Win. Mag. Introduced 1996.
Price: $789.00

Remington Model 700 MTN DM Rifle
Similar to the 700 BDL except weighs 6 3/4 lbs., has a 22" tapered barrel. Redesigned pistol grip, straight comb, contoured cheekpiece, hand-rubbed oil stock finish, deep cut checkering, hinged floorplate and magazine follower, two-position thumb safety. Chambered for 243, 270 Win., 7mm-08, 25-06, 280 Rem., 30-06, 4-shot detachable box magazine. Overall length is 42 1/2". Introduced 1995.
Price: About $636.00

Remington Model 700 APR African Plains Rifle
Similar to the Model 700 BDL except has magnum receiver and specially contoured 26" Custom Shop barrel with satin finish, laminated wood stock with raised cheekpiece, satin finish, black buttpad, 20 lpi cut checkering. Chambered for 7mm Rem. Mag., 300 Win. Mag., 300 Wea. Mag., 338 Win. Mag., 375 H&H. Introduced 1994.
Price: $1,466.00

Remington Model Seven

Remington Model Seven Youth Rifle
Similar to the Model Seven except has hardwood stock with 12 3/16" length of pull and chambered for 243, 7mm-08. Introduced 1993.
Price: $479.00

Remington Model Seven Custom MS Rifle
Similar to the Model Seven except has full-length Mannlicher-style stock of laminated wood with straight comb, solid black recoil pad, black steel forend tip, cut checkering, gloss finish. Barrel length 20", weighs 6 3/4 lbs. Available in 222 Rem., 223, 22-250, 243, 6mm Rem., 7mm-08 Rem., 308, 350 Rem. Mag. Calibers 250 Savage, 257 Roberts, 35 Rem. available on special order. Polished blue finish. Introduced 1993. From Remington Custom Shop.
Price: From $1,114.00

RUGER M77 MARK II RIFLE
Caliber: 223, 243, 6mm Rem., 257 Roberts, 25-06, 6.5x55 Swedish, 270, 280 Rem., 308, 30-06, 7mm Rem. Mag., 300 Win. Mag., 338 Win. Mag., 4-shot magazine.
Barrel: 20", 22"; 24" (magnums).
Weight: About 7 lbs. **Length:** 39 3/4" overall.
Stock: Hand-checkered American walnut; swivel studs, rubber butt pad.
Sights: None furnished. Receiver has Ruger integral scope mount base, comes with Ruger 1" rings. Some models have iron sights.
Features: Short action with new trigger and three-position safety. New trigger guard with redesigned floorplate latch. Left-hand model available. Introduced 1989.
Price: M77RMKII (no sights) $574.00
Price: M77RSMKII (open sights) $635.00
Price: M77LRMKII (left-hand, 270, 30-06, 7mm Rem. Mag., 300 Win. Mag.) $574.00

Ruger M77RSI International Carbine
Same as the standard Model 77 except has 18 1/2" barrel, full-length International-style stock, with steel forend cap, loop-type steel sling swivels. Integral-base receiver, open sights, Ruger 1" steel rings. Improved front sight. Available in 243, 270, 308, 30-06. Weighs 7 lbs. Length overall is 38 3/8".
Price: M77RSIMKII $642.00

REMINGTON MODEL SEVEN BOLT-ACTION RIFLE
Caliber: 223 Rem. (5-shot); 243, 260 Rem., 7mm-08, 308 (4-shot).
Barrel: 18 1/2".
Weight: 6 1/4 lbs. **Length:** 37 1/2" overall.
Stock: Walnut, with modified schnabel forend. Cut checkering.
Sights: Ramp front, adjustable open rear.
Features: Short-action design; silent side safety; free-floated barrel except for single pressure point at forend tip. Introduced 1983.
Price: $583.00

Remington Model Seven SS
Similar to the Model Seven except has stainless steel barreled action and black synthetic stock, 20" barrel. Chambered for 223, 243, 260 Rem., 7mm-08, 308. Introduced 1994.
Price: $641.00

Ruger M77RL Ultra Light
Similar to the standard M77 except weighs only 6 lbs., chambered for 223, 243, 308, 270, 30-06, 257; barrel tapped for target scope blocks; has 20" Ultra Light barrel. Overall length 40". Ruger's steel 1" scope rings supplied. Introduced 1983.
Price: M77RLMKII $610.00

RUGER M77 MARK II MAGNUM RIFLE
Caliber: 375 H&H, 4-shot magazine; 416 Rigby, 3-shot magazine.
Barrel: 26", with integral steel rib; hammer forged.
Weight: 9.25 lbs. (375); 10.25 lbs. (416, 458). **Length:** 40.5" overall.
Stock: Circassian walnut with hand-cut checkering, swivel studs, steel grip cap, rubber butt pad.
Sights: Ramp front, two leaf express on serrated integral steel rib. Rib also serves as base for front scope ring.
Features: Uses an enlarged Mark II action with three-position safety, stainless bolt, steel trigger guard and hinged steel floorplate. Controlled feed. Introduced 1989.
Price: M77RSMMKII $1,550.00

CAUTION: PRICES SHOWN ARE SUPPLIED BY THE MANUFACTURER OR IMPORTER. CHECK YOUR LOCAL GUNSHOP.

CENTERFIRE RIFLES—BOLT ACTION

Ruger M77 All-Weather

Ruger M77 Mark II All-Weather Stainless Rifle
Similar to the wood-stock M77 Mark II except all metal parts are of stainless steel, and has an injection-moulded, glass-fiber-reinforced Du Pont Zytel stock. Also offered with laminated wood stock. Chambered for 223, 243, 270, 308, 30-06, 7mm Rem. Mag., 300 Win. Mag., 338 Win. Mag. Has the fixed-blade-type ejector, three-position safety, and new trigger guard with patented floorplate latch. Comes with integral Scope Base Receiver and 1″ Ruger scope rings, built-in sling swivel loops. Introduced 1990.
Price: KM77RPMKII $574.00
Price: KM77RSPMKII, open sights $635.00
Price: KM77RBZMKII, no sights, laminated wood stock, 223, 243, 270, 280 Rem., 7mm Rem. Mag., 30-06, 308, 300 Win. Mag., 338 Win. Mag. . .$606.00
Price: KM77RSBZMKII, open sights, laminated wood stock, 243, 270, 7mm Rem. Mag., 30-06, 300 Win. Mag., 338 Win. Mag. $667.00

Ruger M77VT Target

RUGER M77 MARK II EXPRESS RIFLE
Caliber: 270, 30-06, 7mm Rem. Mag., 300 Win. Mag., 4-shot magazine.
Barrel: 22″, with integral steel rib; barrel-mounted front swivel stud; hammer forged.
Weight: 7.5 lbs. **Length:** 42.125″ overall.
Stock: Hand-checkered medium quality French walnut with steel grip cap, black rubber butt pad, swivel studs.
Sights: Ramp front, V-notch two-leaf express rear adjustable for windage mounted on rib.
Features: Mark II action with three-position safety, stainless steel bolt, steel trigger guard, hinged steel floorplate. Introduced 1991.
Price: M77RSEXMKII $1,550.00

RUGER M77VT TARGET RIFLE
Caliber: 22-250, 220 Swift, 223, 243, 25-06, 308.
Barrel: 26″ heavy stainless steel with target gray finish.
Weight: Approx. 9.25 lbs. **Length:** Approx. 44″ overall.
Stock: Laminated American hardwood with beavertail forend, steel swivel studs; no checkering or grip cap.
Sights: Integral scope mount bases in receiver.
Features: Ruger diagonal bedding system. Ruger steel 1″ scope rings supplied. Fully adjustable trigger. Steel floorplate and trigger guard. New version introduced 1992.
Price: KM77VTMKII $684.00

Ruger 77/44

RUGER 77/44 BOLT-ACTION RIFLE
Caliber: 44 Magnum, 4-shot magazine.
Barrel: 18 1/2″.
Weight: 6 lbs. **Length:** 38 1/4″ overall.
Stock: American walnut with rubber buttpad and swivel studs.
Sights: Gold bead front, folding leaf rear. Comes with Ruger 1″ scope rings.
Features: Uses same action as the Ruger 77/22. Short bolt stroke; rotary magazine; three-position safety. Introduced 1997. Made in U.S. by Sturm, Ruger & Co.
Price: . . . . .NA

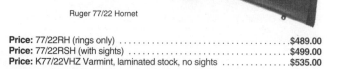
Ruger 77/22 Hornet

RUGER 77/22 HORNET BOLT-ACTION RIFLE
Caliber: 22 Hornet, 6-shot rotary magazine.
Barrel: 20″.
Weight: About 6 lbs. **Length:** 39 3/4″ overall.
Stock: Checkered American walnut, black rubber buttpad.
Sights: Brass bead front, open adjustable rear; also available without sights.
Features: Same basic features as the rimfire model except has slightly lengthened receiver. Uses Ruger rotary magazine. Three-position safety. Comes with 1″ Ruger scope rings. Introduced 1994.
Price: 77/22RH (rings only) $489.00
Price: 77/22RSH (with sights) $499.00
Price: K77/22VHZ Varmint, laminated stock, no sights $535.00

Sako 75 Hunter

Sako 75 Stainless Synthetic Rifle
Similar to the 75 Hunter except all metal is of stainless steel, and the synthetic stock has soft composite panels moulded into the forend and pistol grip. Available in 22-250, 243, 308 Win., 25-06, 270, 30-06 with 22″ barrel, 7mm Rem. Mag., 300 Win. Mag. with 24″ barrel. Introduced 1997. Imported from Finland by Stoeger Industries.
Price: Standard calibers $1,148.00
Price: Magnum calibers $1,178.00

Sako 75 Deluxe Rifle
Similar to the 75 Hunter except has select wood rosewood grip cap and forend tip. Available in 25-06, 270, 280, 30-06; 270 Wea. Mag., 7mm Rem. Mag., 7mm STW, 7mm Wea. Mag., 300 Win. Mag., 338 Win. Mag., 340 Wea. Mag., 375 H&H, 416 Rem. Mag. Introduced 1997. Imported from Finland by Stoeger Industries.
Price: Standard calibers $1,499.00
Price: Magnum Calibers $1,535.00

SAKO 75 HUNTER BOLT-ACTION RIFLE
Caliber: 22-250, 243, 7MM-08, 308 Win., 25-06, 270, 280, 30-06; 270 Wea. Mag., 7mm Rem. Mag., 7mm STW, 7mm Wea. Mag., 300 Win. Mag., 300 Wea. Mag., 338 Win. Mag., 340 Wea. Mag., 375 H&H, 416 Rem. Mag.
Barrel: 22″, standard calibers; 24″, 26″ magnum calibers.
Weight: About 6 lbs. **Length:** NA.
Stock: European walnut with matte lacquer finish.
Sights: None furnished; dovetail scope mount rails.
Features: New design with three locking lugs and a mechanical ejector; cold hammer-forged barrel is free-floating; two-position safety; hinged floorplate or detachable magazine that can be loaded from the top; short 70 degree bolt lift. Available in five action lengths. Introduced 1997. Imported form Finland by Stoeger Industries.
Price: Standard calibers $1,055.00
Price: Magnum Calibers $1,085.00

CENTERFIRE RIFLES—BOLT ACTION

SAKO HUNTER RIFLE
Caliber: 17 Rem., 222, 223 (short action); 22-250, 243, 7mm-08, 308 (medium action); 25-06, 270, 270 Wea. Mag., 7mm Wea. Mag., 30-06, 7mm Rem. Mag., 300 Win. Mag., 338 Win. Mag., 340 Wea. Mag., 375 H&H Mag., 300 Wea. Mag., 416 Rem. Mag. (long action).
Barrel: 22" to 24" depending on caliber.
Weight: 5$\frac{3}{4}$ lbs. (short); 6$\frac{1}{4}$ lbs. (med.); 7$\frac{1}{4}$ lbs. (long).
Stock: Hand-checkered European walnut.
Sights: None furnished.
Features: Adjustable trigger, hinged floorplate. Imported from Finland by Stoeger.
Price: 17 Rem., 222, 223 . . . . $1,050.00
Price: 22-250, 243, 308, 7mm-08 . . . . $1,050.00
Price: Long action cals. (except magnums) $1,085.00
Price: Magnum cals. $1,100.00
Price: 375 H&H, 416 Rem. Mag., from $1,120.00
Price: 300 Wea. $1,120.00

Sako Super Deluxe Sporter
Similar to Hunter except stock is select European walnut with high-gloss finish and deep-cut oak leaf carving. Metal has super high polish, deep blue finish. Special order only.
Price: $3,100.00

Sako Lightweight Deluxe
Same action as Hunter except has select wood, rosewood pistol grip cap and forend tip. Fine checkering on top surfaces of integral dovetail bases, bolt sleeve, bolt handle root and bolt knob. Vent. recoil pad, skip-line checkering, mirror finish bluing.
Price: 17 Rem., 222, 223, 22-250, 243, 308, 7mm-08 . . . . $1,475.00
Price: 25-06, 270, 280 Rem., 30-06 . . . . $1,510.00
Price: 7mm Rem. Mag., 300 Win. Mag., 338 Win. Mag. $1,525.00
Price: 300 Wea., 375 H&H, 416 Rem. Mag. $1,545.00

Sako Long-Range Rifle

Sako Varmint Heavy Barrel
Same as Hunter except has heavy varmint barrel, beavertail forend; available in 17 Rem., 222, 223 (short action), 22 PPC, 6mm PPC (single shot), 22-250, 243, 308, 7mm-08 (medium action). Weight from 8$\frac{1}{4}$ to 8$\frac{1}{2}$ lbs., 5-shot magazine capacity.
Price: 17 Rem., 222, 223 (short action) $1,240.00
Price: 22-250, 243, 308 (medium action) $1,240.00
Price: 22 PPC, 6mm PPC (single shot) $1,475.00

Sako Long-Range Hunting Rifle
Similar to the long action Hunter model except has 26" fluted barrel and is chambered for 25-06, 270 Win., 7mm Rem. Mag., 300 Win. Mag. Introduced 1996. Imported from Finland by Stoeger.
Price: 25-06 . . . . $1,275.00
Price: 7mm Rem. Mag., 300 Win. Mag $1,290.00

Sako TRG-S

SAUER 90 BOLT-ACTION RIFLE
Caliber: 22-250, 243, 6.5x55, 25-06, 270 Win., 308 Win., 30-06, 7mm Rem. Mag., 300 Win. Mag., 300 Wea. Mag., 338 Win. Mag., 375 H&H.
Barrel: 24" (standard calibers); 26" (magnums).
Weight: About 7.5 lbs. **Length:** 42.5" overall (24" barrel).
Stock: Select American Claro walnut with high-gloss epoxy finish, rosewood grip and forend caps; 22 lpi checkering.
Sights: None furnished; drilled and tapped for scope mounting.
Features: Three cam-actuated locking lugs on center of bolt; internal extractor; 65° bolt throw; detachable box magazine; tang safety; loaded chamber indicator; cocking indicator; adjustable trigger. Introduced 1986. Imported from Germany by SIGARMS, Inc.
Price: Standard or magnum $1,450.00

SAKO TRG-S BOLT-ACTION RIFLE
Caliber: 243, 7mm-08, 270, 6.5x55, 30-06, 7mm Rem. Mag., 300 Win. Mag., 338 Win. Mag., 270 Wea. Mag., 7mm Wea. Mag., 340 Wea. Mag., 375 H&H, 416 Rem. Mag., 5-shot magazine (4-shot for 375 H&H).
Barrel: 22", 24" (magnum calibers).
Weight: 7.75 lbs. **Length:** 45.5" overall.
Stock: Reinforced polyurethane with Monte Carlo comb.
Sights: None furnished.
Features: Resistance-free bolt with 60-degree lift. Recoil pad adjustable for length. Free-floating barrel, detachable magazine, fully adjustable trigger. Matte blue metal. Introduced 1993. Imported from Finland by Stoeger.
Price: 243, 7mm-08, 270, 30-06 . . . . $790.00
Price: Magnum calibers $830.00

Sauer Model 202

SAUER 202 BOLT-ACTION RIFLE
Caliber: Standard—243, 6.5x55, 270 Win., 308 Win., 30-06; magnum—7mm Rem. Mag., 300 Win. Mag., 300 Wea. Mag., 375 H&H.
Barrel: 23.6" (standard), 26" (magnum).
Weight: 7.7 lbs. (standard). **Length:** 44.3" overall (23.6" barrel).
Stock: Select American Claro walnut with high-gloss epoxy finish, rosewood grip and forend caps; 22 lpi checkering.
Sights: None furnished; drilled and tapped for scope mounting.
Features: Short 60° bolt throw; detachable box magazine; six-lug bolt; quick-change barrel; tapered bore; adjustable two-stage trigger; firing pin cocking indicator. Introduced 1994. Imported from Germany by SIGARMS, Inc.
Price: Standard or magnum calibers $995.00
Price: Left-hand $1,050.00

Savage 110 FM Sierra

SAVAGE MODEL 110 FM SIERRA LIGHT WEIGHT RIFLE
Caliber: 243, 270, 308, 30-06.
Barrel: 20"
Weight: 6$\frac{1}{4}$ lbs. **Length:** 41$\frac{1}{2}$" overall.
Stock: Graphite/fiberglass-filled composite.
Sights: None furnished; drilled and tapped for scope mounting.
Features: Comes with black nylon sling and quick-detachable swivels. Introduced 1996. Made in U.S. by Savage Arms, Inc.
Price: $410.00

CENTERFIRE RIFLES—BOLT ACTION

SAVAGE MODEL 110GXP3, 110GCXP3 PACKAGE GUNS
Caliber: 223, 22-250, 243, 250 Savage, 25-06, 270, 300 Sav., 30-06, 308, 7mm Rem. Mag., 7mm-08, 300 Win. Mag. (Model 110GXP3); 270, 30-06, 7mm Rem. Mag., 300 Win. Mag. (Model 110GCXP3).
Barrel: 22" (standard calibers), 24" (magnum calibers).
Weight: 7.25-7.5 lbs. **Length:** 43.5" overall (22" barrel).
Stock: Monte Carlo-style hardwood with walnut finish, rubber buttpad, swivel studs.
Sights: None furnished.
Features: Model 110GXP3 has fixed, top-loading magazine, Model 110GCXP3 has detachable box magazine. Rifles come with a factory-mounted and bore-sighted 3-9x32 scope, rings and bases, quick-detachable swivels, sling. Left-hand models available in all calibers. Introduced 1991 (GXP3); 1994 (GCXP3). Made in U.S. by Savage Arms, Inc.
Price: Model 110GXP3, right- or left-hand $420.00
Price: Model 110GCXP3, right- or left-hand $485.00

Savage Model 111FXP3, 111FCXP3 Package Guns
Similar to the Model 110 Series Package Guns except with lightweight, black graphite/fiberglass composite stock with non-glare finish, positive checkering. Same calibers as Model 110 rifles, plus 338 Win. Mag. Model 111FXP3 has fixed top-loading magazine; Model 111FCXP3 has detachable box. Both come with mounted 3-9x32 scope, quick-detachable swivels, sling. Introduced 1994. Made in U.S. by Savage Arms, Inc.
Price: Model 111FXP3, right- or left-hand $450.00
Price: Model 111FCXP3, right- or left-hand $495.00

Savage Model 111 FAK

SAVAGE MODEL 112 VARMINT RIFLES
Caliber: 22-250, 223, 5-shot magazine.
Barrel: 26" heavy.
Weight: 8.8 lbs. **Length:** 47.5" overall.
Stock: Black graphite/fiberglass filled composite with positive checkering.
Sights: None furnished; drilled and tapped for scope mounting.
Features: Pillar-bedded stock. Blued barrel with recessed target-style muzzle. Double front swivel studs for attaching bipod. Introduced 1991. Made in U.S. by Savage Arms, Inc.
Price: Model 112FV $410.00
Price: Model 112FVSS (cals. 223, 22-250, 25-06, 7mm Rem. Mag., 300 Win. Mag., stainless barrel, bolt handle, trigger guard) $515.00
Price: Model 112FVSS-S (as above, single shot) $515.00
Price: Model 112BVSS (heavy-prone laminated stock with high comb, Wundhammer swell, fluted stainless barrel, bolt handle, trigger guard) $540.00
Price: Model 112BVSS-S (as above, single shot) $540.00

Savage Model 114CE Classic European
Similar to the Model 114C except the oil-finished walnut stock has a schnabel forend tip, cheekpiece and skip-line checkering; bead on blade front sight, fully adjustable open rear; solid red buttpad. Chambered for 270, 30-06, 7mm Rem. Mag., 300 Win. Mag. Introduced 1996. Made in U.S. by Savage Arms, Inc.
Price: $600.00

SAVAGE MODEL 110FP TACTICAL RIFLE
Caliber: 223, 25-06, 308, 30-06, 300 Win. Mag., 7mm Rem. Mag., 4-shot magazine.
Barrel: 24", heavy; recessed target muzzle.
Weight: 8 1/2 lbs. **Length:** 45.5" overall.
Stock: Black graphite/fiberglass composition; positive checkering.
Sights: None furnished. Receiver drilled and tapped for scope mounting.
Features: Pillar-bedded stock. Black matte finish on all metal parts. Double swivel studs on the forend for sling and/or bipod mount. Right or left-hand. Introduced 1990. From Savage Arms, Inc.
Price: Right- or left-hand $429.00

> *Manufacturers' addresses in the*
> **Directory of the Arms Trade**
> *page 309, this issue*

SAVAGE MODEL 111 CLASSIC HUNTER RIFLES
Caliber: 223, 22-250, 243, 250 Sav., 25-06, 270, 300 Sav., 30-06, 308, 7mm Rem. Mag., 7mm-08, 300 Win. Mag., 338 Win. Mag. (Models 111G, GL, GNS, F, FL, FNS); 270, 30-06, 7mm Rem. Mag., 300 Win. Mag. (Models 111GC, GLC, FAK, FC, FLC).
Barrel: 22", 24" (magnum calibers).
Weight: 6.3 to 7 lbs. **Length:** 43.5" overall (22" barrel).
Stock: Walnut-finished hardwood (M111G, GC); graphite/fiberglass filled composite.
Sights: Ramp front, open fully adjustable rear; drilled and tapped for scope mounting.
Features: Three-position top tang safety, double front locking lugs, free-floated button-rifled barrel. Comes with trigger lock, target, ear puffs. Introduced 1994. Made in U.S. by Savage Arms, Inc.
Price: Model 111FC (detachable magazine, composite stock, right- or left-hand) $420.00
Price: Model 111F (top-loading magazine, composite stock, right- or left-hand) $380.00
Price: Model 111FNS (as above, no sights, right-hand only) $372.00
Price: Model 111G (wood stock, top-loading magazine, right- or left-hand) $360.00
Price: Model 111GC (as above, detachable magazine) $410.00
Price: Model 111GNS (wood stock, top-loading magzine, no sights, right-hand only) $353.00
Price: Model 111FAK Express (blued, composite stock, top loading magazine, Adjustable Muzzle Brake) $450.00

SAVAGE MODEL 114C CLASSIC RIFLE
Caliber: 270, 30-06, 7mm Rem. Mag., 300 Win. Mag.; 4-shot detachable box magazine in standard calibers, 3-shot for magnums.
Barrel: 22" for standard calibers, 24" for magnums.
Weight: 7 1/8 lbs. **Length:** 45 1/2" overall.
Stock: Oil-finished American walnut; checkered grip and forend.
Sights: None furnished; drilled and tapped for scope mounting.
Features: High polish blue on barrel, receiver and bolt handle; Savage logo laser-etched on bolt body; push-button magazine release. Made in U.S. by Savage Arms, Inc. Introduced 1996.
Price: $525.00

Savage Model 116SE

Savage Model 116US Ultra Stainless Rifle
Similar to the Model 116SE except chambered for 270, 30-06, 7mm Rem. Mag., 300 Win. Mag.; stock has high-gloss finish; no open sights. Stainless steel barreled action with satin finish. Introduced 1995. Made in U.S. by Savage Arms, Inc.
Price: $700.00

SAVAGE MODEL 116SE SAFARI EXPRESS RIFLE
Caliber: 300 Win. Mag., 338 Win. Mag., 425 Express, 458 Win. Mag.
Barrel: 24".
Weight: 8.5 lbs. **Length:** 45.5" overall.
Stock: Classic-style select walnut with ebony forend tip, deluxe cut checkering. Two cross bolts; internally vented recoil pad.
Sights: Bead on ramp front, three-leaf express rear.
Features: Controlled-round feed design; adjustable muzzlebrake; one-piece barrel band stud. Satin-finished stainless steel barreled action. Introduced 1994. Made in U.S. by Savage Arms, Inc.
Price: $900.00

CENTERFIRE RIFLES—BOLT ACTION

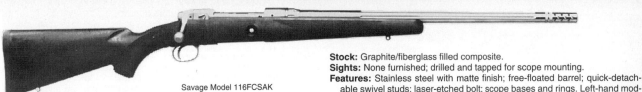

Savage Model 116FCSAK

SAVAGE MODEL 116 WEATHER WARRIORS
Caliber: 223, 243, 270, 30-06, 7mm Rem. Mag., 300 Win. Mag., 338 Win. Mag. (Model 116FSS); 270, 30-06, 7mm Rem. Mag., 300 Win. Mag. (Models 116FCSAK, 116FCS); 270, 30-06, 7mm Rem. Mag., 300 Win. Mag., 338 Win. Mag. (Models 116FSAK, 116FSK).
Barrel: 22", 24" for 7mm Rem. Mag., 300 Win. Mag., 338 Win. Mag. (M116FSS only).
Weight: 6.25 to 6.5 lbs. **Length:** 43.5" overall (22" barrel).
Stock: Graphite/fiberglass filled composite.
Sights: None furnished; drilled and tapped for scope mounting.
Features: Stainless steel with matte finish; free-floated barrel; quick-detachable swivel studs; laser-etched bolt; scope bases and rings. Left-hand models available in all models, calibers at same price. Models 116FCS, 116FSS introduced 1991; Model 116FSK introduced 1993; Model 116FCSAK, 116FSAK introduced 1994. Made in U.S. by Savage Arms, Inc.
Price: Model 116FSS (top-loading magazine) $495.00
Price: Model 116FCS (detachable box magazine) $560.00
Price: Model 116FCSAK (as above with Savage Adjustable Muzzle Brake system) $650.00
Price: Model 116FSAK (top-loading magazine, Savage Adjustable Muzzle Brake system) $585.00
Price: Model 116FSK Kodiak (as above with 22" Shock-Suppressor barrel) $554.00

Steyr-Mannlicher SBS

STEYR MANNLICHER SBS RIFLE
Caliber: 243, 25-06, 308, 6.5x55, 6.5x57, 270, 7x64 Brenneke, 7mm-08, 7.5x55, 30-06, 9.3x62, 6.5x68, 7mm Rem. Mag., 300 Win. Mag., 8x685, 4-shot magazine.
Barrel: 23.6" standard, 26" magnum; 20" full stock standard calibers.
Weight: 7 lbs. **Length:** 40.1" overall.
Stock: Hand-checkered fancy European oiled walnut with schnabel forend.
Sights: Ramp front adjustable for elevation, V-notch rear adjustable for windage.
Features: Single adjustable trigger; 3-position roller safety with "safe-bolt" setting; drilled and tapped for Steyr factory scope mounts. Introduced 1997. Imported from Austria by GSI, Inc.
Price: Half-stock, standard calibers $2,795.00
Price: Half-stock, magnum calibers $2,995.00
Price: Full-stock, standard calibers $2,995.00

Steyr SBS Forester

Steyr SBS Pro-Hunter Rifle
Similar to the SBS Forester except has ABS synthetic stock with adjustable butt spacers, straight comb without cheekpiece, palm swell, Pachmayr 1" swivels. Special 10-round magazine conversion kit available. Introduced 1997. Imported by GSI.
Price Standard calibers $799.00
Price Magnum calibers $899.00

STEYR SBS FORESTER RIFLE
Caliber: 243, 25-06, 270, 7mm-08, 308 Win., 30-06, 7mm Rem. Mag., 300 Win. Mag. Detachable 4-shot magazine.
Barrel: 23.6", standard calibers, 25.6", magnum calibers.
Weight: 7.5 lbs. **Length:** 44.5" overall (23.6" barrel).
Stock: Oil-finished American walnut with Monte Carlo cheekpiece. Pachmayr 1" swivels.
Sights: None furnished. Drilled and tapped for Browning A-Bolt mounts.
Features: Steyr Safe Bolt systems, three-position ambidextrous roller tang safety, for Safe, Loading Fire. Matte finish on barrel and receiver; adjustable trigger. Rotary cold-hammer forged barrel. Introduced 1997. Imported by GSI, In.
Price: Standard caliber $899.00
Price: Magnum calibers $1,045.00

STEYR SSG BOLT-ACTION RIFLES
Caliber: 308 Win., detachable 5-shot rotary magazine.
Barrel: 26"
Weight: 8.5 lbs. **Length:** 44.5" overall.
Stock: Black ABS Cycolac with spacers for length of pull adjustment.
Sights: Hooded ramp front adjustable for elevation, V-notch rear adjustable for windage.
Features: Sliding safety; NATO rail for bipod; 1" swivels; Parkerized finish; single or double-set triggers. Imported from Austria by GSI, Inc.
Price: SSG-PI, iron sights $2,195.00
Price: SSG-PII, heavy barrel, no sights $2,195.00
Price: SSG-PIIK, 20" heavy barrel, no sights $2,195.00
Price: SSG-PIV, 16.75" threaded heavy barrel with flash hider $2,660.00

Tikka Whitetail Hunter

Tikka Whitetail Hunter Synthetic Rifle
Similar to the Whitetail Hunter except has black synthetic stock; calibers 223, 308, 25-06, 270 Win., 30-06, 7mm Rem. Mag., 300 Win. Mag., 338 Win. Mag. Introduced 1996. Imported from Finland by Stoeger.
Price: Standard calibers $589.00
Price: Magnum calibers $619.00

TIKKA WHITETAIL HUNTER BOLT-ACTION RIFLE
Caliber: 22-250, 223, 243, 25-06, 270, 308, 30-06, 7mm Rem. Mag., 300 Win. Mag., 338 Win. Mag.
Barrel: 22 1/2" (std. cals.), 24 1/2" (magnum cals.).
Weight: 7 1/8 lbs. **Length:** 43" overall (std. cals.).
Stock: European walnut with Monte Carlo comb, rubber buttpad, checkered grip and forend.
Sights: None furnished.
Features: Detachable four-shot magazine (standard calibers), three-shot in magnums. Receiver dovetailed for scope mounting. Reintroduced 1996. Imported from Finland by Stoeger Industries.
Price: Standard calibers $589.00
Price: Magnum calibers $619.00

CENTERFIRE RIFLES—BOLT ACTION

Tikka Whitetail Stainless Synthetic

Tikka Continental Varmint Rifle
Similar to the standard Tikka rifle except has 26″ heavy barrel, extra-wide forend. Chambered for 22-250, 223, 308. Reintroduced 1996. Made in Finland by Sako. Imported by Stoeger.
Price: $684.00

Tikka Whitetail Hunter Stainless Synthetic
Similar to the Whitetail Hunter except all metal is of stainless steel, and it has a black synthetic stock. Available in 22-250, 243, 25-06, 308, 30-06, 7mm Rem. Mag., 300 Win. Mag., 338 Win. Mag. Introduced 1997. Imported from Finland by Stoeger.
Price: Standard calibers $649.00
Price: Magnum calibers $679.00

Tikka Continental Long Range Hunting Rifle
Similar to the Whitetail Hunter except has 26″ heavy barrel. Available in 25-06, 270 Win., 7mm Rem. Mag., 300 Win. Mag. Introduced 1996. Imported from Finland by Stoeger.
Price: 25-06, 270 Win. $684.00
Price: 7 Rem. Mag., 300 Win. Mag. $714.00

Ultra Light Arms Model 20

Ultra Light Arms Model 28, Model 40 Rifles
Similar to the Model 20 except in 264, 7mm Rem. Mag., 300 Win. Mag., 338 Win. Mag. (Model 28), 300 Wea. Mag., 416 Rigby (Model 40). Both use 24″ Douglas Premium No. 2 contour barrel. Weighs 5½ lbs., 45″ overall length. KDF or ULA recoil arrestor built in. Any custom feature available on any ULA product can be incorporated.
Price: Right-hand, Model 28 or 40 .... $2,900.00
Price: Left-hand, Model 28 or 40 .... $3,000.00

ULTRA LIGHT ARMS MODEL 20 RIFLE
Caliber: 17 Rem., 22 Hornet, 222 Rem., 223 Rem. (Model 20S); 22-250, 6mm Rem., 243, 257 Roberts, 7x57, 7x57 Ackley, 7mm-08, 284 Win., 308 Savage. Improved and other calibers on request.
Barrel: 22″ Douglas Premium No. 1 contour.
Weight: 4½ lbs. **Length:** 41½″ overall.
Stock: Composite Kevlar, graphite reinforced. DuPont imron paint colors—green, black, brown and camo options. Choice of length of pull.
Sights: None furnished. Scope mount included.
Features: Timney adjustable trigger; two-position three-function safety. Benchrest quality action. Matte or bright stock and metal finish. 3″ magazine length. Shipped in a hard case. From Ultra Light Arms, Inc.
Price: Right-hand $2,500.00
Price: Model 20 Left Hand (left-hand action and stock) $2,600.00
Price: Model 24 (25-06, 270, 280 Rem., 30-06, 3⅜″ magazine length) $2,600.00
Price: Model 24 Left Hand (left-hand action and stock) $2,700.00

Voere VEC-91

Voere VEC-91BR Caseless Rifle
Similar to the VEC-91 except has heavy 20″ barrel, synthetic benchrest stock, and is a single shot. Drilled and tapped for scope mounting. Introduced 1995. Imported from Austria by JagerSport, Ltd.
Price: $1,995.00

Voere VEC-91HB Varmint Special Caseless Rifle
Similar to the VEC-91 except has 22″ heavy sporter barrel, black synthetic or laminated wood stock. Drilled and tapped for scope mounts. Introduced 1995. Imported from Austria by JagerSport, Ltd.
Price: $1,695.00

VOERE VEC-91 LIGHTNING BOLT-ACTION RIFLE
Caliber: 5.56 UCC (223-cal.), 6mm UCC caseless, 5-shot magazine.
Barrel: 20″.
Weight: 6 lbs. **Length:** 39″ overall.
Stock: European walnut with cheekpiece, checkered grip and schnabel forend.
Sights: Blade on ramp front, open adjustable rear.
Features: Fires caseless ammunition via electric ignition; two batteries housed in the pistol grip last for about 5000 shots. Trigger is adjustable from 5 oz. to 7 lbs. Bolt action has twin forward locking lugs. Top tang safety. Drilled and tapped for scope mounting. Ammunition available from importer. Introduced 1991. Imported from Austria by JagerSport, Ltd.
Price: About $1,995.00

Voere VEC-91SS Caseless Rifle
Similar to the VEC-91 except has synthetic stock with straight comb, matte-finished metal. Drilled and tapped for scope mounting. No open sights furnished. Introduced 1995. Imported from Austria by JagerSport, Ltd.
Price: 5.56mm UCC or 6mm UCC $1,495.00

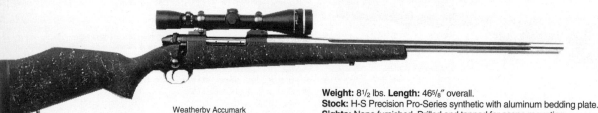
Weatherby Accumark

WEATHERBY ACCUMARK RIFLE
Caliber: 257, 270, 7mm, 300, 340 Wea. Mags., 30-378 Wea. Mag., 7mm STW, 7mm Rem. Mag., 300 Win. Mag.
Barrel: 26″.
Weight: 8½ lbs. **Length:** 46⅝″ overall.
Stock: H-S Precision Pro-Series synthetic with aluminum bedding plate.
Sights: None furnished. Drilled and tapped for scope mounting.
Features: Uses Mark V action with heavy-contour stainless barrel with black oxidized flutes, muzzle diameter of .705″. Introduced 1996. Made in U.S. From Weatherby.
Price: $1,299.00
Price: 30-378 Wea. Mag., 26″, Accubrake $1,427.00

CENTERFIRE RIFLES—BOLT ACTION

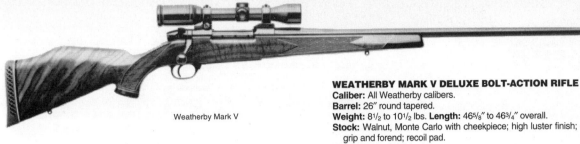

Weatherby Mark V

Weatherby Mark V Eurosport Rifle
Similar to the Mark V Deluxe except has raised-comb Monte Carlo stock with hand-rubbed satin oil finish, low-luster blue metal. No grip cap or forend tip. Right-hand only. Introduced 1995. Made in U.S. From Weatherby.
Price: 257, 270, 7mm, 300, 340 Wea. Mags., 26" barrel $949.00
Price: 7mm Rem. Mag., 300, 338 Win. Mags., 24" barrel $949.00
Price: 375 H&H, 24" barrel $949.00

Weatherby Mark V Stainless Rifle
Similar to the Mark V Deluxe except made of 400-series stainless steel. Has lightweight injection-moulded synthetic stock with raised Monte Carlo comb, checkered grip and forend, custom floorplate release. Right-hand only. Introduced 1995. Made in U.S. From Weatherby.
Price: 257, 270, 7mm, 300, 340 Wea. Mags., 26" barrel $999.00
Price: 7mm Rem. Mag., 300, 338 Win. Mags., 24" barrel $999.00
Price: 375 H&H, 24" barrel $999.00

WEATHERBY MARK V DELUXE BOLT-ACTION RIFLE
Caliber: All Weatherby calibers.
Barrel: 26" round tapered.
Weight: 8 1/2 to 10 1/2 lbs. **Length:** 46 5/8" to 46 3/4" overall.
Stock: Walnut, Monte Carlo with cheekpiece; high luster finish; checkered pistol grip and forend; recoil pad.
Sights: None furnished.
Features: Cocking indicator; adjustable trigger; hinged floorplate, thumb safety; quick detachable sling swivels. Made in U.S. From Weatherby.
Price: 257, 270, 7mm. 300, 340 Wea. Mags 26" barrel $1,499.00
Price: 378 Wea. Mag. 26" barrel $1,586.00
Price: 416 Wea. Mag. with Accubrake, 26" barrel $1,736.00
Price: 460 Wea. Mag. with Accubrake, 26" barrel $2,034.00

Weatherby Mark V Sporter Rifle
Same as the Mark V Deluxe without the embellishments. Metal has low-luster blue, stock is Claro walnut with high-gloss epoxy finish, Monte Carlo comb, recoil pad. Introduced 1993.
Price: 257 270, 7mm, 300, 340 Wea. Mags., 26" $949.00
Price: 375 H&H, 24" $949.00
Price: 7mm Rem. Mag., 300 Win. Mag., 338 Win. Mag., 24", ...$949.00

Weatherby Mark V SLS

Weatherby Euromark Rifle
Similar to the Mark V Deluxe except has raised-comb Monte Carlo stock with hand-rubbed oil finish, fine-line hand-cut checkering, ebony grip and forend tips. All metal has low-luster blue. Right-hand only. Uses Mark V action. Introduced 1995. Made in U.S. From Weatherby.
Price: 257, 270, 7mm, 300, 340 Wea. Mags., 26" barrel $1,459.00
Price: 378 Wea. Mag. $1,586.00
Price: 416 Wea. Mag., 26" barrel $1,736.00
Price: 7mm Rem. Mag., 300 Win. Mag., 338 Win. Mag., 375 H&H, 24" $1,499.00

Weatherby Lazermark V Rifle
Same as Mark V Deluxe except stock has extensive oak leaf pattern laser carving on pistol grip and forend. Introduced 1981.
Price: 257, 270, 7mm Wea. Mag., 300, 340, 26" $1,599.00
Price: 378 Wea. Mag., 26" $1,701.00
Price: 416 Wea. Mag., 26", Accubrake $1,847.00
Price: 460 Wea. Mag., 26", Accubrake $2,174.00

Weatherby Mark V SLS Stainless Laminate Sporter
Similar to the Mark V Stainless except all metalwork is 400 series stainless with a corrosion-resistant black oxide bead-blast matte finish. Action is hand-bedded in a laminated stock with a 1" recoil pad. Weighs 8 1/2 lbs. Introduced 1997. Made in U.S. From Weatherby.
Price: 257, 270, 7mm, 300, 340 Wea. Mag., 26" barrel $1,249.00
Price: 7mm Rem. Mag., 300 Win. Mag., 338 Win. Mag., 24" barrel ..$1,249.00

Weatherby Mark V Fluted Stainless
Similar to the Mark V Stainless except has fluted barrel. Weighs 7 1/2 lbs. Introduced 1997. Made in U.S. from Weatherby.
Price: 257, 270, 7mm, 300 Wea. Mag., 26" barrel $1,149.00
Price: 7mm Rem. Mag., 300 Win. Mag., 24" barrel $1,149.00

Weatherby Mark V Synthetic Rifle
Similar to the Mark V except has synthetic stock with raised Monte Carlo comb, dual-taper checkered forend. Low-luster blued metal. Weighs 8 lbs. Uses Mark V action. Right-hand only. Introduced 1995. Made in U.S. From Weatherby.
Price: 257, 270, 7mm, 300, 340 Wea. Mags., 26" barrel $799.00
Price: 7mm Rem. Mag., 300, 338 Win. Mags., 24" barrel $799.00
Price: 375 H&H, 24" barrel $799.00

Weatherby Mark V Fluted Synthetic
Similar to the Mark V Synthetic except has fluted barrel. Weighs 7 1/2 lbs. Introduced 1997. Made in U.S. from Weatherby.
Price: 257, 270, 7mm, 300 Wea. Mag., 26" barrel $949.00
Price: 7mm Rem. Mag., 300 Win. Mag., 24" barrel $949.00

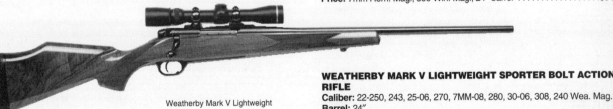

Weatherby Mark V Lightweight

Consult our Directory pages for the location of firms mentioned.

WEATHERBY MARK V LIGHTWEIGHT SPORTER BOLT ACTION RIFLE
Caliber: 22-250, 243, 25-06, 270, 7MM-08, 280, 30-06, 308, 240 Wea. Mag.
Barrel: 24".
Weight: 6 3/4 lbs. **Length:** 44" overall.
Stock: Claro walnut. Monte Carlo with cheekpiece; high luster finish; checkered pistol grip and forend, recoil pad.
Sights: None furnished. Drilled and tapped for scope mounting.
Features: Cocking indicator; adjustable trigger; hinged floorplate, thumb safety; six locking lugs; quick detachable swivels. Introduced 1997. Made in U.S. from Weatherby.
Price: $849.00

CENTERFIRE RIFLES—BOLT ACTION

Weatherby Mark V Carbine

Weatherby Mark V Lightweight Synthetic
Similar to the Mark V Lightweight Stainless except made of matte finished blued steel. Injection moulded synthetic stock. Weighs 6 1/2 lbs., 24" barrel. Available in 22-250, 240 Wea. Mag., 243, 25-06, 270, 7mm-08, 280, 30-06, 308. Introduced 1997. Made in U.S. From Weatherby.
Price: $699.00

WICHITA VARMINT RIFLE
Caliber: 222 Rem., 222 Rem. Mag., 223 Rem., 22 PPC, 6mm PPC, 22-250, 243, 6mm Rem., 308 Win.; other calibers on special order.
Barrel: 20 1/8".
Weight: 9 lbs. **Length:** 40 1/8" overall.
Stock: AAA Fancy American walnut. Hand-rubbed finish, hand checkered, 20 lpi pattern. Hand-inletted, glass bedded, steel grip cap. Pachmayr rubber recoil pad.
Sights: None. Drilled and tapped for scope mounts.
Features: Right- or left-hand Wichita action with three locking lugs. Available as a single shot only. Checkered bolt handle. Bolt is hand fitted, lapped and jeweled. Side thumb safety. Firing pin fall is 3/16". Non-glare blue finish. From Wichita Arms.
Price: Single shot $2,695.00

Weatherby Mark V Lightweight Carbine
Similar to the Mark V Lightweight Synthetic except has 20" barrel; injection moulded synthetic stock. Available in 243, 7mm-08, 308. Weighs 6 lbs.; overall length 40". Introduced 1997. Made in U.S. From Weatherby.
Price: $699.00

Weatherby Mark V Lightweight Stainless
Similar to the lightweight Sporter except made of 400 series stainless steel; injection moulded synthetic stock with Monte Carlo comb, checkered grip and forend. Weighs 6 1/2 lbs. Introduced 1997. Made in U.S. From Weatherby.
Price: $899.00

WICHITA CLASSIC RIFLE
Caliber: 17-222, 17-222 Mag., 222 Rem., 222 Rem. Mag., 223 Rem., 6x47; other calibers on special order.
Barrel: 21 1/8".
Weight: 8 lbs. **Length:** 41" overall.
Stock: AAA Fancy American walnut. Hand-rubbed and checkered (20 lpi). Hand-inletted, glass bedded, steel grip cap. Pachmayr rubber recoil pad.
Sights: None. Drilled and tapped for scope mounting.
Features: Available as single shot only. Octagonal barrel and Wichita action, right- or left-hand. Checkered bolt handle. Bolt is hand-fitted, lapped and jeweled. Adjustable trigger is set at 2 lbs. Side thumb safety. Firing pin fall is 3/16". Non-glare blue finish. From Wichita Arms.
Price: Single shot $3,495.00

Wilderness Explorer

WILDERNESS EXPLORER MULTI-CALIBER CARBINE
Caliber: 22 Hornet, 218 Bee, 44 Magnum, 50 A.E. (interchangeable).
Barrel: 18", match grade.
Weight: 5.5 lbs **Length:** 38 1/2" overall.
Stock: Synthetic or wood.
Sights: None furnished; comes with Weaver-style mount on barrel.
Features: Quick-change barrel and bolt face for caliber switch. Removable box magazine; adjustable trigger with side safety; detachable swivel studs. Introduced 1997. Made in U.S. by Phillips & Rogers, Inc.
Price: $995.00

Winchester Model 70 Classic

Winchester Model 70 Classic Featherweight
Same as the Model 70 Classic except has claw controlled-round feeding system; action is bedded in a standard-grade walnut stock. Available in 22-250, 243, 6.5x55, 308, 7mm-08, 270 Win., 280 Rem., 30-06. Drilled and tapped for scope mounts. Weighs 7.25 lbs. Introduced 1992.
Price: $620.00
Price: Classic Featherweight Stainless, as above except made of stainless steel, and available in 22-250, 243, 270, 308, 30-06, 7mm Rem. Mag., 300 Win. Mag. $716.00

Winchester Model 70 Classic Featherweight All-Terrain
Similar to the Model 70 Classic Featherweight except has black, fiberglass/graphite stock in same style as the Classic Featherweight, barreled action made of stainless steel. Calibers 270 Win., 30-06, 7mm Rem. Mag., 300 Win. Mag. Introduced 1996.
Price: $672.00

WINCHESTER MODEL 70 CLASSIC SPORTER
Caliber: 25-06, 270 Win., 270 Wea., 30-06, 264 Win. Mag., 7mm STW, 7mm Rem. Mag., 300 Win. Mag., 300 Wea. Mag., 338 Win. Mag., 3-shot magazine; 5-shot for 25-06, 270 Win., 30-06.
Barrel: 24", 26" for magnums.
Weight: 7 3/4 lbs. **Length:** 44 3/4" overall.
Stock: American walnut with cut checkering and satin finish. Classic style with straight comb.
Sights: Optional hooded ramp front, adjustable folding leaf rear. Drilled and tapped for scope mounting.
Features: Uses pre-64-type action with controlled round feeding. Three-position safety, stainless steel magazine follower; rubber buttpad; epoxy bedded receiver recoil lug. BOSS barrel vibration modulator and muzzlebrake system optional. From U.S. Repeating Arms Co.
Price: With sights $651.00
Price: Without sights $613.00
Price: With BOSS (25-06, 264 Win. Mag., 270 Win., 270 Wea. Mag., 30-06, 7mm Rem. Mag., 300 Win. Mag., 300 Wea. Mag., 338 Win. Mag.) ...$728.00
Price: Left-hand, 270, 30-06, 7mm Rem. Mag., 7mm STW, 300 Win. Mag., 338 Win. Mag. $641.00
Price: With BOSS, left-hand, 270, 30-06, 7mm Rem. Mag., 7mm STW, 300 Win. Mag. $756.00
Price: Classic Sporter Stainless, 270, 30-06, 7mm Rem. Mag., 300 Win. Mag., 338 Win. Mag. $716.00
Price: As above, left-hand $745.00
Price: Classic Sporter Stainless BOSS $831.00
Price: As above, left-hand $860.00

CENTERFIRE RIFLES—BOLT ACTION

Winchester Model 70 Classic Stainless Rifle
Same as the Model 70 Classic Sporter except has stainless steel barrel and pre-64-style action with controlled round feeding and matte gray finish, black composite stock impregnated with fiberglass and graphite, contoured rubber recoil pad. Available in 22-250, 243, 308, 270 Win., 270 Wea. Mag., 30-06, 7mm Rem. Mag., 300 Win. Mag., 300 Wea. Mag., 338 Win. Mag., 375 H&H Mag. (24" barrel), 3- or 5-shot magazine. Weighs 6.75 lbs. BOSS barrel vibration modulator and muzzlebrake system optional. Introduced 1994.
Price: Without sights $672.00
Price: 375 H&H Mag., with sights $724.00
Price: With BOSS $788.00

Winchester Model 70 Synthetic Heavy Varmint Rifle
Similar to the Model 70 Classic Sporter except has fiberglass/graphite stock, 26" heavy stainless steel barrel, blued receiver. Weighs about 10 3/4 lbs. Available in 220 Swift, 222, 223, 22-250, 243, 308. Uses full-length Pillar Plus Accu Block bedding system. Introduced 1993.
Price: $746.00
Price: With fluted barrel $894.00

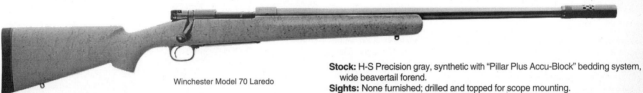
Winchester Model 70 Laredo

WINCHESTER MODEL 70 CLASSIC LAREDO
Caliber: 7mm Rem. Mag., 7mm STW, 300 Win. Mag., 3-shot magazine.
Barrel: 26" heavy; 1:10" (300), 1:9.5" (7mm).
Weight: 9 1/2 lbs. **Length:** 46 3/4" overall.
Stock: H-S Precision gray, synthetic with "Pillar Plus Accu-Block" bedding system, wide beavertail forend.
Sights: None furnished; drilled and topped for scope mounting.
Features: Pre-64-style, controlled round action with claw extractor, receiver-mounted blade ejector; matte blue finish. Introduced 1996. Made in U.S. by U.S. Repeating Arms Co.
Price: $764.00
Price: With BOSS $879.00

Winchester Model 70 Super Grade Classic

WINCHESTER MODEL 70 CLASSIC SUPER GRADE
Caliber: 270, 30-06, 5-shot magazine; 7mm Rem. Mag., 300 Win. Mag., 338 Win. Mag., 3-shot magazine.
Barrel: 24", 26" for magnums.
Weight: About 7 3/4 lbs. to 8 lbs. **Length:** 44 1/2" overall (24" bbl.)
Stock: Walnut with straight comb, sculptured cheekpiece, wrap-around cut checkering, tapered forend, solid rubber buttpad.
Sights: None furnished; comes with scope bases and rings.
Features: Controlled round feeding with stainless steel claw extractor, bolt guide rail, three-position safety; all steel bottom metal, hinged floorplate, stainless magazine follower. BOSS barrel vibration modulator and muzzlebrake system optional. Introduced 1994. From U.S. Repeating Arms Co.
Price: $840.00
Price: With BOSS system $956.00

WINCHESTER MODEL 70 CLASSIC SUPER EXPRESS MAGNUM
Caliber: 375 H&H Mag., 416 Rem. Mag., 458 Win. Mag., 3-shot magazine.
Barrel: 24" (375, 416), 22" (458).
Weight: 8 1/4 to 8 1/2 lbs.
Stock: American walnut with Monte Carlo cheekpiece. Wrap-around checkering and finish.
Sights: Hooded ramp front, open rear.
Features: Controlled round feeding. Two steel cross bolts in stock for added strength. Front sling swivel stud mounted on barrel. Contoured rubber buttpad. From U.S. Repeating Arms Co.
Price: $865.00
Price: Left-hand, 375 H&H only $894.00

WINCHESTER RANGER RIFLE
Caliber: 223, 243, 270, 30-06, 7mm Rem. Mag.
Barrel: 22".
Weight: 7 3/4 lbs. **Length:** 42" overall.
Stock: Stained hardwood.
Sights: Hooded blade front, adjustable open rear.
Features: Three-position safety; push feed bolt with recessed-style bolt face; polished blue finish; drilled and tapped for scope mounting. Introduced 1985. From U.S. Repeating Arms Co.
Price: $482.00
Price: Ranger Ladies/Youth, 223, 243, 7mm-08, 308 only, scaled-down stock $482.00

CENTERFIRE RIFLES—SINGLE SHOT
Classic and modern designs for sporting and competitive use.

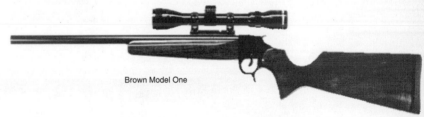
Brown Model One

BROWN MODEL ONE SINGLE SHOT RIFLE
Caliber: 22 LR, 357 Mag., 44 Mag., 7-30 Waters, 30-30 Win., 375 Win., 45-70; custom chamberings from 17 Rem. through 45-caliber available.
Barrel: 22" or custom, bull or tapered.
Weight: 6 lbs. **Length:** NA.
Stock: Smooth walnut; custom takedown design by Woodsmith. Palm swell for right- or left-hand; rubber butt pad.
Sights: Optional. Drilled and tapped for scope mounting.
Features: Rigid barrel/receiver; falling block action with short lock time, automatic case ejection; air-gauged barrels. Muzzle has 11-degree target crown. Matte black oxide finish standard, polished and electroless nickel optional. Introduced 1988. Made in U.S. by E.A. Brown Mfg.
Price: From $1,800.00

ARMSPORT 1866 SHARPS RIFLE, CARBINE
Caliber: 45-70.
Barrel: 28", round or octagonal.
Weight: 8.10 lbs. **Length:** 46" overall.
Stock: Walnut.
Sights: Blade front, folding adjustable rear. Tang sight set optionally available.
Features: Replica of the 1866 Sharps. Color case-hardened frame, rest blued. Imported by Armsport.
Price: $860.00
Price: With octagonal barrel $880.00
Price: Carbine, 22" round barrel $830.00

CENTERFIRE RIFLES—SINGLE SHOT

Browning Model 1885 Traditional Hunter

Browning Model 1885 Traditional Hunter
Similar to the Model 1885 High Wall except chambered for 30-30, 38-55 and 45-70 only; steel crescent buttplate; 1/16" gold bead front sight, adjustable buckhorn rear, and tang-mounted peep sight with barrel-type elevation adjuster and knob-type windage adjustments. Barrel is drilled and tapped for a Browning scope base. Oil-finished select walnut stock with swivel studs. Introduced 1997. Imported for Japan by Browning.
Price: $1,150.00

BROWNING MODEL 1885 HIGH WALL SINGLE SHOT RIFLE
Caliber: 22-250, 30-06, 270, 7mm Rem. Mag., 45-70.
Barrel: 28".
Weight: About 8 1/2 lbs. **Length:** 43 1/2" overall.
Stock: Walnut with straight grip, schnabel forend.
Sights: None furnished; drilled and tapped for scope mounting.
Features: Replica of J.M. Browning's high-wall falling block rifle. Octagon barrel with recessed muzzle. Imported from Japan by Browning. Introduced 1985.
Price: $939.95

Browning Model 1885 Low Wall

Browning Model 1885 Low Wall Rifle
Similar to the Model 1885 High Wall except has trimmer receiver, thinner 24" octagonal barrel. Forend is mounted to the receiver. Adjustable trigger. Walnut pistol grip stock, trim schnabel forend with high-gloss finish. Available in 22 Hornet, 223 Rem., 243 Win. Overall length 39 1/2", weighs 6 lbs., 4 oz. Rifling twist rates: 1:16" (22 Hornet); 1:12" (223); 1:10" (243). Polished blue finish. Introduced 1995. Imported from Japan by Browning.
Price: $939.95

Browning Model 1885 BPCR Rifle
Similar to the 1885 High Wall rifle except the ejector system and shell deflector have been removed; chambered only for 40-65 and 45-70; color case-hardened full-tang receiver, lever, buttplate and grip cap; matte blue 30" part octagon, part round barrel. The Vernier tang sight has indexed elevation, is screw adjustable windage, and has three peep diameters. The hooded front sight has a built-in spirit level and comes with sight interchangeable inserts. Adjustable trigger. Overall length 46 1/8", weighs about 11 lbs. Introduced 1996. Imported from Japan by Browning.
Price: $1,664.95

CABELA'S SHARPS SPORTING RIFLE
Caliber: 45-70.
Barrel: 32", tapered octagon.
Weight: 9 lbs. **Length:** 47 1/4" overall.
Stock: Checkered walnut.
Sights: Blade front, open adjustable rear.
Features: Color case-hardened receiver and hammer, rest blued. Introduced 1995. Imported by Cabela's.
Price: $749.95

CUMBERLAND MOUNTAIN PLATEAU RIFLE
Caliber: 40-65, 45-70.
Barrel: Up to 32"; round.
Weight: About 10 1/2 lbs. (32" barrel). **Length:** 48" overall (32" barrel).
Stock: American walnut.
Sights: Marble's bead front, Marble's open rear.
Features: Falling block action with underlever. Blued barrel and receiver. Stock has lacquer finish, crescent buttplate. Introduced 1995. Made in U.S. by Cumberland Mountain Arms, Inc.
Price: $1,085.00

Dakota Single Shot

DAKOTA SINGLE SHOT RIFLE
Caliber: Most rimmed and rimless commercial calibers.
Barrel: 23".
Weight: 6 lbs. **Length:** 39 1/2" overall.
Stock: Medium fancy grade walnut in classic style. Checkered grip and forend.
Sights: None furnished. Drilled and tapped for scope mounting.
Features: Falling block action with under-lever. Top tang safety. Removable trigger plate for conversion to single set trigger. Introduced 1990. Made in U.S. by Dakota Arms.
Price: $2,995.00
Price: Barreled action $2,000.00
Price: Action only $1,675.00

Dixie 1874 Sharps Silhouette

DIXIE 1874 SHARPS BLACKPOWDER SILHOUETTE RIFLE
Caliber: 45-70.
Barrel: 30"; tapered octagon; blued; 1:18" twist.
Weight: 10 lbs., 3 oz. **Length:** 47 1/2" overall.
Stock: Oiled walnut.
Sights: Blade front, ladder-type hunting rear.
Features: Replica of the Sharps #1 Sporter. Shotgun-style butt with checkered metal buttplate; color case-hardened receiver, hammer, lever and buttplate. Tang is drilled and tapped for tang sight. Double-set triggers. Meets standards for NRA blackpowder cartridge matches. Introduced 1995. Imported from Italy by Dixie Gun Works.
Price: $995.00

Dixie 1874 Sharps Lightweight Hunter/Target Rifle
Same as the Dixie 1874 Sharps Blackpowder Silhouette model except has a straight-grip buttstock with military-style buttplate. Based on the 1874 military model. Introduced 1995. Imported from Italy by Dixie Gun Works.
Price: $995.00

CENTERFIRE RIFLES—SINGLE SHOT

E.M.F. SHARPS RIFLE
Caliber: 45-70.
Barrel: 28", octagon.
Weight: 10¾ lbs. **Length:** NA.
Stock: Oiled walnut.
Sights: Blade front, flip-up open rear.
Features: Replica of the 1874 Sharps Sporting rifle. Color case-hardened lock; double-set trigger; blue finish. Imported by E.M.F.
Price: $950.00
Price: With browned finish $1,000.00
Price: Carbine (round 22" barrel, barrel band) $860.00

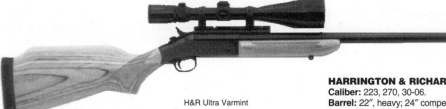

H&R Ultra Varmint

Harrington & Richardson Ultra Hunter Rifle
Similar to the Ultra Varmint rifle except chambered for 25-06 with 26" barrel, or 308 Win., 357 Rem. Max., 7x57 and 7x64 with 22" barrel. Stock and forend are of cinnamon-colored laminate; hand-checkered grip and forend. Introduced 1995. Made in U.S. by H&R 1871, Inc.
Price: $249.95
Price: 7x57, 7x64 ....$219.95

HARRINGTON & RICHARDSON ULTRA VARMINT RIFLE
Caliber: 223, 270, 30-06.
Barrel: 22", heavy; 24" compensated in 270, 30-06.
Weight: About 7.5 lbs. **Length:** NA.
Stock: Hand-checkered laminated birch with Monte Carlo comb.
Sights: None furnished. Drilled and tapped for scope mounting.
Features: Break-open action with side-lever release, positive ejection. Comes with scope mount. Blued receiver and barrel. Swivel studs. Introduced 1993. From H&R 1871, Inc.
Price: $249.95
Price: Compensated, 30-06, 270 ....$289.95

MODEL 1885 HIGH WALL RIFLE
Caliber: 30-40 Krag, 32-40, 38-55, 40-65 WCF, 45-70.
Barrel: 26" (30-40), 28" all others. Douglas Premium #3 tapered octagon.
Weight: NA. **Length:** NA.
Stock: Premium American black walnut.
Sights: Marble's standard ivory bead front, #66 long blade top rear with reversible notch and elevator.
Features: Recreation of early octagon top, thick-wall High Wall with Coil spring action. Tang drilled, tapped for High Wall tang sight. Receiver, lever, hammer and breechblock color case-hardened. Introduced 1991. Available from Montana Armory, Inc.
Price: $1,095.00

HARRIS GUNWORKS ANTIETAM SHARPS RIFLE
Caliber: 40-65, 45-75.
Barrel: 30", 32", octagon or round, hand-lapped stainless or chrome-moly.
Weight: 11.25 lbs. **Length:** 47" overall.
Stock: Choice of straight grip, pistol grip or Creedmoor with schnabel forend; pewter tip optional. Standard wood is A Fancy; higher grades available.
Sights: Montana Vintage Arms #111 Low Profile Spirit Level front, #108 mid-range tang rear with windage adjustments.
Features: Recreation of the 1874 Sharps sidehammer. Action is color case-hardened, barrel satin black. Chrome-moly barrel optionally blued. Optional sights include #112 Spirit Level Globe front with windage, #107 Long Range rear with windage. Introduced 1994. Made in U.S. by Harris Gunworks.
Price: $2,000.00

Navy Arms 1874 Sharps Sniper Rifle
Similar to the Navy Arms Sharps Carbine except has 30" barrel, double-set triggers; weighs 8 lbs., 8 oz., overall length 46¾". Introduced 1984. Imported by Navy Arms.
Price: $1,115.00
Price: 1874 Sharps Infantry Rifle (three-band) $1,060.00

NAVY ARMS 1874 SHARPS CAVALRY CARBINE
Caliber: 45-70.
Barrel: 22".
Weight: 7 lbs., 12 oz. **Length:** 39" overall.
Stock: Walnut.
Sights: Blade front, military ladder-type rear.
Features: Replica of the 1874 Sharps military carbine. Color case-hardened receiver and furniture. Imported by Navy Arms.
Price: $935.00

NAVY ARMS SHARPS BUFFALO RIFLE
Caliber: 45-70, 45-90.
Barrel: 28" heavy octagon.
Weight: 10 lbs., 10 oz. **Length:** 46" overall.
Stock: Walnut; checkered grip and forend.
Sights: Blade front, ladder rear; tang sight optional.
Features: Color case-hardened receiver, blued barrel; double-set triggers. Imported by Navy Arms.
Price: $1,080.00

Navy Arms Sharps Sporting Rifle
Same as the Navy Arms Sharps Plains Rifle except has pistol grip stock. Introduced 1997. Imported by Navy Arms.
Price: $1,080.00

Navy Arms Sharps Plains Rifle
Similar to the Sharps Buffalo rifle except has 32" medium-weight barrel, weighs 9 lbs., 8 oz., and is 49" overall. Imported by Navy Arms.
Price: $1,050.00

Navy Arms 1873 Springfield Infantry Rifle
Same action as the 1873 Springfield Cavalry Carbine except in rifle configuration with 32½" barrel, three-band full-length stock. Introduced 1997. Imported by Navy Arms.
Price: $1,060.00

Navy Arms 1873 Springfield Cavalry

NAVY ARMS ROLLING BLOCK BUFFALO RIFLE
Caliber: 45-70.
Barrel: 26", 30".
Stock: Walnut.
Sights: Blade front, adjustable rear.
Features: Reproduction of classic rolling block action. Available with full-octagon or half-octagon-half-round barrel. Color case-hardened action. From Navy Arms.
Price: $650.00

NAVY ARMS 1873 SPRINGFIELD CAVALRY CARBINE
Caliber: 45-70.
Barrel: 22".
Weight: 7 lbs. **Length:** 40½" overall.
Stock: Walnut.
Sights: Blade front, military ladder rear.
Features: Blued lockplate and barrel; color case-hardened breechblock; saddle ring with bar. Replica of 7th Cavalry gun. Imported by Navy Arms.
Price: $935.00

Navy Arms #2 Creedmoor Rolling Block Rifle
Similar to the Navy Arms Rolling Block Buffalo Rifle except has 30" tapered octagon barrel, checkered full-pistol grip stock, blade front sight, open adjustable rear sight and Creedmoor tang sight. Imported by Navy Arms.
Price: $875.00

CENTERFIRE RIFLES—SINGLE SHOT

New England Firearms Survivor

NEW ENGLAND FIREARMS HANDI-RIFLE
Caliber: 22 Hornet, 223, 243, 30-30, 270, 280 Rem., 30-06, 44 Mag., 45-70.
Barrel: 22", 26" for 280 Rem..
Weight: 7 lbs.
Stock: Walnut-finished hardwood; black rubber recoil pad.
Sights: Ramp front, folding rear (22 Hornet, 30-30, 45-70). Drilled and tapped for scope mount; 223, 243, 270, 280, 30-06 have no open sights, come with scope mounts.
Features: Break-open action with side-lever release. The 223, 243, 270 and 30-06 have recoil pad and Monte Carlo stock for shooting with scope. Swivel studs on all models. Blue finish. Introduced 1989. From New England Firearms.
Price: $209.95
Price: 280 Rem., 26" barrel $214.95

NEW ENGLAND FIREARMS SURVIVOR RIFLE
Caliber: 223, 357 Mag., single shot.
Barrel: 22".
Weight: 6 lbs. **Length:** 36" overall.
Stock: Black polymer, thumbhole design.
Sights: Blade front, fully adjustable open rear.
Features: Receiver drilled and tapped for scope mounting. Stock and forend have storage compartments for ammo, etc.; comes with integral swivels and black nylon sling. Introduced 1996. Made in U.S. by New England Firearms.
Price: Blue $219.95
Price: Electroless nickel $234.95

> *Manufacturers' addresses in the*
> **Directory of the Arms Trade**
> *page 309, this issue*

New England Firearms Super Light

New England Firearms Super Light Rifle
Similar to the Handi-Rifle except has new barrel taper, shorter 20" barrel with recessed muzzle and special lightweight synthetic stock and forend. No sights are furnished on the 223 version, but has a factory-mounted scope base and off-set hammer spur; 22 Hornet has ramp front, fully adjustable open rear. Overall length is 36", weight is 5.5 lbs. Introduced 1997. Made in U.S. by New England Firearms.
Price: 22 Hornet or 223 Rem. $229.95

Remington No. 1 Creedmoor

REMINGTON NO. 1 ROLLING BLOCK CREEDMOOR
Caliber: 45-70.
Barrel: 30".
Weight: $9^7/_8$ lbs. **Length:** $46^1/_2$" overall.
Stock: Semi-fancy American walnut, cut-checkered grip and forend.
Sights: Globe front with four interchangeable inserts and spirit level, tang-mounted Vernier rear.
Features: Recreation of the original pattern using the original stock design. Color case-hardened receiver; blued half-octagon barrel; smooth steel buttplate. Comes in fitted case. Introduced 1997. Made in U.S. by Remington.
Price: $2,799.00

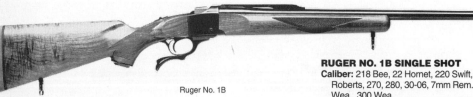

Ruger No. 1B

Ruger No. 1A Light Sporter
Similar to the No. 1B Standard Rifle except has lightweight 22" barrel, Alexander Henry-style forend, adjustable folding leaf rear sight on quarter-rib, dovetailed ramp front with gold bead. Calibers 243, 30-06, 270 and 7x57. Weighs about $7^1/_4$ lbs.
Price: No. 1A $685.00
Price: Barreled action $465.00

Ruger No. 1V Special Varminter
Similar to the No. 1B Standard Rifle except has 24" heavy barrel. Semi-beavertail forend, barrel tapped for target scope block, with 1" Ruger scope rings. Calibers 22-250, 220 Swift, 223, 25-06. Weight about 9 lbs.
Price: No. 1V $685.00
Price: Barreled action $465.00

RUGER NO. 1B SINGLE SHOT
Caliber: 218 Bee, 22 Hornet, 220 Swift, 22-250, 223, 243, 6mm Rem., 25-06, 257 Roberts, 270, 280, 30-06, 7mm Rem. Mag., 300 Win. Mag., 338 Win. Mag., 270 Wea., 300 Wea.
Barrel: 26" round tapered with quarter-rib; with Ruger 1" rings.
Weight: 8 lbs. **Length:** $43^3/_8$" overall.
Stock: Walnut, two-piece, checkered pistol grip and semi-beavertail forend.
Sights: None, 1" scope rings supplied for integral mounts.
Features: Under-lever, hammerless falling block design has auto ejector, top tang safety.
Price: $685.00
Price: Barreled action $465.00

Ruger No. 1H Tropical Rifle
Similar to the No. 1B Standard Rifle except has Alexander Henry forend, adjustable folding leaf rear sight on quarter-rib, ramp front with dovetail gold bead, 24" heavy barrel. Calibers 375 H&H, 404 Jeffery, 416 Rem. Mag. (weighs about $8^1/_4$ lbs.), 416 Rigby, and 458 Win. Mag. (weighs about 9 lbs.).
Price: No. 1H $685.00
Price: Barreled action $465.00

CENTERFIRE RIFLES—SINGLE SHOT

Ruger No. 1 RSI

Ruger No. 1S Medium Sporter
Similar to the No. 1B Standard Rifle except has Alexander Henry-style forend, adjustable folding leaf rear sight on quarter-rib, ramp front sight base and dovetail-type gold bead front sight. Calibers 218 Bee, 7mm Rem. Mag., 338 Win. Mag., 300 Win. Mag. with 26″ barrel, 45-70 with 22″ barrel. Weighs about 7$\frac{1}{2}$ lbs. In 45-70.
Price: No. 1S $685.00
Price: Barreled action $465.00

Ruger No. 1 RSI International
Similar to the No. 1B Standard Rifle except has lightweight 20″ barrel, full-length International-style forend with loop sling swivel, adjustable folding leaf rear sight on quarter-rib, ramp front sight with gold bead. Calibers 243, 30-06, 270 and 7x57. Weight is about 7$\frac{1}{4}$ lbs.
Price: No. 1 RSI $699.00
Price: Barreled action $465.00

C. Sharps New Model 1874

C. SHARPS ARMS NEW MODEL 1874 OLD RELIABLE®
Caliber: 40-50, 40-70, 40-90, 45-70, 45-90, 45-100, 45-110, 45-120, 50-70, 50-90, 50-140.
Barrel: 26″, 28″, 30″ tapered octagon.
Weight: About 10 lbs. **Length:** NA.
Stock: American black walnut; shotgun butt with checkered steel buttplate; straight grip, heavy forend with schnabel tip.
Sights: Blade front, buckhorn rear. Drilled and tapped for tang sight.
Features: Recreation of the Model 1874 Old Reliable Sharps Sporting Rifle. Double set triggers. Reintroduced 1991. Made in U.S. by C. Sharps Arms. Available from Montana Armory, Inc.
Price: $1,175.00

C. Sharps Arms 1875 Classic Sharps
Similar to the New Model 1875 Sporting Rifle except has 26″, 28″ or 30″ full octagon barrel, crescent buttplate with toe plate, Hartford-style forend with cast German silver nose cap. Blade front sight, Rocky Mountain buckhorn rear. Weighs 10 lbs. Introduced 1987. From C. Sharps Arms Co. and Montana Armory, Inc.
Price: $1,185.00

C. Sharps Arms New Model 1875 Target & Long Range
Similar to the New Model 1875 except available in all listed calibers except 22 LR; 34″ tapered octagon barrel; globe with post front sight, Long Range Vernier tang sight with windage adjustments. Pistol grip stock with cheek rest; checkered steel buttplate. Introduced 1991. From C. Sharps Arms Co. and Montana Armory, Inc.
Price: $1,535.00

C. SHARPS ARMS NEW MODEL 1875 OLD RELIABLE® RIFLE
Caliber: 22LR, 32-40 & 38-55 Ballard, 38-56 WCF, 40-65 WCF, 40-90 3$\frac{1}{4}$″, 40-90 2$\frac{5}{8}$″, 40-70 2$\frac{1}{10}$″, 40-70 2$\frac{1}{4}$″, 40-70 2$\frac{1}{2}$″, 40-50 1$\frac{11}{16}$″, 40-50 1$\frac{7}{8}$″, 45-90, 45-70, 45-100, 45-110, 45-120. Also available on special order only in 50-70, 50-90, 50-140.
Barrel: 24″, 26″, 30″ (standard), 32″, 34″ optional.
Weight: 8-12 lbs.
Stock: Walnut, straight grip, shotgun butt with checkered steel buttplate.
Sights: Silver blade front, Rocky Mountain buckhorn rear.
Features: Recreation of the 1875 Sharps rifle. Production guns will have case colored receiver. Available in Custom Sporting and Target versions upon request. Announced 1986. From C. Sharps Arms Co. and Montana Armory, Inc.
Price: 1875 Carbine (24″ tapered round bbl.) $810.00
Price: 1875 Saddle Rifle (26″ tapered oct. bbl.) $910.00
Price: 1875 Sporting Rifle (30″ tapered oct. bbl.) $975.00
Price: 1875 Business Rifle (28″ tapered round bbl.) $860.00

C. Sharps New Model 1885

SHILOH SHARPS 1874 LONG RANGE EXPRESS
Caliber: 40-50 BN, 40-70 BN, 40-90 BN, 45-70 ST, 45-90 ST, 45-110 ST, 50-70 ST, 50-90 ST, 50-110 ST, 32-40, 38-55, 40-70 ST, 40-90 ST.
Barrel: 34″ tapered octagon.
Weight: 10$\frac{1}{2}$ lbs. **Length:** 51″ overall.
Stock: Oil-finished semi-fancy walnut with pistol grip, shotgun-style butt, traditional cheek rest, schnabel forend.
Sights: Globe front, sporting tang rear.
Features: Recreation of the Model 1874 Sharps rifle. Double set triggers. Made in U.S. by Shiloh Rifle Mfg. Co.
Price: $1,334.00
Price: Sporting Rifle No. 1 (similar to above except with 30″ bbl., blade front, buckhorn rear sight) $1,308.00
Price: Sporting Rifle No. 3 (similar to No. 1 except straight-grip stock, standard wood) $1,204.00
Price: 1874 Hartford model $1,374.00

C. SHARPS ARMS NEW MODEL 1885 HIGHWALL RIFLE
Caliber: 22 LR, 22 Hornet, 219 Zipper, 25-35 WCF, 32-40 WCF, 38-55 WCF, 40-65, 30-40-Krag, 40-50 ST or BN, 40-70 ST or BN, 40-90 ST or BN, 45-70 2$\frac{1}{10}$″ ST, 45-90 2$\frac{4}{10}$″ ST, 45-100 2$\frac{6}{10}$″ ST, 45-110 2$\frac{7}{8}$″ ST, 45-120 3$\frac{1}{4}$″ ST.
Barrel: 26″, 28″, 30″, tapered full octagon.
Weight: About 9 lbs., 4 oz. **Length:** 47″ overall.
Stock: Oil-finished American walnut; schnabel-style forend.
Sights: Blade front, buckhorn rear. Drilled and tapped for optional tang sight.
Features: Single trigger; octagonal receiver top; checkered steel buttplate; color case-hardened receiver and buttplate, blued barrel. Many options available. Made in U.S. by C. Sharps Arms Co. Available from Montana Armory, Inc.
Price: From $1,195.00

Shiloh Sharps 1874 Montana Roughrider
Similar to the No. 1 Sporting Rifle except available with half-octagon or full-octagon barrel in 24″, 26″, 28″, 30″, 34″ lengths; standard supreme or semi-fancy wood, shotgun, pistol grip or military-style butt. Weight about 8$\frac{1}{2}$ lbs. Calibers 30-40, 30-30, 40-50x1$\frac{11}{16}$″ BN, 40-70x2$\frac{1}{10}$″ BN, 45-70x2$\frac{1}{10}$″ ST. Globe front and tang sight optional.
Price: Standard supreme $1,204.00
Price: Semi-fancy $1,314.00

CAUTION: PRICES SHOWN ARE SUPPLIED BY THE MANUFACTURER OR IMPORTER. CHECK YOUR LOCAL GUNSHOP.

CENTERFIRE RIFLES—SINGLE SHOT

Shiloh Sharps 1874 Business Rifle
Similar to No. 3 Rifle except has 28" heavy round barrel, military-style buttstock and steel buttplate. Weight about 9½ lbs. Calibers 40-50 BN, 40-70 BN, 40-90 BN, 45-70 ST, 45-90 ST, 50-70 ST, 50-100 ST, 32-40, 38-55, 40-70 ST, 40-90 ST.
Price: $1,210.00
Price: 1874 Saddle Rifle (similar to Carbine except has 26" octagon barrel, semi-fancy shotgun butt) $1,262.00

SHARPS 1874 RIFLE
Caliber: 45-70.
Barrel: 28", octagonal.
Weight: 9¼ lbs. Length: 46" overall.
Stock: Checkered walnut.
Sights: Blade front, adjustable rear.
Features: Double set triggers on rifle. Color case-hardened receiver and buttplate, blued barrel. Imported from Italy by E.M.F.
Price: Rifle or carbine $950.00
Price: Military rifle, carbine $860.00
Price: Sporting rifle $860.00

Thompson/Center Stainless

Thompson/Center Stainless Contender Carbine
Same as the blued Contender Carbine except made of stainless steel with blued sights. Available with walnut or Rynite stock and forend. Chambered for 22 LR, 22 Hornet, 223 Rem., 7-30 Waters, 30-30 Win., 410-bore. Youth model has walnut buttstock with 12" pull length. Introduced 1993.
Price: Rynite stock, forend $509.90

Thompson/Center Contender Carbine Youth Model
Same as the standard Contender Carbine except has 16¼" barrel, shorter buttstock with 12" length of pull. Comes with fully adjustable open sights. Overall length is 29", weight about 4 lbs., 9 oz. Available in 22 LR, 223 Rem.
Price: $479.00
Price: Extra barrels $234.30

THOMPSON/CENTER CONTENDER CARBINE
Caliber: 22 LR, 22 Hornet, 223 Rem., 7mm T.C.U., 7x30 Waters, 30-30 Win., 357 Rem. Maximum, 35 Rem., 44 Mag., 410, single shot.
Barrel: 21".
Weight: 5 lbs., 2 oz. Length: 35" overall.
Stock: Checkered American walnut with rubber buttpad. Also with Rynite stock and forend.
Sights: Blade front, open adjustable rear.
Features: Uses the T/C Contender action. Eleven interchangeable barrels available, all with sights, drilled and tapped for scope mounting. Introduced 1985. Offered as a complete Carbine only.
Price: Rifle calibers $515.00
Price: Extra barrels, rifle calibers, each $234.30
Price: 410 shotgun $535.60
Price: Extra 410 barrel $260.10

Consult our Directory pages for the location of firms mentioned.

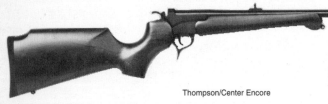
Thompson/Center Encore

WESSON & HARRINGTON BUFFALO CLASSIC RIFLE
Caliber: 45-70.
Barrel: 32" heavy.
Weight: 9 lbs. Length: 52" overall.
Stock: American black walnut.
Sights: None furnished; drilled and tapped for peep sight; barrel dovetailed for front sight.
Features: Color case-hardened Handi-Rifle action with exposed hammer; color case-hardened crescent buttplate; 19th century checkering pattern. Introduced 1995. Made in U.S. by H&R 1871, Inc.
Price: About $349.95

THOMPSON/CENTER ENCORE RIFLE
Caliber: 22-250, 223, 243, 270, 7mm-08, 308, 30-06, 7mm Rem. Mag., 300 Win. Mag.
Barrel: 24", 26".
Weight: 6 lbs., 12 oz. (24" barrel). Length: 38½" (24" barrel).
Stock: American walnut. Monte Carlo style; schnabel forend.
Sights: Ramp-style white bead front, fully adjustable leaf-type rear.
Features: Interchangeable barrels; action opens by squeezing trigger guard; drilled and tapped for T/C scope mounts; polished blue finish. Introduced 1996. Made in U.S. by Thompson/Center Arms.
Price: $535.00
Price: Extra barrels $235.00

UBERTI ROLLING BLOCK BABY CARBINE
Caliber: 22 LR, 22 WMR, 22 Hornet, 357 Mag., single shot.
Barrel: 22".
Weight: 4.8 lbs. Length: 35½" overall.
Stock: Walnut stock and forend.
Sights: Blade front, fully adjustable open rear.
Features: Resembles Remington New Model No. 4 carbine. Brass trigger guard and buttplate; color case-hardened frame, blued barrel. Imported by Uberti USA Inc.
Price: $490.00

DRILLINGS, COMBINATION GUNS, DOUBLE RIFLES

Designs for sporting and utility purposes worldwide.

Beretta 455EELL Express

BERETTA MODEL 455 SxS EXPRESS RIFLE
Caliber: 375 H&H, 458 Win. Mag., 470 NE, 500 NE 3", 416 Rigby.
Barrel: 23½" or 25½".
Weight: 11 lbs.
Stock: European walnut with hand-checkered grip and forend.
Sights: Blade front, folding leaf V-notch rear.
Features: Sidelock action with easily removable sideplates; color case-hardened finish (455), custom big game or floral motif engraving (455EELL). Double triggers, recoil pad. Introduced 1990. Imported from Italy by Beretta U.S.A.
Price: Model 455 ....$36,000.00
Price: Model 455EELL $47,000.00

DRILLINGS, COMBINATION GUNS, DOUBLE RIFLES

BERETTA EXPRESS SSO O/U DOUBLE RIFLES
Caliber: 375 H&H, 458 Win. Mag., 9.3x74R.
Barrel: 25.5".
Weight: 11 lbs.
Stock: European walnut with hand-checkered grip and forend.
Sights: Blade front on ramp, open V-notch rear.
Features: Sidelock action with color case-hardened receiver (gold inlays on SSO6 Gold). Ejectors, double triggers, recoil pad. Introduced 1990. Imported from Italy by Beretta U.S.A.
Price: SSO6 .... $21,000.00
Price: SSO6 Gold $23,500.00

GARBI EXPRESS DOUBLE RIFLE
Caliber: 7x65R, 9.3x74R, 375 H&H.
Barrel: 24 3/4".
Weight: 7 3/4 to 8 1/2 lbs. **Length:** 41 1/2" overall.
Stock: Turkish walnut.
Sights: Quarter-rib with express sight.
Features: Side-by-side double; H&H-pattern sidelock ejector with reinforced action, chopper lump barrels of Boehler steel; double triggers; fine scroll and rosette engraving, or full coverage ornamental; coin-finished action. Introduced 1997. Imported from Spain by Wm. Larkin Moore.
Price: $20,900.00

CHARLES DALY SUPERIOR COMBINATION GUN
Caliber/Gauge: 12 ga. over 22 Hornet, 223 Rem., 22-250, 243 Win., 270 Win., 308 Win., 30-06.
Barrel: 23.5", shotgun choked Imp. Cyl.
Weight: About 7.5 lbs.
Stock: Checkered walnut pistol grip buttstock and semi-beavertail forend.
Features: Silvered, engraved receiver; chrome moly steel barrels; double triggers; extractors; sling swivels; gold bead front sight. Introduced 1997. Imported from Italy by K.B.I. Inc.
Price: $1,229.00

Charles Daly Empire Combination Gun
Same as the Superior grade except has deluxe wood with European-style comb and cheekpiece; slim forend. Introduced 1997. Imported from Italy by K.B.I., Inc.
Price: $1,649.00

Krieghoff Classic Double Rifle

KRIEGHOFF CLASSIC DOUBLE RIFLE
Caliber: 7x65R, 308 Win., 30-06, 30R Blaser, 8x57 JRS, 8x75RS, 9.3x74R.
Barrel: 23.5".
Weight: 7.3 to 8 lbs. **Length:** NA.
Stock: High grade European walnut. Standard has conventional rounded cheekpiece, Bavaria has Bavarian-style cheekpiece.
Sights: Bead front with removable, adjustable wedge (375 H&H and below), standing leaf rear on quarter-rib.
Features: Boxlock action; double triggers; short opening angle for fast loading; quiet extractors; sliding, self-adjusting wedge for secure bolting; Purdey-style barrel extension; horizontal firing pin placement. Many options available. Introduced 1997. Imported from Germany by Krieghoff International.
Price: With small Arabesque engraving $7,850.00
Price: With engraved sideplates $9,800.00
Price: For extra barrels $4,500.00
Price: Extra 20-ga., 28" shotshell barrels $3,200.00

Krieghoff Classic Big Five Double Rifle
Similar to the standard Classic except available in 375 Flanged Mag. N.E., 500/416 N.E., 470 N.E., 500 N.E. 3". Has hinged front trigger, non-removable muzzle wedge (larger than 375-caliber), Universal Trigger System, Combi Cocking Device, steel trigger guard, specially weighted stock bolt for weight and balance. Many options available. Introduced 1997. Imorted from Germany by Krieghott International.
Price: $9,450.00
Price: With engraved sideplates $11,400.00

Merkel Boxlock Double Rifles
Similar to the Model 160 double rifle except with Anson & Deely boxlock action with cocking indicators, double triggers, engraved color case-hardened receiver. Introduced 1995. Imported from Germany by GSI.
Price: Model 140-1 .... $5,895.00
Price: Model 140-1.1 (engraved silver-gray receiver) $6,795.00
Price: Model 150-1 (false sideplates, silver-gray receiver, Arabesque engraving) $7,495.00
Price: Model 150-1.1 (as above with English Arabesque engraving) .. $8,695.00

MERKEL MODEL 160 SIDE-BY-SIDE DOUBLE RIFLE
Caliber: 22 Hornet, 5.6x50R Mag., 5.6x52R, 222 Rem., 243 Win., 6.5x55, 6.5x57R, 7x57R, 7x65R, 308, 30-06, 8x57JRS, 9.3x74R, 375 H&H.
Barrel: 25.6".
Weight: About 7.7 lbs, depending upon caliber. **Length:** NA.
Stock: Oil-finished walnut with pistol grip, cheekpiece.
Sights: Blade front on ramp, fixed rear.
Features: Sidelock action. Double barrel locking lug with Greener cross-bolt; fine engraved hunting scenes on sideplates; Holland & Holland ejectors; double triggers. Imported from Germany by GSI.
Price: From $13,295.00

Merkel Model 210E

MERKEL OVER/UNDER COMBINATION GUNS
Caliber/Gauge: 12, 16, 20 (2 3/4" chamber) over 22 Hornet, 5.6x50R, 5.6x52R, 222 Rem., 243 Win., 6.5x55, 6.5x57R, 7x57R, 7x65R, 308 Win., 30-06, 8x57JRS, 9.3x74R.
Barrel: 25.6".
Weight: About 7.6 lbs. **Length:** NA.
Stock: Oil-finished walnut; pistol grip, cheekpiece.
Sights: Bead front, fixed rear.
Features: Kersten double cross-bolt lock; scroll-engraved, color case-hardened receiver; Blitz action; double triggers. Imported from Germany by GSI.
Price: Model 210E $6,195.00
Price: Model 211E (silver-grayed receivcer, fine hunting scene engraving) $7,495.00
Price: Model 213E (sidelock action, English-style, large scroll Arabesque engraving) $14,795.00
Price: Model 313E (as above, medium-scroll engraving) .. $22,795.00

MERKEL DRILLINGS
Caliber/Gauge: 12, 20, 3" chambers; 16, 2 3/4" chambers; 22 Hornet, 5.6x50R Mag., 5.6x52R, 222 Rem., 243 Win., 6.5x55, 6.5x57R, 7x57R, 7x65R, 308, 30-06, 8x57JRS, 9.3x74R, 375 H&H.
Barrel: 25.6".
Weight: 7.9 to 8.4 lbs. depending upon caliber. **Length:** NA.
Stock: Oil-finished walnut with pistol grip; cheekpiece on 12-, 16-gauge.
Sights: Blade front, fixed rear.
Features: Double barrel locking lug with Greener cross-bolt; scroll-engraved, case-hardened receiver; automatic trigger safety; Blitz action; double triggers. Imported from Germany by GSI.
Price: Model 90S (selective sear safety) $6,795.00
Price: Model 90K (manually cocked rifle system) $7,295.00
Price: Model 95S (selective sear safety) $7,995.00
Price: Model 95K (manually cocked rifle system) $8,595.00

CAUTION: PRICES SHOWN ARE SUPPLIED BY THE MANUFACTURER OR IMPORTER. CHECK YOUR LOCAL GUNSHOP.

DRILLINGS, COMBINATION GUNS, DOUBLE RIFLES

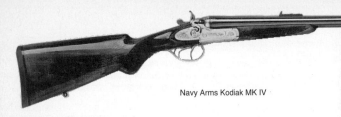
Navy Arms Kodiak MK IV

NAVY ARMS KODIAK MK IV DOUBLE RIFLE
Caliber: 45-70.
Barrel: 24".
Weight: 10 lbs., 3 oz. **Length:** 39 3/4" overall.
Stock: Checkered European walnut.
Sights: Bead front, folding leaf express rear.
Features: Blued, semi-regulated barrels; color case-hardened receiver and hammers; double triggers. Replica of Colt double rifle 1879-1885. Introduced 1996. Imported by Navy Arms.
Price: $3,125.00
Price: Engraved satin-finished receiver, browned barrels $4,000.00

MERKEL OVER/UNDER DOUBLE RIFLES
Caliber: 22 Hornet, 5.6x50R Mag., 5.6x52R, 222 Rem., 243 Win., 6.5x55, 6.5x57R, 7x57R, 7x65R, 308, 30-06, 8x57JRS, 9.3x74R.
Barrel: 25.6".
Weight: About 7.7 lbs, depending upon caliber. **Length:** NA.
Stock: Oil-finished walnut with pistol grip, cheekpiece.
Sights: Blade front, fixed rear.
Features: Kersten double cross-bolt lock; scroll-engraved, case-hardened receiver; Blitz action with double triggers. Imported from Germany by GSI.
Price: Model 221 E (silver-grayed receiver finish, hunting scene engraving) $10,895.00
Price: Model 223E (sidelock action, English-style large-scroll Arabesque engraving) $17,895.00
Price: Model 323E (as above with medium-scroll engraving) $27,195.00

RIZZINI EXPRESS 90L DOUBLE RIFLE
Caliber: 30-06, 7x65R, 9.3x74R.
Barrel: 24".
Weight: NA. **Length:** NA.
Stock: Select European walnut with satin oil finish; English-style cheekpiece.
Sights: Ramp front, quarter-rib with express sight.
Features: Color case-hardened boxlock action; automatic ejectors; single selective trigger; polished blue barrels. Extra 20-gauge shotshell barrels available. Imported for Italy by Wm. Larkin Moore.
Price: With case $4,500.00

Savage 24F Predator

SAVAGE 24F PREDATOR O/U COMBINATION GUN
Caliber/Gauge: 22 Hornet, 223, 30-30 over 12 (24F-12) or 22 LR, 22 Hornet, 223, 30-30 over 20-ga. (24F-20); 3" chambers.
Action: Takedown, low rebounding visible hammer. Single trigger, barrel selector spur on hammer.
Barrel: 24" separated barrels; 12-ga. has Full, Mod., Imp. Cyl. choke tubes, 20-ga. has fixed Mod. choke.
Weight: 8 lbs. **Length:** 40 1/2" overall.
Stock: Black Rynite composition.
Sights: Ramp front, rear open adjustable for elevation. Grooved for tip-off scope mount.
Features: Removable butt cap for storage and accessories. Introduced 1989.
Price: 24F-12 .... $415.00
Price: 24F-20 .... $415.00

SAUER DRILLING
Caliber/Gauge: 12, 2 3/4" chambers/243, 6.5x57R, 7x57R, 7x65R, 30-06, 9.3x74R; 16, 2 3/4" chambers/6.5x57R, 7x57R, 7x65R, 30-06.
Barrel: 25".
Weight: 7.5 lbs. **Length:** 46" overall.
Stock: Fancy French walnut with checkered grip and forend, hog-back comb, sculptured cheekpiece, hand-rubbed oil finish.
Sights: Bead front, automatic pop-up rifle rear.
Features: Greener boxlock cross-bolt action with double underlugs, Greener side safety; separate rifle cartridge extractor. Side-by-side shotgun barrels over rifle barrel. Nitride-coated, hand-engraved receiver available with English Arabesque or relief game animal scene engraving. Lux has profuse relief-engraved game scenes, extra-fancy stump wood. Imported from Germany by SIGARMS, Inc.
Price: Standard $4,800.00
Price: Lux $6,100.00

Springfield M6 Scout

SPRINGFIELD, INC. M6 SCOUT RIFLE/SHOTGUN
Caliber/Gauge: 22 LR or 22 Hornet over 410-bore.
Barrel: 18.25".
Weight: 4 lbs. **Length:** 32" overall.
Stock: Folding detachable with storage for 15 22 LR, four 410 shells.
Sights: Blade front, military aperture for 22; V-notch for 410.
Features: All-metal construction. Designed for quick disassembly and minimum maintenance. Folds for compact storage. Introduced 1982; reintroduced 1996. Imported from the Czech Republic by Springfield, Inc.
Price: Parkerized $167.00
Price: Stainless steel $199.00

TIKKA MODEL 512S COMBINATION GUN
Caliber/Gauge: 12 over 222, 308, 30-06.
Barrel: 24" (Imp. Mod.).
Weight: 7 5/8 lbs.
Stock: American walnut, with recoil pad. Monte Carlo style. Standard measurements 14"x1 3/5"x2"x2 3/5".
Sights: Blade front, flip-up-type open rear.
Features: Barrel selector on trigger. Hand-checkered stock and forend. Barrels are screw-adjustable to change bullet point of impact. Barrels are interchangeable. Introduced 1980. Imported from Italy by Stoeger.
Price: $1,770.00
Price: Extra barrels, from $810.00

TIKKA MODEL 512S DOUBLE RIFLE
Caliber: 300, 30-06, 9.3x74R.
Barrel: 24".
Weight: 8 5/8 lbs.
Stock: American walnut with Monte Carlo style.
Sights: Ramp front, adjustable open rear.
Features: Barrel selector mounted in trigger. Cocking indicators in tang. Recoil pad. Valmet scope mounts available. Introduced 1980. Imported from Italy by Stoeger.
Price: With ejectors $1,890.00

RIMFIRE RIFLES—AUTOLOADERS

Designs for hunting, utility and sporting purposes, including training for competition.

AMT Magnum Hunter

AMT MAGNUM HUNTER AUTO RIFLE
Caliber: 22 WMR, 10-shot magazine.
Barrel: 20".
Weight: 6 lbs. **Length:** 40½" overall.
Stock: Black fiberglass-filled nylon; checkered grip and forend.
Sights: None furnished; drilled and tapped for Weaver mount.
Features: Stainless steel construction. Free-floating target-weight barrel. Introduced 1995. Made in U.S. by AMT.
Price: $409.99

ARMSCOR MODEL AK22 AUTO RIFLE
Caliber: 22 LR, 10-shot magazine.
Barrel: 18.5".
Weight: 7.5 lbs. **Length:** 38" overall.
Stock: Plain mahogany.
Sights: Adjustable post front, leaf rear adjustable for elevation.
Features: Resembles the AK-47. Matte black finish. Introduced 1987. Imported from the Philippines by K.B.I., Inc.
Price: About $215.00

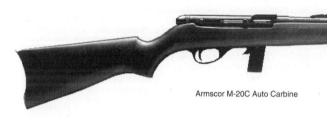

Armscor M-20C Auto Carbine

Armscor M-20P Standard Rifle
Similar to the M-20C except has 20.75" barrel, walnut-finished hardwood stock with Monte Carlo comb. Introduced 1990. Imported from the Philippines by K.B.I., Inc.
Price: $129.00

Armscor M-2000S Classic Auto Rifle
Similar to the M-20C except has 20.75" barrel; hand-checkered stock has Monte Carlo comb and cheekpiece; fully adjustable rear sight. Introduced 1990. Imported from the Philippines by K.B.I., Inc.
Price: $213.00

ARMSCOR M-1600 AUTO RIFLE
Caliber: 22 LR, 10-shot magazine.
Barrel: 18.25".
Weight: 6.2 lbs. **Length:** 38.5" overall.
Stock: Black finished mahogany.
Sights: Post front, aperture rear.
Features: Resembles Colt AR-15. Matte black finish. Introduced 1987. Imported from the Philippines by K.B.I., Inc.
Price: About $196.00

ARMSCOR M-20C AUTO CARBINE
Caliber: 22 LR, 10-shot magazine.
Barrel: 18.25".
Weight: 6.5 lbs. **Length:** 38" overall.
Stock: Walnut-finished mahogany.
Sights: Hooded front, rear adjustable for elevation.
Features: Receiver grooved for scope mounting. Blued finish. Introduced 1990. Imported from the Philippines by K.B.I., Inc.
Price: $161.00

Armscor M-2000SC Super Classic Rifle
Similar to the M-2000S except has oil-finished American walnut stock with 18 lpi hand checkering; black hardwood grip cap and forend tip; highly polished barreled action; jewelled bolt; recoil pad; swivel studs. Imported from the Philippines by K.B.I., Inc.
Price: $340.00

BRNO ZKM 611 AUTO RIFLE
Caliber: 22 WMR, 6-shot magazine.
Barrel: 20".
Weight: 6 lbs., 2 oz. **Length:** 37" overall.
Stock: Walnut; checkered grip and forend.
Sights: Blade front, open rear.
Features: Removable box magazine; polished blue finish; grooved receiver for scope mounting; sling swivels; thumbscrew takedown. Introduced 1995. Imported from the Czech Republic by Magnum Research.
Price: $569.00

Browning Auto-22

CALICO M-100FS SEMI-AUTO CARBINE
Caliber: 22 LR, 100-round helical feed magazine.
Barrel: 16.25".
Weight: 5 lbs. **Length:** 36" overall.
Stock: Glass-filled, impact-resistant polymer butt, grip and forend.
Sights: Post front adjustable for windage and elevation, notch rear.
Features: Easy takedown for storage or transport; ambidextrous safety; aluminum alloy receiver. Introduced 1996. Made in U.S. by Calico.
Price: $504.00

BROWNING AUTO-22 RIFLE
Caliber: 22 LR, 11-shot.
Barrel: 19¼".
Weight: 4¾ lbs. **Length:** 37" overall.
Stock: Checkered select walnut with pistol grip and semi-beavertail forend.
Sights: Gold bead front, folding leaf rear.
Features: Engraved receiver with polished blue finish; cross-bolt safety; tubular magazine in buttstock; easy takedown for carrying or storage. Imported from Japan by Browning.
Price: Grade I $398.95

Browning Auto-22 Grade VI
Same as the Grade I Auto-22 except available with either grayed or blued receiver with extensive engraving with gold-plated animals: right side pictures a fox and squirrel in a woodland scene; left side shows a beagle chasing a rabbit. On top is a portrait of the beagle. Stock and forend are of high-grade walnut with a double-bordered cut checkering design. Introduced 1987.
Price: Grade VI, blue or gray receiver $819.00

CAUTION: PRICES SHOWN ARE SUPPLIED BY THE MANUFACTURER OR IMPORTER. CHECK YOUR LOCAL GUNSHOP.

RIMFIRE RIFLES—AUTOLOADERS

E.A.A./SABATTI MODEL 1822 AUTO RIFLE
Caliber: 22 LR, 10-shot magazine.
Barrel: 18½" round tapered; bull barrel on Heavy and Thumbhole Heavy models.
Weight: 5¼ lbs. (Sporter). **Length:** 37" overall.
Stock: Stained hardwood; Thumbhole model has one-piece stock.
Sights: Bead front, folding leaf rear adjustable for elevation on Sporter model. Heavy and Thumbhole models only dovetailed for scope mount.
Features: Cross-bolt safety. Blue finish. Lifetime warranty. Introduced 1993. Imported from Italy by European American Armory.
Price: Sporter $190.00
Price: Heavy $205.00
Price: Thumbhole Heavy $350.00

KRICO MODEL 260 AUTO RIFLE
Caliber: 22 LR, 5-shot magazine.
Barrel: 19.6".
Weight: 6.6 lbs. **Length:** 38.9" overall.
Stock: Beech.
Sights: Blade on ramp front, open adjustable rear.
Features: Receiver grooved for scope mounting. Sliding safety. Imported from Germany by Mandall Shooting Supplies.
Price: $700.00

ERMA EM1 CARBINE
Caliber: 22 LR, 10-shot magazine.
Barrel: 18".
Weight: 5.6 lbs. **Length:** 35.5" overall.
Stock: Polished beech or oiled walnut.
Sights: Blade front, fully adjustable aperture rear.
Features: Blowback action. Receiver grooved for scope mounting. Imported from Germany by Mandall Shooting Supplies.
Price: $499.95

> *Manufacturers' addresses in the*
> # Directory of the Arms Trade
> *page 309, this issue*

Marlin Model 795

Marlin Model 795 Self-Loading Rifle
Similar to the Model 7000 except has standard-weight 18" barrel with 16-groove Micro-Groove rifling. Comes with ramp front sight with brass bead, screw adjustable open rear. Receiver grooved for scope mount. Introduced 1997. Made in U.S. by Marlin Firearms Co.
Price: $151.00
Price: With 4x scope $157.00

MARLIN MODEL 7000 SELF-LOADING RIFLE
Caliber: 22 LR, 10-shot magazine
Barrel: 18" heavy target with 12-groove Micro-Groove rifling, recessed muzzle.
Weight: 5½ lbs. **Length:** 37" overall.
Stock: Black fiberglass-filled synthetic with Monte Carlo combo, swivel studs, moulded-in checkering.
Sights: None furnished; comes with ring mounts.
Features: Automatic last-shot bolt hold-open, manual bolt hold-open; cross bolt safety; steel charging handle; blue finish, nickel-plated magazine. Introduced 1997. Made in U.S. by Marlin Firearms Co.
Price: $213.00

Marlin Model 60

Marlin Model 60SS Self-Loading Rifle
Same as the Model 60 except breech bolt, barrel and outer magazine tube are made of stainless steel; most other parts are either nickel-plated or coated to match the stainless finish. Monte Carlo stock is of black/gray Maine birch laminate, and has nickel-plated swivel studs, rubber butt pad. Introduced 1993.
Price: $255.00

MARLIN MODEL 60 SELF-LOADING RIFLE
Caliber: 22 LR, 14-shot tubular magazine.
Barrel: 22" round tapered.
Weight: About 5½ lbs. **Length:** 40½" overall.
Stock: Press-checkered, walnut-finished Maine birch with Monte Carlo, full pistol grip; Mar-Shieldr finish.
Sights: Ramp front, open adjustable rear.
Features: Matted receiver is grooved for scope mounting. Manual bolt hold-open; automatic last-shot bolt hold-open.
Price: $158.00
Price: With 4x scope $165.00

Marlin Model 922

MARLIN 70PSS STAINLESS RIFLE
Caliber: 22 LR, 7-shot magazine.
Barrel: 16¼" stainless steel, Micro-Groove® rifling.
Weight: 3¼ lbs. **Length:** 35¼" overall.
Stock: Black fiberglass-filled synthetic with abbreviated forend, nickel-plated swivel studs, moulded-in checkering.
Sights: Ramp front with orange post, cutaway Wide Scan® hood; adjustable open rear. Receiver grooved for scope mounting.
Features: Takedown barrel; cross-bolt safety; manual bolt hold-open; last shot bolt hold-open; comes with padded carrying case. Introduced 1986. Made in U.S. by Marlin.
Price: $259.00

MARLIN MODEL 922 MAGNUM SELF-LOADING RIFLE
Caliber: 22 WMR, 7-shot magazine.
Barrel: 20.5".
Weight: 6.5 lbs. **Length:** 39.75" overall.
Stock: Checkered American black walnut with Monte Carlo comb, swivel studs, rubber buttpad.
Sights: Ramp front with bead and removable Wide-Scan® hood, adjustable folding semi-buckhorn rear.
Features: Action based on the centerfire Model 9 Carbine. Receiver drilled and tapped for scope mounting. Automatic last-shot bolt hold open; magazine safety. Introduced 1993.
Price: $423.00

RIMFIRE RIFLES—AUTOLOADERS

MARLIN MODEL 995SS SELF-LOADING RIFLE
Caliber: 22 LR, 7-shot clip magazine.
Barrel: 18" Micro-Groove®; stainless steel.
Weight: 5 lbs. **Length:** 37" overall.
Stock: Black fiberglass-filled synthetic with nickel-plated swivel studs, moulded-in checkering.
Sights: Ramp front with orange post and cut-away Wide-Scan® hood; screw-adjustable open rear.
Features: Stainless steel breechbolt and barrel. Receiver grooved for scope mount; bolt hold-open device; cross-bolt safety. Introduced 1979.
Price: $247.00

Remington 522 Viper

REMINGTON MODEL 552 BDL SPEEDMASTER RIFLE
Caliber: 22 S (20), L (17) or LR (15) tubular mag.
Barrel: 21" round tapered.
Weight: About 5 3/4 lbs. **Length:** 40" overall.
Stock: Walnut. Checkered grip and forend.
Sights: Bead front, step open rear adjustable for windage and elevation.
Features: Positive cross-bolt safety, receiver grooved for tip-off mount.
Price: About $340.00

REMINGTON MODEL 522 VIPER AUTOLOADING RIFLE
Caliber: 22 LR, 10-shot magazine.
Barrel: 20".
Weight: 4 5/8 lbs. **Length:** 40" overall.
Stock: Black synthetic with positive checkering, beavertail forend.
Sights: Bead on ramp front, fully adjustable open rear. Integral grooved rail for scope mounting.
Features: Synthetic stock and receiver with overall matte black finish. Has magazine safety, cocking indicator; manual and last-shot hold-open; trigger mechanism has primary and secondary sears; integral ejection port shield. Introduced 1993.
Price: $152.00

Remington 597

Remington 597 LSS Auto Rifle
Similar to the Model 597 except has satin-finish stainless barrel, gray-toned alloy receiver with nickle-plated bolt, and laminated wood stock. Receiver is grooved and drilled and tapped for scope mounting. Introduced 1997. Made in U.S. by Remington.
Price: $265.00

REMINGTON 597 AUTO RIFLE
Caliber: 22 LR, 10-shot clip.
Barrel: 20".
Weight: 5 1/2 lbs. **Length:** 40" overall.
Stock: Gray synthetic.
Sights: Bead front, fully adjustable rear.
Features: Matte black finish, nickel-plated bolt. Receiver is grooved and drilled and tapped for scope mounts. Introduced 1997. Made in U.S. by Remington.
Price: $159.00
Price: Model 597 Magnum, 22 WMR, 8-shot clip $292.00

Ruger 10/22T Target

Ruger 10/22T Target Rifle
Similar to the 10/22 except has 20" heavy, hammer-forged barrel with tight chamber dimensions, improved trigger pull, laminated hardwood stock dimensioned for optical sights. No iron sights supplied. Introduced 1996. Made in U.S. by Sturm, Ruger & Co.
Price: $392.50

RUGER 10/22 AUTOLOADING CARBINE
Caliber: 22 LR, 10-shot rotary magazine.
Barrel: 18 1/2" round tapered.
Weight: 5 lbs. **Length:** 37 1/4" overall.
Stock: American hardwood with pistol grip and bbl. band.
Sights: Brass bead front, folding leaf rear adjustable for elevation.
Features: Detachable rotary magazine fits flush into stock, cross-bolt safety, receiver tapped and grooved for scope blocks or tip-off mount. Scope base adaptor furnished with each rifle.
Price: Model 10/22 RB (blue) $213.00
Price: Model K10/22RB (bright finish stainless barrel) $255.00

Ruger 10/22 International

Ruger 10/22 Deluxe Sporter
Same as 10/22 Carbine except walnut stock with hand checkered pistol grip and forend; straight buttplate, no barrel band, has sling swivels.
Price: Model 10/22 DSP $274.00

Ruger 10/22 International Carbine
Similar to the Ruger 10/22 Carbine except has full-length International stock of American hardwood, checkered grip and forend; comes with rubber buttpad, sling swivels. Reintroduced 1994.
Price: Blue (10/22RBI) $262.00
Price: Stainless (K10/22RBI) $282.00

Ruger 10/22RP All-Weather

Ruger K10/22RP All-Weather Rifle
Similar to the stainless K10/22/RP except has black composite stock of thermoplastic polyester resin reinforced with fiberglass; checkered grip and forend. Brushed satin, natural metal finish with clear hardcoat finish. Weighs 5 lbs., measures 36 3/4" overall. Introduced 1997. From Sturm, Ruger & Co.
Price: $255.00

RIMFIRE RIFLES—AUTOLOADERS

Savage Model 64G

SAVAGE MODEL 64G AUTO RIFLE
Caliber: 22 LR, 10-shot magazine.
Barrel: 20".
Weight: 5 1/2 lbs. **Length:** 40" overall.
Stock: Walnut-finished hardwood with Monte Carlo-type comb, checkered grip and forend.
Sights: Bead front, open adjustable rear. Receiver grooved for scope mounting.
Features: Thumb-operated rotating safety. Blue finish. Side ejection, bolt hold-open device. Introduced 1990. Made in Canada, from Savage Arms.
Price: $123.00
Price: Model 64F, black synthetic stock $145.00
Price: Model 64GXP Package Gun includes 4x15 scope and mounts ... $129.00

SURVIVAL ARMS AR-7 EXPLORER RIFLE
Caliber: 22 LR, 8-shot magazine.
Barrel: 16".
Weight: 2.5 lbs. **Length:** 34.5" overall; 16.5" stowed.
Stock: Moulded Cycolac; snap-on rubber butt cap.
Sights: Square blade front, aperture rear adjustable for elevation.
Features: Takedown design stores barrel and action in hollow stock. Light enough to float. Black,. Silvertone or camouflage finish. Reintroduced 1997. Made in U.S. by Survival Arms, Inc.
Price: Black, silver or camo finish $150.00
Price: Sporter (black finish with telescoping stock, 20-shot magazine) ... $200.00

WINCHESTER MODEL 63 AUTO RIFLE
Caliber: 22 LR, 10-shot magazine.
Barrel: 23".
Weight: 6 1/4 lbs. **Length:** 39" overall.
Stock: Walnut.
Sights: Bead front, open adjustable rear.
Features: Recreation of the original Model 63. Magazine tube loads through a port in the buttstock; forward cocking knob at front of forend; easy takedown for cleaning, storage; engraved receiver. Reintroduced 1997. From U.S. Repeating Arms Co.
Price: Grade I $678.00
Price: High grade, select walnut, cut checkering, engraved scenes with gold accents on receiver (made in 1997 only) $1,083.00

RIMFIRE RIFLES—LEVER & SLIDE ACTION
Classic and modern models for sport and utility, including training.

Browning BL-22

BROWNING BL-22 LEVER-ACTION RIFLE
Caliber: 22 S (22), L (17) or LR (15), tubular magazine.
Barrel: 20" round tapered.
Weight: 5 lbs. **Length:** 36 3/4" overall.
Stock: Walnut, two-piece straight grip Western style.
Sights: Bead post front, folding-leaf rear.
Features: Short throw lever, half-cock safety, receiver grooved for tip-off scope mounts. Imported from Japan by Browning.
Price: Grade I $345.95
Price: Grade II (engraved receiver, checkered grip and forend) $395.95

Henry Lever-Action 22

HENRY LEVER-ACTION .22 RIFLE
Caliber: 22 Short (21-shot), 22 Long (17-shot), 22 Long Rifle (15-shot).
Barrel: 18 1/4" round.
Weight: 5 1/2 lbs. **Length:** 36 1/4" overall.
Stock: Walnut.
Sights: Hooded blade front, open adjustabel rear.
Features: Polished blue finish; full-length tubular magazine; receiver grooved for scope mounting. Introduced 1997. From Henry Repeating Arms Co.
Price: $229.95

Marlin Model 39AS

MARLIN MODEL 39AS GOLDEN LEVER-ACTION RIFLE
Caliber: 22 S (26), L (21), LR (19), tubular magazine.
Barrel: 24" Micro-Groove®.
Weight: 6 1/2 lbs. **Length:** 40" overall.
Stock: Checkered American black walnut with white line spacers at pistol grip cap and buttplate; Mar-Shield® finish. Swivel studs; rubber buttpad.
Sights: Bead ramp front with detachable Wide-Scan™ hood, folding rear semi-buckhorn adjustable for windage and elevation.
Features: Hammer-block safety; rebounding hammer. Takedown action, receiver tapped for scope mount (supplied), offset hammer spur; gold-plated steel trigger.
Price: $461.00

RIMFIRE RIFLES—LEVER & SLIDE ACTION

Marlin Model 1897

MARLIN MODEL 1897 CENTURY LIMITED RIFLE
Caliber: 22 Short (19), Long (15), Long Rifle (13 shots).
Barrel: 24" half-round, half-octagon with Micro-Groove rifling.
Weight: 6½ lbs. **Length:** 40" overall.
Stock: Semi-fancy American walnut with cut-checkering.
Sights: Marble brass bead carbine front, adjustable Marble semi-buckhorn rear.
Features: Commemorates 100th anniversary of the Model 1897 rifle. Both sides of the receiver are engraved and gold-accented. Has hard rubber rifle-style buttplate. Introduced 1997. Made in U.S. by Marlin Firearms co.
Price: $1,055.00

Remington 572 BDL

REMINGTON 572 BDL FIELDMASTER PUMP RIFLE
Caliber: 22 S (20), L (17) or LR (14), tubular magazine.
Barrel: 21" round tapered.
Weight: 5½ lbs. **Length:** 42" overall.
Stock: Walnut with checkered pistol grip and slide handle.
Sights: Blade ramp front; sliding ramp rear adjustable for windage and elevation.
Features: Cross-bolt safety; removing inner magazine tube converts rifle to single shot; receiver grooved for tip-off scope mount.
Price: About $353.00

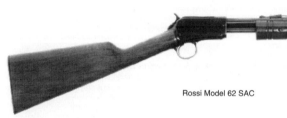

Rossi Model 62 SAC

ROSSI MODEL 62 SA PUMP RIFLE
Caliber: 22 LR, 22 WMR.
Barrel: 23", round or octagonal.
Weight: 5¾ lbs. **Length:** 39¼" overall.
Stock: Walnut, straight grip, grooved forend.
Sights: Fixed front, adjustable rear.
Features: Capacity 20 Short, 16 Long or 14 Long Rifle. Quick takedown. Imported from Brazil by Interarms.
Price: Blue $240.00
Price: Nickel $250.00
Price: Blue, with octagonal barrel $250.00
Price: 22 WMR $280.00

Rossi Model 62 SAC Carbine
Same as standard model except 22 LR, has 16¼" barrel. Magazine holds slightly fewer cartridges.
Price: Blue $240.00
Price: Nickel $250.00

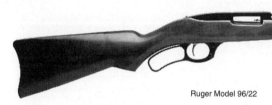

Ruger Model 96/22

RUGER MODEL 96/22 LEVER-ACTION RIFLE
Caliber: 22 LR, 10-shot rotary, magazine; 22 WMR, 9-shot rotary magazine.
Barrel: 18½".
Weight: 5¼ lbs. **Length:** 37¼" overall.
Stock: American hardwood.
Sights: Gold bead front, folding leaf rear.
Features: Cross-bolt safety, visible cocking indicator; short-throw lever action. Screw-on dovetail scope base. Introduced 1996. Made in U.S. by Sturm, Ruger & Co.
Price: 96/22 (22 LR) $327.50
Price: 96/22M (22 WMR) $345.00

Winchester Model 9422 Trapper

WINCHESTER MODEL 9422 LEVER-ACTION RIFLE
Caliber: 22 S (21), L (17), LR (15), tubular magazine.
Barrel: 20½".
Weight: 6¼ lbs. **Length:** 37⅛" overall.
Stock: American walnut, two-piece, straight grip (no pistol grip).
Sights: Hooded ramp front, adjustable semi-buckhorn rear.
Features: Side ejection, receiver grooved for scope mounting, takedown action. From U.S. Repeating Arms Co.
Price: Walnut $407.00
Price: With WinTuff laminated stock $407.00

Winchester Model 9422 Trapper
Similar to the Model 9422 with walnut stock except has 16½" barrel, overall length of 33⅛", weighs 5½ lbs. Magazine holds 15 Shorts, 12 Longs, 11 Long Rifles. Introduced 1996.
Price: $407.00

Winchester Model 9422 Magnum Lever-Action Rifle
Same as the 9422 except chambered for 22 WMR cartridge, has 11-round mag. capacity.
Price: Walnut $424.00
Price: With WinCam green stock $424.00
Price: With WinTuff brown laminated stock $424.00

Winchester Model 9422 25th Anniversary Edition
Similar to the standard Model 9422 except offered in Grade I and High Grade. Grade I has better walnut and engraved receiver sides, high-gloss metal finish. The High Grade has high-grade walnut with high luster finish, and the receiver is engraved and has a silver border. Both produced in limited numbers in 1997—2500 Grade I, 250 High grade. Introduced 1997. From U.S. Repeating Arms Co.
Price: Grade I $606.00
Price: High Grade $1,348.00

CAUTION: PRICES SHOWN ARE SUPPLIED BY THE MANUFACTURER OR IMPORTER. CHECK YOUR LOCAL GUNSHOP.

RIMFIRE RIFLES—BOLT ACTIONS & SINGLE SHOTS

Includes models for a variety of sports, utility and competitive shooting.

Anschutz Achiever

ANSCHUTZ ACHIEVER BOLT-ACTION RIFLE
Caliber: 22 LR, 5-shot clip, single shot adaptor.
Barrel: 19 1/2".
Weight: 5 lbs. **Length:** 35 1/2" to 36 2/3" overall.
Stock: Walnut-finished hardwood with adjustable buttplate, vented forend, stippled pistol grip. Length of pull adjustable from 11 7/8" to 13".
Sights: Hooded front, open rear adjustable for windage and elevation.
Features: Uses Mark 2000-type action with adjustable two-stage trigger. Receiver grooved for scope mounting. Designed for training in junior rifle clubs and for starting young shooters. Introduced 1987. Imported from Germany by AcuSport Corp.
Price: $372.60

ANSCHUTZ 1416D/1516D CLASSIC RIFLES
Caliber: 22 LR (1416D), 5-shot clip; 22 WMR (1516D), 4-shot clip.
Barrel: 22 1/2".
Weight: 6 lbs. **Length:** 41" overall.
Stock: European hardwood with walnut finish; classic style with straight comb, checkered pistol grip and forend.
Sights: Hooded ramp front, folding leaf rear.
Features: Uses Match 64 action. Adjustable single stage trigger. Receiver grooved for scope mounting. Imported from Germany by AcuSport Corp.
Price: 1416D, 22 LR $599.95
Price: 1516D, 22 WMR $629.95
Price: 1416D Classic left-hand $629.95

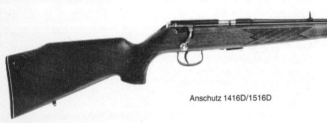
Anschutz 1416D/1516D

Anschutz 1416D/1516D Walnut Luxus Rifles
Similar to the Classic models except have European walnut stocks with Monte Carlo cheekpiece, slim forend with schnabel tip, cut checkering on grip and forend. Introduced 1997. Imported from Germany by AcuSport Corp.
Price: 1416D (22 LR) $689.95
Price: 1516D (22 WMR) $689.95

ANSCHUTZ 1710D CUSTOM RIFLE
Caliber: 22 LR, 5-shot clip.
Barrel: 24 1/4".
Weight: 7 3/8 lbs. **Length:** 42 1/2" overall.
Stock: Select European walnut.
Sights: Hooded ramp front, folding leaf rear; drilled and tapped for scope mounting.
Features: Match 54 action with adjustable single-stage trigger; roll-over Monte Carlo cheekpiece, slim forend with schnabel tip, Wundhammer palm swell on pistol grip, rosewood grip cap with white diamond insert; skip-line checkering on grip and forend. Introduced 1988. Imported from Germany by AcuSport Corp.
Price: $1,159.95

ANSCHUTZ 1518D LUXUS BOLT-ACTION RIFLE
Caliber: 22 WMR, 4-shot magazine.
Barrel: 19 3/4".
Weight: 5 1/2 lbs. **Length:** 37 1/2" overall.
Stock: European walnut.
Sights: Blade on ramp front, folding leaf rear.
Features: Receiver grooved for scopt mounting; single stage trigger; skip-line checkering; rosewood forend tip; sling swivels. Imported from Germany by AcuSport Corp.
Price: $1,089.95

ANSCHUTZ 1712D FEATHERWEIGHT RIFLE
Caliber: 22 LR, 5-shot clip.
Barrel: 22".
Weight: 6.2 lbs. **Length:** 41" overall.
Stock: Black fiberglass with rollover Monte Carlo cheekpiece.
Sights: Hooded front, folding leaf rear.
Features: Uses match 54 action with single stage trigger; claw extractor; receiver grooved for scope mounting; sliding safety. Imported from Germany by AcuSport Corp.
Price: $1,494.95

Manufacturers' addresses in the
Directory of the Arms Trade
page 309, this issue

Armscor M-14P

Armscor M-14P Standard Rifle
Similar to the M-1400S except has short walnut-finished hardwood stock for small shooters. Introduced 1987. Imported from the Philippines by K.B.I., Inc.
Price: $129.00
Price: M-14Y Youth (17.5" barrel) $126.00

ARMSCOR M-1400S CLASSIC BOLT-ACTION RIFLE
Caliber: 22 LR, 10-shot magazine.
Barrel: 22 5/8".
Weight: 6.7 lbs. **Length:** 41.25" overall.
Stock: Walnut-finished mahogany.
Sights: Bead front, rear adjustable for elevation.
Features: Receiver grooved for scope mounting. Blued finish. Introduced 1987. Imported from the Philippines by K.B.I., Inc.
Price: $224.00

RIMFIRE RIFLES—BOLT ACTIONS & SINGLE SHOTS

Armscor M-1400SC Super Classic Rifle
Similar to the M-1400S except has oil-finished American walnut stock with 18 lpi hand checkering; black hardwood grip cap and forend tip; highly polished barreled action; jewelled bolt; recoil pad; swivel studs. Imported from the Philippines by K.B.I., Inc.
Price: $355.00

Armscor M-12Y Youth Rifle
Similar to the M-1400S except has 17.5" barrel, and is a single shot. Weight is 4.1 lbs., overall length 34.4". Imported from the Philippines by K.B.I., Inc.
Price: $122.00

CABANAS MASTER BOLT-ACTION RIFLE
Caliber: 177, round ball or pellet; single shot.
Barrel: 19 1/2".
Weight: 8 lbs. **Length:** 45 1/2" overall.
Stocks: Walnut target-type with Monte Carlo.
Sights: Blade front, fully adjustable rear.
Features: Fires round ball or pellet with 22-cal. blank cartridge. Bolt action. Imported from Mexico by Mandall Shooting Supplies. Introduced 1984.
Price: $189.95
Price: Varmint model (has 21 1/2" barrel, 4 1/2 lbs., 41" overall length, varmint-type stock) $119.95

CABANAS LASER RIFLE
Caliber: 177.
Barrel: 19".
Weight: 6 lbs., 12 oz. **Length:** 42" overall.
Stock: Target-type thumbhole.
Sights: Blade front, open fully adjustable rear.
Features: Fires round ball or pellets with 22 blank cartridge. Imported from Mexico by Mandall Shooting Supplies.
Price: $159.95

Armscor M-1500S Classic Rifle
Similar to the Model 1400S except chambered for 22 WMR. Has 22.6" barrel, double lug bolt, checkered stock, weighs 6.5 lbs. Introduced 1987.
Price: About $236.00

Armscor M-1500SC Super Classic Rifle
Similar to the M-1500S except has oil-finished American walnut stock with 18 lpi hand checkering; black hardwood grip cap and forend tip; highly polished barreled action; jewelled bolt; recoil pad; swivel studs. Imported from the Philippines by K.B.I., Inc.
Price: $379.00

Cabanas Leyre Bolt-Action Rifle
Similar to Master model except 44" overall, has sport/target stock.
Price: $149.95
Price: Model R83 (17" barrel, hardwood stock, 40" o.a.l.) $79.95
Price: Mini 82 Youth (16 1/2" barrel, 33" overall length, 3 1/2 lbs.) $69.95
Price: Pony Youth (16" barrel, 34" overall length, 3.2 lbs.) $69.95

Cabanas Espronceda IV Bolt-Action Rifle
Similar to the Leyre model except has full sporter stock, 18 3/4" barrel, 40" overall length, weighs 5 1/2 lbs.
Price: $134.95

CHIPMUNK SINGLE SHOT RIFLE
Caliber: 22, S, L, LR, single shot.
Barrel: 16 1/8".
Weight: About 2 1/2 lbs. **Length:** 30" overall.
Stocks: American walnut.
Sights: Post on ramp front, peep rear adjustable for windage and elevation.
Features: Drilled and tapped for scope mounting using special Chipmunk base ($13.95). Made in U.S. Introduced 1982. From Rogue Rifle Co., Inc.
Price: Standard $184.95
Price: Deluxe (better wood, checkering) $238.00

Dakota 22 Sporter

DAKOTA 22 SPORTER BOLT-ACTION RIFLE
Caliber: 22 LR, 5-shot magazine.
Barrel: 22" Premium.
Weight: About 6.5 lbs. **Length:** 42 1/2" overall.
Stock: Claro or English walnut in classic design; 13.6" length of pull. Point panel hand checkering. Swivel studs. Black buttpad.
Sights: None furnished.
Features: Combines features of Winchester 52 and Dakota 76 rifles. Full-sized receiver; rear locking lug and bolt machined from bar stock. Trigger and striker-blocking safety; Model 70-style trigger. Introduced 1992. From Dakota Arms, Inc.
Price: $1,795.00

COOPER ARMS MODEL 36 CLASSIC SPORTER RIFLE
Caliber: 22 LR, 5-shot magazine.
Barrel: 22 3/4".
Weight: 7 lbs. **Length:** 42 1/2" overall.
Stock: AAA Claro walnut with 22 lpi checkering, oil finish.
Sights: None furnished.
Features: Action has three mid-bolt locking lugs, 45-degree bolt rotation; fully adjustable single stage match trigger; swivel studs. Pachmayr butt pad. Introduced 1991. Made in U.S. by Cooper Arms.
Price: $1,675.00
Price: Custom Classic (AAA Claro walnut, Monte Carlo beaded cheekpiece, oil finish) $1,960.00
Price: Model 36 Featherweight (black synthetic stock, 6.5 lbs.) $1,740.00
Price: Model 36 Montana Trailblazer (lighter weight, sporter barrel profile) $1,475.00

Kimber Model 82C Classic

Kimber Model 82C Custom Match Bolt-Action Rifle
Same as the Model 82C Classic except has high grade stock of AA French walnut with black ebony forend tip, full coverage 22 lpi borderless checkering, steel Neidner (uncheckered) buttplate, and satin rust blue finish. Reintroduced 1995. Made in U.S. by Kimber of America, Inc.
Price: $2,075.00

Kimber Model 82C SuperAmerica Bolt-Action Rifle
Similar to the Model 82C Classic except has AAA fancy grade Claro walnut with beaded cheekpiece, ebony forend tip; hand-checkered 22 lpi patterns with wrap-around coverage; black rubber buttpad. Reintroduced 1994. Made in U.S. by Kimber of America, Inc.
Price: $1,326.00

KIMBER MODEL 82C CLASSIC BOLT-ACTION RIFLE
Caliber: 22 LR, 4-shot magazine (10-shot available).
Barrel: 22", premium air-gauged, free-floated.
Weight: 6.5 lbs. **Length:** 40.5" overall.
Stock: Classic style of Claro walnut; 13.5" length of pull; hand-checkered; red rubber buttpad; polished steel grip cap.
Sights: None furnished; drilled and tapped for Warne, Leupold or Millett scope mounts.
Features: Action uses aluminum pillar bedding for consistent accuracy; trigger with 2.5-lb. pull is fully adjustable. Reintroduced 1994. Made in U.S. by Kimber of America, Inc.
Price: $810.00

Kimber Model 82C Stainless Classic Bolt-Action Rifle
Similar to the Model 82C except has a match-grade stainless steel barrel and matte-finished receiver. Limited edition of 750 guns. Introduced 1996. Made in U.S. by Kimber of America, Inc.
Price: $899.00

RIMFIRE RIFLES—BOLT ACTIONS & SINGLE SHOTS

Kimber Model 82C HS

Kimber Model 82C HS Rifle
Similar to the Model 82C except has 24" medium-heavy barrel fluted on the back half only; high comb stock. Designed for rimfire silhouette competion and small game hunting. Introduced 1997. Made in U.S. by Kimber of America.
Price: $655.00

Kimber Model 82C SVT Bolt-Action Rifle
Simliar to the Model 82C except has an offhand high comb target-style stock; 18" fluted, stainless steel, target weight match-grade barrel; single shot action; A Claro walnut; weighs 7½ lbs. Designed for off-hand plinking, varmint shooting and competition. Introduced 1996. Made in U.S. by Kimber of America, Inc.
Price: $825.00

KRICO MODEL 300 BOLT-ACTION RIFLE
Caliber: 22 LR, 22 WMR, 22 Hornet.
Barrel: 19.6" (22 RF), 23.6" (Hornet).
Weight: 6.3 lbs. **Length:** 38.5" overall (22 RF).
Stock: Walnut-stained beech.
Sights: Blade on ramp front, open adjustable rear.
Features: Double triggers, sliding safety. Checkered grip and forend. Imported from Germany by Mandall Shooting Supplies.
Price: Model 300 Standard $700.00
Price: Model 300 Deluxe $795.00
Price: Model 300 Stutzen (walnut full-length stock) $825.00
Price: Model 300 SA (walnut Monte Carlo stock) $750.00

Magtech Model 122.2R

MAGTECH MODEL 122.2 BOLT-ACTION RIFLE
Caliber: 22 S, L, LR, 6- and 10-shot magazines.
Barrel: 24" (six-groove).
Weight: 6.5 lbs. **Length:** 43" overall.
Stock: Brazilian hardwood.
Sights: Blade front, open rear adjustable for windage and elevation.
Features: Sliding safety; double extractors; receiver grooved for scope mount. Introduced 1994. Imported from Brazil by Magtech Recreational Products, Inc.
Price: Model 122.2S (no sights) $139.95
Price: Model 122.2R (open sights) $149.95
Price: Model 122.2T (ramp front, micro-type open rear) $169.95

Marlin Model 15YN

MARLIN MODEL 15YN "LITTLE BUCKAROO"
Caliber: 22 S, L, LR, single shot.
Barrel: 16¼" Micro-Groove®.
Weight: 4¼ lbs. **Length:** 33¼" overall.
Stock: One-piece walnut-finished, press-checkered Maine birch with Monte Carlo; Mar-Shield® finish.
Sights: Ramp front, adjustable open rear.
Features: Beginner's rifle with thumb safety, easy-load feed throat, red cocking indicator. Receiver grooved for scope mounting. Introduced 1989.
Price: $179.00

Marlin Model 880

MARLIN MODEL 880 BOLT-ACTION RIFLE
Caliber: 22 LR; 7-shot clip magazine.
Barrel: 22" Micro-Groove®.
Weight: 5½ lbs. **Length:** 41".
Stock: Checkered Monte Carlo American black walnut with checkered p.g. and forend. Rubber buttpad, swivel studs. Mar-Shield® finish.
Sights: Wide-Scan™ ramp front, folding semi-buckhorn rear adjustable for windage and elevation.
Features: Receiver grooved for scope mount. Introduced 1989.
Price: $251.00

Marlin Model 880SQ

Marlin Model 880SQ Squirrel Rifle
Similar to the Model 880 except uses the heavy target barrel of Marlin's Model 2000L target rifle. Black synthetic stock with moulded-in checkering; double bedding screws; matte blue finish. Comes without sights, but has plugged dovetail for a rear sight, filled screw holes for front; receiver grooved for scope mount. Weighs 7 lbs. Introduced 1996. Made in U.S. by Marlin.
Price: $284.00

Marlin Model 881 Bolt-Action Rifle
Same as the Marlin 880 except tubular magazine, holds 17 Long Rifle cartridges. Weighs 6 lbs.
Price: $261.00

RIMFIRE RIFLES—BOLT ACTIONS & SINGLE SHOTS

Marlin Model 882SS Bolt-Action Rifle
Same as the Marlin Model 882 except has stainless steel front breech bolt, barrel, receiver and bolt knob. All other parts are either stainless steel or nickel-plated. Has black Monte Carlo stock of fiberglass-filled polycarbonate with moulded-in checkering, nickel-plated swivel studs. Introduced 1995. Made in U.S. by Marlin Firearms Co.
Price: $294.00

Marlin Model 882 Bolt-Action Rifle
Same as the Marlin 880 except 22 WMR cal. only with 7-shot clip magazine; weight about 6 lbs. Comes with swivel studs.
Price: $277.00
Price: Model 882L (laminated hardwood stock) $293.00

Marlin Model 882SSV

Marlin Model 882SSV Bolt-Action Rifle
Similar to the Model 882SS except has selected heavy 22" stainless steel barrel with recessed muzzle, and comes without sights; receiver is grooved for scope mount and 1" ring mounts are included. Weighs 7 lbs. Introduced 1997. Made in U.S. by Marlin Firearms Co.
Price: $289.00

Marlin Model 883SS Bolt-Action Rifle
Same as the Model 883 except front breech bolt, striker knob, trigger stud, cartridge lifter stud and outer magazine tube are of stainless steel; other parts are nickel-plated. Has two-tone brown laminated Monte Carlo stock with swivel studs, rubber butt pad. Introduced 1993.
Price: $306.00

Marlin Model 883 Bolt-Action Rifle
Same as Marlin 882 except tubular magazine holds 12 rounds of 22 WMR ammunition.
Price: $288.00

Marlin Model 880SS Stainless Steel Bolt-Action Rifle
Same as the Model 880 except barrel, receiver, front breech bolt, striker knob, trigger stud, cartridge lifter stud and outer magazine tube are made of stainless steel. Most other parts are nickel-plated to match the stainless finish. Has black fiberglass-filled AKZO synthetic stock with moulded-in checkering, stainless steel swivel studs. Introduced 1994. Made in U.S. by Marlin Firearms Co.
Price: $270.00

Marlin Model 25N Bolt-Action Repeater
Similar to Marlin 880, except walnut-finished p.g. stock, adjustable open rear sight, ramp front.
Price: $181.00
Price: With 4x scope $188.00

Marlin Model 25MN Bolt-Action Rifle
Similar to the Model 25N except chambered for 22 WMR. Has 7-shot clip magazine, 22" Micro-Groove® barrel, checkered walnut-finished Maine birch stock. Introduced 1989.
Price: $207.00
Price: With 4x scope $273.00

Remington 541-T

Remington 541-T HB Bolt-Action Rifle
Similar to the 541-T except has a heavy target-type barrel without sights. Receiver is drilled and tapped for scope mounting. American walnut stock with straight comb, satin finish, cut checkering, black checkered buttplate, black grip cap and forend tip. Weight is about 6$1/2$ lbs. Introduced 1993.
Price: $481.00

REMINGTON 541-T
Caliber: 22 S, L, LR, 5-shot clip.
Barrel: 24".
Weight: 5$7/8$ lbs. **Length:** 42$1/2$" overall.
Stock: Walnut, cut-checkered p.g. and forend. Satin finish.
Sights: None. Drilled and tapped for scope mounts.
Features: Clip repeater. Thumb safety. Reintroduced 1986.
Price: About $455.00

Ruger K77/22 Varmint

RUGER K77/22 VARMINT RIFLE
Caliber: 22 LR, 10-shot, 22 WMR, 9-shot detachable rotary magazine.
Barrel: 24", heavy.
Weight: 7.25 lbs. **Length:** 43.25" overall.
Stock: Laminated hardwood with rubber butt pad, quick-detachable swivel studs. No checkering or grip cap.
Sights: None furnished. Comes with Ruger 1" scope rings.
Features: Made of stainless steel with target gray finish. Three-position safety, dual extractors. Stock has wide, flat forend. Introduced 1993.
Price: K77/22VBZ, 22 LR $499.00
Price: K77/22VMB, 22 WMR $499.00

RUGER 77/22 RIMFIRE BOLT-ACTION RIFLE
Caliber: 22 LR, 10-shot rotary magazine; 22 WMR, 9-shot rotary magazine.
Barrel: 20".
Weight: About 5$3/4$ lbs. **Length:** 39$3/4$" overall.
Stock: Checkered American walnut or injection-moulded fiberglass-reinforced DuPont Zytel with Xenoy inserts in forend and grip, stainless sling swivels.
Sights: Brass bead front, adjustable folding leaf rear or plain barrel with 1" Ruger rings.
Features: Mauser-type action uses Ruger's 10-shot rotary magazine. Three-position safety, simplified bolt stop, patented bolt locking system. Uses the dual-screw barrel attachment system of the 10/22 rifle. Integral scope mounting system with 1" Ruger rings. Blued model introduced in 1983. Stainless steel model and blued model with the synthetic stock introduced in 1989.
Price: 77/22R (no sights, rings, walnut stock) $473.00
Price: 77/22RS (open sights, rings, walnut stock) $481.00
Price: K77/22RP (stainless, no sights, rings, synthetic stock) ...$473.00
Price: K77/22RSP (stainless, open sights, rings, synthetic stock) ...$481.00
Price: 77/22RM (22 WMR, blue, walnut stock) $473.00
Price: K77/22RSMP (22 WMR, stainless, open sights, rings, synthetic stock) $481.00
Price: K77/22RMP (22 WMR, stainless, synthetic stock) $473.00
Price: 77/22RSM (22 WMR, blue, open sights, rings, walnut stock) $481.00

RIMFIRE RIFLES—BOLT ACTIONS & SINGLE SHOTS

Sako Finnfire

SAKO FINNFIRE BOLT-ACTION RIFLE
Caliber: 22 LR, 5-shot magazine.
Barrel: 22".
Weight: 5.25 lbs. **Length:** 40" overall.
Stock: European walnut with checkered grip and forend.
Sights: Hooded blade front, open adjustable rear.
Features: Adjustable single-stage trigger; has 50-degree bolt lift. Introduced 1994. Imported from Finland by Stoeger Industries.
Price: $732.00
Price: With heavy barrel $815.00

Savage Mark I-G

SAVAGE MARK I-G BOLT-ACTION RIFLE
Caliber: 22 LR, single shot.
Barrel: 20 3/4".
Weight: 5 1/2 lbs. **Length:** 39 1/2" overall.
Stock: Walnut-finished hardwood with Monte Carlo-type comb, checkered grip and forend.
Sights: Bead front, open adjustable rear. Receiver grooved for scope mounting.
Features: Thumb-operated rotating safety. Blue finish. Rifled or smooth bore. Introduced 1990. Made in Canada, from Savage Arms.
Price: Mark I, rifled or smooth bore $119.00
Price: Mark I-GY (Youth), 19" barrel, 37" overall, 5 lbs. $119.00

Savage Mark II-LV

Savage Mark II-LV Heavy Barrel Rifle
Similar to the Mark II-G except has heavy 21" barrel with recessed target-style crown; gray, laminated hardwood stock with cut checkering. No sights furnished, but has dovetailed receiver for scope mounting. Overall length is 39 3/4", weight is 6 1/2 lbs. Comes with 10-shot clip magazine. Introduced 1997. Imported from Canada by Savage Arms, Inc.
Price: $200.00

Savage Mark II-FSS Stainless Rifle
Similar to the Mark II-G except has stainless steel barreled action and graphite/polymer filled stock; free-floated barrel. Weighs 5 lbs. Introduced 1997. Imported from Canada by Savage Arms, Inc.
Price: $150.00

SAVAGE MARK II-G BOLT-ACTION RIFLE
Caliber: 22 LR, 10-shot magazine.
Barrel: 20 1/2".
Weight: 5 1/2 lbs. **Length:** 39 1/2" overall.
Stock: Walnut-finished hardwood with Monte Carlo-type comb, checkered grip and forend.
Sights: Bead front, open adjustable rear. Receiver grooved for scope mounting.
Features: Thumb-operated rotating safety. Blue finish. Introduced 1990. Made in Canada, from Savage Arms.
Price: $126.00
Price: Mark II-GY (youth), 19" barrel, 37" overall, 5 lbs. $126.00
Price: Mark II-GL, left-hand $126.00
Price: Mark II-GLY (youth) left-hand $126.00
Price: Mark II-GXP Package Gun (comes with 4x15 scope) $131.00

> Consult our Directory pages for the location of firms mentioned.

Savage Model 93FSS

SAVAGE MODEL 93G MAGNUM BOLT-ACTION RIFLE
Caliber: 22 WMR, 5-shot magazine.
Barrel: 20 3/4".
Weight: 5 3/4 lbs. **Length:** 39 1/2" overall.
Stock: Walnut-finished hardwood with Monte Carlo-type comb, checkered grip and forend.
Sights: Bead front, adjustable open rear. Receiver grooved for scope mount.
Features: Thumb-operated rotary safety. Blue finish. Introduced 1994. Made in Canada, from Savage Arms.
Price: About $145.00

Savage Model 93FSS Magnum Rifle
Similar to the Model 93G except has stainless steel barreled action and black synthetic stock with positive checkering. Weighs 5 1/2 lbs. Introduced 1997. Imported from Canada by Savage Arms, Inc.
Price: $175.00

TRISTAR PEE-WEE 22 BOLT-ACTION RIFLE
Caliber: 22 Long Rifle, single shot.
Barrel: 16 1/2".
Weight: 2 3/4 lbs. **Length:** 31" overall.
Stock: Hardwood with hard plastic buttplate; Monte Carlo comb; 12" length of pull.
Sights: Ramp front, fully adjustable rear, both removable; receiver grooved for scope mounting; comes with Simmons 4x15 scope and mount.
Features: Manually cocked action; blued barrel and receiver. Comes with hard gun case. Introduced 1997. Made in U.S. From Tristar Sporting Arms.
Price: $199.00

RIMFIRE RIFLES—BOLT ACTIONS & SINGLE SHOTS

Ultra Light Arms Model 20

WINCHESTER MODEL 52B BOLT-ACTION RIFLE
Caliber: 22 Long Rifle, 5-shot magazine.
Barrel: 24".
Weight: 7 lbs. **Length:** 41 3/4" overall.
Stock: Walnut with checkered grip and forend.
Sights: None furnished; grooved receiver and drilled and tapped for scope mounting.
Features: Has Micro Motion trigger adjustable for pull and over-travel; match chamber; detachable magazine. Reintroduced 1997. From U.S. Repeating Arms Co.
Price: $635.00

ULTRA LIGHT ARMS MODEL 20 RF BOLT-ACTION RIFLE
Caliber: 22 LR, single shot or 5-shot repeater.
Barrel: 22" Douglas Premium, #1 contour.
Weight: 5 lbs., 3 oz. **Length:** 41 1/2" overall.
Stock: Composite Kevlar, graphite reinforced. DuPont Imron paint; 13 1/2" length of pull.
Sights: None furnished. Drilled and tapped for scope mounting.
Features: Available as either single shot or repeater with 5-shot removable magazine. Comes with scope mounts. Introduced 1993. Made in U.S. by Ultra Light Arms, Inc.
Price: Single shot $800.00
Price: Repeater $850.00

COMPETITION RIFLES—CENTERFIRE & RIMFIRE

Includes models for classic American and ISU target competition and other sporting and competitive shooting.

AMT 22 TARGET MODEL RIFLE
Caliber: 22 LR, 10-shot magazine.
Barrel: 20" target with .920" diameter.
Weight: 8 lbs., 6 oz. **Length:** 38 1/2" overall.
Stock: Fajen laminated or Hogue synthetic.
Sights: None furnished; comes with Weaver-style scope mount.
Features: Stainless steel, one-piece receiver/scope mount; cryogenic-treated, button-rifled barrel; 2 1/2 lb. target-type trigger. Right or left-hand stock. Introduced 1996. Made in U.S. by AMT.
Price: With Fajen stock $599.99
Price: With Hogue stock $549.99

ANSCHUTZ ACHIEVER ST SUPER TARGET RIFLE
Caliber: 22 LR, single shot.
Barrel: 22", .75" diameter.
Weight: About 6.5 lbs. **Length:** 38.75" to 39.75" overall.
Stock: Walnut-finished European hardwood with hand-stippled panels on grip and forend; 13.5" accessory rail on forend.
Sights: Optional. Receiver grooved for scope mounting.
Features: Designed for the advanced junior shooter with adjustable length of pull from 13.25" to 14.25" via removable butt spacers. Two-stage #5066 adjustable trigger factory set at 2.6 lbs. Introduced 1994. Imported from Germany by Accuracy International, Champion's Choice, Champion Shooter's Supply, Gunsmithing, Inc.
Price: $329.95
Price: Sight Set A $142.75

Anschutz BR-50

ANSCHUTZ 1451R SPORTER TARGET RIFLE
Caliber: 22 LR, 5-shot magazine.
Barrel: 22" heavy match.
Weight: 6.4 lbs. **Length:** 39.75" overall.
Stock: European hardwood with walnut finish.
Sights: None furnished. Grooved receiver for scope mounting or Anschutz micrometer rear sight.
Features: Sliding safety, two-stage trigger. Adjustable buttplate; forend slide rail to accept Anschutz accessories. Imported from Germany by AcuSport Corp.
Price: $484.95
Price: Model 1451E single shot $429.95

ANSCHUTZ 64-MSR, 64-MS LEFT SILHOUETTE RIFLE
Caliber: 22 LR, 5-shot magazine.
Barrel: 21 1/2", medium heavy; 7/8" diameter.
Weight: 8 lbs. **Length:** 39.5" overall.
Stock: Walnut-finished hardwood, silhouette-type.
Sights: None furnished. Receiver drilled and tapped for scope mounting.
Features: Uses Match 64 action. Designed for metallic silhouette competition. Stock has stippled checkering, contoured thumb groove with Wundhammer swell. Two-stage #5091 trigger. Slide safety locks sear and bolt. Introduced 1980. Imported from Germany by AcuSport Corp., Accuracy International, Champion's Choice, Champion Shooter's Supply, Gunsmithing, Inc.
Price: 64-MSR .... $783.70 to $957.95
Price: 64-MS Left .... NA

ANSCHUTZ BR-50 BENCHREST RIFLE
Caliber: 22 LR, single shot.
Barrel: 19.75".
Weight: About 11 lbs. **Length:** 37.75" to 42.5" overall.
Stock: Benchrest style of European hardwood with stippled grip. Cheekpiece vertically adjustable to 1". Stock length adjustable via spacers and buttplate. Finished with glossy blue-black paint.
Sights: None furnished. Receiver grooved for mounts, barrel drilled and tapped for target mounts.
Features: Uses the Anschutz 2013 target action, #5018 two-stage adjustable target trigger factory set at 3.9 oz. Introduced 1994. Imported from Germany by AcuSport Corp., Accuracy International, Champion's Choice, Champion Shooter's Supply, Gunsmithing, Inc.
Price: .... $2,304.00 to $2,879.95

> Consult our Directory pages for the location of firms mentioned.

ANSCHUTZ 1808D-RT SUPER RUNNING TARGET RIFLE
Caliber: 22 LR, single shot.
Barrel: 32 1/2".
Weight: 9.4 lbs. **Length:** 50.5" overall.
Stock: European walnut. Heavy beavertail forend; adjustable cheekpiece and buttplate. Stippled grip and forend.
Sights: None furnished. Grooved for scope mounting.
Features: Designed for Running Target competition. Nine-way adjustable single-stage trigger, slide safety. Introduced 1991. Imported from Germany by Accuracy International, Champion's Choice, Champion Shooter's Supply, Gunsmithing, Inc.
Price: Right-hand $1,430.40

COMPETITION RIFLES—CENTERFIRE & RIMFIRE

ANSCHUTZ 1451 SUPER BOLT-ACTION TARGET RIFLE
Caliber: 22 LR, single shot.
Barrel: 22" heavy match.
Weight: 6.4 lbs. **Length:** 39.75" overall.
Stock: Match-style European hardwood with stippled grip and forend.
Sights: None furnished; receiver grooved for micrometer rear sight or scope mount.
Features: Two-stage match trigger; adjustable buttplate; forend rail accepts Anschutz accessories. Imported from Germany by AcuSport Corp.
Price: $387.95

ANSCHUTZ 1827B BIATHLON RIFLE
Caliber: 22 LR, 5-shot magazine.
Barrel: 21 1/2".
Weight: 8 1/2 lbs. with sights. **Length:** 42 1/2" overall.
Stock: European walnut with cheekpiece, stippled pistol grip and forend.
Sights: Optional globe front specially designed for Biathlon shooting, micrometer rear with hinged snow cap.
Features: Uses Super Match 54 action and nine-way adjustable trigger; adjustable wooden buttplate, Biathlon butthook, adjustable hand-stop rail. Introduced 1982. Imported from Germany by Accuracy International, Champion's Choice, Champion Shooter's Supply, Gunsmithing, Inc.
Price: Right-hand, with sights, about $1,795.00
Price: With laminated stock, with sights, about $1,885.00

Anschutz 1827BT Fortner Biathlon Rifle
Similar to the Anschutz 1827B Biathlon rifle except uses Anschutz/Fortner system straight-pull bolt action, stainless steel barrel. Introduced 1982. Imported from Germany by Accuracy International, Champion's Choice, Champion Shooter's Supply, Gunsmithing, Inc.
Price: Right-hand, with sights $2,895.00
Price: Right-hand, laminated stock, with sights $2,990.00
Price: Left-hand, with sights $3,075.00

Anschutz 1907 ISU Standard Match Rifle
Same action as Model 1913 but with 7/8" diameter 26" barrel. Length is 44.5" overall, weighs 10 lbs. Choice of stock configurations. Vented forend. Designed for prone and position shooting ISU requirements; suitable for NRA matches. Imported from Germany by AcuSport Corp., Accuracy International, Champion's Choice, Champion Shooter's Supply, Gunsmithing, Inc.
Price: Right-hand, no sights, European hardwood stock $1,784.95
Price: With laminated hardwood stock $1,532.10
Price: Right-hand, no sights, walnut stock $1,500.80
Price: M1907-L (true left-hand action and stock) $1,516.90

ANSCHUTZ 1808 MSR SILHOUETTE RIFLE
Caliber: 22 LR, 5-shot magazine.
Barrel: 22.4" match; detachable muzzle tube.
Weight: 7.9 lbs. **Length:** 40.9" overall.
Stock: European walnut, thumbhole design.
Sights: None furnished.
Features: Uses Anschutz 54.18 barreled action with two-stage match trigger. Introduced 1997. Imported from Germany by Acusport Corp.
Price: $1,924.95

ANSCHUTZ 1903 MATCH RIFLE
Caliber: 22 LR, single shot.
Barrel: 25", .75" diameter.
Weight: 8.6 lbs. **Length:** 43.75" overall.
Stock: Walnut-finished hardwood with adjustable cheekpiece; stippled grip and forend.
Sights: None furnished.
Features: Uses Anschutz Match 64 action and #5098 two-stage trigger. A medium weight rifle for intermediate and advanced Junior Match competition. Introduced 1987. Imported from Germany by AcuSport Corp., Accuracy International, Champion's Choice, Champion Shooter's Supply, Gunsmithing, Inc.
Price: Right-hand ....$860.00 to $1,034.95
Price: Left-hand $900.00

ANSCHUTZ 1909 TARGET RIFLE
Caliber: 22 LR, single shot.
Barrel: 26" match.
Weight: 14.3 lbs. **Length:** 44.8" overall.
Stock: European beechwood, thumbhole design with hook buttplate, handstop with swivel.
Sights: None furnished.
Features: Uses Match 54 action. Adjustable comb, buttplate; forend rail accepts Anscuhtz adjustable palm rest. Imported from Germany by AcuSport Corp.
Price: $2,479.95

Anschutz 1913 Super Match Rifle
Same as the Model 1911 except European walnut International-type stock with adjustable cheekpiece, adjustable aluminum hook buttplate, adjustable hand stop, weighs 15.5 lbs., 46" overall. Imported from Germany by AcuSport Corp., Accuracy International, Champion's Choice, Champion Shooter's Supply, Gunsmithing, Inc.
Price: Right-hand, no sights ....$2,475.00 to $3,039.95
Price: M1913 left-hand $2,590.00

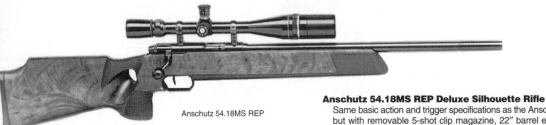
Anschutz 54.18MS REP

Anschutz 54.18MS REP Deluxe Silhouette Rifle
Same basic action and trigger specifications as the Anschutz 1913 Super Match but with removable 5-shot clip magazine, 22" barrel extendable to 30" using optional extension and weight set. Receiver drilled and tapped for scope mounting. Silhouette stock with thumbhole grip is of fiberglass with walnut wood Fibergrain finish. Introduced 1990. Imported from Germany by Accuracy International, Champion's Choice, Champion Shooter's Supply, Gunsmithing, Inc.
Price: 54.18MS Standard with fiberglass stock, barrel extension and weight $1,499.00

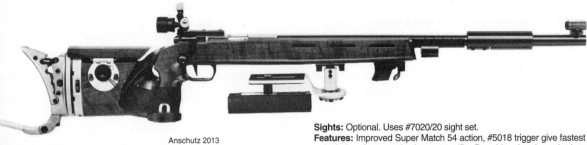
Anschutz 2013

ANSCHUTZ SUPER MATCH MODEL 2013 RIFLE
Caliber: 22 LR, single shot.
Barrel: 19.75" (26" with tube installed).
Weight: 15.5 lbs. **Length:** 43" to 45.5" overall.
Stock: European walnut; target adjustable.
Sights: Optional. Uses #7020/20 sight set.
Features: Improved Super Match 54 action, #5018 trigger give fastest consistent lock time for a production target rifle. Barrel is micro-honed; trigger has nine points of adjustment, two stages. Slide safety. Comes with test target. Introduced 1992. Imported from Germany by AcuSport Corp., Accuracy International, Champion's Choice, Champion Shooter's Supply, Gunsmithing, Inc.
Price: Right-hand ....$2,769.70 to $3,559.95
Price: Left-hand $2,899.20

COMPETITION RIFLES—CENTERFIRE & RIMFIRE

Anschutz Super Match Model 2007 ISU Standard Rifle
Similar to the Model 2013 except has ISU Standard design. European walnut or blonde hardwood stock. Sights optional. Introduced 1992. Imported from Germany by Accuracy International, Champion's Choice, Champion Shooter's Supply, Gunsmithing, Inc.
Price: Left-hand, beech stock $2,102.00
Price: Right-hand, walnut stock $2,003.80

ARMALITE AR-10 (T) RIFLE
Caliber: 308, 10-shot magazine.
Barrel: 24" target-weight Rock 5R custom.
Weight: 10.4 lbs. **Length:** 43.5" overall.
Stock: Green or black compostion; N.M. fiberglass handguard tube.
Sights: Detachable handle, front sight, or scope mount available. Comes with international-style flattop receiver with Picatinny rail.
Features: National Match two-stage trigger. Forged upper receiver. Receivers hard-coat amodized. Introduced 1995. Made in U.S. by ArmaLite,Inc.
Price: $1,995.00
Price: AR-10 (T) LCR, lighter 20" barrel $1,495.00
Price: AR-10 (T) Carbine, lighter 16" barrel, single stage trigger, weighs 8.8 lbs. $1,395.00

ARMALITE M15A4 (T) EAGLE EYE RIFLE
Caliber: 223, 7-shot magazine.
Barrel: 24" heavy stainless; 1:8" twist.
Weight: 9.2 lbs. **Length:** 42 3/8" overall.
Stock: Green or black butt, N.M. fiberglass handguard tube.
Sights: One-piece international-style flat top receiver with Weaver-type rail, including case deflector.
Features: Detachable carry handle, front sight and scope mount (30mm or 1") available. Upper and lower receivers have push-type pivot pin, hard coat anodized. Made in U.S. by ArmaLite, Inc.
Price: $1,325.00

ANSCHUTZ 1911 PRONE MATCH RIFLE
Caliber: 22 LR, single shot.
Barrel: 27 1/4".
Weight: 11 lbs. **Length:** 46" overall.
Stock: Walnut-finished European hardwood; American prone style with Monte Carlo, cast-off cheekpiece, checkered pistol grip, beavertail forend with swivel rail and adjustable swivel, adjustable rubber buttplate.
Sights: None furnished. Receiver grooved for Anschutz sights (extra). Scope blocks.
Features: Two-stage #5018 trigger adjustable from 2.1 to 8.6 oz. Extremely fast lock time. Imported from Germany by Acusport Corp., Accuracy International, Champion's Choice, Champion Shooter's Supply, Gunsmithing, Inc.
Price: Right-hand, no sights . . . . $1,673.50 to $2,094.95

ARMALITE M15A4 ACTION MASTER RIFLE
Caliber: 223, 7-shot magazine.
Barrel: 20" heavy stainless; 1:9" twist.
Weight: 9 lbs. **Length:** 40 1/2" overall.
Stock: Green or black plastic; N.M. fiberglass handguard tube.
Sights: One-piece international-style flattop receiver with Weaver-type rail.
Features: Detachable carry handle, front sight and scope mount available. National Match two-stage trigger group; Picatinny rail; upper and lower receivers have push-type pivot pin; hard coat anodized finish. Made in U.S. by ArmaLite, Inc.
Price: $1,175.00

Bushmaster XM-15-E2S

BUSHMASTER XM-15 E2S V-MATCH RIFLE
Caliber: 223, 10-shot magazine.
Barrel: 20", 24", 26"; 1:9" twist; heavy.
Weight: 8.2 lbs. **Length:** 38.25" overall (20" barrel).
Stock: Black composition.
Sights: None furnished; comes with scope mount base installed.
Features: E2 lower receiver with push-pin-style takedown. Barrel is .950" outside diameter with counter-bored crown; upper receiver has brass deflector; free-floating steel handguard. Made in U.S. by Bushmaster Firearms Co./Quality Parts Co.
Price: 20" barrel $795.00
Price: 24" barrel $805.00
Price: 26" barrel $815.00

BUSHMASTER XM-15 E2S TARGET MODEL RIFLE
Caliber: 223, 10-shot magazine.
Barrel: 20", 24", 26"; 1:9" twist; heavy.
Weight: 8.1 lbs. **Length:** 38.25" overall (20" barrel).
Stock: Black composition.
Sights: Adjustable post front, adjustable aperture rear.
Features: Patterned after Colt M-16A2. Chrome-lined barrel with manganese phosphate exterior. Has E-2 lower receiver with push-pin. Made in U.S. by Bushmaster Firearms Co./Quality Parts Co.
Price: 20" match heavy barrel $740.00
Price: 24" match heavy barrel $750.00
Price: 26" match heavy barrel $760.00

Colt Match Target HBAR

Colt Match Target HBAR Rifle
Similar to the Target Model except has heavy barrel, 800-meter rear sight adjustable for windage and elevation. Introduced 1991.
Price: $1,067.00

Colt Match Target Competition HBAR II Rifle
Similar to the Match Target Competition HBAR except has 16.1" barrel, weighs 7.1 lbs., overall length 34.5"; 1:9" twist barrel. Introduced 1995.
Price: $1,044.00

COLT MATCH TARGET MODEL RIFLE
Caliber: 223 Rem., 5-shot magazine.
Barrel: 20".
Weight: 7.5 lbs. **Length:** 39" overall.
Stock: Composition stock, grip, forend.
Sights: Post front, aperture rear adjustable for windage and elevation.
Features: Five-round detachable box magazine, standard-weight barrel, sling swivels. Has forward bolt assist. Military matte black finish. Model introduced 1991.
Price: $1,019.00

Colt Match Target Competition HBAR Rifle
Similar to the Sporter Target except has flat-top receiver with integral Weaver-type base for scope mounting. Counter-bored muzzle, 1:9" rifling twist. Introduced 1991.
Price: Model R6700 . . . . $1,073.00

COMPETITION RIFLES—CENTERFIRE & RIMFIRE

Cooper Model 36 BR-50

COOPER ARMS MODEL 36 BR-50
Caliber: 22 LR, single shot.
Barrel: 22", .860" straight.
Weight: 6.8 lbs. **Length:** 40.5" overall.
Stock: McMillan Benchrest.
Sights: None furnished.
Features: Action has three mid-bolt locking lugs; fully adjustable match grade trigger; stainless barrel. Introduced 1994. Made in U.S. by Cooper Arms.
Price: $1,850.00

E.A.A./HW 60

E.A.A./WEIHRAUCH HW 60 TARGET RIFLE
Caliber: 22 LR, single shot.
Barrel: 26.8".
Weight: 10.8 lbs. **Length:** 45.7" overall.
Stock: Walnut with adjustable buttplate. Stippled pistol grip and forend. Rail with adjustable swivel.
Sights: Hooded ramp front, match-type aperture rear.
Features: Adjustable match trigger with push-button safety. Left-hand version also available. Introduced 1991. Imported from Germany by European American Armory.
Price: Right-hand $695.00
Price: Left-hand $875.00

E.A.A./HW 660 MATCH RIFLE
Caliber: 22 LR.
Barrel: 26".
Weight: 10.7 lbs. **Length:** 45.3" overall.
Stock: Match-type walnut with adjustable cheekpiece and buttplate.
Sights: Globe front, match aperture rear.
Features: Adjustable match trigger; stippled pistol grip and forend; forend accessory rail. Introduced 1991. Imported from Germany by European American Armory.
Price: About $874.95

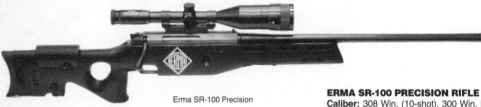

Erma SR-100 Precision

ERMA SR-100 PRECISION RIFLE
Caliber: 308 Win. (10-shot), 300 Win. Mag. (8-shot), 338 Lapua Mag. (5-shot); detachable box magazine.
Barrel: 25.5" (308), 29.5" (300, 338).
Weight: 14.1 lbs. **Length:** 49.6" overall (25.5" barrel).
Stock: Thumbhole style of laminated wood with adjustable recoil pad and comb; aluminum forend rail for bipod or sling swivel.
Sights: None furnished.
Features: Interchangeable barrels; three-lug bolt locks into barrel; 60° bolt rotation; forged aluminum alloy receiver; fully adjustable match trigger; integral muzzlebrake. Inroduced 1996. Imported from Germany by Amtec 2000, Inc.
Price: About $8,000.00

EAGLE ARMS M15A2 GOLDEN EAGLE
Caliber: 223, 30-shot magazine.
Barrel: 20" extra-heavy stainless; 1:8" twist.
Weight: 9.4 lbs. **Length:** 39-9/16" overall.
Stock: Black or green composition.
Sights: N.M. extra-fine front adjustable for elevation, A2-style N.M. rear with 1/2-MOA windage and elevation adjustments, N.M. aperture.
Features: Pre-ban rifle with full front sight housing with bayonet lug; threaded barrel with N.M. compensator; N.M. two-stage trigger; N.M. free-floating barrel sleeve; push-type pivot pin; hard coat anodized; fence-type magazine release. Made in U.S. by ArmaLite, Inc.
Price: $1,300.00

Harris Gunworks M-86

HARRIS GUNWORKS NATIONAL MATCH RIFLE
Caliber: 7mm-08, 308, 5-shot magazine.
Barrel: 24", stainless steel.
Weight: About 11 lbs. (std. bbl.). **Length:** 43" overall.
Stock: Fiberglass with adjustable buttplate.
Sights: Barrel band and Tompkins front; no rear sight furnished.
Features: Gunworks repeating action with clip slot, Canjar trigger. Match-grade barrel. Available in right-hand only. Fiberglass stock, sight installation, special machining and triggers optional. Introduced 1989. From Harris Gunworks, Inc.
Price: $3,300.00

HARRIS GUNWORKS M-86 SNIPER RIFLE
Caliber: 308, 30-06, 4-shot magazine; 300 Win. Mag., 3-shot magazine.
Barrel: 24", Gunworks match-grade in heavy contour.
Weight: 11-1/4 lbs. (308), 11-1/2 lbs. (30-06, 300). **Length:** 43-1/2" overall.
Stock: Specially designed McHale fiberglass stock with textured grip and forend, recoil pad.
Sights: None furnished.
Features: Uses Gunworks repeating action. Comes with bipod. Matte black finish. Sling swivels. Introduced 1989. From Harris Gunworks, Inc.
Price: $2,600.00

COMPETITION RIFLES—CENTERFIRE & RIMFIRE

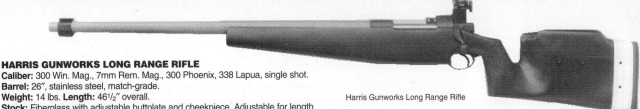
Harris Gunworks Long Range Rifle

HARRIS GUNWORKS LONG RANGE RIFLE
Caliber: 300 Win. Mag., 7mm Rem. Mag., 300 Phoenix, 338 Lapua, single shot.
Barrel: 26", stainless steel, match-grade.
Weight: 14 lbs. **Length:** 46 1/2" overall.
Stock: Fiberglass with adjustable buttplate and cheekpiece. Adjustable for length of pull, drop, cant and cast-off.
Sights: Barrel band and Tompkins front; no rear sight furnished.
Features: Uses Gunworks solid bottom single shot action and Canjar trigger. Barrel twist 1:12". Introduced 1989. From Harris Gunworks, Inc.
Price: $3,300.00

HARRIS GUNWORKS M-89 SNIPER RIFLE
Caliber: 308 Win., 5-shot magazine.
Barrel: 28" (with suppressor).
Weight: 15 lbs., 4 oz.
Stock: Fiberglass; adjustable for length; recoil pad.
Sights: None furnished. Drilled and tapped for scope mounting.
Features: Uses Gunworks repeating action. Comes with bipod. Introduced 1990. From Harris Gunworks, Inc.
Price: Standard (non-suppressed) $3,200.00

HARRIS GUNWORKS COMBO M-87 SERIES 50-CALIBER RIFLES
Caliber: 50 BMG, single shot.
Barrel: 29, with muzzlebrake.
Weight: About 21 1/2 lbs. **Length:** 53" overall.
Stock: Gunworks fiberglass.
Sights: None furnished.
Features: Right-handed Gunworks stainless steel receiver, chrome-moly barrel with 1:15" twist. Introduced 1987. From Harris Gunworks, Inc.
Price: $3,885.00
Price: M87R 5-shot repeater $4,150.00
Price: M-87 (5-shot repeater) "Combo" $4,150.00
Price: M-92 Bullpup (shortened M-87 single shot with bullpup stock) .. $4,770.00
Price: M-93 (10-shot repeater with folding stock, detachable magazine) $4,450.00

Heckler & Koch PSG-1

HECKLER & KOCH PSG-1 MARKSMAN RIFLE
Caliber: 308, 5- and 20-shot magazines.
Barrel: 25.6", heavy.
Weight: 17.8 lbs. **Length:** 47.5" overall.
Stock: Matte black high impact plastic, adjustable for length, pivoting butt cap, vertically-adjustable cheekpiece; target-type pistol grip with adjustable palm shelf.
Sights: Hendsoldt 6x42 scope.
Features: Uses HK-91 action with low-noise bolt closing device, special Marksman trigger group; special forend with T-way rail for sling swivel or tripod. Gun comes in special foam-fitted metal transport case with tripod, two 20-shot and two 5-shot magazines, tripod. Imported from Germany by Heckler & Koch, Inc. Introduced 1986.
Price: $10,811.00

> Consult our Directory pages for the location of firms mentioned.

Krico Model 360S Biathlon

KRICO MODEL 360S BIATHLON RIFLE
Caliber: 22 LR, 5-shot magazine.
Barrel: 21.25".
Weight: 9.26 lbs. **Length:** 40.55" overall.
Stock: Walnut with high comb, adjustable buttplate.
Sights: Globe front, fully adjustable Diana 82 match peep rear.
Features: Straight-pull action with 17.6-oz. match trigger. Comes with five magazines (four stored in stock recess), muzzle/sight snow cap. Introduced 1991. Imported from Germany by Mandall Shooting Supplies.
Price: $1,695.00

KRICO MODEL 360 S2 BIATHLON RIFLE
Caliber: 22 LR, 5-shot magazine.
Barrel: 21.25".
Weight: 9 lbs., 15 oz. **Length:** 40.55" overall.
Stock: Biathlon design of black epoxy-finished walnut with pistol grip.
Sights: Globe front, fully adjustable Diana 82 match peep rear.
Features: Pistol grip-activated action. Comes with five magazines (four stored in stock recess), muzzle/sight snow cap. Introduced 1991. Imported from Germany by Mandall Shooting Supplies.
Price: $1,595.00

KRICO MODEL 400 MATCH RIFLE
Caliber: 22 LR, 22 Hornet, 5-shot magazine.
Barrel: 23.2" (22 LR), 23.6" (22 Hornet).
Weight: 8.8 lbs. **Length:** 42.1" overall (22 RF).
Stock: European walnut, match type.
Sights: None furnished; receiver grooved for scope mounting.
Features: Heavy match barrel. Double-set or match trigger. Imported from Germany by Mandall Shooting Supplies.
Price: $950.00

KRICO MODEL 600 MATCH RIFLE
Caliber: 222, 223, 22-250, 243, 308, 5.6x50 Mag., 4-shot magazine.
Barrel: 23.6".
Weight: 8.8 lbs. **Length:** 43.3" overall.
Stock: Match stock of European walnut with cheekpiece.
Sights: None furnished; drilled and tapped for scope mounting.
Features: Match stock with vents in forend for cooling, rubber recoil pad, sling swivels. Imported from Germany by Mandall Shooting Supplies.
Price: $1,250.00

KRICO MODEL 600 SNIPER RIFLE
Caliber: 222, 223, 22-250, 243, 308, 4-shot magazine.
Barrel: 23.6".
Weight: 9.2 lbs. **Length:** 45.2" overall.
Stock: European walnut with adjustable rubber buttplate.
Sights: None furnished; drilled and tapped for scope mounting.
Features: Match barrel with flash hider; large bolt knob; wide trigger shoe. Parkerized finish. Imported from Germany by Mandall Shooting Supplies.
Price: $2,645.00

CAUTION: PRICES SHOWN ARE SUPPLIED BY THE MANUFACTURER OR IMPORTER. CHECK YOUR LOCAL GUNSHOP.

COMPETITION RIFLES—CENTERFIRE & RIMFIRE

Marlin Model 2000L

MARLIN MODEL 2000L TARGET RIFLE
Caliber: 22 LR, single shot.
Barrel: 22" heavy, Micro-Groove® rifling, match chamber, recessed muzzle.
Weight: 8 lbs. **Length:** 41" overall.
Stock: Laminated black/gray with ambidextrous pistol grip.
Sights: Hooded front with ten aperture inserts, fully adjustable target rear peep.
Features: Buttplate adjustable for length of pull, height and angle. Aluminum forend rail with stop and quick-detachable swivel. Two-stage target trigger; red cocking indicator. Five-shot adaptor kit available. Introduced 1991. From Marlin.
Price: $626.00

MAUSER MODEL SR 86 SNIPER RIFLE
Caliber: 308, detachable 9-shot magazine.
Barrel: 28.75", heavy, with compensator.
Weight: About 13 lbs. **Length:** 50.1" overall.
Stock: Match-style black-finished laminated wood with thumbhole, adjustable cheekpiece and buttplate.
Sights: None furnished. Integral receiver rail.
Features: Adjustable match trigger; muzzle brake; match barrel; forend accessory rail; glass-bedded action. Available with synthetic stock, left-hand action. Introduced 1997. Imported from Germany by GSI, Inc.
Price: $11,795.00

Olympic PCR-1

OLYMPIC ARMS PCR-1 RIFLE
Caliber: 223, 10-shot magazine.
Barrel: 20", 24"; 416 stainless steel.
Weight: 10 lbs., 3 oz. **Length:** 38.25" overall with 20" barrel.
Stock: A2 stowaway grip and trapdoor butt.
Sights: None supplied; flattop upper receiver, cut-down front sight base.
Features: Based on the AR-15 rifle. Broach-cut, free-floating barrel with 1:8.5" or 1:10" twist. No bayonet lug. Crowned barrel; fluting available. Introduced 1994. Made in U.S. by Olympic Arms, Inc.
Price: $1,100.00

Olympic Arms PCR-2, PCR-3 Rifles
Similar to the PCR-1 except has 16" barrel, weighs 8 lbs., 2 oz.; has post front sight, fully adjustable aperture rear. Model PCR-3 has flattop upper receiver, cut-down front sight base. Introduced 1994. Made in U.S. by Olympic Arms, Inc.
Price: $1,035.00

OLYMPIC ARMS PCR-SERVICEMATCH RIFLE
Caliber: 223, 10-shot magazine.
Barrel: 20", broach-cut 416 stainless steel.
Weight: About 10 lbs. **Length:** 39.5" overall.
Stock: A2 stowaway grip and trapdoor buttstock.
Sights: Post front, E2-NM fully adjustable operture rear.
Features: Based on the AR-15. Conforms to all DCM standards. Free-floating 1:8.5" or 1:10" barrel; crowned barrel; no bayonet lug. Introduced 1996. Made in U.S. by Olympic Arms, Inc.
Price: $1,145.00

Remington 40-XB

REMINGTON 40-XB RANGEMASTER TARGET CENTERFIRE
Caliber: 222 Rem., 222 Rem. Mag., 223, 220 Swift, 22-250, 6mm Rem., 243, 25-06, 7mm BR Rem., 7mm Rem. Mag., 30-338 (30-7mm Rem. Mag.), 300 Win. Mag., 7.62 NATO (308 Win.), 30-06, single shot.
Barrel: 27 1/4".
Weight: 11 1/4 lbs. **Length:** 47" overall.
Stock: American walnut or Kevlar with high comb and beavertail forend stop. Rubber non-slip buttplate.
Sights: None. Scope blocks installed.
Features: Adjustable trigger. Stainless barrel and action. Receiver drilled and tapped for sights.
Price: Standard single shot, stainless steel barrel, about $1,360.00
Price: Repeater model $1,462.00
Price: Model 40-XB KS $1,534.00
Price: Repeater model (KS) $1,636.00
Price: Extra for 2-oz. trigger $168.00

REMINGTON 40-XBBR KS
Caliber: 22 BR Rem., 222 Rem., 222 Rem. Mag., 223, 6mmx47, 6mm BR Rem., 7.62 NATO (308 Win.).
Barrel: 20" (light varmint class), 24" (heavy varmint class).
Weight: 7 1/4 lbs. (light varmint class); 12 lbs. (heavy varmint class).
Length: 38" (20" bbl.), 42" (24" bbl.).
Stock: Kevlar.
Sights: None. Supplied with scope blocks.
Features: Unblued stainless steel barrel, trigger adjustable from 1 1/2 lbs. to 3 1/2 lbs. Special 2-oz. trigger at extra cost. Scope and mounts extra.
Price: With Kevlar stock $1,513.00
Price: Extra for 2-oz. trigger, about $168.00

REMINGTON 40-XC KS NATIONAL MATCH COURSE RIFLE
Caliber: 7.62 NATO, 5-shot.
Barrel: 24", stainless steel.
Weight: 11 lbs. without sights. **Length:** 43 1/2" overall.
Stock: Kevlar, position-style, with palm swell, handstop.
Sights: None furnished.
Features: Designed to meet the needs of competitive shooters firing the national match courses. Position-style stock, top loading clip slot magazine, anti-bind bolt and receiver, stainless steel barrel and action. Meets all ISU Army Rifle specifications. Adjustable buttplate, adjustable trigger.
Price: About $1,513.00

REMINGTON 40-XR KS RIMFIRE POSITION RIFLE
Caliber: 22 LR, single shot.
Barrel: 24", heavy target.
Weight: 10 lbs. **Length:** 43" overall.
Stock: Kevlar. Position-style with front swivel block on forend guide rail.
Sights: Drilled and tapped. Furnished with scope blocks.
Features: Meets all ISU specifications. Deep forend, buttplate vertically adjustable, wide adjustable trigger.
Price: About $1,409.00

COMPETITION RIFLES—CENTERFIRE & RIMFIRE

Sako TRG-21

SAKO TRG-21 BOLT-ACTION RIFLE
Caliber: 308 Win., 10-shot magazine.
Barrel: 25.75".
Weight: 10.5 lbs. **Length:** 46.5" overall.
Stock: Reinforced polyurethane with fully adjustable cheekpiece and buttplate.
Sights: None furnished. Optional quick-detachable, one-piece scope mount base, 1" or 30mm rings.
Features: Resistance-free bolt, free-floating heavy stainless barrel, 60-degree bolt lift. Two-stage trigger is adjustable for length, pull, horizontal or vertical pitch. Introduced 1993. Imported from Finland by Stoeger.
Price: $4,265.00
Price: Model TRG-41, as above except in 338 Lapua Mag. $4,825.00

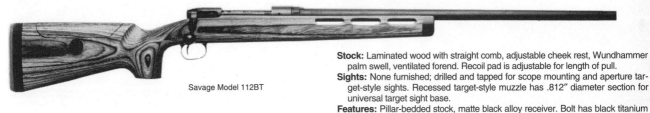
Savage Model 112BT

SAVAGE MODEL 112BT COMPETITION GRADE RIFLE
Caliber: 223, 308, 5-shot magazine, 300 Win. Mag., single shot.
Barrel: 26", heavy contour stainless with black finish; 1:9" twist (223), 1:10" (308).
Weight: 10.8 lbs. **Length:** 47.5" overall.
Stock: Laminated wood with straight comb, adjustable cheek rest, Wundhammer palm swell, ventilated forend. Recoil pad is adjustable for length of pull.
Sights: None furnished; drilled and tapped for scope mounting and aperture target-style sights. Recessed target-style muzzle has .812" diameter section for universal target sight base.
Features: Pillar-bedded stock, matte black alloy receiver. Bolt has black titanium nitride coating, large handle ball. Has alloy accessory rail on forend. Comes with safety gun lock, target and ear puffs. Introduced 1994. Made in U.S. by Savage Arms, Inc.
Price: $1,000.00
Price: 300 Win. Mag. (single shot 112BT-S) $1,000.00

SAVAGE MODEL 900B BIATHALON
Caliber: 22 LR, 5-shot magazine.
Barrel: 21".
Weight: 8 1/4 lbs. **Length:** 39 5/8" overall.
Stock: Natural finish hardwood with clip holder, carrying and shooting rails, butt hook, hand stop.
Sights: Target front with inserts, peep rear with 1/4-minute click adjustments.
Features: Biathlon-style rifle with snow cap muzzle protector. Comes with five magazines. Introduced 1991. Made in Canada, from Savage Arms.
Price: Right- or left-hand $498.00

SAVAGE MODEL 900TR TARGET RIFLE
Caliber: 22 LR, 5-shot magazine.
Barrel: 25".
Weight: 8 lbs. **Length:** 43 5/8" overall.
Stock: Target-type, walnut-finished hardwood.
Sights: Target front with inserts, peep rear with 1/4-minute click adjustments.
Features: Comes with shooting rail and hand stop. Introduced 1991. Made in Canada, from Savage Arms.
Price: Right- or left-hand $415.00

Savage Model 900S Silhouette Rifle
Similar to the Model 900B except has high-comb target-type stock of walnut-finished hardwood, one 5-shot magazine. Comes without sights, but receiver is drilled and tapped for scope base. Weighs about 8 lbs. Introduced 1992. Made in Canada, from Savage Arms.
Price: Right- or left-hand $346.00

Springfield M1A/M-21

Springfield, Inc. M1A/M-21 Tactical Model Rifle
Similar to the M1A Super Match except has special sniper stock with adjustable cheekpiece and rubber recoil pad. Weighs 11.2 lbs. From Springfield, Inc.
Price: $2,204.00

SPRINGFIELD, INC. M1A SUPER MATCH
Caliber: 308 Win.
Barrel: 22", heavy Douglas Premium or National Match.
Weight: About 10 lbs. **Length:** 44.31" overall.
Stock: Heavy walnut competition stock with longer pistol grip, contoured area behind the rear sight, thicker butt and forend, glass bedded.
Sights: National Match front and rear.
Features: Has figure-eight-style operating rod guide. Introduced 1987. From Springfield, Inc.
Price: About $2,050.00

Stoner SR-25 Match

STONER SR-25 MATCH RIFLE
Caliber: 7.62 NATO, 10-shot steel magazine, 5-shot optional.
Barrel: 24" heavy match; 1:11.25" twist.
Weight: 10.75 lbs. **Length:** 44" overall.
Stock: Black synthetic AR-15A2 design. Full floating forend of Mil-spec synthetic attaches to upper receiver at a single point.
Sights: None furnished. Has integral Weaver-style rail. Rings and iron sights optional.
Features: Improved AR-15 trigger; AR-15-style seven-lug rotating bolt. Gas block rail mounts detachable front sight. Introduced 1993. Made in U.S. by Knight's Mfg. Co.
Price: $2,995.00
Price: SR-25 Lightweight Match (20" medium match target contour barrel, 9.5 lbs., 40" overall) $2,995.00

COMPETITION RIFLES—CENTERFIRE & RIMFIRE

TANNER STANDARD UIT RIFLE
Caliber: 308, 7.5mm Swiss, 10-shot.
Barrel: 25.9".
Weight: 10.5 lbs. **Length:** 40.6" overall.
Stock: Match style of seasoned nutwood with accessory rail; coarsely stippled pistol grip; high cheekpiece; vented forend.
Sights: Globe front with interchangeable inserts, Tanner micrometer-diopter rear with adjustable aperture.
Features: Two locking lug revolving bolt encloses case head. Trigger adjustable from 1/2 to 6 1/2 lbs.; match trigger optional. Comes with 300-meter test target. Imported from Switzerland by Mandall Shooting Supplies. Introduced 1984.
Price: About $4,700.00

TANNER 50 METER FREE RIFLE
Caliber: 22 LR, single shot.
Barrel: 27.7".
Weight: 13.9 lbs. **Length:** 44.4" overall.
Stock: Seasoned walnut with palm rest, accessory rail, adjustable hook buttplate.
Sights: Globe front with interchangeable inserts, Tanner micrometer-diopter rear with adjustable aperture.
Features: Bolt action with externally adjustable set trigger. Supplied with 50-meter test target. Imported from Switzerland by Mandall Shooting Supplies. Introduced 1984.
Price: About $3,900.00

Tanner 300 Meter

TANNER 300 METER FREE RIFLE
Caliber: 308 Win., 7.5 Swiss, single shot.
Barrel: 27.58".
Weight: 15 lbs. **Length:** 45.3" overall.
Stock: Seasoned walnut, thumbhole style, with accessory rail, palm rest, adjustable hook butt.
Sights: Globe front with interchangeable inserts, Tanner-design micrometer-diopter rear with adjustable aperture.
Features: Three-lug revolving-lock bolt design; adjustable set trigger; short firing pin travel; supplied with 300-meter test target. Imported from Switzerland by Mandall Shooting Supplies. Introduced 1984.
Price: About $4,900.00

SHOTGUNS—AUTOLOADERS

Includes a wide variety of sporting guns and guns suitable for various competitions.

American Arms/Franchi 610VS

AMERICAN ARMS/FRANCHI 610VS AUTO SHOTGUN
Gauge: 12, 3" chamber.
Barrel: 26", 28", Franchoke tubes.
Weight: 7 lbs., 2 oz. **Length:** 47 1/2" overall.
Stock: 14 1/4"x1 1/2"x2 1/2". European walnut.
Features: Alloy frame with matte black finish; gas-operated with Variopress System; four-lug rotating bolt; loaded chamber indicator. Introduced 1996. Imported from Italy by American Arms.
Price: $750.00
Price: 610VSL, etched receiver, engraved bolt $795.00

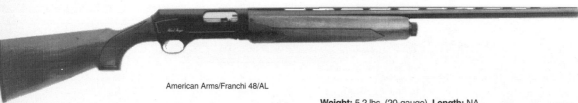

American Arms/Franchi 48/AL

AMERICAN ARMS/FRANCHI 48/AL SHOTGUN
Gauge: 12, 20 or 28, 2 3/4" chamber.
Barrel: 24", 26", 28" (Franchoke Imp. Cyl., Mod., Full choke tubes), 28 ga. has fixed Imp. Cyl. Vent. rib.
Weight: 5.2 lbs. (20-gauge). **Length:** NA
Stock: 14 1/4"x1 5/8"x2 1/2". Walnut with checkered grip and forend.
Features: Recoil-operated action. Chrome-lined bore; cross-bolt safety. Imported from Italy by American Arms, Inc.
Price: 12, 20 . . . . $649.00
Price: 28 ga. $725.00

Benelli Sport

Benelli Sport Shotgun
Similar to the Black Eagle Competition except has matte blue receiver, three carbon fiber interchangeable ventilated ribs, adjustable buttpad, adjustable buttstock, and functions with ultra-light target loads. Walnut stock with satin finish. Introduced 1997. Imported from Italy by Heckler & Koch, Inc.
Price: $1,144.00

BENELLI BLACK EAGLE COMPETITION AUTO SHOTGUN
Gauge: 12, 3" chamber.
Barrel: 26", 28" (Full, Mod., Imp. Cyl., Imp. Mod., Skeet choke tubes). Mid-bead sight.
Weight: 7.1 to 7.6 lbs. **Length:** 49 5/8" overall (26" barrel).
Stock: European walnut with high-gloss finish. Special competition stock comes with drop adjustment kit.
Features: Uses the Montefeltro rotating bolt inertia recoil operating system with a two-piece steel/aluminum etched receiver (bright on lower, blue upper). Drop adjustment kit allows the stock to be custom fitted without modifying the stock. Black lower receiver finish, blued upper. Introduced 1989. Imported from Italy by Heckler & Koch, Inc.
Price: $1,229.00

SHOTGUNS—AUTOLOADERS

Benelli Executive Series Shotguns
Similar to the Black Eagle except has grayed steel lower receiver, hand-engraved and gold inlaid (Type III), and has highest grade of walnut stock with drop adjustment kit. Barrel lengths of 21″, 24″, 26″, 28″; 3″ chamber. **Special order only.** Introduced 1995. Imported from Italy by Heckler & Koch, Inc.
Price: Type I (about two-thirds engraving coverage) $4,950.00
Price: Type II (full coverage engraving) $5,659.00
Price: Type III (full coverage, gold inlays) $6,577.00

Benelli Black Eagle Shotgun
Similar to the Balck Eagle Competition except has standard walnut stock with satin finish, all-blue receiver, comes with five choke tubes, wrench, buttstock drop change kit. Introduced 1997. Imported from Italy by Keckler & Koch, Inc.
Price: $992.00

Benelli M1 Super 90 Camo

Benelli M1 Super 90 Camouflage Field Shotgun
Similar to the M1 Super 90 Field except is covered with Realtree Xtra brown camouflage. Available with 24″, 26″, 28″ barrel, polymer stock. Introduced 1997. Imported from Italy by Heckler & Koch, Inc.
Price: $992.00

Benelli Montefeltro Super 90 Shotgun
Similar to the M1 Super 90 except has checkered walnut stock with high-gloss finish. Uses the Montefeltro rotating bolt system with a simple inertia recoil design. Full, Imp. Mod, Mod., Imp. Cyl. choke tubes. Weighs 7-7½ lbs. Finish is matte black. Introduced 1987.
Price: 21″, 24″, 26″, 28″ $923.00
Price: Left-hand, 26″, 28″ $943.00
Price: 20-ga., Montefeltro Super 90, 24″ 26″, 5¾ lbs. $923.00

BENELLI M1 SPORTING SPECIAL AUTO SHOTGUN
Gauge: 12, 3″ chamber.
Barrel: 18.5″ (Imp. Cyl. Mod., Full choke tubes).
Weight: 6 lbs, 8 oz. **Length:** 39.75″ overall.
Stock: Sporting-style polymer with drop adjustment.

BENELLI M1 SUPER 90 FIELD AUTO SHOTGUN
Gauge: 12, 3″ chamber.
Barrel: 21″, 24″, 26″, 28″ (choke tubes).
Weight: 7 lbs., 4 oz.
Stock: High impact polymer; wood on 26″, 28″.
Sights: Metal bead front.
Features: Sporting version of the military & police gun. Uses the rotating Montefeltro bolt system. Ventilated rib; blue finish. Comes with set of five choke tubes. Imported from Italy by Heckler & Koch, Inc.
Price: $902.00
Price: Wood stock version $916.00

Benelli Montefeltro Super 90 20-Gauge Shotgun
Similar to the 12-gauge Montefeltro Super 90 except chambered for 3″ 20-gauge, 24″ or 26″ barrel (choke tubes), weighs 5 lbs., 12 oz. Has drop-adjustable walnut stock with gloss finish, blued receiver. Overall length 47.5″. Introduced 1993. Imported from Italy by Heckler & Koch, Inc.
Price: 24″ and 26″ barrels $923.00

Sights: Ghost ring.
Features: Uses Montefeltro inertia recoil bolt system. Matte-finish receiver. Introduced 1993. Imported from Italy by Heckler & Koch, Inc.
Price: $924.00

Benelli Super Black Eagle

Benelli Super Black Eagle Custom Slug Gun
Similar to the Benelli Super Black Eagle except has 24″ E.R. Shaw Custom rifled barrel with 3″ chamber, and comes with scope mount base. Uses the Montefeltro inertia recoil bolt system. Matte-finish receiver. Weight is 7.5 lbs., overall length 45.5″. Wood or polymer stocks available. Introduced 1992. Imported from Italy by Heckler & Koch, Inc.
Price: $1,243.00
Price: With polymer stock $1,243.00

Benelli Super Black Eagle Camouflage Shotgun
Similar to the Super Black Eagle except covered with Realtree Xtra Brown camouflage pattern. Available with 24″, 26″, 28″ barrel. Introduced 1997. Imported from Italy by Heckler & Koch, Inc.
Price: $1,302.00

BENELLI SUPER BLACK EAGLE SHOTGUN
Gauge: 12, 3½″ chamber.
Barrel: 24″, 26″, 28″ (Imp. Cyl., Mod., Imp. Mod., Full choke tubes).
Weight: 7 lbs., 5 oz. **Length:** 49⅝″ overall (28″ barrel).
Stock: European walnut with satin finish, or polymer. Adjustable for drop.
Sights: Bead front.
Features: Uses Montfeltro inertia recoil bolt system. Fires all 12-gauge shells from 2¾″ to 3½″ magnums. Introduced 1991. Imported from Italy by Heckler & Koch, Inc.
Price: With 26″ and 28″ barrel, wood stock $1,213.00
Price: With 24″, 26″ barrel, polymer stock $1,199.00

Benelli Super Black Eagle Limited Edition Shotgun
Similar to the Super Black Eagle except has nickel-plated receiver with finely etched scroll and game scenes highlighted with gold; select-grade walnut stock; 26″ vent rib barrel. Limited to 1,000 guns. Introduced 1997. Imported Italy by Keckler & Koch, Inc.
Price: $2,093.00

Beretta Pintail

BERETTA PINTAIL AUTO SHOTGUN
Gauge: 12, 3″ chamber.
Barrel: 24″, 26″ (choke tubes).
Weight: 7 lbs.
Stock: Checkered walnut.
Features: Montefeltro-type short recoil action. Matte finish on wood and metal. Comes with sling swivels. Introduced 1993. Imported from Italy by Beretta U.S.A.
Price: $780.00

SHOTGUNS—AUTOLOADERS

Beretta 390 Silver Mallard

Beretta AL390 Sport Trap/Skeet/Sporting Shotguns
Similar to the AL390 Silver Mallard except has lower-contour, rounded receiver. Available with ported barrel. Trap has 30", 32" barrel (Full, Imp. Mod., Mod. choke tubes); Skeet has 26", 28" barrel (fixed Skeet); Sporting has 28", 30" (Full, Mod., Imp. Cyl., Skeet tubes). Introduced 1995. Imported from Italy by Beretta U.S.A.
Price: AL390 Sport Trap $900.00
Price: As above, fixed Full choke $890.00
Price: AL390 Sport Skeet $890.00
Price: AL390 Sport Sporting $900.00
Price: Ported barrel, above models, add about $100.00

Beretta AL390 Super Trap, Super Skeet Shotguns
Similar to the AL390 Silver Mallard except have adjustable-comb stocks that allow height adjustments via interchangeable comb inserts. Rounded recoil pad system allows adjustments for length of pull. Wide ventilated rib with orange front sight. Factory ported barrels in 28" (fixed Skeet), 30", 32" Trap (Mobilchoke tubes). Weighs 8.3 lbs. In 12-gauge only, with 3" chamber. Introduced 1993. Imported from Italy by Beretta U.S.A.
Price: AL390 Super Trap $1,215.00
Price: AL390 Super Skeet $1,160.00

BERETTA AL390 SILVER MALLARD AUTO SHOTGUN
Gauge: 12, 20, 3" chamber.
Barrel: 24", 26", 28", 30", Mobilchoke choke tubes.
Weight: 6.4 to 7.2 lbs.
Stock: Select walnut or matte black synthetic. Adjustable drop and cast.
Features: Gas-operated action with self-compensating valve allows shooting all loads without adjustment. Alloy receiver, reversible safety; chrome-plated bore; floating vent. rib. Matte-finish models for turkey/waterfowl and Deluxe with gold, engraving; camo models have Advantage finish. Youth models in 20-gauge and slug model available. Introduced 1992. Imported from Italy by Beretta U.S.A.
Price: Walnut or synthetic $860.00
Price: Waterfowl/Turkey (matte finish) $860.00
Price: Gold Mallard, 12 and 20 ....$1,025.00
Price: Slug model $860.00
Price: 20-gauge, 20-gauge Youth $860.00
Price: Camouflage model ....NA

> Consult our Directory pages for the location of firms mentioned.

Browning Auto-5 Stalker

Browning Auto-5 Stalker
Similar to the Auto-5 Light and Magnum models except has matte blue metal finish and black graphite-fiberglass stock and forend. Stock is scratch and impact resistant and has checkered panels. Light Stalker has 2 3/4" chamber, 26" or 28" vent. rib barrel with Invector choke tubes, weighs 8 lbs., 1 oz. (26"). Magnum Stalker has 3" chamber, 28" or 30" back-bored vent. rib barrel with Invector choke tubes, weighs 8 lbs., 11 oz. (28"). Introduced 1992.
Price: Light Stalker $839.95
Price: Magnum Stalker $865.95

Browning Auto-5 Magnum 12
Same as standard Auto-5 except chambered for 3" magnum shells (also handles 2 3/4" magnum and 2 3/4" HV loads). 28" Mod., Full; 30" and 32" (Full) bbls. Back-bored barrel comes with Invector choke tubes. 14"x1 5/8"x2 1/2" stock. Recoil pad. Weighs 8 3/4 lbs.
Price: With back-bored barrel, Invector Plus $865.95
Price: Extra Invector Plus barrel $307.95

BROWNING AUTO-5 LIGHT 12 AND 20
Gauge: 12, 20, 5-shot; 3-shot plug furnished; 2 3/4" or 3" chamber.
Action: Recoil operated autoloader; takedown.
Barrel: 26", 28", 30" Invector (choke tube) barrel; also available with Light 20-ga. 28" (Mod.) or 26" (Imp. Cyl.) barrel.
Weight: 12-, 16-ga. 7 1/4 lbs.; 20-ga. 6 3/8 lbs.
Stock: French walnut, hand checkered half-pistol grip and forend. 14 1/4"x1 5/8"x2 1/2".
Features: Receiver hand engraved with scroll designs and border. Double extractors, extra bbls. Interchangeable without factory fitting; mag. cut-off; cross-bolt safety. All models except Buck Special and game guns have back-bored barrels with Invector Plus choke tubes. Imported from Japan by Browning.
Price: Light 12, 20, vent. rib., Invector Plus $839.95
Price: Extra Invector barrel $307.95
Price: Light 12 Buck Special $828.95
Price: 12, 12 magnum barrel $307.95
Price: Buck Special barrel $296.95

Browning Gold Sporting Clays Auto
Similar to the Gold Hunter except 12-gauge only with 28" or 30" barrel; front and center beads on tapered ventilated rib; ported and back-bored Invector Plus barrel; 2 3/4" chamber; satin-finished stock with solid, radiused recoil pad with hard heel insert; non-glare black alloy receiver has "Sporting Clays" inscribed in gold. Introduced 1996. Imported from Japan by Browning.
Price: $759.95

Browning Gold Deer Hunter

Browning Gold Deer Hunter Auto Shotgun
Similar to the Gold Hunter except 12-gauge only, 22" rifled or smooth Standard Invector barrel with 5" rifled choke tube, cantilever scope mount, extra-thick recoil pad. Weighs 7 lbs., 12 oz., overall length 42 1/2". Sling swivel studs fitted on the magazine cap and butt. Introduced 1997. Imported by Browning.
Price: $799.00

BROWNING GOLD HUNTER AUTO SHOTGUN
Gauge: 12, 20, 3" chamber.
Barrel: 12-ga.—26", 28", 30", Invector Plus choke tubes; 20-ga.—26", 30", Invector choke tubes.
Weight: 7 lbs., 9 oz. (12-ga.), 6 lbs., 12 oz. (20-ga.) **Length:** 46 1/4" overall (20-ga., 26" barrel).
Stock: 14"x1 1/2"x2 1/3"; select walnut with gloss finish; palm swell grip.
Features: Self-regulating, self-cleaning gas system shoots all loads; lightweight receiver with special non-glare deep black finish; large reversible safety button; large rounded trigger guard, gold trigger. The 20-gauge has slightly smaller dimensions; 12-gauge have back-bored barrels, Invector Plus tube system. Introduced 1994. Imported by Browning.
Price: 12- or 20-gauge $734.95
Price: Extra barrels $272.95

SHOTGUNS—AUTOLOADERS

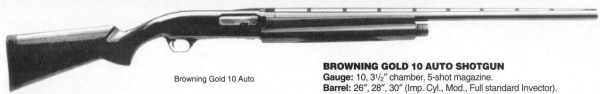
Browning Gold 10 Auto

BROWNING GOLD 10 AUTO SHOTGUN
Gauge: 10, 3½" chamber, 5-shot magazine.
Barrel: 26", 28", 30" (Imp. Cyl., Mod., Full standard Invector).
Weight: 10 lbs, 7 oz. (28" barrel).
Stock: 14³⁄₈"x1½"x2³⁄₈". Select walnut with gloss finish, cut checkering, recoil pad.
Features: Short-stroke, gas-operated action, cross-bolt safety. Forged steel receiver with polished blue finish. Introduced 1993. Imported by Browning.
Price: $1,007.95
Price: Extra barrel $261.95

Browning Gold 10 Stalker Auto Shotgun
Same as the standard Gold 10 except has non-glare metal finish and black graphite-fiberglass composite stock with dull finish and checkering. Introduced 1993. Imported by Browning.
Price: $1,007.95
Price: Extra barrel $261.95

CHURCHILL TURKEY AUTOMATIC SHOTGUN
Gauge: 12, 3" chamber, 5-shot magazine.
Barrel: 25" (Mod., Full, Extra Full choke tubes).
Weight: 7 lbs. **Length:** NA.
Stock: Walnut with satin finish, hand checkering.
Features: Gas-operated action, magazine cut-off, non-glare metal finish. Gold-colored trigger. Introduced 1990. Imported by Ellett Bros.
Price: $569.95

Mossberg Model 9200 Trophy

MOSSBERG MODEL 9200 CROWN GRADE AUTO SHOTGUN
Gauge: 12, 3" chamber.
Barrel: 24" (rifled bore), 24", 28" (Accu-Choke tubes); vent. rib.
Weight: About 7.5 lbs. **Length:** 48" overall (28" bbl.).
Stock: Walnut with high-gloss finish, cut checkering.
Features: Shoots all 2¾" or 3" loads without adjustment. Alloy receiver, ambidextrous top safety. Introduced 1992.
Price: 28", vent. rib $503.00
Price: Trophy, 24" with scope base, rifled bore, Dual-Comb stock $524.00
Price: 24", rifle sights, rifled bore $501.00

Mossberg Model 9200 Persuader
Similar to the Model 9200 Crown Grade except has black synthetic stock and forend, 18½" plain barrel with fixed Mod. choke, swivel studs. Weighs 7 lbs. Made in U.S. by Mossberg. Introduced 1996.
Price: $410.00

Mossberg Model 9200 Viking

Mossberg Model 9200 Viking
Similar to the Model 9200 Crown Grade except has black matte metal finish, moss-green synthetic stock and forend; 28" Accu-Choke vent. rib barrel with Imp. Cyl., Full and Mod. tubes. Made in U.S. by Mossberg. Introduced 1996.
Price: $429.00

Mossberg Model 9200 USST Autoloading Shotgun
Same as the Model 9200 Crown Grade except has "United States Shooting Team" custom engraved receiver. Comes with 26" vent. rib barrel with Accu-Choke tubes (including Skeet), cut-checkered walnut stock and forend. Introduced 1993.
Price: $501.00

Mossberg Model 9200 Bantam
Same as the Model 9200 Crown Grade except has 1" shorter stock, 22" vent. rib barrel with three Accu-Choke tubes. Made in U.S. by Mossberg. Introduced 1996.
Price: $501.00

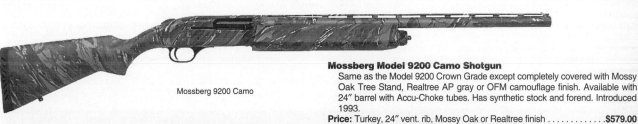
Mossberg 9200 Camo

Mossberg Model 9200 Camo Shotgun
Same as the Model 9200 Crown Grade except completely covered with Mossy Oak Tree Stand, Realtree AP gray or OFM camouflage finish. Available with 24" barrel with Accu-Choke tubes. Has synthetic stock and forend. Introduced 1993.
Price: Turkey, 24" vent. rib, Mossy Oak or Realtree finish $579.00
Price: 28" vent. rib, Accu-Chokes, OFM camo finish $487.00

Remington 11-87 Sporting Clays

REMINGTON MODEL 11-87 SPORTING CLAYS
Gauge: 12, 2¾" chamber
Barrel: 26", 28", vent. rib, Rem Choke (Skeet, Imp. Cyl., Mod., Full); Light Contour barrel. Medium height rib.
Weight: 7.5 lbs. **Length:** 46.5" overall (26" barrel).
Stock: 14³⁄₁₆"x1½"x2¼". Walnut, with cut checkering; sporting clays butt pad.
Features: Top of receiver, barrel and rib have matte finish; shortened magazine tube and forend; lengthened forcing cone; ivory bead front sight; competition trigger. Special no-wrench choke tubes marked on the outside. Comes in two-barrel fitted hard case. Introduced 1992.
Price: $745.00

SHOTGUNS—AUTOLOADERS

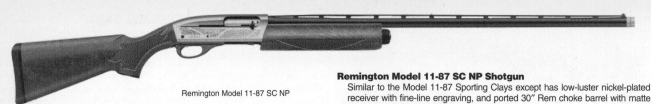

Remington Model 11-87 SC NP

Remington Model 11-87 SC NP Shotgun
Similar to the Model 11-87 Sporting Clays except has low-luster nickel-plated receiver with fine-line engraving, and ported 30" Rem choke barrel with matte finish. Tournament-grade American walnut stock measures $14^{3}/_{16}$" x $2^{1}/_{4}$" x $1^{1}/_{2}$". Sporting Clays choke tubes have knurled extensions. Introduced 1997. Made in U.S. by Remington.
Price: $793.00

Remington Model 11-87 SPS-T Camo

Remington Model 11-87 SPS-T Camo Auto Shotgun
Similar to the 11-87 Special Purpose Magnum except with synthetic stock, 21" vent. rib barrel with Super-Full Turkey (.665" diameter with knurled extension) and Imp. Cyl. Rem Choke tubes. Completely covered with Realtree X-tra Brown camouflage. Bolt body, trigger guard and recoil pad are non-reflective black. Introduced 1993.
Price: $770.00

Remington Model 11-87 Premier Trap
Similar to 11-87 Premier except trap dimension stock with straight or Monte Carlo combs; select walnut with satin finish and Tournament-grade cut checkering; 30" barrel with Rem Chokes (Trap Full, Trap Extra Full, Trap Super Full). Gas system set for $2^{3}/_{4}$" shells only. Introduced 1987.
Price: With Monte Carlo stock $754.00

Remington Model 11-87 Premier Skeet
Similar to 11-87 Premier except Skeet dimension stock with cut checkering, satin finish, two-piece buttplate; 26" barrel with Skeet or Rem Chokes (Skeet, Imp. Skeet). Gas system set for $2^{3}/_{4}$" shells only. Introduced 1987.
Price: $732.00

REMINGTON MODEL 11-87 PREMIER SHOTGUN
Gauge: 12, 3" chamber.
Barrel: 26", 28", 30" Rem Choke tubes. Light Contour barrel.
Weight: About $8^{1}/_{4}$ lbs. **Length:** 46" overall (26" bbl.).
Stock: Walnut with satin or high-gloss finish; cut checkering; solid brown buttpad; no white spacers.
Sights: Bradley-type white-faced front, metal bead middle.
Features: Pressure compensating gas system allows shooting $2^{3}/_{4}$" or 3" loads interchangeably with no adjustments. Stainless magazine tube; redesigned feed latch, barrel support ring on operating bars; pinned forend. Introduced 1987.
Price: $684.00
Price: Left-hand $734.00
Price: Premier Cantilever Deer Barrel, sling, swivels, Monte Carlo stock $749.00

Remington Model 11-87 Special Purpose Magnum
Similar to the 11-87 Premier except has dull stock finish, Parkerized exposed metal surfaces. Bolt and carrier have dull blackened coloring. Comes with 26" or 28" barrel with Rem Chokes, padded Cordura nylon sling and quick detachable swivels. Introduced 1987.
Price: $670.00
Price: With synthetic stock and forend (SPS) $670.00
Price: Magnum-Turkey with synthetic stock (SPS-T) $684.00

Remington Model 11-87 SPS Camo

Remington Model 11-87 SPS-Deer Shotgun
Similar to the 11-87 Special Purpose Camo except has fully-rifled 21" barrel with rifle sights, black non-reflective, synthetic stock and forend, black carrying sling. Introduced 1993.
Price: $692.00

Remington Model 11-87 SPS Special Purpose Synthetic Camo
Similar to the 11-87 Special Purpose Magnum except has synthetic stock and all metal (except bolt and trigger guard) and stock covered with Mossy Oak Break-Up camo finish. In 12-gauge only, 26", Rem Choke. Comes with camo sling, swivels. Introduced 1992.
Price: $757.00

Remington Model 11-87 SPS Cantilever Shotgun
Similar to the 11-87 SPS except has fully rifled barrel; synthetic stock with Monte Carlo comb; cantilever scope mount deer barrel. Comes with sling and swivels. Introduced 1994.
Price: $752.00

Remington 1100 Synthetic

Remington Model 1100 Synthetic
Similar to the 1100 LT magnum except in 12- or 20-gauge, and has black synthetic stock; vent. rib 28" barrel on 12-gauge, 26" on 20, both with Mod. Rem Choke tube. Weighs about $7^{1}/_{2}$ lbs. Introduced 1996.
Price: $492.00

REMINGTON MODEL 1100 LT-20 AUTO
Gauge: 20.
Barrel: 25" (Full, Mod.), 26", 28" with Rem Chokes.
Weight: $7^{1}/_{2}$ lbs.
Stock: 14"x$1^{1}/_{2}$"x$2^{1}/_{2}$". American walnut, checkered pistol grip and forend.
Features: Quickly interchangeable barrels. Matted receiver top with scroll work on both sides of receiver. Cross-bolt safety.
Price: With Rem Chokes, 20-ga. about $651.00
Price: Youth Gun LT-20 (21" Rem Choke) $651.00
Price: 20-ga., 3" magnum $651.00
Price: Skeet, 26", cut checkering, Rem. Choke $732.00

SHOTGUNS—AUTOLOADERS

Remington 1100 Sporting 28

Remington Model 110V Synthetic FR CL Shotgun
Similar to the Model 1100 LT-20 except 12-gauge, has 21" fully rifled barrel with cantilever scope mount and fiberglass-reinforced synthetic stock with Monte Carlo comb. Introduced 1997. Made in U.S. by Remington.
Price: $585.00

Remington Model 1100 LT-20 Synthetic FR RS Shotgun
Similar to the Model 1100 LT-20 except has 21" fully rifled barrel with rifle sights, $2^{3}/_{4}$" chamber, and fiberglasss-reinforced synthetic stock. Introduced 1997. Made in U.S. by Remington.
Price: $475.00

Remington Model 1100 Sporting 28
Similar to the 1100 LT-20 except in 28-gauge with 25" barrel; comes with Skeet, Imp. Cyl., Light Mod., Mod. Rem Choke tubes. Fancy walnut with gloss finish, Sporting rubber buttpad. Made in U.S. by Remington. Introduced 1996.
Price: $748.00

Remington Model 1100 Special Field
Similar to Standard Model 1100 except 12- and 20-ga. only, comes with 23" Rem Choke barrel. LT-20 version $6^{1}/_{2}$ lbs.; has straight-grip stock, shorter forend, both with cut checkering. Comes with vent. rib only; matte finish receiver without engraving. Introduced 1983.
Price: 12- and 20-ga., 23" Rem Choke, about $651.00

Remington Model SP-10 Magnum-Camo

Remington Model SP-10 Magnum-Camo Auto Shotgun
Similar to the SP-10 Magnum except buttstock, forend, receiver, barrel and magazine cap are covered with Mossy Oak Break-Up camo finish; bolt body and trigger guard have matte black finish. Comes with Extra-Full Turkey Rem Choke tube, 23" vent. rib barrel with mid-rib bead and Bradley-style front sight, swivel studs and quick-detachable swivels, and a non-slip Cordura carrying sling in the same camo pattern. Introduced 1993.
Price: $1,145.00

REMINGTON MODEL SP-10 MAGNUM AUTO SHOTGUN
Gauge: 10, $3^{1}/_{2}$" chamber, 3-shot magazine.
Barrel: 26", 30" (Full and Mod. Rem Chokes).
Weight: 11 to $11^{1}/_{4}$ lbs. **Length:** $47^{1}/_{2}$" overall (26" barrel).
Stock: Walnut with satin finish. Checkered grip and forend.
Sights: Metal bead front.
Features: Stainless steel gas system with moving cylinder; $^{3}/_{8}$" ventilated rib. Receiver and barrel have matte finish. Brown recoil pad. Comes with padded Cordura nylon sling. Introduced 1989.
Price: $1,054.00

Remington Model 11-96 Euro

REMINGTON MODEL 11-96 EURO LIGHTWEIGHT AUTO SHOTGUN
Gauge: 12, 3" chamber.
Barrel: 26", 28", Rem Chokes.
Weight: $6^{7}/_{8}$ lbs. (26" barrel). **Length:** 46" overall (26" barrel).
Stock: Semi-fancy Claro walnut with cut checkering; solid rubber butt pad.
Features: Pressure-compensating gas system allows shooting $2^{3}/_{4}$" or 3" loads interchangeably with no adjustments. Lightweight steel receiver with scroll-engraved panels; stainless steel magazine tube; 6mm ventilated rib on light contour barrel. Introduced 1996. Made in U.S. by Remington.
Price: $862.00

SHOTGUNS—SLIDE ACTIONS
Includes a wide variety of sporting guns and guns suitable for competitive shooting.

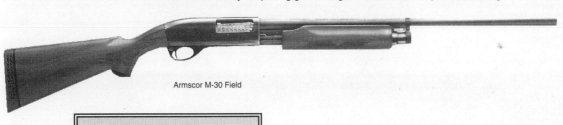

Armscor M-30 Field

Manufacturers' addresses in the
Directory of the Arms Trade
page 309, this issue

ARMSCOR M-30 FIELD PUMP SHOTGUN
Gauge: 12, 3" chamber.
Barrel: 28" fixed Mod., or with Mod. and Full choke tubes.
Weight: 7.6 lbs.
Stock: Walnut-finished hardwood.
Features: Double action slide bars; blued steel receiver; damascened bolt. Introduced 1996. Imported from the Philippines by K.B.I., Inc.
Price: With fixed choke $254.00
Price: With choke tubes $289.00

SHOTGUNS—SLIDE ACTIONS

Browning BPS 10-Ga.

BROWNING BPS PUMP SHOTGUN
Gauge: 10, 12, 3½" chamber; 12 or 20, 3" chamber (2¾" in target guns), 28, 2¾" chamber, 5-shot magazine.
Barrel: 10-ga.—24" Buck Special, 28", 30", 32" Invector; 12-, 20- ga.—22", 24", 26", 28", 30", 32" (Imp. Cyl., Mod. or Full). Also available with Invector choke tubes, 12- or 20-ga.; Upland Special has 22" barrel with Invector tubes. BPS 3" and 3½" have back-bored barrel.
Weight: 7 lbs. 8 oz. (28" barrel). **Length:** 48¾" overall (28" barrel).
Stock: 14¼"x1½"x2½". Select walnut, semi-beavertail forend, full pistol grip stock.
Features: All 12-gauge 3" guns except Buck Special and game guns have back-bored barrels with Invector Plus choke tubes. Bottom feeding and ejection, receiver top safety, high post vent. rib. Double action bars eliminate binding. Vent. rib barrels only. All 12- and 20-gauge guns with 3" chamber available with fully engraved receiver flats at no extra cost. Each gauge has its own unique game scene. Introduced 1977. Imported from Japan by Browning.
Price: 10-ga., Hunting, Invector $671.95
Price: 12-ga., 3½" Mag., Hunting, Invector Plus $671.95
Price: 12-, 20-ga., Hunting, Invector Plus $534.95
Price: 12-, 20-ga., Upland Special, Invector Plus $534.95
Price: 10-ga. Buck Special $676.95
Price: 12-ga. Buck Special $519.95
Price: 28-ga., Hunting, Invector $534.95

Browning BPS Game Gun Turkey Special
Similar to the standard BPS except has satin-finished walnut stock and dull-finished barrel and receiver. Receiver is drilled and tapped for scope mounting. Rifle-style stock dimensions and swivel studs. Has Extra-Full Turkey choke tube. Introduced 1992.
Price: $571.95

Browning BPS Pigeon Grade Pump Shotgun
Same as the standard BPS except has select high grade walnut stock and forend, and gold-trimmed receiver. Available in 12-gauge only with 26" or 28" vent. rib barrels. Introduced 1992.
Price: $713.95
Price: 10-gauge Waterfowl Model $860.95

Browning BPS Stalker

Browning BPS Stalker Pump Shotgun
Same gun as the standard BPS except all exposed metal parts have a matte blued finish and the stock has a durable black finish with a black recoil pad. Available in 10-ga. (3½") and 12-ga. with 3" or 3½" chamber, 22", 28", 30" barrel with Invector choke system. Introduced 1987.
Price: 12-ga., 3" chamber, Invector Plus $534.95
Price: 10-, 12-ga., 3½" chamber $671.95

Browning BPS Pump Shotgun Ladies and Youth Model
Same as BPS Upland Special except 20-ga. only, 22" Invector barrel, stock has pistol grip with recoil pad. Length of pull is 13¼". Introduced 1986.
Price: $534.95

Browning BPS Game Gun Deer Special
Similar to the standard BPS except has newly designed receiver/magazine tube/barrel mounting system to eliminate play, heavy 20.5" barrel with rifle-type sights with adjustable rear, solid receiver scope mount, "rifle" stock dimensions for scope or open sights, sling swivel studs. Gloss or matte finished wood with checkering, polished blue metal. Introduced 1992.
Price: $603.95

Hawk HP9 Combo

HAWK HP9 FIELD PUMP SHOTGUN
Gauge: 12, 3" chamber.
Barrel: 24", 28" (Mod. choke tube); vent. rib.
Weight: 7.3 lbs. **Length:** 44" overall.
Stock: 14" x 1½" x 2½". Black polymer or oil-finished hardwood.
Sights: Bead front.
Features: Twin action bars; steel receiver; removable sling swivels; cross-bolt safety; matte blue finish. Introduced 1997. Imported by Brolin Industries, Inc.
Price: Synthetic or wood stock, 24" or 28" barrel $269.95

Hawk HP9 Combo Pump Shotgun
Same as the HP9 Field except comes with 18½" Cyl.-bored barrel with bead or rifle sights, and 28" vent. rib. barrel with Mod. choke tube. Black synthetic or hardwood stock; pistol grip comes with bead sight on short barrel. Introduced 1997. Imported by Brolin Industries, Inc.
Price: Combo with bead sight $299.95
Price: Combo with rifle sight $319.95

Ithaca Model 37 Turkeyslayer

ITHACA MODEL 37 FIELD PUMP SHOTGUN
Gauge: 12, 20, 3" chamber.
Barrel: 26", 28", 30" (12-gauge), 26", 28" (20-gauge), choke tubes.
Weight: 7 lbs. (12-gauge).
Stock: Walnut with cut-checkered grip and forend.
Features: Steel receiver; bottom ejection; brushed blue finish, vent rib barrels. Reintroduced 1996. Made in U.S. by Ithaca Gun Co.
Price: $529.95

Ithaca Model 37 Turkeyslayer Pump Shotgun
Similar to the Model 37 Field except has 22" barrel with rifle sights, extended choke tube and full-coverage, Realtree Advantage, Realtree All-Purpose Brown, All-Purpose Grey, or Xtra Brown camouflage finish. Introduced 1996. Made in U.S. by Ithaca Gun Co.
Price: $549.95

SHOTGUNS—SLIDE ACTIONS

Ithaca Deerslayer II

Weight: 7 lbs.
Stock: Cut-checkered American walnut with Monte Carlo comb.
Sights: Rifle-type.
Features: Integral barrel and receiver. Bottom ejection. Brushed blue finish. Reintroduced 1997. Made in U.S. by Ithaca Gun Co.
Price: $549.95
Price: Smooth Bore Deluxe $499.95
Price: Rifled Deluxe $499.95

ITHACA DEERSLAYER II PUMP SHOTGUN
Gauge: 12, 20, 3" chamber.
Barrel: 20", 25", fully rifled.

Magtech Model 586.2

Weight: 7 1/4 lbs. **Length:** 46.5" overall (26" barrel).
Stock: Brazilian Embuia hardwood.
Features: Double action slide bars. Ventilated rib with bead front sight. Polished blue finish. Introduced 1995. Imported from Brazil by Magtech Recreational Products.
Price: About $255.00

MAGTECH MODEL 586.2-VR PUMP SHOTGUN
Gauge: 12, 3" chamber.
Barrel: 26", 28", choke tubes.

Maverick Model 88

Weight: 7 1/4 lbs. **Length:** 48" overall with 28" bbl.
Stock: Black synthetic with ribbed synthetic forend.
Sights: Bead front.
Features: Alloy receiver with blue finish; dual slide bars; cross-bolt safety in trigger guard; interchangeable barrels. Rubber recoil pad. Mossberg Cablelock included. Introduced 1989. From Maverick Arms, Inc.
Price: Model 88, synthetic stock, 28" Mod. $221.00
Price: Model 88, synthetic stock, 28" ACCU-TUBE, Mod. $235.00
Price: Model 88, synthetic stock, 24" with rifle sights $235.00

MAVERICK MODEL 88 PUMP SHOTGUN
Gauge: 12, 3" chamber.
Barrel: 18 1/2" (Cyl.), 28" (Mod.).

Mossberg Model 500 Sporting

MOSSBERG MODEL 500 SPORTING PUMP
Gauge: 12, 20, 410, 3" chamber.
Barrel: 18 1/2" to 28" with fixed or Accu-Choke, plain or vent. rib.
Weight: 6 1/4 lbs. (410), 7 1/4 lbs. (12). **Length:** 48" overall (28" barrel).
Stock: 14"x1 1/2"x2 1/2". Walnut-stained hardwood. Cut-checkered grip and forend.
Sights: White bead front, brass mid-bead.
Features: Ambidextrous thumb safety, twin extractors, disconnecting safety, dual action bars. Quiet Carry forend. Many barrels are ported. Mossberg Cablelock included. From Mossberg.
Price: From about $281.00
Price: Sporting Combos (field barrel and Slugster barrel), from $353.00

Mossberg Model 500 Bantam Pump
Same as the Model 500 Sporting Pump except 20-gauge only, 22" vent. rib Accu-Choke barrel with Mod. choke tube; has 1" shorter stock, reduced length from pistol grip to trigger, reduced forend reach. Introduced 1992.
Price: $301.00
Price: With full camouflage finish $309.00

Mossberg Model 500 Viking
Similar to the Model 500 Sporting except in 12-gauge with 24" ported rifled bore, rifle sights or 28" vent. rib with Mod. Accu-Choke tube, or 20-gauge 26" vent. rib with Mod. Accu-Choke tube; moss-green synthetic stock and forend, matte metal finish. Made in U.S. by Mossberg. Introduced 1996.
Price: $287.00
Price: With rifled barrel $326.00
Price: Scope and case combo $394.00

Mossberg Model 500 Camo Pump
Same as the Model 500 Sporting Pump except 12-gauge only and entire gun is covered with special camouflage finish. Receiver drilled and tapped for scope mounting. Comes with quick detachable swivel studs, swivels, camouflage sling, Mossberg Cablelock.
Price: From about $309.00
Price: Camo Combo (as above with extra Slugster barrel), from ...$409.00

Mossberg Model 500 Trophy Slugster

MOSSBERG MODEL 500 TROPHY SLUGSTER
Gauge: 12, 20, 3" chamber.
Barrel: 24", ported rifled bore. Integral scope mount.
Weight: 7 1/4 lbs. **Length:** 44" overall.
Stock: 14" pull, 1 3/8" drop at heel. Walnut; Dual Comb design for proper eye positioning with or without scoped barrels. Recoil pad and swivel studs.
Features: Ambidextrous thumb safety, twin extractors, dual slide bars. Comes with scope mount. Mossberg Cablelock included. Introduced 1988.
Price: Rifled bore, with integral scope mount, Dual-Comb stock, 12 or 20 **$369.00**
Price: Cyl. bore, rifle sights $303.00
Price: Rifled bore, rifle sights $341.00
Price: 20 ga., Standard or Bantam $341.00

SHOTGUNS—SLIDE ACTIONS

Mossberg Model 835 Turkey

MOSSBERG MODEL 835 CROWN GRADE ULTI-MAG PUMP
Gauge: 12, 3½″ chamber.
Barrel: Ported 24″ rifled bore, 24″, 28″, Accu-Mag with four choke tubes for steel or lead shot.
Weight: 7¾ lbs. **Length:** 48½″ overall.
Stock: 14″x1½″x2½″. Dual Comb. Cut-checkered walnut or camo synthetic; both have recoil pad.
Sights: White bead front, brass mid-bead.
Features: Shoots 2¾″, 3″ or 3½″ shells. Backbored and ported barrel to reduce recoil, improve patterns. Ambidextrous thumb safety, twin extractors, dual slide bars. Mossberg Cablelock included. Introduced 1988.

Price: 28″ vent. rib, Dual-Comb walnut stock $428.00
Price: As above, standard stock $421.00
Price: 24″ Trophy Slugster, rifled bore, scope base, Dual-Comb stock . .$381.00
Price: Combo, 24″ rifled bore, rifle sights, 28″ vent. rib, Accu-Mag choke tubes $410.00
Price: Combo, 24″ Trophy Slugster rifled bore, 28″ vent. rib, Accu-Mag Mod. tube, Dual-Comb stock $440.00
Price: Realtree or Mossy Oak Camo Turkey, 24″ vent. rib, Accu-Mag Extra-Full tube, synthetic stock $515.00
Price: Realtree Camo, 28″ vent. rib, Accu-Mag tubes, synthetic stock . . .$515.00
Price: Realtree Camo Combo, 24″ rifled bore, rifle sights, 24″ vent. rib, Accu-Mag choke tubes, synthetic stock, hard case $640.00
Price: OFM Camo, 28″ vent. rib, Accu-Mag tubes, synthetic stock $359.00
Price: OFM Camo Combo, 24″ rifled bore, rifle sights, 28″ vent. rib, Accu-Mag tubes, synthetic stock $554.00

Mossberg American Field Model 835 Pump Shotgun
Same as the Model 835 Crown Grade except has walnut-stained hardwood stock and comes only with Modified choke tube, 28″ barrel. Introduced 1992.
Price: $325.00

Mossberg Model 835 Viking
Similar to the Model 835 Crown Grade except has moss-green synthetic stock and forend, matte metal finish, 28″ vent. rib Accu-Mag. barrel with Mod. tube. Made in U.S. by Mossberg. Introduced 1996.
Price: $316.00

Remington 870 Wingmaster

Remington Model 870 Express
Similar to the 870 Wingmaster except has a walnut-toned hardwood stock with solid, black recoil pad and pressed checkering on grip and forend. Outside metal surfaces have a black oxide finish. Comes with 26″ or 28″ vent. rib barrel with a Mod. Rem Choke tube. Introduced 1987.
Price: 12 or 20 ....$299.00
Price: Express Combo, 26″ vent rib with Mod. Rem Choke and 20″ fully rifled barrel with rifle sights $421.00
Price: Express 20-ga., 28″ with Mod. Rem Choke tubes $292.00

Remington Model 870 Marine Magnum
Similar to the 870 Wingmaster except all metal is plated with electroless nickel and has black synthetic stock and forend. Has 18″ plain barrel (Cyl.), bead front sight, 7-shot magazine. Introduced 1992.
Price: $500.00

Remington Model 870 Express Rifle-Sighted Deer Gun
Same as the Model 870 Express except comes with 20″ barrel with fixed Imp. Cyl. choke, open iron sights, Monte Carlo stock. Introduced 1991.
Price: $287.00
Price: With fully rifled barrel $325.00

Remington Model 870 Express Synthetic
Similar to the 870 Express with 26″, 28″ barrel except has synthetic stock and forend. Introduced 1994.
Price: $299.00

REMINGTON MODEL 870 WINGMASTER
Gauge: 12, 3″ chamber.
Barrel: 26″, 28″, 30″ (Rem Chokes). Light Contour barrel.
Weight: 7¼ lbs. **Length:** 46½″ overall (26″ bbl.).
Stock: 14″x2½″x1″. American walnut with satin or high-gloss finish, cut-checkered pistol grip and forend. Rubber buttpad.
Sights: Ivory bead front, metal mid-bead.
Features: Double action bars; cross-bolt safety; blue finish. Available in right- or left-hand style. Introduced 1986.
Price: $519.00
Price: Fully rifled Cantilever, 20″ $599.00

Remington Model 870 TC Trap Gun
Similar to the 870 Wingmaster except has tournament-grade, satin-finished American walnut stock with or Monte Carlo comb, over-bored 30″ vent. rib barrel with 2¾″ chamber, over-bore-matched Rem Choke tubes. Made in U.S. by Remington. Reintroduced 1996.
Price: With Monte Carlo stock $647.00

Remington Model 870 Express Turkey
Same as the Model 870 Express except comes with 3″ chamber, 21″ vent. rib turkey barrel and Extra-Full Rem Choke Turkey tube; 12-ga. only. Introduced 1991.
Price: $305.00

Remington Model 870 Express Youth Gun
Same as the Model 870 Express except comes with 12½″ length of pull, 21″ barrel with Mod. Rem Choke tube. Hardwood stock with low-luster finish. Introduced 1991.
Price: 20-ga. Express Youth (1″ shorter stock) $292.00
Price: 20-ga. Express Youth Deer (rifle sights, fully rifled barrel) $325.00

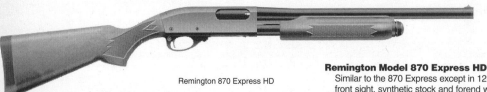
Remington 870 Express HD

Remington Model 870 Express Small Gauge
Similar to the 870 Express except is scaled down for 28-gauge and 410-bore. Has 25″ vent. rib barrel with fixed Mod. choke; solid black rubber buttpad. Reintroduced 1996.
Price: $319.00

Remington Model 870 Express HD
Similar to the 870 Express except in 12-gauge only, 18″ (Cyl.) barrel with bead front sight, synthetic stock and forend with non-reflective black finish and positive checkering. Introduced 1995.
Price: $292.00

SHOTGUNS—SLIDE ACTIONS

Remington Model 870 SPS-Deer Shotgun
Has fully-rifled 20″ barrel with rifle sights, black non-reflective, synthetic stock and forend, black carrying sling. Introduced 1993.
Price: $436.00

Remington Model 870 SPS Cantilever Shotgun
Similar to the 870 SPS-Deer except has rifled barrel; synthetic stock with Monte Carlo comb; cantilever scope mount deer barrel. Comes with sling and swivels. Introduced 1994.
Price: With fully rifled barrel $496.00

Remington Model 870 SPS-T Camo

Remington Model 870 SPS-T Camo Pump Shotgun
Similar to the 870 Special Purpose Magnum except with synthetic stock, 21″ vent. rib barrel with Super-Full Turkey (.665″ diameter with knurled extension) and Imp. Cyl. Rem Choke tubes. Completely covered with Realtree X-tra Brown camouflage. Bolt body, trigger guard and recoil pad are non-reflective black. Introduced 1993.
Price: $511.00

Remington Model 870 SPS-T Special Purpose Magnum
Similar to the Model 870 except chambered only for 12-ga., 3″ shells, 26″ or 28″ Rem Choke barrel. All exposed metal surfaces are finished in dull, non-reflective black. Black synthetic stock and forend. Comes with padded Cordura 2″ wide sling, quick-detachable swivels. Chrome-lined bores. Dark recoil pad. Introduced 1985.
Price: $425.00

Remington Model 870 Special Purpose Synthetic Camo
Similar to the 870 Special Purpose Magnum except has synthetic stock and all metal (except bolt and trigger guard) and stock covered with Mossy Oak Break-Up camo finish, In 12-gauge only, 26″ vent. rib, Rem Choke. Comes with camo sling, swivels. Introduced 1992.
Price: $511.00

Remington Model 870 High Grades
Same as 870 except better walnut, hand checkering. Engraved receiver and barrel. Vent. rib. Stock dimensions to order.
Price: 870D, about $2,610.00
Price: 870F, about $5,377.00
Price: 870F with gold inlay, about $8,062.00

TRISTAR MODEL 1887 LEVER-ACTION SHOTGUN
Gauge: 12, 2³/₄″ chamber, 5-shot magazine.
Barrel: 30″ (Full).
Weight: 8 lbs. **Length:** 48″ overall.
Stocks: 12³/₄″ pull. Rounded-knob pistol grip; walnut with oil finish; blued, checkered steel buttplate. Dimensions duplicate original WRA Co. specifications.
Sights: Brass, bead front.
Features: Recreation of Browning's original 1885 patents and design as made by Winchester Repeating Arms. External hammer with half- and full-cock positions; has original-type WRA Co. logo on left side of receiver; two-piece walnut forend. Introduced 1997. Imported by Tristar Sporting Arms.
Price: $599.00

SARSILMAZ COBRA PUMP SHOTGUN
Gauge: 12, 3″ chamber, 7-shot magazine.
Barrel: 18″ (Cyl.).
Weight: 6¹/₄ lbs. **Length:** 38″ overall.
Stock: European hardwood.
Sights: Bead front.
Features: Comes with extra pistol grip. Imported from Turkey by Armsport.
Price: $350.00

Winchester Model 1300 Black Shadow Field Gun
Similar to the Model 1300 Walnut except has black composite stock and forend, matte black finish. Has vent. rib 26″ or 28″ barrel, 3″ chamber, comes with Mod. Winchoke tube. Introduced 1995. From U.S. Repeating Arms Co., Inc.
Price: 12- or 20-gauge $296.00

WINCHESTER MODEL 1300 WALNUT PUMP
Gauge: 12, 20, 3″ chamber, 5-shot capacity.
Barrel: 26″, 28″, vent. rib, with Full, Mod., Imp. Cyl. Winchoke tubes.
Weight: 6³/₈ lbs. **Length:** 42⁵/₈″ overall.
Stock: American walnut, with deep cut checkering on pistol grip, traditional ribbed forend; high luster finish.
Sights: Metal bead front.
Features: Twin action slide bars; front-locking rotary bolt; roll-engraved receiver; blued, highly polished metal; cross-bolt safety with red indicator. Introduced 1984. From U.S. Repeating Arms Co., Inc.
Price: $340.00

Winchester Model 1300 Black Shadow Deer Gun
Similar to the Model 1300 Black Shadow Turkey Gun except has ramp-type front sight, fully adjustable rear, drilled and tapped for scope mounting. Black composite stock and forend, matte black metal. Smoothbore 22″ barrel with one Imp. Cyl. WinChoke tube; 12-gauge only, 3″ chamber. Weighs 7¹/₄ lbs. Introduced 1994. From U.S. Repeating Arms Co., Inc.
Price: $296.00
Price: With rifled barrel $317.00

Winchester 1300 Advantage

Price: Combo, 22″ Cyl. barrel with 28″ Mod. vent rib barrel $366.00

Winchester Model 1300 Advantage Camo Deer Gun
Similar to the Model 1300 Black Shadow Deer Gun except has full coverage Advantage camouflage. Has 22″ rifled or smoothbore barrel, padded camouflage sling, swivels and swivel posts, rifle sights. Receiver drilled and tapped for scope mounting. Introduced 1995. From U.S. Repeating Arms Co., Inc.
Price: Rifled bore $432.00
Price: Smoothbore $410.00

Winchester Model 1300 Black Shadow Turkey Gun
Similar to the Model 1300 Realtree® Turkey except synthetic stock and forend are matte black, and all metal surfaces finished matte black. Drilled and tapped for scope mounting. In 12- or 20-gauge, 3″ chamber, 22″ vent. rib barrel; comes with one Extra-Full Winchoke tube (20-gauge has Full). Introduced 1994. From U.S. Repeating Arms Co., Inc.
Price: $296.00

Winchester 1300 Realtree Turkey

Price: Camouflage stock and forend only $370.00

Winchester Model 1300 Realtree® Turkey Gun
Similar to the standard Model 1300 except has synthetic Realtree® camo stock and forend, matte finished barrel and receiver, 22″ barrel with Extra Full, Full and Mod. Winchoke tubes. Drilled and tapped for scope mounting. Comes with padded, adjustable sling. In 12-gauge only, 3″ chamber; weighs about 7 lbs. Introduced 1994. From U.S. Repeating Arms Co., Inc.
Price: $370.00

SHOTGUNS—SLIDE ACTIONS

Winchester 1300 Ranger

Winchester Model 1300 Ranger Pump Gun Combo & Deer Gun
Similar to the standard Ranger except comes with two barrels: 22" (Cyl.) deer barrel with rifle-type sights and an interchangeable 28" vent. rib Winchoke barrel with Full, Mod. and Imp. Cyl. choke tubes. Drilled and tapped; comes with rings and bases. Available in 12- and 20-gauge 3" only, with recoil pad. Introduced 1983.
Price: Deer Combo with two barrels $379.00
Price: 12-ga., 22" rifled barrel $401.00
Price: Rifled Deer Combo (22" rifled and 28" vent. rib barrels, 12 or 20-ga.) $379.00

WINCHESTER MODEL 1300 RANGER PUMP GUN
Gauge: 12, 20, 3" chamber, 5-shot magazine.
Barrel: 26", 28" vent. rib with Full, Mod., Imp. Cyl. Winchoke tubes.
Weight: 7 to 7 1/4 lbs. **Length:** 48 5/8" to 50 5/8" overall.
Stock: Walnut-finished hardwood with ribbed forend.
Sights: Metal bead front.
Features: Cross-bolt safety, black rubber recoil pad, twin action slide bars, front-locking rotating bolt. From U.S. Repeating Arms Co., Inc.
Price: Vent. rib barrel, Winchoke $309.00
Price: Model 1300 Ladies/Youth, 20-ga., 22" vent. rib $309.00

SHOTGUNS—OVER/UNDERS

Includes a variety of game guns and guns for competitive shooting.

American Arms Silver I

American Arms Silver II Shotgun
Similar to the Silver I except 26" barrel (Imp. Cyl., Mod., Full choke tubes, 12- and 20-ga.), 28" (Imp. Cyl., Mod., Full choke tubes, 12-ga. only), 26" (Imp. Cyl. & Mod. fixed chokes, 28 and 410), automatic selective ejectors. Weight is about 6 lbs., 15 oz. (12-ga., 26").
Price: $750.00
Price: 28, 410 ....$775.00
Price: Two-barrel sets $1,150.00

AMERICAN ARMS SILVER SPORTING O/U
Gauge: 12, 2 3/4" chambers, 20 3" chambers.
Barrel: 28", 30" (Skeet, Imp. Cyl., Mod., Full choke tubes).
Weight: 7 3/8 lbs. **Length:** 45 1/2" overall.
Stock: 14 3/8"x1 1/2"x2 3/8". Figured walnut, cut checkering; Sporting Clays quick-mount buttpad.
Sights: Target bead front.
Features: Boxlock action with single selective mechanical trigger, automatic selective ejectors; special broadway channeled rib; vented barrel rib; chrome bores. Chrome-nickel finish on frame, with engraving. Introduced 1990. Imported from Italy by American Arms, Inc.
Price: $925.00

AMERICAN ARMS SILVER I O/U
Gauge: 12, 20, 28, 410, 3" chamber (28 has 2 3/4").
Barrel: 26" (Imp. Cyl. & Mod., all gauges), 28" (Mod. & Full, 12, 20).
Weight: About 6 3/4 lbs.
Stock: 14 1/8"x1 3/8"x2 3/8". Checkered walnut.
Sights: Metal bead front.
Features: Boxlock action with scroll engraving, silver finish. Single selective trigger, extractors. Chrome-lined barrels. Manual safety. Rubber recoil pad. Introduced 1987. Imported from Italy by American Arms, Inc.
Price: 12- or 20-gauge $625.00
Price: 28 or 410 ....$650.00

American Arms Silver Upland Lite
Similar to the Silver I except weighs 6 lbs., 4 oz. (12-gauge), 5 lbs., 12 oz. (20-gauge). Single selective trigger, automatic selective ejectors. Franchoke tubes, vent. rib, engraved frame with antique silver finish. Introduced 1994. Imported by American Arms, Inc.
Price: 12- , 20-ga., 3" chambers, 26" $925.00

American Arms WS/OU 12

AMERICAN ARMS/FRANCHI ALCIONE 2000 SX O/U
Gauge: 12, 3" chambers.
Barrel: 28", Franchoke tubes.
Weight: 7 lbs., 4 oz. **Length:** 45 1/8" overall.
Stock: 14 1/4"x1 1/2"x2 1/2". Select European walnut.
Features: Extensively engraved silvered action; single selective trigger, automatic selective ejectors; chrome-lined barrels vent. top rib, separated barrels. Imported from Italy by American Arms, Inc.
Price: $1,895.00

AMERICAN ARMS WS/OU 12, TS/OU 12 SHOTGUNS
Gauge: 12, 3 1/2" chambers.
Barrel: WS/OU—28" (Imp. Cyl., Mod., Full choke tubes); TS/OU—24" (Imp. Cyl., Mod., Full choke tubes).
Weight: 6 lbs., 15 oz. **Length:** 46" overall.
Stock: 14 1/8"x1 1/8"x2 3/8". European walnut with cut checkering, black vented recoil pad, matte finish.
Features: Boxlock action with single selective trigger, automatic selective ejectors; chrome bores. Matte metal finish. Imported by American Arms, Inc.
Price: $775.00
Price: With Mossy Oak Break-Up camo $850.00

American Arms WT/OU 10 Shotgun
Similar to the WS/OU 12 except chambered for 10-gauge 3 1/2" shell, 26" (Full & Full, choke tubes) barrel. Single selective trigger, extractors. Non-reflective finish on wood and metal. Imported by American Arms, Inc.
Price: $995.00

SHOTGUNS—OVER/UNDERS

American Arms/Franchi Falconet

AMERICAN ARMS/FRANCHI SPORTING 2000 O/U
Gauge: 12, 2³/₄″ chambers.
Barrel: 28″, Franchoke tubes.
Weight: 7 lbs., 12 oz. **Length:** 45³/₈″ overall.
Stock: 14¹/₄″x1³/₈″x2¹/₄″. Select European walnut.
Features: Wide single selective mechanical trigger, automatic selective ejectors; chrome-lined barrels; ventilated 10mm top rib, separated barrels. Imported from Italy by American Arms, Inc.
Price: $1,495.00

BABY BRETTON OVER/UNDER SHOTGUN
Gauge: 12 or 20, 2³/₄″ chambers.
Barrel: 27¹/₂″ (Cyl., Imp. Cyl., Mod., Full choke tubes).
Weight: About 5 lbs.
Stock: Walnut, checkered pistol grip and forend, oil finish.
Features: Receiver slides open on two guide rods, is locked by a large thumb lever on the right side. Extractors only. Light alloy barrels. Imported from France by Mandall Shooting Supplies.
Price: Sprint Standard $895.00
Price: Sprint Deluxe $975.00
Price: Model Fairplay $1,025.00

BAIKAL TOZ-34P OVER/UNDER SHOTGUN
Gauge: 12, 2³/₄″ chambers.
Barrel: 28″ (Full & Imp. Cyl.).
Weight: 7.5 lbs. **Length:** 44″ overall.
Stock: European walnut.
Features: Engraved, blued receiver; cocking indicator; double triggers. Ventilated rib, ventilated rubber buttpad. Imported from Russia by Century International Arms.
Price: About $329.00
Price: With ejectors, about $379.00

Beretta 682 Gold Skeet

BERETTA MODEL 686 SILVER ESSENTIAL O/U
Gauge: 12, 3″ chambers.
Barrel: 26″, 28″, Mobilchoke tubes (Imp. Cyl., Mod., Full).
Weight: 6.7 lbs. **Length:** 45.7″ overall (28″ barrels).
Stock: 14.5″x2.2″x1.4″. American walnut; radiused black buttplate.
Features: Matte chrome finish on receiver, matte barrels; hard-chrome bores; low-profile receiver with dual conical locking lugs; single selective trigger, ejectors. Introduced 1994. Imported from Italy by Beretta U.S.A.
Price: $1,005.00

BERETTA MODEL SO5, SO6, SO9 SHOTGUNS
Gauge: 12, 2³/₄″ chambers.
Barrel: To customer specs.
Stock: To customer specs.
Features: SO5—Trap, Skeet and Sporting Clays models SO5; SO6—SO6 and SO6 EELL are field models. SO6 has a case-hardened or silver receiver with contour hand engraving. SO6 EELL has hand-engraved receiver in a fine floral or "fine English" pattern or game scene, with bas-relief chisel work and gold inlays. SO6 and SO6 EELL are available with sidelocks removable by hand. Imported from Italy by Beretta U.S.A.
Price: SO5 Trap, Skeet, Sporting $13,000.00
Price: SO6 Trap, Skeet, Sporting $17,500.00
Price: SO6 EELL Field, custom specs $28,000.00
Price: SO9 (12, 20, 28, 410, 26″, 28″, 30″, any choke) ...$31,000.00

AMERICAN ARMS/FRANCHI FALCONET 2000 OVER/UNDER
Gauge: 12, 2³/₄″ chambers.
Barrel: 26″, Franchoke tubes.
Weight: 6 lbs., 2 oz. **Length:** 43¹/₈″ overall.
Stock: 14¹/₄″ x 1¹/₂″x2¹/₂″. European walnut.
Features: Alloy frame with engraving, gold-plated game bird scene engraving; chrome-lined barrels; single selective trigger, automatic selective ejectors; vent. top rib, separated barrels. Imported from Italy by American Arms, Inc.
Price: $1,375.00

> *Manufacturers' addresses in the*
> **Directory of the Arms Trade**
> *page 309, this issue*

BAIKAL IJ-27M OVER/UNDER SHOTGUN
Gauge: 12, 2³/₄″ chambers.
Barrel: 28.5″ (Mod. & Full).
Weight: 7.5 lbs. **Length:** 44.5″ overall.
Stock: European hardwood.
Features: Engraved boxlock action with double triggers, extractors; chrome-lined barrels; sling swivels. Imported from Russia by Century International Arms.
Price: About $340.00

BERETTA 686 ONYX SPORTING O/U SHOTGUN
Gauge: 12, 3″ chambers.
Barrel: 28″, 30″ (Mobilchoke tubes).
Weight: 7.7 lbs.
Stock: Checkered American walnut.
Features: Intended for the beginning sporting clays shooter. Has wide, vented 12.5mm target rib, radiused recoil pad. Matte black finish on receiver and barrels. Introduced 1993. Imported from Italy by Beretta U.S.A.
Price: $1,450.00

BERETTA SERIES 682 GOLD SKEET, TRAP OVER/UNDERS
Gauge: 12, 2³/₄″ chambers.
Barrel: Skeet—28″; trap—30″ and 32″, Imp. Mod. & Full and Mobilchoke; trap mono shotguns—32″ and 34″ Mobilchoke; trap top single guns—32″ and 34″ Full and Mobilchoke; trap combo sets—from 30″ O/U, to 32″ O/U, 34″ top single.
Stock: Close-grained walnut, hand checkered.
Sights: White Bradley bead front sight and center bead.
Features: Receiver has Greystone gunmetal gray finish with gold accents. Trap Monte Carlo stock has deluxe trap recoil pad. Various grades available; contact Beretta U.S.A. for details. Imported from Italy by Beretta U.S.A.
Price: 682 Gold Skeet $2,850.00
Price: 682 Gold Trap $2,910.00
Price: 682 Gold Trap Top Combo $3,845.00
Price: 682 Gold Super Trap Top Combo $4,040.00
Price: 686 Silver Pigion Skeet (28″) $1,760.00
Price: 687 EELL Diamond Pigeon Trap $4,815.00
Price: 687 EELL Diamond Pigeon Skeet (4-bbl. set) ...$8,405.00
Price: 687 EELL Diamond Pigeon Trap Top Mono .....$5,055.00 to $5,105.00
Price: ASE Gold Skeet $12,060.00
Price: ASE Gold Trap $12,145.00
Price: ASE Gold Trap Combo $16,055.00

BERETTA ULTRALIGHT OVER/UNDER
Gauge: 12, 2³/₄″ chambers.
Barrel: 28″, Mobilchoke choke tubes.
Weight: About 5 lbs., 13 oz.
Stock: Select American walnut with checkered grip and forend.
Features: Low-profile aluminum alloy receiver with titanium breech face insert. Silvered receiver with game scene engraving. Single selective trigger; automatic safety. Introduced 1992. Imported from Italy by Beretta U.S.A.
Price: $1,795.00

SHOTGUNS—OVER/UNDERS

Beretta 682 Gold Sporting

BERETTA SPORTING CLAYS SHOTGUNS
Gauge: 12 and 20, 2³⁄₄" and 3" chambers.
Barrel: 28", 30", 32" Mobilchoke.
Stock: Close-grained walnut.
Features: Equipped with Beretta Mobilchoke flush-mounted screw-in choke tube system. Dual-purpose O/U for hunting and Sporting Clays.12- or 20-gauge, 28", 30" Mobilchoke tubes (four, Skeet, Imp. Cyl., Mod., Full). Wide 12.5mm top rib with 2.5mm center groove; 686 Silver Pigeon has silver receiver with scroll engraving; 687 Silver Pigeon Sporting has silver receiver, highly figured walnut; 687 EL Pigeon Sporting has game scene engraving with gold inlaid animals on full sideplate. Introduced 1994. Imported from Italy by Beretta U.S.A.

Price: 682 Gold Sporting, 28", 30", 31" (with case) $2,910.00
Price: 682 Gold Sporting, 28", 30", ported, adj. l.o.p. $3,035.00
Price: 682 Continental Course Sporting, 2³⁄₄" chambers, 28" $2,345.00
Price: 686 Silver Pigeon Sporting $1,780.00
Price: 686 Silver Pigeon Sporting (20-gauge) $1,760.00
Price: 687 Silver Pigeon Sporting $2,575.00
Price: 687 Silver Pigeon Sporting (20 gauge) $2,575.00
Price: 687 Diamond Pigeon EELL Sporter (hand engraved sideplates, deluxe wood) $5,310.00
Price: 687 Silver Pigeon Sporting Combo, 28" and 30" $3,395.00
Price: ASE Gold Sporting Clay $12,145.00

Beretta 687EL Gold Pigeon Sporting O/U
Similar to the 687 Silver Pigeon Sporting except has sideplates with gold inlay game scene, vent. side and top ribs, bright orange front sight. Stock and forend are of high grade walnut with fine-line checkering. Available in 12-gauge only with 28" or 30" barrels and Mobilchoke tubes. Weight is 6 lbs., 13 oz. Introduced 1993. Imported from Italy by Beretta U.S.A.
Price: $3,365.00

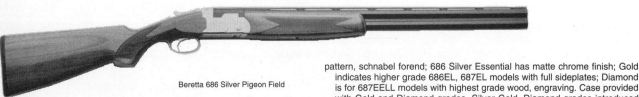

Beretta 686 Silver Pigeon Field

BERETTA OVER/UNDER FIELD SHOTGUNS
Gauge: 12, 20, 28, and 410 bore, 2³⁄₄", 3" and 3¹⁄₂" chambers.
Barrel: 26" and 28" (Mobilchoke tubes).
Stock: Close-grained walnut.
Features: Highly-figured, American walnut stocks and forends, and a unique, weather-resistant finish on barrels. The 686 Onyx bears a gold P. Beretta signature on each side of the receiver. Silver designates standard 686, 687 models with silver receivers; 686 Silver Pigeon has enhanced engraving pattern, schnabel forend; 686 Silver Essential has matte chrome finish; Gold indicates higher grade 686EL, 687EL models with full sideplates; Diamond is for 687EELL models with highest grade wood, engraving. Case provided with Gold and Diamond grades. Silver Gold, Diamond grades introduced 1994. Imported from Italy by Beretta U.S.A.

Price: 686 Silver Essential $1,005.00
Price: 686 Onyx $1,420.00
Price: 686 Silver Pigeon two-bbl. set $2,350.00
Price: 686 Silver Pigeon $1,895.00
Price: 686EL Gold Perdiz (engraved sideplates, hard case) $1,930.00
Price: 687L Silver Pigeon $2,115.00
Price: 687EL Gold Pigeon (gold inlays, sideplates) $3,595.00
Price: 687EL Gold Pigeon, 410, 26", 28-ga., 28" $3,760.00
Price: 687EELL Diamond Pigeon (engraved sideplates) $5,215.00
Price: 687EELL Diamond Pigeon Combo, 20- and 28-ga., 26" $5,815.00

Browning Citori Gran Lightning

Browning Citori Special Trap Models
Similar to standard Citori except 12 gauge only; 30", 32" ported or non-ported (Invector Plus); Monte Carlo cheek piece (14³⁄₈"x1³⁄₈"x1³⁄₈"x2"); fitted with trap-style recoil pad; high post target rib, ventilated side ribs.
Price: Grade I, Invector Plus, ported bbls. $1,586.00
Price: Grade III, Invector Plus Ported $2,179.00
Price: Golden Clays $3,239.00
Price: Adjustable comb stock, add $210.00

Browning Superlight Citori Over/Under
Similar to the standard Citori except available in 12, 20 with 24", 26" or 28" Invector barrels, 28 or 410 with 26" barrels choked Imp. Cyl. & Mod. or 28" choked Mod. & Full. Has straight grip stock, schnabel forend tip. Superlight 12 weighs 6 lbs., 9 oz. (26" barrels); Superlight 20, 5 lbs., 12 oz. (26" barrels). Introduced 1982.
Price: Grade I only, 28 or 410, Invector $1,439.00
Price: Grade III, Invector, 12 or 20 . . . . $2,006.00
Price: Grade III, 28 or 410, Invector $2,242.00
Price: Grade VI, Invector, 12 or 20 . . . . $2,919.00
Price: Grade VI, 28 or 410, Invector $3,145.00
Price: Grade I Invector, 12 or 20 . . . . $1,386.00
Price: Grade I Invector, Upland Special (24" bbls.), 12 or 20 . . . . $1,386.00

BROWNING CITORI O/U SHOTGUN
Gauge: 12, 20, 28 and 410.
Barrel: 26", 28" in 28 and 410. Offered with Invector choke tubes. All 12- and 20-gauge models have back-bored barrels and Invector Plus choke system.
Weight: 6 lbs., 8 oz. (26" 410) to 7 lbs., 13 oz. (30" 12-ga.).
Length: 43" overall (26" bbl.).
Stock: Dense walnut, hand checkered, full pistol grip, beavertail forend. Field-type recoil pad on 12-ga. field guns and trap and Skeet models.
Sights: Medium raised beads, German nickel silver.
Features: Barrel selector integral with safety, automatic ejectors, three-piece takedown. Imported from Japan by Browning. Contact Browning for complete list of models and prices.
Price: Grade I, Hunting, Invector, 12 and 20 . . . . $1,334.00
Price: Grade I, Lightning, 28 and 410, Invector $1,418.00
Price: Grade III, Lightning, 28 and 410, Invector $2,242.00
Price: Grade VI, 28 and 410 Lightning, Invector $3,145.00
Price: Grade I, Lightning, Invector Plus, 12, 20 . . . . $1,376.00
Price: Grade I, Hunting, 28", 30" only, 3¹⁄₂", Invector Plus $1,418.00
Price: Grade III, Lightning, Invector, 12, 20 . . . . $2,006.00
Price: Grade VI, Lightning, Invector, 12, 20 . . . . $2,919.00
Price: Gran Lightning, 26", 28", Invector, 12 ,20 . . . . $1,869.00
Price: Gran Lightning, 28, 410 . . . . $1,969.00

Browning Micro Citori Lightning
Similar to the standard Citori 20-ga. Lightning except scaled down for smaller shooter. Comes with 24" Invector Plus back-bored barrels, 13³⁄₄" length of pull. Weighs about 6 lbs., 3 oz. Introduced 1991.
Price: Grade I $1,428.00

SHOTGUNS—OVER/UNDERS

Browning Lightning Sporting Clays
Similar to the Citori Lightning with rounded pistol grip and classic forend. Has high post tapered rib or lower hunting-style rib with 30" back-bored Invector Plus barrels, ported or non-ported, 3" chambers. Gloss stock finish, radiused recoil pad. Has "Lightning Sporting Clays Edition" engraved and gold filled on receiver. Introduced 1989.
Price: Low-rib, ported $1,496.00
Price: High-rib, ported $1,565.00
Price: Golden Clays, low rib, ported $3,092.00
Price: Golden Clays, high rib, ported $3,203.00
Price: Adjustable comb stock, all models, add $210.00

Browning Citori Ultra Sporter
Similar to the Citori Hunting except has slightly grooved, semi-beavertail forend, satin-finish stock, radiused rubber buttpad. Has three interchangeable trigger shoes, trigger has three length of pull adjustments. Ventilated rib tapers from 13mm to 10mm, 28" or 30" barrels (ported or non-ported) with Invector Plus choke tubes. Ventilated side ribs. Introduced 1989.
Price: With ported barrels, gray or blue receiver $1,722.00
Price: Golden Clays $3,203.00

Browning Citori O/U Special Skeet
Similar to standard Citori except 26", 28" barrels, ventilated side ribs, Invector choke tubes; stock dimensions of 14 3/8"x1 1/2"x2", fitted with Skeet-style recoil pad; conventional target rib and high post target rib.
Price: Grade I Invector, 12-, 20-ga., Invector Plus (high post rib) $1,586.00
Price: Grade I, 28 and 410 (high post rib) $1,549.00
Price: Grade III, 28, 410 (high post rib) $2,184.00
Price: Golden Clays $3,239.00
Price: Grade III, 12-ga. Invector Plus $2,179.00
Price: Adjustable comb stock, add $210.00

Browning Special Sporting Clays
Similar to the Citori Ultra Sporter except has full pistol grip stock with palm swell, gloss finish, 28", 30" or 32" barrels with back-bored Invector Plus chokes (ported or non-ported); high post tapered rib. Also available as 28" and 30" two-barrel set. Introduced 1989.
Price: With ported barrels $1,565.00
Price: As above, adjustable comb $1,775.00
Price: Golden Clays $3,203.00
Price: With adjustable comb stock $3,413.00

Browning 802 ES

BROWNING LIGHT SPORTING 802 ES O/U
Gauge: 12, 2 3/4" chambers.
Barrel: 28", back-bored Invector Plus. Comes with flush-mounted Imp. Cyl. and Skeet; 2" extended Imp. Cyl. and Mod.; and 4" extended Imp. Cyl. and Mod. tubes.
Weight: 7 lbs., 5 oz. **Length:** 45" overall.
Stock: 14 3/8"±1/8"x1 9/16"x1 3/4". Select walnut with radiused solid recoil pad, schnabel-type forend.
Features: Trigger adjustable for length of pull; narrow 6.2mm ventilated rib; ventilated barrel side rib; blued receiver. Introduced 1996. Imported from Japan from Browning.
Price: $1,880.00

Browning 425 Sporting Clays

BROWNING 425 SPORTING CLAYS
Gauge: 12, 20, 2 3/4" chambers.
Barrel: 12-ga.—28", 30", 32" (Invector Plus tubes), back-bored; 20-ga.—28", 30" (Invector Plus tubes).
Weight: 7 lbs., 13 oz. (12-ga., 28").
Stock: 14 13/16" (±1/8")x1 7/16"x2 3/16" (12-ga.). Select walnut with gloss finish, cut checkering, schnabel forend.
Features: Grayed receiver with engraving, blued barrels. Barrels are ported on 12-gauge guns. Has low 10mm wide vent rib. Comes with three interchangeable trigger shoes to adjust length of pull. Introduced in U.S. 1993. Imported by Browning.
Price: Grade I, 12-, 20-ga., Invector Plus $1,775.00
Price: Golden Clays, 12-, 20-ga., Invector Plus $3,308.00
Price: Adjustable comb stock, add $210.00

Browning 425 WSSF Shotgun
Similar to the 425 Sporting Clays except in 12-gauge only, 28" barrels, has stock dimensions specifically tailored to women shooters (14 1/4"x1 1/2"x1 1/2"); top lever and takedown lever are easier to operate. Stock and forend have teal-colored finish or natural walnut with Women's Shooting Sports Foundation logo. Weighs 7 lbs., 4 oz. Introduced 1995. Imported by Browning.
Price: $1,775.00

CENTURY CENTURION OVER/UNDER SHOTGUN
Gauge: 12, 2 3/4" chambers.
Barrel: 28" (Mod. & Full).
Weight: 7.3 lbs. **Length:** 44.5" overall.
Stock: European walnut.
Features: Double triggers; extractors. Polished blue finish. Introduced 1993. Imported by Century International Arms.
Price: About $380.00

Connecticut Valley Classics Sporter

CONNECTICUT VALLEY CLASSICS CLASSIC SPORTER O/U
Gauge: 12, 3" chambers.
Barrel: 28", 30", 32" (Skeet, Imp. Cyl. Mod., Full CV choke tubes); elongated forcing cones.
Weight: 7 3/4 lbs. **Length:** 44 7/8" overall (28" barrels).
Stock: 14 1/2"x1 1/2"x2 1/8". AA grade semi-fancy American black walnut with 20 lpi hand-checkered grip and forend; hand rubbed oil finish.
Features: Receiver duplicates Classic Doubles M101 specifications. Stainless receiver with fine engraving. Bores and chambers suitable for steel shot. Optionally available are CV Plus (2 3/8" tubes) choke tubes. Introduced 1993. Made in U.S. by Connecticut Valley Classics.
Price: Grade I $3,195.00
Price: Grade II, AAA fancy walnut, 22 lpi checkering, enhanced engraving $3,795.00
Price: Grade III, AAA select fancy walnut, 22 lpi Fleur de lis checkering, enhanced engraving, bird scenes, gold inlay, gold-plated trigger ... $4,195.00
Price: Women's Classic Sporter, 28" only, shorter stock, different stock dimensions $3,195.00

Connecticut Valley Classics Classic Field Waterfowler
Similar to the Classic Sporter except with 30" barrel only, blued, non-reflective overall finish. Interchangeable CV choke tube system includes Skeet, Imp. Cyl., Mod. Full tubes. Introduced 1995. Made in U.S. by Connecticut Valley Classics.
Price: $2,995.00

Connecticut Valley Classics Classic Flyer
Similar to the Classic Sporter except has AAA American black or Claro walnut, Premier Grade 22 lpi checkering, choice of standard or schnabel forend, special engraving pattern, gold inlay, gold-plated trigger, 11mm tapered top rib. Introduced 1995. Made in U.S. by Century International Arms.
Price: $3,995.00

CAUTION: PRICES SHOWN ARE SUPPLIED BY THE MANUFACTURER OR IMPORTER. CHECK YOUR LOCAL GUNSHOP.

SHOTGUNS—OVER/UNDERS

Connecticut Valley Classics Classic Field O/U
Similar to the Classic Sporter except 27½" barrels with standard choke tubes, slightly different stock shape and dimensions for hunting. Over-bored barrels with lengthened forcing cones. Introduced 1995. Made in U.S. by Century International Arms.
Price: Grade I $3,195.00
Price: Grade II (see Sporter prices) $3,595.00
Price: Grade III $4,195.00

Connecticut Valley Classics Classic Skeet
Similar to the Classic Sporter except has 29" barrel with 9mm Skeet rib. AA American black or Claro walnut, 20 lpi checkering. Introduced 1995. Made in U.S. by Century International Arms.
Price: $3,195.00

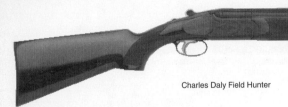
Charles Daly Field Hunter

Charles Daly Field Hunter AE Shotgun
Similar to the Field Hunter except 28-gauge and 410-bore only; 26" (Imp. Cyl. & Mod., 28-gauge), 26" (Full & Full, 410); automatic; ejectors. Introduced 1997. Imported from Italy by K.B.I., Inc.
Price: 28 .... $919.00
Price: 410 .... $959.00

CHARLES DALY FIELD HUNTER OVER/UNDER SHOTGUN
Gauge: 12, 20, 28 and 410 bore 3" chambers, 28-ga. use has 2¾".
Barrel: 28" Mod. & Full, 26" Imp. Cyl. & Mod. (410 is Full & Full).
Weight: About 7 lbs.
Stock: Checkered walnut pistol grip and forend.
Features: Blued, engraved receiver, chrome moly steel barrels; gold single selectvie trigger; automatic safety; extractors; gold bead front sight. Introduced 1997. Imported from Italy by K.B.I., Inc.
Price: 12- or 20- $769.00
Price: 28 .... $834.00
Price: 410-bore $874.00

CHARLES DALY SUPERIOR TRAP
Gauge: 12, 2¾" chambers.
Barrel: 30" (Full & Full).
Weight: About 7 lbs.
Stock: Checkered walnut; pistol grip, semi-beavertail forend.
Features: Silver engraved receiver, chrome moly steel barrels; gold single selective trigger; automatic safety, automatic ejectors; red bead front sight, metal bead center; recoil pad. Introduced 1997. Imported from Italy by K.B.I., Inc.
Price: $1,179.00
Price: Superior Trap MC model (choke tubes) $1,299.00

Charles Daly Field Hunter AE-MC
Similar to the Field Hunter except in 12 or 20 only, 26" or 28" barrels with five multichoke tubes; automatic ejectors. Introduced 1997. Imported from Italy by K.B.I., Inc.
Price: 12 or 20 .... $979.00
Price: With choke tubes $1,199.00

Charles Daly Superior Hunter AE Shotgun
Similar to the Field Hunter AE except has silvered, engraved receiver. Introduced 1997. Imported from Italy by F.B.I., Inc.
Price: 28 .... $1,099.00
Price: 410 .... $1,139.00

CHARLES DALY SUPERIOR SKEET
Gauge: 12 and 20, 3" chambers.
Barrel: 26" (Skeet 1 & Skeet 2)
Weight: About 7 lbs.
Stock: Checkered walnut; pistol grip, semi-beavertail forend.
Features: Silvered engraved receiver, chrome moly steel barrels; gold single selective trigger; automatic safety; automatic-ejectors; red bead front sight, metal bead center; recoil pad. Introduced 1997. Imported from Italy by K.B.I., Inc.
Price: $1,109.00
Price: Superior Skeet-MC (choke tubes) $1,239.00

Charles Daly Superior Sporting O/U
Similar to the Field Hunter AE-MC except 28" or 30" barrels; silvered, engraved receiver; five choke tubes; ported barrels; red bead front sight. Introduced 1997. Imported from Italy by K.B.I., Inc.
Price: $1,179.00

Charles Daly Empire EDL

Charles Daly Empire Sporting O/U
Similar to the Empire DL Hunter except 12- or 20-gauge only, 28", 30" barrels with choke tubes; ported barrels; special stock dimensions. Introduced 1997. Imported from Italy by K.B.I., Inc.
Price: $1,389.00

CHARLES DALY EMPIRE DL HUNTER O/U
Gauge: 12, 20, 410, 3" chambers, 28-ga., 2¾".
Barrel: 26", 28" (12, 20, choke tubes), 26" (Imp. Cyl. & Mod., 28-ga.), 26" (Full & Full, 410).
Weight: About 7 lbs.
Stock: Checkered walnut pistol grip buttstock, semi-beavertail forend; recoil pad.
Features: Silvered, engraved receiver; chrome moly barrels; gold single selective trigger; automatic safety; automatic ejectors; red bead front sight, metal bead middle sight. Introduced 1997. imported from Italy by K.B.I., Inc.
Price: 12 or 20 .... $1,229.00
Price: 28 .... $1,299.00
Price: 410 .... $1,349.00
Price: Empire EDL (dummy sideplates) 12 or 20 ... $1,399.00
Price: Empire EDL, 28 .... $1,469.00
Price: Empire EDL, 410 .... $1,519.00

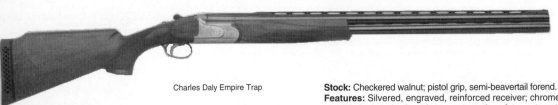
Charles Daly Empire Trap

CHARLES DALY EMPIRE TRAP
Gauge: 12, 2¾" chambers.
Barrel: 30" (Full & Full).
Weight: About 7 lbs.
Stock: Checkered walnut; pistol grip, semi-beavertail forend.
Features: Silvered, engraved, reinforced receiver; chrome moly steel barrels; gold single selective trigger; automatic safety; automatic ejector; red bead front sight, metal bead center; recoil pad. Introduced 1997. Imported from Italy by K.B.I., Inc.
Price: $1,349.00
Price: Empire Trap-MC (choke tubes) $1,469.00

SHOTGUNS—OVER/UNDERS

CHARLES DALY EMPIRE SKEET O/U SHOTGUN
Gauge: 12, 20, 3" chambers.
Barrel: 26" (Skeet 1 & Skeet 2).
Weight: About 7 lbs.
Stock: Checkered walnut; pistol grip; semi-beavertail forend. Skeet dimensions.
Features: Silvered, engraved, reinforced receiver; chrome steel barrels; gold single selective trigger; automatic safety; automatic-ejectors; red bead front sight, metal bead middle sight; recoil pad. Introduced 1997. Imported from Italy by K.B.I. Inc.
Price: $1,249.00
Price: Empire Skeet-MC (choke tubes) $1,399.00

CHARLES DALY DIAMOND GTX DL HUNTER O/U
Gauge: 12, 20, 410, 3" chambers, 28, 2 3/4"chambers.
Barrel: 26, 28", choke tubes in 12 and 20 ga., 26" (Imp. Cyl. & Mod.), 26" (Full & Full) in 410-bore.
Weight: About 8.5 lbs.
Stock: Select fancy European walnut stock, with 24 lpi hand checkering; hand-rubbed oil finish.
Features: Boss-type action with internal side lugs, hand-engraved scrollwork and game scene. GTX detachable single selective trigger system with coil springs; chrome moly steel barrels, automatic safety, automatic-ejectors, red bead front sight, recoil pad. Introduced 1997. Imported from Italy by K.B.I., Inc.
Price: 12 or 20 ....$12,399.00
Price: 28 ....$12,489.00
Price: 410 ....$12,529.00
Price: GTX EDL Hunter (with gold inlays), 12, 20 ....$15,999.00
Price: As above, 28 ....$16,179.00
Price: As above, 410 ....$16,219.00

CHARLES DALY DIAMOND GTX SKEET O/U SHOTGUN
Gauge: 12, 20, 3" chambers.
Barrel: 26" (SK1 & SK2).
Weight: About 8.5 lbs.
Stock: Checkered deluxed walnut; pistol grip; Skeet dimensions; semi-beavertail forend; hand-rubbed oil finish.
Features: Silvered, hand-engraved receiver; chrome moly steel barrels; GTX detachable single selective trigger system with coil springs, automatic safety; automatic ejectors; red bead front sight, metal bead center sight; recoil pad. Introduced 1997. Imported from Italy by K.B.I., Inc.
Price: $5,479.00
Price: Diamond GTX Skeet-MC (choke tubes) $5,629.00

CHARLES DALY DIAMOND GXT SPORTING O/U SHOTGUN
Gauge: 12, 20, 3" chambers.
Barrel: 28", 30" with choke tubes.
Weight: About 8.5 lbs.
Stock: Checkered deluxe walnut; Sporting Clays dimensions. Pistol grip; semi-beavertail forend; hand rubbed oil finish.
Features: Chromed, hand-engraved receiver; chrome moly steel barrels; GTX detachabel single selective trigger system with coil springs, automatic safety; automatic ejectors; red bead front sight; ported barrels. Introduced 1997. Imported from Italy by K.B.I., Inc.
Price: $5,799.00

CHARLES DALY DIAMOND GXT TRAP O/U SHOTGUN
Gauge: 12, 2 3/4" chambers.
Barrel: 30" (Full & Full).
Weight: About 8.5 lbs.
Stock: Checkered deluxe walnut; pistol grip; trap dimensions; semi-beavertail forend; hand-rubbed oil finish.
Features: Silvered, hand-engraved receiver; chrome moly steel barrels; GTX detachable single selective trigger system with coil springs, automatic safety, automatic-ejectors, red bead front sight, metal bead middle; recoil pad. Introduced 1997. Imported from Italy by K.B.I., Inc.
Price: $5,629.00
Price: Diamond GTX Trap-MC (choke tubes) $5,799.00

CHARLES DALY DIAMOND REGENT GTX DL HUNTER O/U
Gauge: 12, 20, 410, 3" chambers, 28, 2 3/4" chambers.
Barrel: 26", 28", 30" (choke tubes), 26"(Imp. Cyl. & Mod. in 28, 26" (Full & Full) in 410.
Weight: About 7 lbs.
Stock: Extra select fancy European walnut with 24" hand checkering, hand-rubbed oil finish.
Features: Boss-type action with internal side lumps. Deep cut hand-engraved scrollwork and game scene set in full sideplates. GTX detachable single selective trigger system with coil springs; chrome moly steel barrels; automatic safety; automatic ejectors, white bead front sight, metal bead center sight. Introduced 1997. Imported from Italy by K.B.I., Inc.
Price: 12 or 20 ....$22,299.00
Price: 28 ....$22,369.00
Price: 410 ....$22,419.00
Price: Diamond Regent GTX EDL Hunter (as above with engraved scroll and birds, 10 gold inlays), 12 or 20 ....$26,249.00
Price: As above, 28 ....$26,499.00
Price: As above, 410 ....$26,549.00

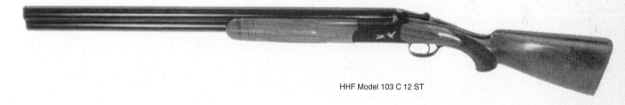

HHF Model 103 C 12 ST

HHF MODEL 103 F 12 ST OVER/UNDER
Gauge: 12, 20, 3" chambers.
Barrel: 28", choke tubes or fixed chokes.
Weight: About 7 1/2 lbs.
Stock: Circassian walnut.
Features: Boxlock action with dummy sideplates. Single selective trigger; manual safety; extractors. Can be ordered with many custom options. Has 100 percent engraving coverage, inlaid animals on blackened sideplates. Introduced 1995. Imported from Turkey by Turkish Firearms Corp.
Price: With extractors $1,120.00
Price: With automatic ejectors $1,750.00
Price: Model 103 C 12 ST (black receiver, 50 percent engraving coverage, extractors) $1,050.00
Price: As above, ejectors $1,680.00
Price: Model 103 D 12 ST (standard boxlock with 80 percent engraving coverage, extractors) $1,050.00
Price: As above, ejectors $1,680.00
Price: Model 103 B 12 ST (double triggers, extractors, 80 percent engraving coverage, fixed chokes) $995.00
Price: As above, 28, 410 ....$1,550.00
Price: With choke tubes, extractors (12, 20) $1,050.00

HHF MODEL 104 A 12 ST OVER/UNDER
Gauge: 12, 3" chambers.
Barrel: 28", fixed chokes or choke tubes.
Weight: About 7 1/2 lbs.
Stock: Circassian walnut, field dimensions.
Features: Boxlock action with manual safety, extractors, double triggers. Silvered,

> Consult our Directory pages for the location of firms mentioned.

HHF MODEL 101 B 12 ST TRAP O/U
Gauge: 12, 3" chambers.
Barrel: 30", fixed chokes or choke tubes; 16mm rib.
Weight: About 8 lbs.
Stock: Circassian walnut to trap dimensions; Monte Carlo comb, palm swell grip, recoil pad.
Features: Single selective trigger; manual safety; automatic ejectors or extractors. Many custom features available. Silvered frame with 50 percent envgraving coverage. Introduced 1995. Imported from Turkey by Turkish Firearms Corp.
Price: With extractors $1,050.00
Price: With ejectors $1,680.00
Price: Model 101 B 12 AT-DT (trap combo, 32" barrels) $2,295.00

engraved receiver. Has 15 percent engraving coverage. Introduced 1995. Imported from Turkey by Turkish Firearms Corp.
Price: Fixed chokes, extractors $925.00
Price: As above, 28, 410 ....$1,295.00
Price: Choke tubes, ejectors (12, 20) $925.00

SHOTGUNS—OVER/UNDERS

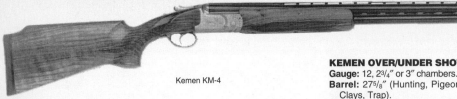
Kemen KM-4

KRIEGHOFF K-80 SKEET SHOTGUN
Gauge: 12, 2 3/4" chambers.
Barrel: 28" (Skeet & Skeet, optional Tula or choke tubes).
Weight: About 7 3/4 lbs.
Stock: American Skeet or straight Skeet stocks, with palm-swell grips. Walnut.
Features: Satin gray receiver finish. Selective mechanical trigger adjustable for position. Choice of ventilated 8mm parallel flat rib or ventilated 8-12mm tapered flat rib. Introduced 1980. Imported from Germany by Krieghoff International, Inc.
Price: Standard, Skeet chokes $6,900.00
Price: As above, Tula chokes $7,825.00
Price: Lightweight model (weighs 7 lbs.), Standard $6,900.00
Price: Two-Barrel Set (tube concept), 12-ga., Standard $11,840.00
Price: Skeet Special (28", tapered flat rib, Skeet & Skeet choke tubes) $7,575.00

KEMEN OVER/UNDER SHOTGUNS
Gauge: 12, 2 3/4" or 3" chambers.
Barrel: 27 5/8" (Hunting, Pigeon, Sporting Clays, Skeet), 30", 32" (Sporting Clays, Trap).
Weight: 7.25 to 8.5 lbs.
Stock: Dimensions to customer specs. High grade walnut.
Features: Drop-out trigger assembly; ventilated flat or step top rib, ventilated, solid or no side ribs. Low-profile receiver with black finish on Standard model, antique silver on sideplate models and all engraved, gold inlaid models. Barrels, forend, trigger parts interchangeable with Perazzi. Comes with hard case, accessory tools, spares. Introduced 1989. Imported from Spain by U.S.A. Sporting Clays.
Price: KM-4 Standard $6,179.00
Price: KM-4 Luxe-A (engraved scroll), Luxe-B (game scenes) ... $10,644.00
Price: KM-4 Super Luxe (engraved game scene) $12,064.00
Price: KM-4 Extra Luxe-A (scroll engraved sideplates) $13,960.00
Price: KM-4 Extra Luxe-B (game scene sideplates) $16,030.00
Price: KM-4 Extra Gold (inlays, game scene) $19,607.00

Krieghoff K-80 Trap

Krieghoff K-80 Four-Barrel Skeet Set
Similar to the Standard Skeet except comes with barrels for 12, 20, 28, 410. Comes with fitted aluminum case.
Price: Standard grade $16,950.00

Krieghoff K-80 International Skeet
Similar to the Standard Skeet except has 1/2" ventilated Broadway-style rib, special Tula chokes with gas release holes at muzzle. International Skeet stock. Comes in fitted aluminum case.
Price: Standard grade $7,825.00

LAURONA SUPER MODEL OVER/UNDERS
Gauge: 12, 20, 2 3/4" or 3" chambers.
Barrel: 26", 28" (Multichoke), 29" (Multichokes and Full).
Weight: About 7 lbs.
Stock: European walnut. Dimensions may vary according to model. Full pistol grip.
Features: Boxlock action, silvered with engraving. Automatic selective ejectors; choke tubes available on most models; single selective or twin single triggers; black chrome barrels. Has 5-year warranty, including metal finish. Imported from Spain by Galaxy Imports.
Price: Model 83 MG, 12- or 20-ga. $1,215.00
Price: Model 84S Super Trap (fixed chokes) $1,340.00
Price: Model 85 Super Game, 12- or 20-ga. $1,215.00
Price: Model 85 MS Super Trap (Full/Multichoke) $1,390.00
Price: Model 85 MS Super Pigeon $1,370.00
Price: Model 85 S Super Skeet, 12-ga. $1,300.00

Laurona Silhouette 300 Trap
Same gun as the Silhouette 300 Sporting Clays except has 29" barrels, trap stock dimensions of 14 3/8"x17/16"x1 5/8", weighs 7 lbs., 15 oz. Available with flush or knurled Multichokes.
Price: $1,310.00

KRIEGHOFF K-80 O/U TRAP SHOTGUN
Gauge: 12, 2 3/4" chambers.
Barrel: 30", 32" (Imp. Mod. & Full or choke tubes).
Weight: About 8 1/2 lbs.
Stock: Four stock dimensions or adjustable stock available; all have palm-swell grips. Checkered European walnut.
Features: Satin nickel receiver. Selective mechanical trigger, adjustable for position. Ventilated step rib. Introduced 1980. Imported from Germany by Krieghoff International, Inc.
Price: K-80 O/U (30", 32", Imp. Mod. & Full), from $7,375.00
Price: K-80 Unsingle (32", 34", Full), Standard, from $7,950.00
Price: K-80 Combo (two-barrel set), Standard, from $9,975.00

KRIEGHOFF K-80 SPORTING CLAYS O/U
Gauge: 12.
Barrel: 28", 30" or 32" with choke tubes.
Weight: About 8 lbs.
Stock: #3 Sporting stock designed for gun-down shooting.
Features: Choice of standard or lightweight receiver with satin nickel finish and classic scroll engraving. Selective mechanical trigger adjustable for position. Choice of tapered flat or 8mm parallel flat barrel rib. Free-floating barrels. Aluminum case. Imported from Germany by Krieghoff International, Inc.
Price: Standard grade with five choke tubes, from $8,150.00

LAURONA SILHOUETTE 300 SPORTING CLAYS
Gauge: 12, 2 3/4" or 3" chambers.
Barrel: 28", 29" (Multichoke tubes, flush-type or knurled).
Weight: 7 lbs., 12 oz.
Stock: 14 3/8"x1 3/8"x2 1/2". European walnut with full pistol grip, beavertail forend. Rubber buttpad.
Features: Selective single trigger, automatic selective ejectors. Introduced 1988. Imported from Spain by Galaxy Imports.
Price: $1,250.00
Price: Silhouette Ultra-Magnum, 3 1/2" chambers $1,265.00

Ljutic LM-6 Super Deluxe

LJUTIC LM-6 SUPER DELUXE O/U SHOTGUN
Gauge: 12.
Barrel: 28" to 34", choked to customer specs for live birds, trap, International Trap.
Weight: To customer specs.
Stock: To customer specs. Oil finish, hand checkered.
Features: Custom-made gun. Hollow-milled rib, pull or release trigger, pushbutton opener in front of trigger guard. From Ljutic Industries.
Price: Super Deluxe LM-6 O/U $19,995.00
Price: Over/under Combo (interchangeable single barrel, two trigger guards, one for single trigger, one for doubles) $26,995.00
Price: Extra over/under barrel sets, 29"-32" $5,995.00

SHOTGUNS—OVER/UNDERS

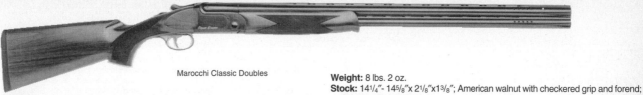
Marocchi Classic Doubles

MAROCCHI CLASSIC DOUBLES MODEL 92 SPORTING CLAYS O/U SHOTGUN
Gauge: 12, 3″ chambers.
Barrel: 30″; backbored, ported (ContreChoke Plus tubes); 10 mm concave ventilated top rib, ventilated middle rib.
Weight: 8 lbs. 2 oz.
Stock: 14 1/4″- 14 5/8″x 2 1/8″x1 3/8″; American walnut with checkered grip and forend; Sporting Clays buttpad.
Features: Low profile frame; fast lock time; automatic selective ejectors; blued receiver and barrels. Comes with three choke tubes. Ergonomically shaped trigger adjustable for pull length without tools. Barrels are backbored and ported. Introduced 1996. Imported from Italy by Precision Sales International.
Price: $1,598.00

Marocchi Lady Sport

Marocchi Lady Sport O/U Shotgun
Ergonomically designed specifically for women shooters. Similar to the Conquista Sporting Clays model except has 28″ or 30″ barrels with five Contrechoke tubes, stock dimensions of 13 7/8″-14 1/4″x1 11/32″x2 9/32″; weighs about 7 1/2 lbs. Also available as left-hand model—opening lever operates from left to right; stock has left-hand cast. Also available with colored graphics finish on frame and opening lever. Introduced 1995. Imported from Italy by Precision Sales International.
Price: Grade I, right-hand $2,120.00
Price: Left-hand, add (all grades) $101.00
Price: Colored graphics frame (Grade I only), add $79.00

Marocchi Conquista Trap Over/Under Shotgun
Similar to the Conquista Sporting Clays model except has 30″ or 32″ barrels choked Full & Full, stock dimensions of 14 1/2″-14 7/8″x1 11/16″x1 9/32″; weighs about 8 1/4 lbs. Introduced 1994. Imported from Italy by Precision Sales International.
Price: Grade I, right-hand $1,995.00
Price: Grade II, right-hand $2,330.00
Price: Grade III, right-hand, from $3,599.00

MAROCCHI CONQUISTA SPORTING CLAYS O/U SHOTGUNS
Gauge: 12, 2 3/4″ chambers.
Barrel: 28″, 30″, 32″ (Contrechoke tubes); 10mm concave vent. rib.
Weight: About 8 lbs.
Stock: 14 1/2″-14 7/8″x2 3/16″x1 7/16″; American walnut with checkered grip and forend; Sporting Clays butt pad.
Sights: 16mm luminescent front.
Features: Has lower monoblock and frame profile. Fast lock time. Ergonomically-shaped trigger is adjustable for pull length. Automatic selective ejectors. Coin-finished receiver, blued barrels. Comes with five choke tubes, hard case. Also available as true left-hand model—opening lever operates from left to right; stock has left-hand cast. Introduced 1994. Imported from Italy by Precision Sales International.
Price: Grade I, right-hand $1,995.00
Price: Grade I, left-hand $2,120.00
Price: Grade II, right-hand $2,330.00
Price: Grade II, left-hand $2,685.00
Price: Grade III, right-hand, from $3,599.00
Price: Grade III, left-hand, from $3,995.00

Marocchi Conquista Skeet Over/Under Shotgun
Similar to the Conquista Sporting Clays except has 28″ (Skeet & Skeet) barrels, stock dimensions of 14 3/8″-14 3/4″x2 3/16″x1 1/2″. Weighs about 7 3/4 lbs. Introduced 1994. Imported from Italy by Precision Sales International.
Price: Grade I, right-hand $1,995.00
Price: Grade II, right-hand $2,330.00
Price: Grade III, right-hand, from $3,599.00

Merkel Model 201E

MERKEL MODEL 201E O/U SHOTGUN
Gauge: 12, 20, 3″ chambers, 16, 28, 2 3/4″ chambers.
Barrel: 12, 16-ga.—28″; 20, 28-ga. —26 3/4″.
Weight: About 7 lbs. (12-ga.).
Stock: Oil-finished walnut; English or pistol grip.
Features: Self-cocking Blitz boxlock action with cocking indicators; Kersten double cross-bolt lock; silver-grayed receiver with engraved hunting scenes; coil spring ejectors; single selective or doulbe triggers. Imported form Germany by GSI, Inc.
Price: 12, 16, 20 .... $5,895.00
Price: 28-ga. $6,495.00
Price: Model 201E Skeet or Trap $8,495.00
Price: Model 200ET Trap $5,195.00

Merkel Model 202E O/U Shotgun
Similar to the Model 201E except has dummy sideplates, Arabesque engraving with hunting scenes; 12, 16, 20-gauge. Imported from Germany by GSI, Inc.
Price: $9,995.00

Merkel Model 203E, 303E O/U Shotgun
Similar to the Model 201E except has Holland & Holland-style sidelock action with cocking indicators (203E has quick-detachable sideplates with cranked screw, 303E has integral hook); English-style Arabesque engraving. Available in 12, 16, 20 gauge. Imported from Germany by GSI, Inc.
Price: Model 203E $11,995.00
Price: Model 303E $19,995.00
Price: Model 203E Skeet or Trap $14,595.00

Perazzi Sporting Classic

Perazzi Sporting Classic O/U
Same as the Mirage Special Sporting except is deluxe version with select wood and engraving, Available with flush mount choke tubes, 29.5″ barrels. Introduced 1993.
Price: From $10,510.00

PERAZZI MIRAGE SPECIAL SPORTING O/U
Gauge: 12, 2 3/4″ chambers.
Barrel: 28 3/8″ (Imp. Mod. & Extra Full), 29 1/2″ (choke tubes).
Weight: 7 lbs., 12 oz.
Stock: Special specifications.
Features: Has single selective trigger; flat 7/16″x 5/16″ vent. rib. Many options available. Imported from Italy by Perazzi U.S.A., Inc.
Price: $9,430.00

SHOTGUNS—OVER/UNDERS

Perazzi Mirage Special Four-Gauge Skeet
Similar to the Mirage Sporting model except has Skeet dimensions, interchangeable, adjustable four-position trigger assembly. Comes with four barrel sets in 12, 20, 28, 410, flat 5/16"x5/16" rib.
Price: From $20,720.00

Perazzi Mirage Special Skeet Over/Under
Similar to the MX8 Skeet except has adjustable four-position trigger, Skeet stock dimensions.
Price: From $8,830.00

Perazzi MX8/20 Over/Under Shotgun
Similar to the MX8 except has smaller frame and has a removable trigger mechanism. Available in trap, Skeet, sporting or game models with fixed chokes or choke tubes. Stock is made to customer specifications. Introduced 1993.
Price: From $8,330.00

PERAZZI MX10 OVER/UNDER SHOTGUN
Gauge: 12, 2¾" chambers.
Barrel: 29.5", 31.5" (fixed chokes).
Weight: NA.
Stock: Walnut; cheekpiece adjustable for elevation and cast.
Features: Comes with six pattern adjustment rib inserts. Vent. side rib. Externally selective trigger. Available in single barrel, combo, over/under trap, Skeet, pigeon and sporting models. Introduced 1993. Imported from Italy by Perazzi U.S.A., Inc.
Price: From $10,610.00

PERAZZI MX28, MX410 GAME O/U SHOTGUNS
Gauge: 28, 2¾" chambers, 410, 3" chambers.
Barrel: 26" (Imp. Cyl. & Full).
Weight: NA.

PERAZZI MX8/MX8 SPECIAL TRAP, SKEET
Gauge: 12, 2¾" chambers.
Barrel: Trap—29½" (Imp. Mod. & Extra Full), 31½" (Full & Extra Full). Choke tubes optional. Skeet—27⅝" (Skeet & Skeet).
Weight: About 8½ lbs. (Trap); 7 lbs., 15 oz. (Skeet).
Stock: Interchangeable and custom made to customer specs.
Features: Has detachable and interchangeable trigger group with flat V springs. Flat 7/16" ventilated rib. Many options available. Imported from Italy by Perazzi U.S.A., Inc.
Price: From $8,330.00
Price: MX8 Special (adj. four-position trigger), from $8,330.00
Price: MX8 Special Combo (o/u and single barrel sets), from $11,620.00

PERAZZI MX12 HUNTING OVER/UNDER
Gauge: 12, 2¾" chambers.
Barrel: 26", 27⅝", 28⅜", 29½" (Mod. & Full); choke tubes available in 27⅝", 29½" only (MX12C).
Weight: 7 lbs., 4 oz.
Stock: To customer specs; Interchangeable.
Features: Single selective trigger; coil springs used in action; schnabel forend tip. Imported from Italy by Perazzi U.S.A., Inc.
Price: From $8,330.00
Price: MX12C (with choke tubes), from $8,935.00

Perazzi MX20 Hunting Over/Under
Similar to the MX12 except 20-ga. frame size. Available in 20, 28, 410 with 2¾" or 3" chambers. 26" standard, and choked Mod. & Full. Weight is 6 lbs., 6 oz.
Price: From $8,330.00
Price: MX20C (as above, 20-ga. only, choke tubes), from $8,935.00

Stock: To customer specifications.
Features: Made on scaled-down frames proportioned to the gauge. Introduced 1993. Imported from Italy by Perazzi U.S.A., Inc.
Price: From $16,650.00

Piotti Boss

PIOTTI BOSS OVER/UNDER SHOTGUN
Gauge: 12, 20.
Barrel: 26" to 32", chokes as specified.
Weight: 6.5 to 8 lbs.
Stock: Dimensions to customer specs. Best quality figured walnut.
Features: Essentially a custom-made gun with many options. Introduced 1993. Imported from Italy by Wm. Larkin Moore.
Price: From $38,400.00

Remington Peerless

REMINGTON PEERLESS OVER/UNDER SHOTGUN
Gauge: 12, 3" chambers.
Barrel: 26", 28", 30" (Imp. Cyl., Mod., Full Rem Chokes).
Weight: 7¼ lbs. (26" barrels). **Length:** 43" overall (26" barrels).
Stock: 14 3/16"x1½"x2¼". American walnut with Imron gloss finish, cut-checkered grip and forend. Black, ventilated recoil pad.
Features: Boxlock action with removable sideplates. Gold-plated, single selective trigger, automatic safety, automatic ejectors. Fast lock time. Mid-rib bead, Bradley-type front. Polished blue finish with light scrollwork on sideplates, Remington logo on bottom of receiver. Introduced 1993.
Price: $1,225.00

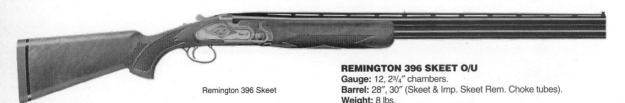
Remington 396 Skeet

REMINGTON 396 SKEET O/U
Gauge: 12, 2¾" chambers.
Barrel: 28", 30" (Skeet & Imp. Skeet Rem. Choke tubes).
Weight: 8 lbs.
Stock: 14 3/16"x1½"x2¼". Fancy, figured American walnut. Target-style forend, larger-radius comb, grip palm swell.
Features: Boxlock action with removable sideplates. Barrels have lengthened forcing cones; 10mm non-stepped, parallel rib; engraved receiver, sideplates, trigger guard, top lever, forend iron are finished with gray nitride. Made in U.S. by Remington. Introduced 1996.
Price: $1,859.00

Remington 396 Sporting O/U
Similar to the 396 Skeet except the 28", 30" barrels are factory ported, and come with Skeet, Imp. Skeet, Imp. Cyl. and Mod. Rem Choke tubes. Made in U.S. by Remington. Introduced 1996.
Price: $1,993.00

SHOTGUNS—OVER/UNDERS

Rizzini Artemis

Rizzini Artemis Over/Under Shotgun
Same as the Aurum model except has dummy sideplates with extensive game scene engraving. Fancy European walnut stock. Comes with fitted case. Introduced 1996. Imported from Italy by Wm. Larkin Moore & Co.
Price: From $2,375.00

RIZZINI AURUM OVER/UNDER SHOTGUN
Gauge: 12, 16, 20, 28, 410.
Barrel: 26", 27 1/2", Mod. & Full, Imp. Cyl. & Imp. Mod. choke tubes.
Weight: About 6.6 lbs.
Stock: 14"x1 1/2"x2 1/8".
Features: Boxlock action; single selective trigger; ejectors; profuse engraving on silvered receiver. Comes with fitted case. Introduced 1996. Imported from Italy by Wm. Larkin Moore & Co.
Price: From $2,125.00

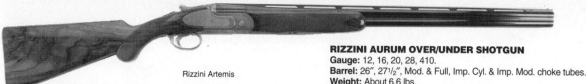
Rizzini S782 EMEL

RIZZINI S790 SPORTING EL OVER/UNDER
Gauge: 12, 2 3/4" chambers.
Barrel: 28", 29.5", Imp. Mod., Mod., Full choke tubes.
Weight: 8.1 lbs.
Stock: 14"x1 1/2"x2". Extra-fancy select walnut.
Features: Boxlock action; automatic ejectors; single selective trigger; 10mm top rib. Comes with case. Introduced 1996. Imported from Italy by Wm. Larkin Moore & Co.
Price: $6,190.00

Rizzini S792 EMEL Over/Under Shotgun
Similar to the S790 EMEL except has dummy sideplates with extensive engraving coverage. Comes with Nizzoli leather case. Introduced 1996. Imported from Italy by Wm. Larkin Moore & Co.
Price: From $10,000.00

ROTTWEIL PARAGON OVER/UNDER
Gauge: 12, 2 3/4" chambers.
Barrel: 28", 30", five choke tubes.
Weight: 7 lbs.
Stock: 14 1/2"x1 1/2"x2 1/2"; European walnut.

RIZZINI S782 EMEL OVER/UNDER SHOTGUN
Gauge: 12, 2 3/4" chambers.
Barrel: 26", 27.5" (Imp. Cyl. & Imp. Mod.).
Weight: About 6.75 lbs.
Stock: 14"x1 1/2"x2 1/8". Extra fancy select walnut.
Features: Boxlock action with dummy sideplates; extensive engraving with gold inlaid game birds; silvered receiver; automatic ejectors; single selective trigger. Comes with Nizzoli leather case. Introduced 1996. Imported from Italy by Wm. Larkin Moore & Co.
Price: From $12,000.00

RIZZINI S790 EMEL OVER/UNDER SHOTGUN
Gauge: 20, 28, 410.
Barrel: 26", 27.5" (Imp. Cyl. & Imp. Mod.).
Weight: About 6 lbs.
Stock: 14"x1 1/2"x2 1/8". Extra-fancy select walnut.
Features: Boxlock action with profuse engraving; automatic ejectors; single selective trigger; silvered receiver. Comes with Nizzoli leather case. Introduced 1996. Imported from Italy by Wm. Larkin Moore & Co.
Price: From $10,500.00

Features: Boxlock action. Detachable trigger assembly; ejectors can be deactivated; convertible top lever for right- or left-hand use; trigger adjustable for position. Imported from Germany by Dynamit Nobel-RWS, Inc.
Price: $5,995.00

Ruger English Field

RUGER RED LABEL O/U SHOTGUN
Gauge: 12, 20, 3" chambers; 28 2 3/4" chambers.
Barrel: 26", 28" (Skeet, Imp. Cyl., Full, Mod. screw-in choke tubes). Proved for steel shot.
Weight: About 7 lbs. (20-ga.); 7 1/2 lbs. (12-ga.). **Length:** 43" overall (26" barrels).
Stock: 14"x1 1/2"x2 1/2". Straight grain American walnut. Checkered pistol grip and forend, rubber butt pad.
Features: Choice of blue or stainless receiver. Single selective mechanical trigger, selective automatic ejectors; serrated free-floating vent. rib. Comes with two Skeet, one Imp. Cyl., one Mod., one Full choke tube and wrench. Made in U.S. by Sturm, Ruger & Co.
Price: Red Label with pistol grip stock $1,215.00
Price: English Field with straight-grip stock $1,215.00

Ruger Sporting Clays O/U Shotgun
Similar to the Red Label except 30" back-bored barrels, stainless steel choke tubes. Weighs 7.75 lbs., overall length 47". Stock dimensions of 14 1/8"x1 1/2"x2 1/2". Free-floating serrated vent. rib with brass front and mid-rib beads. No barrel side spacers. Comes with two Skeet, one Imp. Cyl., one Mod. choke tubes. Full and Extra-Full available at extra cost. 12 ga. introduced 1992, 20 ga. introduced 1994.
Price: 12 or 20 . . . . $1,349.00

Ruger Woodside

RUGER WOODSIDE OVER/UNDER SHOTGUN
Gauge: 12, 3" chambers.
Barrel: 26", 28" (Full, Mod., Imp. Cyl. and two Skeet tubes). 30" (Mod., Imp. Cyl. and two Skeet tubes).
Weight: 7 1/2 to 8 lbs.
Stock: 14 1/8"x1 1/2"x2 1/2". Select Circassian walnut; pistol grip or straight English grip.
Features: Has a newly patented Ruger cocking mechanism for easier, smoother opening. Buttstock extends forward into action as two side panels. Single selective mechanical trigger, selective automatic ejectors; serrated free-floating rib; back-bored barrels with stainless steel choke tubes. Blued barrels, stainless steel receiver. Engraved action available. Introduced 1995. Made in U.S. by Sturm, Ruger & Co.
Price: $1,675.00
Price: Woodside Sporting Clays (30" barrels) $1,675.00

SHOTGUNS—OVER/UNDERS

SARSILMAZ OVER/UNDER SHOTGUN
Gauge: 12, 3" chambers.
Barrel: 28" (Mod. & Full).
Weight: About 7½ lbs.
Stock: European walnut; checkered grip and forend.
Features: Silvered, engraved action, blued barrels. Imported from Turkey by Armsport.
Price: 26" or 28", double triggers $475.00
Price: 26" or 28", single selective trigger $550.00

Sigarms SA3

SIGARMS SA3 OVER/UNDER SHOTGUN
Gauge: 12, 3" chambers.
Barrel: 26", 27" (Full, Mod, Imp. Cyl. choke tubes).
Weight: 7 lbs.
Stock: Select grade walnut; checkered grip and forend.
Features: Chrome-lined bores; single selective trigger, automatic ejectors; satin nickel receiver finish, rest blued. Introduced 1997. Imported by SIGARMS, Inc.
Price: $1,325.00

SIGARMS SA5 OVER/UNDER SHOTGUN
Gauge: 12, 20, 3" chamber.
Barrel: 26½", 27" (Full, Imp. Mod., Mod., Imp. Cyl., Cyl. choke tubes).
Weight: 6.9 lbs. (12-gauge), 5.9 lbs. (20-gauge).
Stock: 14½" x 1½" x 2½". Select grade walnut; checkered 20 l.p.i. at grip and forend.
Features: Single selective trigger, automatic ejectors; hand-engraved detachable sideplated; matte nickel receiver, rest blued; tapered bolt lock-up. Introduced 1997. Imported by SIGARMS, inc.
Price: With fitted case $3,000.00

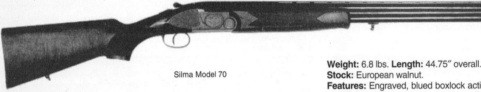

Silma Model 70

SILMA MODEL 70 OVER/UNDER SHOTGUN
Gauge: 12, 3" chambers.
Barrel: 27.5" (Mod. & Imp. Cyl.).
Weight: 6.8 lbs. **Length:** 44.75" overall.
Stock: European walnut.
Features: Engraved, blued boxlock action with single trigger; sling swivels. Introduced 1995. Imported from Italy by Century International Arms.
Price: About $540.00
Price: With selective trigger $599.00
Price: With selective trigger, auto ejectors, tubes $959.00

SKB Model 585

SKB Model 505 Shotguns
Similar to the Model 585 except blued receiver, standard bore diameter, standard Inter-Choke system on 12, 20, 28, diffrent receiver engraving. Imported from Japan by G.U. Inc.
Price: Field, 12 (26", 28"), 20 (26" only) $1,049.00
Price: Sporting Clays, 12 (28", 30") $1,149.00

SKB MODEL 585 OVER/UNDER SHOTGUN
Gauge: 12 or 20, 3"; 28, 2¾"; 410, 3".
Barrel: 12-ga.—26", 28", 30", 32", 34" (Inter-Choke tube); 20-ga.—26", 28" (Inter-Choke tube); 28—26", 28" (Inter-Choke tube); 410—26", 28" (Imp. Cyl. & Mod., Mod. & Full). Ventilated side ribs.
Weight: 6.6 to 8.5 lbs. **Length:** 43" to 51⅜" overall.
Stock: 14⅛"x1½"x2³⁄₁₆". Hand checkered walnut with high-gloss finish. Target stocks available in standard and Monte Carlo.
Sights: Metal bead front (field), target style on Skeet, trap, Sporting Clays.
Features: Boxlock action; silver nitride finish with Field or Target pattern engraving; manual safety, automatic ejectors, single selective trigger. All 12-gauge barrels are back-bored, have lengthened forcing cones and longer choke tube system. Sporting Clays models in 12-gauge with 28" or 30" barrels available with optional ⅜" step-up target-style rib, matte finish, nickel center bead, white front bead. Introduced 1992. Imported from Japan by G.U., Inc.
Price: Field $1,329.00
Price: Two-barrel Field Set, 12 & 20 .... $2,129.00
Price: Two-barrel Field Set, 20 & 28 or 28 & 410) $2,179.00
Price: Trap, Skeet $1,429.00
Price: Two-barrel trap combo $2,129.00
Price: Sporting Clays model .... $1,149.00 to $1,529.00
Price: Skeet Set (20, 28, 410) $3,329.00

SKB 785 Sporting Clays

SKB MODEL 785 OVER/UNDER SHOTGUN
Gauge: 12, 20, 3"; 28, 2¾"; 410, 3".
Barrel: 26", 28", 30", 32" (Inter-Choke tubes).
Weight: 6 lbs., 10 oz. to 8 lbs.
Stock: 14⅛"x1½"x2³⁄₁₆" (Field). Hand-checkered American black walnut with high-gloss finish; semi-beavertail forend. Target stocks available in standard or Monte Carlo styles.
Sights: Metal bead front (Field), target style on Skeet, trap, Sporting Clays models.
Features: Boxlock action with Greener-style cross bolt; single selective chrome-plated trigger, chrome-plated selective ejectors; manual safety. Chrome-plated, over-size, back-bored barrels with lengthened forcing cones. Introduced 1995. Imported from Japan by G.U. Inc.
Price: Field, 12 or 20 .... $1,949.00
Price: Field, 28 or 410 .... $2,029.00
Price: Field set, 12 and 20 .... $2,829.00
Price: Field set, 20 and 28 or 28 and 410 .... $2,929.00
Price: Sporting Clays, 12 or 20 .... $2,099.00
Price: Sporting Clays, 28 .... $2,169.00
Price: Sporting Clays set, 12 and 20 .... $2,999.00
Price: Skeet, 12 or 20 .... $2,029.00
Price: Skeet, 28 or 410 .... $2,069.00
Price: Skeet, three-barrel set, 20, 28, 410 .... $4,089.00
Price: Trap, standard or Monte Carlo $2,029.00
Price: Trap combo, standard or Monte Carlo $2,829.00

SHOTGUNS—OVER/UNDERS

Stoeger/IGA Deluxe Hunter Clays

Stoeger/IGA Deluxe Hunter Clays O/U
Similar to the Condor Supreme except 12-gauge only with 28" choke tube barrels, select semi-fancy American walnut stock with black Pachmayr target-style recoil pad, high luster blued barrels, gold-plated trigger, red bead front and mid-rib sights. Introduced 1997. Imported for Brazil by Stoeger.
Price: $699.00

STOEGER/IGA CONDOR I OVER/UNDER SHOTGUN
Gauge: 12, 20, 3" chambers.
Barrel: 26" (Imp. Cyl. & Mod. choke tubes), 28" (Mod. & Full choke tubes).
Weight: $6^3/_4$ to 7 lbs.
Stock: $14^1/_2$"x$1^1/_2$"x$2^1/_2$". Oil-finished hardwood with checkered pistol grip and forend.
Features: Manual safety, single trigger, extractors only, ventilated top rib. Introduced 1983. Imported from Brazil by Stoeger Industries.
Price: With choke tubes $559.00
Price: Condor II (sames as Condor I except has double triggers, moulded buttplate) $459.00
Price: Condor Supreme (same as Condor I with single trigger, choke tubes, but with auto. ejectors), 12- or 20-ga., 26", 28" $629.00

Stoeger/IGA Condor Waterfowl

Stoeger/IGA Condor Waterfowl O/U
Similar to the Condor I except has Advantage camouflage on the barrels, stock and forend; all other metal has matte black finish. Comes only with 28" choke tube barrels, 3" chambers, automatic ejectors, single trigger and manual safety. Designed for steel shot. Introduced 1997. Imported for Brazil by Stoeger.
Price: $729.00

Stoeger/IGA Turkey Model O/U
Similar to the Condor I model except has Advantage camouflage on the barrels stock and forend. All exposed metal and recoil pad are matte black. Has 26" (Full & Full) barrels, single trigger, manual safety, 3" chambers. Introduced 1997. Imported from Brazil by Stoeger.
Price: $729.00

TIKKA MODEL 512S FIELD GRADE OVER/UNDER
Gauge: 12, 20, 3" chambers.
Barrel: 26", 28", with stainless steel screw-in chokes (Imp. Cyl, Mod., Imp. Mod., Full); 20-ga. 28" only.
Weight: About $7^1/_4$ lbs.
Stock: American walnut. Standard dimensions—$13^9/_{10}$"x$1^1/_2$"x$2^2/_5$". Checkered pistol grip and forend.
Features: Free interchangeability of barrels, stocks and forends into double rifle model, combination gun, etc. Barrel selector in trigger; auto. top tang safety; barrel cocking indicators. Introduced 1980. Imported from Italy by Stoeger.
Price: Model 512S (ejectors), Field Grade $1,325.00
Price: Model 512S Sporting Clays, 12-ga., 28", choke tubes $1,360.00

Tristar Model 333

Tristar Model 330 Over/Under
Similar to the Model 333 except has standard grade walnut, etched engraving, fixed chokes, extractors only. Introduced 1996. Imported from Turkey by Tristar Sporting Arms, Ltd.
Price: $579.00
Price: Model 330D (as above except with three choke tubes, ejectors) . . $709.00

TRISTAR MODEL 333 OVER/UNDER
Gauge: 12, 20, 3" chambers.
Barrel: 12 ga.—26", 28", 30"; 20 ga.—26", 28"; five choke tubes.
Weight: $7^1/_2$-$7^3/_4$ lbs. Length: 45" overall.
Stock: Hand-checkered fancy grade Turkish walnut; full pistol grip, semi-beavertail forend; black recoil pad.
Features: Boxlock action with slivered, hand-engraved receiver; automatic selective ejectors, mechanical single selective trigger; stainless steel firing pins; auto safety; hard chrome bores. Introduced 1995. Imported from Turkey by Tristar Sporting Arms, Ltd.
Price: $799.95

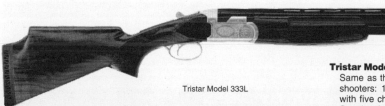
Tristar Model 333L

Tristar Model 333SC Over/Under
Same as the Model 333 except has 11mm rib with target sight beads, elongated forcing cones, ported barrels, stainless extended Sporting choke tubes (Skeet, Imp. Cyl., Imp. Cyl., Mod.), Sporting Clays recoil pad. Introduced 1996. Imported from Turkey by Tristar Sporting Arms, Ltd.
Price: $899.95

Tristar Model 333SCL Over/Under
Same as the Model 333SC except has special stock dimensions for female shooters: $13^1/_2$"x$1^1/_2$"x3"x$1/_4$". Introduced 1996. Imported from Turkey by Tristar Sporting Arms, Ltd.
Price: $899.95

Tristar Model 333L Over/Under
Same as the Model 333 except has special stock dimensions for female shooters: $13^1/_2$"x$1^1/_2$"x3"x$1/_4$". Available in 12-ga. with 26", 28" or 20 ga. 26", with five choke tubes. Introduced 1996. Imported from Turkey by Tristar Sporting Arms, Ltd.
Price: $799.95

> Consult our Directory pages for the location of firms mentioned.

Tristar Model 300 Over/Under
Similar to the Model 333 except has standard grade walnut, extractors, etched frame, double triggers, manual safety, plastic buttplate. Available in 12-ga. only with 26" (Imp. Cyl. & Mod.) or 28" (Mod. & Full) barrels. Introduced 1996. Imported from Turkey by Tristar Sporting Arms, Ltd.
Price: $429.95

SHOTGUNS—OVER/UNDERS

Weatherby Athena Grade V Classic

Weatherby Athena Grade V Classic Field O/U
Similar to the Athena Grade IV except has rounded pistol grip, slender forend, oil-finished Claro walnut stock with fine-line checkering, Old English recoil pad. Sideplate receiver has rose and scroll engraving. Available in 12-gauge, 26", 28", 20-gauge, 26", 28", all with 3" chambers. Introduced 1993.
Price: $2,599.00

WEATHERBY ORION O/U SHOTGUNS
Gauge: 12, 20, 3" chambers.
Barrel: 12-gauge—26", 28", 30"; 20-gauge— 26", 30"; IMC Multi-Choke tubes.
Weight: 6½ to 9 lbs.
Stock: American walnut, checkered grip and forend. Rubber recoil pad. Dimensions for Field and Skeet models, 14¼"x1½"x2½".
Features: Selective automatic ejectors, single selective mechanical trigger. Top tang safety, Greener cross bolt. Orion I has plain blued receiver, no engraving; Orion III has silver-gray receiver with engraving. Imported from Japan by Weatherby.
Price: Orion I, Field, 12, IMC, 26", 28", 30" $1,329.00
Price: Orion I, Field, 20, IMC, 26", 28" $1,329.00
Price: Orion III, Field, 12, IMC, 26", 28" $1,699.00
Price: Orion III, Field, 20, IMC, 26", 28" $1,699.00

WEATHERBY ATHENA GRADE IV O/U SHOTGUNS
Gauge: 12, 20, 3" chambers.
Action: Boxlock (simulated sidelock) top lever break-open. Selective auto ejectors, single selective trigger (selector inside trigger guard).
Barrel: 26", 28", IMC Multi-Choke tubes.
Weight: 12-ga., 7⅜ lbs.; 20-ga. 6⅞ lbs.
Stock: American walnut, checkered pistol grip and forend (14¼"x1½"x2½").
Features: Mechanically operated trigger. Top tang safety, Greener cross bolt, fully engraved receiver, recoil pad installed. IMC models furnished with three interchangeable flush-fitting choke tubes. Imported from Japan by Weatherby. Introduced 1982.
Price: 12-ga., IMC, 26", 28" $2,259.00
Price: 20-ga., IMC, 26", 28" $2,259.00

Weatherby Orion Grade II Classic Sporting O/U
Similar to the Orion II Classic Field except in 12 gauge only with 3" chambers, 28", 30" barrels with Skeet, SC1, SC2, Imp. Cyl., Mod. chokes. Weighs 7.5-8 lbs. Competition center vent rib; middle barrel and enlarge front beads. Rounded grip; high gloss stock. Radiused heel recoil pad. Receiver finished in silver nitride with acid-etched, gold plate clay pigeon monogram. Barrels have lengthened forcing cones. Introduced 1993. Imported by Weatherby.
Price: $1,499.00

Weatherby Orion Grade II Sporting Clays
Similar to the Orion II Classic Sporting has traditional pistol grip with diamond inlay, and standard full-size forend. Available in 12-gauge only, 28", 30" barrels with Skeet, Imp. Cyl., SC2, Mod. Has lengthened forcing cones, backboring, stepped competition rib, radius heel recoil pad, hand-engraved, silver/nitride receiver. Introduced 1992. Imported by Weatherby.
Price: $1,499.00

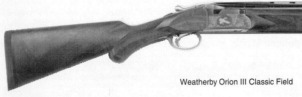
Weatherby Orion III Classic Field

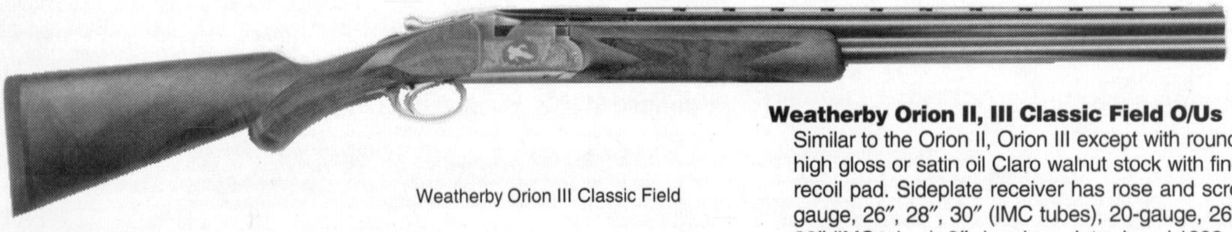

Weatherby Orion II Sporting Clays O/U
Similar to the Orion II Field except in 12-gauge only with 2¾" chambers, 28", 30" barrels with Imp. Cyl., Mod., Full chokes. High-gloss stock finish. Stock dimensions are 14¼"x1½"x2¼"; weighs 7.5 to 8 lbs. Matte finish, competition center vent. rib, mid-barrel and enlarged front beads. Rounded recoil pad. Receiver finished in silver nitride with acid-etched, gold-plate clay pigeon monogram. Barrels have lengthened forcing cones. Introduced 1992.
Price: $1,460.00

Weatherby Orion II, III Classic Field O/Us
Similar to the Orion II, Orion III except with rounded pistol grip, slender forend, high gloss or satin oil Claro walnut stock with fine-line checkering, Old English recoil pad. Sideplate receiver has rose and scroll engraving. Available in 12-gauge, 26", 28", 30" (IMC tubes), 20-gauge, 26", 28" (IMC tubes), 28-gauge, 26" (IMC tubes), 3" chambers. Introduced 1993.
Price: Orion II Classic Field $1,399.00
Price: Orion III Classic Field (12 and 20 only) $1,699.00

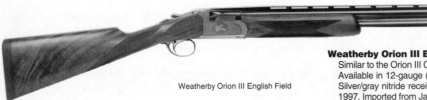
Weatherby Orion III English Field

Weatherby Orion III English Field O/U
Similar to the Orion III Classic Field except has straight grip English-style stock. Available in 12-gauge (28"), 20-gauge (26", 28") with IMC Multi-Choke tubes. Silver/gray nitride receiver is engraved and has gold-plate overlay. Introduced 1997. Imported from Japan by Weatherby.
Price: $1,699.00

SHOTGUNS—SIDE BY SIDES
Variety of models for utility and sporting use, including some competitive shooting.

American Arms Brittany

AMERICAN ARMS BRITTANY SHOTGUN
Gauge: 12, 20, 3" chambers.
Barrel: 12-ga.—27"; 20-ga.—25" (Imp. Cyl., Mod., Full choke tubes).
Weight: 6 lbs., 7 oz. (20-ga.).
Stock: 14⅛"x1⅜"x2⅜". Hand-checkered walnut with oil finish, straight English-style with semi-beavertail forend.
Features: Boxlock action with case-color finish, engraving; single selective trigger, automatic selective ejectors; rubber recoil pad. Introduced 1989. Imported from Spain by American Arms, Inc.
Price: $860.00

SHOTGUNS—SIDE BY SIDES

American Arms Gentry

AMERICAN ARMS TS/SS 12 DOUBLE
Gauge: 12, 3½" chambers.
Barrel: 26", choke tubes; solid raised rib.
Weight: 7 lbs., 6 oz.
Stock: Walnut; cut-checked grip and forend.
Features: Non-reflective metal and wood finishes; boxlock action; single trigger; extractors. Imported by American Arms, Inc.
Price: $785.00

ARIZAGA MODEL 31 DOUBLE SHOTGUN
Gauge: 12, 16, 20, 28, 410.
Barrel: 26", 28" (standard chokes).
Weight: 6 lbs., 9 oz. **Length:** 45" overall.
Stock: Straight English style or pistol grip.
Features: Boxlock action with double triggers; blued, engraved receiver. Imported by Mandall Shooting Supplies.
Price: $550.00

BERETTA MODEL 452 SIDELOCK SHOTGUN
Gauge: 12, 2¾" or 3" chambers.
Barrel: 26", 28", 30", choked to customer specs.
Weight: 6 lbs., 13 oz.
Stock: Dimensions to customer specs. Highly figured walnut; Model 452 EELL has walnut briar.
Features: Full sidelock action with English-type double bolting; automatic selective ejectors, manual safety; double triggers, single or single non-selective trigger on request. Essentially custom made to specifications. Model 452 is coin finished without engraving; 452 EELL is fully engraved. Imported from Italy by Beretta U.S.A.
Price: 452 ....$31,500.00
Price: 452 EELL $43,500.00

AMERICAN ARMS GENTRY DOUBLE SHOTGUN
Gauge: 12, 20, 410, 3" chambers; 28 ga. 2¾" chambers.
Barrel: 26" (Imp. Cyl. & Mod., all gauges), 28" (Mod. & Full, 12 and 20 gauges).
Weight: 6¼ to 6¾ lbs.
Stock: 14⅛"x1⅜"x2⅜". Hand-checkered walnut with semi-gloss finish.
Sights: Metal bead front.
Features: Boxlock action with English-style scroll engraving, color case-hardened finish. Double triggers, extractors. Independent floating firing pins. Manual safety. Five-year warranty. Introduced 1987. Imported from Spain by American Arms, Inc.
Price: 12 or 20 ....$735.00
Price: 28 or 410 ....$775.00

ARRIETA SIDELOCK DOUBLE SHOTGUNS
Gauge: 12, 16, 20, 28, 410.
Barrel: Length and chokes to customer specs.
Weight: To customer specs.
Stock: 14½"x1½"x2½" (standard dimensions), or to customer specs. Straight English with checkered butt (standard), or pistol grip. Select European walnut with oil finish.
Features: Essentially a custom gun with myriad options. Holland & Holland-pattern hand-detachable sidelocks, selective automatic ejectors, double triggers (hinged front) standard. Some have self-opening action. Finish and engraving to customer specs. Imported from Spain by Wingshooting Adventures.
Price: Model 557, auto ejectors, from $2,750.00
Price: Model 570, auto ejectors, from $3,380.00
Price: Model 578, auto ejectors, from $3,740.00
Price: Model 600 Imperial, self-opening, from $4,990.00
Price: Model 601 Imperial Tiro, self-opening, from $5,750.00
Price: Model 801, from $7,950.00
Price: Model 802, from $7,950.00
Price: Model 803, from $5,850.00
Price: Model 871, auto ejectors, from $4,290.00
Price: Model 872, self-opening, from $9,790.00
Price: Model 873, self-opening, from $6,850.00
Price: Model 874, self-opening, from $7,950.00
Price: Model 875, self-opening, from $12,950.00

Beretta Model 470 Silver Hawk

BERETTA MODEL 470 SILVER HAWK SHOTGUN
Gauge: 12, 20, 3" chambers.
Barrel: 26" (Imp. Cyl. & Imp. Mod.), 28" (Mod. & Full).
Weight: 5.9 lbs. (20-gauge).
Stock: Select European walnut, straight English grip.
Features: Boxlock action with single selective trigger; selector provides automatic ejection or extraction; silver-chrome action and forend iron with fine engraving; top lever highlighted with gold inlaid hawk's head. Comes with ABS case. Introduced 1997. Imported from Italy by Beretta U.S.A.
Price: ....NA

Charles Daly Field Hunter

CHARLES DALY DIAMOND DL DOUBLE SHOTGUN
Gauge: 12, 20, 410, 3" chambers, 28, 2¾" chambers.
Barrel: 28" (Mod. & Full), 26" (Imp. Cyl. & Mod.), 26" (Full & Full, 410).
Weight: About 5-7 lbs.
Stock: Select fancy European walnut, English-style butt, beavertail forend; hand-checkered, hand-rubbed oil finish.
Features: Drop-forged action with gas escape valves; demiblock barrels with concave rib; selective automatic ejectors; hand-detachable double safety sidelocks with hand-engraved rose and scrollwork. Hinged front trigger. Color case-hardened receiver. Introduced 1997. Imported from Spain by K.B.I., Inc.
Price: 12 or 20 ....$6,799.00
Price: 28 ....$7,229.00
Price: 410 ....$7,229.00

Charles Daly Empire Hunter Double Shotgun
Similar to the Superior Hunter except has deluxe wood, game scene engraving, automatic ejectors. Introduced 1997. Imported from Italy by K.B.I., Inc.
Price: 12 or 20 ....$1,339.00

CHARLES DALY FIELD HUNTER DOUBLE SHOTGUN
Gauge: 10, 12, 20, 28, 410 (3" chambers; 28 has 2¾").
Barrel: 32" (Mod. & Mod.), 28, 30" (Mod. & Full), 26" (Imp. Cyl. & Mod.) 410 (Full & Full).
Weight: 6 lbs. to 11.4 lbs.
Stock: Checkered walnut pistol grip and forend.
Features: Silvered, engraved receiver; gold single selective trigger in 10–, 12–, and 20–ga.; double triggers in 28 and 410; automatic safety; extractors; gold bead front sight. Introduced 1997. Imported from Spain by K.B.I., Inc.
Price: 10-ga. $979.00
Price: 12-ga. or 20-ga. $799.00
Price: 28-ga. $849.00
Price: 410-bore $849.00
Price: Field Hunter-MC (choke tubes, 10 ga.) $1,089.00
Price: As above, 12 or 20 ....$929.00

CHARLES DALY SUPERIOR HUNTER DOUBLE SHOTGUN
Gauge: 12, 20, 3" chambers.
Barrel: 28" (Mod. & Full) 26" (Imp. Cyl. & Mod.).
Weight: About 7 lbs.
Stock: Checkered walnut pistol grip buttstock, splinter forend.
Features: Silvered, engraved receiver; chrome-lined barrels; gold single trigger; automatic safety; extractors; gold bead front sight. Introduced 1997. Imported from Italy by K.B.I., Inc.
Price: $989.00

SHOTGUNS—SIDE BY SIDES

A.H. Fox DE Grade

CHARLES DALY DIAMOND REGENT DL DOUBLE SHOTGUN
Gauge: 12, 20, 410 3" chambers, 28, 2 3/4" chambers.
Barrel: 28" (Mod. & Full), 26" (Imp. Cyl. & Mod.), 26" (Full & Full, 410).
Weight: About 5-7 lbs.
Stock: Special select fancy European walnut, English-style butt, splinter forend; hand-checkered; hand-rubbed oil finish.
Features: Drop-forged action with gas escape valves; demiblock barrels of chrome-nickel steel with concave rib; selective automatic-ejectors; hand-detachable, double-safety H&H sidelocks with demi-relief hand engraving; H&H pattern easy-opening feature; hinged trigger; coin finished action. Introduced 1997. Imported from Spain by K.B.I., Inc.
Price: 12 or 20 .... $21,659.00
Price: 28 .... $22,081.00
Price: 410 .... $22,081.00

A.H. FOX SIDE-BY-SIDE SHOTGUNS
Gauge: 16, 20, 28, 410.
Barrel: Length and chokes to customer specifications. Rust-blued Chromox or Krupp steel.
Weight: 5 1/2 to 6 3/4 lbs.
Stock: Dimensions to customer specifications. Hand-checkered Turkish Circassian walnut with hand-rubbed oil finish. Straight, semi- or full pistol grip; splinter, schnabel or beavertail forend; traditional pad, hard rubber buttplate or skeleton butt.
Features: Boxlock action with automatic ejectors; double or Fox single selective trigger. Scalloped, rebated and color case-hardened receiver; hand finished and hand-engraved. Grades differ in engraving, inlays, grade of wood, amount of hand finishing. Add $1,000 for 28 or 410-bore. Introduced 1993. Made in U.S. by Connecticut Shotgun Mfg.
Price: CE Grade $9,500.00
Price: XE Grade $11,000.00
Price: DE Grade $13,500.00
Price: FE Grade $18,500.00
Price: Exhibition Grade $26,000.000
Price: 28/410 CE Grade $8,200.00
Price: 28/410 XE Grade $9,700.00
Price: 28/410 DE Grade $13,800.00
Price: 28/410 FE Grade $14,700.00
Price: 28/410 Exhibition Grade $25,000.00

Garbi Model 100

Garbi Model 101 Side-by-Side
Similar to the Garbi Model 100 except is hand engraved with scroll engraving, select walnut stock. Better overall quality than the Model 100. Imported from Spain by Wm. Larkin Moore.
Price: From $6,000.00

Garbi Model 200 Side-by-Side
Similar to the Garbi Model 100 except has heavy-duty locks, magnum proofed. Very fine Continental-style floral and scroll engraving, well figured walnut stock. Other mechanical features remain the same. Imported from Spain by Wm. Larkin Moore.
Price: $9,500.00

BILL HANUS CLASSIC 600 SERIES SHOTGUN
Gauge: 20, 3" chambers.
Barrel: 27".
Weight: 5 lbs., 10 oz. **Length:** 43 3/4" overall.
Stock: 14 7/16" x 1 9/16" x 2 5/16". Select walnut.
Features: Boxlock action with ejectors; tapered solid rib. Imported by Bill Hanus Birdguns.
Price: About $1,895.00

HHF MODEL 200 A 12 ST SIDE-BY-SIDE
Gauge: 12, 3" chambers.
Barrel: 28", fixed chokes or choke tubes.
Weight: About 7 1/2 lbs.
Stock: Circassian walnut, field dimensions.
Features: Boxlock action with single selective trigger, extractors, manual safety. Silvered receiver with 15 percent engraving coverage. Many options available. Introduced 1995. Imported from Turkey by Turkish Firearms Corp.
Price: Fixed chokes, extractors $1,050.00
Price: As above, 28, 410 .... $1,495.00
Price: Choke tubes, extractors $1,050.00
Price: Model 202 A 12 ST (double triggers, 30 percent engraving coverage $1,025.00
Price: As above, 28, 410 .... $1,495.00
Price: With extractors, choke tubes, 12, 20 .... $1,025.00

GARBI MODEL 100 DOUBLE
Gauge: 12, 16, 20, 28.
Barrel: 26", 28", choked to customer specs.
Weight: 5 1/2 to 7 1/2 lbs.
Stock: 14 1/2" x 2 1/4" x 1 1/2". European walnut. Straight grip, checkered butt, classic forend.
Features: Sidelock action, automatic ejectors, double triggers standard. Color case-hardened action, coin finish optional. Single trigger; beavertail forend; etc. optional. Five other models are available. Imported from Spain by Wm. Larkin Moore.
Price: From $4,700.00

Garbi Model 103A, B Side-by-Side
Similar to the Garbi Model 100 except has Purdey-type fine scroll and rosette engraving. Better overall quality than the Model 101. Model 103B has nickel-chrome steel barrels, H&H-type easy opening mechanism; other mechanical details remain the same. Imported from Spain by Wm. Larkin Moore.
Price: Model 103A, from $7,250.00
Price: Model 103B, from $9,950.00

> Consult our Directory pages for the location of firms mentioned.

CRUCELEGUI HERMANOS MODEL 150 DOUBLE
Gauge: 12, 16 or 20, 2 3/4" chambers.
Action: Greener triple cross bolt.
Barrel: 20", 26", 28", 30", 32" (Cyl. & Cyl., Full & Full, Mod. & Full, Mod. & Imp. Cyl., Imp. Cyl. & Full, Mod. & Mod.).
Weight: 5 to 7 1/4 lbs.
Stock: Hand-checkered walnut, beavertail forend.
Features: Double triggers; color case-hardened receiver; sling swivels; chrome-lined bores. Imported from Spain by Mandall Shooting Supplies.
Price: $450.00

IAR COWBOY SHOTGUNS
Gauge: 12.
Barrel: 20", 28".
Weight: 7 lbs. (20" barrel). **Length:** 36 7/8" overall (20" barrel).
Stock: Walnut.
Features: Exposed hammers; blued or brown barrels; double triggers. Introduced 1997. Imported from Italy by IAR, Inc.
Price: Gentry model, 20" or 28", engraved, bright-finished locks, blue barrels $1,895.00
Price: Cowboy model, 20" or 28", no engraving on color case-hardened locks, brown patina barrels $1,895.00

SHOTGUNS—SIDE BY SIDES

Merkel Model 47E

Merkel Model 47S, 147S Side-by-Sides
Similar to the Model 122 except with Holland & Holland-style sidelock action with cocking indicators, ejectors. Silver-grayed receiver and sideplates have Arabesque engraving, engraved border and screws (Model 47S), or fine hunting scene engraving (Model 147S). Imported from Germany by GSI.
Price: Model 47S $5,295.00
Price: Model 147S $6,695.00
Price: Model 247S (English-style engraving, large scrolls) $6,995.00
Price: Model 347S (English-style engraving, medium scrolls) $7,895.00
Price: Model 447S (English-style engraving, small scrolls) $8,995.00

PARKER REPRODUCTIONS SIDE-BY-SIDE SHOTGUN
Gauge: 12, 16/20 combo, 20, 28, 2¾" and 3" chambers.
Barrel: 26" (Skeet 1 & 2, Imp. Cyl. & Mod.), 28" (Mod. & Full, 2¾" and 3", 12, 20, 28; Skeet 1 & 2, Imp. Cyl. & Mod., Mod. & Full 16-ga. only).
Weight: 6¾ lbs. (12-ga.)
Stock: Checkered (26 lpi) AAA fancy California English or Claro walnut, skeleton steel and checkered butt. Straight or pistol grip, splinter or beavertail forend.
Features: Exact reproduction of the original Parker—parts interchange. Double or single selective trigger, selective ejectors, hard-chromed bores,

MERKEL MODEL, 47E SIDE-BY-SIDE SHOTGUNS
Gauge: 12, 3" chambers, 16, 2¾" chambers, 20, 3" chambers.
Barrel: 12-, 16-ga.—28"; 20-ga.—26¾" (Imp. Cyl. & Mod., Mod. & Full).
Weight: About 6¾ lbs. (12-ga.).
Stock: Oil-finished walnut; straight English or pistol grip.
Features: Anson & Deeley-type boxlock action with single selective or double triggers, automatic safety, cocking indicators. Color case-hardened receiver with standard Arabesque engraving. Imported from Germany by GSI.
Price: Model 47E (H&H ejectors) $2,695.00
Price: Model 147 (extractors, silver-grayed receiver with hunting scenes) $2,995.00
Price: Model 147E (as above with ejectors) $3,195.00
Price: Model 122 (as above with false sideplates, fine engraving) $4,995.00

designed for steel shot. One, two or three (16-20, 20) barrel sets available. Hand-engraved snap caps included. Introduced 1984. Made by Winchester. Imported from Japan by Parker Division, Reagent Chemical.
Price: D Grade, one-barrel set $3,370.00
Price: Two-barrel set, same gauge $4,200.00
Price: Two-barrel set, 16/20 . . . . $4,870.00
Price: Three-barrel set, 16/20/20 . . . . $5,630.00
Price: A-1 Special two-barrel set $11,200.00
Price: A-1 Special three-barrel set $13,200.00

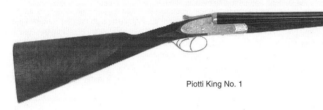
Piotti King No. 1

PIOTTI KING NO. 1 SIDE-BY-SIDE
Gauge: 12, 16, 20, 28, 410.
Barrel: 25" to 30" (12-ga.), 25" to 28" (16, 20, 28, 410). To customer specs. Chokes as specified.
Weight: 6½ lbs. to 8 lbs. (12-ga. to customer specs.).
Stock: Dimensions to customer specs. Finely figured walnut; straight grip with checkered butt with classic splinter forend and hand-rubbed oil finish standard. Pistol grip, beavertail forend, satin luster finish optional.
Features: Holland & Holland pattern sidelock action, automatic ejectors. Double trigger with front trigger hinged standard; non-selective single trigger optional. Coin finish standard; color case-hardened optional. Top rib; level, file-cut standard; concave, ventilated optional. Very fine, full coverage scroll engraving with small floral bouquets, gold crown in top lever, name in gold, and gold crest in forend. Imported from Italy by Wm. Larkin Moore.
Price: From $22,400.00

Piotti Lunik Side-by-Side
Similar to the Piotti King No. 1 except better overall quality. Has Renaissance-style large scroll engraving in relief, gold crown in top lever, gold name and gold crest in forend. Best quality Holland & Holland-pattern sidelock ejector double with chopper lump (demi-bloc) barrels. Other mechanical specifications remain the same. Imported from Italy by Wm. Larkin Moore.
Price: From $22,400.00

Piotti King Extra Side-by-Side
Similar to the Piotti King No. 1 except highest quality wood and metal work. Choice of either bulino game scene engraving or game scene engraving with gold inlays. Engraved and signed by a master engraver. Exhibition grade wood. Other mechanical specifications remain the same. Imported from Italy by Wm. Larkin Moore.
Price: From $26,900.00

RIZZINI SIDELOCK SIDE-BY-SIDE
Gauge: 12, 16, 20, 28, 410.
Barrel: 25" to 30" (12-, 16-, 20-ga.), 25" to 28" (28, 410). To customer specs. Chokes as specified.
Weight: 6½ lbs. to 8 lbs. (12-ga. to customer specs).
Stock: Dimensions to customer specs. Finely figured walnut; straight grip with checkered butt with classic splinter forend and hand-rubbed oil finish standard.

PIOTTI PIUMA SIDE-BY-SIDE
Gauge: 12, 16, 20, 28, 410.
Barrel: 25" to 30" (12-ga.), 25" to 28" (16, 20, 28, 410).
Weight: 5½ to 6¼ lbs. (20-ga.).
Stock: Dimensions to customer specs. Straight grip stock with walnut checkered butt, classic splinter forend, hand-rubbed oil finish are standard; pistol grip, beavertail forend, satin luster finish optional.
Features: Anson & Deeley boxlock ejector double with chopper lump barrels. Level, file-cut rib, light scroll and rosette engraving, scalloped frame. Double triggers with hinged front standard, single non-selective optional. Coin finish standard, color case-hardened optional. Imported from Italy by Wm. Larkin Moore.
Price: From $12,900.00

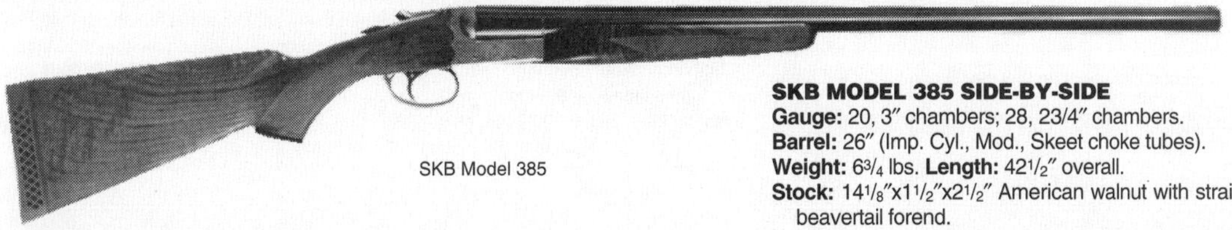

Pistol grip, beavertail forend, satin luster finish optional.
Features: Holland & Holland pattern sidelock action, auto ejectors. Double triggers with front trigger hinged optional; non-selective single trigger standard. Coin finish standard. Top rib level, file cut standard; concave optional. Imported from Italy by Wm. Larkin Moore.
Price: 12-, 20-ga., from $43,700.00
Price: 28, 410 bore, from $48,700.00

SKB Model 385

SKB Model 485 Side-by-Side
Similar to the Model 385 except has dummy sideplates, extensive upland game scene engraving, semi-fancy American walnut English or pistol grip stock. Imported from Japan by G.U. Inc.
Price: $2,369.00

SKB MODEL 385 SIDE-BY-SIDE
Gauge: 20, 3" chambers; 28, 2¾" chambers.
Barrel: 26" (Imp. Cyl., Mod., Skeet choke tubes).
Weight: 6¾ lbs. **Length:** 42½" overall.
Stock: 14⅛"x1½"x2½" American walnut with straight or pistol grip stock, semi-beavertail forend.
Features: Boxlock action. Silver nitrided receiver with engraving; solid barrel rib; single selective trigger, selective automatic ejectors, automatic safety. Introduced 1996. Imported from Japan by G.U. Inc.
Price: $1,769.00
Price: Field Set, 20-, 28-ga., 26", English or pistol grip $2,499.00

SHOTGUNS—SIDE BY SIDES

Stoeger/IGA Deluxe Uplander

Stoeger/IGA Deluxe Uplander Shotgun
Similar to the Uplander except with semi-fancy American walnut with thin black Pachmayr rubber recoil pad, matte lacquer finish. Choke tubes and 3" chambers standard on 12– and 20– gauge; 28-gauge has 26", 3" chokes, fixed Mod. & Full. Double gold-plated triggers; extractors. Introduced 1997. Imported from Brazil by Stoeger.
Price: 12, 20 ....$559.00
Price: 28, 410 ....$519.00

STOEGER/IGA UPLANDER SIDE-BY-SIDE SHOTGUN
Gauge: 12, 20, 28, 2¾" chambers; 410, 3" chambers.
Barrel: 26" (Full & Full, 410 only, Imp. Cyl. & Mod.), 28" (Mod. & Full).
Weight: 6¾ to 7 lbs.
Stock: 14½"x1½"x2½". Oil-finished hardwood. Checkered pistol grip and forend.
Features: Automatic safety, extractors only, solid matted barrel rib. Double triggers only. Introduced 1983. Imported from Brazil by Stoeger Industries.
Price: $414.00
Price: With choke tubes $454.00
Price: Coach Gun, 12-, 20, 410, 20" bbls. $399.00
Price: Coach Gun, nickel finish, black stock $444.00
Price: Coach Gun, engraved stock $459.00

Stoeger/IGA Turkey

Stoeger/IGA English Stock Side-by-Side
Similar to the Uplander except in 410-bore only with 24" barrels (Mod. & Mod.), straight English stock and beavertail forend. Has automatic safety, extractors, double triggers. Intro 1996. Imported from Brazil by Stoeger.
Price: $414.00

Stoeger/IGA Ladies Side-by-Side
Similar to the Uplander except in 20-ga. only with 24" barrels (Imp. Cyl. & Mod. choke tubes), 13" length of pull, ventilated rubber recoil pad. Has extractors, double triggers, automatic safety. Introduced 1996. Imported from Brazil by Stoeger.
Price: $464.00

Stoeger/IGA Deluxe Coach Gun
Similar to the Uplander except 12 or 20-gauge, 20" barrels, choked Imp. Cyl. & Mod., 3" chambers; select semi-fancy American walnut pistol grip stock with checkering; double triggers; extractors. Introduced 1997. Imported form Brazil by Stoeger.
Price: $499.00

Stoeger/IGA Turkey Side-by-Side
Similar to the Uplander Model except has Advantage camouflage on stock, forend and barrels; 12-gauge only with 3" chambers, and has 24" choke tube barrels. Overall length 40". Introduced 1997. Imported from Brazil by Stoeger.
Price: $559.00

Stoeger/IGA Youth Side-by-Side
Similar to the Uplander except in 410-bore with 24" barrels (Mod. & Full), 13" length of pull, ventilated recoil pad. Has double triggers, extractors, auto safety. Intro 1996. Imported from Brazil by Stoeger.
Price: $424.00

Tristar Model 311R

TRISTAR MODEL 311 DOUBLE
Gauge: 12, 20, 3" chambers.
Barrel: 26", 28", five choke tubes.
Weight: About 7 lbs.
Stock: 14⅜"x1⅜"x2⅜"x⅜"; hand-checkered Turkish walnut; recoil pad.
Features: Boxlock action; underlug and Greener bolt lockup; extractors, manual safety, double triggers. Black chrome finish. Introduced 1996. Imported from Turkey by Tristar Sporting Arms, Ltd.
Price: $599.00
Price: Model 311R (20" Cyl. & Cyl. barrels) $429.00

SHOTGUNS—BOLT ACTIONS & SINGLE SHOTS
Variety of designs for utility and sporting purposes, as well as for competitive shooting.

Browning A-Bolt Stalker

BROWNING A-BOLT SHOTGUN
Gauge: 12, 3" chamber, 2-shot detachable magazine.
Barrel: 22" (fully rifled), 23" (5" Invector choke tubes).
Weight: 7 lbs., 2 oz. Length: 44¾" overall.
Stock: 14"x⅝"x½". Walnut with satin finish on Hunter; Stalker has black graphite fiberglass composite. Swivel studs.
Sights: Blade front with red insert, open adjustable rear or none. Drilled and tapped for scope mounting.
Features: Uses same bolt system as A-Bolt rifle with 60° bolt throw; front-locking bolt with claw extractor; hinged floorplate. Matte finish on barrel and receiver. Introduced 1995. Imported by Browning.
Price: Hunter, rifled choke tube, open sights $828.95
Price: As above, no sights $804.95
Price: Hunter, rifled barrel, open sights $881.95
Price: As above, no sights $856.95
Price: Stalker, rifled, choke tube, open sights $744.95
Price: As above, no sights $719.95
Price: Stalker, rifled barrel, no sights $772.95

Manufacturers' addresses in the
Directory of the Arms Trade
page 309, this issue

SHOTGUNS—BOLT ACTIONS & SINGLE SHOTS

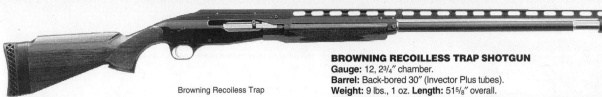

Browning Recoiless Trap

BROWNING RECOILLESS TRAP SHOTGUN
Gauge: 12, 2 3/4" chamber.
Barrel: Back-bored 30" (Invector Plus tubes).
Weight: 9 lbs., 1 oz. **Length:** 51 5/8" overall.
Stock: 14"-14 3/4"x1 3/8"-1 3/4"x1 1/8"-1 3/4". Select walnut with high gloss finish, cut checkering.
Features: Eliminates up to 72 percent of recoil. Mass of the inner mechansim (barrel, receiver and inner bolt) is driven forward when trigger is pulled, cancelling most recoil. Forend is used to cock action when the action is forward. Ventilated rib adjusts to move point of impact; drop at comb and length of pull adjustable. Introduced 1993. Imported by Browning.
Price: $1,995.00

Browning Micro Recoilless Trap Shotgun
Same as the standard Recoilless Trap except has 27" barrel, weighs 8 lbs., 10 oz., and stock length of pull adjustable from 13" to 13 3/4", Overall length 47 5/8". Introduced 1993. Imported by Browning.
Price: $1,995.00

Browning BT-100 Trap

BROWNING BT-100 TRAP SHOTGUN
Gauge: 12, 2 3/4" chamber.
Barrel: 32", 34" (Invector Plus); back-bored; also with fixed Full choke.
Weight: 8 lbs., 9 oz. **Length:** 48 1/2" overall (32" barrel).
Stock: 14 3/8"x1 9/16"x1 7/16"x2" (Monte Carlo); 14 3/8"x1 3/4"x1 1/4"x2 1/8" (thumbhole). Walnut with high gloss finish; cut checkering. Wedge-shaped forend with finger groove.

Features: Available in stainless steel or blue. Has drop-out trigger adjustable for weight of pull from 3 1/2 to 5 1/2 lbs., and for three length postions; Ejector-Selector allows ejection or extraction of shells. Available with adjustable comb stock and thumbhole style. Introduced 1995. Imported from Japan by Browning.
Price: Grade I, blue, Monte Carlo, Invector Plus $1,995.00
Price: As above, fixed Full choke $1,948.00
Price: Stainless steel, Monte Carlo, Invector Plus $2,415.00
Price: As above, fixed Full choke $2,368.00
Price: Thumbhole stock, blue, Invector Plus $2,270.00
Price: Thumbhole stock, stainless, Invector Plus $2,690.00
Price: Adjustable comb stock, add $210.00
Price: Replacement trigger assembly $525.00

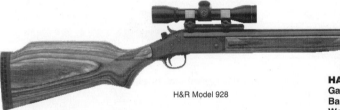

H&R Model 928

Harrington & Richardson Model 928 Ultra Slug Hunter Deluxe
Similar to the SB2-980 Ultra Slug except uses 12-gauge action and 12-gauge barrel blank bored to 20-gauge, then fully rifled with 1:35" twist. Has hand-checkered camo laminate Monte Carlo stock and forend. Comes with Weaver-style scope base, offset hammer extension, ventilated recoil pad, sling swivels and camo nylon sling. Introduced 1997. Made in U.S. by H&R 1871 Inc.
Price: $239.95

HARRINGTON & RICHARDSON SB2-980 ULTRA SLUG
Gauge: 12, 20, 3" chamber.
Barrel: 22" (20 ga. Youth) 24", fully rifled.
Weight: 9 lbs. **Length:** NA.
Stock: Walnut-stained hardwood.
Sights: None furnished; comes with scope mount.
Features: Uses the H&R 10-gauge action with heavy-wall barrel. Monte Carlo stock has sling swivels; comes with black nylon sling. Introduced 1995. Made in U.S. by H&R 1871, Inc.
Price: $209.95

H&R Topper 098

Harrington & Richardson Topper Deluxe Model 098
Similar to the standard Topper 098 except 12-gauge only with 3 1/2" chamber, 28" barrel with choke tube (comes with Mod. tube, others optional). Satin nickel frame, blued barrel, black-finished wood. Introduced 1992. From H&R 1871, Inc.
Price: $134.95

HARRINGTON & RICHARDSON TOPPER MODEL 098
Gauge: 12, 16, 20, 28 (2 3/4"), 410, 3" chamber.
Barrel: 12 ga.—28" (Mod., Full); 16 ga.— 28" (Mod.); 20 ga.—26" (Mod.); 28 ga.—26" (Mod.); 410 bore—26" (Full).
Weight: 5-6 lbs.
Stock: Black-finish hardwood with full pistol grip; semi-beavertail forend.
Sights: Gold bead front.
Features: Break-open action with side-lever release, automatic ejector. Satin nickel frame, blued barrel. Reintroduced 1992. From H&R 1871, Inc.
Price: $114.95
Price: Topper Junior 098 (as above except 22" barrel, 20-ga. (Mod.), 410-bore (Full), 12 1/2" length of pull) $119.95

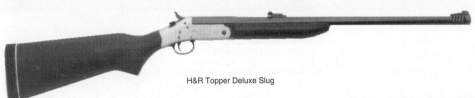

H&R Topper Deluxe Slug

Harrington & Richardson Topper Deluxe Rifled Slug Gun
Similar to the 12-gauge Topper Model 098 except has fully rifled and ported barrel, ramp front sight and fully adjustable rear. Barrel twist is 1:35". Nickel-plated frame, blued barrel, black-finished stock and forend. Introduced 1995. Made in U.S. by H&R 1871, Inc.
Price: $169.95

Harrington & Richardson Topper Junior Classic Shotgun
Similar to the Topper Junior 098 except available in 20-gauge (3", Mod.), 410-bore (Full) with 3" chamber; 28-gauge, 2 3/4" chamber (Mod.); all have 22" barrel. Stock is American black walnut with cut-checkered pistol grip and forend. Ventilated rubber recoil pad with white line spacers. Blued barrel, blued frame. Introduced 1992. From H&R 1871, Inc.
Price: $144.95

SHOTGUNS—BOLT ACTIONS & SINGLE SHOTS

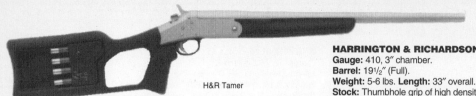
H&R Tamer

HARRINGTON & RICHARDSON TAMER SHOTGUN
Gauge: 410, 3" chamber.
Barrel: 19½" (Full).
Weight: 5-6 lbs. **Length:** 33" overall.
Stock: Thumbhole grip of high density black polymer.
Features: Uses H&R Topper action with matte electroless nickel finish. Stock holds four spare shotshells. Introduced 1994. From H&R 1871, Inc.
Price: $124.95

Krieghoff KS-5 Trap

Krieghoff KS-5 Special
Same as the KS-5 except the barrel has a fully adjustable rib and adjustable stock. Rib allows shooter to adjust point of impact from 50%/50% to nearly 90%/10%. Introduced 1990.
Price: $4,695.00

KRIEGHOFF K-80 SINGLE BARREL TRAP GUN
Gauge: 12, 2¾" chamber.
Barrel: 32" or 34" Unsingle; 34" Top Single. Fixed Full or choke tubes.
Weight: About 8¾ lbs.
Stock: Four stock dimensions or adjustable stock available. All hand-checkered European walnut.

KRIEGHOFF KS-5 TRAP GUN
Gauge: 12, 2¾" chamber.
Barrel: 32", 34"; Full choke or choke tubes.
Weight: About 8½ lbs.
Stock: Choice of high Monte Carlo (1½"), low Monte Carlo (1⅜") or factory adjustable stock. European walnut.
Features: Ventilated tapered step rib. Adjustable trigger or optional release trigger. Satin gray electroless nickel receiver. Comes with fitted aluminum case. Introduced 1988. Imported from Germany by Krieghoff International, Inc.
Price: Fixed choke, cased $3,695.00
Price: With choke tubes $4,120.00

Features: Satin nickel finish with K-80 logo. Selective mechanical trigger adjustable for finger position. Tapered step vent. rib. Adjustable point of impact on Unsingle.
Price: Standard grade full Unsingle, from $7,950.00
Price: Standard grade full Top Single combo (special order), from $9,975.00
Price: RT (removable trigger) option, add $1,000.00

Ljutic Mono Gun

Ljutic LTX Super Deluxe Mono Gun
Super Deluxe version of the standard Mono Gun with high quality wood, extra-fancy checkering pattern in 24 lpi, double recessed choking. Available in two weights: 8¼ lbs. or 8¾ lbs. Extra light 33" barrel; medium-height rib. Introduced 1984. From Ljutic Industries.
Price: $5,995.00
Price: With three screw-in choke tubes $6,395.00

LJUTIC MONO GUN SINGLE BARREL
Gauge: 12 only.
Barrel: 34", choked to customer specs; hollow-milled rib, 35½" sight plane.
Weight: Approx. 9 lbs.
Stock: To customer specs. Oil finish, hand checkered.
Features: Totally custom made. Pull or release trigger; removable trigger guard contains trigger and hammer mechanism; Ljutic pushbutton opener on front of trigger guard. From Ljutic Industries.
Price: With standard, medium or Olympic rib, custom 32"-34" bbls. ...$4,995.00
Price: As above with screw-in choke barrel $5,300.00

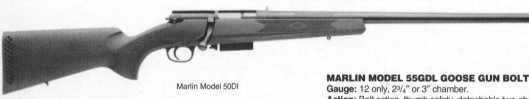
Marlin Model 50DL

Marlin Model 50DL Bolt-Action Shotgun
Similar to the Model 55DL except has 28" barrel with Mod. choke. Weighs 7½ lbs., measures 48¾" overall. Introduced 1997. Made in U.S. by Marlin Firearms Co.
Price: $322.00

Consult our Directory pages for the location of firms mentioned.

Marlin Model 512DL Slugmaster Shotgun
Similar to the Model 512 except has black fiberglass-filled synthetic stock with moulded-in checkering, swivel studs; ventilated recoil pad; padded black nylon sling. Has 21" fully rifled barrel with 1:28" rifling twist. Introduced 1997. Made in U.S. by Marlin Firearms Co.
Price: $372.00

MARLIN MODEL 55GDL GOOSE GUN BOLT-ACTION SHOTGUN
Gauge: 12 only, 2¾" or 3" chamber.
Action: Bolt action, thumb safety, detachable two-shot clip. Red cocking indicator.
Barrel: 36" (Full) with burnished bore for lead or steel shot.
Weight: 8 lbs. **Length:** 56¾" overall.
Stock: Black fiberglass-filled synthetic with moulded-in checkering and swivel studs; ventilated recoil pad.
Sights: Brass bead front, U-groove rear.
Features: Brushed blue finish; thumb safety; red cocking indicator; 2-shot detachable box magazine. Introduced 1997. Made in U.S. by Marlin Firearms Co.
Price: $372.00

MARLIN MODEL 512 SLUGMASTER SHOTGUN
Gauge: 12, 3" chamber; 2-shot detachable box magazine.
Barrel: 21", rifled (1:28" twist).
Weight: 8 lbs. **Length:** 44¾" overall.
Stock: Walnut-finished, press-checkered Maine birch with Mar-Shield® finish, ventilated recoil pad.
Sights: Ramp front with brass bead and removable Wide-Scan™ hood, adjustable folding semi-buckhorn rear. Drilled and tapped for scope mounting.
Features: Uses Model 55 action with thumb safety. Designed for shooting sabot-ed slugs. Comes with special Weaver scope mount. Introduced 1994. Made in U.S. by Marlin Firearms Co.
Price: $356.00

SHOTGUNS—BOLT ACTIONS & SINGLE SHOTS

Maverick 95 Bolt-Action

MAVERICK MODEL 95 BOLT-ACTION SHOTGUN
Gauge: 12, 3" chamber, 2-shot magazine.
Barrel: 25" (Mod.).
Weight: 6.5 lbs.
Stock: Textured black synthetic.
Sights: Bead front.
Features: Full-length stock with integral magazine; ambidextrous rotating safety; twin extractors; rubber recoil pad. Blue finish. Introduced 1995. From Maverick Arms.
Price: $184.00

Mossberg Model 695 Slugster

Mossberg Model 695 Turkey
Same as the Model 695 Slugster except has smoothbore 22" barrel with Extra-Full Turkey Accu-Choke tube, full OFM camouflage finish, fixed U-notch rear sight, bead front. Made in U.S. by Mossberg. Introduced 1996.
Price: $276.00

MOSSBERG MODEL 695 SLUGSTER
Gauge: 12, 3" chamber.
Barrel: 22"; fully rifled, ported.
Weight: 7 1/2 lbs.
Stock: Black synthetic, with swivel studs and rubber recoil pad.
Sights: Blade front, folding rifle-style leaf rear. Comes with Weaver-style scope bases.
Features: Matte metal finish; rotating thumb safety; detachable 2-shot magazine. Mossberg Cablelock. Made in U.S. by Mossberg. Introduced 1996.
Price: $307.00
Price: Scope Combo Model includes Protecto case and Bushnell 1.5-4.5x scope $429.00

New England Turkey and Goose

NEW ENGLAND FIREARMS TURKEY AND GOOSE GUN
Gauge: 10, 3 1/2" chamber.
Barrel: 28" (Full), 32" (Mod.).
Weight: 9.5 lbs. **Length:** 44" overall.
Stock: American hardwood with walnut, or matte camo finish; ventilated rubber recoil pad.
Sights: Bead front.
Features: Break-open action with side-lever release; ejector. Matte finish on metal. Introduced 1992. From New England Firearms.
Price: Walnut-finish wood $149.95
Price: Camo finish, sling and swivels $159.95
Price: Camo finish, 32", sling and swivels $179.95
Price: Black matte finish, 24", Turkey Full choke tube, sling and swivels .$184.95

NEW ENGLAND FIREARMS TRACKER SLUG GUN
Gauge: 12, 20, 3" chamber.
Barrel: 24" (Cyl.).
Weight: 6 lbs. **Length:** 40" overall.
Stock: Walnut-finished hardwood with full pistol grip, recoil pad.
Sights: Blade front, fully adjustable rifle-type rear.
Features: Break-open action with side-lever release; blued barrel, color case-hardened frame. Introduced 1992. From New England Firearms.
Price: Tracker $129.95
Price: Tracker II (as above except fully rifled bore) $139.95

NEW ENGLAND FIREARMS SURVIVOR
Gauge: 12, 20, 410/45 Colt, 3" chamber.
Barrel: 22" (Mod.); 20" (410/45 Colt, rifled barrel, choke tube).
Weight: 6 lbs. **Length:** 36" overall.
Stock: Black polymer with thumbhole/pistol grip, sling swivels; beavertail forend.
Sights: Bead front.
Features: Buttplate removes to expose storage for extra ammunition; forend also holds extra ammunition. Black or nickel finish. Introduced 1993. From New England Firearms.
Price: Black $129.95
Price: Nickel $145.95
Price: 410/45 Colt, black $145.95
Price: 410/45 Colt, nickel $164.95

NEW ENGLAND FIREARMS STANDARD PARDNER
Gauge: 12, 20, 410, 3" chamber; 16, 28, 2 3/4" chamber.
Barrel: 12-ga.—28" (Full, Mod.), 32" (Full); 16-ga.—28" (Full); 20-ga.—26" (Full, Mod.); 28-ga.—26" (Mod.); 410-bore—26" (Full).
Weight: 5-6 lbs. **Length:** 43" overall (28" barrel).
Stock: Walnut-finished hardwood with full pistol grip.
Sights: Bead front.
Features: Transfer bar ignition; break-open action with side-lever release. Introduced 1987. From New England Firearms.
Price: $99.95
Price: Youth model (20-, 28-ga., 410, 22" barrel, recoil pad) $109.95
Price: 12-ga., 32" (Full) $104.95

Perazzi TMX Special

PERAZZI TMX SPECIAL SINGLE TRAP
Gauge: 12, 2 3/4" chamber.
Barrel: 32" or 34" (Extra Full).
Weight: 8 lbs., 6 oz.
Stock: To customer specs; interchangeable.
Features: Special high rib; adjustable four-position trigger. Also available with choke tubes. Imported from Italy by Perazzi U.S.A., Inc.
Price: From $6,790.00

CAUTION: PRICES SHOWN ARE SUPPLIED BY THE MANUFACTURER OR IMPORTER. CHECK YOUR LOCAL GUNSHOP.

SHOTGUNS—BOLT ACTIONS & SINGLE SHOTS

REMINGTON 90-T SUPER SINGLE SHOTGUN
Gauge: 12, 2 3/4" chamber.
Barrel: 30", 32", 34", fixed choke or Rem Choke tubes; ported or non-ported. Medium-high tapered, ventilated rib; white Bradley-type front bead, stainless center bead.
Weight: About 8 3/4 lbs.
Stock: 14 3/8"x1 3/8" (or 1 1/2" or 1 1/4")x1 1/2". Choice of drops at comb, pull length available plus or minus 1". Figured American walnut with low-luster finish, checkered 18 lpi; black vented rubber recoil pad. Cavity in forend and butt stock for added weight.
Features: Barrel is over-bored with elongated forcing cones. Removable sideplates can be ordered with engraving; drop-out trigger assembly. Metal has non-glare matte finish. Available with extra barrels in different lengths, chokes, extra trigger assemblies and sideplates, porting, stocks. Introduced 1990. From Remington.
Price: Depending on options $3,199.00
Price: With high post adjustable rib $3,992.00

Savage Model 210FT

Savage Model 210FT Master Shot Shotgun
Similar to the Model 210F except has smoothbore barrel threaded for Winchoke-style choke tubes (comes with one Full tube); Advantage camo pattern covers the stock; pillar-bedded synthetic stock; bead front sight, U-notch rear. Introduced 1997. Made in U.S. by Savage Arms, Inc.
Price: $440.00

SAVAGE MODEL 210F MASTER SHOT SLUG GUN
Gauge: 12, 3" chamber; 2-shot magazine.
Barrel: 24" 1:35" rifling twist.
Weight: 7 1/2 lbs. **Length:** 43.5" overall.
Stock: Glass-filled polymer with positive checkering.
Features: Based on the Savage Model 110 action; 60° bolt lift; controlled round feed; comes with scope mount. Introduced 1996. Made in U.S. by Savage Arms.
Price: $380.00

SNAKE CHARMER II SHOTGUN
Gauge: 410, 3" chamber.
Barrel: 18 1/4".
Weight: About 3 1/2 lbs. **Length:** 28 5/8" overall.
Stock: ABS grade impact resistant plastic.
Features: Thumbhole-type stock holds four extra rounds. Stainless steel barrel and frame. Reintroduced 1989. From Sporting Arms Mfg., Inc.
Price: $149.00
Price: New Generation Snake Charmer (as above except with black carbon steel bbl.) $139.00

Tar-Hunt RSG-20 Mountaineer

Tar-Hunt RSG-20 Mountaineer Rifled Slug Gun
Similar to the RSG-12 Professional Slug gun except chambered only for 20-gauge 2 3/4" shells; 21" fully rifled barrel with muzzlebrake; weighs 6 1/2 lbs. Right and left-hand models at same price. Introduced 1997. Made in U.S. by Tar-Hunt Rifles, Inc.
Price: $1,195.00

TAR-HUNT RSG-12 PROFESSIONAL RIFLED SLUG GUN
Gauge: 12, 2 3/4" chamber.
Barrel: 21 1/2"; fully rifled, with muzzlebrake.
Weight: 7 3/4 lbs. **Length:** 41 1/2" overall.
Stock: Matte black McMillan fiberglass with Pachmayr Decelerator pad.
Sights: None furnished; comes with Leupold windage bases only.
Features: Uses rifle-style action with two locking lugs; two-position safety; single-stage, trigger; muzzlebrake. Many options available. Right- and left-hand models at same prices. Introduced 1991. Made in U.S. by Tar-Hunt Custom Rifles, Inc.
Price: Professional model, right- or left hand $1,395.00
Price: Matchless model (400-grit gloss metal finish, McMillan Fibergrain stock), right- or left-hand $1,783.50
Price: Peerless model NP-3 nickel/teflon metal finish, McMillan Fibergrain stock), right- or left-hand $1,973.25

SHOTGUNS—MILITARY & POLICE
Designs for utility, suitable for and adaptable to competitions and other sporting purposes.

Armscor M-30SAS Special Purpose

ARMSCOR M-30 SECURITY SHOTGUNS
Gauge: 12, 3" chamber.
Barrel: 18.5", 20" (Cyl.).
Weight: About 7 lbs.
Stock: Walnut-finished hardwood.
Sights: Metal bead front.
Features: Dual action slide bars; damascened bolt; blued steel receiver. Imported from the Philippines by K.B.I., Inc.
Price: M-30R6 (5-shot) $229.00
Price: M-30R8 (7-shot) $245.00

ARMSCOR M-30 SPECIAL PURPOSE SHOTGUNS
Gauge: 12, 3" chamber.
Barrel: 20" (Cyl.).
Weight: 7.5 lbs.
Stock: Walnut-finished hardwood, or synthetic speedfeed.
Sights: Rifle sights on M-30DG, metal bead front on M-305AS.
Features: M-30DB has 7-shot magazine, polished blue receiver; M-305AS based on Special Air Services gun with 7-shot magazine, ventilated barrel shroud, Parkerized finish. Introduced 1996. Imported from the Philippines by K.B.I., Inc.
Price: M-30DG $279.00
Price: M-30SAS $319.00

SHOTGUNS—MILITARY & POLICE

Benelli M1 Super 90 Tactical

BENELLI M1 SUPER 90 TACTICAL SHOTGUN
Gauge: 12, 3", 5-shot magazine.
Barrel: 18.5", choke tubes.
Weight: 6.5 lbs. **Length:** 39.75" overall.
Stock: Black polymer.
Sights: Rifle type with Ghost Ring system, tritium night sights optional.
Features: Semi-auto intertia recoil action. Cross-bolt safety; bolt release button; matte-finish metal. Introduced 1993. Imported from Italy by Heckler & Koch, Inc.
Price: With rifle sights, standard stock $873.00
Price: As above with pistol grip stock $909.00
Price: With Ghost Ring rifle sights, standard stock $916.00
Price: As above with pistol grip stock $951.00

BENELLI M3 SUPER 90 PUMP/AUTO SHOTGUN
Gauge: 12, 3" chamber, 7-shot magazine.
Barrel: 19 3/4" (Cyl.).
Weight: 7 lbs., 8 oz. **Length:** 41" overall.
Stock: High-impact polymer with sling loop in side of butt; rubberized pistol grip on stock.
Sights: Post front, buckhorn rear adjustable for windage. Ghost ring system available.
Features: Combination pump/auto action. Alloy receiver with inertia recoil rotating locking lug bolt; matte finish; automatic shell release lever. Introduced 1989. Imported by Heckler & Koch, Inc.
Price: With standard stock $1,040.00
Price: With Ghost Ring sight system, standard stock $1,081.00

Beretta Model 1201FP

BERETTA MODEL 1201FP AUTO SHOTGUN
Gauge: 12, 3" chamber.
Barrel: 18" (Cyl.).
Weight: 6.3 lbs.
Stock: Special strengthened technopolymer, matte black finish.
Stock: Fixed rifle type.
Features: Has 6-shot magazine. Introduced 1988. Imported from Italy by Beretta U.S.A.
Price: $745.00
Price: With tritium sights $830.00

Crossfire Shotgun/Rifle

CROSSFIRE SHOTGUN/RIFLE
Gauge/Caliber: 12, 2 3/4" chamber 4-shot/223 Rem. (10-shot).
Barrel: 22" (shotgun), 20" (rifle).
Weight: About 9 lbs. **Length:** 40 1/4" overall.
Stock: Composite.
Sights: Meprolight night sights. Integral scope rail.
Features: Combination pump-action shotgun, autoloading rifle; single selector, single trigger; dual action bars for both upper and lower actions; ambidextrous selector and safety. Introduced 1997. Made in U.S. From Hesco.
Price: $1,500.00

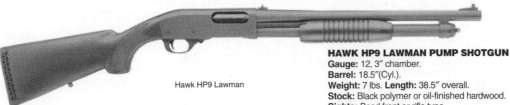
Hawk HP9 Lawman

HAWK HP9 LAWMAN PUMP SHOTGUN
Gauge: 12, 3" chamber.
Barrel: 18.5"(Cyl.).
Weight: 7 lbs. **Length:** 38.5" overall.
Stock: Black polymer or oil-finished hardwood.
Sights: Bead front or rifle type.
Features: Twin action bars; steel receiver; cross-bolt safety. Introduced 1997. Imported by Brolin Industries, Inc.
Price: Matte blue finish, synthetic stock $249.95
Price: Stainless finish synthetic stock $269.95
Price: Matte blue finish synthetic stock, rifle sights $259.95
Price: Matte blue finish, wood stock $249.95
Price: Matte blue finish, wood stock, rifle sights $259.95

> Consult our Directory pages for the location of firms mentioned.

Magtech MT 586P

MAGTECH MT 586P PUMP SHOTGUN
Gauge: 12, 3" chamber, 7-shot magazine (8-shot with 2 3/4" shells).
Barrel: 19" (Cyl.).
Weight: 7.3 lbs. **Length:** 39.5" overall.
Stock: Brazilian hardwood.
Sights: Bead front.
Features: Dual action slide bars, cross-bolt safety. Blue finish. Introduced 1991. Imported from Brazil by Magtech Recreational Products.
Price: About $219.00

SHOTGUNS—MILITARY & POLICE

Mossberg Model 500

Mossberg Model HS410 Shotgun
Similar to the Model 500 Security pump except chambered for 20 gauge or 410 with 3" chamber; has pistol grip forend, thick recoil pad, muzzlebrake and has special spreader choke on the 18.5" barrel. Overall length is 37.5", weight is 6.25 lbs. Blue finish; synthetic field stock. Mossberg Cablelock and video included. Introduced 1990.
Price: HS 410 .... $294.00

MOSSBERG MODEL 500 PERSUADER SECURITY SHOTGUNS
Gauge: 12, 20, 410, 3" chamber.
Barrel: 18 1/2", 20" (Cyl.).
Weight: 7 lbs.
Stock: Walnut-finished hardwood or black synthetic.
Sights: Metal bead front.
Features: Available in 6- or 8-shot models. Top-mounted safety, double action slide bars, swivel studs, rubber recoil pad. Blue, Parkerized, Marinecote finishes. Mossberg Cablelock included. From Mossberg.
Price: 12- or 20-ga., 18 1/2", blue, wood or synthetic stock, 6-shot $282.00
Price: Cruiser, 12- or 20-ga., 18 1/2", blue, pistol grip, heat shield $274.00
Price: As above, 410-bore $281.00
Price: 12-ga., 8-shot, blue, wood or synthetic stock $282.00
Price: 6-shot Mil-Spec, 20" barrel, Mil-packaging $478.00

Mossberg Model 590 Ghost-Ring

Mossberg Model 500, 590 Ghost-Ring Shotguns
Similar to the Model 500 Security except has adjustable blade front, adjustable Ghost-Ring rear sight with protective "ears." Model 500 has 18.5" (Cyl.) barrel, 6-shot capacity; Model 590 has 20" (Cyl.) barrel, 9-shot capacity. Both have synthetic field stock. Mossberg Cablelock included. Introduced 1990. From Mossberg.
Price: Model 500, blue $332.00
Price: As above, Parkerized $385.00
Price: Model 590, blue $381.00
Price: As above, Parkerized $434.00
Price: Parkerized Speedfeed stock $466.00

MOSSBERG MODEL 590 SHOTGUN
Gauge: 12, 3" chamber.
Barrel: 20" (Cyl.).
Weight: 7 1/4 lbs.
Stock: Synthetic field or Speedfeed.
Sights: Metal bead front.
Features: Top-mounted safety, double slide action bars. Comes with heat shield, bayonet lug, swivel studs, rubber recoil pad. Blue, Parkerized or Marinecote finish. Mossberg Cablelock included. From Mossberg.
Price: Blue, synthetic stock $331.00
Price: Parkerized, synthetic stock $381.00
Price: Blue, Speedfeed stock $363.00
Price: Parkerized, Speedfeed stock $413.00

Mossberg Model 500, 590 Mariner Pump
Similar to the Model 500 or 590 Security except all metal parts finished with Marinecote metal finish to resist rust and corrosion. Synthetic field stock; pistol grip kit included. Mossberg Cablelock included.
Price: 6-shot, 18 1/2" barrel $404.00
Price: As above with Ghost-Ring sights $460.00
Price: 9-shot, 20" barrel $416.00
Price: As above with Ghost-Ring sights $472.00

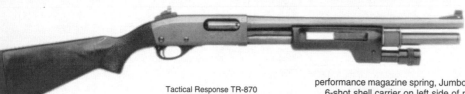
Tactical Response TR-870

TACTICAL RESPONSE TR-870 STANDARD MODEL SHOTGUN
Gauge: 12, 3" chamber, 7-shot magazine.
Barrel: 18" (Cyl.).
Weight: 9 lbs. **Length:** 38" overall.
Stock: Fiberglass-filled polypropolene with non-snag recoil absorbing butt pad. Nylon tactical forend houses flashlight.
Sights: Trak-Lock ghost ring sight system. Front sight has tritium insert.
Features: Highly modified Remington 870P with Parkerized finish. Comes with nylon three-way adjustable sling, high visibility non-binding follower, high performance magazine spring, Jumbo Head safety, and Side Saddle extended 6-shot shell carrier on left side of receiver. Introduced 1991. From Scattergun Technologies, Inc.
Price: Standard model $815.00
Price: FBI model $770.00
Price: Patrol model $595.00
Price: Border Patrol model $605.00
Price: Military model $690.00
Price: K-9 model (Rem. 11-87 action) $860.00
Price: Urban Sniper, Rem. 11-87 action $1,290.00
Price: Louis Awerbuck model $705.00
Price: Practical Turkey model $725.00
Price: Expert model $1,350.00
Price: Professional model $815.00
Price: Entry model $840.00
Price: Compact model $635.00
Price: SWAT model $1,050.00

Winchester Model 1300 Defender

WINCHESTER MODEL 1300 DEFENDER PUMP GUN
Gauge: 12, 20, 3" chamber, 5- or 8-shot capacity.
Barrel: 18" (Cyl.).
Weight: 6 3/4 lbs. **Length:** 38 5/8" overall.
Stock: Walnut-finished hardwood stock and ribbed forend, or synthetic; or pistol grip.
Sights: Metal bead front.
Features: Cross-bolt safety, front-locking rotary bolt, twin action slide bars. Black rubber buttpad. From U.S. Repeating Arms Co.
Price: 8-shot, wood or synthetic stock $290.00
Price: 5-shot, wood stock $290.00
Price: Defender Field Combo with pistol grip $393.00

SHOTGUNS—MILITARY & POLICE

Winchester 8-Shot Pistol Grip Pump Security Shotgun
Same as regular Defender Pump but with pistol grip and forend of high-impact resistant ABS plastic with non-glare black finish. Introduced 1984.
Price: Pistol Grip Defender $290.00

Winchester Model 1300 Stainless Marine Pump Gun
Same as the Defender except has bright chrome finish, stainless steel barrel, rifle-type sights only. Phosphate coated receiver for corrosion resistance. Pistol grip optional.
Price: $460.00

Winchester Model 1300 Lady Defender Pump Gun
Similar to the Model 1300 defender except in 20-gauge only, weighs 6 1/4 lbs. Available with synthetic full stock or synthetic pistol grip only. Introduced 1997. From U.S. Repeating Arms Co.
Price: $290.00

BLACKPOWDER SINGLE SHOT PISTOLS—FLINT & PERCUSSION

CVA Hawken

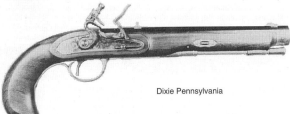
Dixie Pennsylvania

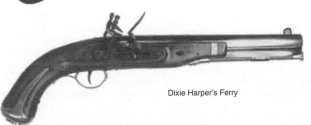
Dixie Harper's Ferry

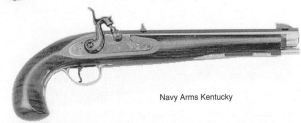
Navy Arms Kentucky

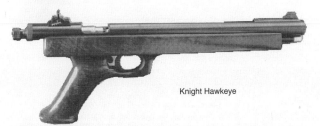
Knight Hawkeye

CVA HAWKEN PISTOL
Caliber: 50.
Barrel: 9 3/4"; 15/16" flats.
Weight: 50 oz. **Length:** 16 1/2" overall.
Stock: Select hardwood.
Sights: Beaded blade front, fully adjustable open rear.
Features: Color case-hardened lock, polished brass wedge plate, nose cap, ramrod thimble, trigger guard, grip cap. Imported by CVA.
Price: $149.95
Price: Kit $109.95

DIXIE PENNSYLVANIA PISTOL
Caliber: 44 (.430" round ball).
Barrel: 10" (7/8" octagon).
Weight: 2 1/2 lbs.
Stock: Walnut-stained hardwood.
Sights: Blade front, open rear drift-adjustable for windage; brass.
Features: Available in flint only. Brass trigger guard, thimbles, nosecap, wedge-plates; high-luster blue barrel. Imported from Italy by Dixie Gun Works.
Price: Finished $183.75
Price: Kit $174.95

FRENCH-STYLE DUELING PISTOL
Caliber: 44.
Barrel: 10".
Weight: 35 oz. **Length:** 15 3/4" overall.
Stock: Carved walnut.
Sights: Fixed.
Features: Comes with velvet-lined case and accessories. Imported by Mandall Shooting Supplies.
Price: $295.00

HARPER'S FERRY 1806 PISTOL
Caliber: 58 (.570" round ball).
Barrel: 10".
Weight: 40 oz. **Length:** 16" overall.
Stock: Walnut.
Sights: Fixed.
Features: Case-hardened lock, brass-mounted browned barrel. Replica of the first U.S. Gov't.-made flintlock pistol. Imported by Navy Arms, Dixie Gun Works.
Price: .... $275.00 to $405.00
Price: Kit (Dixie) $237.00
Price: Cased set (Navy Arms) $355.00

KENTUCKY FLINTLOCK PISTOL
Caliber: 44, 45.
Barrel: 10 1/8".
Weight: 32 oz. **Length:** 15 1/2" overall.
Stock: Walnut.
Sights: Fixed.
Features: Specifications, including caliber, weight and length may vary with importer. Case-hardened lock, blued barrel; available also as brass barrel flint Model 1821. Imported by Navy Arms (44 only), The Armoury.
Price: .... $145.00 to $225.00
Price: In kit form, from .... $90.00 to $112.00
Price: Single cased set (Navy Arms) $350.00
Price: Double cased set (Navy Arms) $580.00

Kentucky Percussion Pistol
Similar to flint version but percussion lock. Imported by The Armoury, Navy Arms, CVA (50-cal.).
Price: .... $129.95 to $250.00
Price: Steel barrel (Armoury) $179.00
Price: Single cased set (Navy Arms) $335.00
Price: Double cased set (Navy Arms) $550.00

KNIGHT HAWKEYE PISTOL
Caliber: 50.
Barrel: 12", 1:20" twist.
Weight: 3 1/4 lbs. **Length:** 20" overall.
Stock: Black composite, autumn brown or shadow black laminate.
Sights: Bead front on ramp, open fully adjustable rear.
Features: In-line ignitiion design; patented double safety system; removable breech plug; fully adjustable trigger; receiver drilled and tapped for scope mounting. Made in U.S. by Modern Muzzle Loading, Inc.
Price: Blued $359.95
Price: Stainless $429.95

CAUTION: PRICES SHOWN ARE SUPPLIED BY THE MANUFACTURER OR IMPORTER. CHECK YOUR LOCAL GUNSHOP.

BLACKPOWDER SINGLE SHOT PISTOLS—FLINT & PERCUSSION

Lyman Plains Pistol

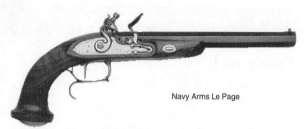

Navy Arms Le Page

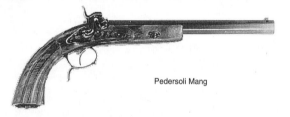

Pedersoli Mang

Dixie Queen Anne

LE PAGE PERCUSSION DUELING PISTOL
Caliber: 44.
Barrel: 10", rifled.
Weight: 40 oz. **Length:** 16" overall.
Stock: Walnut, fluted butt.
Sights: Blade front, notch rear.
Features: Double-set triggers. Blued barrel; trigger guard and buttcap are polished silver. Imported by Dixie Gun Works.
Price: $259.95

LYMAN PLAINS PISTOL
Caliber: 50 or 54.
Barrel: 8", 1:30" twist, both calibers.
Weight: 50 oz. **Length:** 15" overall.
Stock: Walnut half-stock.
Sights: Blade front, square notch rear adjustable for windage.
Features: Polished brass trigger guard and ramrod tip, color case-hardened coil spring lock, spring-loaded trigger, stainless steel nipple, blackened iron furniture. Hooked patent breech, detachable belt hook. Introduced 1981. From Lyman Products.
Price: Finished $224.95
Price: Kit $179.95

NAVY ARMS LE PAGE DUELING PISTOL
Caliber: 44.
Barrel: 9", octagon, rifled.
Weight: 34 oz. **Length:** 15" overall.
Stock: European walnut.
Sights: Adjustable rear.
Features: Single-set trigger. Polished metal finish. From Navy Arms.
Price: Percussion $500.00
Price: Single cased set, percussion $775.00
Price: Double cased set, percussion $1,300.00
Price: Flintlock, rifled $625.00
Price: Flintlock, smoothbore (45-cal.) $625.00
Price: Flintlock, single cased set $900.00
Price: Flintlock, double cased set $1,575.00

PEDERSOLI MANG TARGET PISTOL
Caliber: 38.
Barrel: 10.5", octagonal; 1:15" twist,
Weight: 2.5 lbs. **Length:** 17.25" overall.
Stock: Walnut with fluted grip.
Sights: Blade front, open rear adjustable for windage.
Features: Browned barrel, polished breech plug, rest color case-hardened. Imported from Italy by Dixie Gun Works.
Price: $786.00

QUEEN ANNE FLINTLOCK PISTOL
Caliber: 50 (.490" round ball).
Barrel: 7 1/2", smoothbore.
Stock: Walnut.
Sights: None.
Features: Browned steel barrel, fluted brass trigger guard, brass mask on butt. Lockplate left in the white. Made by Pedersoli in Italy. Introduced 1983. Imported by Dixie Gun Works.
Price: $195.00
Price: Kit $170.00

THOMPSON/CENTER SCOUT PISTOL
Caliber: 45, 50 and 54.
Barrel: 12", interchangeable.
Weight: 4 lbs., 6 oz. **Length:** NA.
Stocks: American black walnut stocks and forend.
Sights: Blade on ramp front, fully adjustable Patridge rear.
Features: Patented in-line ignition system with special vented breech plug. Patented trigger mechanism consists of only two moving parts. Interchangeable barrels. Wide grooved hammer. Brass trigger guard assembly. Introduced 1990. From Thompson/Center.
Price: 45-, 50- or 54-cal. $350.00

TRADITIONS BUCKHUNTER PRO IN-LINE PISTOL
Caliber: 50, 54.
Barrel: 10" round.
Weight: 48 oz. **Length:** 14" overall.
Stocks: Smooth walnut or black epoxy coated grip and forend.
Sights: Beaded blade front, folding adjustable rear.
Features: Thumb safety; removable stainless steel breech plug; adjustable trigger, barrel drilled and tapped for scope mounting. From Traditions.
Price: With walnut grip $219.00
Price: Nickel with black grip $233.75

Thompson/Center Scout

Traditions Buckhunter

BLACKPOWDER SINGLE SHOT PISTOLS—FLINT & PERCUSSION

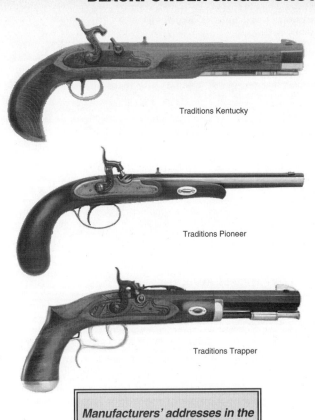

Traditions Kentucky

Traditions Pioneer

Traditions Trapper

Manufacturers' addresses in the
Directory of the Arms Trade
page 309, this issue

TRADITIONS BUCKSKINNER PISTOL
Caliber: 50.
Barrel: 10" octagonal, $7/8$" flats, 1:20" twist.
Weight: 40 oz. **Length:** 15" overall.
Stocks: Stained beech.
Sights: Blade front, fixed rear.
Features: Percussion ignition. Blackened furniture. Imported by Traditions.
Price: Beech stocks $145.50

TRADITIONS KENTUCKY PISTOL
Caliber: 50.
Barrel: 10"; octagon with $7/8$" flats; 1:20" twist.
Weight: 40 oz. **Length:** 15" overall.
Stock: Stained beech.
Sights: Blade front, fixed rear.
Features: Birds-head grip; brass thimbles; color case-hardened lock. Percussion only. Introduced 1995. From Traditions.
Price: Finished $131.00
Price: Kit $101.25

TRADITIONS PIONEER PISTOL
Caliber: 45.
Barrel: $9^5/8$", $13/16$" flats, 1:16" twist.
Weight: 31 oz. **Length:** 15" overall.
Stock: Beech.
Sights: Blade front, fixed rear.
Features: V-type mainspring. Single trigger. German silver furniture, blackened hardware. From Traditions.
Price: $140.00
Price: Kit $116.25

TRADITIONS TRAPPER PISTOL
Caliber: 50.
Barrel: $9^3/4$", $7/8$" flats, 1:20" twist.
Weight: $2^3/4$ lbs. **Length:** 16" overall.
Stock: Beech.
Sights: Blade front, adjustable rear.
Features: Double-set triggers; brass buttcap, trigger guard, wedge plate, forend tip, thimble. From Traditions.
Price: Percussion $175.00
Price: Flintlock $189.50
Price: Kit $131.00

TRADITIONS WILLIAM PARKER PISTOL
Caliber: 50.
Barrel: $10^3/8$", $15/16$" flats; polished steel.
Weight: 37 oz. **Length:** $17^1/2$" overall.
Stock: Walnut with checkered grip.
Sights: Brass blade front, fixed rear.
Features: Replica dueling pistol with 1:20" twist, hooked breech. Brass wedge plate, trigger guard, cap guard; separate ramrod. Double-set triggers. Polished steel barrel, lock. Imported by Traditions.
Price: $250.00

WHITE MUZZLELOADING SYSTEMS JAVELINA PISTOL
Caliber: 41, 45, 50.
Barrel: 14".
Weight: NA. **Length:** NA.
Stock: Black composite with rear grip.
Sights: Blade front, fully adjustable rear. Drilled and tapped for scope mounting.
Features: Stainless steel construction. Action based on the G-Series rifle (Whitetail, Bison) system. Introduced 1996. Made in U.S. by White Muzzleloading Systems.
Price: $499.95

BLACKPOWDER REVOLVERS

ARMY 1860 PERCUSSION REVOLVER
Caliber: 44, 6-shot.
Barrel: 8".
Weight: 40 oz. **Length:** $13^5/8$" overall.
Stocks: Walnut.
Sights: Fixed.
Features: Engraved Navy scene on cylinder; brass trigger guard; case-hardened frame, loading lever and hammer. Some importers supply pistol cut for detachable shoulder stock, have accessory stock available. Imported by American Arms, Cabela's (1860 Lawman), E.M.F., Navy Arms, The Armoury, Cimarron, Dixie Gun Works (half-fluted cylinder, not roll engraved), Euroarms of America (brass or steel model), Armsport, Traditions (brass or steel), Uberti U.S.A. Inc., United States Patent Fire-Arms.
Price: About ....$92.95 to $395.00
Price: Hartford model, steel frame, German silver trim, cartouches (E.M.F.) $215.00
Price: Single cased set (Navy Arms) $300.00
Price: Double cased set (Navy Arms) $490.00

American Arms 1860 Army

Price: 1861 Navy: Same as Army except 36-cal., $7^1/2$" bbl., weighs 41 oz., cut for shoulder stock; round cylinder (fluted available), from CVA (brass frame, 44-cal.), United States Patent Fire-Arms ....$99.95 to $385.00
Price: Steel frame kit (E.M.F., Euroarms) ....$125.00 to $216.25
Price: Colt Army Police, fluted cyl., $5^1/2$", 36-cal. (Cabela's) $124.95

CAUTION: PRICES SHOWN ARE SUPPLIED BY THE MANUFACTURER OR IMPORTER. CHECK YOUR LOCAL GUNSHOP.

BLACKPOWDER REVOLVERS

Colt 1847 Walker

COLT 1847 WALKER PERCUSSION REVOLVER
Caliber: 44.
Barrel: 9", 7 groove, right-hand twist.
Weight: 73 oz.
Stocks: One-piece walnut.
Sights: German silver front sight, hammer notch rear.
Features: Made in U.S. Faithful reproduction of the original gun, including markings. Color case-hardened frame, hammer, loading lever and plunger. Blue steel backstrap, brass square-back trigger guard. Blue barrel, cylinder, trigger and wedge. From Colt Blackpowder Arms Co.
Price: $442.50

Colt Walker 150th Anniversary Revolver
Similar to the standard Walker except has original-type "A Company No. 1" markings embellished in gold. Serial numbers begin with 221, a continuation of A Company numbers. Imported by Colt Blackpowder Arms Co.
Price: $675.00

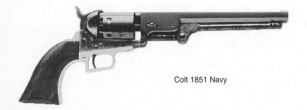
Colt 1851 Navy

COLT 1851 NAVY PERCUSSION REVOLVER
Caliber: 36.
Barrel: 7 1/2", octagonal, 7 groove left-hand twist.
Weight: 40 1/2 oz.
Stocks: One-piece oiled American walnut.
Sights: Brass pin front, hammer notch rear.
Features: Faithful reproduction of the original gun. Color case-hardened frame, loading lever, plunger, hammer and latch. Blue cylinder, trigger, barrel, screws, wedge. Silver-plated brass backstrap and square-back trigger guard. From Colt Blackpowder Arms Co.
Price: $427.50

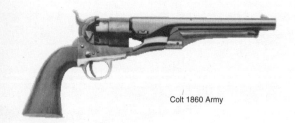
Colt 1860 Army

COLT 1861 NAVY PERCUSSION REVOLVER
Caliber: 36.
Barrel: 7 1/2".
Weight: 42 oz. **Length:** 13 1/8" overall.
Stocks: One-piece walnut.
Sights: Blade front, hammer notch rear.
Features: Color case-hardened frame, loading lever, plunger; blued barrel, backstrap, trigger guard; roll-engraved cylinder and barrel. From Colt Blackpowder Arms Co.
Price: $465.00

ARMY 1851 PERCUSSION REVOLVER
Caliber: 44, 6-shot.
Barrel: 7 1/2".
Weight: 45 oz. **Length:** 13" overall.
Stocks: Walnut finish.
Sights: Fixed.
Features: 44-caliber version of the 1851 Navy. Imported by The Armoury, Armsport.
Price: $129.00

BABY DRAGOON 1848, 1849 POCKET, WELLS FARGO
Caliber: 31.
Barrel: 3", 4", 5", 6"; seven-groove, RH twist.
Weight: About 21 oz.
Stocks: Varnished walnut.
Sights: Brass pin front, hammer notch rear.
Features: No loading lever on Baby Dragoon or Wells Fargo models. Unfluted cylinder with stagecoach holdup scene; cupped cylinder pin; no grease grooves; one safety pin on cylinder and slot in hammer face; straight (flat) mainspring. From Armsport, Dixie Gun Works, Uberti USA Inc., Cabela's.
Price: 6" barrel, with loading lever (Dixie Gun Works) $254.95
Price: 4" (Cabela's, Uberti USA Inc.) $335.00

CABELA'S PATERSON REVOLVER
Caliber: 36, 5-shot cylinder.
Barrel: 7 1/2".
Weight: 24 oz. **Length:** 11 1/2" overall.
Stocks: One-piece walnut.
Sights: Fixed.
Features: Recreation of the 1836 gun. Color case-hardened frame, steel backstrap; roll-engraved cylinder scene. Imported by Cabela's.
Price: $229.95

COLT 1849 POCKET DRAGOON REVOLVER
Caliber: 31.
Barrel: 4".
Weight: 24 oz. **Length:** 9 1/2" overall.
Stocks: One-piece walnut.
Sights: Fixed. Brass pin front, hammer notch rear.
Features: Color case-hardened frame. No loading lever. Unfluted cylinder with engraved scene. Exact reproduction of original. From Colt Blackpowder Arms Co.
Price: $390.00

Uberti 1861 Navy Percussion Revolver
Similar to 1851 Navy except has round 7 1/2" barrel, rounded trigger guard, German silver blade front sight, "creeping" loading lever. Available with fluted or round cylinder. Imported by Uberti USA Inc.
Price: Steel backstrap, trigger guard, cut for stock $300.00

COLT 1860 ARMY PERCUSSION REVOLVER
Caliber: 44.
Barrel: 8", 7 groove, left-hand twist.
Weight: 42 oz.
Stocks: One-piece walnut.
Sights: German silver front sight, hammer notch rear.
Features: Steel backstrap cut for shoulder stock; brass trigger guard. Cylinder has Navy scene. Color case-hardened frame, hammer, loading lever. Reproduction of original gun with all original markings. From Colt Blackpowder Arms Co.
Price: $427.50

Colt 1860 "Cavalry Model" Percussion Revolver
Similar to the 1860 Army except has fluted cylinder. Color case-hardened frame, hammer, loading lever and plunger; blued barrel, backstrap and cylinder, brass trigger guard. Has four-screw frame cut for optional shoulder stock. From Colt Blackpowder Arms Co.
Price: $465.00

COLT 1862 POCKET POLICE "TRAPPER MODEL" REVOLVER
Caliber: 36.
Barrel: 3 1/2".
Weight: 20 oz. **Length:** 8 1/2" overall.
Stocks: One-piece walnut.
Sights: Blade front, hammer notch rear.
Features: Has separate 4 5/8" brass ramrod. Color case-hardened frame and hammer; silver-plated backstrap and trigger guard; blued semi-fluted cylinder, blued barrel. From Colt Blackpowder Arms Co.
Price: $442.50

BLACKPOWDER REVOLVERS

Griswold & Gunnison

GRISWOLD & GUNNISON PERCUSSION REVOLVER
Caliber: 36 or 44, 6-shot.
Barrel: 7½".
Weight: 44 oz. (36-cal.). **Length:** 13" overall.
Stocks: Walnut.
Sights: Fixed.
Features: Replica of famous Confederate pistol. Brass frame, backstrap and trigger guard; case-hardened loading lever; rebated cylinder (44-cal. only). Rounded Dragoon-type barrel. Imported by Navy Arms as Reb Model 1860.
Price: $115.00
Price: Kit $90.00
Price: Single cased set $235.00
Price: Double cased set $365.00

LE MAT REVOLVER
Caliber: 44/65.
Barrel: 6¾" (revolver); 4⅞" (single shot).
Weight: 3 lbs., 7 oz.
Stocks: Hand-checkered walnut.
Sights: Post front, hammer notch rear.
Features: Exact reproduction with all-steel construction; 44-cal. 9-shot cylinder, 65-cal. single barrel; color case-hardened hammer with selector; spur trigger guard; ring at butt; lever-type barrel release. From Navy Arms.
Price: Cavalry model (lanyard ring, spur trigger guard) $595.00
Price: Army model (round trigger guard, pin-type barrel release) $595.00
Price: Naval-style (thumb selector on hammer) $595.00
Price: Engraved 18th Georgia cased set $795.00
Price: Engraved Beauregard cased set $1,000.00

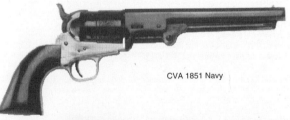
CVA 1851 Navy

Navy Arms 1858 Remington

North American Companion

COLT THIRD MODEL DRAGOON
Caliber: 44.
Barrel: 7½".
Weight: 66 oz. **Length:** 13¾" overall.
Stocks: One-piece walnut.
Sights: Blade front, hammer notch rear.
Features: Color case-hardened frame, hammer, lever and plunger; round trigger guard; flat mainspring; hammer roller; rectangular bolt cuts. From Colt Blackpowder Arms Co.
Price: Three-screw frame with brass grip straps $487.50
Price: Four-screw frame with blued steel grip straps, shoulder stock cuts, dovetailed folding leaf rear sight $502.50

DIXIE WYATT EARP REVOLVER
Caliber: 44.
Barrel: 12" octagon.
Weight: 46 oz. **Length:** 18" overall.
Stocks: Two-piece walnut.
Sights: Fixed.
Features: Highly polished brass frame, backstrap and trigger guard; blued barrel and cylinder; case-hardened hammer, trigger and loading lever. Navy-size shoulder stock ($45) will fit with minor fitting. From Dixie Gun Works.
Price: $130.00

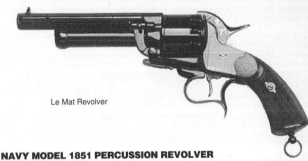
Le Mat Revolver

NAVY MODEL 1851 PERCUSSION REVOLVER
Caliber: 36, 44, 6-shot.
Barrel: 7½".
Weight: 44 oz. **Length:** 13" overall.
Stocks: Walnut finish.
Sights: Post front, hammer notch rear.
Features: Brass backstrap and trigger guard; some have 1st Model squareback trigger guard, engraved cylinder with navy battle scene; case-hardened frame, hammer, loading lever. Imported by American Arms, The Armoury, Cabela's, Navy Arms, E.M.F., Dixie Gun Works, Euroarms of America, Armsport, CVA (44-cal. only), Traditions (44 only), Uberti USA Inc., United States Patent Fire-Arms.
Price: Brass frame ....$99.95 to $385.00
Price: Steel frame ....$130.00 to $285.00
Price: Kit form ....$110.00 to $123.95
Price: Engraved model (Dixie Gun Works) $139.95
Price: Single cased set, steel frame (Navy Arms) $280.00
Price: Double cased set, steel frame (Navy Arms) $455.00
Price: Confederate Navy (Cabela's) $69.95
Price: Hartford model, steel frame, German silver trim, cartouche (E.M.F.) $190.00

NAVY ARMS DELUXE 1858 REMINGTON-STYLE REVOLVER
Caliber: 44.
Barrel: 8".
Weight: 2 lbs., 13 oz.
Stocks: Smooth walnut.
Sights: Dovetailed blade front.
Features: First exact reproduction—correct in size and weight to the original, with progressive rifling; highly polished with blue finish. From Navy Arms.
Price: Deluxe model $415.00

NORTH AMERICAN COMPANION PERCUSSION REVOLVER
Caliber: 22.
Barrel: 1⅛".
Weight: 5.1 oz. **Length:** 4⁵⁄₁₀" overall.
Stocks: Laminated wood.
Sights: Blade front, notch fixed rear.
Features: All stainless steel construction. Uses standard #11 percussion caps. Comes with bullets, powder measure, bullet seater, leather clip holster, gun rag. Long Rifle or Magnum frame size. Introduced 1996. Made in U.S. by North American Arms.
Price: Long Rifle frame $160.00
Price: Magnum frame (1⅝" barrel) $180.00

North American Magnum Companion Percussion Revolver
Similar to the Companion except has larger frame. Weighs 7.2 oz., has 1⅝" barrel, measures 5⁷⁄₁₆" overall. Comes with bullets, powder measure, bullet seater, leather clip holster, gun rag. Introduced 1996. Made in U.S. by North American Arms.
Price: $180.00

BLACKPOWDER REVOLVERS

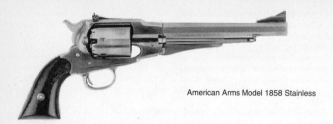

American Arms Model 1858 Stainless

NEW MODEL 1858 ARMY PERCUSSION REVOLVER
Caliber: 36 or 44, 6-shot.
Barrel: 6 1/2" or 8".
Weight: 38 oz. **Length:** 13 1/2" overall.
Stocks: Walnut.
Sights: Blade front, groove-in-frame rear.
Features: Replica of Remington Model 1858. Also available from some importers as Army Model Belt Revolver in 36-cal., a shortened and lightened version of the 44. Target Model (Uberti USA Inc., Navy Arms) has fully adjustable target rear sight, target front, 36 or 44. Imported by American Arms, Cabela's, CVA (as 1858 Army, brass frame, 44 only), Dixie Gun Works, Navy Arms, The Armoury, E.M.F., Euroarms of America (engraved, stainless and plain), Armsport, Traditions (44 only), Uberti USA Inc.
Price: Steel frame, about . . . . $99.95 to $280.00
Price: Steel frame kit (Euroarms, Navy Arms) . . . . $115.95 to $150.00
Price: Single cased set (Navy Arms) $290.00
Price: Double cased set (Navy Arms) $480.00
Price: Stainless steel Model 1858 (American Arms, Euroarms, Uberti USA Inc., Cabela's, Navy Arms, Armsport, Traditions) . . . . . $169.95 to $380.00
Price: Target Model, adjustable rear sight (Cabela's, Euroarms, Uberti USA Inc., Navy Arms, Stone Mountain Arms) . . . . $95.95 to $399.00
Price: Brass frame (CVA, Cabela's, Traditions, Navy Arms) . . . $79.95 to $125.00
Price: As above, kit (Dixie Gun Works, Navy Arms) . . . . $145.00 to $188.95
Price: Buffalo model, 44-cal. (Cabela's) $129.95
Price: Hartford model, steel frame, German silver trim, cartouche (E.M.F.) $215.00

POCKET POLICE 1862 PERCUSSION REVOLVER
Caliber: 36, 5-shot.
Barrel: 4 1/2", 5 1/2", 6 1/2", 7 1/2".
Weight: 26 oz. **Length:** 12" overall (6 1/2" bbl.).
Stocks: Walnut.
Sights: Fixed.
Features: Round tapered barrel; half-fluted and rebated cylinder; case-hardened frame, loading lever and hammer; silver or brass trigger guard and backstrap. Imported by CVA (7 1/2" only), Dixie Gun Works, Navy Arms (5 1/2" only), Uberti USA Inc. (5 1/2", 6 1/2" only), United States Patent Fire-Arms.
Price: About . . . . $139.95 to $335.00
Price: Single cased set with accessories (Navy Arms) $365.00
Price: Hartford model, steel frame, German silver trim, cartouche (E.M.F.) $215.00

ROGERS & SPENCER PERCUSSION REVOLVER
Caliber: 44.
Barrel: 7 1/2".
Weight: 47 oz. **Length:** 13 3/4" overall.
Stocks: Walnut.
Sights: Cone front, integral groove in frame for rear.
Features: Accurate reproduction of a Civil War design. Solid frame; extra large nipple cut-out on rear of cylinder; loading lever and cylinder easily removed for cleaning. From Dixie Gun Works, Euroarms of America (standard blue, engraved, burnished, target models), Navy Arms.
Price: . . . . $160.00 to $289.00
Price: Nickel-plated $215.00
Price: Engraved (Euroarms) $287.00
Price: Kit version . . . . $245.00 to $252.00
Price: Target version (Euroarms, Navy Arms) . . . . $239.00 to $270.00
Price: Burnished London Gray (Euroarms, Navy Arms) . . . . $245.00 to $270.00

Euroarms Rogers & Spencer

RUGER OLD ARMY PERCUSSION REVOLVER
Caliber: 45, 6-shot. Uses .457" dia. lead bullets.
Barrel: 7 1/2" (6-groove, 16" twist).
Weight: 46 oz. **Length:** 13 3/4" overall.
Stocks: Smooth walnut.
Sights: Ramp front, rear adjustable for windage and elevation; or fixed (groove).
Features: Stainless steel; standard size nipples, chrome-moly steel cylinder and frame, same lockwork as in original Super Blackhawk. Also available in stainless steel. Made in USA. From Sturm, Ruger & Co.
Price: Stainless steel (Model KBP-7) $465.00
Price: Blued steel (Model BP-7) $413.00
Price: Stainless steel with high-gloss finish, simulated ivory grips, fixed sight (GKBPI-7F) $495.00
Price: Blued steel, fixed sight (BP-7F) $413.00

Ruger Old Army

SHERIFF MODEL 1851 PERCUSSION REVOLVER
Caliber: 36, 44, 6-shot.
Barrel: 5".
Weight: 40 oz. **Length:** 10 1/2" overall.
Stocks: Walnut.
Sights: Fixed.
Features: Brass backstrap and trigger guard; engraved navy scene; case-hardened frame, hammer, loading lever. Imported by E.M.F.
Price: Steel frame $172.00
Price: Brass frame $140.00

Consult our Directory pages for the location of firms mentioned.

SPILLER & BURR REVOLVER
Caliber: 36 (.375" round ball).
Barrel: 7", octagon.
Weight: 2 1/2 lbs. **Length:** 12 1/2" overall.
Stocks: Two-piece walnut.
Sights: Fixed.
Features: Reproduction of the C.S.A. revolver. Brass frame and trigger guard. Also available as a kit. From Cabela's, Dixie Gun Works, Navy Arms.
Price: . . . . $89.95 to $199.00
Price: Kit form (Dixie) $129.95
Price: Single cased set (Navy Arms) $270.00
Price: Double cased set (Navy Arms) $430.00

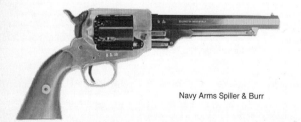

Navy Arms Spiller & Burr

BLACKPOWDER REVOLVERS

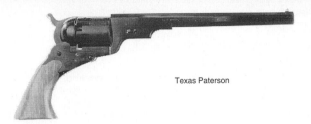

Texas Paterson

TEXAS PATERSON 1836 REVOLVER
Caliber: 36 (.375" round ball).
Barrel: 7 1/2".
Weight: 42 oz.
Stocks: One-piece walnut.
Sights: Fixed.
Features: Copy of Sam Colt's first commercially-made revolving pistol. Has no loading lever but comes with loading tool. From Dixie Gun Works, Navy Arms, Uberti USA Inc.
Price: About . . . .$310.00 to $395.00
Price: With loading lever (Uberti USA Inc.) $450.00
Price: Engraved (Navy Arms) $485.00

UBERTI 1st MODEL DRAGOON
Caliber: 44.
Barrel: 7 1/2", part round, part octagon.
Weight: 64 oz.
Stocks: One-piece walnut.
Sights: German silver blade front, hammer notch rear.
Features: First model has oval bolt cuts in cylinder, square-back flared trigger guard, V-type mainspring, short trigger. Ranger and Indian scene roll-engraved on cylinder. Color case-hardened frame, loading lever, plunger and hammer; blue barrel, cylinder, trigger and wedge. Available with old-time charcoal blue or standard blue-black finish. Polished brass backstrap and trigger guard. From Uberti USA Inc., United States Patent Fire-Arms.
Price: . . . .$325.00 to $435.00

Uberti 2nd Model Dragoon Revolver
Similar to the 1st Model except distinguished by rectangular bolt cuts in the cylinder. From Uberti USA, United States Patent Fire-Arms.
Price: . . . .$325.00 to $435.00

Uberti 3rd Model Dragoon Revolver
Similar to the 2nd Model except for oval trigger guard, long trigger, modifications to the loading lever and latch. Imported by Uberti USA Inc., United States Patent Fire-Arms.
Price: Military model (frame cut for shoulder stock, steel backstrap) . . . .$330.00 to $435.00
Price: Civilian (brass backstrap, trigger guard) $325.00

UBERTI 1862 POCKET NAVY PERCUSSION REVOLVER
Caliber: 36, 5-shot.
Barrel: 5 1/2", 6 1/2", octagonal, 7-groove, LH twist.
Weight: 27 oz. (5 1/2" barrel). **Length:** 10 1/2" overall (5 1/2" bbl.).
Stocks: One-piece varnished walnut.
Sights: Brass pin front, hammer notch rear.
Features: Rebated cylinder, hinged loading lever, brass or silver-plated backstrap and trigger guard, color-cased frame, hammer, loading lever, plunger and latch, rest blued. Has original-type markings. From Uberti USA Inc.
Price: With brass backstrap, trigger guard $310.00

WALKER 1847 PERCUSSION REVOLVER
Caliber: 44, 6-shot.
Barrel: 9".
Weight: 84 oz. **Length:** 15 1/2" overall.
Stocks: Walnut.
Sights: Fixed.
Features: Case-hardened frame, loading lever and hammer; iron backstrap; brass trigger guard; engraved cylinder. Imported by Cabela's, Navy Arms, Dixie Gun Works, Uberti USA Inc., E.M.F., Cimarron, Traditions, United States Patent Fire-Arms.
Price: About . . . .$225.00 to $445.00
Price: Single cased set (Navy Arms) $405.00
Price: Deluxe Walker with French fitted case (Navy Arms) $540.00
Price: Hartford model, steel frame, German silver trim, cartouche (E.M.F.) $295.00

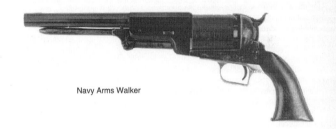

Navy Arms Walker

U.S. PATENT FIRE-ARMS 1862 POCKET NAVY
Caliber: 36.
Barrel: 4 1/2", 5 1/2", 6 1/2".
Weight: 27 oz. (5 1/2" barrel). **Length:** 10 1/2" overall (5 1/2" barrel).
Stocks: Smooth walnut.
Sights: Brass pin front, hammer notch rear.
Features: Blued barrel and cylinder, color case-hardened frame, hammer, lever; silver-plated backstrap and trigger guard. Imported from Italy; available from United States Paten Fire-Arms Mfg. Co.
Price: $335.00

BLACKPOWDER MUSKETS & RIFLES

Armoury R140 Hawken

ARMOURY R140 HAWKEN RIFLE
Caliber: 45, 50 or 54.
Barrel: 29".
Weight: 8 3/4 to 9 lbs. **Length:** 45 3/4" overall.
Stock: Walnut, with cheekpiece.
Sights: Dovetail front, fully adjustable rear.
Features: Octagon barrel, removable breech plug; double set triggers; blued barrel, brass stock fittings, color case-hardened percussion lock. From Armsport, The Armoury.
Price: . . . .$225.00 to $245.00

ARMSPORT 1863 SHARPS RIFLE, CARBINE
Caliber: 45, 54.
Barrel: 28", round.
Weight: 8.4 lbs. **Length:** 46" overall.
Stock: Walnut.
Sights: Blade front, folding adjustable rear. Tang sight set optionally available.
Features: Replica of the 1863 Sharps. Color case-hardened frame, rest blued. Imported by Armsport.
Price: $900.00
Price: Carbine, 54 caliber, 22" barrel $750.00

BOSTONIAN PERCUSSION RIFLE
Caliber: 45.
Barrel: 30", octagonal.
Weight: 7 1/4 lbs. **Length:** 46" overall.
Stock: Walnut.
Sights: Blade front, fixed notch rear.
Features: Color case-hardened lock, brass trigger guard, buttplate, patchbox. Imported from Italy by E.M.F.
Price: $285.00

BLACKPOWDER MUSKETS & RIFLES

CABELA'S BLUE RIDGE RIFLE
Caliber: 32, 36, 45, 50, 54.
Barrel: 39", octagonal.
Weight: About 7 3/4 lbs. **Length:** 55" overall.
Stock: American black walnut.
Sights: Blade front, rear drift adjustable for windage.
Features: Color case-hardened lockplate and cock/hammer, brass trigger guard and buttplate, double set, double-phased triggers. From Cabela's.
Price: Percussion $299.95
Price: Flintlock $319.95
Price: Percussion carbine (28" barrel) $259.95

CABELA'S SHARPS SPORTING RIFLE
Caliber: 45, 54.
Barrel: 31", octagonal.
Weight: About 10 lbs. **Length:** 49" overall.
Stock: American walnut with checkered grip and forend.
Sights: Blade front, ladder-type adjustable rear.
Features: Color case-hardened lock and buttplate. Adjustable double set, double-phased triggers. From Cabela's.
Price: $649.00

CABELA'S TRADITIONAL HAWKEN
Caliber: 45, 50, 54, 58.
Barrel: 29".
Weight: About 9 lbs.
Stock: Walnut.
Sights: Blade front, open adjustable rear.
Features: Flintlock or percussion. Adjustable double-set triggers. Polished brass furniture, color case-hardened lock. Imported by Cabela's.
Price: Percussion, right-hand $159.95
Price: Percussion, left-hand $169.95
Price: Flintlock, right-hand $184.95

Cabela's Sporterized Hawken Hunter Rifle
Similar to the Traditional Hawken's except has more modern stock style with rubber recoil pad, blued furniture, sling swivels. Percussion only, in 45-, 50-, 54- or 58-caliber.
Price: Carbine or rifle, right-hand $179.95
Price: Carbine or rifle, left-hand $189.95

COLT MODEL 1861 MUSKET
Caliber: 58.
Barrel: 40".
Weight: 9 lbs., 3 oz. **Length:** 56" overall.
Stock: Oil-finished walnut.
Sights: Blade front, adjustable folding leaf rear.
Features: Made to original specifications and has authentic Civil War Colt markings. Bright-finished metal, blued nipple and rear sight. Bayonet and accessories available. From Colt Blackpowder Arms Co.
Price: $615.00

CABELA'S RED RIVER RIFLE
Caliber: 45, 50, 54, 58.
Barrel: NA.
Weight: About 7 lbs. **Length:** 45" overall.
Stock: Walnut-stained hardwood.
Sights: Blade front, adjustable buckhorn rear.
Features: Brass trigger guard, forend cap, thimbles; color case-hardened lock and hammer; rubber recoil pad. Introduced 1995. Imported by Cabela's.
Price: $119.95

CABELA'S ROLLING BLOCK MUZZLELOADER
Caliber: 50, 54.
Barrel: 26 1/2" octagonal; 1:32" (50), 1:48" (54) twist.
Weight: About 9 1/4 lbs. **Length:** 43 1/2" overall.
Stock: American walnut, rubber butt pad.
Sights: Blade front, adjustable buckhorn rear.
Features: Uses in-line ignition system, Brass trigger guard, color case-hardened hammer, block and buttplate; black-finished, engraved receiver; easily removable screw-in breech plug; black ramrod and thimble. From Cabela's.
Price: $289.95

Cabela's Rolling Block Muzzleloader Carbine
Similar to the rifle version except has 22 1/4" barrel, weighs 8 1/4 lbs. Has bead on ramp front sight, modern fully adjustable rear. From Cabela's.
Price: $269.95

Manufacturers' addresses in the
Directory of the Arms Trade
page 309, this issue

COLT GAME MASTER 50 RIFLE
Caliber: 50.
Barrel: 32", 1:28" twist.
Weight: About 8 lbs. **Length:** NA.
Stock: European walnut.
Sights: Bead on ramp front, fully adjustable aperture rear.
Features: Polished blue finish. Introduced 1997. Imported from Italy by Colt Blackpowder Arms Co.
Price: ....NA

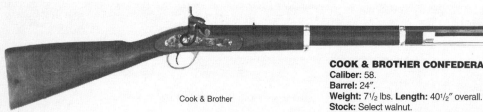

Cook & Brother

COOK & BROTHER CONFEDERATE CARBINE
Caliber: 58.
Barrel: 24".
Weight: 7 1/2 lbs. **Length:** 40 1/2" overall.
Stock: Select walnut.
Features: Recreation of the 1861 New Orleans-made artillery carbine. Color case-hardened lock, browned barrel. Buttplate, trigger guard, barrel bands, sling swivels and nose cap of polished brass. From Euroarms of America.
Price: $447.00
Price: Cook & Brother rifle (33" barrel) $480.00

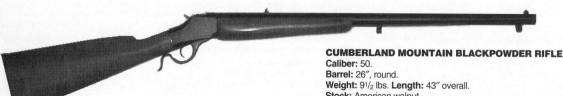

Cumberland Mountain

CUMBERLAND MOUNTAIN BLACKPOWDER RIFLE
Caliber: 50.
Barrel: 26", round.
Weight: 9 1/2 lbs. **Length:** 43" overall.
Stock: American walnut.
Sights: Bead front, open rear adjustable for windage.
Features: Falling block action fires with shotshell primer. Blued receiver and barrel. Introduced 1993. Made in U.S. by Cumberland Mountain Arms, Inc.
Price: $931.50

BLACKPOWDER MUSKETS & RIFLES

CVA Accubolt Pro

CVA AccuBolt Pro In-Line Rifle
Similar to the standard AccuBolt except has 24" Badger barrel, Bell & Carlson composite thumbhole stock, and comes with a hard case. No iron sights. Introduced 1997. From CVA.
Price: $449.95

CVA Staghorn Rifle
Similar to the Apollo Shadow except has blued barrel and action, and is available in 50, 54 caliber. Drilled and tapped receiver for scope mount or aperture sight. Introduced 1996.
Price: $184.95

CVA Buckmaster Rifle
Similar to the Apollo Shadow except has Dura-Grip synthetic stock with Advantage camouflage pattern, and blued barrel and action. Introduced 1996. From CVA.
Price: $246.95

CVA BOBCAT RIFLE
Caliber: 50 and 54.
Barrel: 26"; 1:48" twist.
Weight: 6 1/2 lbs. **Length:** 40" overall.
Stock: Dura-Grip synthetic.
Sights: Blade front, open rear.
Features: Oversize trigger guard; wood ramrod; matte black finish. Introduced 1995. From CVA.
Price: $125.95

CVA ACCUBOLT IN-LINE RIFLE
Caliber: 50.
Barrel: 24" hammer-forged; 1:32" twist.
Weight: 7 1/2 lbs. **Length:** NA.
Stock: Dura-Grip synthetic, with Monte Carlo comb.
Sights: CVA Illuminator Fiber Optic Sight system. Drilled and tapped for scope mounting.
Features: Uses CVA's AccuSystem copper-coated bullets and special bullet sizer (included). Synthetic ramrod, removable breech plug, swivel studs, rubber recoil pad. Introduced 1997. From CVA.
Price: $334.95

CVA APOLLO SHADOW, BROWN BEAR RIFLES
Caliber: 50, 54.
Barrel: 24"; round with octagon integral receiver; 1:32" twist.
Weight: 7-7 1/2 lbs. **Length:** 42" overall.
Stock: Synthetic Dura-Grip (Shadow); select hardwood (Brown Bear); pistol grip, solid rubber buttpad.
Sights: Blade on ramp front, fully adjustable rear; drilled and tapped for scope mounting.
Features: In-line ignition, modern-style trigger with automatic safety; oversize trigger guard; synthetic ramrod. From CVA.
Price: Shadow $239.95
Price: Brown Bear $223.95
Price: Brown Bear left-hand $237.95
Price: Eclipse $207.95

CVA Bobcat Hunter
Similar to the Bobcat except has black synthetic stock with checkered wrist and forend, drilled and tapped for scope mounting, engraved, blued lockplate and offset hammer, and has sporter adjustable rear sight. Available in 36 caliber. Introduced 1995. From CVA.
Price: $189.95

CVA Firebolt

CVA GREY WOLF RIFLES
Caliber: 50, 54.
Barrel: 26" octagonal; 1:32" twist; 15/16" flats; blue finish.
Weight: 6 1/2 lbs. **Length:** 40" overall.
Stock: Tuff-Lite polymer—gray finish, solid buttplate (Grey Wolf); Realtree All Purpose® camo finish, solid buttplate (Timber Wolf); checkered grip.
Sights: Blade front on ramp, fully adjustable open rear; drilled and tapped for scope mounting.
Features: Oversize trigger guard; synthetic ramrod; offset hammer. From CVA.
Price: Grey Wolf $169.95

CVA FIREBOLT BOLT-ACTION IN-LINE RIFLES
Caliber: 50, 54.
Barrel: 24".
Weight: 7 lbs. **Length:** NA.
Stock: Dura Grip synthetic; thumbhole, traditional, camo.
Sights: CVA Illuminator Fiber Optic Sight System.
Features: Bolt-action, in-line ignition system. Stainless steel or matte blue barrel; removable breech plug; trigger-block safety. Introduced 1997. From CVA.
Price: Stainless barrel, traditional stock $329.95
Price: Matte blue barrel, Advantage camo stock $329.95
Price: Matte blue barrel, thumbhole stock $299.95
Price: Matte blue barrel, traditional stock $269.95

CVA RMEF Elk Master

CVA RMEF SERIES RIFLES
Caliber: 54.
Barrel: 24".
Weight: 7 1/2 lbs. **Length:** NA.
Stock: Bell & Carlson composite; Elk Master has thumbhole style, Elk Hunter has traditional design.
Sights: CVA Illuminator Fiber Optic Sight System.
Features: In-line ignition system. Stocks have Realtree X-tra Grey camouflage, bronze Rocky Mountain Elk Foundation medallion inset into butt. Parkerized barrel; sling swivel studs; oversize trigger guard; recoil pad. Introduced 1997. From CVA.
Price: Elk Master (thumbhold stock) $416.95
Price: Elk Hunter (traditional stock) $317.95

CVA ST. LOUIS HAWKEN RIFLE
Caliber: 50, 54.
Barrel: 28", octagon; 15/16" across flats; 1:48" twist.
Weight: 8 lbs. **Length:** 44" overall.
Stock: Select hardwood.
Sights: Beaded blade front, fully adjustable open rear.
Features: Fully adjustable double-set triggers; synthetic ramrod (kits have wood); brass patch box, wedge plates, nosecap, thimbles, trigger guard and buttplate; blued barrel; color case-hardened, engraved lockplate. V-type mainspring. Button breech. Introduced 1981. From CVA.
Price: St. Louis Hawken, finished (50-, 54-cal.) $219.95
Price: Left-hand, percussion $234.95
Price: Flintlock, 50-cal. only $234.95
Price: Flintlock, left-hand $249.95
Price: Percussion kit (50-cal., blued, wood ramrod) $169.95

BLACKPOWDER MUSKETS & RIFLES

CVA VARMINT RIFLE
Caliber: 32.
Barrel: 24" octagonal; 7/8" flats; 1:48" rifling.
Weight: 6 3/4 lbs. **Length:** 40" overall.
Stock: Select hardwood.
Sights: Blade front, Patridge-style click adjustable rear.
Features: Brass trigger guard, nose cap, wedge plate, thimble and buttplate. Drilled and tapped for scope mounting. Color case-hardened lock. Single trigger. Aluminum ramrod. Imported by CVA.
Price: $219.95

DIXIE DELUX CUB RIFLE
Caliber: 40.
Barrel: 28".
Weight: 6 1/2 lbs.
Stock: Walnut.
Sights: Fixed.
Features: Short rifle for small game and beginning shooters. Brass patchbox and furniture. Flint or percussion. From Dixie Gun Works.
Price: Finished $395.00
Price: Kit $350.00
Price: Super Cub (50-caliber) $350.00

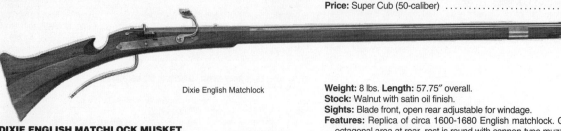
Dixie English Matchlock

Weight: 8 lbs. **Length:** 57.75" overall.
Stock: Walnut with satin oil finish.
Sights: Blade front, open rear adjustable for windage.
Features: Replica of circa 1600-1680 English matchlock. Getz barrel with 11" octagonal area at rear, rest is round with cannon-type muzzle. All steel finished in the white. Imported by Dixie Gun Works.
Price: $895.00

DIXIE ENGLISH MATCHLOCK MUSKET
Caliber: 72.
Barrel: 44".

Dixie Inline Carbine

Weight: 6.5 lbs. **Length:** 41" overall.
Stock: Walnut-finished hardwood with Monte Carlo comb.
Sights: Ramp front with red insert, open fully adjustable rear.
Features: Sliding "bolt" fully encloses cap and nipple. Fully adjustable trigger, automatic safety. Aluminum ramrod. Imported from Italy by Dixie Gun Works.
Price: $349.95

DIXIE INLINE CARBINE
Caliber: 50, 54.
Barrel: 24"; 1:32" twist.

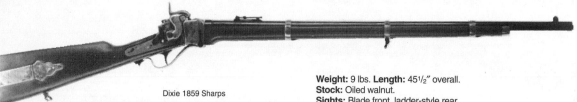
Dixie 1859 Sharps

Weight: 9 lbs. **Length:** 45 1/2" overall.
Stock: Oiled walnut.
Sights: Blade front, ladder-style rear.
Features: Blued barrel, color case-hardened barrel bands, receiver, hammer, nose cap, lever, patchbox cover and buttplate. Introduced 1995. Imported from Italy by Dixie Gun Works.
Price: $895.00

DIXIE SHARPS NEW MODEL 1859 MILITARY RIFLE
Caliber: 54.
Barrel: 30", 6-groove; 1:48" twist.

Dixie Model 1816

DIXIE U.S. MODEL 1816 FLINTLOCK MUSKET
Caliber: 69.
Barrel: 42", smoothbore.
Weight: 9.75 lbs. **Length:** 56.5" overall.
Stock: Walnut with oil finish.
Sights: Blade front.
Features: All metal finished "National Armory Bright"; three barrel bands with springs; steel ramrod with button-shaped head. Imported by Dixie Gun Works.
Price: $725.00

DIXIE TENNESSEE MOUNTAIN RIFLE
Caliber: 32 or 50.
Barrel: 41 1/2", 6-groove rifling, brown finish. **Length:** 56" overall.
Stock: Walnut, oil finish; Kentucky-style.
Sights: Silver blade front, open buckhorn rear.
Features: Recreation of the original mountain rifles. Early Schultz lock, interchangeable flint or percussion with vent plug or drum and nipple. Tumbler has fly. Double-set triggers. All metal parts browned. From Dixie Gun Works.
Price: Flint or percussion, finished rifle, 50-cal. $575.00
Price: Kit, 50-cal. $495.00
Price: Left-hand model, flint or percussion $575.00
Price: Left-hand kit, flint or perc., 50-cal. $495.00
Price: Squirrel Rifle (as above except in 32-cal. with 13/16" barrel flats), flint or percussion $575.00
Price: Kit, 32-cal., flint or percussion $495.00

DIXIE U.S. MODEL 1861 SPRINGFIELD
Caliber: 58.
Barrel: 40".
Weight: About 8 lbs. **Length:** 55 13/16" overall.
Stock: Oil-finished walnut.
Sights: Blade front, step adjustable rear.
Features: Exact recreation of original rifle. Sling swivels attached to trigger guard bow and middle barrel band. Lockplate marked "1861" with eagle motif and "U.S. Springfield" in front of hammer; "U.S." stamped on top of buttplate. From Dixie Gun Works.
Price: $595.00
Price: From Stone Mountain Arms $599.00
Price: Kit $525.00

DIXIE 1863 SPRINGFIELD MUSKET
Caliber: 58 (.570" patched ball or .575" Minie).
Barrel: 50", rifled.
Stocks: Walnut stained.
Sights: Blade front, adjustable ladder-type rear.
Features: Bright-finish lock, barrel, furniture. Reproduction of the last of the regulation muzzleloaders. Imported from Japan by Dixie Gun Works.
Price: Finished $595.00
Price: Kit $525.00

BLACKPOWDER MUSKETS & RIFLES

E.M.F. 1863 SHARPS MILITARY CARBINE
Caliber: 54.
Barrel: 22", round.
Weight: 8 lbs. **Length:** 39" overall.
Stock: Oiled walnut.
Sights: Blade front, military ladder-type rear.
Features: Color case-hardened lock, rest blued. Imported by E.M.F.
Price: $860.00

EUROARMS BUFFALO CARBINE
Caliber: 58.
Barrel: 26", round.
Weight: 7 3/4 lbs. **Length:** 42" overall.
Stock: Walnut.
Sights: Blade front, open adjustable rear.
Features: Shoots .575" round ball. Color case-hardened lock, blue hammer, barrel, trigger; brass furniture. Brass patchbox. Imported by Euroarms of America.
Price: $440.00

Euroarms Volunteer

EUROARMS VOLUNTEER TARGET RIFLE
Caliber: .451.
Barrel: 33" (two-band), 36" (three-band).
Weight: 11 lbs. (two-band). **Length:** 48.75" overall (two-band).
Stock: European walnut with checkered wrist and forend.
Sights: Hooded bead front, adjustable rear with interchangeable leaves.
Features: Alexander Henry-type rifling with 1:20" twist. Color case-hardened hammer and lockplate, brass trigger guard and nose cap, rest blued. Imported by Euroarms of America.
Price: Two-band $720.00
Price: Three-band $773.00

Euroarms 1861

FORT WORTH FIREARMS SABINE RIFLE
Caliber: 22.
Barrel: 16 1/4".
Weight: 3 1/2 lbs. **Length:** 32" overall.
Stocks: Walnut-finished hardwood.
Sights: Hooded blade front, open adjustable rear.
Features: In-line design with side cocking lever. Positive click safety; blued finish. Introduced 1997. From Fort Worth Firearms.
Price: $189.95

EUROARMS 1861 SPRINGFIELD RIFLE
Caliber: 58.
Barrel: 40".
Weight: About 10 lbs. **Length:** 55.5" overall.
Stock: European walnut.
Sights: Blade front, three-leaf military rear.
Features: Reproduction of the original three-band rifle. Lockplate marked "1861" with eagle and "U.S. Springfield." Metal left in the white. Imported by Euroarms of America.
Price: $530.00

> Consult our Directory pages for the location of firms mentioned.

FORT WORTH FIREARMS PECOS RIFLE
Caliber: 50.
Barrel: 22", stainless; 1:24" twist.
Weight: 6 1/2 lbs. **Length:** 39" overall.
Stock: Black or camo composite with checkered grip and forend.
Sights: Ramped blade front, open adjustable rear. Drilled and tapped for scope mounting.
Features: In-line design with stainless steel barrel and receiver; fully adjustable trigger; synthetic Delron ramrod. Introduced 1997. From Fort Worht Firearms.
Price: $457.95

FORT WORTH FIREARMS RIO GRANDE RIFLE
Caliber: 45, 50.
Barrel: 22"; 1:22" twist.
Weight: 6 1/2 lbs. **Length:** 39" overall.
Stock: Black composite; checkered grip and forend; swivel studs; recoil pad.
Sights: Ramped blade front, open adjustable rear. Drilled and tapped for scope mounting.
Features: Bolt-action design with stainless barrel and receiver. Flash diffuser protects optics from blow-by. Fully adjustable trigger with safety; synthetic Delron ramrod. Introduced 1996. From Fort Worth Firearms.
Price: $457.95

Gonic GA-87

Gonic GA-93 Magnum M/L Rifle
Similar to the GA-87 except has open bolt mechanism, single safety, 22" barrel and comes only in 50-caliber. Stock is black wrinkle-finish wood or gray or brown, standard or thumbhole laminate. **Partial listing shown.** Introduced 1993. From Gonic Arms, Inc.
Price: Black stock, blue, no sights $483.30
Price: As above, stainless $562.25
Price: Black stock, blue, open sights $500.57
Price: As above, stainless $603.04

HARPER'S FERRY 1803 FLINTLOCK RIFLE
Caliber: 54 or 58.
Barrel: 35".
Weight: 9 lbs. **Length:** 59 1/2" overall.
Stock: Walnut with cheekpiece.

GONIC GA-87 M/L RIFLE
Caliber: 45, 50.
Barrel: 26".
Weight: 6 to 6 1/2 lbs. **Length:** 43" overall (Carbine).
Stock: American walnut with checkered grip and forend, or laminated stock.
Sights: Optional bead front, open or peep rear adjustable for windage and elevation; drilled and tapped for scope bases (included).
Features: Closed-breech action with straight-line ignition. Modern trigger mechanism with ambidextrous safety. Satin blue finish on metal, satin stock finish. Introduced 1989. From Gonic Arms, Inc.
Price: Standard rifle, no sights $800.41
Price: As above, with sights, from $869.93
Price: Walnut stock, peep sight $880.47

Sights: Brass blade front, fixed steel rear.
Features: Brass trigger guard, sideplate, buttplate; steel patch box. Imported by Euroarms of America, Navy Arms (54-cal. only), Cabela's.
Price: . . . .$495.95 to $729.00
Price: 54-cal. (Navy Arms) $625.00

BLACKPOWDER MUSKETS & RIFLES

HAWKEN RIFLE
Caliber: 45, 50, 54 or 58.
Barrel: 28", blued, 6-groove rifling.
Weight: 8 3/4 lbs. **Length:** 44" overall.
Stock: Walnut with cheekpiece.
Sights: Blade front, fully adjustable rear.
Features: Coil mainspring, double-set triggers, polished brass furniture. From Armsport, Navy Arms, E.M.F.
Price: ....$220.00 to $345.00

J.P. HENRY TRADE RIFLE
Caliber: 54.
Barrel: 34", 1" flats.
Weight: 8 1/2 lbs. **Length:** 45" overall.
Stock: Premium curly maple.
Sights: Silver blade front, fixed buckhorn rear.
Features: Brass buttplate, side plate, trigger guard and nose cap; browned barrel and lock; L&R Large English percussion lock; single trigger. Made in U.S. by October Country.
Price: $965.50

Ithaca-Navy Hawken

ITHACA-NAVY HAWKEN RIFLE
Caliber: 50.
Barrel: 32" octagonal, 1" dia.
Weight: About 9 lbs.
Stocks: Walnut.
Sights: Blade front, rear adjustable for windage.
Features: Hooked breech, 1 7/8" throw percussion lock. Attached twin thimbles and under-rib. German silver barrel key inlays, Hawken-style toe and buttplates, lock bolt inlays, barrel wedges, entry thimble, trigger guard, ramrod and cleaning jag, nipple and nipple wrench. Introduced 1977. From Navy Arms.
Price: Complete, percussion $445.00

KENTUCKIAN RIFLE & CARBINE
Caliber: 44.
Barrel: 35" (Rifle), 27 1/2" (Carbine).
Weight: 7 lbs. (Rifle), 5 1/2 lbs. (Carbine). **Length:** 51" overall (Rifle), 43" (Carbine).
Stock: Walnut stain.
Sights: Brass blade front, steel V-ramp rear.
Features: Octagon barrel, case-hardened and engraved lockplates. Brass furniture. Imported by Dixie Gun Works.
Price: Rifle or carbine, flint, about $269.95
Price: As above, percussion, about $259.95

Navy Arms Kentucky Flintlock

Kentucky Percussion Rifle
Similar to flintlock except percussion lock. Finish and features vary with importer. Imported by Navy Arms, The Armoury, CVA.
Price: About $259.95
Price: 45- or 50-cal. (Navy Arms) $400.00
Price: Kit, 50-cal. (CVA) $189.95

KENTUCKY FLINTLOCK RIFLE
Caliber: 44, 45, or 50.
Barrel: 35".
Weight: 7 lbs. **Length:** 50" overall.
Stock: Walnut stained, brass fittings.
Sights: Fixed.
Features: Available in carbine model also, 28" bbl. Some variations in detail, finish. Kits also available from some importers. Imported by Navy Arms, The Armoury.
Price: About ....$217.95 to $345.00
Price: Flintlock, 45 or 50-cal. (Navy Arms) $410.00

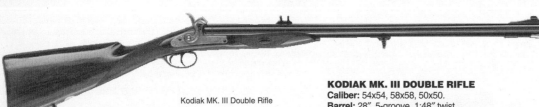

Kodiak MK. III Double Rifle

KODIAK MK. III DOUBLE RIFLE
Caliber: 54x54, 58x58, 50x50.
Barrel: 28", 5-groove, 1:48" twist.
Weight: 9 1/2 lbs. **Length:** 43 1/4" overall.
Stock: Czechoslovakian walnut, hand-checkered.
Sights: Adjustable bead front, adjustable open rear.
Features: Hooked breech allows interchangeability of barrels. Comes with sling, swivels, bullet mould and bullet starter. Engraved lockplates, top tang and trigger guard. Locks and top tang polished, rest browned. Introduced 1976. Imported from Italy by Navy Arms.
Price: 50-, 54-, 58-cal. SxS $775.00

KNIGHT BK-92 BLACK KNIGHT RIFLE
Caliber: 50, 54.
Barrel: 24", blued.
Weight: 6 1/2 lbs.
Stock: Black composition.
Sights: Blade front on ramp, open adjustable rear.
Features: Patented double safety system; removable breech plug for cleaning; adjustable Accu-Lite trigger; Green Mountain barrel; receiver drilled and tapped for scope bases. Made in U.S. by Modern Muzzleloading, Inc.
Price: With composition stock $399.95

KNIGHT LK-93 WOLVERINE RIFLE
Caliber: 50.
Barrel: 22", blued.
Weight: 6 lbs.
Stock: Black Fiber-Lite synthetic.
Sights: Blade front on ramp, open adjustable rear.
Features: Patented double safety system; removable breech plug; Sure-Fire in-line percussion ignition system. Made in U.S. by Modern Muzzleloading, Inc.
Price: $319.95
Price: LK-93 Stainless $399.95
Price: LK-93 Thumbhole $409.95

KNIGHT MK-85 RIFLE
Caliber: 50, 54.
Barrel: 24".
Weight: 6 3/4 lbs.
Stock: Walnut, laminated or composition.
Sights: Hooded blade front on ramp, open adjustable rear.
Features: Patented double safety; Sure-Fire in-line percussion ignition; Timney Featherweight adjustable trigger; aluminum ramrod; receiver drilled and tapped for scope bases. Made in U.S. by Modern Muzzleloading, Inc.
Price: Hunter, walnut stock $539.95
Price: Stalker, laminated or composition stock $679.95
Price: Predator (stainless steel), laminated or composition stock $759.95
Price: Knight Hawk, blued, composition thumbhole stock $779.95
Price: As above, stainless steel $869.95

BLACKPOWDER MUSKETS & RIFLES

Knight MK-95 Magnum

LONDON ARMORY 2-BAND 1858 ENFIELD
Caliber: .577" Minie, .575" round ball.
Barrel: 33".
Weight: 10 lbs. **Length:** 49" overall.
Stock: Walnut.
Sights: Folding leaf rear adjustable for elevation.
Features: Blued barrel, color case-hardened lock and hammer, polished brass buttplate, trigger guard, nosecap. From Navy Arms, Euroarms of America, Dixie Gun Works.
Price: ....$385.00 to $531.00

KNIGHT MK-95 MAGNUM ELITE RIFLE
Caliber: 50, 54.
Barrel: 24", stainless.
Weight: 6 3/4 lbs.
Stock: Composition; black or Realtree All-Purpose camouflage.
Sights: Hooded blade front on ramp, open adjustable rear.
Features: Enclosed Posi-Fire ignition system uses large rifle primers; Timney Featherweight adjustable trigger; Green Mountain barrel; receiver drilled and tapped for scope bases. Made in U.S. by Modern Muzzleloading, Inc.
Price: Black composition stock $839.95

London Armory 1861

LONDON ARMORY 1861 ENFIELD MUSKETOON
Caliber: 58, Minie ball.
Barrel: 24", round.
Weight: 7-7 1/2 lbs. **Length:** 40 1/2" overall.
Stock: Walnut, with sling swivels.
Sights: Blade front, graduated military-leaf rear.
Features: Brass trigger guard, nose cap, buttplate; blued barrel, bands, lockplate, swivels. Imported by Euroarms of America, Navy Arms.
Price: ....$300.00 to $427.00
Price: Kit ....$365.00 to $373.00

LONDON ARMORY 3-BAND 1853 ENFIELD
Caliber: 58 (.577" Minie, .575" round ball, .580" maxi ball).
Barrel: 39".
Weight: 9 1/2 lbs. **Length:** 54" overall.
Stock: European walnut.
Sights: Inverted "V" front, traditional Enfield folding ladder rear.
Features: Recreation of the famed London Armory Company Pattern 1853 Enfield Musket. One-piece walnut stock, brass buttplate, trigger guard and nose cap. Lockplate marked "London Armoury Co." and with a British crown. Blued Baddeley barrel bands. From Dixie Gun Works, Euroarms of America, Navy Arms.
Price: About ....$350.00 to $495.00
Price: Assembled kit (Dixie, Euroarms of America) ....$425.00 to $431.00

LYMAN DEERSTALKER RIFLE
Caliber: 50, 54.
Barrel: 24", octagonal; 1:48" rifling.
Weight: 7 1/2 lbs.
Stock: Walnut with black rubber buttpad.
Sights: Lyman #37MA beaded front, fully adjustable fold-down Lyman #16A rear.
Features: Stock has less drop for quick sighting. All metal parts are blackened, with color case-hardened lock; single trigger. Comes with sling and swivels. Available in flint or percussion. Introduced 1990. From Lyman.
Price: 50- or 54-cal., percussion $299.95
Price: 50- or 54-cal., flintlock $324.95
Price: 50- or 54-cal., percussion, left-hand $314.95
Price: 50-cal., flintlock, left-hand $329.95
Price: Stainless steel ....NA

LYMAN COUGAR IN-LINE RIFLE
Caliber: 50 or 54.
Barrel: 22"; 1:24" twist.
Weight: NA. **Length:** NA.
Stock: Smooth walnut; swivel studs.
Sights: Bead on ramp front, folding adjustable rear. Drilled and tapped for Lyman 57WTR receiver sight and Weaver scope bases.
Features: Blued barrel and receiver. Has bolt safety notch and trigger safety. Rubber recoil pad. Delrin ramrod. Introduced 1996. From Lyman.
Price: $299.95
Price: Stainless steel $382.95

Lyman Deerstalker Custom Carbine
Similar to the Deerstalker rifle except in 50-caliber only with 21" stepped octagon barrel; 1:24" twist for optimum performance with conical projectiles. Comes with Lyman 37MA front sight, Lyman 16A folding rear. Weighs 6 3/4 lbs., measures 38 1/2" overall. Percussion or flintlock. Comes with Delrin ramrod, modern sling and swivels. Introduced 1991.
Price: Percussion $324.95
Price: Percussion, left-hand $329.95

Lyman Great Plains Hunter

Lyman Great Plains Hunter Rifle
Similar to the Great Plains model except has 1:32" twist shallow-groove barrel and comes drilled and tapped for the Lyman 57GPR peep sight.
Price: $424.95

LYMAN GREAT PLAINS RIFLE
Caliber: 50- or 54-cal.
Barrel: 32", 1:66" twist.
Weight: 9 lbs.
Stock: Walnut.
Sights: Steel blade front, buckhorn rear adjustable for windage and elevation and fixed notch primitive sight included.
Features: Blued steel furniture. Stainless steel nipple. Coil spring lock, Hawken-style trigger guard and double-set triggers. Round thimbles recessed and sweated into rib. Steel wedge plates and toe plate. Introduced 1979. From Lyman.
Price: Percussion $424.95
Price: Flintlock $449.95
Price: Percussion kit $344.95
Price: Flintlock kit $369.95
Price: Left-hand percussion $434.95
Price: Left-hand flintlock $459.95

BLACKPOWDER MUSKETS & RIFLES

Marlin Model MLS-50

MARLIN MODELS MLS-50, MLS-54 IN-LINE RIFLES
Caliber: 50, 54.
Barrel: 22", 1:28" twist.
Weight: 6½ lbs. **Length:** 41" overall.
Stock: Black fiberglass-reinforced Rynite with moulded-in checkering, rubber buttpad, swivel studs.
Sights: Ramp front with brass bead, adjustable Marble open rear. Receiver drilled and tapped for scope mounting.
Features: All stainless steel construction. Reversible cocking handle for right- or left-hand shooters; automatic tang safety; one-piece barrel/receiver. Introduced 1997. Made in U.S. by Marlin Firearms Co.
Price: $411.00

LYMAN TRADE RIFLE
Caliber: 50, 54.
Barrel: 28" octagon, 1:48" twist.
Weight: 8¾ lbs. **Length:** 45" overall.
Stock: European walnut.
Sights: Blade front, open rear adjustable for windage or optional fixed sights.
Features: Fast twist rifling for conical bullets. Polished brass furniture with blue steel parts, stainless steel nipple. Hook breech, single trigger, coil spring percussion lock. Steel barrel rib and ramrod ferrules. Introduced 1980. From Lyman.
Price: Percussion $299.95
Price: Flintlock $324.95

Navy Arms J.P. Murray

J.P. MURRAY 1862-1864 CAVALRY CARBINE
Caliber: 58 (.577" Minie).
Barrel: 23".
Weight: 7 lbs., 9 oz. **Length:** 39" overall.
Stock: Walnut.
Sights: Blade front, rear drift adjustable for windage.
Features: Browned barrel, color case-hardened lock, blued swivel and band springs, polished brass buttplate, trigger guard, barrel bands. From Navy Arms, Euroarms of America.
Price: ....$405.00 to $453.00

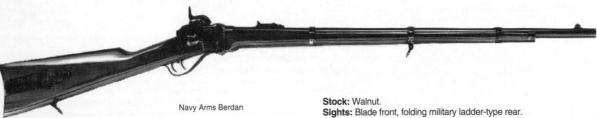

Navy Arms Berdan

NAVY ARMS BERDAN 1859 SHARPS RIFLE
Caliber: 54.
Barrel: 30".
Weight: 8 lbs., 8 oz. **Length:** 46¾" overall.
Stock: Walnut.
Sights: Blade front, folding military ladder-type rear.
Features: Replica of the Union sniper rifle used by Berdan's 1st and 2nd Sharpshooter regiments. Color case-hardened receiver, patch box, furniture. Double-set triggers. Imported by Navy Arms.
Price: $1,095.00
Price: 1859 Sharps Infantry Rifle (three-band) $1,030.00

Navy Arms Country Boy

NAVY ARMS COUNTRY BOY IN-LINE RIFLE
Caliber: 50.
Barrel: 24".
Weight: 8 lbs. **Length:** 41" overall.
Stock: Black composition.
Sights: Bead front, fully adjustable open rear.
Features: Chrome-lined barrel; receiver drilled and tapped for scope mount; buttstock has trap containing takedown tool for nipple and breech plug removal. Introduced 1996. From Navy Arms.
Price: $165.00
Price: With satin chrome finish $175.00

NAVY ARMS HAWKEN HUNTER RIFLE/CARBINE
Caliber: 50, 54, 58.
Barrel: 22½" or 28"; 1:48" twist.
Weight: 6 lbs., 12 oz. **Length:** 39" overall.
Stock: Walnut with cheekpiece.
Sights: Blade front, fully adjustable rear.
Features: Double-set triggers; all metal has matte black finish; rubber recoil pad; detachable sling swivels. Imported by Navy Arms.
Price: Rifle or Carbine $240.00

Navy Arms Mortimer Match

NAVY ARMS MORTIMER FLINTLOCK RIFLE
Caliber: 54.
Barrel: 36".
Weight: 9 lbs. **Length:** 52¼" overall.
Stock: Checkered walnut.
Sights: Bead front, rear adjustable for windage.
Features: Waterproof pan, roller frizzen; sling swivels; browned barrel; external safety. Introduced 1991. Imported by Navy Arms.
Price: $780.00
Price: Mortimer Match Rifle (hooded globe front sight, fully adjustable target aperture rear, color case-hardened lock) $905.00

BLACKPOWDER MUSKETS & RIFLES

Navy Arms Whitworth

NAVY ARMS PARKER-HALE VOLUNTEER RIFLE
Caliber: .451".
Barrel: 32".
Weight: 9 1/2 lbs. **Length:** 49" overall.
Stock: Walnut, checkered wrist and forend.
Sights: Globe front, adjustable ladder-type rear.
Features: Recreation of the type of gun issued to volunteer regiments during the 1860s. Rigby-pattern rifling, patent breech, detented lock. Stock is glass bedded for accuracy. Imported by Navy Arms.
Price: $850.00

NAVY ARMS PENNSYLVANIA LONG RIFLE
Caliber: 32, 45.
Barrel: 40 1/2".
Weight: 7 1/2 lbs. **Length:** 56 1/2" overall.
Stock: Walnut.
Sights: Blade front, fully adjustable rear.
Features: Browned barrel, brass furniture, polished lock with double-set triggers. Imported by Navy Arms.
Price: Percussion $460.00
Price: Flintlock $475.00

NAVY ARMS 1777 CHARLEVILLE MUSKET
Caliber: 69.
Barrel: 44 5/8".
Weight: 10 lbs., 4 oz. **Length:** 59 3/4" overall.
Stock: Walnut.
Sights: Brass blade front.
Features: Exact copy of the musket used in the French Revolution. All steel is polished, in the white. Brass flashpan. Introduced 1991. Imported by Navy Arms.
Price: $810.00
Price: 1816 M.T. Wickham Musket $810.00

NAVY ARMS PARKER-HALE WHITWORTH MILITARY TARGET RIFLE
Caliber: 45.
Barrel: 36".
Weight: 9 1/4 lbs. **Length:** 52 1/2" overall.
Stock: Walnut. Checkered at wrist and forend.
Sights: Hooded post front, open step-adjustable rear.
Features: Faithful reproduction of the Whitworth rifle, only bored for 45-cal. Trigger has a detented lock, capable of being adjusted very finely without risk of the sear nose catching on the half-cock bent and damaging both parts. Introduced 1978. Imported by Navy Arms.
Price: $875.00

NAVY ARMS SMITH CARBINE
Caliber: 50.
Barrel: 21 1/2".
Weight: 7 3/4 lbs. **Length:** 39" overall.
Stock: American walnut.
Sights: Brass blade front, folding ladder-type rear.
Features: Replica of the breech-loading Civil War carbine. Color case-hardened receiver, rest blued. Cavalry model has saddle ring and bar, Artillery model has sling swivels. Imported by Navy Arms.
Price: Cavalry model $600.00
Price: Artillery model $600.00

NAVY ARMS 1859 SHARPS CAVALRY CARBINE
Caliber: 54.
Barrel: 22".
Weight: 7 3/4 lbs. **Length:** 39" overall.
Stock: Walnut.
Sights: Blade front, military ladder-type rear.
Features: Color case-hardened action, blued barrel. Has saddle ring. Introduced 1991. Imported from Navy Arms.
Price: $885.00

Navy Arms 1863

NAVY ARMS 1861 SPRINGFIELD RIFLE
Caliber: 58.
Barrel: 40".
Weight: 10 lbs., 4 oz. **Length:** 56" overall.
Stock: Walnut.
Sights: Blade front, military leaf rear.
Features: Steel barrel, lock and all furniture have polished bright finish. Has 1855-style hammer. Imported by Navy Arms.
Price: $550.00

NAVY ARMS 1863 SPRINGFIELD
Caliber: 58, uses .575" Minie.
Barrel: 40", rifled.
Weight: 9 1/2 lbs. **Length:** 56" overall.
Stock: Walnut.
Sights: Open rear adjustable for elevation.
Features: Full-size three-band musket. Polished bright metal, including lock. From Navy Arms.
Price: Finished rifle $550.00

NAVY ARMS 1863 C.S. RICHMOND RIFLE
Caliber: 58.
Barrel: 40".
Weight: 10 lbs. **Length:** NA.
Stock: Walnut.
Sights: Blade front, adjustable rear.
Features: Copy of the three-band rifle musket made at Richmond Armory for the Confederacy. All steel polished bright. Imported by Navy Arms.
Price: $550.00

> Consult our Directory pages for the location of firms mentioned.

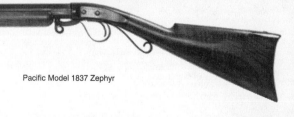

Pacific Model 1837 Zephyr

PACIFIC RIFLE MODEL 1837 ZEPHYR
Caliber: 62.
Barrel: 30", tapered octagon.
Weight: 7 3/4 lbs. **Length:** NA.
Stocks: Oil-finished fancy walnut.
Sights: German silver blade front, semi-buckhorn rear. Options available.
Features: Improved underhammer action. First production rifle to offer Forsyth rifle, with narrow lands and shallow rifling with 1:144" pitch for high velocity round balls. Metal finish is slow rust brown with nitre blue accents. Optional sights, finishes and integral muzzle brake available. Introduced 1995. Made in U.S. by Pacific Rifle Co.
Price: From $795.00

BLACKPOWDER MUSKETS & RIFLES

PENNSYLVANIA FULL-STOCK RIFLE
Caliber: 45 or 50.
Barrel: 32″ rifled, 15/16″ dia.
Weight: 8 1/2 lbs.
Stock: Walnut.
Sights: Fixed.
Features: Available in flint or percussion. Blued lock and barrel, brass furniture. Offered complete or in kit form. From The Armoury.
Price: Flint $250.00
Price: Percussion $225.00

PRAIRIE RIVER ARMS PRA CLASSIC RIFLE
Caliber: 50, 54.
Barrel: 26″; 1:28″ twist.
Weight: 7 1/2 lbs. **Length:** 40 1/2″ overall.
Stock: Hardwood or black all-weather.
Sights: Blade front, open adjustable rear.
Features: Patented internal percussion ignition system. Drilled and tapped for scope mount. Introduced 1995. Made in U.S. by Prairie River Arms, Ltd.
Price: 4140 alloy barrel, hardwood stock $375.00
Price: As above, stainless barrel $425.00
Price: 4140 alloy barrel, black all-weather stock $390.00
Price: As above, stainless barrel $440.00

STONE MOUNTAIN SILVER EAGLE RIFLE
Caliber: 50.
Barrel: 26″, octagonal; 15/16″ flats; 1:48″ twist.
Weight: About 6 1/2 lbs. **Length:** 40″ overall.
Stock: Dura-Grip synthetic; checkered grip and forend.
Sights: Blade front, fixed rear.
Features: Weatherguard nickel finish on metal; oversize trigger guard. Introduced 1995. From Stone Mountain Arms.
Price: $139.95

PEIFER MODEL TS-93 RIFLE
Caliber: 45, 50.
Barrel: 24″ Douglas premium; 1:20″ twist in 45, 1:28″ in 50.
Weight: 7 lbs. **Length:** 43 1/4″ overall.
Stock: Bell & Carlson solid composite, with recoil pad, swivel studs.
Sights: Williams bead front on ramp, fully adjustable open rear. Drilled and tapped for Weaver scope mounts with dovetail for rear peep.
Features: In-line ignition uses #209 shotshell primer; extremely fast lock time; fully enclosed breech; adjustable trigger; automatic safety; removal primer holder. Blue or stainless. Made in U.S. by Peifer Rifle Co. Introduced 1996.
Price: Blue, black stock $663.00
Price: Blue, wood or camouflage composite stock, or stainless with black composite stock $728.75
Price: Stainless, wood or camouflage composite stock $795.00

PRAIRIE RIVER ARMS PRA BULLPUP RIFLE
Caliber: 50, 54.
Barrel: 28″; 1:28″ twist.
Weight: 7 1/2 lbs. **Length:** 31 1/2″ overall.
Stock: Hardwood or black all-weather.
Sights: Blade front, open adjustable rear.
Features: Bullpup design thumbhole stock. Patented internal percussion ignition system. Left-hand model available. Dovetailed for scope mount. Introduced 1995. Made in U.S. by Prairie River Arms.
Price: 4140 alloy barrel, hardwood stock $375.00
Price: As above, black stock $390.00
Price: Stainless barrel, hardwood stock $425.00
Price: As above, black stock $440.00

STONE MOUNTAIN SILVER WOLF RIFLE
Caliber: 50.
Barrel: 26″.
Weight: 6 1/2 lbs. **Length:** 40″ overall.
Stock: Tuff Lite synthetic with checkered grip and forend.
Sights: Bead front, open fully adjustable rear.
Features: Oversize trigger guard; synthetic ramrod; nickeled barrel and lock; adjustable sear. Introduced 1997. From Stone Mountain Arms.
Price: $189.95

Remington Model 700 ML

REMINGTON MODEL 700 ML, MLS RIFLE
Caliber: 50, 54.
Barrel: 24″; 1:28″ twist.
Weight: 7 3/4 lbs. **Length:** 44 1/2″ overall.
Stock: Black fiberglass-reinforced synthetic with checkered grip and forend; magnum-style buttpad.
Sights: Ramped bead front, open fully adjustable rear. Drilled and tapped for scope mounts.
Features: Uses the Remington 700 bolt action, stock design, safety and trigger mechanisms; removable stainelss steel breech plug, No. 11 nipple; solid aluminum ramrod. Comes with cleaning tools and accessories.
Price: ML, blued, 50-caliber only $372.00
Price: MLS, stainless, 50- or 54-caliber $469.00
Price: ML, blued, Mossy Oak Break-Up camo stock $405.00
Price: MLS, stainless, Mossy Oak Break-Up camo stock $503.00

Ruger 77/50 In-Line

RUGER 77/50 IN-LINE PERCUSSION RIFLE
Caliber: 50.
Barrel: 22″, 1:28″ twist.
Weight: 6 1/2 lbs. **Length:** 41 1/2″ overall.
Stocks: Birch with rubber buttpad and swivel studs.
Sights: Gold bead front, folding leaf rear. Comes with Ruger scope mounts.
Features: Shares design features with the Ruger 77/22 rifle. Stainless steel bolt and nipple/breech plug; uses #11 caps; three-position safety; blued steel ramrod. Introduced 1997. Made in U.S. by Sturm, Ruger & Co.
Price: ....NA

SECOND MODEL BROWN BESS MUSKET
Caliber: 75, uses .735″ round ball.
Barrel: 42″, smoothbore.
Weight: 9 1/2 lbs. **Length:** 59″ overall.
Stock: Walnut (Navy); walnut-stained hardwood (Dixie).
Sights: Fixed.
Features: Polished barrel and lock with brass trigger guard and buttplate. Bayonet and scabbard available. From Navy Arms, Dixie Gun Works, Cabela's.
Price: Finished ....$475.00 to $850.00
Price: Kit (Dixie Gun Works, Navy Arms) ....$575.00 to $625.00
Price: Carbine (Navy Arms) $750.00

C.S. RICHMOND 1863 MUSKET
Caliber: 58.
Barrel: 40″.
Weight: 11 lbs. **Length:** 56 1/4″ overall.
Stock: European walnut with oil finish.
Sights: Blade front, adjustable folding leaf rear.
Features: Reproduction of the three-band Civil War musket. Sling swivels attached to trigger guard and middle barrel band. Lock plate marked "1863" and "C.S. Richmond." All metal left in the white. Brass buttplate and forend cap. Imported by Euroarms of America.
Price: $530.00

BLACKPOWDER MUSKETS & RIFLES

Thompson/Center Fire Hawk

THOMPSON/CENTER BIG BOAR RIFLE
Caliber: 58.
Barrel: 26" octagon; 1:48" twist.
Weight: 7 3/4 lbs. **Length:** 42 1/2" overall.
Stock: American black walnut; rubber buttpad; swivels.
Sights: Bead front, fullt adjustable open rear.
Features: Percussion lock; single trigger with wide bow trigger guard. Comes with soft leather sling. Introduced 1991. From Thompson/Center.
Price: $355.00

THOMPSON/CENTER GREY HAWK PERCUSSION RIFLE
Caliber: 50, 54.
Barrel: 24"; 1:48" twist.
Weight: 7 lbs. **Length:** 41" overall.
Stock: Black Rynite with rubber recoil pad.
Sights: Bead front, fully adjustable open hunting rear.
Features: Stainless steel barrel, lock, hammer, trigger guard, thimbles; blued sights. Percussion only. Introduced 1993. From Thompson/Center Arms.
Price: $330.00

THOMPSON/CENTER FIRE HAWK RIFLE
Caliber: 50, 54.
Barrel: 24"; 1:38" twist.
Weight: 7 lbs. **Length:** 41 3/4" overall.
Stock: American black walnut or black Rynite; Rynite thumbhole style; all with cheekpiece and swivel studs.
Sights: Ramp front with bead, adjustable leaf-style rear.
Features: In-line ignition with sliding thumb safety; free-floated barrel; exposed nipple; adjustable trigger. Available in blue or stainless. Comes with Weaver-style scope mount bases. Introduced 1995. Made in U.S. by Thompson/Center Arms.
Price: Blue, walnut stock, 50, 54 ....$365.00
Price: Stainless, walnut stock, 50, 54 ....$405.00
Price: Stainless, Rynite stock, 50, 54 ....$395.00
Price: Blue, thumbhole stock, 50, 54 ....$385.00
Price: Stainless, thumbhole stock, 50, 54 ....$425.00
Price: Bantam model with 13 1/4" pull, 21" barrel $365.00
Price: Blue, Advantage camo stock, 50, 54 ....$395.00

Manufacturers' addresses in the
Directory of the Arms Trade
page 309, this issue

T/C Hawken

Thompson/Center Hawken Silver Elite Rifle
Similar to the 50-caliber Hawken except all metal is satin-finished stainless steel. Has semi-fancy American walnut stock without patchbox. Percussion only. Introduced 1996. Made in U.S. by Thompson/Center Arms.
Price: $495.00

THOMPSON/CENTER HAWKEN RIFLE
Caliber: 45, 50 or 54.
Barrel: 28" octagon, hooked breech.
Stock: American walnut.
Sights: Blade front, rear adjustable for windage and elevation.
Features: Solid brass furniture, double-set triggers, button rifled barrel, coil-type mainspring. From Thompson/Center.
Price: Percussion model (45-, 50- or 54-cal.) $455.00
Price: Flintlock model (50-cal.) $425.00
Price: Percussion kit $315.00
Price: Flintlock kit $335.00

T/C Renegade

THOMPSON/CENTER RENEGADE RIFLE
Caliber: 50 and 54.
Barrel: 26", 1" across the flats.
Weight: 8 lbs.
Stock: American walnut.
Sights: Open hunting (Patridge) style, fully adjustable for windage and elevation.
Features: Coil spring lock, double-set triggers, blued steel trim. From Thompson/Center.
Price: Percussion only $360.00

THOMPSON/CENTER PENNSYLVANIA HUNTER RIFLE
Caliber: 50.
Barrel: 31", half-octagon, half-round.
Weight: About 7 1/2 lbs. **Length:** 48" overall.
Stock: Black walnut.
Sights: Open, adjustable.
Features: Rifled 1:66" for round ball shooting. Available in flintlock or percussion. From Thompson/Center.
Price: Flintlock $375.00

Thompson/Center Pennsylvania Hunter Carbine
Similar to the Pennsylvania Hunter except has 21" barrel, weighs 6.5 lbs., and has an overall length of 38". Designed for shooting patched round balls. Available in flintlock only. Introduced 1992. From Thompson/Center.
Price: Flintlock $365.00

THOMPSON/CENTER NEW ENGLANDER RIFLE
Caliber: 50, 54.
Barrel: 28", round.
Weight: 7 lbs., 15 oz.
Stock: American walnut or Rynite.
Sights: Open, adjustable.
Features: Color case-hardened percussion lock with engraving, rest blued. Also accepts 12-ga. shotgun barrel. Introduced 1987. From Thompson/Center.
Price: Right-hand model $310.00
Price: Left-hand model $330.00
Price: Accessory 12-ga. barrel, right-hand $170.00

Thompson/Center Pennsylvania Match Rifle
Similar to the Pennsylvania Hunter except has a tang peep sight, globe front with Seven interchangeable inserts. Introduced 1996. Made in U.S. by Thompson/Center Arms.
Price: $400.00

BLACKPOWDER MUSKETS & RIFLES

T/C Scout Rifle

THOMPSON/CENTER SCOUT CARBINE
Caliber: 50 and 54.
Barrel: 21", interchangeable, 1:38" twist.
Weight: 7 lbs., 4 oz. **Length:** 38 5/8" overall.
Stocks: American black walnut stock and forend.
Sights: Bead front, adjustable semi-buckhorn rear.
Features: Patented in-line ignition system with special vented breech plug. Patented trigger mechanism consists of only two moving parts. Interchangeable barrels. Wide grooved hammer. Brass trigger guard assembly, brass barrel band and buttplate. Ramrod has blued hardware. Comes with quick detachable swivels and suede leather carrying sling. Drilled and tapped for standard scope mounts. Introduced 1990. From Thompson/Center.
Price: 50- or 54-cal. $425.00
Price: With black Rynite stock $345.00

Thompson/Center Scout Rifle
Similar to the Scout Carbine except has 24" part octagon, part round barrel (round only on Rynite-stocked model), solid brass forend cap on walnut-stocked gun. Barrel twist is 1:38". Available in 50- and 54-caliber. Introduced 1995. Made in U.S. by Thompson/Center Arms.
Price: With walnut stock $435.00
Price: With Rynite stock $360.00

T/C System 1

Sights: Ramp front with white bead, adjustable leaf rear.
Features: In-line ignition. Interchangeable barrels; removable breech plug allows cleaning from the breech; fully adjustable trigger; sliding thumb safety; QLA muzzle system; rubber recoil pad; sling swivel studs. Introduced 1997. Made in U.S. by Thompson/Center Arms.
Price: Blue, walnut stock $370.00
Price: Stainless, composite stock, 50-, 54- caliber $400.00
Price: Stainless, camo composite stock, 50-caliber $435.00
Price: Extra barrels, blue $162.00
Price: Extra barrels, stainless, 50-, 54-caliber $206.00

THOMPSON/CENTER SYSTEM 1 IN-LINE RIFLE
Caliber: 32, 50, 54, 58; 12-gauge.
Barrel: 26" round; 1:38" twist.
Weight: About 7 1/2 lbs. **Length:** 44" overall.
Stock: American black walnut or composite.

THOMPSON/CENTER THUNDERHAWK CARBINE
Caliber: 50, 54.
Barrel: 21", 24"; 1:38" twist.
Weight: 6.75 lbs. **Length:** 38.75" overall.
Stock: American walnut or black Rynite with rubber recoil pad.
Sights: Bead on ramp front, adjustable leaf rear.
Features: Uses modern in-line ignition system, adjustable trigger. Knurled striker handle indicators for Safe and Fire. Black wood ramrod. Drilled and tapped for T/C scope mounts. Introduced 1993. From Thompson/Center Arms.
Price: Blue with walnut stock $315.00

Thompson/Center ThunderHawk Shadow
Similar to the ThunderHawk except 24" barrel only, blued or stainless barrel and receiver, composite stock, polycarbonate adjustable rear sight. Available in 50- or 54-caliber. Introduced 1996. Made in U.S. by Thompson/Center Arms.
Price: Blued $275.00
Price: Stainless $306.00
Price: Blue with camo stock $379.90

Traditions Buckhunter

TRADITIONS IN-LINE BUCKHUNTER SERIES RIFLES
Caliber: 50, 54.
Barrel: 24", round; 1:32" (50), 1:48" (54) twist.
Weight: 7 lbs., 6 oz. to 8 lbs. **Length:** 41" overall.
Stock: All-Weather composite.
Sights: Beaded blade front, click adjustable rear. Drilled and tapped for scope mounting.
Features: Removable breech plug; PVC ramrod; sling swivels. Introduced 1995. From Traditions.
Price: $160.00
Price: With RS Redi-Pak (powder measure, powder flask, two fast loaders, 5-in-1 loader, capper, ball starter, ball puller, cleaning jag, nipple wrench, bullets) $219.00
Price: With RP Redi-Pak (as above with Pyrodex and percussion caps ..$233.75

TRADITIONS BUCKSKINNER CARBINE
Caliber: 50.
Barrel: 21", 15/16" flats, half octagon, half round; 1:20" or 1:66" twist.
Weight: 6 lbs. **Length:** 37" overall.
Stock: Beech or black laminated.
Sights: Beaded blade front, hunting-style open rear click adjustable for windage and elevation.
Features: Uses V-type mainspring, single trigger. Non-glare hardware. From Traditions.
Price: Flintlock $204.00
Price: Flintlock, laminated stock $277.50
Price: Percussion, 50 ....$189.50
Price: Percussion, laminated stock, 50 ....$256.00
Price: Percussion, left-hand $212.00

Traditions Buckhunter Pro In-Line

TRADITIONS BUCKHUNTER PRO IN-LINE RIFLES
Caliber: 50 (1:32" twist), 54 (1:48" twist).
Barrel: 24" tapered round.
Weight: 7 1/2 lbs. **Length:** 42" overall.
Stock: Beech, composite or laminated; thumbhole available in black, Break-Up or Realtree® Advantage camouflage.
Sights: Beaded blade front, fully adjustable open rear. Drilled and tapped for scope mounting.
Features: In-line percussion ignition system; adjustable trigger; manual thumb safety; removable stainless steel breech plug. Seventeen models available. Introduced 1996. From Traditions.
Price: ....$196.00 to $262.50

BLACKPOWDER MUSKETS & RIFLES

Traditions Deerhunter

Traditions Deerhunter Scout Rifle
Similar to the Deerhunter except in 50-caliber percussion only with 22" octagon barrel; 1:48" twist; weighs 5 lbs., 10 oz.; 36 1/2" overall length; beech stock; drilled and tapped for scope mounting; hooked breech; PVC ramrod. Introduced 1996. Imported by Traditions.
Price: $152.50

Traditions Deerhunter Composite Rifle
Similar to the Deerhunter except has black composite stock with checkered grip and forend. Blued barrel, C-Nickel or Advantage camouflage finish, 50, 54 percussion, 50-caliber flintlock. Introduced 1996. Imported by Traditions.
Price: Blued, flintlock, 50-cal. $160.00
Price: Blued, percussion, 50- or 54-cal. $131.00
Price: Blued or Advantage camo, percussion, 50-cal. $175.00
Price: C-Nickel, percussion, 50-cal. $152.50
Price: C-Nickel, percussion, 54-cal. $152.50

TRADITIONS DEERHUNTER RIFLE SERIES
Caliber: 32, 50 or 54.
Barrel: 24", octagonal, 15/16" flats; 1:48" or 1:66" twist.
Weight: 6 lbs. **Length:** 40" overall.
Stock: Stained beech and All-Weather Composite with rubber buttpad, sling swivels.
Sights: Blade front, fixed rear.
Features: Flint or percussion with color case-hardened lock. Hooked breech, oversized trigger guard, blackened furniture, PVC ramrod. All-Weather has composite stock and C-Nickel barrel. Drilled and tapped for scope mounting. Imported by Traditions, Inc.
Price: Percussion, 50 or 54, 1:48" twist $152.50
Price: Flintlock, 50-caliber only, 1:66" twist $167.50
Price: Percussion kit, 50 or 54 .... $135.00
Price: Flintlock, All-Weather, 50-cal. $160.00
Price: Percussion, All-Weather, 50 or 54 .... $131.00
Price: Small Game, 32-cal., percussion $152.50

TRADITIONS HAWKEN WOODSMAN RIFLE
Caliber: 50 and 54.
Barrel: 28"; 15/16" flats.
Weight: 7 lbs., 11 oz. **Length:** 44 1/2" overall.
Stock: Walnut-stained hardwood.
Sights: Beaded blade front, hunting-style open rear adjustable for windage and elevation.
Features: Percussion only. Brass patchbox and furniture. Double triggers. From Traditions.
Price: 50 or 54 .... $219.00
Price: 50-cal., left-hand $233.75

Traditions Lightning

TRADITIONS KENTUCKY RIFLE
Caliber: 50.
Barrel: 33 1/2"; 7/8" flats; 1:66" twist.
Weight: 7 lbs. **Length:** 49" overall.
Stock: Beech; inletted toe plate.
Sights: Blade front, fixed rear.
Features: Full-length, two-piece stock; brass furniture; color case-hardened lock. Introduced 1995. From Traditions.
Price: Finished $219.00
Price: Kit $175.00

TRADITIONS PENNSYLVANIA RIFLE
Caliber: 50.
Barrel: 40 1/4", 7/8" flats; 1:66" twist, octagon.
Weight: 9 lbs. **Length:** 57 1/2" overall.
Stock: Walnut.
Sights: Blade front, adjustable rear.
Features: Brass patchbox and ornamentation. Double-set triggers. From Traditions.
Price: Flintlock $463.00
Price: Percussion $454.00

TRADITIONS LIGHTNING BOLT-ACTION MUZZLELOADER
Caliber: 50, 54.
Barrel: 24" round; blued, stainless, C-Nickel or Ultra Coat.
Weight: 7 lbs. **Length:** 43" overall.
Stock: Beech, brown laminated, All-Weather Composite, Advantage, X-tra Brown or Break-Up camouflage.
Sights: Beaded blade front, fully adjustabel open rear.
Features: Twenty-one variations available. Field-removable stainless steel bolt; silent thumb safety; adjustable trigger; drilled and tapped for scope mounting. Introduced 1997. Imported by Traditions.
Price: Beech stock $277.50
Price: Laminated stock, stainless steel barrel $380.00
Price: All-Weather composite stock, blue finish $262.50
Price: All-weather composite, C-Nickel finish $277.50
Price: All-weather composite, stainless steel $307.50
Price: Camouflage composite $307.50
Price: All-weather composite, Teflon finish $307.50
Price: Camouflage composite, Teflon finish $351.25

TRADITIONS PANTHER RIFLE
Caliber: 50.
Barrel: 24" octagon (1:48" twist); 15/16" flats.
Weight: 6 lbs. **Length:** 40" overall.
Stock: All-Weather composite.
Sights: Brass blade front, fixed rear.
Features: Percussion only; color case-hardened lock; blackened furniture; sling swivels; PVC ramrod. Introduced 1996. Imported by Traditions.
Price: $116.25
Price: With RS Redi-Pak (powder measure, flask, fast loaders, 5-in-1 loader, capper, ball starter, ball puller, cleaning jag, nipple wrench) $175.00
Price: With RP Redi-Pak (as above with Pyrodex and percussion caps) .$189.75

Traditions Shenandoah

TRADITIONS SHENANDOAH RIFLE
Caliber: 50.
Barrel: 33 1/2" octagon, 1:66" twist.
Weight: 7 lbs., 3 oz. **Length:** 49 1/2" overall.
Stock: Walnut.
Sights: Blade front, buckhorn rear.
Features: V-type mainspring; double-set trigger; solid brass buttplate, patchbox, nose cap, thimbles, rigger guard. Introduced 1996. From Traditions.
Price: Flintlock $336.25
Price: Percussion $322.00

BLACKPOWDER MUSKETS & RIFLES

Traditions Tennessee

TRADITIONS TENNESSEE RIFLE
Caliber: 50.
Barrel: 24", octagon with 15/16" flats; 1:32" twist.
Weight: 6 lbs. **Length:** 40 1/2" overall.
Stock: Stained beech.
Sights: Blade front, fixed rear.
Features: One-piece stock has inletted brass furniture, cheekpiece; double-set trigger; V-type mainspring. Flint or percussion. Introduced 1995. From Traditions.
Price: Percussion $270.00
Price: Flintlock $285.00

Traditions Model 1861

TRADITIONS 1861 U.S. SPRINGFIELD RIFLE
Caliber: 58.
Barrel: 40"; 1:66" twist.
Weight: 10 lbs. **Length:** 56" overall.
Stock: Walnut.
Sights: Military front, adjustable ladder-type rear.
Features: Full-length stock with white steel barrel, buttplate, ramrod, trigger guard, barrel bands, swivels, lockplate. Introduced 1995. From Traditions.
Price: $513.00

TRADITIONS 1853 THREE-BAND ENFIELD
Caliber: 58.
Barrel: 39"; 1:48" twist.
Weight: 10 lbs. **Length:** 55" overall.
Stock: Walnut.
Sights: Military front, adjustable ladder-type rear.
Features: Color case-hardened lock; brass buttplate, trigger guard, nose cap. Has V-type mainspring; steel ramrod; sling swivels. Introduced 1995. From Traditions.
Price: $483.75

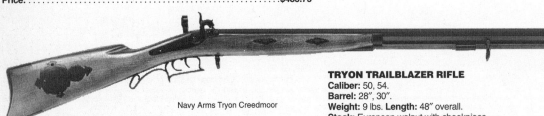
Navy Arms Tryon Creedmoor

TRYON TRAILBLAZER RIFLE
Caliber: 50, 54.
Barrel: 28", 30".
Weight: 9 lbs. **Length:** 48" overall.
Stock: European walnut with cheekpiece.
Sights: Blade front, semi-buckhorn rear.
Features: Reproduction of a rifle made by George Tryon about 1820. Double-set triggers, back action lock, hooked breech with long tang. From Armsport.
Price: About $825.00

Navy Arms Tryon Creedmoor Target Model
Similar to the standard Tryon rifle except 45-caliber only, 33" octagon barrel, globe front sight with inserts, fully adjustable match rear. Has double-set triggers, sling swivels. Imported by Navy Arms.
Price: $780.00

UFA Grand Teton Rifle
Similar to the Teton model except has 30" tapered octagon barrel in 45- or 50-caliber only. Available in blue or stainless steel with brushed or matte finish, brown or black laminated wood stock and forend. Weighs 9 lbs., overall length 46". Introduced 1994. Made in U.S. by UFA, Inc.
Price: $995.00
Price: With premium walnut or maple $1,145.00

UFA TETON RIFLE
Caliber: 45, 50, 12-bore (rifled, 72-cal.), 12-gauge.
Barrel: 26".
Weight: 8 lbs. **Length:** 42" overall.
Stock: Black or brown laminated wood; 1" recoil pad.
Sights: Marble's bead front, Marble's fully adjustable rear.
Features: Removable, interchangeable barrel; removable one-piece breech plug/nipple, hammer/trigger assembly; hammer blowback block; glass-bedded stock and forend. Introduced 1994. Made in U.S. by UFA, Inc.
Price: Stainless or blued $834.00
Price: With premium walnut or maple $984.00
Price: Extra barrels $165.00

UFA Teton Blackstone Rifle
Similar to the Teton model except in 50-caliber only, 26" barrel with shallow groove 1:26" rifling. Available only in stainless steel with matte finish. Has hardwood stock with black epoxy coating, 1" recoil pad. Weighs 7 1/2 lbs., overall length 42". Introduced 1994. Made in U.S. by UFA, Inc.
Price: $534.00

White Shooting Systems Super 91

WHITE MUZZLELOADING SYSTEMS SUPER 91 BLACKPOWDER RIFLE
Caliber: 41, 45 or 50.
Barrel: 26".
Weight: 7 1/2 lbs. **Length:** 43.5" overall.
Stock: Black laminate or black composite; recoil pad, swivel studs.
Sights: Bead front on ramp, fully adjustable open rear.
Features: Insta-Fire straight-line ignition system; all stainless steel construction; side-swing safety; fully adjustable trigger; full barrel under-rib with two ramrod thimbles. Introduced 1991. Made in U.S. by White Muzzleloading Systems, Inc.
Price: Stainless $659.95
Price: Stainless, laminate stock $699.95

White Muzzleloading Systems Super Safari Rifle
Same as the stainless Super 91 except has Mannlicher-style stock of black composite. Introduced 1993. From White Muzzleloading Systems, Inc.
Price: $799.00

BLACKPOWDER MUSKETS & RIFLES

WHITE MUZZLELOADING SYSTEMS WHITETAIL RIFLE
Caliber: 41, 45 or 50.
Barrel: 22".
Weight: 6.5 lbs. **Length:** 39.5" overall.
Stock: Black composite; classic style; recoil pad, swivel studs.
Sights: Bead front on ramp, fully adjustable open rear.
Features: Insta-Fire straight-line ignition; action and trigger safeties; adjustable trigger; stainless steel. Introduced 1992. Made in U.S. by White Muzzleloading Systems, Inc.
Price: Blue, wood stock, bull bbl., 50-cal. $399.00
Price: Stainless, composite stock $499.95
Price: Stainless, laminate stock $529.95

White Muzzleloading Systems Bison Blackpowder Rifle
Similar to the blued Whitetail model except in 54-caliber (1:28" twist) with 22" ball barrel. Uses Insta-Fire in-line percussion system, double safety. Adjustable sight, black-finished hardwood stock, matte blue metal finish, Delron ramrod, swivel studs. Drilled and tapped for scope mounting. Weighs 7 1/4 lbs. Introduced 1993. From White Muzzleloading Systems, Inc.
Price: $399.95

White Muzzleloading Systems White Lightning Rifle
Similar to the Whitetail stainless rifle except uses smaller action with cocking lever and secondary safety on right side, primary safety on the left. Available only in 50-caliber with 22" barrel. Weighs 6.4 lbs, 40" overall. Has black hardwood stock. Introduced 1995. From White Muzzleloading Systems, Inc.
Price: $299.95

Navy Arms 1841 Mississippi

Mississippi 1841 Percussion Rifle
Similar to Zouave rifle but patterned after U.S. Model 1841. Imported by Dixie Gun Works, Euroarms of America, Navy Arms.
Price: About ....$430.00 to $500.00

ZOUAVE PERCUSSION RIFLE
Caliber: 58, 59.
Barrel: 32 1/2".
Weight: 9 1/2 lbs. **Length:** 48 1/2" overall.
Stock: Walnut finish, brass patchbox and buttplate.
Sights: Fixed front, rear adjustable for elevation.
Features: Color case-hardened lockplate, blued barrel. From Navy Arms, Dixie Gun Works, Euroarms of America (M1863), E.M.F., Cabela's.
Price: About ....$325.00 to $465.00
Price: Kit (Euroarms 58-cal. only) $331.00

BLACKPOWDER SHOTGUNS

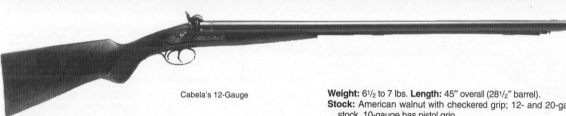

Cabela's 12-Gauge

CABELA'S BLACKPOWDER SHOTGUNS
Gauge: 10, 12, 20.
Barrel: 28 1/2" (10-, 12-ga.), Imp. Cyl., Mod., Full choke tubes; 27 1/2" (20-ga.), Imp. Cyl., Mod. choke tubes.
Weight: 6 1/2 to 7 lbs. **Length:** 45" overall (28 1/2" barrel).
Stock: American walnut with checkered grip; 12- and 20-gauge have straight stock, 10-gauge has pistol grip.
Features: Blued barrels, engraved, color case-hardened locks and hammers, brass ramrod tip. From Cabela's.
Price: 10-gauge $379.95
Price: 12-gauge $359.95
Price: 20-gauge $329.95

CVA TRAPPER PERCUSSION
Gauge: 12.
Barrel: 28".
Weight: 6 lbs. **Length:** 46" overall.
Stock: English-style checkered straight grip of walnut-finished hardwood.
Sights: Brass bead front.
Features: Single blued barrel; color case-hardened lockplate and hammer; screw adjustable sear engagements, V-type mainspring; brass wedge plates; color case-hardened and engraved trigger guard and tang. From CVA.
Price: Finished $239.95

CVA CLASSIC DOUBLE SHOTGUN
Gauge: 12.
Barrel: 28".
Weight: 9 lbs. **Length:** 45" overall.
Stock: European walnut; classic English style with checkered straight grip, wrap-around forend with bottom screw attachment.
Sights: Bead front.
Features: Hinged double triggers; color case-hardened and engraved lockplates, trigger guard and tang. Polymer-coated fiberglass ramrod. Rubber recoil pad. Not suitable for steel shot. Introduced 1990. Imported by CVA.
Price: $459.95

Dixie Magnum

DIXIE MAGNUM PERCUSSION SHOTGUN
Gauge: 10, 12, 20.
Barrel: 30" (Imp. Cyl. & Mod.) in 10-gauge; 28" in 12-gauge.
Weight: 6 1/4 lbs. **Length:** 45" overall.
Stock: Hand-checkered walnut, 14" pull.
Features: Double triggers; light hand engraving; case-hardened locks in 12-gauge, polished steel in 10-gauge; sling swivels. From Dixie Gun Works.
Price: Upland $449.00
Price: 12-ga. kit $375.00
Price: 20-ga. $495.00
Price: 10-ga. $495.00
Price: 10-ga. kit $395.00

CAUTION: PRICES SHOWN ARE SUPPLIED BY THE MANUFACTURER OR IMPORTER. CHECK YOUR LOCAL GUNSHOP.

BLACKPOWDER SHOTGUNS

Navy Arms Fowler

NAVY ARMS FOWLER SHOTGUN
Gauge: 10, 12.
Barrel: 28".
Weight: 7 lbs., 12 oz. **Length:** 45" overall.
Stock: Walnut-stained hardwood.
Features: Color case-hardened lockplates and hammers; checkered stock. Imported by Navy Arms.
Price: $340.00

NAVY ARMS MORTIMER FLINTLOCK SHOTGUN
Gauge: 12.
Barrel: 36".
Weight: 7 lbs. **Length:** 53" overall.
Stock: Walnut, with cheekpiece.
Features: Waterproof pan, roller frizzen, external safety. Color case-hardened lock, rest blued. Imported by Navy Arms.
Price: $735.00

NAVY ARMS STEEL SHOT MAGNUM SHOTGUN
Gauge: 10.
Barrel: 28" (Cyl. & Cyl.).
Weight: 7 lbs., 9 oz. **Length:** 45 1/2" overall.
Stock: Walnut, with cheekpiece.
Features: Designed specifically for steel shot. Engraved, polished locks; sling swivels; blued barrels. Imported by Navy Arms.
Price: $560.00

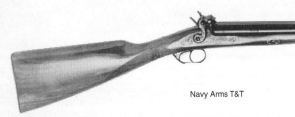
Navy Arms T&T

NAVY ARMS T&T SHOTGUN
Gauge: 12.
Barrel: 28" (Full & Full).
Weight: 7 1/2 lbs.
Stock: Walnut.
Sights: Bead front.
Features: Color case-hardened locks, double triggers, blued steel furniture. From Navy Arms.
Price: $540.00

T/C New Englander

THOMPSON/CENTER NEW ENGLANDER SHOTGUN
Gauge: 12.
Barrel: 28" (Imp. Cyl.), round.
Weight: 5 lbs., 2 oz.
Stock: Select American black walnut with straight grip.
Features: Percussion lock is color case-hardened, rest blued. Also accepts 26" round 50- and 54-cal. rifle barrel. Introduced 1986. From Thompson/Center.
Price: Right-hand $330.00

Traditions Buckhunter Pro

TRADITIONS BUCKHUNTER PRO SHOTGUN
Gauge: 12.
Barrel: 24"; choke tube.
Weight: 6 lbs., 4 oz. **Length:** 43" overall.
Stock: Composite matte black, Break-Up or Advantage camouflage.
Features: In-line action with removable stainless steel breech plug; thumb safety; adjustable trigger; rubber buttpad. Introduced 1996. From Traditions.
Price: $385.00
Price: With Advantage or Break-Up camouflage stock $329.00

White Shooting Systems White Thunder

WHITE MUZZLELOADING SYSTEMS WHITE THUNDER SHOTGUN
Gauge: 12.
Barrel: 26" (Imp. Cyl., Mod., Full choke tubes); ventilated rib.
Weight: About 5 3/4 lbs.
Stock: Black hardwood.
Features: InstaFire in-line ignition; double safeties; match-grade trigger; Delron ramrod. Introduced 1995. From White Muzzleloading Systems, Inc.
Price: $399.95

White Muzzleloading Systems "Tominator" Shotgun
Similar to the White Thunder except has Imp. Cyl., Mod., Full and Super Full Turkey choke tubes; black laminate stock. Introduced 1995. From White Shooting Systems, Inc.
Price: $499.95

AIRGUNS—HANDGUNS

Anics A-101 Magnum

ANICS A-201 AIR REVOLVER
Caliber: 177, 4.5mm, BB; 36-shot cylinder.
Barrel: 4″, 6″ steel smoothbore.
Weight: 36 oz. **Length:** 9.75″ overall.
Power: CO_2
Stocks: Checkered plastic.
Sights: Blade front, fully adjustable rear.
Features: Velocity about 425 fps. Fixed barrel; single/double action; rotating cylinder; manual cross-bolt safety; blue and silver finish. Introduced 1996. Imported by Anics, Inc.
Price: $75.00

BEEMAN P1 MAGNUM AIR PISTOL
Caliber: 177, 5mm, single shot.
Barrel: 8.4″.
Weight: 2.5 lbs. **Length:** 11″ overall.
Power: Top lever cocking; spring-piston.
Stocks: Checkered walnut.
Sights: Blade front, square notch rear with click micrometer adjustments for windage and elevation. Grooved for scope mounting.
Features: Dual power for 177 and 20-cal.: low setting gives 350-400 fps; high setting 500-600 fps. Rearward expanding mainspring simulates firearm recoil. All Colt 45 auto grips fit gun. Dry-firing feature for practice. Optional wooden shoulder stock. Introduced 1985. Imported by Beeman.
Price: 177, 5mm $415.00

Beeman/Feinwerkbau 103

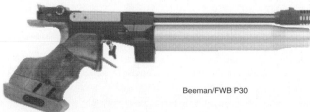
Beeman/FWB P30

BEEMAN/FWB P30 MATCH AIR PISTOL
Caliber: 177, single shot.
Barrel: $10^{5}/_{16}$″, with muzzlebrake.
Weight: 2.4 lbs. **Length:** 16.5″ overall.
Power: Pre-charged pneumatic.
Stocks: Stippled walnut; adjustable match type.
Sights: Undercut blade front, fully adjustable match rear.
Features: Velocity to 525 fps; up to 200 shots per CO_2 cartridge. Fully adjustable trigger; built-in muzzlebrake. Introduced 1995. Imported from Germany by Beeman.
Price: Right-hand $1,465.00
Price: Left-hand $1,520.00

ANICS A-101 AIR PISTOL
Caliber: 177, 4.5mm, BB; 15-shot magazine.
Barrel: 4.5″ steel smoothbore.
Weight: 35 oz. **Length:** 7″ overall.
Power: CO_2
Stocks: Checkered plastic.
Sights: Blade front, fixed rear.
Features: Velocity to 460 fps. Semi-automatic action; double action only; cross-bolt safety; black and silver finish. Comes with two 15-shot magazines. Introduced 1996. Imported by Anics, Inc.
Price: With case, about $65.00

Anics A-101 Magnum Air Pistol
Similar to the A-101 except has 6″ barrel with compensator, gives about 490 fps. Introduced 1996. Imported by Anics, Inc.
Price: With case, about $72.00

Beeman P1

Beeman P2 Match Air Pistol
Similar to the Beeman P1 Magnum except shoots only 177 pellets; completely recoilless single-stroke pnuematic action. Weighs 2.2 lbs. Choice of thumbrest match grips or standard style. Introduced 1990.
Price: 177, 5mm, standard grip $445.00
Price: 177, match grip $480.00

BEEMAN/FEINWERKBAU 103 PISTOL
Caliber: 177, single shot.
Barrel: 10.1″, 12-groove rifling.
Weight: 2.5 lbs. **Length:** 16.5″ overall.
Power: Single-stroke pneumatic, underlever cocking.
Stocks: Stippled walnut with adjustable palm shelf.
Sights: Blade front, open rear adjustable for windage and elevation. Notch size adjustable for width. Interchangeable front blades.
Features: Velocity 510 fps. Fully adjustable trigger. Cocking effort of 2 lbs. Imported by Beeman.
Price: Right-hand $1,520.00
Price: Left-hand $1,580.00

> Consult our Directory pages for the location of firms mentioned.

BEEMAN/FEINWERKBAU 65 MKII AIR PISTOL
Caliber: 177, single shot.
Barrel: 6.1″; removable bbl. wgt. available.
Weight: 42 oz. **Length:** 13.3″ overall.
Power: Spring, sidelever cocking.
Stocks: Walnut, stippled thumbrest; adjustable or fixed.
Sights: Front, interchangeable post element system, open rear, click adjustable for windage and elevation and for sighting notch width. Scope mount available.
Features: New shorter barrel for better balance and control. Cocking effort 9 lbs. Two-stage trigger, four adjustments. Quiet firing, 525 fps. Programs instantly for recoil or recoilless operation. Permanently lubricated. Steel piston ring. Imported by Beeman.
Price: Right-hand $1,170.00
Price: Left-hand $1,220.00

AIRGUNS—HANDGUNS

Beeman/FWB C55

BEEMAN HW70A AIR PISTOL
Caliber: 177, single shot.
Barrel: 6 1/4", rifled.
Weight: 38 oz. **Length:** 12 3/4" overall.
Power: Spring, barrel cocking.
Stocks: Plastic, with thumbrest.
Sights: Hooded post front, square notch rear adjustable for windage and elevation. Comes with scope base.
Features: Adjustable trigger, 31-lb. cocking effort, 440 fps MV; automatic barrel safety. Imported by Beeman.
Price: $225.00
Price: HW70S, black grip, silver finish $240.00

BEEMAN/WEBLEY NEMESIS AIR PISTOL
Caliber: 177, single shot.
Barrel: 7".
Weight: 2.2 lbs. **Length:** 9.8" overall.
Power: Single-stroke pneumatic.
Stocks: Checkered black composition.
Sights: Blade on ramp front, fully adjustable rear. Integral scope rail.
Features: Velocity to 400 fps. Adjustable two-stage trigger, manual safety. Recoilless action. Introduced 1995. Imported from England by Beeman.
Price: $200.00

BEEMAN/WEBLEY TEMPEST AIR PISTOL
Caliber: 177, 22, single shot.
Barrel: 6 7/8".
Weight: 32 oz. **Length:** 8.9" overall.
Power: Spring-piston, break barrel.
Stocks: Checkered black plastic with thumbrest.
Sights: Blade front, adjustable rear.
Features: Velocity to 500 fps (177), 400 fps (22). Aluminum frame; black epoxy finish; manual safety. Imported from England by Beeman.
Price: $210.00

Beeman/Webley Hurricane Air Pistol
Similar to the Tempest except has extended frame in the rear for a click-adjustable rear sight; hooded front sight; comes with scope mount. Imported from England by Beeman.
Price: $240.00

Benjamin Sheridan CO2

Benjamin Sheridan Pneumatic

BEEMAN/FWB C55 CO$_2$ RAPID FIRE PISTOL
Caliber: 177, single shot or 5-shot magazine.
Barrel: 7.3".
Weight: 2.5 lbs. **Length:** 15" overall.
Power: Special CO$_2$ cylinder.
Stocks: Anatomical, adjustable.
Sights: Interchangeable front, fully adjustable open micro-click rear with adjustable notch size.
Features: Velocity 510 fps. Has 11.75" sight radius. Built-in muzzlebrake. Introduced 1993. Imported by Beeman Precision Airguns.
Price: Right-hand $1,705.00
Price: Left-hand $1,755.00

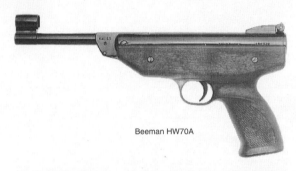
Beeman HW70A

Beeman/Webley Nemesis

Beeman/Webley Tempest

BENJAMIN SHERIDAN CO$_2$ PELLET PISTOLS
Caliber: 177, 20, 22, single shot.
Barrel: 6 3/8", rifled brass.
Weight: 29 oz. **Length:** 9.8" overall.
Power: 12-gram CO$_2$ cylinder.
Stocks: Walnut.
Sights: High ramp front, fully adjustable notch rear.
Features: Velocity to 500 fps. Turn-bolt action with cross-bolt safety. Gives about 40 shots per CO$_2$ cylinder. Black or nickel finish. Made in U.S. by Benjamin Sheridan Co.
Price: Black finish, EB17 (177), EB20 (20), EB22 (22), about $105.00
Price: Nickel finish, E17 (177), E20 (20), E22 (22), about $120.00

BENJAMIN SHERIDAN PNEUMATIC PELLET PISTOLS
Caliber: 177, 20, 22, single shot.
Barrel: 9 3/8", rifled brass.
Weight: 38 oz. **Length:** 13 1/8" overall.
Power: Underlever pneumatic, hand pumped.
Stocks: Walnut stocks and pump handle.
Sights: High ramp front, fully adjustable notch rear.
Features: Velocity to 525 fps (variable). Bolt action with cross-bolt safety. Choice of black or nickel finish. Made in U.S. by Benjamin Sheridan Co.
Price: Black finish, HB17 (177), HB20 (20), HB22 (22), about $115.00
Price: Nickel finish, H17 (177), H20 (20), H22 (22), about $120.00

AIRGUNS—HANDGUNS

BRNO TAU-7

BSA 240 MAGNUM AIR PISTOL
Caliber: 177, 22, single shot
Barrel: 6".
Weight: 2 lbs. **Length:** 9" overall.
Power: Spring-air, top-lever cocking.
Stocks: Walnut.
Sights: Blade front, micrometer adjustable rear.
Features: Velocity 510 fps (177), 420 fps (22); crossbolt safety. Combat autoloader styling. Imported from U.K. by Precision Sales International, Inc.
Price: $293.00

COPPERHEAD BLACK VENOM PISTOL
Caliber: 177 pellets, BB, 17-shot magazine; darts, single shot.
Barrel: 4.75" smoothbore.
Weight: 16 oz. **Length:** 10.8" overall.
Power: Spring.
Stocks: Checkered.
Sights: Blade front, adjustable rear.
Features: Velocity to 270 fps (BBs), 250 fps (pellets). Spring-fed magazine; cross-bolt safety. Introduced 1996. Made in U.S. by Crosman Corp.
Price: About $20.00

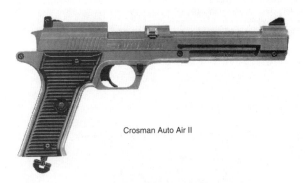
Crosman Auto Air II

Manufacturers' addresses in the
Directory of the Arms Trade
page 309, this issue

Crosman Model 1008

BRNO TAU-7 CO$_2$ MATCH PISTOL
Caliber: 177.
Barrel: 10.24".
Weight: 37 oz. **Length:** 15.75" overall.
Power: 12.5-gram CO$_2$ cartridge.
Stocks: Stippled hardwood with adjustable palm rest.
Sights: Blade front, open fully adjustable rear.
Features: Comes with extra seals and counterweight. Blue finish. Imported by Great Lakes Airguns.
Price: About $369.50

BSA 240 Magnum

COPPERHEAD BLACK FANG PISTOL
Caliber: 177 BB, 17-shot magazine.
Barrel: 4.75" smoothbore.
Weight: 10 oz. **Length:** 10.8" overall.
Power: Spring.
Stocks: Checkered.
Sights: Blade front, fixed notch rear.
Features: Velocity to 250 fps. Spring-fed magazine; cross-bolt safety. Introduced 1996. Made in U.S. by Crosman Corp.
Price: About $16.00

CROSMAN AUTO AIR II PISTOL
Caliber: BB, 17-shot magazine, 177 pellet, single shot.
Barrel: 8$5/8$" steel, smoothbore.
Weight: 13 oz. **Length:** 10$3/4$" overall.
Power: CO$_2$ Powerlet.
Stocks: Grooved plastic.
Sights: Blade front, adjustable rear; highlighted system.
Features: Velocity to 480 fps (BBs), 430 fps (pellets). Semi-automatic action with BBs, single shot with pellets. Silvered finish. Introduced 1991. From Crosman.
Price: About $35.00

CROSMAN MODEL 357 SERIES AIR PISTOL
Caliber: 177, 6- and 10-shot pellet clips.
Barrel: 4" (Model 3574GT), 6" (Model 3576GT).
Weight: 32 oz. (6"). **Length:** 11$3/8$" overall (357-6).
Power: CO$_2$ Powerlet.
Stocks: Checkered wood-grain plastic.
Sights: Ramp front, fully adjustable rear.
Features: Average 430 fps (Model 3574GT). Break-open barrel for easy loading. Single or double action. Vent. rib barrel. Wide, smooth trigger. Two cylinders come with each gun. Black and gold finish. From Crosman.
Price: 4" or 6", about $60.00

CROSMAN MODEL 1008 REPEAT AIR
Caliber: 177, 8-shot pellet clip
Barrel: 4.25", rifled steel.
Weight: 17 oz. **Length:** 8.625" overall.
Power: CO$_2$ Powerlet.
Stocks: Checkered black plastic.
Sights: Post front, adjustable rear.
Features: Velocity about 430 fps. Break-open barrel for easy loading; single or double semi-automatic action; two 8-shot clips included. Optional carrying case available. Introduced 1992. From Crosman.
Price: About $60.00
Price: With case, about $65.00
Price: Model 1008SB (silver and black finish), about $60.00

CAUTION: PRICES SHOWN ARE SUPPLIED BY THE MANUFACTURER OR IMPORTER. CHECK YOUR LOCAL GUNSHOP.

AIRGUNS—HANDGUNS

Crosman Model 1377

CROSMAN MODEL 1322, 1377 AIR PISTOLS
Caliber: 177 (M1377), 22 (M1322), single shot.
Barrel: 8", rifled steel.
Weight: 39 oz. **Length:** 13 5/8".
Power: Hand pumped.
Sights: Blade front, rear adjustable for windage and elevation.
Features: Moulded plastic grip, hand size pump forearm. Cross-bolt safety. Model 1377 also shoots BBs. From Crosman.
Price: About $60.00

DAISY MODEL 91 MATCH PISTOL
Caliber: 177, single shot.
Barrel: 10.25", rifled steel.
Weight: 2.5 lbs. **Length:** 16.5" overall.
Power: CO_2, 12-gram cylinder.
Stocks: Stippled hardwood; anatomically shaped and adjustable.
Sights: Blade and ramp front, changeable-width rear notch with full micrometer adjustments.
Features: Velocity to 476 fps. Gives 55 shots per cylinder. Fully adjustable trigger. Imported by Daisy Mfg. Co.
Price: About $670.00

Daisy Model 91

DAISY MODEL 288 AIR PISTOL
Caliber: 177 pellets, 24-shot.
Barrel: Smoothbore steel.
Weight: .8 lb. **Length:** 12.1" overall.
Power: Single stroke spring-air.
Stocks: Moulded resin with checkering and thumbrest.
Sights: Blade and ramp front, open fixed rear.
Features: Velocity to 215 fps. Cross-bolt trigger block safety. Black finish. From Daisy Mfg. Co.
Price: About $26.00

DAISY/POWER LINE 44 REVOLVER
Caliber: 177 pellets, 6-shot.
Barrel: 6", rifled steel; interchangeable 4" and 8".
Weight: 2.7 lbs.
Power: CO_2.
Stocks: Moulded plastic with checkering.
Sights: Blade on ramp front, fully adjustable notch rear.
Features: Velocity up to 400 fps. Replica of 44 Magnum revolver. Has swingout cylinder and interchangeable barrels. Introduced 1987. From Daisy Mfg. Co.
Price: $73.95

DAISY/POWER LINE 93 PISTOL
Caliber: 177, BB, 15-shot clip.
Barrel: 5", steel.
Weight: 17 oz. **Length:** NA.
Power: CO_2.
Stocks: Checkered plastic.
Sights: Fixed.
Features: Velocity to 400 fps. Semi-automatic repeater. Manual lever-type trigger-block safety. Introduced 1991. From Daisy Mfg. Co.
Price: About $80.00
Price: Model 693 (nickel-chrome plated), about $85.00

Daisy/Power Line 93

DAISY/POWER LINE MATCH 777 PELLET PISTOL
Caliber: 177, single shot.
Barrel: 9.61" rifled steel by Lothar Walther.
Weight: 32 oz. **Length:** 13 1/2" overall.
Power: Sidelever, single-pump pneumatic.
Stocks: Smooth hardwood, fully contoured with palm and thumbrest.
Sights: Blade and ramp front, match-grade open rear with adjustable width notch, micro. click adjustments.
Features: Adjustable trigger; manual cross-bolt safety. MV of 385 fps. Comes with cleaning kit, adjustment tool and pellets. From Daisy Mfg. Co.
Price: About $345.00

DAISY/POWER LINE 717 PELLET PISTOL
Caliber: 177, single shot.
Barrel: 9.61".
Weight: 2.8 lbs. **Length:** 13 1/2" overall.
Stocks: Moulded wood-grain plastic, with thumbrest.
Sights: Blade and ramp front, micro-adjustable notch rear.
Features: Single pump pneumatic pistol. Rifled steel barrel. Cross-bolt trigger block. Muzzle velocity 385 fps. From Daisy Mfg. Co. Introduced 1979.
Price: About $84.95

Daisy/Power Line 747 Pistol
Similar to the 717 pistol except has a 12-groove rifled steel barrel by Lothar Walther, and adjustable trigger pull weight. Velocity of 360 fps. Manual cross-bolt safety.
Price: About $165.95

Daisy/Power Line 717

Daisy/Power Line 1140

DAISY/POWER LINE 1140 PELLET PISTOL
Caliber: 177, single shot.
Barrel: Rifled steel.
Weight: 1.3 lbs. **Length:** 11.7" overall.
Power: Single-stroke barrel cocking.
Stocks: Checkered resin.
Sights: Hooded post front, open adjustable rear.
Features: Velocity to 325 fps. Made of black lightweight engineering resin. Introduced 1995. From Daisy.
Price: About $45.50

AIRGUNS—HANDGUNS

Daisy/PowerLine 1270

DAISY/POWER LINE 1700 AIR PISTOL
Caliber: 177 BB, 60-shot magazine.
Barrel: Smoothbore steel.
Weight: 1.4 lbs. **Length:** 11.2″ overall.
Power: CO_2.
Stocks: Moulded checkered plastic.
Sights: Blade front, adjustable rear.
Features: Velocity to 420 fps. Cross-bolt trigger block safety; matte finish. Has ³⁄₈″ dovetail mount for scope or point sight. Introduced 1994. From Daisy Mfg. Co.
Price: About $42.50

"GAT" AIR PISTOL
Caliber: 177, single shot.
Barrel: 7¹⁄₂″ cocked, 9¹⁄₂″ extended.
Weight: 22 oz.
Power: Spring-piston.
Stocks: Cast checkered metal.
Sights: Fixed.
Features: Shoots pellets, corks or darts. Matte black finish. Imported from England by Stone Enterprises, Inc.
Price: $24.95

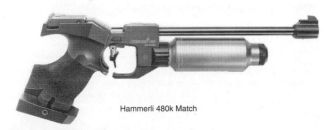
Hammerli 480k Match

MARKSMAN 1010 REPEATER PISTOL
Caliber: 177, 18-shot repeater.
Barrel: 2¹⁄₂″, smoothbore.
Weight: 24 oz. **Length:** 8¹⁄₄″ overall.
Power: Spring.
Features: Velocity to 200 fps. Thumb safety. Black finish. Uses BBs, darts, bolts or pellets. Repeats with BBs only. From Marksman Products.
Price: Matte black finish $26.00
Price: Model 2000 (as above except silver-chrome finish) $27.00

MARKSMAN 2005 LASERHAWK SPECIAL EDITION AIR PISTOL
Caliber: 177, 24-shot magazine.
Barrel: 3.8″, smoothbore.
Weight: 22 oz. **Length:** 10.3″ overall.
Power: Spring-air.
Stocks: Checkered.
Sights: Fixed fiber optic front sight.
Features: Velocity to 300 fps with Hyper-Velocity pellets. Square trigger guard with skeletonized trigger; extended barrel for greater velocity and accuracy. Shoots BBs, pellets, darts or bolts. Made in the U.S. From Marksman Products.
Price: $32.00

PARDINI K58 MATCH AIR PISTOL
Caliber: 177, single shot.
Barrel: 9.0″.
Weight: 37.7 oz. **Length:** 15.5″ overall.
Power: Pre-charged compressed air; single-stroke cocking.
Stocks: Adjustable match type; stippled walnut.
Sights: Interchangeable post front, fully adjustable match rear.
Features: Fully adjustable trigger. Introduced 1995. Imported from Italy by Nygord Precision Products.
Price: $650.00
Price: K60 model (CO_2) $650.00

DAISY/POWER LINE CO_2 1200 PISTOL
Caliber: BB, 177.
Barrel: 10¹⁄₂″, smooth.
Weight: 1.6 lbs. **Length:** 11.1″ overall.
Power: Daisy CO_2 cylinder.
Stocks: Contoured, checkered moulded wood-grain plastic.
Sights: Blade ramp front, fully adjustable square notch rear.
Features: 60-shot BB reservoir, gravity feed. Cross-bolt safety. Velocity of 420-450 fps for more than 100 shots. From Daisy Mfg. Co.
Price: About $47.00

DAISY/POWER LINE 1270 CO_2 AIR PISTOL
Caliber: BB, 60-shot magazine.
Barrel: Smoothbore steel.
Weight: 17 oz. **Length:** 11″ overall.
Power: CO_2 pump action.
Stocks: Moulded black polymer.
Sights: Blade on ramp front, adjustable rear.
Features: Velocity to 420 fps. Crossbolt trigger block safety; plated finish. Introduced 1997. Made in U.S. by Daisy Mfg. Co.
Price: About $49.00

DRULOV DU-10 CONDOR TARGET PISTOL
Caliber: 177, 5-shot magazine.
Barrel: 7.09″.
Weight: 2.32 lbs. **Length:** 11.81″ overall.
Power: CO_2.
Stocks: Target-type walnut with stippling.
Sights: Blade front, fully adjustable open rear.
Features: Velocity to 472 fps. Developed for Olympic Rapid Fire and 10-meter Sport Pistol events. Introduced 1997. Imported from the Czech Republic by Great Lakes Airguns.
Price: About $500.00

HAMMERLI 480 MATCH AIR PISTOL
Caliber: 177, single shot.
Barrel: 9.8″.
Weight: 37 oz. **Length:** 16.5″ overall.
Power: Air or CO_2.
Stocks: Walnut with 7-degree rake adjustment. Stippled grip area.
Sights: Undercut blade front, fully adjustable open match rear.
Features: Under-barrel cannister charges with air or CO_2 for power supply; gives 320 shots per filling. Trigger adjustable for position. Introduced 1994. Imported from Switzerland by Hammerli Pistols U.S.A.
Price: $1,325.00

Hammerli 480k Match Air Pistol
Similar to the 480 except has a short, detachable aluminum air cylinder for use only with compressed air; can be filled while on the gun or off; special adjustable barrel weights. Muzzle velocity of 470 fps, gives about 180 shots. Has stippled black composition grip with adjustable palm shelf and rake angle. Comes with air pressure gauge. Introduced 1996. Imported from Switzerland by SIGARMS, Inc.
Price: $1,155.00

Marksman 2005 Laserhawk

MORINI 162E MATCH AIR PISTOL
Caliber: 177, single shot.
Barrel: 9.4″.
Weight: 32 oz. **Length:** 16.1″ overall.
Power: Pre-charged CO_2.
Stocks: Adjustable match type.
Sights: Interchangeable blade front, fully adjustable match-type rear.
Features: Power mechanism shuts down when pressure drops to a pre-set level. Adjustable electronic trigger. Introduced 1995. Imported from Switzerland by Nygord Precision Products.
Price: $950.00

CAUTION: PRICES SHOWN ARE SUPPLIED BY THE MANUFACTURER OR IMPORTER. CHECK YOUR LOCAL GUNSHOP.

AIRGUNS—HANDGUNS

RECORD CHAMPION REPEATER PISTOL
Caliber: 177, 12-shot magazine.
Barrel: 7.6", rifled.
Weight: 2.8", rifled. **Length:** 10.2" overall.
Power: Spring-air.
Stocks: Oil-finished walnut.
Sights: Post front, fully adjustable rear.
Features: Velocity about 420 fps. Magazine loads through bottom of the grip. Full-length dovetail for scope mounting. Manual safety. Introduced 1996. Imported from Germany by Great Lakes Airguns.
Price: $161.50

Record Champion Repeater

RECORD JUMBO DELUXE AIR PISTOL
Caliber: 177, single shot.
Barrel: 6", rifled.
Weight: 1.9 lbs. **Length:** 7.25" overall.
Power: Spring-air, lateral cocking lever.
Stocks: Smooth walnut.
Sights: Blade front, fully adjustable open rear.
Features: Velocity to 322 fps. Thumb safety. Grip magazine compartment for extra pellet storage. Introduced 1983. Imported from Germany by Great Lakes Airguns.
Price: $121.34

RWS/DIANA MODEL 5G AIR PISTOL
Caliber: 177, single shot.
Barrel: 7".
Weight: $2^{3}/_{4}$ lbs. **Length:** 15" overall.
Power: Spring-air, barrel cocking.
Stocks: Plastic, thumbrest design.
Sights: Tunnel front, micro-click open rear.
Features: Velocity of 450 fps. Adjustable two-stage trigger with automatic safety. Imported from Germany by Dynamit Nobel-RWS, Inc.
Price: $260.00

Record Jumbo

RWS/DIANA MODEL 6M MATCH AIR PISTOL
Caliber: 177, single shot.
Barrel: 7".
Weight: 3 lbs. **Length:** 15" overall.
Power: Spring-air, barrel cocking.
Stocks: Walnut-finished hardwood with thumbrest.
Sights: Adjustable front, micro. click open rear.
Features: Velocity of 410 fps. Recoilless double piston system, movable barrel shroud to protect from sight during cocking. Imported from Germany by Dynamit Nobel-RWS, Inc.
Price: Right-hand $585.00
Price: Left-hand $640.00

RWS/Diana Model 6M

RWS/Diana Model 6G Air Pistols
Similar to the Model 6M except does not have the movable barrel shroud. Has click micrometer rear sight, two-stage adjustable trigger, interchangeable tunnel front sight. Available in right- or left-hand models.
Price: Right-hand $450.00
Price: Left-hand $490.00

Consult our Directory pages for the location of firms mentioned.

STEYR CO$_2$ MATCH LP1C PISTOL
Caliber: 177, single shot.
Barrel: 9".
Weight: 38.7 oz. **Length:** 15.3" overall.
Power: Refillable CO$_2$ cylinders.
Stocks: Fully adjustable Morini match with palm shelf; stippled walnut.
Sights: Interchangeable blade in 4mm, 4.5mm or 5mm widths, fully adjustable open rear with interchangeable 3.5mm or 4mm leaves.
Features: Velocity about 500 fps. Adjustable trigger, adjustable sight radius from 12.4" to 13.2". With compensator. Imported from Austria by Nygord Precision Products.
Price: $1,250.00

Steyr Match LP1C

STEYR LP 5CP MATCH AIR PISTOL
Caliber: 177, 5-shot magazine.
Barrel: NA.
Weight: 40.7 oz. **Length:** 15.2" overall.
Power: Pre-charged air cylinder.
Stocks: Adjustable match type.
Sights: Interchangeable blade front, fully adjustable match rear.
Features: Adjustable sight radius; fully adjustable trigger. Has barrel compensator. Introduced 1995. Imported from Austria by Nygord Precision Products.
Price: $1,395.00

STEYR LP5C MATCH PISTOL
Caliber: 177, 5-shot magazine.
Barrel: NA.
Weight: 40.2 oz. **Length:** 13.39" overall.
Power: Refillable CO$_2$ cylinders.
Stocks: Adjustable Morini match with palm shelf; stippled walnut.
Sights: Movable 2.5mm blade front; 2-3mm interchangeable in .2mm increments; fully adjustable open match rear.
Features: Velocity about 500 fps. Fully adjustable trigger; compensator; has dry-fire feature. Barrel and grip weights available. Introduced 1993. Imported from Austria by Nygord Precision Products.
Price: $1,350.00

AIRGUNS—HANDGUNS

WALTHER CPM-1 CO$_2$ MATCH PISTOL
Caliber: 177, single shot.
Barrel: 8.66".
Weight: NA. **Length:** 15.1" overall.
Power: CO$_2$.
Stocks: Orthopaedic target type.
Sights: Undercut blade front, open match rear fully adjustable for windage and elevation.
Features: Adjustable velocity; matte finish. Introduced 1995. Imported from Germany by Nygord Precision Products.
Price: $950.00

Walther CP88

WALTHER CP88 PELLET PISTOL
Caliber: 177, 8-shot rotary magazine.
Barrel: 4", 6".
Weight: 37 oz. (4" barrel) **Length:** 7" (4" barrel).
Power: CO$_2$.
Stocks: Checkered plastic.
Sights: Blade front, fully adjustable rear.
Features: Faithfully replicates size, weight and trigger pull of the 9mm Walther P88 compact pistol. Has SA/DA trigger mechanism; ambidextrous safety, levers. Comes with two magazines, 500 pellets, one CO$_2$ cartridge. Introduced 1997. Imported from Germany by Interarms.
Price: Blue $179.00
Price: Nickel $189.00

Walther CP88 Competition Pellet Pistol
Similar to the standard CP88 except has 6" match-grade barrel, muzzle weight, wood or plastic stocks. Weighs 41 oz., has overall length of 9". Introduced 1997. Imported from Germany by Interarms.
Price: Blue, plastic stocks $189.00
Price: Nickel, plastic stocks $199.00
Price: Blue, wood stocks $220.00
Price: Nickel, wood stocks $232.00

Walther CP88 Competition

AIRGUNS—LONG GUNS

Air Arms S-300

AIR ARMS S-300 HI-POWER AIR RIFLE
Caliber: 22.
Barrel: 19 1/2".
Weight: 6.4 lbs. **Length:** 38.5" overall.
Stock: Stained European hardwood; ambidextrous.
Power: Precharged pneumatic.
Sights: None furnished.
Features: Velocity about 750 fps. Two-stage trigger. Blue finish. Introduced 1997. Imported from England by Great Lakes Airarms.
Price: $764.07

AIR ARMS TX 200 AIR RIFLE
Caliber: 177; single shot.
Barrel: 15.7".
Weight: 9.3 lbs. **Length:** 41.5" overall.
Power: Spring-air; underlever cocking.
Stock: Oil-finished hardwood; checkered grip and forend; rubber buttpad.
Sights: None furnished.
Features: Velocity about 900 fps. Automatic safety; adjustable two-stage trigger. Imported from England by Great Lakes Airguns.
Price: $578.65

Consult our Directory pages for the location of firms mentioned.

Airrow A-8S1P

AIRROW MODEL A-8S1P STEALTH AIR GUN
Caliber: #2512 16" arrow.
Barrel: 16".
Weight: 4.4 lbs. **Length:** 30.1" overall.
Power: CO$_2$ or compressed air; variable power.
Stock: Telescoping CAR-15-type.
Sights: Scope rings only.
Features: Velocity to 650 fps with 260-grain arrow. Pneumatic air trigger. All aircraft aluminum and stainless steel construction. Mil-spec materials and finishes. Waterproof case. Introduced 1991. From Swivel Machine Works, Inc.
Price: About $1,699.00

AIRGUNS—LONG GUNS

Anschutz 2002

ANSCHUTZ 2002 MATCH AIR RIFLE
Caliber: 177, single shot.
Barrel: 26".
Weight: 10 1/2 lbs. **Length:** 44.5" overall.
Stock: European walnut, blonde hardwood or colored laminated hardwood; stippled grip and forend.
Sights: Optional sight set #6834.
Features: Muzzle velocity 575 fps. Balance, weight match the 1907 ISU smallbore rifle. Uses #5021 match trigger. Recoil and vibration free. Fully adjustable cheekpiece and buttplate; accessory rail under forend. Introduced 1988. Imported from Germany by Gunsmithing, Inc., Champion's Choice, Champion Shooter's Supply, Accuracy International, AcuSport Corp.
Price: Right-hand, blonde hardwood stock ....$1,684.95 to $1,789.95
Price: Left-hand, blonde hardwood stock $1,272.80
Price: Right-hand, walnut stock $1,261.40
Price: Right-hand, color laminated stock $1,291.60
Price: Left-hand, color laminated stock $1,355.30
Price: Model 2002D-RT Running Target, right-hand, no sights $1,419.20
Price: #6834 Sight Set $245.90

AIRROW MODEL A-8SRB STEALTH AIR GUN
Caliber: 177, 22, 25, 38, 9-shot.
Barrel: 19.7"; rifled.
Weight: 6 lbs. **Length:** 34" overall.
Power: CO_2 or compressed air; variable power.
Stock: Telescoping CAR-15-type.
Sights: Variable 3.5-10x scope.
Features: Velocity 1100 fps in all calibers. Pneumatic air trigger. All aircraft aluminum and stainless steel construction. Mil-spec materials and finishes. Introduced 1992. From Swivel Machine Works, Inc.
Price: About $2,599.00

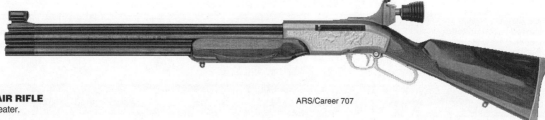
ARS/Career 707

ARS/CAREER 707 AIR RIFLE
Caliber: 22, 6-shot repeater.
Barrel: 23".
Weight: 7.75 lbs. **Length:** 40.5" overall.
Power: Pre-compressed air; variable power.
Stock: Indonesian walnut with checkered grip, gloss finish.
Sights: Hooded post front with interchangeable inserts, fully adjustable diopter rear.
Features: Velocity to 1000 fps. Lever-action with straight feed magazine; pressure gauge in lower front air reservoir; scope mounting rail included. Introduced 1996. Imported from the Philippines by Air Rifle Specialists.
Price: $580.00

ARS/FARCO FP SURVIVAL AIR RIFLE
Caliber: 22, 25, single shot.
Barrel: 22 3/4".
Weight: 5 3/4 lbs. **Length:** 42 3/4" overall.
Power: Multi-pump foot pump.
Stock: Philippine hardwood.
Sights: Blade front, fixed rear.
Features: Velocity to 850 fps (22 or 25). Receiver grooved for scope mounting. Imported from the Philippines by Air Rifle Specialists.
Price: $295.00

ARS/FARCO CO₂ AIR SHOTGUN
Caliber: 51 (28-gauge).
Barrel: 30".
Weight: 7 lbs. **Length:** 48 1/2" overall.
Power: 10-oz. refillable CO_2 tank.
Stock: Hardwood.
Sights: Blade front, fixed rear.
Features: Gives over 100 ft. lbs. energy for taking small game. Imported from the Philippines by Air Rifle Specialists.
Price: $460.00

ARS/Farco CO₂ Stainless Steel Air Rifle
Similar to the ARS/Farco CO_2 shotgun except in 22- or 25-caliber with 21 1/2" barrel; weighs 6 3/4 lbs., 42 1/2" overall; Philippine hardwood stock with stippled grip and forend; blade front sight, adjustable rear, grooved for scope mount. Uses 10-oz. refillable CO_2 cylinder. Made of stainless steel. Imported from the Philippines by Air Rifle Specialists.
Price: Including CO_2 cylinder $460.00

ARS HUNTING MASTER AR6 AIR RIFLE
Caliber: 22, 6-shot repeater.
Barrel: 25 1/2".
Weight: 7 lbs. **Length:** 41 1/4" overall.
Power: Pre-compressed air from 3000 psi diving tank.
Stock: Indonesian walnut with checkered grip; rubber buttpad.
Sights: Blade front, adjustable peep rear.
Features: Velocity over 1000 fps with 32-grain pellet. Receiver grooved for scope mounting. Has 6-shot rotary magazine. Imported by Air Rifle Specialists.
Price: $580.00

ARS/KING HUNTING MASTER AIR RIFLE
Caliber: 22, 5-shot repeater.
Barrel: 22 3/4".
Weight: 7 3/4 lbs. **Length:** 42" overall.
Power: Pre-compressed air from 3000 psi diving tank.
Stock: Indonesian walnut with checkered grip and forend; rubber buttpad.
Sights: Blade front, fully adjustable open rear. Receiver grooved for scope mounting.
Features: Velocity over 1000 fps with 32-grain pellet. High and low power switch for hunting or target velocities. Side lever cocks action and inserts pellet. Rotary magazine. Imported from Korea by Air Rifle Specialists.
Price: $580.00
Price: Hunting Master 900 (9mm, limited production) $1,000.00

ARS/QB77 DELUXE AIR RIFLE
Caliber: 177, 22, single shot.
Barrel: 21 1/2".
Weight: 5 1/2 lbs. **Length:** 40" overall.
Power: Two 12-oz. CO_2 cylinders.
Stock: Walnut-stained hardwood.
Sights: Blade front, adjustable rear.
Features: Velocity to 625 fps (22), 725 fps (177). Receiver grooved for scope mounting. Comes with bulk-fill valve. Imported by Air Rifle Specialists.
Price: $195.00

ARS/Magnum 6 Air Rifle
Similar to the King Hunting Master except is 6-shot repeater with 23 3/4" barrel, weighs 8 1/4 lbs. Stock is walnut-stained hardwood with checkered grip and forend; rubber buttpad. Velocity of 1000+ fps with 32-grain pellet. Imported from Korea by Air Rifle Specialists.
Price: $500.00

AIRGUNS—LONG GUNS

BEEMAN BEARCUB AIR RIFLE
Caliber: 177, single shot.
Barrel: 13".
Weight: 7.2 lbs. **Length:** 37.8" overall.
Power: Spring-piston, barrel cocking.
Stock: Stained hardwood.
Sights: Hooded post front, open fully adjustable rear.
Features: Velocity to 915 fps. Polished blue finish; receiver dovetailed for scope mounting. Imported from England by Beeman Precision Airguns.
Price: $325.00

BEEMAN CROW MAGNUM AIR RIFLE
Caliber: 20, 22, 25, single shot.
Barrel: 16"; 10-groove rifling.
Weight: 8.5 lbs. **Length:** 46" overall.
Power: Gas-spring; adjustable power to 32 foot pounds muzzle energy. Barrel-cocking.
Stock: Classic-style hardwood; hand checkered.
Sights: For scope use only; built-in base and 1" rings included.
Features: Adjustable two-stage trigger. Automatic safety. Also available in 22-caliber on special order. Introduced 1992. Imported by Beeman.
Price: $1,220.00

Beeman/Feinwerkbau P70

BEEMAN/FEINWERKBAU 300-S SERIES MATCH RIFLE
Caliber: 177, single shot.
Barrel: 19.9", fixed solid with receiver.
Weight: Approx. 10 lbs. with optional bbl. sleeve. **Length:** 42.8" overall.
Power: Spring-piston, single stroke sidelever.
Stock: Match model—walnut, deep forend, adjustable buttplate.
Sights: Globe front with interchangeable inserts. Click micro. adjustable match aperture rear. Front and rear sights move as a single unit.
Features: Recoilless, vibration free. Five-way adjustable match trigger. Grooved for scope mounts. Permanent lubrication, steel piston ring. Cocking effort 9 lbs. Optional 10-oz. barrel sleeve. Available from Beeman.
Price: Right-hand $1,270.00
Price: Left-hand $1,370.00

BEEMAN/FEINWERKBAU 300-S MINI-MATCH
Caliber: 177, single shot.
Barrel: 17 1/8".
Weight: 8.8 lbs. **Length:** 40" overall.
Power: Spring-piston, single stroke sidelever cocking.
Stock: Walnut. Stippled grip, adjustable buttplate. Scaled-down for youthful or slightly built shooters.
Sights: Globe front with interchangeable inserts, micro. adjustable rear. Front and rear sights move as a single unit.
Features: Recoilless, vibration free. Grooved for scope mounts. Steel piston ring. Cocking effort about 9 1/2 lbs. Barrel sleeve optional. Left-hand model available. Introduced 1978. Imported by Beeman.
Price: Right-hand $1,270.00
Price: Left-hand $1,370.00

BEEMAN/FEINWERKBAU P70 AIR RIFLE
Caliber: 177, single shot.
Barrel: 16.6".
Weight: 10.6 lbs. **Length:** 42.6" overall.
Power: Precharged pneumatic.
Stock: Laminated hardwoods and hard rubber for stability. Multi-colored stock also available.
Sights: Tunnel front with interchangeable inserts, click micrometer match aperture rear.
Features: Velocity to 570 fps. Recoilless action; double supported barrel; special short rifled area frees pellet from barrel faster so shooter's motion has minimum effect on accuracy. Fully adjustable match trigger with separately adjustable trigger and trigger slack weight. Trigger and sights blocked when loading latch is open. Introduced 1997. Imported by Beeman.
Price: $1,975.00

BEEMAN/FEINWERKBAU 603 AIR RIFLE
Caliber: 177, single shot.
Barrel: 16.6".
Weight: 10.8 lbs. **Length:** 43" overall.
Power: Single stroke pneumatic.
Stock: Special laminated hardwoods and hard rubber for stability. Multi-colored stock also available.
Sights: Tunnel front with interchangeable inserts, click micrometer match aperture rear.
Features: Velocity to 570 fps. Recoilless action; double supported barrel; special, short rifled area frees pellet form barrel faster so shooter's motion has minimum effect on accuracy. Fully adjustable match trigger with separately adjustable trigger and trigger slack weight. Trigger and sights blocked when loading latch is open. Introduced 1997. Imported by Beeman.
Price: $1,975.00

Beeman/HW 97

BEEMAN/HW 97 AIR RIFLE
Caliber: 177, 20, single shot.
Barrel: 17.75".
Weight: 9.2 lbs. **Length:** 44.1" overall.
Power: Spring-piston, underlever cocking.
Stock: Walnut-stained beech; rubber buttpad.
Sights: None. Receiver grooved for scope mounting.
Features: Velocity 830 fps (177). Fixed barrel with fully opening, direct loading breech. Adjustable trigger. Introduced 1994. Imported by Beeman Precision Airguns.
Price: Right-hand only $550.00

Beeman Mako

BEEMAN MAKO AIR RIFLE
Caliber: 177, single shot.
Barrel: 20", with compensator.
Weight: 7.3 lbs. **Length:** 38.5" overall.
Power: Pre-charged pneumatic.
Stock: Stained beech; Monte Carlo cheekpiece; checkered grip.
Sights: None furnished.
Features: Velocity to 930 fps. Gives over 50 shots per charge. Manual safety; brass trigger blade; vented rubber butt pad. Requires scuba tank for air. Introduced 1994. Imported from England by Beeman.
Price: $875.00
Price: Mako FT (thumbhole stock) $1,250.00

CAUTION: PRICES SHOWN ARE SUPPLIED BY THE MANUFACTURER OR IMPORTER. CHECK YOUR LOCAL GUNSHOP.

AIRGUNS—LONG GUNS

BEEMAN KODIAK AIR RIFLE
Caliber: 25, single shot.
Barrel: 17.6".
Weight: 9 lbs. Length: 45.6" overall.
Power: Spring-piston, barrel cocking.
Stock: Stained hardwood.
Sights: Blade front, open fully adjustable rear.
Features: Velocity to 820 fps. Up to 30 foot pounds muzzle energy. Introduced 1993. Imported by Beeman.
Price: $625.00

BEEMAN R1 AIR RIFLE
Caliber: 177, 20 or 22, single shot.
Barrel: 19.6", 12-groove rifling.
Weight: 8.5 lbs. Length: 45.2" overall.
Power: Spring-piston, barrel cocking.
Stock: Walnut-stained beech; cut-checkered pistol grip; Monte Carlo comb and cheekpiece; rubber buttpad.
Sights: Tunnel front with interchangeable inserts, open rear click-adjustable for windage and elevation. Grooved for scope mounting.
Features: Velocity of 940-1000 fps (177), 860 fps (20), 800 fps (22). Non-drying nylon piston and breech seals. Adjustable metal trigger. Milled steel safety. Right- or left-hand stock. Available with adjustable cheekpiece and buttplate at extra cost. Custom and Super Laser versions available. Imported by Beeman.
Price: Right-hand, 177, 20, 22 ....$540.00
Price: Left-hand, 177, 20, 22 ....$575.00

BEEMAN R1 CARBINE
Caliber: 177, 20, 22, 25, single shot.
Barrel: 16.1".
Weight: 8.6 lbs. Length: 41.7" overall.
Power: Spring-piston, barrel cocking.
Stock: Stained beech; Monte Carlo comb and checkpiece; cut checkered pistol grip; rubber buttpad.
Sights: Tunnel front with interchangeable inserts, open adjustable rear; receiver grooved for scope mounting.
Features: Velocity up to 1000 fps (177). Non-drying nylon piston and breech seals. Adjustable metal trigger. Machined steel receiver end cap and safety. Right- or left-hand stock. Imported by Beeman.
Price: 177, 20, 22, 25, right-hand $540.00
Price: As above, left-hand $575.00
Price: R1-AW (synthetic stock, nickel plating) $650.00

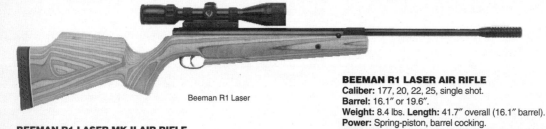
Beeman R1 Laser

BEEMAN R1 LASER MK II AIR RIFLE
Caliber: 177, 20, 22, 25, single shot.
Barrel: 16.1" or 19.6".
Weight: 8.4 lbs. Length: 41.7" overall.
Power: Spring-piston, barrel cocking.
Stock: Laminated wood with high cheekpiece, ventilated recoil pad.
Sights: Tunnel front with interchangeable inserts, open adjustable rear; receiver grooved for scope mounting.
Features: Velocity to 1150 fps (177). Special powerplant components. Built from the Beeman R1 rifle by Beeman.
Price: $995.00

BEEMAN R1 LASER AIR RIFLE
Caliber: 177, 20, 22, 25, single shot.
Barrel: 16.1" or 19.6".
Weight: 8.4 lbs. Length: 41.7" overall (16.1" barrel).
Power: Spring-piston, barrel cocking.
Stock: Laminated wood with Monte Carlo comb and cheekpiece; stippled pistol grip and forend; ambidextrous ventilated recoil pad.
Sights: Tunnel front with interchangeable inserts, open adjustable rear.
Features: Velocity up to 1150 fps (177). Special powerplant components. Built from the Beeman R1 rifle by Beeman.
Price: 177, 20, 22, 25 ....$995.00

BEEMAN R6 AIR RIFLE
Caliber: 177, single shot.
Barrel: NA.
Weight: 7.1 lbs. Length: 41.8" overall.
Power: Spring-piston, barrel cocking.
Stock: Stained hardwood.
Sights: Tunnel post front, open fully adjustable rear.
Features: Velocity to 815 fps. Two-stage Rekord adjustable trigger; receiver dovetailed for scope mounting; automatic safety. Introduced 1996. Imported from Germany by Beeman Precision Airguns.
Price: $325.00

Manufacturers' addresses in the
Directory of the Arms Trade
page 309, this issue

BEEMAN R9 AIR RIFLE
Caliber: 177, 20, single shot.
Barrel: NA.
Weight: 7.3 lbs. Length: 43" overall.
Power: Spring-piston, barrel cocking.
Stock: Stained hardwood.
Sights: Tunnel post front, fully adjustable open rear.
Features: Velocity to 1000 fps (177), 800 fps (20). Adjustable Rekord trigger; automatic safety; receiver dovetailed for scope mounting. Introduced 1996. Imported from Germany by Beeman Precision Airguns.
Price: $350.00

BEEMAN R7 AIR RIFLE
Caliber: 177, 20, single shot.
Barrel: 17".
Weight: 6.1 lbs. Length: 40.2" overall.
Power: Spring piston.
Stock: Stained beech.
Sights: Hooded front, fully adjustable micrometer click open rear.
Features: Velocity to 700 fps (177), 620 fps (20). Receiver grooved for scope mounting; double-jointed cocking lever; fully adjustable trigger; checkered grip. Imported by Beeman.
Price: $325.00

Beeman R9 Deluxe

Beeman R9 Deluxe Air Rifle
Same as the R9 except has an extended forend stock, checkered pistol grip, grip cap, carved Monte Carlo cheekpiece. Globe front sight with inserts. Introduced 1997. Imported by Beeman.
Price: $400.00

AIRGUNS—LONG GUNS

Beeman R11

BEEMAN RX-1 GAS-SPRING MAGNUM AIR RIFLE
Caliber: 177, 20, 22, 25, single shot.
Barrel: 19.6", 12-groove rifling.
Weight: 8.8 lbs.
Power: Gas-spring piston air; single stroke barrel cocking.
Stock: Walnut-finished hardwood, hand checkered, with cheekpiece. Adjustable cheekpiece and buttplate.
Sights: Tunnel front, click-adjustable rear.
Features: Velocity adjustable to about 1200 fps. Uses special sealed chamber of air as a mainspring. Gas-spring cannot take a set. Introduced 1990. Imported by Beeman.
Price: 177, 20, 22 or 25 regular, right-hand $590.00
Price: 177, 20, 22, 25, left-hand $625.00

BEEMAN SUPER 12 AIR RIFLE
Caliber: 22, 25, 12-shot magazine.
Barrel: 19", 12-groove rifling.
Weight: 7.8 lbs. **Length:** 41.7" overall.
Power: Pre-charged pneumatic; external air reservoir.
Stock: European walnut.
Sights: None furnished; drilled and tapped for scope mounting; scope mount included.
Features: Velocity to 850 fps (25-caliber). Adjustable power setting gives 30-70 shots per 400 cc air bottle. Requires scuba tank for air. Introduced 1995. Imported by Beeman.
Price: $1,675.00

BEEMAN R11 AIR RIFLE
Caliber: 177, single shot.
Barrel: 19.6".
Weight: 8.8 lbs. **Length:** 47" overall.
Power: Spring-piston, barrel cocking.
Stock: Walnut-stained beech; adjustable buttplate and cheekpiece.
Sights: None furnished. Has dovetail for scope mounting.
Features: Velocity 910-940 fps. All-steel barrel sleeve. Imported by Beeman.
Price: $560.00

BEEMAN S1 MAGNUM AIR RIFLE
Caliber: 177, single shot.
Barrel: 19".
Weight: 7.1 lbs. **Length:** 45.5" overall.
Power: Spring-piston, barrel cocking.
Stock: Stained beech with Monte Carlo cheekpiece; checkered grip.
Sights: Hooded post front, fully adjustable micrometer click rear.
Features: Velocity to 900 fps. Automatic safety; receiver grooved for scope mounting; two-stage adjustable trigger; curved rubber buttpad. Introduced 1995. Imported by Beeman.
Price: $210.00

BENJAMIN SHERIDAN CO₂ AIR RIFLES
Caliber: 177, 20 or 22, single shot.
Barrel: 19 3/8", rifled brass.
Weight: 5 lbs. **Length:** 36 1/2" overall.
Power: 12-gram CO₂ cylinder.
Stock: American walnut with buttplate.
Sights: High ramp front, fully adjustable notch rear.
Features: Velocity to 680 fps (177). Bolt action with ambidextrous push-pull safety. Gives about 40 shots per cylinder. Black or nickel finish. Introduced 1991. Made in the U.S. by Benjamin Sheridan Co.
Price: Black finish, Model G397 (177), Model G392 (22), about $130.00
Price: Black finish, Model FB9 (20), about $124.50

Benjamin Sheridan Pneumatic

Benjamin Sheridan 397C Pneumatic Carbine
Similar to the standard Model 397 except has 16 3/4" barrel, weighs 4 lbs., 3 oz. Velocity about 650 fps. Introduced 1995. Made in U.S. by Benjamin Sheridan Co.
Price: About $125.00

BENJAMIN SHERIDAN PNEUMATIC (PUMP-UP) AIR RIFLES
Caliber: 177 or 22, single shot.
Barrel: 19 3/8", rifled brass.
Weight: 5 1/2 lbs. **Length:** 36 1/4" overall.
Power: Underlever pneumatic, hand pumped.
Stock: American walnut stock and forend.
Sights: High ramp front, fully adjustable notch rear.
Features: Variable velocity to 800 fps. Bolt action with ambidextrous push-pull safety. Black or nickel finish. Introduced 1991. Made in the U.S. by Benjamin Sheridan Co.
Price: Black finish, Model 397 (177), Model 392 (22), about $130.00
Price: Nickel finish, Model S397 (177), Model S392 (22), about $140.00

BRNO TAU-200

BRNO TAU-200 AIR RIFLE
Caliber: 177, single shot
Barrel: 19", rifled.
Weight: 8 lbs. **Length:** 42" overall.
Power: 6-oz. CO₂ cartridge.
Stock: Wood match style with adjustable comb and buttplate.
Sights: Globe front with interchangeable inserts, fully adjustable open rear.
Features: Adjustable trigger. Comes with sling, extra seals, CO₂ cartridges, large CO₂ bottle, counterweight. Introduced 1993. Imported by Century International Arms, Great Lakes Airguns..
Price: About $446.50
Price: Junior Match (synthetic stock, 7 lbs.) $259.95

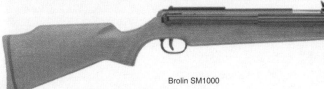
Brolin SM1000

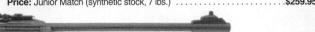

BROLIN MAX VELOCITY SM1000 AIR RIFLE
Caliber: 177 or 22, single shot.
Barrel: 19.5".
Weight: 9.2 lbs. **Length:** 46" overall.
Stock: Walnut-stained hardwood; smooth, checkered or adjustable styles.
Sights: Adjustable blade front, fully adjustable rear. Comes with Weaver-style base mount.
Features: Velocity 1100+ fps (177) 900+ fps (22). Telescoping side cocking lever; double locking safety system. Introduced 1997. Imported by Brolin Industries, Inc.
Price: 177 or 22, smooth stock $189.95
Price: 177 or 22, checkered stock $199.95
Price: 177 or 22, adjustable stock $239.95

AIRGUNS—LONG GUNS

BSA Airsporter Crown Grade

BSA MAGNUM AIRSPORTER RB2 AIR RIFLE, CARBINE
Caliber: 177, 22, 25, single shot.
Barrel: 18".
Weight: 8 lbs., 4 oz. **Length:** 44½" overall.
Power: Spring-air, underlever cocking.
Stock: Oil-finished hardwood; Monte Carlo with cheekpiece, checkered at grip; recoil pad.
Sights: Ramp front, micrometer adjustable rear, comes with Maxi-Grip scope rail.
Features: Velocity 1020 fps (177), 800 fps (22), 675 fps (25). Maxi-Grip scope rail protects optics from recoil; automatic anti-beartrap plus manual tang safety. Imported from U.K. by Precision Sales International, Inc.
Price: $437.00
Price: RB2 Carbine (14" barrel, 41" overall, muzzle brake) $508.00

BSA MAGNUM SUPERSTAR™ MKII AIR RIFLE, CARBINE
Caliber: 177, 22, 25, single shot.
Barrel: 18".
Weight: 8 lbs., 8 oz. **Length:** 43" overall.
Power: Spring-air, underlever cocking.
Stock: Oil-finished hardwood; Monte Carlo with cheekpiece, checkered at grip; recoil pad.
Sights: Ramp front, micrometer adjustable rear. Maxi-Grip scope rail.
Features: Velocity 1020 fps (177), 800 fps (22), 675 fps (25). Patented rotating breech design. Maxi-Grip scope rail protects optics from recoil; automatic anti-beartrap plus manual safety. Imported from U.K. by Precision Sales International, Inc.
Price: $585.00
Price: MKII Carbine (14" barrel, 39½" overall) $585.00

BSA MAGNUM GOLDSTAR AIR RIFLE
Caliber: 177, 22, 10-shot repeater.
Barrel: 18".
Weight: 8 lbs., 8 oz. **Length:** 42.5" overall.
Power: Spring-air, underlever cocking.
Stock: Oil-finished hardwood; Monte Carlo with cheekpiece, checkered at grip; recoil pad.
Sights: Ramp front, micrometer adjustable rear; comes with Maxi-Grip scope rail.
Features: Velocity 1020 fps (177), 800 fps (22). Patented 10-shot indexing magazine; Maxi-Grip scope rail protects optics from recoil; automatic anti-beartrap plus manual safety; muzzlebreak standard. Imported from U.K. by Precision Sales International, Inc.
Price: $847.00

BSA Magnum SuperTEN

BSA MAGNUM SUPERTEN AIR RIFLE
Caliber: 177, 22 10-shot repeater.
Barrel: 17½".
Weight: 7 lbs., 8 oz. **Length:** 37" overall.
Power: Precharged pneumatic via buddy bottle.
Stock: Oil-finished hardwood; Monte Carlo with cheekpiece, cut checkering at grip; adjustable recoil pad.
Sights: No sights; intended for scope use.
Features: Velocity 1300+ fps (177), 1000+ fps (22). Patented 10-shot indexing magazine, bolt-action loading. Left-hand version also available. Imported from U.K. by Precision Sales International, Inc.
Price: $999.00
Price: Left-hand $1,069.00

BSA MAGNUM AIRSPORTER CROWN GRADE AIR RIFLE
Caliber: 177, 22, 25, single shot.
Barrel: 18".
Weight: 8 lbs., 6 oz. **Length:** 44½" overall.
Power: Spring-air, underlever cocking.
Stock: Laminated hardwood; Monte Carlo with rollover cheekpiece, recoil pad.
Sights: No sights; intended for scope use; comes with Maxi-Grip scope rail.
Features: Velocity 1020 fps (177), 800 fps (22), 675 fps (25). Maxi-Grip scope rail protects optics from recoil; automatic anti-beartrap plus manual tang safety. Imported from U.K. by Precision Sales International, Inc.
Price: $536.00

BSA MAGNUM STUTZEN RB2 AIR RIFLE
Caliber: 177, 22, 25, single shot.
Barrel: 14".
Weight: 8 lbs., 8 oz. **Length:** 37" overall.
Power: Spring-air, underlever cocking.
Stock: Oil-finished hardwood; Monte Carlo with cheekpiece; checkered at grip, recoil pad.
Sights: Ramp front, adjustable leaf rear. Comes with Maxi-Grip scope rail.
Features: Velocity 1020 fps (177), 800 fps (22), 675 fps (25). Maxi-Grip scope rail protects optics from recoil; automatic anti-beartrap plus manual tang safety. Imported from U.K. by Precision Sales International, Inc.
Price: $698.00

BSA MAGNUM SUPERSPORT™ AIR RIFLE
Caliber: 177, 22, 25, single shot.
Barrel: 18".
Weight: 6 lbs., 8 oz. **Length:** 41" overall.
Power: Spring-air, barrel cocking.
Stock: Oil-finished hardwood; Monte Carlo with cheekpiece, recoil pad.
Sights: Ramp front, micrometer adjustable rear. Maxi-Grip scope rail.
Features: Velocity 1020 fps (177), 800 fps (22), 675 fps (25). Patented Maxi-Grip scope rail protects optics from recoil; automatic anti-beartrap plus manual tang safety. Muzzle brake standard. Imported for U.K. by Precision Sales International, Inc.
Price: $279.00
Price: Carbine, 14" barrel, muzzle brake $308.00

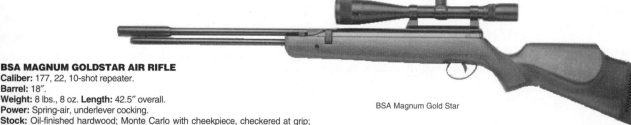

BSA Magnum Gold Star

BSA METEOR MK6 AIR RIFLE
Caliber: 177, 22, single shot.
Barrel: 18".
Weight: 6 lbs. **Length:** 41" overall.
Power: Spring-air, barrel cocking.
Stock: Oil-finished hardwood.
Sights: Ramp front, micrometer adjustable rear.
Features: Velocity 650 fps (177), 500 fps (22). Automatic anti-beartrap; manual tang safety. Receiver grooved for scope mounting. Imported from U.K. by Precision Sales International, Inc.
Price: $210.00

AIRGUNS—LONG GUNS

COPPERHEAD BLACK FIRE RIFLE
Caliber: 177 BB only.
Barrel: 14" smoothbore steel.
Weight: 2 lbs., 7 oz. **Length:** 31 1/2" overall.
Power: Pneumatic, hand pumped.
Stock: Textured plastic.
Sights: Blade front, open adjustable rear.
Features: Velocity to 437 fps. Introduced 1996. Made in U.S. by Crosman Corp.
Price: About $25.00

CROSMAN MODEL 66 POWERMASTER
Caliber: 177 (single shot pellet) or BB, 200-shot reservoir.
Barrel: 20", rifled steel.
Weight: 3 lbs. **Length:** 38 1/2" overall.
Power: Pneumatic; hand pumped.
Stock: Wood-grained ABS plastic; checkered pistol grip and forend.
Sights: Ramp front, fully adjustable open rear.
Features: Velocity about 645 fps. Bolt action, cross-bolt safety. Introduced 1983. From Crosman.
Price: About $55.00
Price: Model 66RT (as above with Realtree® camo finish), about $65.00
Price: Model 664X (as above, with 4x scope) $70.00
Price: Model 664SB (as above with silver and black finish), about $70.00
Price: Model 664GT (black and gold finish, 4x scope) about $70.00

COPPERHEAD BLACK LIGHTNING RIFLE
Caliber: 177 BB, 15-shot magazine.
Barrel: 14" smoothbore.
Weight: 2 lbs. **Length:** 32" overall.
Power: Single-stroke pneumatic.
Stock: Textured plastic.
Sights: Bead front.
Features: Velocity to 350 fps. Cross-bolt safety. Introduced 1996. Made in U.S. by Crosman Corp.
Price: About $25.00

COPPERHEAD BLACK SERPENT RIFLE
Caliber: 177 pellets, 5-shot, on BB, 195-shot magazine.
Barrel: 19 1/2" smoothbore steel.
Weight: 2 lbs., 14 oz. **Length:** 35 7/8" overall.
Power: Pneumatic, single pump.
Stock: Textured plastic.
Sights: Blade front, open adjustable rear.
Features: Velocity to 405 fps. Introduced 1996. Made in U.S. by Crosman Corp.
Price: About $40.00

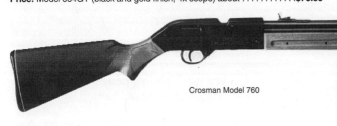

Crosman Model 760

CROSMAN MODEL 782 BLACK DIAMOND AIR RIFLE
Caliber: 177 pellets (5-shot clip) or BB (195-shot reservoir).
Barrel: 18", rifled steel.
Weight: 3 lbs.
Power: CO_2 Powerlet.
Stock: Wood-grained ABS plastic; checkered grip and forend.
Sights: Blade front, open adjustable rear.
Features: Velocity up to 595 fps (pellets), 650 fps (BB). Black finish with white diamonds. Introduced 1990. From Crosman.
Price: About $60.00

CROSMAN MODEL 760 PUMPMASTER
Caliber: 177 pellets (single shot) or BB (200-shot reservoir).
Barrel: 19 1/2", rifled steel.
Weight: 2 lbs., 12 oz. **Length:** 33.5" overall.
Power: Pneumatic, hand pumped.
Stock: Walnut-finished ABS plastic stock and forend.
Features: Velocity to 590 fps (BBs, 10 pumps). Short stroke, power determined by number of strokes. Post front sight and adjustable rear sight. Cross-bolt safety. Introduced 1966. From Crosman.
Price: About $36.00
Price: Model 760SB (silver and black finish), about $50.00

CROSMAN MODEL 795 SPRING MASTER RIFLE
Caliber: 177, single shot.
Barrel: Rifled steel.
Weight: 4 lbs., 8 oz. **Length:** 42" overall.
Power: Spring-piston.
Stock: Black synthetic.
Sights: Hooded front, fully adjustable rear.
Features: Velocity about 550 fps. Introduced 1995. From Crosman.
Price: About $90.00

Crosman Model 1077 Repeatair

CROSMAN MODEL 1389 BACKPACKER RIFLE
Caliber: 177, single shot.
Barrel: 14", rifled steel.
Weight: 3 lbs. 3 oz. **Length:** 31" overall.
Power: Hand pumped, pneumatic.
Stock: Composition, skeletal type.
Sights: Blade front, rear adjustable for windage and elevation.
Features: Velocity to 560 fps. Detachable stock. Receiver grooved for scope mounting. Metal parts blued. From Crosman.
Price: About $70.00

CROSMAN MODEL 2100 CLASSIC AIR RIFLE
Caliber: 177 pellets (single shot), or BB (200-shot BB reservoir).
Barrel: 21", rifled.
Weight: 4 lbs., 13 oz. **Length:** 39 3/4" overall.
Power: Pump-up, pneumatic.
Stock: Wood-grained checkered ABS plastic.
Features: Three pumps give about 450 fps, 10 pumps about 755 fps (BBs). Cross-bolt safety; concealed reservoir holds over 200 BBs. From Crosman.
Price: About $70.00
Price: Model 2100SB (silver and black finish), about $80.00
Price: Model 2104GT (black and gold finish, 4x scope), about $80.00
Price: Model 2100W (walnut stock, pellets only), about $100.00

CROSMAN MODEL 1077 REPEATAIR RIFLE
Caliber: 177 pellets, 12-shot clip
Barrel: 20.3", rifled steel.
Weight: 3 lbs., 11 oz. **Length:** 38.8" overall.
Power: CO_2 Powerlet.
Stock: Textured synthetic or American walnut.
Sights: Blade front, fully adjustable rear.
Features: Velocity 590 fps. Removable 12-shot clip. True semi-automatic action. Introduced 1993. From Crosman.
Price: About $62.75
Price: 1077SB Silver Series (black stock, silver bbl.) $65.00
Price: 1077W (walnut stock) $100.00

CROSMAN MODEL 2200 MAGNUM AIR RIFLE
Caliber: 22, single shot.
Barrel: 19", rifled steel.
Weight: 4 lbs., 12 oz. **Length:** 39" overall.
Stock: Full-size, wood-grained ABS plastic with checkered grip and forend or American walnut.
Sights: Ramp front, open step-adjustable rear.
Features: Variable pump power—three pumps give 395 fps, six pumps 530 fps, 10 pumps 595 fps (average). Full-size adult air rifle. Has white line spacers at pistol grip and buttplate. Introduced 1978. From Crosman.
Price: About $70.00
Price: 2200W, about $100.00

CAUTION: PRICES SHOWN ARE SUPPLIED BY THE MANUFACTURER OR IMPORTER. CHECK YOUR LOCAL GUNSHOP.

AIRGUNS—LONG GUNS

Daisy Red Ryder

DAISY 1938 RED RYDER CLASSIC
Caliber: BB, 650-shot repeating action.
Barrel: Smoothbore steel with shroud.
Weight: 2.2 lbs. **Length:** 35.4" overall.
Stock: Walnut stock burned with Red Ryder lariat signature.
Sights: Post front, adjustable V-slot rear.
Features: Walnut forend. Saddle ring with leather thong. Lever cocking. Gravity feed. Controlled velocity. One of Daisy's most popular guns. From Daisy Mfg. Co.
Price: About $45.00

DAISY MODEL 840
Caliber: 177 pellet single shot; or BB 350-shot.
Barrel: 19", smoothbore, steel.
Weight: 2.7 lbs. **Length:** 36.8" overall.
Power: Pneumatic, single pump.
Stock: Moulded wood-grain stock and forend.
Sights: Ramp front, open, adjustable rear.
Features: Muzzle velocity 335 fps (BB), 300 fps (pellet). Steel buttplate; straight pull bolt action; cross-bolt safety. Forend forms pump lever. Introduced 1978. From Daisy Mfg. Co.
Price: About $40.00

DAISY/POWER LINE 853
Caliber: 177 pellets.
Barrel: 20.9"; 12-groove rifling, high-grade solid steel by Lothar Walther™, precision crowned; bore size for precision match pellets.
Weight: 5.08 lbs. **Length:** 38.9" overall.
Power: Single-pump pneumatic.
Stock: Full-length, select American hardwood, stained and finished; black buttplate with white spacers.
Sights: Globe front with four aperture inserts; precision micrometer adjustable rear peep sight mounted on a standard 3/8" dovetail receiver mount.
Features: Single shot. From Daisy Mfg. Co.
Price: About $270.00

DAISY/POWER LINE 856 PUMP-UP AIRGUN
Caliber: 177 pellets (single shot) or BB (100-shot reservoir).
Barrel: Rifled steel with shroud.
Weight: 2.7 lbs. **Length:** 37.4" overall.
Power: Pneumatic pump-up.
Stock: Moulded wood-grain with Monte Carlo cheekpiece.
Sights: Ramp and blade front, open rear adjustable for elevation.
Features: Velocity from 315 fps (two pumps) to 650 fps (10 pumps). Shoots BBs or pellets. Heavy die-cast metal receiver. Cross-bolt trigger-block safety. Introduced 1984. From Daisy Mfg. Co.
Price: About $45.00

> Consult our Directory pages for the location of firms mentioned.

Daisy/Power Line 880 Anniversary

DAISY/POWER LINE 880 SILVER ANNIVERSARY
Caliber: 177 pellet or BB, 50-shot BB magazine, single shot for pellets.
Barrel: Rifled steel, plated finish.
Weight: 3.7 lbs. **Length:** 37.6" overall.
Power: Multi-pump pneumatic.
Stock: Moulded black wood grain; Monte Carlo comb.
Sights: Hooded front, adjustable rear.
Features: Celebrates 25th year of production. Velocity to 685 fps. (BB). Variable power (velocity and range) increase with pump strokes; resin receiver with dovetail scope mount. Introduced 1997. Made in U.S. by Daisy Mfg. Co.
Price: About $63.50

DAISY MODEL 990 DUAL-POWER AIR RIFLE
Caliber: 177 pellets (single shot) or BB (100-shot magazine).
Barrel: Rifled steel.
Weight: 4.1 lbs. **Length:** 37.4" overall.
Power: Pneumatic pump-up and 12-gram CO_2.
Stock: Moulded woodgrain.
Sights: Ramp and blade front, adjustable open rear.
Features: Velocity to 650 fps (BB), 630 fps (pellet). Choice of pump or CO_2 power. Shoots BBs or pellets. Heavy die-cast receiver dovetailed for scope mount. Cross-bolt trigger block safety. Introduced 1993. From Daisy Mfg. Co.
Price: About $70.00

> *Manufacturers' addresses in the*
> **Directory of the Arms Trade**
> *page 309, this issue*

Daisy/Power Line 1000

DAISY/POWER LINE 1170 PELLET RIFLE
Caliber: 177, single shot.
Barrel: Rifled steel.
Weight: 5.5 lbs. **Length:** 42.5" overall.
Power: Spring-air, barrel cocking.
Stock: Hardwood.
Sights: Hooded post front, micrometer adjustable open rear.
Features: Velocity to 800 fps. Monte Carlo comb. Introduced 1995. From Daisy Mfg. Co.
Price: About $162.50

DAISY/POWER LINE 1000 AIR RIFLE
Caliber: 177, single shot.
Barrel: NA.
Weight: 6 lbs. **Length:** 43" overall.
Power: Spring-air, barrel cocking.
Stock: Stained hardwood.
Sights: Hooded blade front on ramp, fully adjustable micrometer rear.
Features: Velocity to 1000 fps. Blued finish; trigger block safety. Introduced 1997. From Daisy Mfg. Co.
Price: About $260.00

AIRGUNS—LONG GUNS

DAISY/POWER LINE 2002 PELLET RIFLE
Caliber: 177, 35-shot magazine.
Barrel: Rifled steel.
Weight: 3.6 lbs. **Length:** 37.5" overall.
Power: 12-gram CO_2.
Stock: Moulded polymer.
Sights: Ramped blade front, open fully adjustable rear.
Features: Velocity to 630 fps. Continuous feed helical design Mag Clip. Cross-bolt trigger block safety. Introduced 1995. From Daisy Mfg. Co.
Price: About $83.50

DAISY/YOUTH LINE RIFLES

Model:	95	111	105
Caliber:	BB	BB	BB
Barrel:	18"	18"	13$\frac{1}{2}$"
Length:	35.2"	34.3"	29.8"
Power:	Spring	Spring	Spring
Capacity:	700	650	400
Price: About	$45.00	$35.00	$29.00

Features: Model 95 stock and forend are wood; 105 and 111 have plastic stocks. From Daisy Mfg. Co.

DAISY/POWER LINE EAGLE 7856 PUMP-UP AIRGUN
Caliber: 177 (pellets), BB, 100-shot BB magazine.
Barrel: Rifled steel with shroud.
Weight: 2$\frac{3}{4}$ lbs. **Length:** 37.4" overall.
Power: Pneumatic pump-up.
Stock: Moulded wood-grain plastic.
Sights: Ramp and blade front, open rear adjustable for elevation.
Features: Velocity from 315 fps (two pumps) to 650 fps (10 pumps). Finger grooved forend. Cross-bolt trigger-block safety. Introduced 1985. From Daisy Mfg. Co.
Price: With 4x scope, about $60.00

"GAT" AIR RIFLE
Caliber: 177, single shot.
Barrel: 17$\frac{1}{4}$" cocked, 23$\frac{1}{4}$" extended.
Weight: 3 lbs.
Power: Spring-piston.
Stock: Composition.
Sights: Fixed.
Features: Velocity about 450 fps. Shoots pellets, darts, corks. Imported from England by Stone Enterprises, Inc.
Price: $38.95

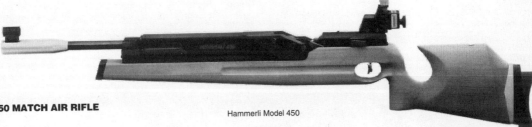
Hammerli Model 450

HAMMERLI MODEL 450 MATCH AIR RIFLE
Caliber: 177, single shot.
Barrel: 19.5".
Weight: 9.8 lbs. **Length:** 43.3" overall.
Power: Pneumatic.
Stock: Match style with stippled grip, rubber buttpad. Beach or walnut.
Sights: Match tunnel front, Hammerli diopter rear.
Features: Velocity about 560 fps. Removable sights; forend sling rail; adjustable trigger; adjustable comb. Introduced 1994. Imported from Switzerland by SIGARMS, Inc.
Price: Beech stock $1,355.00
Price: Walnut stock $1,395.00

Marksman 1710

MARKSMAN 1710 PLAINSMAN AIR RIFLE
Caliber: 177, BB, 20-shot repeater.
Barrel: Smoothbore steel with shroud.
Weight: 2.25 lbs. **Length:** 34" overall.
Power: Spring-air.
Stock: Stained hardwood.
Sights: Blade on ramp front, adjustable V-slot rear.
Features: Velocity about 275 fps. Positive feed; automatic safety. Introduced 1994. Made in U.S. From Marksman Products.
Price: $36.45

Marksman 1745S

MARKSMAN 1745 BB REPEATER AIR RIFLE
Caliber: 177 BB or pellet, 18-shot BB reservoir.
Barrel: 15$\frac{1}{2}$", rifled.
Weight: 4.75 lbs. **Length:** 36" overall.
Power: Spring-air.
Stock: Moulded composition with ambidextrous Monte Carlo cheekpiece and rubber recoil pad.
Sights: Hooded front, adjustable rear.
Features: Velocity about 450 fps. Break-barrel action; automatic safety. Uses BBs, pellets, darts or bolts. Introduced 1997. Made in the U.S. From Marksman Products.
Price: $58.00
Price: Model 1745S (same as above except comes with #1804 4x20 scope) $73.00

MARKSMAN 1750 BB BIATHLON REPEATER RIFLE
Caliber: BB, 18-shot magazine.
Barrel: 15", smoothbore.
Weight: 4.7 lbs.
Power: Spring-piston, barrel cocking.
Stock: Moulded composition.
Sights: Tunnel front, open adjustable rear.
Features: Velocity of 450 fps. Automatic safety. Positive Feed System loads a BB each time gun is cocked. Introduced 1990. From Marksman Products.
Price: $59.75

MARKSMAN 1792 COMPETITION TRAINER AIR RIFLE
Caliber: 177, single shot.
Barrel: 15", rifled.
Weight: 4.7 lbs.
Power: Spring-air, barrel cocking.
Stock: Synthetic.
Sights: Hooded front, match-style diopter rear.
Features: Velocity about 450 fps. Automatic safety. Introduced 1993. More economical version of the 1790 Biathlon Trainer. Made in U.S. From Marksman Products.
Price: $62.75

AIRGUNS—LONG GUNS

Marksman 1790

MARKSMAN 1790 BIATHLON TRAINER
Caliber: 177, single shot.
Barrel: 15″, rifled.
Weight: 4.7 lbs.
Power: Spring-air, barrel cocking.
Stock: Synthetic.
Sights: Hooded front, match-style diopter rear.
Features: Velocity of 450 fps. Endorsed by the U.S. Shooting Team. Introduced 1989. From Marksman Products.
Price: $70.00

MARKSMAN 1795 BOLT-ACTION REPEATER AIR RIFLE
Caliber: 177 pellet or BB, 10-shot clip.
Barrel: 15″ rifled.
Weight: 3.9 lbs. **Length:** 38″ overall.
Power: Spring-air.
Stock: Moulded composition with rubber recoil pad.
Sights: Blade front, fully adjustable rear sight.
Features: Velocity about 500 fps. Bolt-action; positive feed; automatic safety. Uses BBs or pellets. Made in the U.S. From Marksman Products.
Price: $50.00
Price: Model 1795S (same as above except comes wit #1804 4x20 scope $65.00

MARKSMAN 2015 LASERHAWK™ BB REPEATER AIR RIFLE
Caliber: 177 BB, 20-shot magazine.
Barrel: 10.5″ smoothbore.
Weight: 1.6 lbs. **Length:** Adjustable to 33″, 34″ or 35″ overall.
Power: Spring-air.
Stock: Moulded composition.
Sights: Fixed fiber optic front sight, adjustable elevation V-slot rear.
Features: Velocity about 275 fps. Positive feed; automatic safety. Adjustable stock. Introduced 1997. Made in the U.S. From Marksman Products.
Price: $33.00

RWS Model 75S T01

RWS MODEL 75S T01 MATCH
Caliber: 177, single shot.
Barrel: 19″.
Weight: 11 lbs. **Length:** 43.7″ overall.
Power: Dual spring piston.
Stock: Oil-finished beech with stippled grip; adjustable cheekpiece, buttplate.
Sights: Globe front, fully adjustable match peep rear.
Features: Velocity of 580 fps. Fully adjustable trigger; recoilless action. Introduced 1990. Imported from Germany by Dynamit Nobel-RWS.
Price: $1,650.00

RWS MODEL CA 100 AIR RIFLE
Caliber: 177, single shot.
Barrel: 22″.
Weight: 11.4 lbs. **Length:** 44″ overall.
Power: Compressed air; interchangeable cylinders.
Stock: Laminated hardwood with adjustable cheekpiece and buttplate.
Sights: Optional.
Features: Gives 250 shots per full charge. Double-sided power regulator. Introduced 1995. Imported from England by Dynamit Nobel-RWS, Inc.
Price: $2,100.00

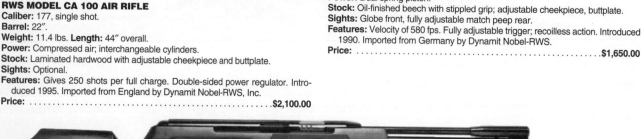

RWS TX200SR

RWS TX200 MAGNUM AIR RIFLE
Caliber: 177, 22, single shot.
Barrel: 14 3/4″; 12-groove Walther with choke.
Weight: 8 1/2 lbs. **Length:** 42″ overall.
Power: Spring-air, underlever cocking.
Stock: Beech or walnut (177 only) with Monte Carlo cheekpiece; checkered grip and forend; rubber recoil pad.
Sights: None furnished; scope rail.
Features: Adjustable two-stage match trigger; automatic safety; floating piston. Made by Air Arms. Introduced 1995. Imported from England by Dynamit Nobel-RWS, Inc.
Price: $560.00
Price: TX200SR (recoilless version of above, slightly different stock), from $660.00

RWS/DIANA MODEL 24 AIR RIFLE
Caliber: 177, 22, single shot.
Barrel: 17″, rifled.
Weight: 6 lbs. **Length:** 42″ overall.
Power: Spring-air, barrel cocking.
Stock: Beech.
Sights: Hooded front, adjustable rear.
Features: Velocity of 700 fps (177). Easy cocking effort; blue finish. Imported from Germany by Dynamit Nobel-RWS, Inc.
Price: $205.00
Price: Model 24C $205.00

RWS/Diana Model 34 Air Rifle
Similar to the Model 24 except has 19″ barrel, weighs 7.5 lbs. Gives velocity of 1000 fps (177), 800 fps (22). Adjustable trigger, synthetic seals. Comes with scope rail.
Price: 177 or 22 .... $285.00
Price: Model 34N (nickel-plated metal, black epoxy-coated wood stock) .$330.00
Price: Model 34BC (matte black metal, black stock, 4x32 scope, mounts) $485.00

Manufacturers' addresses in the
Directory of the Arms Trade
page 309, this issue

AIRGUNS—LONG GUNS

RWS/Diana Model 36

RWS/DIANA MODEL 36 AIR RIFLE
Caliber: 177, 22, single shot.
Barrel: 19", rifled.
Weight: 8 lbs. **Length:** 45" overall.
Power: Spring-air, barrel cocking.
Stock: Beech.
Sights: Hooded front (interchangeable inserts available), adjustable rear.
Features: Velocity of 1000 fps (177-cal.). Comes with scope mount; two-stage adjustable trigger. Imported from Germany by Dynamit Nobel-RWS, Inc.
Price: $415.00
Price: Model 36 Carbine (same as Model 36 except has 15" barrel) $415.00

RWS/Diana Model 52 Deluxe

RWS/DIANA MODEL 45 AIR RIFLE
Caliber: 177, single shot.
Weight: 8 lbs. **Length:** 45" overall.
Power: Spring-air, barrel cocking.
Stock: Walnut-finished hardwood with rubber recoil pad.
Sights: Globe front with interchangeable inserts, micro. click open rear with four-way blade.
Features: Velocity of 820 fps. Dovetail base for either micrometer peep sight or scope mounting. Automatic safety. Imported from Germany by Dynamit Nobel-RWS, Inc.
Price: $330.00

RWS/DIANA MODEL 52 AIR RIFLE
Caliber: 177, 22, single shot.
Barrel: 17", rifled.
Weight: 8½ lbs. **Length:** 43" overall.
Power: Spring-air, sidelever cocking.
Stock: Beech, with Monte Carlo, cheekpiece, checkered grip and forend.
Sights: Ramp front, adjustable rear.
Features: Velocity of 1100 fps (177). Blue finish. Solid rubber buttpad. Imported from Germany by Dynamit Nobel-RWS, Inc.
Price: $535.00
Price: Model 52 Deluxe (select walnut stock, rosewood grip and forend caps, palm swell grip) $775.00
Price: Model 48B (as above except matte black metal, black stock) $535.00
Price: Model 48 (same as Model 52 except no Monte Carlo, cheekpiece or checkering) $530.00

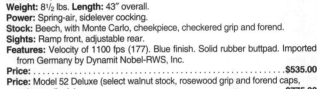
RWS/Diana Model 54 Air King

Consult our Directory pages for the location of firms mentioned.

RWS/DIANA MODEL 54 AIR KING RIFLE
Caliber: 177, 22, single shot.
Barrel: 17".
Weight: 9 lbs. **Length:** 43" overall.
Power: Spring-air, sidelever cocking.
Stock: Walnut with Monte Carlo cheekpiece, checkered grip and forend.
Sights: Ramp front, fully adjustable rear.
Features: Velocity to 1000 fps (177), 900 fps (22). Totally recoilless system; floating action absorbs recoil. Imported from Germany by Dynamit Nobel-RWS, Inc.
Price: $750.00

RWS/Diana Model 100

RWS/DIANA MODEL 100 MATCH AIR RIFLE
Caliber: 177, single shot.
Barrel: 19".
Weight: 11 lbs. **Length:** 43" overall.
Power: Spring-air, sidelever cocking.
Stock: Walnut.
Sights: Tunnel front, fully adjustable match rear.
Features: Velocity of 580 fps. Single-stroke cocking; cheekpiece adjustable for height and length; recoilless operation. Cocking lever secured against rebound. Introduced 1990. Imported from Germany by Dynamit Nobel-RWS, Inc.
Price: Right-hand only $1,650.00

CAUTION: PRICES SHOWN ARE SUPPLIED BY THE MANUFACTURER OR IMPORTER. CHECK YOUR LOCAL GUNSHOP.

AIRGUNS—LONG GUNS

Slavia Model 631

SLAVIA MODEL 631 AIR RIFLE
Caliber: 177, single shot.
Barrel: 21".
Weight: 6.8 lbs. **Length:** 45.5" overall.
Power: Spring-air; barrel cocking.
Stock: Oil-finished European hardwood; checkered forend.
Sights: Hooded post front, fully adjustable open rear.
Features: Velocity to 720 fps. Adjustable two-stage trigger; receiver grooved for scope mounting; automatic safety. Introduced 1996. Imported from the Czech Republic by Great Lakes Airguns.
Price: $114.40

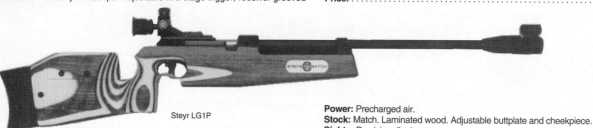
Steyr LG1P

STEYR LG1P AIR RIFLE
Caliber: 177, single shot.
Barrel: 23.75", (13.75" rifled).
Weight: 10.5 lbs. **Length:** 51.7" overall.
Power: Precharged air.
Stock: Match. Laminated wood. Adjustable buttplate and cheekpiece.
Sights: Precision diopter.
Features: Velocity 577 fps. Air cylinders are refillable; about 320 shots per cylinder. Designed for 10-meter shooting. Introduced 1996. Imported from Austria by Nygord Precision Products.
Price: About $1,295.00
Price: Left-hand, about $1,350.00

Webley Patriot

WEBLEY PATRIOT AIR RIFLE
Caliber: 22, single shot.
Barrel: 17.5".
Weight: 9 lbs. **Length:** 45.6" overall.
Power: Spring-air; barrel cocking.
Stock: Walnut-stained beech; checkered grip; rubber buttpad.
Sights: Post front, fully adjustable open rear.
Features: Velocity to 932 fps. Automatic safety; receiver grooved for scope mounting. Imported from England by Great Lakes Airguns.
Price: $518.00

Webley Tracker

WEBLEY TRACKER AIR RIFLE
Caliber: 177, single shot.
Barrel: 11 3/8", rifled.
Weight: 7 lbs. **Length:** 36 1/2" overall.
Power: Spring-air, sidelever cocking.
Stock: European hardwood.
Sights: Adjustable field-type.
Features: Velocity to 772 fps. Receiver grooved for scope mounting. Imported from England by Great Lakes Airguns.
Price: $398.65

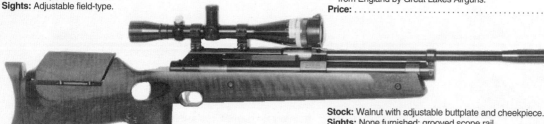
Whiscombe JW70 FB

WHISCOMBE JW SERIES AIR RIFLES
Caliber: 177, 20, 22, 25.
Barrel: 17", Lothar Walther. Polygonal rifling.
Weight: 9 lbs., 8 oz. **Length:** 39" overall.
Power: Dual spring-piston, multi-stroke; underlever cocking.
Stock: Walnut with adjustable buttplate and cheekpiece.
Sights: None furnished; grooved scope rail.
Features: Velocity 660-890 fps (22-caliber, fixed barrel) depending upon model. Interchangeable barrels; automatic safety; muzzle weight; semi-floating action; twin opposed pistons with counter-wound springs; adjustable trigger. All models include H.O.T. System (Harmonic Optimization Tunable System). Introduced 1995. Imported from England by Pelaire Products.
Price: JW50, MKII fixed barrel only $1,820.00
Price: JW60, MKII fixed barrel only $1,870.00
Price: JW70, MKII fixed barrel only $1,920.00
Price: JW75, MKII fixed barrel only $1,950.00
Price: JW80, MKII $1,975.00

SCOPES & MOUNTS

Maker and Model	Magn.	Field at 100 Yds. (feet)	Eye Relief (in.)	Length (in.)	Tube Dia. (in.)	W&E Adjustments	Weight (ozs.)	Price	Other Data
ADCO									[1]Multi-Color Dot system changes from red to green. [2]For airguns, paintball, rimfires. Uses common lithium wafer battery. [3]Comes with standard dovetail mount. [4]3/8" dovetail mount; poly body; adj. intensity diode. [5]10 MOA dot; black or nickel. [6]Square format; with mount; battery.
Magnum 45mm[5]	0	—	—	4.1	45mm	Int.	6.8	$289.00	
MiRAGE Ranger 1"	0	—	—	5.2	1	Int.	3.9	159.00	
MiRAGE Ranger 30mm	0	—	—	5.5	30mm	Int.	5.0	179.00	
MiRAGE Sportsman[1]	0	—	—	5.2	1	Int.	4.5	249.00	
MiRAGE Competitor[1]	0	—	—	5.5	30mm	Int.	5.5	269.00	
IMP Sight[2]	0	—	—	4.5	—	Int.	1.3	19.95	
Square Shooter[3]	0	—	—	5.0	—	Int.	5	129.00	
MiRAGE Eclipse[1]	0	—	—	5.5	30mm	Int.	5.5	249.00	
MiRAGE Champ Red Dot	0	—	—	4.5	—	Int.	2	39.95	
Vantage 1"	0	—	—	3.9	1	Int.	3.9	129.00	
Vantage 30mm	0	—	—	4.2	30mm	Int.	4.9	132.00	
Vision 2000[6]	0	60	—	4.7	—	Int.	6.2	99.00	
AIMPOINT									Illuminates red dot in field of view. Noparallax (dot does not need to be centered). Unlimited field of view and eye relief. On/off, adj. intensity. Dot covers 3" @ 100 yds. [1]Comes with 30mm rings, battery, lens cloth. [2]Requires 1" rings. Black or stainless finish. 3x scope attachment (for rifles only), **$129.95**. [3]Black finish (AP 5000-B) or stainless (AP 5000-S) avail. with regular 3-min. or 10-min. Mag Dot as B2 or S2. From Aimpoint U.S.A.
Comp	0	—	—	4.6	30mm	Int.	4.3	308.00	
Series 5000[3]	0	—	—	5.75	30mm	Int.	5.8	277.00	
Series 3000 Universal[2]	0	—	—	5.5	1	Int.	5.5	232.00	
Series 5000/2x[1]	2	—	—	7	30mm	Int.	9	367.00	
ARMSON O.E.G.									Shows red dot aiming point. No batteries needed. Standard model fits 1" ring mounts (not incl.). Other O.E.G. models for shotguns and rifles can be special ordered. [1]Daylight Only Sight with 3/8" dovetail mount for 22s. Does not contain tritium. From Trijicon, Inc.
Standard	0	—	—	5 1/8	1	Int.	4.3	202.00	
22 DOS[1]	0	—	—	3 3/4	—	Int.	3.0	127.00	
22 Day/Night	0	—	—	3 3/4	—	Int.	3.0	169.00	
M16/AR-15	0	—	—	5 1/8	—	Int.	5.5	226.00	
BAUSCH & LOMB									[1]Adj. objective, sunshade. [2]Also in matte and silver finish. [3]Also in matte finish. [4]Also in matte and silver finish. [5]Also in matte finish. [6]50mm objective; also in matte finish. [9]Also in silver finish. **Partial listing shown. Contact Bausch & Lomb Sports Optics Div. for details.**
Elite 4000									
40-6244A[1]	6-24	18-4.5	3	16.9	1	Int.	20.2	640.95	
40-2104G[2]	2.5-10	41.5-10.8	3	13.5	1	Int.	16	560.95	
40-1636G[3]	1.5-6	61.8-16.1	3	12.8	1	Int.	15.4	528.95	
40-1040	10	10.5	3.6	13.8	1	Int.	22.1	1,858.00	
Elite 3000									
30-4124A[1]	4-12	26.9-9	3	13.2	1	Int.	15.0	417.95	
30-3940G[4]	3-9	33.8-11.5	3	12.6	1	Int.	13.0	319.95	
30-2732G[5]	2-7	44.6-12.7	3	11.6	1	Int.	12.0	303.95	
30-3950G[6]	3-9	31.5-10.5	3	15.7	1	Int.	19	382.95	
30-1545M	1.5-4.5	63-20	3.3	12.5	1	Int.	13	433.95	
30-3955E	3-9	31.5-10.5	3	15.6	30mm	Int.	22	592.95	
Elite 3000 Handgun									
30-2632G[7]	2-6	10-4	20	9.0	1	Int.	10.0	417.95	
BEEMAN									All scopes have 5 point reticle, all glass, fully coated lenses. [1]Includes mount. [2]Also as 66RL with lighted color reticle, **$355.00**. [3]Also as SS-2L 3x with color 4pt. reticle. Imported by Beeman
Blue Ribbon SS-3[1]	1.5-4	42-25	3	5.8	7/8	Int.	8.5	300.00	
Blue Ribbon 66R[2]	2-7	62-16	3	11.4	1	Int.	14.9	315.00	
Blue Ribbon SS-2[1,3]	4	25	3.5	7.0	1.4	Int.	13.7	305.00	
Blue Ribbon 25 Pistol	2	19	10-24	9.1	1	Int.	7.4	155.00	
BURRIS									All scopes avail. in Plex reticle. Steel-on-steel click adjustments. [1]Dot reticle on some models. [2]Post crosshair reticle extra. [3]Matte satin finish. [4]Available with parallax adjustment (standard on 10x, 12x, 4-12x, 6-12x, 6-18x, 6x HBR and 3-12x Signature). [5]Silver matte finish. [6]Target knobs extra, standard on silhouette models, LER and XER with P.A., 6x HBR. Sunshade avail. [7]Avail. with Fine Plex reticle. [8]Available with Heavy Plex reticle. [9]Available with Posi-Lock. [10]Available with Peep Plex reticle. [11]Also avail. for rimfires, airguns. Selected models available with camo finish. **Partial listing shown.** Contact Burris for complete details.
Fullfield									
1x XER[3]	1	51	4.5-20	8.8	1	Int.	7.9	275.00	
1 1/2x[9]	1.6	62	3.5-3.75	10 1/4	1	Int.	9.0	284.00	
2 1/2x[9]	2.5	55	3.5-3.75	10 1/4	1	Int.	9.0	293.00	
4x[1,2,3]	3.75	36	3.5-3.75	11 1/4	1	Int.	11.5	293.00	
6x[1,3]	5.8	23	3.5-3.75	13	1	Int.	12.0	319.00	
12x[1,4,6,7,8]	11.8	10.5	3.5-3.75	15	1	Int.	15	433.00	
1-4x XER[3]	1.0-3.8	53-15	4.25-30	8.8	1	Int.	10.3	360.00	
1 3/4-5x[1,2,9,10]	1.7-4.6	66-25	3.5-3.75	10 7/8	1	Int.	13	348.00	
2-7x[1,2,3]	2.5-6.8	47-18	3.5-3.75	12	1	Int.	14	374.00	
3-9x[1,2,3,10]	3.3-8.7	38-15	3.5-3.75	12 5/8	1	Int.	15	340.00	
3.5-10x50mm[3,5,10]	3.7-9.7	29.5-11	3.5-3.75	14	1	Int.	19	462.00	
4-12x[1,4,8,11]	4.4-11.8	27-10	3.5-3.75	15	1	Int.	18	469.00	
6-18x[1,3,4,6,7,8]	6.5-17.6	16-7	3.5-3.75	15.8	1	Int.	18.5	492.00	
Compact Scopes									
4x[4,5]	3.6	24	3 3/4-5	8 1/4	1	Int.	7.8	246.00	
6x[1,4]	5.5	17	3 3/4-5	9	1	Int.	8.2	260.00	
6x HBR[1,5,8]	6.0	13	4.5	11 1/4	1	Int.	13.0	337.00	
2-7x	2.5-6.9	32-14	3 3/4-5	12	1	Int.	10.5	336.00	
3-9x[5]	3.6-8.8	25-11	3 3/4-5	12 5/8	1	Int.	11.5	343.00	
4-12x[1,4,6]	4.5-11.6	19-8	3 3/4-4	15	1	Int.	15	453.00	

CAUTION: PRICES SHOWN ARE SUPPLIED BY THE MANUFACTURER OR IMPORTER. CHECK YOUR LOCAL GUNSHOP.

HUNTING, TARGET & VARMINT SCOPES

Maker and Model	Magn.	Field at 100 Yds. (feet)	Eye Relief (in.)	Length (in.)	Tube Dia. (in.)	W&E Adjustments	Weight (ozs.)	Price	Other Data
Signature Series									LER=Long Eye Relief; IER=Intermediate Eye Relief; XER=Extra Eye Relief. Partial listing shown, contact maker for complete data. From Burris.
1.5-6x[2,3,5,9,10]	1.7-5.8	70-20	3.5-4.0	10.8	1	Int.	13.0	440.00	
4x[3]	4.0	30	3.5-4.0	12 1/8	1	Int.	14	359.00	
6x[3]	6.0	20	3.5-4.0	12 1/8	1	Int.	14	378.00	
2-8x[3,5,11]	2.1-7.7	53-17	3.5-4.0	11.75	1	Int.	14	512.00	
3-9x[3,5,10]	3.3-8.8	36-14	3.5-4.0	12 7/8	1	Int.	15.5	523.00	
2 1/2-10x[3,5,10]	2.7-9.5	37-10.5	3.5-4.0	14	1	Int.	19.0	636.00	
3-12x[3,10]	3.3-11.7	34-9	3.5-4.0	14 1/4	1	Int.	21	653.00	
4-16x[1,3,5,6,8,10]	4.3-15.7	33-9	3.5-4.0	15.4	1	Int.	23.7	666.00	
6-24x[1,3,5,6,8,10]	6.6-23.8	17-6	3.5-4.0	16.0	1	Int.	22.7	680.00	
8-32x[8,10,12]	8.6-31.4	13-3.8	3.5-4.0	17	1	Int.	24	746.00	
Handgun									
1 1/2-4x LER[1,5,10]	1.6-3.	16-11	11-25	10 1/4	1	Int.	11	352.00	
2-7x LER[3,4,5,10]	2-6.5	21-7	7-27	9.5	1	Int.	12.6	411.00	
3-9x LER[4,5,10]	3.4-8.4	12-5	22-14	11	1	Int.	14	456.00	
1x LER[1]	1.1	27	10-24	8 3/4	1	Int.	6.8	275.00	
2x LER[4,5,6]	1.7	21	10-24	8 3/4	1	Int.	6.8	275.00	
4x LER[1,4,5,6,10]	3.7	11	10-22	9 5/8	1	Int.	9.0	302.00	
10x IER[1,4,6]	9.5	4	8-12	13 1/2	1	Int.	14	398.00	
Scout Scope									
1 1/2x XER[3,9]	1.5	22	7-18	9	1	Int.	7.3	264.00	
2 3/4x XER[3,9]	2.7	15	7-14	9 3/8	1	Int.	7.5	240.00	
BUSHNELL									
Trophy									[1]Wide angle. [2]Also silver finish. [3]Also silver finish. [4]56mm objective. [5]Selective red L.E.D. dot for low light hunting. [6]Also silver finish. [7]Adj. obj. [8]Variable intensity; interchangeable extra reticles (Dual Rings, Open Cross Hairs, Rising Dot) **$136.95**; fits Weaver-style base. Comp model 430 with diamond reticle and 1911 No-hole or 5-hole pattern mount, or STI mount, **$800.95**. [9]Blackpowder scope; extended eye relief, Circle-X reticle. [10]50mm objective.
73-1420	1.75-4	73-30	3.5	10.8	1	Int.	10.9	237.95	
73-1500[1]	1.75-5	68-23	3.5	10.8	1	Int.	12.3	234.95	
73-4124[1]	4-12	32-11	3	12.5	1	Int.	16.1	285.95	
73-3940	3-9	42-14	3	11.7	1	Int.	13.2	159.95	
73-6184	6-18	17.3-6	3	14.8	1	Int.	17.9	360.95	
HOLOsight Model 400[8]	1	—	—	6	—	Int.	8.7	638.95	
Trophy Handgun									
73-0232[2]	2	20	9-26	8.7	1	Int.	7.7	218.95	
73-2632[3]	2-6	21-7	9-26	9.1	1	Int.	9.6	268.95	
Banner									
71-1545	1.5-4.5	67-23	3.5	10.5	1	Int.	10.5	116.95	
71-3944[9]	3-9	36-13	4	11.5	1	Int.	12.5	120.95	
71-3950[10]	3-9	31-10	3	16	1	Int.	19	205.95	
71-4124[7]	4-12	29-11	3	12	1	Int.	15	157.95	
71-6185[10]	6-18	17-6	3	16	1	Int.	18	209.95	
Sportview									
79-0428	4	25	3	7.6	1	Int.	8.5	75.95	
79-0004	4	31	4	11.7	1	Int.	11.2	97.95	
79-0039	3-9	38-13	3.5	10.75	1	Int.	11.2	116.95	
79-0412[7]	4-12	27-9	3.2	13.1	1	Int.	14.6	141.95	
79-1393[6]	3-9	35-12	3.5	11.75	1	Int.	10	68.95	
79-1545	1.5-4.5	69-24	3	10.7	1	Int.	8.6	86.95	
79-3145	3.5-10	36-13	3	12.75	1	Int.	13.9	154.95	
79-1403	4	29	4	11.75	1	Int.	9.2	56.95	
79-6184	6-18	19.1-6.8	3	14.5	1	Int.	15.9	170.95	
79-3938	3-9	42-14	3	12.7	1	Int.	12.5	88.95	
Turkey & Brush									
73-1420	1.75-4	73-30	3.5	10.8	32mm	Int.	10.9	237.95	
CHARLES DALY									Waterproof, fog-proof. [1]Shotgun scope. From Outdoor Sports Headquarters.
4x32	4	28	3.25	11.75	1	Int.	9.5	70.00	
4x32[1]	4	16	6	8.8	1	Int.	9.2	90.00	
4x40 WA	4	36	3.25	13	1	Int.	11.5	98.00	
2-7x32 WA	2-7	56-17	3	11.5	1	Int.	12	125.00	
3-9x40	3-9	35-14	3	12.5	1	Int.	11.25	110.00	
3-9x40 WA	3-9	36-13	3	12.75	1	Int.	12.5	125.00	
4-12x40 WA	4-12	30-11	3	13.75	1	Int.	14.5	133.00	
DOCTER OPTIC									Matte black and matte silver finish available. All lenses multi-coated. Illuminated reticle avail., choice of reticles. Rail mount, aspherical lenses avail. Aspherical lens model, **$1,375.00**. Imported from Germany by Docter Optic Technologies, Inc.
Fixed Power									
4x32	4	31	3	10.7	26mm	Int.	10.0	898.00	
6x42	6	20	3	12.8	26mm	Int.	12.7	1,004.00	
8x56[1]	8	15	3	14.7	26mm	Int.	15.6	1,240.00	
Variables									
1-4x24	1-4	79.7-31.3	3	10.8	30mm	Int.	13	1,300.00	
1.2-5x32	1.2-5	65-25	3	11.6	30mm	Int.	15.4	1,345.00	
1.5-6x42	1.5-6	41.3-20.6	3	12.7	30mm	Int.	16.8	1,378.00	
2.5-10x48	2.5-10	36.6-12.4	3	13.7	30mm	Int.	18.6	1,378.00	
2-12x56	3-12	44.2-13.8	3	14.8	30mm	Int.	20.3	1,425.00	
3-10x40	3-10	34.4-11.7	3	13	1	Int.	18	795.00	
EUROPTIK SUPREME									[1]Military scope with adjustable parallax. Fixed powers have 26mm tubes, variables have 30mm tubes. Some models avail. with steel tubes. All lenses multi-coated. Dust and water tight. From Europtik.
4x36K	4	39	3.5	11.6	26mm	Int.	14	795.00	
6x42K	6	21	3.5	13	26mm	Int.	15	875.00	
8x56K	8	18	3.5	14.4	26mm	Int.	20	925.00	
1.5-6x42K	1.5-6	61.7-23	3.5	12.6	30mm	Int.	17	1,095.00	
2-8x42K	2-8	52-17	3.5	13.3	30mm	Int.	17	1,150.00	
2.5-10x56K	2.5-10	40-13.6	3.5	15	30mm	Int.	21	1,295.00	
3-12x56 Super	3-12	10.8-34.7	3.5-2.5	15.2	30mm	Int.	24	1,495.00	
4-16x56 Super	4-16	9.8-3.9	3.1	18	30mm	Int.	26	1,575.00	

HUNTING, TARGET & VARMINT SCOPES

Maker and Model	Magn.	Field at 100 Yds. (feet)	Eye Relief (in.)	Length (in.)	Tube Dia. (in.)	W&E Adjustments	Weight (ozs.)	Price	Other Data
Europtic Sup. (cont.)									
3-9x40 Micro	3-9	3.2-12.1	2.7	13	1	Int.	14	1,450.00	
2.5-10x46 Micro	2.5-10	13.7-33.4	2.7	14	30mm	Int.	20	1,395.00	
4-16x56 EDP[1]	4-16	22.3-7.5	3.1	18	30mm	Int.	29	1,995.00	
7-12x50 Target	7-12	8.8-5.5	3.5	15	30mm	Int.	21	1,495.00	
KAHLES									[1]Steel tube. [2]Ballistic cam system with military rangefinder. Waterproof, fogproof, nitrogen filled. Choice of reticles. Imported from Austria by Swarovski Optic NA.
K1.5-6x42-L	1.5-6	61-21	—	12.5	30mm	Int.	15.8	721.12	
K2.2-9x42-L	2.2-9	39.5-15	—	13.3	30mm	Int.	15.5	887.78	
K3-12x56-L	3-12	30-11	—	15.2	30mm	Int.	18	943.33	
KZF84-6[1,2]	6	23	—	12.5	1	Int.	17.6	1,245.00	
KZF84-10[1,2]	10	13	—	13.25	1	Int.	18	1,245.00	
KILHAM									Unlimited eye relief; internal click adjustments; crosshair reticle. Fits Thompson/Center rail mounts, for S&W K, N, Ruger Blackhawk, Super, Super Single-Six, Contender.
Hutson Handgunner II	1.7	8	—	5½	7/8	Int.	5.1	119.95	
Hutson Handgunner	3	8	10-12	6	7/8	Int.	5.3	119.95	
LEUPOLD									Constantly centered reticles, choice of Duplex, tapered CPC, Leupold Dot, Crosshair and Dot. CPC and Dot reticles extra. [1]2x and 4x scopes have from 12"-24" of eye relief and are suitable for handguns, top ejection arms and muzzleloaders. [2]3x9 Compact, 6x Compact, 12x, 3x9, 3.5x10 and 6.5x20 come with adjustable objective. Sunshade available for all adjustable objective scopes, **$23.20-41.10**. [3]Silver finish about **$15.00** extra. [4]Gloss finish. Partial listing shown. **Contact Leupold for complete details.**
Vari-X III 3.5x10 STD Tactical	3.5-10	29.5-10.7	3.6-4.6	12.5	1	Int.	13.5	741.10	
M8-2X EER[1]	1.7	21.2	12-24	7.9	1	Int.	6.0	282.10	
M8-2X EER Silver[1]	1.7	21.2	12-24	7.9	1	Int.	6.0	303.60	
M8-2.5x28 IER Scout	2.3	22	9.3	10.1	1	Int.	7.5	367.90	
M8-4X EER[1]	3.7	9	12-24	8.4	1	Int.	7.0	382.10	
M8-4X EER Silver[1]	3.7	9	12-24	8.4	1	Int.	7.0	382.10	
Vari-X 2.5-8 EER	2.5-8.0	13-4.3	11.7-12	9.7	1	Int.	10.9	551.80	
M8-4X Compact	3.6	25.5	4.5	9.2	1	Int.	7.5	350.00	
Vari-X 2-7x Compact	2.5-6.6	41.7-16.5	5-3.7	9.9	1	Int.	8.5	437.50	
Vari-X 3-9x Compact	3.2-8.6	34-13.5	4.0-3.0	11-11.3	1	Int.	11.0	475.00	
M8-4X	4.0	24	4.0	10.7	1	Int.	9.3	350.00	
M8-6X[6]	5.9	17.7	4.3	11.4	1	Int.	10.0	373.20	
M8-6x 42mm	6.0	17	4.5	12	1	Int.	11.3	462.50	
M8-6x42 A.O. Tactical	6	17	4.2	12.1	1	Int.	11.3	600.00	
M8-12x A.O. Varmint	11.6	9.1	4.2	13.0	1	Int.	13.5	517.90	
Vari-X 3-9x Compact EFR A.O.	3.8-8.6	34.0-13.5	4.0-3.0	11.0	1	Int.	11	508.90	
Vari-X-II 1x4	1.6-4.2	70.5-28.5	4.3-3.8	9.2	1	Int.	9.0	396.40	
Vari-X-II 2x7	2.5-6.6	42.5-17.8	4.9-3.8	11.0	1	Int.	10.5	428.60	
Vari-X-II 3x9[1,3]	3.3-8.6	32.3-14.0	4.1-3.7	12.3	1	Int.	13.5	410.70	
Vari-X-II 3-9x50mm	3.3-8.6	32.3-14	4.7-3.7	12	1	Int.	13.6	489.30	
Vari-X II 3-9x40 Tactical	3-9	32.3-14.0	4.7-3.7	12.2	1	Int.	13	510.70	
Vari-X-II 4-12 A.O. Matte	4.4-11.6	22.8-11.0	5.0-3.3	12.3	1	Int.	13.5	564.30	
Vari-X-III 1.5-5x20	1.5-4.5	66.0-23.0	5.3-3.7	9.4	1	Int.	9.5	573.20	
Vari-X-III 1.75-6x 32	1.9-5.6	47-18	4.8-3.7	9.8	1	Int.	11	598.20	
Vari-X-III 2.5x8	2.6-7.8	37.0-13.5	4.7-3.7	11.3	1	Int.	11.5	617.90	
Vari-X-III 3.5-10x50	3.3-9.7	29.5-10.7	4.6-3.6	12.4	1	Int.	13.0	814.30	
Vari-X-III 3.5-10x50[2]	3.3-9.7	29.5-10.7	4.6-3.6	12.4	1	Int.	14.4	739.30	
Vari-X-III 4.5-14 A.O.	4.7-13.7	20.8-7.4	5.0-3.7	12.4	1	Int.	14.5	717.90	
Vari-X-III 4.5-14x50 A.O.	4.7-13.7	20.8-7.4	5.0-3.7	12.4	1	Int.	14.5	837.50	
Vari-X-III 6.5-20 A.O. Varmint	6.5-19.2	14.2-5.5	5.3-3.6	14.2	1	Int.	17.5	850.00	
Vari-X-III 6.5-20x Target EFR A.O.	6.5-19.2	—	5.3-3.6	14.2	1	Int.	16.5	841.10	
Vari-X III 8.5-20x 40 A.O. Target	8.5-25	10.86-4.2	5.3	14.3	1	Int.	17.5	855.40	
Mark 4 M3-6x	6	17.7	4.5	13.1	30mm	Int.	21	1,676.80	
Mark 4 M1-10x[4]	10	11.1	3.6	13 1/8	1	Int.	21	1,676.80	
Mark 4 M1-16x[4]	16	6.6	4.1	12 7/8	1	Int.	22	1,676.80	
Mark 4 M3-10x[4]	10	11.1	3.6	13 1/8	1	Int.	21	1,676.80	
Vari-X-III 6.5x20[2] A.O.	6.5-19.2	14.2-5.5	5.3-3.6	14.2	1	Int.	16.0	748.20	
BR-D 24x40 A.O. Target	24	4.7	3.2	13.6	1	Int.	15.3	932.10	
BR-D 36x-40 A.O. Target	36	3.2	3.4	14.1	1	Int.	15.6	975.00	
LPS 1.5-6x42	1.5-6	58.7-15.7	4	11.2	30mm	Int.	16	1,426.80	
LPS 3.5-14x52 A.O.	3.5-14	28-7.2	4	13.1	30mm	Int.	22	1,516.10	
Rimfire									
Vari-X-II 2-7x RF Special	3.6	25.5	4.5	9.2	1	Int.	7.5	437.50	
Shotgun									
M8 4x	3.7	9.0	12-24	8.4	1	Int.	6.0	371.40	
Vari-X-II 1x4	1.6-4.2	70.5-28.5	4.3-3.8	9.2	1	Int.	9.0	396.40	
Vari-X-II 2x7	2.5-6.6	42.5-17.8	4.9-3.8	11.0	1	Int.	9.0	428.60	
LYMAN									Made under license from Lyman to Lyman's orig. specs. Blue steel. Three-point suspension rear mount with ¼-min. click adj. Data listed are for 20x model. [1]Price approximate. Made in U.S. by Parsons Optical Mfg. Co.
Super TargetSpot[1]	10,12,15,20,25,30	5.5	2	24.3	.75	Int.	27.5	685.00	
McMILLAN									42mm obj. lens; ¼-MOA clicks; nitrogen filled, waterproof; etched duplex-type reticle. [1]Tactical Scope with external adj. knobs, military reticle; 60+ min. adj.
Vision Master 2.5-10x	2.5-10	14.2-4.4	4.3-3.3	13.3	30mm	Int.	17.0	1,250.00	
Vision Master Model I[1]	2.5-10	14.2-4.4	4.3-3.3	13.3	30mm	Int.	17.0	1,250.00	
MILLETT									Full coated lenses; parallax-free; three lenses; 30mm has 10-min. dot, 1-Inch has 3-min. dot. Black or silver finish. From Millett Sights.
Red Dot 1 Inch	1	36.65	—	NA	1	Int.	NA	189.95	
Red Dot 30mm	1	58	—	NA	30mm	Int.	NA	289.95	
MIRADOR									[1]Wide Angle scope. Multi-coated objective lens. Nitrogen filled; waterproof; shockproof. From Mirador Optical Corp.
RXW 4x40[1]	4	37	3.8	12.4	1	Int.	12	179.95	
RXW 1.5-5x20[1]	1.5-5	46-17.4	4.3	11.1	1	Int.	10	188.95	
RXW 3-9x40	3-9	43-14.5	3.1	12.9	1	Int.	13.4	251.95	

CAUTION: PRICES SHOWN ARE SUPPLIED BY THE MANUFACTURER OR IMPORTER. CHECK YOUR LOCAL GUNSHOP.

HUNTING, TARGET & VARMINT SCOPES

Maker and Model	Magn.	Field at 100 Yds. (feet)	Eye Relief (in.)	Length (in.)	Tube Dia. (in.)	W&E Adjustments	Weight (ozs.)	Price	Other Data
NIGHTFORCE									Lighted reticles with eleven intensity levels. Most scopes have choice of reticles. From Lightforce U.S.A.
Fixed Power									
36x56	36	2.6	3	17	30mm	Int.	35	795.95	
Variable Power									
1.75-6x42	1.75-6	47.2-15.7	4.1	12.5	30mm	Int.	22	685.95	
2.5-10x50	2.5-10	31.4-9.4	3.3	13.9	30mm	Int.	28	735.95	
3.5-15x56	3.5-15	24.5-6.9	3	15.8	30mm	Int.	32	775.95	
5.5-22x56	5.5-22	15.7-4.4	3	19.4	30mm	Int.	38.5	825.95	
8-32x56	8-32	9.4-3.1	3	16.6	30mm	Int.	36	855.95	
12-42x56	12-42	6.7-2.3	3	17	30mm	Int.	36	885.95	
NIKON									Super multi-coated lenses and blackening of all internal metal parts for maximum light gathering capability; positive 1/4-MOA; fogproof; waterproof; shockproof; luster and matte finish. [1] Also available in matte silver finish. [2] Available in silver matte finish. From Nikon, Inc.
4x40[2]	4	26.7	3.5	11.7	1	Int.	11.7	284.00	
1.5-4.5x20	1.5-4.5	67.8-22.5	3.7-3.2	10.1	1	Int.	9.5	358.00	
1.5-4.5x24 EER	1.5-4.4	13.7-5.8	24-18	8.9	1	Int.	9.3	352.00	
2-7x32	2-7	46.7-13.7	3.9-3.3	11.3	1	Int.	11.3	367.00	
3-9x40[1]	3-9	33.8-11.3	3.6-3.2	12.5	1	Int.	12.5	371.00	
3.5-10x50	3.5-10	25.5-8.9	3.9-3.8	13.7	1	Int.	15.5	489.00	
4-12x40 A.O.	4-12	25.7-8.6	3.6-3.2	14	1	Int.	16.6	476.00	
4-12x50 A.O.	4-12	25.4-8.5	3.6-3.5	14.0	1	Int.	18.3	578.00	
6.5-20x44	6.5-19.4	16.2-5.4	3.5-3.1	14.8	1	Int.	19.6	591.00	
2x20 EER	2	22	26.4	8.1	1	Int.	6.3	213.00	
NORINCO									Partial listing shown. Some with Ruby Lens coating, blue/black and matte finish. Imported by Nic Max, Inc.
N2520	2.5	44.1	4	—	1	Int.	—	52.28	
N420	4	29.3	3.7	—	1	Int.	—	52.70	
N640	6	20	3.1	—	1	Int.	—	67.88	
N154520	1.5-4.5	63.9-23.6	4.1-3.2	—	1	Int.	—	80.14	
N251042	2.5-10	27-11	3.5-2.8	—	1	Int.	—	206.60	
N3956	3-9	35.1-6.3	3.7-2.6	—	1	Int.	—	231.88	
N31256	3-12	26-10	3.5-2.8	—	1	Int.	—	290.92	
NC2836M	2-8	50.8-14.8	3.6-2.7	—	1	Int.	—	255.60	
PARSONS									Adjustable for parallax, focus. Micrometer rear mount with 1/4-min. click adjustments. Price is approximate. Made in U.S. by Parsons Optical Mfg. Co.
Parsons Long Scope	6	10	2	28-34+	3/4	Ext.	13	475.00-525.00	
PENTAX									[1] Glossy finish; matte finish, $530.00; satin chrome, $550.00. [2] Glossy finish; matte finish, $560.00; satin chrome, $580.00. [3] Glossy finish; matte finish, $580.00; satin chrome, $600.00. [4] Glossy-XL finish; matte-XL finish, $720.00; satin chrome-XL, $740.00. [5] Glossy finish; matte finish, $770.00. [6] Glossy finish, Fine Plex; matte finish, Fine Plex, $810.00; dot reticle, add $10.00. [7] Glossy finish; matte finish, $504.00; satin chrome, $524.00. [8] Glossy finish; matte finish, $420.00; satin chrome $440.00. [9] Lightseeker II $624.00 glossy; $648.00 matte. [10] Lightseeker II $804.00 glossy; $828.00 matte. [11] Glossy finish; matte finish, $360.00. [12] Glossy finish; matte finish, $440.00. [13] Glossy finish; matte finish, $310.00; Mossy Oak, $330.00. [14] Glossy finish; satin chrome, $260.00. [15] Glossy finish; satin chrome, $380.00. [16] Glossy finish; satin chrome, $390.00. [17] Lightseeker II $836.00 glossy, $844.00 satin chrome. Imported by Pentax Corp.
Lightseeker 2-8x[1]	2-8	53-17	3-3.5	11.7	1	Int.	14.0	530.00	
Lightseeker 3-9x[2,9]	3-9	36-14	3-3.5	12.7	1	Int.	15.0	560.00	
Lightseeker 1.75-6x[7]	1.75-6	71-20	3.5-4	10.75	1	Int.	13.0	484.00	
Lightseeker 3.5-10x[3]	3.5-10	29.5-11	3-3.25	14.0	1	Int.	19.5	588.00	
Lightseeker 3-11x[4]	3-11	38.5-13	3-3.25	13.3	1	Int.	19	700.00	
Lightseeker 4-16x AO[5,10]	4-16	3-3.5	33-9	15.4	1	Int.	23.7	760.00	
Lightseeker 6-24 AO[6,17]	6-24	18-5.5	3-3.25	16	1	Int.	22.7	800.00	
3-9x[8]	3-9	38-14.7	3-3.25	13.0	1	Int.	15.0	400.00	
Shotgun									
Lightseeker Zero-X SG Plus[11]	0	51	4.5-15	8.9	1	Int.	7.9	340.00	
Lightseeker Zero-X/V SG Plus[12]	0-4	53.8-15	3.5-7	8.9	1	Int.	10.3	420.00	
Lightseeker 2.5x SG Plus[13]	2.5	55	3-3.5	10.0	1	Int.	9.0	346.00	
Pistol									
2x[14]	2	21	10-24	8.8	1	Int.	6.8	230.00	
1.5-4x[15]	1.5-4	16-11	11-25, 11-18	10.0	1	Int.	11.0	350.00	
2.5-7x[16]	2.5-7	12-7.5	11-28, 9-14	12.0	1	Int.	12.5	370.00	
RWS									Air gun scopes. All have Dyna-Plex reticle. Model 800 is for air pistols. [1] M450, 3-9x40mm, $200.00. Imported from Japan by Dynamit Nobel-RWS.
300	4	36	3.5	11 3/4	1	Int.	13.2	170.00	
400[1]	2-7	55-16	3.5	11 3/4	1	Int.	13.2	190.00	
450	3-9	43-14	3.5	12	1	Int.	14.3	215.00	
500	4	36	3.5	12 1/4	1	Int.	13.9	225.00	
550	2-7	55-16	3.5	12 3/4	1	Int.	14.3	235.00	
600	3-9	43-14	3.5	13	1	Int.	16.5	260.00	
REDFIELD									*Accutrac feature avail. on these scopes at extra cost. Traditionals have round lenses. 4-Plex reticle is standard. [1] Magnum proof. Specially designed for magnum and auto pistols. Uses Double Dovetail mounts. Also in nickel-plated finish, 2x, $239.95, 4x, $239.95, 2 1/2-7x, $322.95, 2 1/2-7x matte black, $322.95. [2] With matte finish $616.95. [3] Also available with matte finish at extra cost. [4] All Golden Five Star scopes come with Butler Creek flip-up lens covers. [5] 56mm adj. objective; European #4 or 4-Plex reticle; comes with 30mm steel rings with Rotary Dovetail System. 1/4-min. click adj. Also in matte finish, $810.95. [6] Also available nickel-plated $377.95. [7] With target knob, $439.95; black matte finish, $493.95; matte black with target knob, $446.95. [8] Black matte finish $400.95. [9] Also avail. in black matte, $246.95. [10] Also avail. in black $460.95; black matte with target knobs, $480.95; with Accu-Trac, black matte, $512.95. [11] Fine crosshair, black finish; $681.95 dot reticle or black or fine crosshair and matte finish; $737.95 with dot, matte; Quick-Zero target knobs, 1/8-MOA reticle, adj. obj. [12] Price shown for variable dot sight with 4, 8, 12, 16 MOA dots. Multi-reticle sight has four dial-in reticle patterns, $399.50. Selected models shown. Contact Redfield for full data.
Ultimate Illuminator 3-12x[5]	2.9-11.7	27-10.5	3-3 1/2	15.4	30mm	Int.	23	801.95	
Widefield Illuminator 2-7x	2.0-6.8	56-17	3-3.5	11.7	1	Int.	13.5	536.95	
Widefield Illuminator 3-9x*[2]	2.9-8.7	38-13	3 1/2	12 3/4	1	Int.	17	605.95	
Widefield Illuminator 3-10x	3-10.1	29-10.5	3-3.5	14.75	1	Int.	18.0	685.95	
Target 8-32x	8-32	13.3-3.4	2.9	16.4	1	Int.	16.1	712.95	
Tracker 4x[3]	3.9	28.9	3 1/2	11.02	1	Int.	9.8	188.95	
Tracker 6x[3]	6.2	18	3.5	12.4	1	Int.	11.1	216.95	
Tracker 8x	8.1	13.5	3.5	12.4	1	Int.	11.1	237.95	
Tracker 2-7x[3]	2.3-6.9	36.6-12.2	3 1/2	12.20	1	Int.	11.6	240.95	
Tracker 3-9x[3]	3.0-9.0	34.4-11.3	3 1/2	14.96	1	Int.	13.4	268.95	
Traditional 4x 3/4"	4	24 1/2	3 1/2	9 3/8	3/4	Int.	—	233.95	
Traditional 2 1/2x	2 1/2	43	3 1/2	10 1/4	1	Int.	8 1/2	175.95	
Golden Five Star 4x[4]	4	28.5	3.75	11.3	1	Int.	9.75	257.95	
Golden Five Star 6x[4]	6	18	3.75	12.2	1	Int.	11.5	281.95	
Golden Five Star 2-7x[4]	2.4-7.4	42-14	3-3.75	11.25	1	Int.	12	331.95	
Golden Five Star 3-9x[4,6]	3.0-9.1	34-11	3-3.75	12.50	1	Int.	13	356.95	

CAUTION: PRICES SHOWN ARE SUPPLIED BY THE MANUFACTURER OR IMPORTER. CHECK YOUR LOCAL GUNSHOP.

HUNTING, TARGET & VARMINT SCOPES

Maker and Model	Magn.	Field at 100 Yds. (feet)	Eye Relief (in.)	Length (in.)	Tube Dia. (in.)	W&E Adjustments	Weight (ozs.)	Price	Other Data
Golden Five Star 3-9x 50mm[4]	3.0-9.1	36.0-11.5	3-3.5	12.8	1	Int.	16	426.95	
Golden Five Star 4-12x A.O.*[4,10]	3.9-11.4	27-9	3-3.75	13.8	1	Int.	16	453.95	
Golden Five Star 6-18x A.O.*[4,7]	6.1-18.1	18.6	3-3.75	14.3	1	Int.	18	481.95	
6-24x Varmint[11]	5.9-23.8	15-5.5	3-3.5	15.75	1	Int.	26	659.95	
I.E.R. 1-4x Shotgun	1.3-3.8	48-16	6	10.2	1	Int.	12	330.95	
Handgun Scopes									
Golden Five Star 2x	2	24	9.5-20	7.88	1	Int.	6	222.95	
Golden Five Star 4x	4	75	13-19	8.63	1	Int.	6.1	222.95	
Golden Five Star 2½-7x	2½-7	11-3.75	11-26	9.4	1	Int.	9.3	303.95	
Widefield Low Profile Compact									
Widefield 3-9x LP Compact	3.3-9	37.0-13.7	3-3.5	10.20	1	Int.	13	386.95	
ESD[12]	1	14.9	—	5.25	30mm	Int.	6.1	365.95	
Low Profile Scopes									
Widefield 2¾xLP	2¾	55½	3½	10½	1	Int.	8	282.95	
Widefield 4xLP	3.6	37½	3½	11½	1	Int.	10	316.95	
Widefield 6xLP	5.5	23	3½	12¾	1	Int.	11	340.95	
Widefield 1¾x-5xLP[8]	1¾-5	70-27	3½	10¾	1	Int.	11½	388.95	
Widefield 2x-7xLP*	2-7	49-19	3½	11¾	1	Int.	13	399.95	
Widefield 3x-9xLP*	3-9	39-15	3½	12½	1	Int.	14	442.95	
SCHMIDT & BENDER									
Fixed									All scopes have 30-yr. warranty, click adjustments, centered reticles, rotation indicators. [1]Glass reticle; aluminum. Available in aluminum with mounting rail. [2]Aluminum only. [3]Aluminum tube. Choice of two bullet drop compensators, choice of two sunshades, two rangefinding reticles. From Schmidt & Bender, Inc. [4]Parallax adjustment in third turret; extremely fine crosshairs. [5]Available with illuminated reticle that glows red; third turret houses on/off switch, dimmer and battery.
4x36	4	30	3.25	11	1	Int.	14	725.00	
6x42	6	21	3.25	13	1	Int.	17	795.00	
8x56	8	16.5	3.25	14	1	Int.	22	915.00	
10x42	10	10.5	3.25	13	1	Int.	18	910.00	
Variables									
1.25-4x20[1]	1.25-4	96-16	3.25	10	30mm	Int.	15.5	980.00	
1.25-4x20 Safari[5]	1.25-4	96-16	3.75	10	30mm	Int.	15.5	980.00	
1.5-6x42[1,5]	1.5-6	60-19.5	3.25	12	30mm	Int.	19.7	1,073.00	
2.5-10x56[1,5]	2.5-10	37.5-12	3.25	14	30mm	Int.	24.6	1,298.00	
3-12x42[2]	3-12	34.5-11.5	3.25	13.5	30mm	Int.	19.0	1,222.00	
3-12x50[1,5]	3-12	33.3-12.6	3.25	13.5	30mm	Int.	22.9	1,262.00	
Police/Marksman									
Fixed									
6x42[1]	6	21	3.25	13.0	30mm	Int.	17.0	900.00	
10x42[3]	10	10.5	3.25	13.0	30mm	Int.	18	950.00	
Variables									
1.5-6x42[3]	1.5-6	60-19.5	3.25	12.0	30mm	Int.	NA	1,200.00	
3-12x42[3]	3-12	34.5-11.5	3.25	13.5	30mm	Int.	NA	1,360.00	
3-12x50[3]	3-12	33.3-12.6	3.25	13.5	30mm	Int.	NA	1,400.00	
4-16x50 Varmint[4]	4-16	22.5-7.5	3.25	14	30mm	Int.	26	1,485.00	
SHEPHERD									[1]Also avail. as 310-P, 310-PE, **$524.25**. [2]Also avail. as 310-P1, 310-P2, 310-P3, 310-P1a, 310-PE1, 310-P22, 310-P22 Mag., 310-PE, **$524.95**. All have patented Dual Reticle system with rangefinder bullet drop compensation; multi-coated lenses, waterproof, shockproof, nitrogen filled, matte finish. From Shepherd Scope, Ltd.
3940-E	3-9	43.5-15	3.3	13	1	Int.	17	1,039.40	
310-2[1,2]	3-10	35.3-11.6	3-3.75	12.8	1	Int.	18	524.25	
SIGHTRON									
Electronic Red Dot									[1]Adjustable objective. [2]3 MOA dot; also with 5 or 10 MOA dot. [3]Variable 3, 5, 10 MOA dot; black finish; also stainless. [4]Satin black; also stainless. Electronic Red Dot scopes come with ring mounts, front and rear extension tubes, polarizing filter, battery, haze filter caps, wrench. Rifle, pistol, shotgun scopes have aluminum tubes, Exac Trak adjustments. Lifetime warranty. From Sightron, Inc.
S33-3[2,4]	1	58	—	5.15	33mm	Int.	5.43	248.99	
S33-3D[3,4]	1	58	—	5.74	33mm	Int.	6.27	369.99	
Riflescopes									
Variables									
SII 1.56x42	1.5-6	50-15	3.8-4.0	11.69	1	Int.	15.35	377.99	
SII2.58x42	2.5-8	36-12	3.6-4.2	11.89	1	Int.	12.82	339.99	
SII 39x42[4]	3-9	34-12	3.6-4.2	12.00	1	Int.	13.22	358.99	
SII312x42	3-12	32-9	3.6-4.2	11.89	1	Int.	12.99	379.99	
SII3.510x42[6]	3.5-10	32-11	3.6	11.89	1	Int.	13.16	379.99	
SII4.514x42[7]	4.5-14	22-7.9	3.6	13.88	1	Int.	16.07	426.99	
SII6.520x50[1]	6.5-20	15.7-5.3	3.8	14.32	1	Int.	18.46	489.99	
Fixed									
SII 4x42	4	31	4.0	12.48	1	Int.	12.34	249.99	
SII 6x42[4]	6	20	4.0	12.48	1	Int.	12.34	249.99	
SII 8x42[4]	8	16	4.0	12.28	1	Int.	12.34	249.99	
Target									
SII 24x44	24	4.1	4.33	13.30	1	Int.	15.87	406.99	
SII 416x42[1,4,5]	4-16	26-7	3.6	13.62	1	Int.	16.0	466.99	
SII 624-42[1,4,5]	6-24	16-5	3.6	14.6	1	Int.	18.7	489.99	
Compact									
SII 4x32	4	25	4.5	9.69	1	Int.	9.34	179.99	
SII2.5-10x32	2.5-10	25	4.5	10.9	1	Int.	10.39	339.99	
Shotgun									
SII 2.5x20SG	2.5	41	4.3	10.28	1	Int.	8.46	193.99	
Pistol									
SII 1x28P[4]	1	30	9.0-24.0	9.49	1	Int.	8.46	197.99	
SII 2x28P[4]	2	16-10	9.0-24.0	9.49	1	Int.	8.28	197.99	
Rimfire									
SII4x32RF	4	25	4.5	9.69	1	Int.	9.34	179.99	

CAUTION: PRICES SHOWN ARE SUPPLIED BY THE MANUFACTURER OR IMPORTER. CHECK YOUR LOCAL GUNSHOP.

HUNTING, TARGET & VARMINT SCOPES

Maker and Model	Magn.	Field at 100 Yds. (feet)	Eye Relief (in.)	Length (in.)	Tube Dia. (in.)	W&E Adjustments	Weight (ozs.)	Price	Other Data
SIMMONS									[1]Matte; also polished finish. [2]Silver; also black matte or polished. [3]Black matte finish. [4]Granite finish; black polish **$216.95**; silver $218.95; also with 50mm obj., black granite **$336.95**. [5]Camouflage. [6]Black polish. [7]With ring mounts. [8]Black polished; also black or silver matte. [9]Lighted reticle, Black Granite finish. [10]50mm obj.; black matte. [11]Black or silver matte. [12]75-yd. parallax; black or silver matte. [13]TV view. [14]Adj. obj. [15]V-TAC reticle in 1st focal plane; 4" sunshade; flat black. [16]Adj. objective; 4" sunshade; black matte. [17]Octagon body; rings included; black matter or silver finish. [18]Black matte finish; also available in silver. **Only selected models shown.** Contact Simmons Outdoor Corp. for complete details. [19]Smart reticle. [20]44mm A.O.; also available with 50mm A.O., **$399.95**.
AETEC									
2100[8]	2.8-10	44-14	5	11.9	1	Int.	15.5	349.95	
2104[16]	3.8-12	33-11	4	13.5	1	Int.	20.0	364.95	
2107[14,20]	6-24	20-6.4	3.2	17.3	1	Int.	22.1	389.95	
V-TAC									
3006[15]	3-9	33-11	4.1-3.0	12 3/8	1	Int.	17.0	699.95	
3007	4.5-14	22-9.4	4.1-2.8	15.2	1	Int.	20.6	749.95	
44 Mag									
M-1044[11]	3-10	34-10.5	3.4-3.3	12.75	1	Int.	15.5	259.95	
M-1045	4-12	29.5-9.5	3	13.2	1	Int.	18.2	279.95	
M-1047	6.5-20	14-.5	2.6-3.4	12.8	1	Int.	19.5	289.95	
M-1050M[19]	3.8-12	30-9.5	3	13.2	1	Int.	18.25	299.95	
Prohunter									
7700[1]	2-7	53-16.25	3.25	11.6	1	Int.	12.4	169.95	
7710[2]	3-9	36-13	3	12.6	1	Int.	13.4	179.95	
7716	4-12	29.6-10.0	3	13.6	1	Int.	14	199.95	
7720	6-18	18.5-6	2.5	12.5	1	Int.	12	224.95	
7740[3]	6	21.75	3	12.6	1	Int.	12	144.95	
Prohunter Handgun									
7737[18]	2	21.5	10.5-26.4	8.3	1	Int.	6.5	179.95	
7738[18]	4	15	10.5-26.4	8.5	1	Int.	8	189.95	
7744[18]	2.5-7	11.5-7	15.7-19.7	9.3	1	Int.	9.75	229.95	
Whitetail Classic									
WTC9[9]	3	11.5	11-20	9.0	1	Int.	9.2	329.95	
WTC11	1.5-5	75-23.5	3.4-3.2	9.3	1	Int.	9.7	184.95	
WTC12	2.5-8	46.5-14.5	3.2-3	12.6	1	Int.	11.25	199.95	
WTC13	3.5-10	35-12	3.2-3	12.4	1	Int.	13.5	219.95	
WTC16	4	36.8	4	9.9	1	Int.	12	149.95	
Pro50									
8800[10]	4-12	27-9	3.5	13.2	1	Int.	18.25	179.95	
8810[10]	6-18	17-5.8	3.6	13.2	1	Int.	18.25	199.95	
Deerfield									
21006	4	29.5	4	11.5	1	Int.	10	74.95	
21029	3-9	37-13	3.4	12.1	1	Int.	12.3	104.95	
21031	4-12	28-11	3-2.8	13.25	1	Int.	14.6	139.95	
Gold Medal Silhouette									
23002	6-20	18-5.4	3	14.5	1	Int.	19.75	529.95	
Gold Medal Handgun									
22002[6]	2.5-7	10.5-4.0	8.9-19.4	9.25	1	Int.	9.0	329.95	
22004[6]	2	21.5	10.5-26.4	7.8	1	Int.	5.75	229.95	
22008	1.5-4	14-6.3	10-26	8.7	1	Int.	7.25	229.95	
Shotgun									
21005	2.5	24	6	7.1	1	Int.	7.0	99.95	
7789D	2	31	5.5	8.5	1	Int.	8.75	129.95	
7790D	4	17	5.5	8.8	1	Int.	8.75	139.95	
7791D	1.5-5	75-23	3.4	9.5	1	Int.	10.75	139.95	
Rimfire									
1022[7]	4	29.5	3.0	11.5	1	Int.	11	74.95	
1022T	3-9	42-14	3.5	11.5	1	Int.	12	199.95	
Blackpowder									
BP0420M[17]	4	19.5	4	7.5	1	Int.	8.3	139.95	
BP2732M[12]	2-7	57.7-16.6	3	11.6	1	Int.	12.4	129.95	
Competition Air Gun									
21612[14]	4-12	25-9	3.1-2.9	13.1	1	Int.	15.8	179.95	
21618[14]	6-18	18-7	2.9-2.7	13.8	1	Int.	18.2	189.95	
STEINER									Waterproof, fogproof, nitrogen filled, accordion-type eye cup. Heavy-duplex or European #4 reticle. Aluminum tubes; matte black finish. From Pioneer Research.
Penetrator									
6x42	6	20.4	3.1	14.8	26mm	Int.	14	1,099.00	
8x56	8	15	3.1	14.8	26mm	Int.	17	1,299.00	
Hunting Z									
1.5-5x20[1]	1.5-5	32-12	4.3	9.6	30mm	Int.	11.7	1,499.00	
2.5-8x36[1]	2.5-8	40-15	4	11.6	30mm	Int.	13.4	1,799.00	
3.5-10x50[1]	3.5-10	77-25	4	12.4	30mm	Int.	16.9	1,899.00	
SWAROVSKI HABICHT									All models offered in either steel or lightweight alloy tubes. Weights shown are for lightweight versions. Choice of nine constantly centered reticles. Eyepiece recoil mechanism and rubber ring shield to protect face. American-style plex reticle available in 2.2-9x42 and 3-12x56 traditional European scopes. [1]Alloy weighs 12.3 oz. [2]Alloy weighs 15.9 oz. [3]Alloy weighs 14.8 oz. [4]Alloy weighs 18.3 oz. [5]Alloy weighs 16.6 oz. Imported by Swarovski Optik North America Ltd.
PH Series									
1.25-4x24[1]	1.25-4	86-27	4.5	10.6	30mm	Int.	15.9	998.89	
1.5-6x42[2]	1.5-6	65.4-21	3.75	13	30mm	Int.	20.5	1,132.22	
2.5-10x42[3]	2.5-10	39.6-12.3	3.75	13.2	30mm	Int.	19.4	1,298.89	
2.5-10x56[4]	2.5-10	39.6-12.3	3.75	14.7	30mm	Int.	24.3	1,398.89	
3-12x50[5]	3-12	33-10.5	3.75	14.3	30mm	Int.	22.0	1,376.67	
6-24x50	6-24	18.6-5.4	3.1	15.4	30mm	Int.	22.6	1,665.56	
6x42	6	23	3.25	12.6	1	Int.	17.9	921.11	
8x50	8	17	3.25	14.4	30mm	Int.	19.9	954.44	
8x56	8	17	3 1/4	14.4	30mm	Int.	23	998.89	
AL Series									
4x32A	4	30	3.2	11.5	1	Int.	10.8	554.44	
6x36A	6	21	3.2	11.9	1	Int.	11.5	610.00	
1.5-4.5x20A	1.5-4.5	75-25.8	3.5	9.53	1	Int.	10.6	665.56	
3-9x36	3-9	39-13.5	3.3	11.9	1	Int.	13	698.89	
3-10x42	3-10	33-11.7	3.3	12.5	1	Int.	13.7	776.67	

HUNTING, TARGET & VARMINT SCOPES

Maker and Model	Magn.	Field at 100 Yds. (feet)	Eye Relief (in.)	Length (in.)	Tube Dia. (in.)	W&E Adjustments	Weight (ozs.)	Price	Other Data
SWIFT									All Swift scopes, with the exception of the 4x15, have Quadraplex reticles and are fogproof and waterproof. The 4x15 has crosshair reticle and is non-waterproof. [1]Available in black or silver finish—same price. [2]Comes with ring mounts, wrench, lens caps, extension tubes, filter, battery. From Swift Instruments.
600 4x15	4	16.2	2.4	11	3/4	Int.	4.7	24.00	
601 3-7x20	3-7	25-12	3-2.9	11	1	Int.	5.6	53.00	
649 4-12x50	4-12	30-10	3-2.8	13.2	1	Int.	14.6	216.00	
650 4x32	4	29	3.5	12	1	Int.	9	80.00	
653 4x40WA[1]	4	35.5	3.75	12.25	1	Int.	12	98.00	
654 3-9x32	3-9	35.75-12.75	3	12.75	1	Int.	13.75	95.00	
656 3-9x40WA[1]	3-9	42.5-13.5	2.75	12.75	1	Int.	14	103.00	
657 6x40	6	18	3.75	13	1	Int.	10	99.50	
660 4x20	4	25	4	11.8	1	Int.	9	80.00	
664 4-12x40[1]	4-12	27-9	3-2.8	13.3	1	Int.	14.8	143.00	
665 1.5-4.5x21	1.5-4.5	69-24.5	3.5-3	10.9	1	Int.	9.6	98.00	
666 Shotgun 1x20	1	113	3.2	7.5	1	Int.	9.6	102.00	
667 Fire-Fly[2]	1	—	—	5.3	30mm	Int.	5	215.00	
668M 4x32	4	25	4	10	1	Int.	8.9	95.00	
Pistol Scopes									
661 4x32	4	90	10-22	9.2	1	Int.	9.5	115.00	
662 2.5x32	2.5	14.3	9-22	8.9	1	Int.	9.3	110.00	
663 2x20[1]	2	18.3	9-21	7.2	1	Int.	8.4	115.00	
TASCO									[1]Water, fog & shockproof; fully coated optics; 1/4-min. click stops; haze filter caps; 30-day/limited lifetime warranty. [2]30/30 range finding reticle. [3]World Class Wide Angle; Supercon multi-coated optics; Opti-Centered® 30/30 range finding reticle; lifetime warranty. [4]1/3 greater zoom range. [5]Trajectory compensating scopes, Opti-Centered® stadia reticle. [6]Anodized finish. [7]True one-power scope. [8]Coated optics; crosshair reticle; ring mounts included to fit most 22, 10mm receivers. [9]Fits Remington 870, 1100, 11-87. [10]Electronic dot reticle with rheostat; coated optics; adj. for windage and elevation; waterproof, shockproof, fogproof; Lithium battery; 3x power booster avail.; matte black or matte aluminum finish; dot or T-3 reticle. [11]TV view. [12]Also matte aluminum finish. [13]Also with crosshair reticle. [14]Also 30/30 reticle. [15]Also in stainless finish. [16]Black matte or stainless finish. [17]Also with stainless finish. [18]Also in matte black. [19]Available with 5-min. or 10-min. dot. [20]Available with 10, 15, 20-min. dot. [21]20mm; also 32mm. [22]20mm; black matte; also stainless steel; also 32mm. [23]Choice of 30/30 or Pro-Shot reticle. [24]Has 4, 8, 12, 16 MOA dots (switchable). [25]Available with BDC. **Contact Tasco for details on complete line.**
Titan									
T1.254.5x26NG	1.25-4.5	59-20	3.5	12	30mm	Int.	16.4	594.00	
T1.56x42N	1.5-6	59-20	3.5	12	30mm	Int.	16.4	680.00	
T39x42N	3-9	37-13	3.5	12.5	30mm	Int.	16.8	645.00	
T312x52N	3-12	27-10	4.5	14	30mm	Int.	20.7	764.00	
Big Horn									
BH2.510x50	2.5-10	44-11	4	13.5	1	Int.	18.7	611.00	
BH4.510x50	4.5-10	30-7.3	4	13.5	1	Int.	18.9	679.00	
World Class									
WA4x40	4	36	3	13	1	Int.	11.5	135.00	
WA6x40	6	23	3	12.75	1	Int.	11.5	144.00	
WA13.5x20[1,3,10]	1-3.5	115-31	3.5	9.75	1	Int.	10.2	161.00	
WA1.75-5x20[1,3]	1.75-5	72-24	3	10 5/8	1	Int.	10.0	152.00	
WA2.58x40[18]	2.5-8	44-14	3	11.75	1	Int.	14.25	178.00	
WA27x32[1,3,9]	2-7	56-17	3.25	11.5	1	Int.	12	161.00	
WA39x40[1,3,6,11,18]	3-9	43.5-15	3	12.75	1	Int.	13.0	199.00	
World Class Airgun									
AG4x40WA	4	36	3	13	1	Int.	14	374.00	
AG39x50WA	3-9	41-14	3	15	1	Int.	17.5	509.00	
World Class Electronic									
ERD39x40WA	3-9	41-14	3	12.75	1	Int.	16	323.00	
World Class Mag IV-44									
WC2510x44[6,19]	2.5-10	41-11	3.5	12.5	1	Int.	14.4	305.00	
World Class TS									
TS24x44[19]	24	4.5	3	14	1	Int.	17.9	407.00	
TS36x44[19]	36	3	3	14	1	Int.	17.9	441.00	
TS832x44[19]	8-24	11-3.5	3	14	1	Int.	19.5	492.00	
TS624x44[19]	6-24	15-4.5	3	14	1	Int.	18.5	475.00	
World Class Pistol									
PWC2x22[12]	2	25	11-20	8.75	1	Int.	7.3	288.00	
PWC4x28[12]	4	8	12-19	9.45	1	Int.	7.9	340.00	
P1.254x28[12]	1.25-4	23-9	15-23	9.25	1	Int.	8.2	339.00	
Mag IV									
W312x40[1,2,4]	3-12	35-9	3	12.25	1	Int.	12	152.00	
W416x40[1,2,4,15,16]	4-16	26-7	3	14.25	1	Int.	15.6	203.00	
W416x50	4-16	31-8	4	13.5	1	Int.	16	350.00	
W520x50[25]	5-20	24-6	4	13.5	1	Int.	16	NA	
W624x40	6-24	17-4	3	15.25	1	Int.	16.8	255.00	
Golden Antler									
GA4x32TV	4	32	3	13	1	Int.	12.7	79.00	
GA4x40TV	4	32	3	12	1	Int.	12.5	85.00	
GA39x32TV[11]	3-9	39-13	3	—	1	Int.	12.2	102.00	
GA39x40TV	3-9	39-13	3	12.5	1	Int.	13	135.00	
GA39x40WA	3-9	41-15	3	12.75	1	Int.	13	152.00	
Silver Antler									
SA2.5x32	2.5	42	3 1/4	11	1	Int.	10	99.00	
SA4x40	4	32	3	12	1	Int.	12.5	85.00	
SA39x32	3-9	39-13	3	13.25	1	Int.	12.2	101.00	
SA39x40[12]	3-9	41-15	3	12.75	1	Int.	13	152.00	
SA39x40	3-9	39-13	3	12.5	1	Int.	13	135.00	
SA4x32[12]	4	32	3	13	1	Int.	12.7	79.00	
Pronghorn									
PH4x32	4	32	3	12	1	Int.	12.5	61.00	
PH39x32	3-9	39-13	3	12	1	Int.	11	83.00	
PH39x40	3-9	39-13	3	13	1	Int.	12.1	110.00	
High Country									
HC416x40	4-16	26-7	3.25	14.25	1	Int.	15.6	254.00	
HC624x10	6-24	17-4	3	15.25	1	Int.	16.8	280.00	
HC39x40	3-9	41-15	3	12.75	1	Int.	13.0	195.00	
HC3.510x40	3.5-10	30-10.5	3	11.75	1	Int.	14.25	220.00	
Rubber Armored									
RC39x40A	3-9	35-12	3.25	12.5	1	Int.	14.3	255.00	

CAUTION: PRICES SHOWN ARE SUPPLIED BY THE MANUFACTURER OR IMPORTER. CHECK YOUR LOCAL GUNSHOP.

HUNTING, TARGET & VARMINT SCOPES

Maker and Model	Magn.	Field at 100 Yds. (feet)	Eye Relief (in.)	Length (in.)	Tube Dia. (in.)	W&E Adjustments	Weight (ozs.)	Price	Other Data
Tasco (cont.)									
World Class TR Scopes									
TR39x40WA	3-9	41-14	3	13	1	Int.	12.5	305.00	
TR416x40	4-16	26-7	3	14.25	1	Int.	16.8	373.00	
TR624x40	6-24	17-4	3	15.5	1	Int.	17.5	407.00	
Bantam									
S1.5-45x20[21,23]	1.5-4.5	69.5-23	4	10.25	1	Int.	10	102.00	
S1.54x32[23]	1.5-4.5	69.5-23	4	11.25	1	Int.	12	110.00	
S2.5x20[22,23]	2.5	22	6	7.5	1	Int.	7.5	80.00	
Airgun									
AG4x20	4	20	2.5	10.75	.75	Int.	5	40.00	
AG4x40WA	4	36	3	13.0	1	Int.	14	373.00	
AG4x32N	4	30	3	—	1	Int.	12.25	144.00	
AG27x32	2-7	48-17	3	12.25	1	Int.	14	178.00	
AG37x20	3-7	24-11	3	11.5	1	Int.	6.5	73.00	
AG39x50WA	3-9	41-14	3	15	1	Int.	17.5	475.00	
Rimfire									
RF4x15[8]	4	22.5	2.5	11	.75	Int.	4	17.00	
RF4x20WA	4	23	2.5	10.5	.75	Int.	3.8	24.00	
RF4x32[18]	4	31	3	12.25	1	Int.	12.6	86.00	
RF37x20	3-7	24-11	2.5	11.5	.75	Int.	5.7	45.00	
P1.5x15	1.5	22.5	9.5-20.75	8.75	.75	Int.	3.25	37.00	
Propoint									
PDP2[10,12,19]	1	40	Unltd.	5	30mm	Int.	5	254.00	
PDP3[10,12,19]	1	52	Unltd.	5	30mm	Int.	5	367.00	
PDP4[16,20]	1	82	Unltd.	—	45mm	Int.	6.1	458.00	
PB1[13]	3	35	Unltd.	5.5	30mm	Int.	6.0	183.00	
PB3	2	30	Unltd.	1.25	30mm	Int.	2.6	214.00	
PDP3CMP	1	68	Unltd.	4.75	33mm	Int.	—	390.00	
PDP5CMP[24]	1	82	Unltd.	4	47mm	Int.	8	407.00	
World Class Plus									
WCP4x44	4	32	3¼	12.75	1	Int.	13.5	271.00	
WCP3.510x50[18]	3.5-10	30-10.5	3¾	13	1	Int.	17.1	407.00	
WCP6x44	6	21	3.25	12.75	1	Int.	13.6	288.00	
WCP39x44[1,16]	3-9	39-14	3.5	12.75	1	Int.	15.8	305.00	
DWC832x50 Target	8-32	13-4	3	14.5	1	Int.	25.1	560.00	
DWCP1040x50 Target	10-40	11-2.5	3	14.5	1	Int.	25.3	611.00	
THOMPSON/CENTER RECOIL PROOF SCOPES									
Pistol Scopes									
8356[1]	2	22.1	10.5-26.4	7⁴⁄₅	1	Int.	6.4	264.00	[1]Black finish; silver, **$269.00**. [2]Black finish; silver, **$357.00**. [3]Black; silver, **$305.00**. [4]Lighted reticle, black. [5]Red dot scope. [6]Adj. obj. [7]adj. obj. [8]Matte black; silver finish **$165.00**. From Thompson/Center.
8315[2]	2.5-7	15-5	8-21, 8-11	9¼	1	Int.	9.2	324.00	
8352[3]	4	22.1	10.5-26.4	7⁴⁄₅	1	Int.	6.4	300.00	
8322	2.5	15	9-21	7²⁄₅	1	Int.	7.2	314.00	
8326[4]	2.5-7	15-5	8-21, 8-11	9¼	1	Int.	10.5	389.00	
8650[5]	1	40	—	5¼	30mm	Int.	4.8	265.00	
Muzzleloader Scopes									
8658	1	60	3.8	9⅛	1	Int.	10.2	128.00	
8656[8]	1.5-5	53-16	3	11½	1	Int.	12.5	160.00	
8664[6]	6-18	18.8-6.2	3	14⅓	1	Int.	13.5	210.00	
8666[7]	4-12	26.7-9	3	12⁴⁄₅	1	Int.	19.5	263.00	
TRIJICON									
Reflex 1x24	1	—	—	4.25	—	Int.	4.6	299.00	[1]Advanced Combat Optical Gunsight for AR-15, M-16, with integral mount. Other mounts available. From Trijicon, Inc.
TA44 1.5x16[1]	1.5	43.8	2.4	4.1	—	Int.	3.5	595.00	
TA45 1.5x24[1]	1.5	28.9	3.6	5.6	—	Int.	3.9	595.00	
TA47 2x20[1]	2	33.1	2.1	4.5	—	Int.	3.8	595.00	
TA50 3x24[1]	3	28.9	1.4	4.8	—	Int.	3.9	619.00	
TA11 3.5x35[1]	3.5	28.9	2.4	8.0	—	Int.	14.0	1,295.00	
TA01 4x32[1]	4	36.8	1.5	5.8	—	Int.	9.9	895.00	
Variable AccuPoint									
3-9x40	3-9	—	3.2-3.6	12.2	1	Int.	12.8	649.00	
ULTRA DOT									
Micro-Dot Scopes[1]									
1.5-4.5x20 Rifle	1.5-4.5	80-26	3	9.8	1	Int.	10.5	297.00	[1]Brightness-adjustable fiber optic red dot reticle. Waterproof, nitrogen-filled one-piece tube tube. Tinted see-through lens covers and battery included. [2]Parallax adjustable. [3]Ultra Dot sights include rings, battery, polarized filter, and 5-year warranty. All models available in black or satin finish. [4]Illuminated red dot has eleven brightness settings. Shock-proof aluminum tube. From Ultra Dot Distribution.
2-7x32	2-7	54-18	3	11.0	1	Int.	12.1	308.00	
3-9x40	3-9	40-14	3	12.2	1	Int.	13.3	327.00	
4x-12x56[2]	4-12	30-10	3	14.3	1	Int.	18.3	417.00	
Ultra-Dot Sights[3]									
Ultra-Dot 25[4]	1	—	—	5.1	1	Int.	3.9	159.00	
Ultra-Dot 30[4]	1	—	—	5.1	30mm	Int.	4.0	179.00	
UNERTL									
1" Target	6,8,10	16-10	2	21½	¾	Ext.	21	358.00	[1]Dural ¼-MOA click mounts. Hard coated lenses. Non-rotating objective lens focusing. [2]¼-MOA click mounts. [3]With target mounts. [4]With calibrated head. [5]Same as 1" Target but without objective lens focusing. [6]With new Posa mounts. [7]Range focus unit near rear of tube. Price is with Posa or standard mounts. Magnum clamp. From Unertl.
1¼" Target[1]	8,10,12,14	12-16	2	25	¾	Ext.	21	466.00	
1½" Target	10,12,14,16,18,20	11.5-3.2	2¼	25½	¾	Ext.	31	487.00	
2" Target[2]	10,12,14,16,18,24,30,32,36	8	2¼	26¼	1	Ext.	44	642.00	
Varmint, 1¼"[3]	6,8,10,12	1-7	2½	19½	⅞	Ext.	26	466.00	
Ultra Varmint, 2"[4]	8,10,12,15	12.6-7	2½	24	1	Ext.	34	630.00	
Small Game[5]	3,4,6	25-17	2¼	18	¾	Ext.	16	284.00	

HUNTING, TARGET & VARMINT SCOPES

Maker and Model	Magn.	Field at 100 Yds. (feet)	Eye Relief (in.)	Length (in.)	Tube Dia. (in.)	W&E Adjustments	Weight (ozs.)	Price	Other Data
Unertl (cont.)									
Programmer 200[7]	10,12,14, 16,18,20, 24,30,36	11.3-4	—	26½	1	Ext.	45	805.00	
BV-20[8]	20	8	4.4	17⅞	1	Ext.	21¼	595.00	
Tube Sight	—	—	—	17	—	Ext.	—	262.50	
U.S. OPTICS									
SN-1/TAR Fixed Power System									
9.6x	10	11.3	3.8	14.5	30mm	Int.	24	1,600.00	Prices shown are estimates; scopes built as ordered, to order; choice of reticles; choice of front or rear focal plane; extra-heavy MIL-SPEC construction; extra-long turrets; individual w&e rebound springs; up to 88mm dia. objectives; up to 50mm tubes; all lenses multi-coated. Made in U.S. by U.S. Optics.
16.2x	15	8.6	4.3	16.5	30mm	Int.	27	1,700.00	
22.4x	20	5.8	3.8	18.0	30mm	Int.	29	1,800.00	
26x	24	5.0	3.4	18.0	30mm	Int.	31	1,900.00	
31x	30	4.6	3.5	18.0	30mm	Int.	32	2,100.00	
37x	36	4.0	3.6	18.0	30mm	Int.	32	2,300.00	
48x	50	3.0	3.8	18.0	30mm	Int.	32	2,500.00	
Variables									
SN-2	4-22	26.8-5.8	5.4-3.8	18.0	30mm	Int.	24	1,762.00	
SN-3	1.6-8	—	4.4-4.8	18.4	30mm	Int.	36	1,435.00	
SN-4	1-4	116-31.2	4.6-4.9	18.0	30mm	Int.	35	1,065.00	
Fixed Power									
SN-6	4,6,8,10	—	4.2-4.8	9.2	30mm	Int.	18	1,195.00	
SN-8	4, 10, 20, 40	32	3.3	7.5	30mm	Int.	11.1	890.00	
WEAVER									
Riflescopes									
K2.5[1]	2.5	35	3.7	9.5	1	Int.	7.3	176.93	[1]Gloss black, [2]Matte black, [3]Silver, [4]Satin, [5]Silver and black (slightly higher in price). [6]Field of view measured at 18" eye relief. ¼ MOA click adjustments, except T-Series which vary from ⅛ to ¼ clicks. One-piece tubes with multi-coated lenses. All scopes are shock-proof, waterproof, and fogproof. Dual-X reticle available in all except V24 which has a fine X-hair and ot; T-Series in which certain models are available in fine X-hair and dots; Qwik-Point red dot scopes which are available in fixed 4 or 12 MOA, or variable 4-8-12 MOA. V16 also available with fine X-hair, dot or Dual-X reticle. T-Series scopes have Micro-Trac® adjustments.
K4[1-2]	3.7	26.5	3.3	11.3	1	Int.	10	185.72	
K6[1]	5.7	18.5	3.3	11.4	1	Int.	10	198.40	
KT15[1]	14.6	7.5	3.2	12.9	1	Int.	14.7	374.90	
V3[1-2]	1.1-2.8	88-32	3.9-3.7	9.2	1	Int.	8.5	223.57	
V9[1-2]	2.8-8.7	33-11	3.5-3.4	12.1	1	Int.	11.1	255.05	
V98x50[1-2]	3-9	29.4-9.9	3.6-3.0	13.1	1	Int.	14.5	332.36	
V10[1-2-3]	2.2-9.6	38.5-9.5	3.4-3.3	12.2	1	Int.	11.2	260.62	
V10-50[1-2-3]	2.3-9.7	40.2-9.2	2.9-2.8	13.75	1	Int.	15.2	357.58	
V16 MDX[2-3]	3.8-15.5	26.8-6.8	3.1	13.9	1	Int.	16.5	437.22	
V16 MFC[2-3]	3.8-15.5	26.8-6.8	3.1	13.9	1	Int.	16.5	437.22	
V16 MDT[2-3]	3.8-15.5	26.8-6.8	3.1	13.9	1	Int.	16.5	437.22	
V24 Varm.[2]	6-24	15.3-4	3.15	14.3	1	Int.	17.5	509.19	
Handgun									
H2[1-3]	2	21	4-29	8.5	1	Int.	6.7	225.23	
H4[1-3]	4	18	11.5-18	8.5	1	Int.	6.7	248.80	
VH4[1-3]	1.5-4	13.6-5.8	11-17	8.6	1	Int.	8.1	300.47	
VH8[1-2-3]	2.5-8	8.5-3.7	12.16	9.3	1	Int.	8.3	314.23	
Shotgun									
SV6[2]	1.5-6	58-13.2	4.5	10.6	1	Int.	10.8	248.34	
Rimfire									
R4[2-3]	3.9	29	3.9	9.7	1	Int.	8.8	153.74	
RV7[2-3]	2.5-7	37-13	3.7-3.3	10.75	1	Int.	10.7	176.63	
T-Series									
T-6[4]	6	14	3.58	12.75	1	Int.	14.9	458.51	
T-10[4]	10	9.3	3.0	15.1	1	Int.	16.7	836.32	
T16[4]	16	6.5	3.0	15.1	1	Int.	16.7	842.89	
T-24[4]	24	4.4	3.0	15.1	1	Int.	16.7	849.49	
T-36[3-4]	36	3.0	3.0	15.1	1	Int.	16.7	856.07	
Qwik-Point[6]									
QP 30mm 4 or 12 MOA[5]	1	12.6	—	5.39	30mm	Int.	5.3	235.79	
QP 33mm VCariable[2-5]	1	14.4	—	5.74	33mm	Int.	6.3	382.29	
QP 45mm 4 or 12 MOA[5]	1	21.8	—	4.8	45mm	Int.	8.46	296.17	
QP 45mm 12MOA[2]	1	21.8	—	4.8	45mm	Int.	8.46	383.08	
QP 45mm Variable[2,5]	1	21.8	—	4.8	45mm	Int.	8.46	383.18	
ZEISS									
Diatal Z 6x42	6	22.9	3.2	12.7	1	Int.	13.4	955.00	All scopes have ¼-minute click-stop adjustments. Choice of Z-Plex or fine crosshair reticles. Rubber armored objective bell, rubber eyepiece ring. Lenses have T-Star coating for highest light transmission. Z-Series scopes offered in non-rail,tubes with duplex reticles only; 1" and 30mm. [1]Black matte finish. [2]Also in stainless matte finish. [3]Also with illuminated reticle, $1,810.00. [4]Matte finish. Bullet Drop Compensator avail. for all Z-Series scopes. Imported from Germany by Carl Zeiss Optical, Inc.
Diatal Z 8x56	8	18	3.2	13.8	1	Int.	17.6	1,135.00	
Diavari 1.25-4x24	1.25-4	105-33	3.2	11.46	30mm	Int.	17.3	1,085.00	
Diavari Z 2.5x10x48[1,2]	2.5-10	33-11.7	3.2	14.5	30mm	Int.	24	1,465.00	
Diavari C 3-9x36MC[4]	3-9	36-13	3.5	11.9	1	Int.	15	625.00	
Diavari Z 1.5-6x42[1,2]	1.5-6	65.5-22.9	3.2	12.4	1.18 (30mm)	Int.	18.5	1,240.00	
Diavari Z 3-12x56[1,2,3]	3-12	27.6-9.9	3.2	15.3	1.18 (30mm)	Int.	25.8	1,575.00	

Hunting scopes in general are furnished with a choice of reticle—crosshairs, post with crosshairs, tapered or blunt post, or dot crosshairs, etc. The great majority of target and varmint scopes have medium or fine crosshairs but post or dot reticles may be ordered. W—Windage E—Elevation MOA—Minute of angle or 1" (approx.) at 100 yards, etc.

CAUTION: PRICES SHOWN ARE SUPPLIED BY THE MANUFACTURER OR IMPORTER. CHECK YOUR LOCAL GUNSHOP.

LASER SIGHTS

Maker and Model	Wavelength (nm)	Beam Color	Lens	Operating Temp. (degrees F.)	Weight (ozs.)	Price	Other Data
ALPEC							[1]Range 1000 yards, [2]Range 500 yards, [3]Range 300 yards; Laser Shot II 500 yards; Super Laser Shot 1000 yards. Black or stainless finish aluminum; removable pressure or push-button switch. Mounts for most handguns, many rifles and shotguns. From Alpec Team, Inc.
Power Shot[1]	635	Red	Glass	NA	2.5	NA	
Mini Shot[2]	650	Red	Glass	NA	2.5	NA	
Laser Shot[3]	670	Red	Glass	NA	3.0	NA	
LASERAIM							[1]Red dot/laser combo; 300-yd. range; LA3XHD Hotdot has 500-yd. range **$249.00**; 4 MOA dot size, laser gives 2" dot size at 100 yds. [3]30mm obj. lens; 4 MOA dot at 100 yds.; fits Weaver base. [4]300-yd. range; 2" dot at 100 yds.; rechargeable Nicad battery. [4]1.5-mile range; 1" dot at 100 yds.; 20+ hrs. batt. life. [5]1.5-mile range; 1" dot at 100 yds.; rechargeable Nicad battery (comes with in-field charger); [6]Black or satin finish. With mount, **$169.00**. [7]Laser projects 2" dot at 100 yds.; with rotary switch; with Hotdot **$237.00**; with Hotdot, touch switch **$357.00**. [8]For Glock 17-27; G1 Hotdot **$299.00**; price installed. [10]Fits std. Weaver base, no rings required; 6-MOA dot; seven brightness settings. All have w&e adj.; black or satin silver finish. From Laseraim Technologies, Inc.
LA3X Dualdot[1]	—	—	—	—	12	199.00	
LA5[3]	—	—	—	—	1.2	236.00	
LA10 Hotdot[4]	—	—	—	—	NA	396.00	
LA11 Hotdot[5]	—	—	—	—	NA	292.00	
LA14	—	—	—	—	NA	314.00	
LA16 Hotdot Mighty Sight[6]	—	—	—	—	1.5	169.00	
Red Dot Sights							
LA93 Illusion III[2]	—	—	—	—	5.0	139.00	
LA9750 Grand Illusion[10]	—	—	—	—	7.0	199.00	
Lasers							
MA3 Mini Aimer[7]	—	—	—	—	1.0	155.00	
G1 Laser[8]	—	—	—	—	2.0	289.00	
LASER DEVICES							Projects high intensity beam of laser light onto target as an aiming point. Adj. for w. & e. [1]Diode laser system. From Laser Devices, Inc.
He Ne FA-6	—	—	—	—	11	229.50	
He Ne FA-9	—	—	—	—	16	299.00	
He Ne FA-9P	—	—	—	—	14	299.00	
FA-4[1]	—	—	—	—	3.5	299.00	
LASERGRIPS							Replaces existing grips with built-in laser high in the right grip panel. Integrated pressure sensitive pad in grip activates the laser. Also has master on/off switch. [1]For Beretta 92, 96, Colt 1911/Commander, Ruger MkII, S&W J-frames, SIG Sauer P228, P229. [2]For all Glock models. Option on/off switch. Requires factory installation. From Crimson Trace Corp.
LG-201[1]	633	Red-Orange	Glass	NA	—	349.00	
GLS-630[2]	633	Red-Orange	Glass	NA	—	595.00	
LASERMAX							Replaces the recoil spring guide rod; includes a customized takedown lever that serves as the laser's instant on/off switch. For Glock, Smith & Wesson, Sigarms and Beretta. From LaserMax.
Guide Rod	650/635	Red-Orange	Glass	40-120	.25	NA	
TACSTAR LASERLYTE							[1]Dot/circle or dot/crosshair projection; black or stainless. [2]Also 635/645mm model. From TacStar Laserlyte.
LLX-0006-140/090[1]	635/645	Red	—	—	1.4	112.50	
WPL-0004-140/090[2]	670	Red	—	—	1.2	76.00	
TPL-0004-140/090[2]	670	Red	—	—	1.2	76.00	
T7S-0004-140[2]	670	Red	—	—	0.8	76.00	

Lasergrip Beretta

Lasergrip Colt

Laseraim G1 laser

TacStar LaserLyte LLX-0006-140/090

CAUTION: PRICES SHOWN ARE SUPPLIED BY THE MANUFACTURER OR IMPORTER. CHECK YOUR LOCAL GUNSHOP.

SCOPE MOUNTS

Maker, Model, Type	Adjust.	Scopes	Price
AIMTECH			
Handguns			
AMT Auto Mag II 22 Mag	No	Weaver rail	$56.99
AMT Auto Mag III 30 Carb.	No	Weaver rail	64.95
Auto Mag IV 45WM	No	Weaver rail	64.95
Astra 44 Mag Revolver	No	Weaver rail	63.25
Beretta/Taurus 92/99	No	Weaver rail	63.25
Browning Buckmark/Challenger II	No	Weaver rail	56.99
Browning Hi-Power	No	Weaver rail	63.25
CZ75	No	Weaver rail	63.25
EA9/P9 Tanfoglio frame	No	Weaver rail	63.25
Glock 17, 17L, 19, 22, 23	No	Weaver rail	63.25
Glock 20, 21	No	Weaver rail	63.25
Govt. 45 Autos/38 Super	No	Weaver rail	63.25
Hi-Standard 22 all makes	No	Weaver rail	63.25
Rossi 85/851/951 Revolvers	No	Weaver rail	63.25
Ruger Mk I, Mk II	No	Weaver rail	49.95
Ruger P89	No	Weaver rail	63.25
S&W K,L,N frames	No	Weaver rail	63.25
S&W K,L,N with tapped top strap[1]	No	Weaver rail	69.95
S&W Model 41 Target 22	No	Weaver rail	63.25
S&W Model 52 Target 38	No	Weaver rail	63.25
S&W 2nd Gen. 59/459/659	No	Weaver rail	56.99
S&W 3rd Gen. 59 Series	No	Weaver rail	69.95
S&W 422/622/2206/2206TGT	No	Weaver rail	56.99
S&W 645/745	No	Weaver rail	56.99
S&W Sigma	No	Weaver rail	64.95
Taurus PT908	No	Weaver rail	63.25
Taurus 44 6.5" bbl.	No	Weaver rail	69.95
Shotguns			
Benelli M-1 Super 90	No	Weaver rail	40.95
Benelli Montefeltro 12-ga.	No	Weaver rail	40.95
Benelli Super Black Eagle	No	Weaver rail	49.95
Browning Auto-5 12-ga.	No	Weaver rail	40.95
Browning BPS	No	Weaver rail	40.95
Ithaca 37/87 12-ga.	No	Weaver rail	40.95
Mossberg 500/Maverick 12-ga.[2]	No	Weaver rail	40.95
Mossberg 500/Vaverick 20-ga.[2]	No	Weaver rail	40.95
Mossberg 835 Ulti-Mag[2]	No	Weaver rail	40.95
Mossberg 5500/9200[2]	No	Weaver rail	40.95
Remington 1100/1187 12-ga.[2]	No	Weaver rail	40.95
Remington 1100/1187 12-ga. LH	No	Weaver rail	40.95
Remington 1100/1187 20-ga.	No	Weaver rail	40.95
Remington 1100/1187 20-ga. LH	No	Weaver rail	40.95
Remington 870 12-ga.[2]	No	Weaver rail	40.95
Remington 870 12-ga. LH	No	Weaver rail	40.95
Remington 870 20-ga.	No	Weaver rail	40.95
Remington 870 20-ga. LH	No	Weaver rail	40.95
Winchester 1300[2]	No	Weaver rail	40.95
Winchester 1400[2]	No	Weaver rail	40.95
Rifles			
AR-15/M16	No	Weaver rail	21.95
Browning A-Bolt	No	Weaver rail	21.95
Browning BAR	No	Weaver rail	21.95
Browning BLR	No	Weaver rail	21.95
CVA Apollo	No	Weaver rail	21.95
Marlin 336	No	Weaver rail	21.95
Mauser Mark X	No	Weaver rail	21.95
Modern Muzzleloading MK85	No	Weaver rail	21.95
Remington 700 Short	No	Weaver rail	21.95
Remington 700 Long	No	Weaver rail	21.95
Remington 7400/7600	No	Weaver rail	21.95
Ruger 10/22	No	Weaver rail	21.95
Savage 110, 111, 113, 114, 115, 116	No	Weaver rail	21.95
Thompson/Center Thunderhawk	No	Weaver rail	21.95
Traditions Buckhunter	No	Weaver rail	21.95
White W Series	No	Weaver rail	21.95
White G Series	No	Weaver rail	21.95
White WG Series	No	Weaver rail	21.95
Winchester Model 70	No	Weaver rail	21.95
Winchester 94AE	No	Weaver rail	21.95

All mounts no-gunsmithing, see-through/iron sight usable. Rifle mounts are solid see-through bases. All mounts accomodate standard split rings of all makes. From Aimtech, L&S Technologies, Inc. [1]3 blade sight and mount combination. [2]These models also available in RSP camouflage.

Maker, Model, Type	Adjust.	Scopes	Price
A.R.M.S.			
M16A1/A2/AR-15	No	Weaver rail	59.95
Multibase	No	Weaver rail	59.95
#19 Weaver/STANAG Throw Lever Rail	No	Weaver rail	140.00
STANAG Rings	No	30mm	75.00
Throw Lever Rings	No	Weaver rail	88.00
Ring Inserts	No	1", 30mm	29.00
#38 Std. Swan Sleeve[1]	No	—	150.00

[1]Avail in three lengths. From A.R.M.S., Inc.

Maker, Model, Type	Adjust.	Scopes	Price
ARMSON			
AR-15[1]	No	1"	45.00
Mini-14[2]	No	1"	66.00
H&K[3]	No	1"	82.00

[1]Fastens with one nut. [2]Models 181, 182, 183, 184, etc. [3]Claw mount. From Trijicon, Inc.

Maker, Model, Type	Adjust.	Scopes	Price
ARMSPORT			
100 Series[1]	No	1" rings. Low, med., high	10.75
104 22-cal.	No	1"	10.75
201 See-Thru	No	1"	13.00
1-Piece Base[2]	No	—	5.50
2-Piece Base[2]	No	—	2.75

[1]Weaver-type rings. [2]Weaver-type base; most poular rifles. Made in U.S. From Armsport.

Maker, Model, Type	Adjust.	Scopes	Price
B-SQUARE			
Pistols			
Beretta/Taurus 92/99[6]	—	1"	69.95
Browning Buck Mark[6]	No	1"	54.95
Colt 45 Auto	E only	1"	69.95
Colt Python/MkIV, 4",6",8"[1,6]	E	1"	69.95
Dan Wesson Clamp-On[2,6]	E	1"	69.95
Ruger Single-Six[4]	No	1"	69.95
Ruger Blackhawk, Super B'hwk[7]	W&E	1"	69.95
Ruger GP-100[8]	No	1"	69.95
Ruger Redhawk[7]	W&E	1"	69.95
Taurus 66[9]	No	1"	69.95
S&W K, L, N frame[2,6]	No	1"	69.95
T/C Contender (Dovetail Base)	W&E	1"	39.95
Rifles			
Charter AR-7	No	1"	39.95
Mini-14 (dovetail/NATO Stanag)[5,6]	W&E	1"	64.95
M-94 Side Mount	W&E	1"	54.95
RWS, Beeman/FWB, Anschutz, Diana, Walther Air Rifles	E only	—	59.95
SMLE Side Mount with rings	W&E	1"	69.95
Military			
AK-47, SKS-56[10]	No	1"	64.95
M1-A[7]	W&E	1"	99.95
AR-15/16[7]	W&E	1"	64.95
FN-LAR/FAL[6,7]	E only	1"	99.95
HK-91/93/94[6,7]	E only	1"	99.95
Shotguns[6]			
Ithaca 37[6]	No	1"	49.95
Mossberg 500, 712, 5500[6]	No	1"	49.95
Rem. 870/1100 (12 & 20 ga.)[6]	No	1"	49.95
Rem. 870, 1100 (and L.H.)[6]	No	1"	49.95
Interlock Bases			
One-Piece[11]	No	Standard dovetail rings	9.95-10.95

[1]Clamp-on, blue finish; stainless finish $59.95. [2]Blue finish; stainless finish $59.95. [3]Clamp-on, blue; stainless finish $59.95. [4]Dovetail; stainless finish $59.95. [5]No gunsmithing, no sight removal; blue; stainless finish $79.95. [6]Weaver-style rings. Rings not included with Weaver-type bases; stainless finish add $10. [7]Blue; stainless $79.95. [9]Blue; stainless $69.95. [10]Receiver mounts. [11]Most popular sporting rifles. Mounts for many shotguns, airguns, military and law enforcement guns also available. **Partial listing of mounts shown here. Contact B-Square for more data.** B-Square makes mounts for the following military rifles: AK47/AKS, Egyptian Hakim, French MAS 1936, M91 Argentine Mauser, Model 98 Brazilian and German Mausers, Model 93, Spanish Mauser (long and short), Model 1916 Mauser, Model 38 and 96 Swedish Mausers, Model 91 Russian (round and octagon receivers), Chinese SKS 56, SMLE No. 1, Mk. III, 1903 Springfield, U.S. 30-cal. Carbine, and others. Those following replace gun's rear sight: AK47/AKS, P14/1917 Enfield, FN49, M1 Garand, M1-A/M14 (no sight removal), SMLE No. 1, Mk III/No. 4 & 5, Mk. 1, 1903/1903-A3 Springfield, Beretta AR 70 (no sight removal).

Maker, Model, Type	Adjust.	Scopes	Price
BEEMAN			
Two-Piece, Med.	No	1"	31.50
Deluxe Two-Piece, High	No	1"	33.00
Deluxe Two-Piece	No	30mm	41.00
Deluxe One-Piece	No	1"	50.00
Dampamount	No	1"	110.00

All grooved receivers and scope bases on all known air rifles and 22-cal. rimfire rifles (½" to ⅝"—6mm to 15mm).

Maker, Model, Type	Adjust.	Scopes	Price
BOCK			
Swing ALK[1]	W&E	1", 26mm, 30mm	349.00
Safari KEMEL[2]	W&E	1", 26mm, 30mm	149.00

CAUTION: PRICES SHOWN ARE SUPPLIED BY THE MANUFACTURER OR IMPORTER. CHECK YOU LOCAL GUNSHOP.

SCOPE MOUNTS

Maker, Model, Type	Adjust.	Scopes	Price
Bock (cont.)			
Claw KEMKA[3]	W&E	1", 26mm, 30mm	224.00
ProHunter Fixed[4]	No	1", 26mm, 30mm	95.00

[1]Q.D.; pivots right for removal. For Steyr-Mannlicher, Win. 70, Rem. 700, Mauser 98, Dakota, Sako, Sauer 80, 90. Magnum has extra-wide rings, same price. [2]Heavy-duty claw-type; reversible for front or rear removal. For Steyr-Mannlicher rifles. [3]True claw mount for bolt-action rifles. Also in extended model. For Steyr-Mannlicher, Win. 70, Rem. 700. Also avail. as Gunsmith Bases—bases not drilled or contoured—same price. [4]Extra-wide rings. Imported from Germany by GSI, Inc.

Maker, Model, Type	Adjust.	Scopes	Price
BURRIS			
Supreme (SU) One Piece (T)[1]	W only	1" split rings, 3 heights	1 piece base—28.00-34.00
Trumount (TU) Two Piece (T)	W only	1" split rings, 3 heights	2 piece base—26.00-40.00
Trumount (TU) Two Piece Ext.	W only	1" split rings	32.00
Browning 22-cal. Auto Mount[2]	No	1" split rings	20.00
1" 22-cal. Ring Mounts[3]	No	1" split rings	1" rings—24.00-47.00
L.E.R. (LU) Mount Bases[4]	W only	1" split rings	26.00-67.00
L.E.R. No Drill-No Tap Bases[4,7,8]	W only	1" split rings	47.00-53.00
Extension Rings[5]	No	1" scopes	28.00-53.00
Ruger Ring Mount[6,9]	W only	1" split rings	53.00-75.00
Std. 1" Rings[9]	—	Low, medium, high heights	36.00-45.00
Zee Rings[9]	—	Fit Weaver bases; medium and high heights	34.00-47.00

[1]Most popular rifles. Universal rings, mounts fit Burris, Universal, Redfield, Leupold and Browning bases. Comparable prices. [2]Browning Standard 22 Auto rifle. [3]Grooved receivers. [4]Universal dovetail; accept Burris, Universal, Redfield, Leupold rings. Prices for Dan Wesson, S&W, Virginian, Ruger Blackhawk, Win. 94. [5]Medium standard front, extension rear, per pair. Low standard front, extension rear, per pair. [6]Compact scopes, scopes with 2" bell, for M77R. [7]Selected rings and bases available with matte Safari or silver finish. [8]For S&W K,L,N frames, Colt Python, Dan Wesson with 6" or longer barrels. [9]Also in 30mm.

Maker, Model, Type	Adjust.	Scopes	Price
CAPE OUTFITTERS			
Quick Detachable	No	1" split rings, lever quick detachable	99.95

Double rifles; Rem. 700-721, Colt Sauer, Sauer 200, Kimber, Win. 61-63-07-100-70, Browning High Power, 22, BLR, BAR, BBR, A-Bolt; Wea. Mark V, Vanguard; Modern Muzzle Loading, Knight, Thompson/Center, CVA rifles, Dixie rifles. All steel; returns to zero. From Cape Outfitters.

Maker, Model, Type	Adjust.	Scopes	Price
CLEAR VIEW			
Universal Rings, Mod. 101[1]	No	1" split rings	21.95
Standard Model[2]	No	1" split rings	21.95
Broad View[3]	No	1"	21.95
22 Model[4]	No	3/4", 7/8", 1"	13.95
SM-94 Winchester[5]	No	1"	23.95
94 EJ[6]	No	1" split rings	21.95

[1]Most rifles by using Weaver-type base; allows use of iron sights. [2]Most popular rifles; allows use of iron sights. [3]Most popular rifles; low profile, wide field of view. [4]22 rifles with grooved receiver. [5]Side mount. [6]For Win. A.E. From Clear View Mfg.

Maker, Model, Type	Adjust.	Scopes	Price
CONETROL			
Huntur[1]	W only	1", 26mm, 26.5mm solid or split rings, 3 heights	69.96
Gunnur[2]	W only	1", 26mm, 26.5mm solid or split rings, 3 heights	79.92
Custum[3]	W only	1", 26mm, 26.5mm solid or split rings, 3 heights	99.96
One Piece Side Mount Base[4]	W only	1", 26mm, 26.5mm solid or split rings, 3 heights	—
DapTar Bases[5]	W only	1", 26mm, 26.5mm solid or split rings, 3 heights	—
Pistol Bases, 2 or 3-ring[6]	W only	1" scopes	—
Fluted Bases[7]	W only	Standard Conetrol rings	99.96
30mm Rings[8]	W only	30mm	69.96-89.94

[1]All popular rifles, including metric-drilled foreign guns. Price shown for base, two rings. Matte finish. [2]Gunnur grade has mirror-finished rings, satin-finish base. Price shown for base, two rings. [3]Custum grade has mirror-finished rings and mirror-finished, streamlined base. Price shown for base, two rings. [4]Win. 94, Krag, older split-bridge Mannlicher-Schoenauer, Mini-14, etc. Prices same as above. [5]For all popular guns with integral mounting provision, including Sako, BSA, Ithacagun, Ruger, Tikka, H&K, BRNO—$34.98-$49.98—and many others. Also for grooved-receiver rimfires and air rifles. Prices same as above. [6]For XP-100, T/C Contender, Colt SAA, Ruger Blackhawk, S&W. [7]Sculptured two-piece bases as found on fine custom rifles. Price shown is for base alone. Also available unfinished—$79.92, or finished but unblued—$89.94. [8]30mm rings made in projectionless style, medium height only. Three-ring mount available for T/C Contender and other pistols in Conetrol's three grades. Any Conetrol mount available in stainless or Teflon for double regular cost of grade.

Maker, Model, Type	Adjust.	Scopes	Price
EAW			
Quick-Loc Mount	W&E	1", 26mm	253.00
	W&E	30mm	291.00
Magnum Fixed Mount	W&E	1", 26mm	198.00
	W&E	30mm	215.00

Fit most popular rifles. Avail. in 4 heights, 4 extensions. Reliable return to zero. Stress-free mounting. Imported by New England Custom Gun Svc.

Maker, Model, Type	Adjust.	Scopes	Price
GENTRY			
Feather-Light Bases	No	—	39.00-59.00
Feather-Light Rings	No	1", 30mm	48.00-65.00

Bases for Rem. Seven, 700, Mauser 98, Browning A-Bolt, Weatherby Mk. V, Win. 70, HVA, Dakota. Two-piece base for Rem. Seven, chrome moly or stainless. Rings in matte or regular blue, or stainless gray; four heights. From David Gentry.

Maker, Model, Type	Adjust.	Scopes	Price
GRIFFIN & HOWE			
Topmount[1]	No	1", 30mm	625.00
Sidemount[2]	No	1", 30mm	255.00
Garand Mount[3]	No	1", 30mm	255.00

[1]Quick-detachable, double-lever mount with 1" rings, installed; with 30mm rings $875.00. [2]Quick-detachable, double-lever mount with 1" rings; with 30mm rings $375.00; installed, 1" rings $405.00; installed, 30mm rings $525.00. [3]Price installed, with 1" rings $405.00. From Griffin & Howe.

Maker, Model, Type	Adjust.	Scopes	Price
G. G. & G.			
Swan G-3[1]	No	Weaver rail	229.00
FN FAL[2]	No	Weaver rail	149.00
Remington 700	No	Weaver base	95.00
Sniper Grade Rings	No	1", 30mm	135.00
M-14 Mount	No	1", 30mm	NA
M16/AR15 IRS Delta	—	—	175.00
M16/AR15 IRS Delta Plus	—	—	225.00
M16/AR15 F.I.R.E. System	—	—	595.00

[1]Universal top claw lock. [2]Paratrooper model, $159.00. From G,G&G.

Maker, Model, Type	Adjust.	Scopes	Price
IRONSIGHTER			
Wide Ironsighter™	No	1" split rings	35.98
Ironsighter Center Fire[1]	No	1" split rings	32.95
Ironsighter S-94	No	1" split rings	39.95
Ironsighter AR-15/M-16[8]	No	1", 30mm	$103.95
Ironsighter 22-Cal.Rimfire			
Model #570[9]	No	1" split rings	32.95
Model #573[9]	No	30mm split rings	32.95
Model #722	No	1" split rings	17.75
Model #727	No	7/8" split rings	17.75
Series #700[5]	No	1" split rings	32.95
Ruger Base Mounts[6]	No	1" split rings	83.95
Ironsighter Handguns[4]	No	1" split rings	38.95
Blackpowder Mount[7]	No	1"	32.95-76.95

[1]Most popular rifles, including Ruger Mini-14, H&R M700, and muzzleloaders. Rings have oval holes to permit use of iron sights. [2]For 1" dia. scopes. [3]For 7/8" dia. scopes. [4]For 1" dia. extended eye relief scopes. [5]702—Browning A-Bolt; 709—Marlin 39A. [6]732—Ruger 77/22 R&RS, No. 1, Ranch Rifle; 778 fits Ruger 77R, RS. Both 733, 778 fit Ruger integral bases. [7]Fits most popular blackpowder rifles; one model for Holden Ironsighter mounts, one for Weaver rings. [8]Model 716 with 1" #540 rings; Model 717 with 30mm #530 rings. [9]Fits mount rail on Rem. 522 Viper. Adj. rear sight is integral. Some models in stainless finish. From Ironsighter Co.

Maker, Model, Type	Adjust.	Scopes	Price
K MOUNT By KENPATABLE			
Shotgun Mount	No	1", laser or red dot device	49.95
SKS[1]	No	1"	39.95

Wrap-around design; no gunsmithing required. Models for Browning BPS, A-5 12-ga., Sweet 16, 20, Rem. 870/1100 (LTW and L.H.), S&W 916, Mossberg 500, Ithaca 37 & 51 12-ga., S&W 1000/3000, Win. 1400. [1]Requires simple modification to gun. From KenPatable Ent.

Maker, Model, Type	Adjust.	Scopes	Price
KRIS MOUNTS			
Side-Saddle[1]	No	1", 26mm split rings	12.98
Two Piece (T)[2]	No	1", 26mm split rings	8.98
One Piece (T)[3]	No	1", 26mm split rings	12.98

[1]One-piece mount for Win. 94. [2]Most popular rifles and Ruger. [3]Blackhawk revolver. Mounts have oval hole to permit use of iron sights.

Maker, Model, Type	Adjust.	Scopes	Price
KWIK-SITE			
KS-See-Thru[1]	No	1"	31.95
KS-22 See-Thru[2]	No	1"	23.95
KS-W94[3]	No	1"	39.95
Bench Rest	No	1"	31.95
KS-WEV	No	1"	31.95
KS-WEV-HIGH	No	1"	37.95
KS-T22 1"[4]	No	1"	23.95
KS-FL Flashlite[5]	No	Mini or C cell flashlight	49.95
KS-T88[6]	No	1"	11.95
KS-T89	No	30mm	14.95
KSN 22 See-Thru	No	1", 7/8"	20.95
KSN-T22	No	1", 7/8"	20.95
KSN-M16 See-Thru	No	1"	99.95
KS-202[1]	No	1"	31.95
KS-203	No	30mm	43.95
KSBP[7]	No	Integral	76.95

SCOPE MOUNTS

Maker, Model, Type	Adjust.	Scopes	Price
Kwik-Site (cont.)			
KSSM[8]	No	1"	31.95
KSB Base Set	—		5.95
Combo Bases & Rings	No	1"	31.95

Bases interchangeable with Weaver bases. [1]Most rifles. Allows use of iron sights. [2]22-cal. rifles with grooved receivers. Allows use of iron sights. [3]Model 94, 94 Big Bore. No drilling or tapping. Also in adjustable model $49.95. [4]Non-see-through model for grooved receivers. [5]Allows Mag Lite or C or D, Mini Mag Lites to be mounted atop See-Thru mounts. [6]Fits any Redfield, Tasco, Weaver or universal-style Kwik-Site dovetail base. [7]Blackpowder mount with integral rings and sights. [8]Shotgun side mount. Bright blue, black matte or satin finish. Standard, high heights.

Maker, Model, Type	Adjust.	Scopes	Price
LASER AIM	No	Laser Aim	19.00-69.00

Mounts Laser Aim above or below barrel. Avail. for most popular handguns, rifles, shotguns, including militaries. From Laseraim Technologies, Inc.

Maker, Model, Type	Adjust.	Scopes	Price
LEUPOLD			
STD Bases[1]	W only	One- or two-piece bases	23.80
STD Rings[2]	—	1" super low, low, medium, high	31.40
STD Handgun mounts[3]	No	—	57.00
Dual Dovetail Bases[1,4]	No		23.80
Dual Dovetail Rings[9]	—	1", super low, low	31.40
Ring Mounts[5,6,7]	No	7/8", 1"	79.80
22 Rimfire[9]	No	7/8", 1"	58.20
Gunmaker Base[8]	W only	1"	16.50
Quick Release Rings	—	1", low, med., high	31.90-68.90
Quick Release Bases[10]	No	1", one- or two-piece	69.30

[1]Rev. front and rear combinations; matte finish $22.90. [2]Avail. polished, matte or silver (low, med. only) finish. [3]Base and two rings; Casull, Ruger, S&W, T/C; add $5.00 for silver finish. [4]Rem. 700, Win. 70-type actions. [5]For Ruger No. 1, 77, 77/22; interchangeable with Ruger units. [6]For dovetailed rimfire rifles. [7]Sako; high, medium, low. [8]Must be drilled, tapped for each action. [9]Most dovetail-receiver 22s. [10]BSA Monarch, Rem. 40X, 700, 721, 725, Ruger M77, S&W 1500, Weatherby Mark V, Vanguard, Win M70.

Maker, Model, Type	Adjust.	Scopes	Price
MARLIN			
One Piece QD (T)	No	1" split rings	10.10

Most Marlin lever actions.

Maker, Model, Type	Adjust.	Scopes	Price
MILLETT			
Black Onyx Smooth	—	1", low, medium, high	31.15
Chaparral Engraved	—	1" engraved	46.15
One-Piece Bases[6]	Yes	1"	23.95
Universal Two-Piece Bases			
700 Series	W only	Two-piece bases	25.15
FN Series	W only	Two-piece bases	25.15
70 Series[1]	W only	1", two-piece bases	25.15
Angle-Loc Rings[2]	W only	1", low, medium, high	32.20-47.20
Ruger 77 Rings[3]	—	1"	47.20
Shotgun Rings[4]	—	1"	28.29
Handgun Bases, Rings[5]	—	1"	34.60-69.15
30mm Rings[7]	—	30mm	37.75-42.95
Extension Rings[8]	—	1"	35.65
See-Thru Mounts[9]	No	1"	27.95-32.95
Shotgun Mounts[10]	No	1"	49.95

[1]Rem. 40X, 700, 722, 725, Ruger 77 (round top), Weatherby, FN Mauser, FN Brownings, Colt 57, Interarms Mark X, Parker-Hale, Sako (round receiver), many others. [2]Fits Win. M70, 70XTR, 670, Browning BBR, BAR, BLR, A-Bolt, Rem. 7400/7600, Four, Six, Marlin 336, Win. 94 A.E., Sav. 110. [3]To fit Weaver-type bases. [4]Engraved. Smooth $34.60. [5]For Rem. 870, 1100; smooth. [6]Two and three-ring sets for Colt Python, Trooper, Diamondback, Peacekeeper, Dan Wesson, Ruger Redhawk, Super Redhawk. [7]Turn-in bases and Weaver-style for most popular rifles and T/C Contender, XP-100 pistols. [8]Both Weaver and turn-in styles; three heights. [9]Med. or high; ext. front—std. rear, ext. front—ext. rear—std. front, ext. rear. For double extension. [10]Many popular rifles, Knight MK-85, T/C Hawken, Renegade, Mossberg 500 Slugster, 835 Slug. [10]For Rem. 870/1100, Win. 1200, 1300/1400, 1500, Mossberg 500. Some models available in nickel at extra cost. From Millett Sights.

Maker, Model, Type	Adjust.	Scopes	Price
MMC			
AK[1]	No	—	39.95
FN FAL/LAR[2]	No	—	59.95

[1]Fits all AK derivative receivers; Weaver-style base; low-profile scope position. [2]Fits all FAL versions; Weaver-style base. From MMC.

Maker, Model, Type	Adjust.	Scopes	Price
PEM'S			
22T Mount[1]	No	1"	17.95
The Mount[2]	Yes	1"	29.50

[1]Fit all 3/8" dovetail on rimfire rifles. [2]Base and ring set; for over 100 popular rifles; low, medium rings. From Pem's.

Maker, Model, Type	Adjust.	Scopes	Price
RAM-LINE			
Mini-14 Mount	Yes	1"	24.97

No drilling or tapping. Use std. dovetail rings. Has built-in shell deflector. Made of solid black polymer. From Ram-Line, Inc.

Maker, Model, Type	Adjust.	Scopes	Price
REDFIELD			
NGS	No	Weaver rail	30.95-78.95

Maker, Model, Type	Adjust.	Scopes	Price
American Rings[6]	No	1", low, med., high	27.95-37.95
American Bases[6]	No		4.95-10.95
American Widefield See-Thru[7]	No	1"	15.95
JR-SR (T)[1]	W only	3/4", 1", 26mm, 30mm	JR—23.95-52.95 SR—18.95-22.95
Ring (T)[2]	No	3/4" and 1"	27.95-29.95
Three-Ring Pistol System SMP[3]	No	1" split rings (three)	49.95-52.95
Widefield See-Thru Mounts	No	1"	15.95
Ruger Rings[4]	No	1", med., high	34.95-36.95
Ruger 30mm[5]	No	1"	42.95
Midline Ext. Rings	No	1"	24.95

[1]Low, med. & high, split rings. Reversible extension front rings for 1". 2-piece bases for Sako. Colt Sauer bases $39.95. Med. Top Access JR rings nickel-plated, $28.95. SR two-piece ABN mount nickel-plated, $22.95. [2]Split rings for grooved 22s; 30mm, black matte $42.95. [3]Used with MP scopes for: S&W K, L or N frame, XP-100, T/C Contender, Ruger receivers. [4]For Ruger Model 77 rifles, medium and high; medium only for M77/22. [5]For Model 77. Also in matte finish, $45.95. [6]Aluminum 22 groove mount $14.95; base and medium rings $18.95. [7]Fits American or Weaver-style base. Non-Gunsmithing mount system. For many popular shotguns, rifles, handguns and blackpowder rifles. Uses existing screw holes.

Maker, Model, Type	Adjust.	Scopes	Price
S&K			
Insta-Mount (T) bases and rings[1]	W only	Use S&K rings only	47.00-117.00
Conventional rings and bases[2]	W only	1" split rings	From 65.00
SKulptured Bases, Rings[2]	W only	1", 26mm, 30mm	From 65.00
Smooth Kontoured Rings[3]	Yes	1", 26mm, 30mm	90.00-120.00

[1]1903, A3, M1 Carbine, Lee Enfield #1, Mk. III, #4, #5, M1917, M98 Mauser, AR-15, AR-180, M-14, M-1, Ger. K-43, Mini-14, M1-A, Krag, AKM, Win. 94, SKS Type 56, Daewoo, H&K. [2]Most popular rifles already drilled and tapped. [3]No projections; weigh 1/2-oz. each; matte or gloss finish. Horizontally and vertically split rings, matte or high gloss.

Maker, Model, Type	Adjust.	Scopes	Price
SSK INDUSTRIES			
T'SOB	No	1"	65.00-145.00
Quick Detachable	No	1"	From 160.00

Custom installation using from two to four rings (included). For T/C Contender, most 22 auto pistols, Ruger and other S.A. revolvers, Ruger, Dan Wesson, S&W, Colt DA revolvers. Black or white finish. Uses Kimber rings in two- or three-ring sets. In blue or SSK Khrome. For T/C Contender or most popular revolvers. Standard, non-detachable model also available, from $65.00.

Maker, Model, Type	Adjust.	Scopes	Price
SAKO			
QD Dovetail	W only	1" only	70.00-155.00

Sako, or any rifle using Sako action, 3 heights available. Stoeger, importer.

Maker, Model, Type	Adjust.	Scopes	Price
SPRINGFIELD, INC.			
M1A Third Generation	No	1" or 30mm	123.00
M1A Standard	No	1" or 30mm	77.00
SAR-4800 Mount	No	—	96.00
M6 Scount Mount	No	—	29.00

Weaver-style bases. From Springfield, Inc.

Maker, Model, Type	Adjust.	Scopes	Price
TALBOT			
QD Bases	No	—	180.00-190.00
Rings	No	1", 30mm	40.00-60.00

Blue or stainless steel; standard or extended bases; rings in three heights. For most popular rifles. From Talbot QD Mounts.

Maker, Model, Type	Adjust.	Scopes	Price
TASCO			
World Class			
Universal "W" Ringmount[1]	No	1", 30mm	25.50-30.00
22, Air Rifle[2]	No	1", 30mm	18.00-82.00
Ringsets[3]	No	1", 26mm, 30mm	39.00-51.00
Aluminum Ringsets	Yes	1", 30mm	12.00-17.00
See-Thru	No	1"	19.00
Shotgun Bases	Yes		34.00

[1]Steel; low, high only; also high-profile see-through; fit Tasco, Weaver, other universal bases; black gloss or satin chrome. [2]Low, med., high; 3/8" grooved receivers; black or satin chrome. [3]Low, med., high; black gloss, matte satin chrome; also Traditional Ringsets $31.00 (1"), $42.00 (26mm), $53.00 (30mm).

Maker, Model, Type	Adjust.	Scopes	Price
THOMPSON/CENTER			
Duo-Ring Mount[1]	No	1"	65.00
Weaver-Style Bases[2]	No	—	13.00
Weaver-Style Rings[3]	No	1"	29.00-41.00
Weaver-Style See-Through Rings[4]	No	1"	29.00
Quick Release System[5]	No	1"	Rings 56.00 Base 33.00

[1]Attaches directly to T/C Contender bbl., no drilling/tapping; also for T/C M/L rifles, needs base adapter; blue or stainless; for M/L guns, $59.80. [2]For T/C ThunderHawk, FireHawk rifles; blue; silver, $37.00. [3]Medium and high; blue or silver finish. [4]For T/C FireHawk, ThunderHawk; blue; silver, $25.00. [5]For Contender pistol, Carbine, Scout, all M/L long guns. From Thompson/Center.

Maker, Model, Type	Adjust.	Scopes	Price
UNERTL			
1/4 Click[1]	Yes	3/4", 1" target scopes	Per set 186.00

[1]Unertl target or varmint scopes. Posa or standard mounts, less bases. From Unertl.

CAUTION: PRICES SHOWN ARE SUPPLIED BY THE MANUFACTURER OR IMPORTER. CHECK YOU LOCAL GUNSHOP.

SCOPE MOUNTS

Maker, Model, Type	Adjust.	Scopes	Price
WARNE			
Deluxe Series (all steel non-Q.D. rings)			
Standard	No	1", 4 heights	95.50
		30mm, 2 heights	107.50
Sako	No	1", 4 heights	95.50
		30mm, 3 heights	107.50
Deluxe Series Rings fit Premier Series Bases			
Premier Series (all-steel Q.D. rings)			
Adjustable Double Levers	No	1", 4 heights	105.50
		26mm, 2 heights	117.50
		30mm, 3 heights	117.50
Thumb Knob	No	1", 4 heights	95.50
		26mm, 2 heights	107.50
		30mm, 3 heights	107.50
Brno 19mm	No	1", 3 heights	105.50
		30mm, 2 heights	117.50
Brno 16mm		1" 2 heights	105.50
Ruger	No	1", 4 heights	105.50
		30mm, 3 heights	117.50
Ruger M77	No	1", 3 heights	105.50
		30mm, 2 heights	117.50
Sako Medium & Long Action	No	1", 4 heights	105.50
		30mm, 3 heights	117.50
Sako Short Action	No	1", 3 heights	105.50
All-Steel One-Piece Base, ea.			32.00
All-Steel Two-Piece Base, ea.			12.50
Maxima Series (fits all Weaver-style bases)			
Permanently Attached[1]	No	1", 3 heights	31.40
		30mm, 3 heights	35.40
Adjustable Double Lever[2]	No	1", 3 heights	65.50
		30mm, 3 heights	69.50
Thumb Knob	No	1", 3 heights	55.50
		30mm, 3 heights	59.50
All-Steel Two-Piece Base, ea.			12.50

Vertically split rings with dovetail clamp, precise return to zero. Fit most popular rifles, handguns. Regular blue, matte blue, silver finish. [1]All-steel, non-q.d. rings. [2]All-steel, q.d. rings. From Warne Mfg. Co.

Maker, Model, Type	Adjust.	Scopes	Price
WEAVER			
Detachable Mounts			
Top Mount	No	7/8", 1", 30mm, 33mm	25.00-42.00
Side Mount	No	1", 1" Long	30.00-35.00
Tip-Off Rings	No	7/8", 1"	22.00-28.00
Pivot Mounts	No	1"	39.00
Complete Mount Systems			
Pistol	No	1"	75.00-105.00
Rifle	No	1"	20.00
SKS Mount System	No	1"	49.75
Pro-View (no base required)	No	1"	15.00-17.00
Converta-Mount, 12-ga. (Rem. 870, Moss. 500)	No	1", 30mm	75.00
See-Thru Mounts			
Detachable	No	1"	27.00-32.00
System (no base required)	No	1"	15.00-35.00
Tip-Off	No	1"	15.00

Nearly all modern rifles, pistols, and shotguns. Detachable rings in standard, See-Thru, and extension styles, in Low, Medium, High or X-High heights; gloss (blued), silver and matte finishes to match scopes. Extension rings are only available in 1" High style and See-Thru X-tensions only in gloss finish. Tip-Off rings only for 3/8" grooved receivers or 3/8" grooved adaptor bases; no base required. See-Thru & Pro-View mounts for most modern big bore rifles, some in silver. No Drill & Tap Pistol systems in gloss or silver for: Colt Python, Trooper, 357, Officer's Model; Ruger Single-Six, Security-Six (gloss finish only), Blackhawk, Super Blackhawk, Blackhawk SRM 357, Redhawk, Mini-14 Series (not Ranch), Ruger 22 Auto Pistols, Mark II; Smith & Wesson I- and current K-frames with adj. rear sights. Converta-Mount Systems in Standard and See-Under for: Mossberg 500 (12- and 20-ga.); Remington 870, 11-87 (12- and 20-ga. lightweight); Winchester 1200, 1300, 1400, 1500. Converta Brackets, Bases Rings also avail. for Beretta A303 and A390; Browning A-5, BPS Pump; Ithaca 37, 87. From Weaver.

Maker, Model, Type	Adjust.	Scopes	Price
WEIGAND			
Browning Buck Mark[1]	No	—	29.95
Colt 22 Automatic[1]	No	—	19.95
Integramounts[2]	No	—	39.95-69.00
S&W Revolver[3]	No	—	29.95
Ruger 10/22[4]	No	—	14.95-39.95
Ruger Revolver[5]	No	—	29.95
Taurus Revolver[4]	No	—	29.95-65.00
T/C Encore Monster Mount	No	—	69.00
T/C Contender Monster Mount	No	—	69.00
Lightweight Rings	No	1", 30mm	29.95-39.95
1911, P-9 Scopemounts			
SM3[6]	No	Weaver rail	99.95
SRS1911[7]	No	30mm	99.95
APCMNT[8]	No	—	69.95

[1]No gunsmithing. [2]S&W K, L, N frames; Taurus vent rib models; Colt Anaconda/Python; Ruger Redhawk; Ruger 10/22. [3]K, L, N frames. [4]Three models. [5]Redhawk, Blackhawk, GP-100. [6]3rd Gen.; drill and tap; without slots, $59.95. [7]Ringless design, silver only; SRS 1911-2, $79.95; for P-9, $59.95. [8]For Aimpoint Comp Red Dot scope, silver only. From Weigand Combat Handguns, In.c

Maker, Model, Type	Adjust.	Scopes	Price
WIDEVIEW			
Premium 94 Angle Eject	No	1"	24.00
Premium See-Thru	No	1"	22.00
22 Premium See-Thru	No	3/4", 1"	16.00
Universal Ring Angle Cut	No	1"	24.00
Universal Ring Straight Cut	No	1"	22.00
Solid Mounts			
Lo Ring Solid[1]	No	1"	16.00
Hi Ring Solid[1]	No	1"	16.00
SR Rings	—	1", 30mm	18.64
22 Grooved Receiver	No	1"	16.00
94 Side Mount	No	1"	26.00
Blackpowder Mounts[2]	No	1"	22.00-44.00

[1]For Weaver-type bases. Models for many popular rifles. Low ring, high ring and grooved receiver types. [2]No drilling, tapping; for T/C Renegade, Hawken, CVA, Knight Traditions guns. From Wideview Scope Mount Corp.

Maker, Model, Type	Adjust.	Scopes	Price
WILLIAMS			
Sidemount with HCO Rings[1]	No	1", split or extension rings.	74.21
Sidemount, offset rings[2]	No	Same	61.08
Sight-Thru Mounts[3]	No	1", 7/8" sleeves	18.95
Streamline Mounts	No	1" (bases form rings).	25.70
Guideline Handgun[4]	No	1" split rings.	61.75

[1]Most rifles, Br. S.M.L.E. (round rec.) **$14.41** extra. [2]Most rifles including Win. 94 Big Bore. [3]Many modern rifles, including CVA Apollo, others with 1" octagon barrels. [4]No drilling, tapping required; heat treated alloy. For Ruger MkII Bull Barrel (**$61.75**); Streamline Top Mount for T/C Contender (**$41.15**), Scout Rifle, (**$24.00**), High Top Mount with sub-base (**$51.45**). From Williams Gunsight Co.

Maker, Model, Type	Adjust.	Scopes	Price
YORK			
M-1 Garand	Yes	1"	39.95

Centers scope over the action. No drilling, tapping or gunsmithing. Uses standard dovetail rings. From York M-1 Conversions.

NOTES

(S)—Side Mount (T)—Top Mount; 22mm=.866"; 25.4mm=1.024"; 26.5mm=1.045"; 30mm=1.81"

DIRECTORY OF THE ARMS TRADE

The Directory of the Arms Trade is divided into two sections to help the reader more easily find a listing. The **Product Directory** contains fifty-three product categories. Each entry is cross-referenced to the **Manufacturer's Directory,** an alphabetical listing giving the address, telephone number and, where available, the FAX number to allow faster access to the company.

DIRECTORY OF THE ARMS TRADE INDEX

PRODUCT DIRECTORY 292-308

- AMMUNITION, COMMERCIAL .... .292
- AMMUNITION, CUSTOM .... .292
- AMMUNITION, FOREIGN .... .292
- AMMUNITION COMPONENTS—BULLETS, POWDER, PRIMERS .... .292
- ANTIQUE ARMS DEALERS .... .293
- APPRAISERS—GUNS, ETC. .... .293
- AUCTIONEERS—GUNS, ETC .... .294
- BOOKS (Publishers and Dealers) .... .294
- BULLET AND CASE LUBRICANTS .... .294
- BULLET SWAGE DIES AND TOOLS .... .294
- CARTRIDGES FOR COLLECTORS .... .294
- CASES, CABINETS, RACKS AND SAFES—GUN .294
- CHOKE DEVICES, RECOIL ABSORBERS AND RECOIL PADS .... .295
- CHRONOGRAPHS AND PRESSURE TOOLS ...295
- CLEANING AND REFINISHING SUPPLIES ..... .295
- COMPUTER SOFTWARE—BALLISTICS .... .295
- CUSTOM GUNSMITHS .... .295
- CUSTOM METALSMITHS .... .297
- DECOYS .... .297
- ENGRAVERS, ENGRAVING TOOLS .... .297
- GAME CALLS .... .298
- GUN PARTS, U.S. AND FOREIGN .... .298
- GUNS, AIR .... .298
- GUNS, FOREIGN—IMPORTERS .... .298
- GUNS, FOREIGN—MANUFACTURERS .... .299
- GUNS, U.S.-MADE .... .300
- GUNS AND GUN PARTS, REPLICA AND ANTIQUE .... .300
- GUNS, SURPLUS—PARTS AND AMMUNITION ..300
- GUNSMITHS, CUSTOM (See Custom Gunsmiths)
- GUNSMITHS, HANDGUN (See Pistolsmiths)
- GUNSMITH SCHOOLS .... .300
- GUNSMITH SUPPLIES, TOOLS, SERVICES .... .301
- HANDGUN ACCESSORIES .... .301
- HANDGUN GRIPS .... .301
- HEARING PROTECTORS .... .302
- HOLSTERS AND LEATHER GOODS .... .302
- HUNTING AND CAMP GEAR, CLOTHING, ETC ..302
- KNIVES AND KNIFEMAKER'S SUPPLIES— FACTORY AND MAIL ORDER .... .302
- LABELS, BOXES, CARTRIDGE HOLDERS ..... .303
- LOAD TESTING AND PRODUCT TESTING (Chronographing, Ballistic Studies) .... .303
- MISCELLANEOUS .... .303
- MUZZLE-LOADING GUNS, BARRELS AND EQUIPMENT .... .303
- PISTOLSMITHS .... .304
- REBORING AND RERIFLING .... .304
- RELOADING TOOLS AND ACCESSORIES ..... .304
- RESTS—BENCH, PORTABLE—AND ACCESSORIES .... .305
- RIFLE BARREL MAKERS (See also Muzzle-Loading Guns, Barrels and Equipment) .... .305
- SCOPES, MOUNTS, ACCESSORIES, OPTICAL EQUIPMENT .... .306
- SHOOTING/TRAINING SCHOOLS .... .306
- SIGHTS, METALLIC .... .306
- STOCKS (Commercial and Custom) .... .307
- TARGETS, BULLET AND CLAYBIRD TRAPS 307
- TAXIDERMY .... .308
- TRAP AND SKEET SHOOTER'S EQUIPMENT ...308
- TRIGGERS, RELATED EQUIPMENT .... .308

MANUFACTURERS' DIRECTORY 309-336

30th EDITION, 1998

PRODUCT DIRECTORY

AMMUNITION, COMMERCIAL

ACTIV Industries, Inc.
AFSCO Ammunition
American Ammunition
Arizona Ammunition, Inc.
Arms Corporation of the Philippines
A-Square Co., Inc.
Atlantic Rose, Inc.
Bergman & Williams
Berger Bullets, Ltd.
Big Bear Arms & Sporting Goods, Inc.
Black Hills Ammunition, Inc.
Blammo Ammo
Blount, Inc., Sporting Equipment Div.
Brenneke KG, Wilhelm
Brown Dog Ent.
Buffalo Bullet Co., Inc.
BulletMakers Workshop, The
Bull-X, Inc.
California Magnum
CBC
Colorado Sutlers Arsenal
Cor-Bon Bullet & Ammo Co.
Cumberland States Arsenal
Daisy Mfg. Co.
Dead Eye's Sport Center
Delta Frangible Ammunition, LLC
Denver Bullets, Inc.
Diana
Dynamit Nobel-RWS, Inc.
Effebi SNC, Dr. Franco Beretta
Eldorado Cartridge Corp.
Eley Ltd.
Elite Ammunition
Estate Cartridge, Inc.
Federal Cartridge Co.
Fiocchi of America, Inc.
4W Ammunition
Hunters Supply
Garrett Cartridges, Inc.
Gibbs Rifle Co., Inc.
GOEX, Inc.
Goldcoast Reloaders, Inc.
Grand Falls Bullets, Inc.
Hansen & Co.
Hansen Cartridge Co.
Hart & Son, Inc., Robert W.
Hirtenberger Aktiengesellschaft
Hornady Mfg. Co.
ICI-America
IMI
Israel Military Industries Ltd.
Jones, J.D.
Keng's Firearms Specialty, Inc.
Kent Cartridge Mfg. Co. Ltd.
Lapua Ltd.
Lightfield Ammunition Corp.
M&D Munitions Ltd.
Mac-1 Distributors
MagSafe Ammo Co.
Maionchi-L.M.I.
Markell, Inc.
Mathews & Son, Inc., George E.
McBros Rifle Co.
Men—Metallwerk Elisenhuette, GmbH
Mullins Ammunition
NECO
New England Ammunition Co.
Oklahoma Ammunition Co.
Old Western Scrounger, Inc.
Omark Industries
Pacific Cartridge, Inc.
PMC/Eldorado Cartridge Corp.
Polywad, Inc.
Pony Express Reloaders
Precision Delta Corp.
Pro Load Ammunition, Inc.
Remington Arms Co., Inc.
Rocky Fork Enterprises
Rucker Dist. Inc.
RWS
Shooting Components Marketing
Slug Group, Inc.
Spence, George W.
SSK Industries
Starr Trading Co., Jedediah
Talon Mfg. Co., Inc.
Taylor & Robbins
TCCI
Thompson Bullet Lube Co.
3-D Ammunition & Bullets
3-Ten Corp.
USAC
Valor Corp.
Victory USA
Vihtavuori Oy/Kaltron-Pettibone
Voere-KGH m.b.H.
Vom Hoffe
Weatherby, Inc.
Westley Richards & Co.
Widener's Reloading & Shooting Supply, Inc.
Winchester Div., Olin Corp.
Zero Ammunition Co., Inc.
Zonie Bullets
McKillen & Heyer, Inc.
McMurdo, Lynn
Men-Metallwerk Elisenhuette, GmbH
Milstor Corp.
Mountain Rifles Inc.
Mullins Ammunition
Naval Ordnance Works
NECO
Northern Precision Custom Swaged Bullets
Old Western Scrounger, Inc.
Oklahoma Ammunition Company
Oregon Trail Bullet Company
Parts & Surplus
Personal Protection Systems
Precision Delta Corp.
Precision Munitions, Inc.
Precision Reloading, Inc.
Professional Hunter Supplies
Recoilless Technologies, Inc.
Sanders Custom Gun Service
Sandia Die & Cartridge Co.
SOS Products Co.
Specialty Gunsmithing
Spence, George W.
Spencer's Custom Guns
Star Custom Bullets
State Arms Gun Co.
Stewart's Gunsmithing
Talon Mfg. Co., Inc.
3-D Ammunition & Bullets
3-Ten Corp.
Unmussig Bullets, D.L.
Vitt/Boos
Vom Hoffe
Vulpes Ventures, Inc.
Warren Muzzleloading Co., Inc.
Weaver Arms Corp. Gun Shop
Wells Custom Gunsmith, R.A.
Worthy Products, Inc.
Yukon Arms Classic Ammunition
Zonie Bullets

AMMUNITION, FOREIGN

AFSCO Ammunition
Armscorp USA, Inc.
A-Square Co., Inc.
Atlantic Rose, Inc.
Berger Bullets, Ltd.
BulletMakers Workshop, The
B-West Imports, Inc.
CBC
Cheddite France, S.A.
Cubic Shot Shell Co., Inc.
Dead Eye's Sport Center
Diana
DKT, Inc.
Dynamit Nobel-RWS, Inc.
Fiocchi of America, Inc.
First, Inc., Jack
Fisher Enterprises, Inc.
Fisher, R. Kermit
FN Herstal
Forgott Jr., Valmore J.
Gibbs Rifle Co., Inc.
GOEX, Inc.
Hansen & Co.
Hansen Cartridge Co.
Heidenstrom Bullets
Hirtenberger Aktiengesellschaft
Hornady Mfg. Co.
IMI
IMI Services USA, Inc.
Israel Military Industries Ltd.
JägerSport, Ltd.
K.B.I., Inc.
Keng's Firearms Specialty, Inc.
Magnum Research, Inc.
MagSafe Ammo Co.
MagTech Recreational Products, Inc.
Maionchi-L.M.I.
MAST Technology
Merkuria Ltd.
Mullins Ammunition
Naval Ordnance Works
Oklahoma Ammunition Co.
Old Western Scrounger, Inc.
Petro-Explo, Inc.
Precision Delta Corp.
R.E.T. Enterprises
Rocky Fork Enterprises
RWS
Sentinel Arms
Southern Ammunition Co., Inc.
Spence, George W.
Stratco, Inc.
SwaroSports, Inc.
T.F.C. S.p.A.
USA Sporting Inc.
Vom Hoffe
Vihtavuori Oy/Kaltron-Pettibone
Yukon Arms Classic Ammunition

AMMUNITION, CUSTOM

Accuracy Unlimited (Littleton, CO)
AFSCO Ammunition
Allred Bullet Co.
American Derringer Corp.
Arizona Ammunition, Inc.
Arms Corporation of the Philippines
A-Square Co., Inc.
Atlantic Rose, Inc.
Ballistica Maximus North
Berger Bullets, Ltd.
Black Hills Ammunition, Inc.
Blue Mountain Bullets
Bruno Shooters Supply
Brynin, Milton
Buck Stix—SOS Products Co.
Buckskin Bullet Co.
BulletMakers Workshop, The
Bull-X, Inc.
Calhoon Varmint Bullets, James
Carroll Bullets
CBC
CHAA, Ltd.
Country Armourer, The
Cubic Shot Shell Co., Inc.
Custom Tackle and Ammo
Dakota Arms, Inc.
Dead Eye's Sport Center
Delta Frangible Ammunition, LLC
DKT, Inc.
Elite Ammunition
Estate Cartridge, Inc.
4W Ammunition
Freedom Arms, Inc.
GDL Enterprises
Glaser Safety Slug, Inc.
GOEX, Inc.
"Gramps" Antique Cartridges
Grand Falls Bullets, Inc.
Granite Custom Bullets
Gun Accessories
Heidenstrom Bullets
Hirtenberger Aktiengesellschaft
Hoelscher, Virgil
Horizons Unlimited
Hornady Mfg. Co.
Hunters Supply
IMI
Israel Military Industries Ltd.
Jensen Bullets
Jensen's Custom Ammunition
Jensen's Firearms Academy
Kaswer Custom, Inc.
Keeler, R.H.
Kent Cartridge Mfg. Co. Ltd.
KJM Fabritek, Inc.
Lindsley Arms Cartridge Co.
Loch Leven Industries
MagSafe Ammo Co.
MAST Technology
McBros Rifle Co.

AMMUNITION COMPONENTS—BULLETS, POWDER, PRIMERS, CASES

Acadian Ballistic Specialties
Accuracy Unlimited (Littleton, CO)
Accurate Arms Co., Inc.
Accurate Bullet Co.
Action Bullets, Inc.
ACTIV Industries, Inc.
Alaska Bullet Works, Inc.
Alliant Techsystems
Allred Bullet Co.
Alpha LaFranck Enterprises
American Products Inc.
Arco Powder
Armfield Custom Bullets
A-Square Co., Inc.
Atlantic Rose, Inc.
Baer's Hollows
Ballard Built
Ballistic Products, Inc.
Barnes Bullets, Inc.
Beartooth Bullets
Beeline Custom Bullets Limited
Bell Reloading, Inc.
Belt MTN Arms
Berger Bullets, Ltd.
Bergman & Williams
Berry's Mfg., Inc.
Bertram Bullet Co.
Big Bore Bullets of Alaska
Big Bore Express
Bitterroot Bullet Co.
Black Belt Bullets
Black Hills Shooters Supply
Black Powder Products
Blount, Inc., Sporting Equipment Div.
Brenneke KG, Wilhelm
Briese Bullet Co., Inc.
Brown Co., E. Arthur
Brown Dog Ent.
Brownells, Inc.
BRP, Inc.
Bruno Shooters Supply
Buck Stix
Buckeye Custom Bullets
Buckskin Bullet Co.
Buffalo Arms Co.
Buffalo Rock Shooters Supply
Bullet, Inc.
Bullseye Bullets
Bull-X, Inc.
Butler Enterprises
Buzztail Brass
Calhoon Varmint Bullets, James
Canyon Cartridge Corp.
Carnahan Bullets
Cascade Bullet Co., Inc.
Cast Performance Bullet Company
CCI
Champion's Choice, Inc.

PRODUCT DIRECTORY

Cheddite France, S.A.
CheVron Bullets
C.J. Ballistics, Inc.
Colorado Sutlers Arsenal
Competitor Corp., Inc.
Cook Engineering Service
Copperhead Bullets, Inc.
Cor-Bon Bullet & Ammo Co.
Cumberland States Arsenal
Cummings Bullets
Curtis Cast Bullets
Curtis Gun Shop
Custom Bullets by Hoffman
Cutsinger Bench Rest Bullets
D&J Bullet Co. & Custom Gun Shop, Inc.
Dakota Arms, Inc.
Dixie Gun Works, Inc.
DKT, Inc.
Dohring Bullets
Double A Ltd.
DuPont
Eichelberger Bullets, Wm.
Eldorado Cartridge Corp.
Elkhorn Bullets
Epps, Ellwood
Federal Cartridge Co.
Finch Custom Bullets
Fiocchi of America, Inc.
Forkin, Ben
4W Ammunition
Fowler, Bob
Fowler Bullets
Foy Custom Bullets
Freedom Arms, Inc.
Fusilier Bullets
G&C Bullet Co., Inc.
Gain Twist Barrel Co.
Gander Mountain, Inc.
Gehmann, Walter
GOEX, Inc.
Golden Bear Bullets
Gonic Bullet Works
Gotz Bullets
"Gramps" Antique Cartridges
Granite Custom Bullets
Grayback Wildcats
Green Mountain Rifle Barrel Co., Inc.
Grier's Hard Cast Bullets
Group Tight Bullets
Gun City
Hammets VLD Bullets
Hardin Specialty Dist.
Harris Enterprises
Harrison Bullets
Hart & Son, Inc., Robert W.
Haselbauer Products, Jerry
Hawk, Inc.
Hawk Laboratories, Inc.
Haydon Shooters' Supply, Russ
Heidenstrom Bullets
Hercules, Inc.
Hi-Performance Ammunition Company
Hirtenberger Aktiengesellschaft
Hobson Precision Mfg. Co.
Hodgdon Powder Co.
Hornady Mfg. Co.
HT Bullets
Huntington Die Specialties
Hunters Supply
IMI Services USA, Inc.
Imperial Magnum Corp.
IMR Powder Co.
J-4, Inc.
J&D Components
J&L Superior Bullets
Jensen Bullets
Jensen's Firearms Academy
Jericho Tool & Die Co. Inc.
Jester Bullets
JLK Bullets
JRP Custom Bullets
Ka Pu Kapili
Kasmarsik Bullets
Kaswer Custom, Inc.
Keith's Bullets
Ken's Kustom Kartridge
Keng's Firearms Specialty, Inc.
Kent Cartridge Mfg. Co. Ltd.
KJM Fabritek, Inc.
KLA Enterprises
Knight Rifles
Kodiak Custom Bullets
Lage Uniwad
Lapua Ltd.
Lawrence Brand Shot
Legend Products Corp.
Liberty Shooting Supplies
Lightfield Ammunition Corp.
Slug Group, Inc.
Lightning Performance Innovations, Inc.
Lindsley Arms Cartridge Co.
Littleton, J.F.
Loweth, Richard H.R.
Lomont Precision Bullets
M&D Munitions Ltd.
Magnus Bullets
Maine Custom Bullets
Maionchi-L.M.I.
Marchmon Bullets
Markesbery Muzzle Loaders, Inc.
MarMik, Inc.
Marple & Associates, Dick
MAST Technology
Mathews & Son, Inc., George E.
McMurdo, Lynn
Meister Bullets
Men—Metallwerk Elisenhuette, GmbH
Merkuria Ltd.
Michael's Antiques
Miller Enterprises, Inc., R.P.
Mitchell Bullets, R.F.
MI-TE Bullets
Modern Muzzleloading, Inc.
MoLoc Bullets
Montana Armory, Inc.
Montana Precision Swaging
Mountain State Muzzleloading Supplies, Inc.
Mt. Baldy Bullet Co.
Mulhern, Rick
Murmur Corp.
Mushroom Express Bullet Co.
Nagel's Bullets
National Bullet Co.
Naval Ordnance Works
Navy Arms Co.
Necromancer Industries, Inc.
Norma
North American Shooting Systems
North Devon Firearms Services
Northern Precision Custom Swaged Bullets
Nosler, Inc.
Oklahoma Ammunition Co.
Old Wagon Bullets
Old Western Scrounger, Inc.
Omark Industries
Ordnance Works, The
Oregon Trail Bullet Company
Pacific Cartridge, Inc.
Page Custom Bullets
Patrick Bullets
Pattern Control
Pease Accuracy, Bob
Petro-Explo, Inc.
Phillippi Custom Bullets, Justin
Pinetree Bullets
PMC/Eldorado Cartridge Corp.
Polywad, Inc.
Pomeroy, Robert
Precision Components
Precision Components and Guns
Precision Delta Corp.
Precision Munitions, Inc.
Prescott Projectile Co.
Price Bullets, Patrick W.
PRL Bullets
Professional Hunter Supplies
Rainier Ballistics Corp.
Ranger Products
Red Cedar Precision Mfg.
Redwood Bullet Works
Reloading Specialties, Inc.
Remington Arms Co., Inc.
Redding Reloading Equipment
Rhino
Rifle Works & Armory
R.I.S. Co., Inc.
R.M. Precision, Inc.
Robinson H.V. Bullets
Rolston, Inc., Fred W.
Rubright Bullets
SAECO
Scharch Mfg., Inc.
Schmidtman Custom Ammunition
Schneider Bullets
Schroeder Bullets
Scot Powder
Seebeck Assoc., R.E.
Shappy Bullets
Shilen, Inc.
Sharps Arms Co. Inc., C.
Shooting Components Marketing
Sierra Bullets
Silhouette, The
SOS Products Co.
Specialty Gunsmithing
Speer Products
Spencer's Custom Guns
Stanley Bullets
Star Ammunition, Inc.
Star Custom Bullets
Stark's Bullet Mfg.
Starke Bullet Company
Stewart's Gunsmithing
Swift Bullet Co.
Talon Mfg. Co., Inc.
Taracorp Industries
TCCI
TCSR
T.F.C. S.p.A.
Thompson Precision
3-D Ammunition & Bullets
TMI Products
Traditions, Inc.
Trico Plastics
Trophy Bonded Bullets, Inc.
True Flight Bullet Co.
Tucson Mold, Inc.
Unmussig Bullets, D.L.
USAC
Vann Custom Bullets
Vihtavuori Oy/Kaltron-Pettibone
Vincent's Shop
Viper Bullet and Brass Works
Vom Hoffe
Warren Muzzleloading Co., Inc.
Watson Trophy Match Bullets
Weatherby, Inc.
Western Nevada West Coast Bullets
Widener's Reloading & Shooting Supply
Williams Bullet Co., J.R.
Winchester Div., Olin Corp.
Windjammer Tournament Wads, Inc.
Winkle Bullets
Woodleigh
Worthy Products, Inc.
Wosenitz VHP, Inc.
Wyant Bullets
Wyoming Bonded Bullets
Wyoming Custom Bullets
Yukon Arms Classic Ammunition
Zero Ammunition Co., Inc.
Zonie Bullets

ANTIQUE ARMS DEALERS

Ackerman & Co.
Ad Hominem
Antique American Firearms
Antique Arms Co.
Aplan Antiques & Art, James O.
Armoury, Inc., The
Bear Mountain Gun & Tool
Bob's Tactical Indoor Shooting Range & Gun Shop
Boggs, Wm.
British Antiques
Buckskin Machine Works
Buffalo Arms Co.
Cape Outfitters
Carlson, Douglas R.
Chadick's Ltd.
Chambers Flintlocks Ltd., Jim
Champlin Firearms, Inc.
Chuck's Gun Shop
Classic Guns, Inc.
Clements' Custom Leathercraft, Chas
Cole's Gun Works
Colonial Arms, Inc.
D&D Gunsmiths, Ltd.
Dixie Gun Works, Inc.
Dixon Muzzleloading Shop, Inc.
Duffy, Charles E.
Dyson & Son Ltd., Peter
Ed's Gun House
Enguix Import-Export
Fagan & Co., William
Fish Mfg. Gunsmith Sptg. Co., Marshall F.
Flayderman & Co., N.
Forgett Jr., Valmore J.
Frielich Police Equipment
Fulmer's Antique Firearms, Chet
Getz Barrel Co.
Glass, Herb
Goergen's Gun Shop, Inc.
Golden Age Arms Co.
Greenwald, Leon E. "Bud"
Gun Room, The
Gun Room Press, The
Guncraft Sports, Inc.
Gun Works, The
Guns Antique & Modern DBA/Charles E. Duffy
Hallowell & Co.
HandiCrafts Unltd.
Hansen & Co.
Hunkeler, A.
Johns Master Engraver, Bill
Kelley's
Ledbetter Airguns, Riley
LeFever Arms Co., Inc.
Lever Arms Service Ltd.
Lock's Philadelphia Gun Exchange
Log Cabin Sport Shop
Madis, George
Mandall Shooting Supplies, Inc.
Martin's Gun Shop
Montana Outfitters
Mountain Bear Rifle Works, Inc.
Museum of Historical Arms, Inc.
Muzzleloaders Etcetera, Inc.
New England Arms Co.
Pioneer Guns
Pony Express Sport Shop, Inc.
Retting, Inc., Martin B.
R.G.-G., Inc.
S&S Firearms
Scott Fine Guns, Inc., Thad
Shootin' Shack, Inc.
Steves House of Guns
Stott's Creek Armory, Inc.
Strawbridge, Victor W.
Vic's Gun Refinishing
Vintage Arms, Inc.
Westley Richards & Co.
Wiest, M.C.
Winchester Sutler, Inc., The
Wood, Frank
Yearout, Lewis E.

APPRAISERS—GUNS, ETC.

Accuracy Gun Shop
Antique Arms Co.
Armoury, Inc., The
Arundel Arms & Ammunition, Inc., A.
Barsotti, Bruce
Beitzinger, George
Billings Gunsmiths, Inc.
Blue Book Publications, Inc.
Bob's Tactical Indoor Shooting Range & Gun Shop
British Antiques
Bustani, Leo
Butterfield & Butterfield
Camilli, Lou
Cannon's, Andy Cannons
Cape Outfitters
Chadick's Ltd.
Champlin Firearms, Inc.
Christie's East
Clark Firearms Engraving
Classic Guns, Inc.
Clements' Custom Leathercraft, Chas
Cole's Gun Works
Colonial Arms, Inc.
Colonial Repair
Corry, John
Custom Tackle and Ammo
D&D Gunsmiths, Ltd.
DGR Custom Rifles
Dixon Muzzleloading Shop, Inc.
Duane's Gun Repair
Ed's Gun House
Epps, Ellwood
Eversull Co., Inc., K.
Fagan & Co., William
Ferris Firearms
Fish Mfg. Gunsmith Sptg. Co, Marshall F.
Flayderman & Co., Inc., N.
Forgett, Valmore J., Jr.
Forty Five Ranch Enterprises
Francotte & Cie S.A., Auguste

30th EDITION, 1998

DIRECTORY OF THE ARMS TRADE

Frontier Arms Co., Inc.
Getz Barrel Co.
Gillmann, Edwin
Golden Age Arms Co.
Gonzalez Guns, Ramon B.
Greenwald, Leon E. "Bud"
Griffin & Howe, Inc.
Gun City
Gun Hunter Trading Co.
Gun Room Press, The
Gun Shop, The
Guncraft Sports, Inc.
Guns
Hallowell & Co.
Hammans, Charles E.
HandiCrafts Unltd.
Hank's Gun Shop
Hansen & Co.
Hughes, Steven Dodd
Irwin, Campbell H.
Island Pond Gun Shop
Jackalope Gun Shop
Jaeger, Inc., Paul/Dunn's
Jensen's Custom Ammunition
Kelley's
LaRocca Gun Works, Inc.
Ledbetter Airguns, Riley
LeFever Arms Co., Inc.
L.L. Bean, Inc.
Lock's Philadelphia Gun Exchange
Long, George F.
Mac's .45 Shop
Madis, George
Mandall Shooting Supplies, Inc.
Martin's Gun Shop
McCann's Muzzle-Gun Works
Montana Outfitters

Museum of Historical Arms, Inc.
Muzzleloaders Etcetera, Inc.
Navy Arms Co.
New England Arms Co.
Nitex, Inc.
Orvis Co., The
Pasadena Gun Center
Pentheny de Pentheny
Perazzi USA, Inc.
Peterson Gun Shop, Inc., A.W.
Pettinger Books, Gerald
Pioneer Guns
Pony Express Sport Shop, Inc.
R.E.T. Enterprises
Retting, Inc., Martin B.
Richards, John
River Road Sporting Clays
Safari Outfitters Ltd.
Scott Fine Guns, Inc., Thad
Shootin' Shack, Inc.
Steger, James R.
Stratco, Inc.
Strawbridge, Victor W.
Swampfire Shop, The
Thurston Sports, Inc.
Vic's Gun Refinishing
Wayne Firearms for Collectors and Investors, James
Wells Custom Gunsmith, R.A.
Whildin & Sons Ltd., E.H.
Wiest, M.C.
Williams Shootin' Iron Service
Winchester Sutler, Inc., The
Wood, Frank
Yearout, Lewis E.
Yee, Mike

Wiest, M.C.
Wilderness Sound Products Ltd.
Williams Gun Sight Co.

Winchester Press
Wolfe Publishing Co.
Wolf's Western Traders

BULLET AND CASE LUBRICANTS

Blackhawk West
Bonanza
Brown Co., E. Arthur
Camp-Cap Products
Chem-Pak, Inc.
C-H Tool & Die Corp.
Cooper-Woodward
CVA
Elkhorn Bullets
E-Z-Way Systems
Forster Products
4-D Custom Die Co.
Guardsman Products
HEBB Resources
Hollywood Engineering
Hornady Mfg. Co.
Imperial
Le Clear Industries
Lee Precision, Inc.

Lestrom Laboratories, Inc.
Lithi Bee Bullet Lube
M&N Bullet Lube
Michaels of Oregon Co.
MI-TE Bullets
NECO
Paco's
RCBS
Reardon Products
Rooster Laboratories
Shay's Gunsmithing
Small Custom Mould & Bullet Co.
Tamarack Products, Inc.
Uncle Mike's
Warren Muzzleloading Co., Inc.
Widener's Reloading & Shooting Supply, Inc.
Young Country Arms

BULLET SWAGE DIES AND TOOLS

Brynin, Milton
Bullet Swaging Supply, Inc.
Camdex, Inc.
Corbin Mfg. & Supply, Inc.
Cumberland Arms
Eagan, Donald V.
Heidenstrom Bullets

Holland's
Hollywood Engineering
Necromancer Industries, Inc.
Niemi Engineering, W.B.
North Devon Firearms Services
Rorschach Precision Products
Sport Flite Manufacturing Co.

AUCTIONEERS—GUNS, ETC.

Butterfield & Butterfield
Christie's East
Kelley's

"Little John's" Antique Arms
Sotheby's

BOOKS (Publishers and Dealers)

Action Direct, Inc.
American Handgunner Magazine
Armory Publications, Inc.
Arms & Armour Press
Ballistic Products, Inc.
Barnes Bullets, Inc.
Blackhawk West
Blacksmith Corp.
Blacktail Mountain Books
Blue Book Publications, Inc.
Blue Ridge Machinery & Tools, Inc.
Brown Co., E. Arthur
Brownell Checkering Tools, W.E.
Brownell's, Inc.
Bullet'n Press
Calibre Press, Inc.
Cape Outfitters
Colonial Repair
Colorado Sutlers Arsenal
Corbin Mfg. & Supply, Inc.
Cumberland States Arsenal
DBI Books
Flores Publications, Inc., J.
Forgett Jr., Valmore J.
Golden Age Arms Co.
Gun City
Gun Hunter Books
Gun Hunter Trading Co.
Gun List
Gun Parts Corp., The
Gun Room Press, The
Gun Works, The
Guncraft Books
Guncraft Sports, Inc.
Gunnerman Books
GUNS Magazine
H&P Publishing
Handgun Press
Harris Publications
Hawk Laboratories, Inc.
Hawk, Inc.
Heritage/VSP Gun Books
Hodgdon Powder Co., Inc.
Home Shop Machinist, The
Hornady Mfg. Co.
Hungry Horse Books
I.D.S.A. Books
Info-Arm
Ironside International Publishers, Inc.
Koval Knives

Krause Publications, Inc.
Lane Publishing
Lapua Ltd.
Lethal Force Institute
Liberty Shooting Supplies
Lyman Products Corp.
Madis Books
Martin Bookseller, J.
McKee Publications
MI-TE Bullets
Montana Armory, Inc.
Mountain South
New Win Publishing, Inc.
NgraveR Co., The
OK Weber, Inc.
Outdoorsman's Bookstore, The
Paintball Games International Magazine (Aceville Publications)
Paintball Sports Magazine
Pejsa Ballistics
Petersen Publishing Co.
Pettinger Books, Gerald
Police Bookshelf
Precision Shooting, Inc.
PWL Gunleather
Remington Double Shotguns
R.G.-G., Inc.
Riling Arms Books Co., Ray
Rocky Mountain Wildlife Products
Rutgers Book Center
S&S Firearms
Safari Press, Inc.
Saunders Gun & Machine Shop
Semmer, Charles
Sharps Arms Co. Inc., C.
Shootin' Accessories, Ltd.
Sierra Bullets
SPG, Inc.
Stackpole Books
Stewart Game Calls, Inc., Johnny
Stoeger Industries
Stoeger Publishing Co.
"Su-Press-On," Inc.
Thomas, Charles C.
Track of the Wolf, Inc.
Trafalgar Square
Trotman, Ken
Vintage Industries, Inc.
VSP Publishers
WAMCO—New Mexico

CARTRIDGES FOR COLLECTORS

Ad Hominem
Buck Stix—SOS Products Co.
Cameron's
Campbell, Dick
Cartridge Transfer Group
Cole's Gun Works
Colonial Repair
Country Armourer, The
de Coux, Pete
DGR Custom Rifles
Duane's Gun Repair
Ed's Gun House
Enguix Import-Export
Epps, Ellwood
First, Inc., Jack
Fitz Pistol Grip Co.
Forty Five Ranch Enterprises
Goergen's Gun Shop, Inc.

"Gramps" Antique Cartridges
Gun City
Gun Parts Corp., The
Gun Room Press, The
Mandall Shooting Supplies, Inc.
MAST Technology
Michael's Antiques
Montana Outfitters
Mountain Bear Rifle Works, Inc.
Pasadena Gun Center
San Francisco Gun Exchange
Samco Global Arms, Inc.
Scott Fine Guns, Inc., Thad
SOS Products Co.
Stone Enterprises, Ltd.
Ward & Van Valkenburg
Yearout, Lewis E.

CASES, CABINETS, RACKS AND SAFES—GUN

Abel Safe & File, Inc.
Alco Carrying Cases
All Rite Products, Inc.
Allen Co., Bob
Allen Co., Inc.
Allen Sportswear, Bob
Alumna Sport by Dee Zee
American Display Co.
American Security Products Co.
Americase
Ansen Enterprises
Arizona Custom Case
Arkfeld Mfg. & Dist. Co., Inc.
Art Jewel Enterprises Ltd.
Ashby Turkey Calls
Bagmaster Mfg., Inc.
Barramundi Corp.
BEC, Inc.
Berry's Mfg., Inc.
Big Sky Racks, Inc.
Big Spring Enterprises "Bore Stores"
Bill's Custom Cases
Bison Studios
Black Sheep Brand
Boyt
Brauer Bros. Mfg. Co.
Brown, H.R.
Browning Arms Co.
Bucheimer, J.M.
Bushmaster Hunting & Fishing
Cannon Safe, Inc.
Chipmunk
Cobalt Mfg., Inc.
CONKKO
Connecticut Shotgun Mfg. Co.
D&L Industries

Dara-Nes, Inc.
Deepeeka Exports Pvt. Ltd.
D.J. Marketing
Doskocil Mfg. Co., Inc.
DTM International, Inc.
Elk River, Inc.
English, Inc., A.G.
Enhanced Presentations, Inc.
Eutaw Co., Inc.,, The
Eversull Co., Inc. K.
Fort Knox Security Products
Frontier Safe Co.
Galati Internationl
GALCO International Ltd.
Granite Custom Bullets
Gun Locker
Gun-Ho Sports Cases
Gusdorf Corp.
Hafner Creations, Inc.
Hall Plastics, Inc., John
Harrison-Hurtz Enterprises, Inc.
Hastings Barrels
Homak
Hoppe's Div.
Huey Gun Cases
Hugger Hooks Co.
Hunter Co., Inc.
Impact Case Co.
Johanssons Vapentillbehor, Bert
Johnston Bros.
Jumbo Sports Products
Kalispel Case Line
Kane Products, Inc.
KK Air International
Knock on Wood Antiques
Kolpin Mfg., Inc.
Lakewood Products, LLC

PRODUCT DIRECTORY

Liberty Safe
Marsh, Mike
Maximum Security Corp.
McWelco Products
Morton Booth Co.
MPC
MTM Molded Products Co., Inc.
Nalpak
National Security Safe Co., Inc.
Necessary Concepts, Inc.
Nesci Enterprises, Inc.
Oregon Arms, Inc.
Outa-Site Gun Carriers
Outdoor Connection, Inc., The
Pachmayr Ltd.
Palmer Security Products
Penguin Industries, Inc.
Perazzi USA, Inc.
Pflumm Mfg. Co.
Poburka, Philip
Powell & Son (Gunmakers) Ltd., William
Protecto Plastics
Prototech Industries, Inc.
Quality Arms, Inc.
Rogue Rifle Co., Inc.
Schulz Industries
Silhouette Leathers
Southern Security
Sportsman's Communicators
Sun Welding Safe Co.
Surecase Co., The
Sweet Home, Inc.
Tinks & Ben Lee Hunting Products
Waller & Son, Inc., W.
WAMCO, Inc.
Wilson Case, Inc.
Woodstream
Zanotti Armor, Inc.
Ziegel Engineering

CHOKE DEVICES, RECOIL ABSORBERS AND RECOIL PADS

Accuright
Action Products, Inc.
Allen Co., Bob
Allen Sportswear, Bob
Answer Products Co.
Arms Ingenuity Co.
Baer Custom, Inc., Les
Baker, Stan
Bansner's Gunsmithing Specialties
Bartlett Engineering
Briley Mfg., Inc.
B-Square Co., Inc.
Buffer Technologies
Bull Mountain Rifle Co.
C&H Research
Cape Outfitters
Cation
Chuck's Gun Shop
Colonial Arms, Inc.
Connecticut Shotgun Mfg. Co.
Crane Sales Co., George S.
CRR, Inc./Marble's Inc.
Danuser Machine Co.
Dayson Arms Ltd.
Dever Co., Jack
Dina Arms Corporation
D-Max, Inc.
Elsen, Inc., Pete
Galazan
Gentry Custom Gunmaker, David
Graybill's Gun Shop
Hastings Barrels
Holland's
I.N.C., Inc.
Jackalope Gun Shop
Jaeger, Inc., Paul/Dunn's
Jenkins Recoil Pads, Inc.
J.P. Enterprises, Inc.
Kick Eez
London Guns Ltd.
Lyman Instant Targets, Inc.
Lyman Products Corp.
Mag-Na-Port International, Inc.
Marble Arms
Mathews & Son, Inc., George E.
Meadow Industries
Menck, Gunsmith Inc., T.W.
Michaels of Oregon Co.
Middlebrooks Custom Shop
Morrow, Bud
Nelson/Weather-rite, Inc.
One Of A Kind
Original Box, Inc.
Pachmayr Ltd.
Palsa Outdoor Products
PAST Sporting Goods, Inc.
Pro-Port Ltd.
Protektor Model
Que Industries
R.M. Precision, Inc.
Shell Shack
Shotguns Unlimited
Simmons Gun Repair, Inc.
Spencer's Custom Guns
Stone Enterprises Ltd.
3-Ten Corp.
Trulock Tool
Uncle Mike's
Wise Guns, Dale

CHRONOGRAPHS AND PRESSURE TOOLS

Brown Co., E. Arthur
Canons Delcour
Competition Electronics, Inc.
Custom Chronograph, Inc.
D&H Precision Tooling
Hege Jagd-u. Sporthandels, GmbH
Hornady Mfg. Co.
Kent Cartridge Mfg. Co. Ltd.
Oehler Research, Inc.
P.A.C.T., Inc.
Shooting Chrony, Inc.
SKAN A.R.
Stratco, Inc.
Tepeco

CLEANING AND REFINISHING SUPPLIES

AC Dyna-tite Corp.
Acculube II, Inc.
Accupro Gun Care
American Gas & Chemical Co., Ltd.
Answer Products Co.
Armite Laboratories
Armsport, Inc.
Atlantic Mills, Inc.
Atsko/Sno-Seal, Inc.
Barnes Bullets, Inc.
Birchwood Casey
Blackhawk East
Blount, Inc., Sporting Equipment Div.
Blue and Gray Products, Inc.
Break-Free, Inc.
Bridgers Best
Brown Co., E. Arthur
Camp-Cap Products
Cape Outfitters
Chem-Pak, Inc.
CONKKO
Connecticut Shotgun Mfg. Co
Crane & Crane Ltd.
Creedmoor Sports, Inc.
CRR, Inc./Marble's Inc.
Custom Products
D&H Prods. Co., Inc.
Dara-Nes, Inc.
Decker Shooting Products
Deepeeka Exports Pvt. Ltd.
Dewey Mfg. Co., Inc., J.
Du-Lite Corp.
Dutchman's Firearms, Inc., The
Dykstra, Doug
E&L Mfg., Inc.
Eezox, Inc.
Ekol Leather Care
Faith Associates, Inc.
Flitz International Ltd.
Fluoramics, Inc.
Frontier Products Co.
G96 Products Co., Inc.
Goddard, Allen
Golden Age Arms Co.
Gozon Corp., U.S.A.
Great Lakes Airguns
Guardsman Products
Half Moon Rifle Shop
Heatbath Corp.
Hoppe's Div.
Hornady Mfg. Co.
Hydrosorbent Products
Iosso Products
Johnston Bros.
Kellogg's Professional Products
Kent Cartridge Mfg. Co. Ltd.
Kesselring Gun Shop
Kleen-Bore, Inc.
Laurel Mountain Forge
Lee Supplies, Mark
LEM Gun Specialties, Inc.
Lewis Lead Remover, The
List Precision Engineering
LPS Laboratories, Inc.
Marble Arms
Micro Sight Co.
Minute Man High Tech Industries
Mountain View Sports, Inc.
MTM Molded Products Co., Inc.
Muscle Products Corp.
Nesci Enterprises, Inc.
Northern Precision Custom Swaged Bullets
Now Products, Inc.
Old World Oil Products
Omark Industries
Original Mink Oil, Inc.
Outers Laboratories, Div. of Blount
Ox-Yoke Originals, Inc.
P&M Sales and Service
Pachmayr Ltd.
PanaVise Products, Inc.
Parker Gun Finishes
Pendleton Royal
Penguin Industries, Inc.
Precision Reloading, Inc.
Prolix® Lubricants
Pro-Shot Products, Inc.
R&S Industries Corp.
Radiator Specialty Co.
Rickard, Inc., Pete
RIG Products Co.
Rod Guide Co.
Rooster Laboratories
Rusteprufe Laboratories
Rusty Duck Premium Gun Care Products
Saunders Gun & Machine Shop
Shiloh Creek
Shooter's Choice
Shootin' Accessories, Ltd.
Silencio/Safety Direct
Sno-Seal, Inc.
Spencer's Custom Guns
Stoney Point Products, Inc.
Svon Corp.
Tag Distributors
TDP Industries, Inc.
Tetra Gun Lubricants
Texas Platers Supply Co.
T.F.C. S.p.A.
Thompson Bullet Lube Co.
Thompson/Center Arms
Track of the Wolf, Inc.
United States Products Co.
Van Gorden & Son, Inc., C.S.
Venco Industries, Inc.
VibraShine, Inc.
Warren Muzzleloading Co., Inc.
WD-40 Co.
Wick, David E.
Willow Bend
Young Country Arms
Z-Coat Industrial Coatings, Inc.

COMPUTER SOFTWARE—BALLISTICS

ADC, Inc.
Action Target, Inc.
AmBr Software Group Ltd.
Arms, Programming Solutions
Arms Software
Ballistic Engineering & Software, Inc.
Ballistic Program Co., Inc., The
Barnes Bullets, Inc.
Beartooth Bullets
Blackwell, W.
Canons Delcour
Corbin Mfg. & Supply, Inc.
Country Armourer, The
Data Tech Software Systems
Exe, Inc.
FlashTek, Inc.
Hodgdon Powder Co., Inc.
Hutton Rifle Ranch
Jensen Bullets
J.I.T. Ltd.
JWH: Software
Kent Cartridge Mfg. Co. Ltd.
Load From A Disk
Maionchi-L.M.I.
Oehler Research, Inc.
P.A.C.T., Inc.
PC Bullet/ADC, Inc.
Pejsa Ballistics
Powley Computer
RCBS
Sierra Bullets
Tioga Engineering Co., Inc.
Vancini, Carl
W. Square Enterprises

CUSTOM GUNSMITHS

A&W Repair
A.A. Arms, Inc.
Acadian Ballistic Specialties
Ackerman & Co.
Ace Custom 45's, Inc.
Ad Hominem
Accuracy Gun Shop
Accuracy Unlimited (Glendale, AZ)
Actions by "T"
Adair Custom Shop, Bill
Ahlman Guns
Aldis Gunsmithing & Shooting Supply
Alpha Gunsmith Division
Alpine's Precision Gunsmithing & Indoor Shooting Range
Amrine's Gun Shop
Answer Products Co.
Antique Arms Co.
Armament Gunsmithing Co., Inc.
Arms Craft Gunsmithing
Arms Ingenuity Co.
Arnold Arms Co., Inc.
Aro-Tek, Ltd.
Art's Gun & Sport Shop, Inc.
Artistry in Wood
Arundel Arms & Ammunition, Inc., A.
Baelder, Harry
Baer Custom, Inc., Les
Bain & Davis, Inc.
Baity's Custom Gunworks
Bansner's Gunsmithing Specialties
Barnes Bullets, Inc.
Barsotti, Bruce
Barta's Gunsmithing
Bear Arms
Bear Mountain Gun & Tool
Beaver Lodge
Behlert Precision, Inc.
Beitzinger, George
Belding's Custom Gun Shop
Bellm Contenders
Belt MTN Arms
Benchmark Guns
Bengtson Arms Co., L.
Biesen, Al
Biesen, Roger
Billeb, Stephen L.
Billings Gunsmiths, Inc.
BlackStar AccuMax Barrels
BlackStar Barrel Accurizing
Bond Custom Firearms
Borden's Accuracy
Borovnik KG, Ludwig
Brace, Larry D.
Brgoch, Frank
Briese Bullet Co., Inc.
Briganti, A.J.
Briley Mfg., Inc.
Broad Creek Rifle Works
Brockman's Custom Gunsmithing
Broken Gun Ranch
Broughton Rifle Barrels
Brown Precision, Inc.
Buckhorn Gun Works
Buckskin Machine Works
Budin, Dave
Bull Mountain Rifle Co.
Bullberry Barrel Works, Ltd.
Burkhart Gunsmithing, Don
C&J Enterprises, Inc.
Cache La Poudre Rifleworks
CAM Enterprises
Camilli, Lou
Campbell, Dick
Cannon's, Andy Cannon
Carolina Precision Rifles

DIRECTORY OF THE ARMS TRADE

Carter's Gun Shop
Caywood, Shane J.
Chambers Flintlocks Ltd., Jim
Champlin Firearms, Inc.
Chicasaw Gun Works
Christman Jr., Gunmaker, David
Chuck's Gun Shop
Clark Custom Guns, Inc.
Clark Firearms Engraving
Classic Arms Corp.
Classic Guns, Inc.
Clearview Products
Cloward's Gun Shop
Cochran, Oliver
Cogar's Gunsmithing
Cole's Gun Works
Coleman's Custom Repair
Colonial Repair
Colorado Gunsmithing Academy
Colorado School of Trades
Colt's Mfg. Co., Inc.
Competitive Pistol Shop, The
Conrad, C.A.
Corkys Gun Clinic
Cox, C. Ed
Craig Custom Ltd.
Creekside Gun Shop, Inc.
Cullity Restoration, Daniel
Cumberland Knife & Gun Works
Curtis Custom Shop
Custom Checkering Service
Custom Gun Products
Custom Gun Stocks
Custom Gunsmiths
Custom Shop, The
Cylinder & Slide, Inc.
D&D Gunsmiths, Ltd.
D&J Bullet Co. & Custom
 Gun Shop, Inc.
Dangler, Homer L.
Darlington Gun Works, Inc.
Dave's Gun Shop
Davis, Don
Davis Service Center, Bill
Dayton Traister
Delorge, Ed
Dever Co., Jack
DGS, Inc.
DGR Custom Rifles
Dietz Gun Shop & Range, Inc.
Dilliott Gunsmithing, Inc.
Donnelly, C.P.
Dowtin Gunworks
Duane's Gun Repair
Duffy, Charles E.
Duncan's Gun Works, Inc.
Dyson & Son Ltd., Peter
Echols & Co., D'Arcy
Eckelman Gunsmithing
Eggleston, Jere D.
EGW Evolution Gun Works
Erhardt, Dennis
Eskridge Rifles, Steven Eskridge
Eversull Co., Inc., K.
Eyster Heritage Gunsmiths, Inc., Ken
Fanzoj GmbH
Ferris Firearms
Fish Mfg. Gunsmith Spt. Co.,
 Marshall F.
Fisher, Jerry A.
Flaig's
Fleming Firearms
Flynn's Custom Guns
Forkin, Ben
Forster, Kathy
Forster, Larry L.
Forthofer's Gunsmithing
 & Knifemaking
Francesca, Inc.
Francotte & Cie S.A., Auguste
Frank Custom Classic Arms, Ron
Frazier Brothers Enterprises
Frontier Arms Co., Inc.
Fullmer, Geo. M.
G.G. & G.
Gator Guns & Repair
Genecco Gun Works, K.
Gentry Custom Gunmaker, David
Gilkes, Anthony W.
Gillmann, Edwin
Gilman-Mayfield, Inc.
Giron, Robert E.
Goens, Dale W.
Gonzalez Guns, Ramon B.
Goodling's Gunsmithing
Goodwin, Fred
Gordie's Gun Shop
Grace, Charles E.
Graybill's Gun Shop
Green, Roger M.
Greg Gunsmithing Repair
Gr,-Tan Rifles
Griffin & Howe, Inc.
Gun Shop, The
Guncraft Sports, Inc.
Guns
Guns Antique & Modern
 DBA/Charles E. Duffy
Gunsite Custom Shop
Gunsite Gunsmithy
Gunsite Training Center
Gunsmithing Ltd.
Hagn Rifles & Actions, Martin
Hallberg Gunsmith, Fritz
Halstead, Rick
Hamilton, Alex B.
Hamilton, Jim
Hammans, Charles E.
Hammond Custom Guns Ltd.
Hank's Gun Shop
Hanus, Bill Birdguns
Hanson's Gun Center, Dick
Harold's Custom Gun Shop, Inc.
Harris Gunworks
Hart & Son, Inc., Robert W.
Hart Rifle Barrels, Inc.
Hartmann & Weiss GmbH
Harwood, Jack O.
Hawken Shop, The
Hecht, Hubert J.
Heilmann, Stephen
Heinie Specialty Products
Hendricks Gun Works
Hensler, Jerry
Hensley, Gunmaker, Darwin
Heppler, Keith M.
Heydenberk, Warren R.
High Bridge Arms, Inc.
High Performance International
Highline Machine Co.
Hill, Loring F.
Hiptmayer, Armurier
Hiptmayer, Klaus
Hoag, James W.
Hobbie Gunsmithing, Duane A.
Hodgson, Richard
Hoehn Sales, Inc.
Hoelscher, Virgil
Hoenig & Rodman
Hofer Jagdwaffen, P.
Holland, Dick
Holland's
Hollis Gun Shop
Horst, Alan K.
Huebner, Corey O.
Hughes, Steven Dodd
Hunkeler, Al
Hyper-Single, Inc.
Ide, Kenneth G.
Imperial Magnum Corp.
Irwin, Campbell H.
Island Pond Gun Shop
Ivanoff, Thomas G.
J&S Heat Treat
Jackalope Gun Shop
Jaeger, Inc., Paul/Dunn's
Jamison's Forge Works
Jarrett Rifles, Inc.
Jarvis, Inc.
Jensen's Custom Ammunition
Jim's Gun Shop
Jim's Precision
Johnston, James
Jones, J.D.
Juenke, Vern
Jurras, L.E.
K-D, Inc.
KDF, Inc.
Keith's Custom Gunstocks
Ken's Gun Specialties
Ketchum, Jim
Kilham & Co.
Kimball, Gary
King's Gun Works
KLA Enterprises
Klein Custom Guns, Don
Kleinendorst, K.W.
Kneiper Custom Guns, Jim
Knippel, Richard
Kopp, Terry K.
Korzinek Riflesmith, J.
LaFrance Specialties
Lair, Sam
Lampert, Ron
LaRocca Gun Works, Inc.
Lathrop's, Inc.
Laughridge, William R.
Lawson Co., Harry
Lee's Red Ramps
LeFever Arms Co., Inc.
Liberty Antique Gunworks
Lind Custom Guns, Al
Linebaugh Custom Sixguns
 & Rifle Works
List Precision Engineering
Ljutic Industries, Inc.
Lock's Philadelphia Gun Exchange
London Guns Ltd.
Long, George F.
Lyons Gunworks, Larry
Mac-1 Distributors
Mac's .45 Shop
Mag-Na-Port International, Inc.
Mahony, Philip Bruce
Makinson, Nicholas
Mandall Shooting Supplies, Inc.
Martin's Gun Shop
Martz, John V.
Masker, Seely
Mathews & Son, Inc., George E.
Mazur Restoration, Pete
McCament, Jay
McCann's Machine & Gun Shop
McCann's Muzzle-Gun Works
McCluskey Precision Rifles
McFarland, Stan
McGowen Rifle Barrels
McKinney, R.P.
McMillan Rifle Barrels
MCS, Inc.
Mercer Custom Stocks, R.M.
Michael's Antiques
Mid-America Recreation, Inc.
Middlebrooks Custom Shop
Miller Co., David
Miller Arms, Inc.
Miller Custom
Mills Jr., Hugh B.
Mo's Competitor Supplies
Moeller, Steve
Monell Custom Guns
Morrison Custom Rifles, J.W.
Morrow, Bud
Mountain Bear Rifle Works, Inc.
Mowrey's Guns & Gunsmithing
Mullis Guncraft
Nastoff's 45 Shop, Inc., Steve
Nelson, Stephen
Nettestad Gun Works
New England Arms Co.
New England Custom Gun Service
Newman Gunshop
NCP Products, Inc.
Nicholson Custom
Nickels, Paul R.
Nicklas, Ted
Nitex, Inc.
Norman Custom Gunstocks, Jim
Norrell Arms, John
North American Shooting Systems
North Fork Custom Gunsmithing
Nu-Line Guns, Inc.
Nygord Precision Products
Oakland Custom Arms, Inc.
Old World Gunsmithing
Olson, Vic
Orvis Co., The
Ottmar, Maurice
Ozark Gun Works
P&S Gun Service
Pagel Gun Works, Inc.
Parker Gun Finishes
Pasadena Gun Center
Paterson Gunsmithing
PEM's Mfg. Co.
Pence Precision Barrels
Penrod Precision
Pentheny de Pentheny
Perazone, Brian
Performance Specialists
Peterson Gun Shop, Inc., A.W.
Powell & Son (Gunmakers) Ltd.,
 William
Power Custom, Inc.
Professional Hunter Supplies
P.S.M.G. Gun Co.
Quality Firearms of Idaho, Inc.
R&J Gun Shop
Ray's Gunsmith Shop
Renfrew Guns & Supplies
Ridgetop Sporting Goods
Ries, Chuck
Rifles Inc.
Rigby & Co., John
River Road Sporting Clays
RMS Custom Gunsmithing
Robar Co.'s, Inc., The
Robinson, Don
Rocky Mountain Arms, Inc.
Rocky Mountain Rifle Works Ltd.
Rogers Gunsmithing, Bob
Romain's Custom Guns, Inc.
Rudnicky, Susan
Rupert's Gun Shop
Ryan, Chad L.
Sanders Custom Gun Service
Schiffman, Curt
Schiffman, Mike
Schiffman, Norman
Schumakers Gun Shop
Schwartz Custom Guns, Wayne E.
Score High Gunsmithing
Scott, Dwight
Scott, McDougall & Associates
Shaw, Inc., E.R.
Shay's Gunsmithing
Shell Shack
Shockley, Harold H.
Shooten' Haus, The
Shooter Shop, The
Shooters Supply
Shootin' Shack, Inc.
Shooting Specialties
Shotgun Shop, The
Shotguns Unlimited
Silver Ridge Gun Shop
Simmons Gun Repair, Inc.
Singletary, Kent
Sipes Gun Shop
Siskiyou Gun Works
Skeoch, Brian R.
Slezak, Jerome F.
Small Arms Mfg. Co.
Smith, Art
Smith, Sharmon
Snapp's Gunshop
Speiser, Fred D.
Spencer's Custom Guns
Spencer Reblue Service
Sportsmen's Exchange & Western
 Gun Traders, Inc.
SSK Industries
Star Custom Bullets
Starnes Gunmaker, Ken
Steelman's Gun Shop
Steffens, Ron
Steger, James R.
Stiles Custom Guns
Storey, Dale A.
Stott's Creek Armory, Inc.
Strawbridge, Victor W.
Sturgeon Valley Sporters
Sullivan, David S.
Swampfire Shop, The
Swann, D.J.
Swenson's 45 Shop, A.D.
Swift River Gunworks
Szweda, Robert
Talmage, William G.
Tank's Rifle Shop
Tarnhelm Supply Co., Inc.
Taylor & Robbins
Ten-Ring Precision, Inc.
Thompson, Randall
Tom's Gunshop
300 Gunsmith Service, Inc.
Thurston Sports, Inc.
Time Precision, Inc.
Titus, Daniel
Tom's Gun Repair
Tooley Custom Rifles
Trevallion Gunstocks
Trulock Tool
Tucker, James C.
Unmussig Bullets, D.L.
Upper Missouri Trading Co.
USA Sporting Inc.
Van Horn, Gil
Van Patten, J.W.
Van's Gunsmith Service
Vest, John
Vic's Gun Refinishing
Vintage Arms, Inc.
Volquartsen Custom Ltd.
Von Minden Gunsmithing Services
Vortek Products, Inc.
Walker Arms Co., Inc.
Wasmundt, Jim

PRODUCT DIRECTORY

Weaver Arms Corp. Gun Shop
Weber & Markin Custom Gunsmiths
Weems, Cecil
Weigand Combat Handguns, Inc.
Wells, Fred F.
Wells Custom Gunsmith, R.A.
Welsh, Bud
Wenig Custom Gunstocks, Inc.
Werth, T.W.
Wessinger Custom Guns & Engraving
West, Robert G.
Western Design
Westley Richards & Co.
Westwind Rifles, Inc.
White Muzzleloading Systems
White Shooting Systems, Inc.

Wichita Arms, Inc.
Wiebe, Duane
Wild West Guns
Williams Gun Sight Co.
Williams Shootin' Iron Service
Williamson Precision Gunsmithing
Wilson Gun Shop
Winter, Robert M.
Wise Guns, Dale
Wiseman and Co., Bill
Wood, Frank
Wright's Hardwood Gunstock Blanks
Yankee Gunsmith
Yee, Mike
Zeeryp, Russ

Waldron, Herman
Weber & Markin Custom Gunsmiths
Wells, Fred F.
Wells Custom Gunsmith, R.A.
Welsh, Bud
Werth, T.W.
Wessinger Custom Guns & Engraving
West, Robert G.
Westrom, John

White Rock Tool & Die
Wiebe, Duane
Williams Gun Sight Co.
Williams Shootin' Iron Service
Williamson Precision Gunsmithing
Wise Guns, Dale
Wood, Frank
Zufall, Joseph F.

CUSTOM METALSMITHS

Ahlman Guns
Aldis Gunsmithing & Shooting Supply
Amrine's Gun Shop
Answer Products Co.
Arnold Arms Co., Inc.
Arundel Arms & Ammunition, Inc., A.
Baer Custom, Inc., Les
Bansner's Gunsmithing Specialties
Baron Technology
Barsotti, Bruce
Bear Mountain Gun & Tool
Behlert Precision, Inc.
Beitzinger, George
Bell, Sid
Benchmark Guns
Bengtson Arms Co., L.
Biesen, Al
Billeb, Stephen L.
Billingsley & Brownell
Brace, Larry D.
Briganti, A.J.
Broad Creek Rifle Works
Brown Precision, Inc.
Buckhorn Gun Works
Bull Mountain Rifle Co.
Bullberry Barrel Works, Ltd.
Campbell, Dick
Carter's Gun Shop
Champlin Firearms, Inc.
Checkmate Refinishing
Chicasaw Gun Works
Christman Jr., Gunmaker, David
Classic Guns, Inc.
Cochran, Oliver
Colonial Repair
Colorado Gunsmithing Academy
Craftguard
Crandall Tool & Machine Co.
Cullity Restoration, Daniel
Custom Gun Products
Custom Gunsmiths
Custom Shop, The
D&D Gunsmiths, Ltd.
D&H Precision Tooling
DAMASCUS-U.S.A.
Delorge, Ed
DGR Custom Rifles
DGS, Inc.
Dietz Gun Shop & Range, Inc.
Duane's Gun Repair
Duncan's Gunworks, Inc.
Eversull Co., Inc., K.
Eyster Heritage Gunsmiths, Inc., Ken
Ferris Firearms
Forster, Larry L.
Forthofer's Gunsmithing
  & Knifemaking
Francesca, Inc.
Frank Custom Classic Arms, Ron
Fullmer, Geo. M.
Gentry Custom Gunmaker, David
Gilkes, Anthony W.
Gordie's Gun Shop
Grace, Charles E.
Graybill's Gun Shop
Green, Roger M.
Griffin & Howe, Inc.
Gun Shop, The
Guns
Hagn Rifles & Actions, Martin
Hamilton, Alex B.
Hart & Son, Inc., Robert W.
Hartmann & Weiss GmbH
Harwood, Jack O.
Hecht, Hubert J.
Heilmann, Stephen
Heritage Wildlife Carvings
Highline Machine Co.

Hiptmayer, Armurier
Hiptmayer, Klaus
Hoag, James W.
Hoelscher, Virgil
Holland's
Hollis Gun Shop
Hyper-Single, Inc.
Island Pond Gun Shop
Ivanoff, Thomas G.
J&S Heat Treat
Jaeger, Inc., Paul/Dunn's
Jamison's Forge Works
Jeffredo Gunsight
Johnston, James
KDF, Inc.
Ken's Gun Specialties
Kilham & Co.
Klein Custom Guns, Don
Kleinendorst, K.W.
Kopp, Terry K.
Lampert, Ron
Lawson Co., Harry
List Precision Engineering
Mac's .45 Shop
Makinson, Nicholas
McCament, Jay
McCann's Machine & Gun Shop
McFarland, Stan
Morrison Custom Rifles, J.W.
Morrow, Bud
Mullis Guncraft
Nelson, Stephen
Nettestad Gun Works
New England Custom Gun Service
Nicholson Custom
Nitex, Inc.
Noreen, Peter H.
North Fork Custom Gunsmithing
Nu-Line Guns, Inc.
Olson, Vic
Ozark Gun Works
P&S Gun Service
Pagel Gun Works, Inc.
Parker Gun Finishes
Pasadena Gun Center
Penrod Precision
Precision Metal Finishing
Precise Metalsmithing Enterprises
Precision Metal Finishing,
  John Westrom
Precision Specialties
Rice, Keith
Rifles Inc.
River Road Sporting Clays
Robar Co.'s, Inc., The
Rocky Mountain Arms, Inc.
Score High Gunsmithing
Simmons Gun Repair, Inc.
Sipes Gun Shop
Skeoch, Brian R.
Smith, Art
Snapp's Gunshop
Spencer's Custom Guns
Sportsmen's Exchange & Western
  Gun Traders, Inc.
Starnes Gunmaker, Ken
Steffens, Ron
Steger, James R.
Stiles Custom Guns
Storey, Dale A.
Strawbridge, Victor W.
Ten-Ring Precision, Inc.
Thompson, Randall
Tom's Gun Repair
Tooley Custom Rifles
Van Horn, Gil
Van Patten, J.W.
Von Minden Gunsmithing Services

DECOYS

A&M Waterfowl, Inc.
Baekgaard Ltd.
Boyds' Gunstock Industries, Inc.
Carry-Lite, Inc.
Deer Me Products Co.
Fair Game International
Farm Form Decoys, Inc.
Feather Flex Decoys
Flambeau Products Corp.
G&H Decoys, Inc.
Herter's Manufacturing, Inc.
Hiti-Schuch, Atelier Wilma

Klingler Woodcarving
L.L. Bean, Inc.
Molin Industries
North Wind Decoy Co.
Penn's Woods Products, Inc.
Russ Trading Post
Quack Decoy & Sporting Clays
Sports Innovations, Inc.
Tanglefree Industries
Waterfield Sports, Inc.
Woods Wise Products

ENGRAVERS, ENGRAVING TOOLS

Ackerman & Co.
Adair Custom Shop, Bill
Adams, John J. & Son Engravers
Adams Jr., John J.
Ahlman Guns
Alfano, Sam
Allard, Gary
Allen Firearm Engraving
Altamont Co.
American Pioneer Video
Anthony and George Ltd.
Baron Technology
Barraclough, John K.
Bates Engraving, Billy
Bell, Sid
Blair Engraving, J.R.
Bleile, C. Roger
Boessler, Erich
Bone Engraving, Ralph
Bratcher, Dan
Brgoch, Frank
Brooker, Dennis
Brownell Checkering Tools, W.E.
Burgess, Byron
CAM Enterprises
Churchill, Winston
Clark Firearms Engraving
Collings, Ronald
Creek Side Metal & Woodcrafters
Cullity Restoration, Daniel
Cupp, Custom Engraver, Alana
Custom Gun Engraving
DAMASCUS-U.S.A.
Davidson, Jere
Dayton Traister
Delorge, Ed
Desquesnes, Gerald
Dixon Muzzleloading Shop, Inc.
Dolbare, Elizabeth
Drain, Mark
Dubber, Michael W.
Engraving Artistry
Evans Engraving, Robert
Eversull Co., Inc., K.
Eyster Heritage Gunsmiths, Inc., Ken
Fanzoj GmbH
Firearms Engraver's Guild of America
Flannery Engraving Co., Jeff W.
Forty Five Ranch Enterprises
Fountain Products
Francotte & Cie S.A., Auguste
Frank Knives
French, Artistic Engraving, J.R.
Gene's Custom Guns
George, Tim
Glimm, Jerome C.
Golden Age Arms Co.
Gournet, Geoffroy
Grant, Howard V.
Griffin & Howe, Inc.
GRS Corp., Glendo
Gun Room, The
Guns
Gurney, F.R.
Gwinnell, Bryson J.
Hale/Engraver, Peter
Half Moon Rifle Shop
Hands Engraving, Barry Lee
Harris Gunworks
Harris Hand Engraving, Paul A.

Harwood, Jack O.
Hawken Shop, The
Hendricks, Frank E.
Heritage Wildlife Carvings
Hiptmayer, Armurier
Hiptmayer, Heidemarie
Horst, Alan K.
Ingle, Engraver, Ralph W.
Jaeger, Inc., Paul/Dunn's
Jantz Supply
Johns Master Engraver, Bill
Kamyk Engraving Co., Steve
Kane, Edward
Kehr, Roger
Kelly, Lance
Klingler Woodcarving
Koevenig's Engraving Service
Kudlas, John M.
LeFever Arms Co., Inc.
Leibowitz, Leonard
Lindsay, Steve
Little Trees Ramble
Lutz Engraving, Ron
Master Engravers, Inc.
McCombs, Leo
McDonald, Dennis
McKenzie, Lynton
Mele, Frank
Metals Hand Engraver
Mittermeier, Inc., Frank
Montgomery Community College
Moschetti, Mitchell R.
Mountain States Engraving
Nelson, Gary K.
New England Custom Gun Service
New Orleans Jewelers Supply Co.
NgraveR Co., The
Oker's Engraving
P&S Gun Service
Pedersen, C.R.
Pedersen, Rex C.
Pilgrim Pewter, Inc.
Pilkington, Scott
Piquette, Paul R.
Potts, Wayne E.
Rabeno, Martin
Reed, Dave
Reno, Wayne
Riggs, Jim
Roberts, J.J.
Rohner, Hans
Rohner, John
Rosser, Bob
Rundell's Gun Shop
Runge, Robert P.
Sampson, Roger
Schiffman, Mike
Sherwood, George
Singletary, Kent
Smith, Mark A.
Smith, Ron
Smokey Valley Rifles
Theis, Terry
Thiewes, George W.
Thirion Gun Engraving, Denise
Thompson/Center Arms
Valade Engraving, Robert
Vest, John
Viramontez, Ray
Vorhes, David

DIRECTORY OF THE ARMS TRADE

Wagoner, Vernon G.
Wallace, Terry
Warenski, Julie
Warren, Kenneth W.
Weber & Markin Custom Gunsmiths
Welch, Sam
Wells, Rachel
Wessinger Custom Guns & Engraving
Wood, Mel
Yee, Mike

GAME CALLS

Adventure Game Calls
Arkansas Mallard Duck Calls
Ashby Turkey Calls
Bostick Wildlife Calls, Inc.
Carter's Wildlife Calls, Inc., Garth
Cedar Hill Game Calls, Inc.
Crit'R Call
Custom Calls
D&H Prods. Co., Inc.
D-Boone Ent., Inc.
Deepeeka Exports Pvt. Ltd.
Dr. O's Products Ltd.
Duck Call Specialists
Faulhaber Wildlocker
Faulk's Game Call Co., Inc.
Flow-Rite of Tennessee, Inc.
Gander Mountain, Inc.
Green Head Game Call Co.
Hally Caller
Haydel's Game Calls, Inc.
Herter's Manufacturing, Inc.
Hunter's Specialties, Inc.
Keowee Game Calls
Kingyon, Paul L.
Knight & Hale Game Calls
Lohman Mfg. Co., Inc.
Mallardtone Game Calls
Marsh, Johnny
Moss Double Tone, Inc.
Mountain Hollow Game Calls
Oakman Turkey Calls
Olt Co., Philip S.
Penn's Woods Products, Inc.
Primos, Inc.
Quaker Boy, Inc.
Rickard, Inc., Pete
Rocky Mountain Wildlife Products
Russ Trading Post
Salter Calls, Inc., Eddie
Sceery Game Calls
Scobey Duck & Goose Calls, Glynn
Scruggs' Game Calls, Stanley
Simmons Outdoor Corp.
Sports Innovations, Inc.
Stewart Game Calls, Inc., Johnny
Sure-Shot Game Calls, Inc.
Tanglefree Industries
Tink's & Ben Lee Hunting Products
Tink's Safariland Hunting Corp.
Wellington Outdoors
Wilderness Sound Products Ltd.
Woods Wise Products
Wyant's Outdoor Products, Inc.

GUN PARTS, U.S. AND FOREIGN

A.A. Arms, Inc.
Accuracy Gun Shop
Actions by "T"
Ahlman Guns
Amherst Arms
Armscorp USA, Inc.
Aro-Tek, Ltd.
Auto-Ordnance Corp.
Badger Shooters Supply, Inc.
Bear Mountain Gun & Tool
Billings Gunsmiths, Inc.
Bob's Gun Shop
Bowen Classic Arms Corp.
Briese Bullet Co., Inc.
British Antiques
Buffer Technologies
Bushmaster Firearms
Bustani, Leo
Cape Outfitters
Caspian Arms Ltd.
Chicasaw Gun Works
Clark Custom Guns, Inc.
Cochran, Oliver
Cole's Gun Works
Colonial Repair
Cylinder & Slide, Inc.
Dayton Traister
Delta Arms Ltd.
DGR Custom Rifles
Dibble, Derek A.
Duane's Gun Repair
Duffy, Charles E.
Dyson & Son Ltd., Peter
E&L Mfg., Inc.
E.A.A. Corp.
EGW Evolution Gun Works
Elliott Inc., G.W.
EMF Co., Inc.
Enguix Import-Export
European American Armory Corp.
Fleming Firearms
Forrest, Inc., TomGalati International
Gentry Custom Gunmaker, David
Glimm, Jerome C.
Goodwin, Fred
Greider Precision
Groenewold, John
Gun Parts Corp., The
Gun Shop, The
Guns Antique & Modern DBA/Charles E. Duffy
Gunsmithing, Inc.
Gun-Tec
Hastings Barrels
Hawken Shop, The
High Performance International
Irwin, Campbell H.
I.S.S.
Jaeger, Inc., Paul/Dunn's
Jamison's Forge Works
Johnson's Gunsmithing, Inc., Neal G
J.R. Distributing (Wolf competiition guns)
K&T Co.
Kimber of America, Inc.
K.K. Arms Co.
Krico Jagd-und Sportwaffen GmbH
Laughridge, William R.
List Precision Engineering
Ljutic Industries, Inc.
Lodewick, Walter H.
Long, George F.
Lothar Walther Precision Tool, Inc.
Mac's .45 Shop
Mandall Shooting Supplies, Inc.
Markell, Inc.
Martin's Gun Shop
Martz, John V.
McCormick Corp., Chip
MCS, Inc.
Merkuria Ltd.
Mid-America Recreation, Inc.
Mo's Competitor Supplies
Morrow, Bud
Mountain Bear Rifle Works, Inc.
NCP Products, Inc.
North Star West
Nu-Line Guns, Inc.
Olympic Arms
Pachmayr Ltd.
Parts & Surplus
Pennsylvania Gun Parts
Perazone, Brian
Perazzi USA, Inc.
Performance Specialists
Peterson Gun Shop, Inc., A.W.
Pre-Winchester 92-90-62 Parts Co.
P.S.M.G. Gun Co.
Quality Firearms of Idaho, Inc.
Quality Parts Co.
Randco UK
Ravell Ltd.
Retting, Inc., Martin B.
R.G.-G., Inc.
Ruger
S&S Firearms
Sabatti S.R.L.
Sarco, Inc.
Scherer
Shockley, Harold H.
Shootin' Shack, Inc.
Silver Ridge Gun Shop
Simmons Gun Repair, Inc.
Sipes Gun Shop
Smires, C.L.
Smith & Wesson
Southern Ammunition Co., Inc.
Southern Armory, The
Sportsmen's Exchange & Western Gun Traders, Inc.
Springfield, Inc.
Springfield Sporters, Inc.
Starr Trading Co., Jedediah
Steyr Mannlicher AG & CO KG
Sturm, Ruger & Co., Inc.
"Su-Press-On," Inc.
Swampfire Shop, The
Tank's Rifle Shop
Tarnhelm Supply Co., Inc.
Triple-K Mfg. Co., Inc.
Twin Pine Armory
USA Sporting Inc.
Vintage Arms, Inc.
Volquartsen Custom Ltd.
Walker Arms Co., Inc.
Waller & Son, Inc. W.
Weaver Arrms Corp. Gun Shop
Wescombe, Bill
Westfield Engineering
Whitestone Lumber Corp.
Williams Mfg. of Oregon
Winchester Sutler, Inc., The
Wise Guns, Dale
Wolff Co., W.C.

GUNS, AIR

Air Arms
Air Venture
Airrow
Allred Bullet Co.
Anschutz GmbH
Arms Corporation of the Philippines
Arms United Corp.
Baikal
Beeman Precision Airguns
Benjamin/Sheridan Co.
Brass Eagle, Inc.
Brocock Ltd.
BSA Guns Ltd.
Compasseco, Ltd.
Component Concepts, Inc.
Creedmoor Sports, Inc.
Crosman Airguns
Crosman Products of Canada Ltd.
Daisy Mfg. Co.
Daystate Ltd.
Diana
Dynamit Nobel-RWS, Inc.
FAS
Frankonia Jagd
FWB
Gamo USA, Inc.
Gaucher Armes, S.A.
Great Lakes Airguns
Hebard Guns, Gil
Hofmann & Co.
Interarms
Labanu, Inc.
List Precision Engineering
Mac-1 Distributors
Marksman Products
Maryland Paintball Supply
Merkuria Ltd.
Pardini Armi Srl
Park Rifle Co., Inc.
Penguin Industries, Inc.
Precision Airgun Sales, Inc.
Precision Sales Int'l., Inc.
Ripley Rifles
Robinson, Don
RWS
S.G.S. Sporting Guns Srl
SKAN A.R.
Smart Parts
Steyr Mannlicher AG & CO KG
Stone Enterprises Ltd.
Swivel Machine Works, Inc.
Theoben Engineering
Tippman Pneumatics, Inc.
Tristar Sporting Arms, Ltd.
Trooper Walsh
UltraSport Arms, Inc.
Valor Corp.
Vortek Products
Walther GmbH, Carl
Webley and Scott Ltd.
Weihrauch KG, Hermann
Whiscombe
World Class Airguns

GUNS, FOREIGN—IMPORTERS (Manufacturers)

Accuracy International (Anschutz GmbH target rifles)
AcuSport Corporation (Anschutz GmbH)
Air Rifle Specialists (airguns)
Air Venture (airguns)
American Arms, Inc. (Fausti Cav. Stefano & Figlie snc; Franchi S.p.A.; Grulla Armes; Uberti, Aldo; Zabala Hermanos S.A.; blackpowder arms)
American Frontier Firearms Mfg. Inc. (single-action revolvers)
Amtec 2000, Inc. (Erma Werke GmbH)
Anics Firm, Inc. (Anics)
Arms United Corp. (Gamo)
Armsport, Inc. (Bernardelli S.p.A., Vincenzo)
Aspen Outfitting Co. (Ugartechea S.A., Ignacio)
Auto-Ordnance Corp. (Techno Arms)
Autumn Sales, Inc. (Blaser Jagdwaffen GmbH)
Beauchamp & Son, Inc. (Pedersoli and Co., Davide)
Beeman Precision Airguns (Beeman Precision Airguns, Inc.; FWB; Webley & Scott Ltd.; Weihrauch KG, Hermann)
Bell's Legendary Country Wear (Miroku, B.C/Daly, Charles)
Beretta U.S.A. Corp. (Beretta S.p.A., Pietro)
Big Bear Arms & Sporting Goods, Inc. (Russian/Big Bear Arms)
Bohemia Arms Co. (BRNO)
British Sporting Arms
Browning Arms Co. (Browning Arms Co.)
B-West Imports, Inc.
Cabela's (Pedersoli and Co., Davide; Uberti, Aldo; blackpowder arms)
Cape Outfitters (Armi Sport; Pedersoli and Co., Davide; San Marco; Societa Armi Bresciane Srl.; blackpowder arms)
Century International Arms, Inc. (FEG)
Champion Shooters' Supply (Anschutz GmbH)
Champion's Choice (Anschutz GmbH, Walther GmbH, Carl; target rifles)
Chapuis USA (Chapuis Armes)
Champlin Firearms, Inc. (Chapuis Armes; M.Thys)
Christopher Firearms Co., Inc., E.
Cimarron Arms (Uberti, Aldo; Armi San Marco; Pedersoli)
CVA (blackpowder arms)
CZ USA
Daisy Mfg. Co. (Daisy Mfg. Co.; Gamo)
Dixie Gun Works, Inc. (Pedersoli and Co., Davide; Uberti, Aldo; blackpowder arms)
Dynamit Nobel-RWS, Inc. (Brenneke KG, Wilhelm; Diana; Gamo; Norma Precision AB; RWS)
E.A.A. Corp. (Astra-Sport, S.A.; Sabatti S.r.l.; Tanfoglio Fratelli S.r.l.; Weihrauch KG, Hermann; Star Bonifacio Echeverria S.A.)
Eagle Imports, Inc. (Bersa S.A.)
Ellett Bros. (Churchill)
EMF Co., Inc. (Dakota; Hartford; Pedersoli and Co., Davide; San Marco; Uberti, Aldo; blackpowder arms)
Euroarms of America, Inc. (blackpowder arms)
Eversull Co., Inc., K.
Fiocchi of America, Inc. (Fiocchi Munizioni S.p.A.)
Forgett Jr., Valmore J. (Navy Arms Co.; Uberti, Aldo)
Franzen International, Inc. (Peters Stahl GmbH)
Galaxy Imports Ltd., Inc. (Hanus Birdguns, Bill; Ignacio Ugartechea S.A.; Laurona Armas Eibar, S.A.D.)
Gamba, USA (Societa Armi Bresciane Srl.)
Gamo USA, Inc. (Gamo airguns)
Giacomo Sporting, Inc.
Glock, Inc. (Glock GmbH)

PRODUCT DIRECTORY

Great Lakes Airguns (air pistols & rifles)
Griffin & Howe, Inc. (Arrieta, S.L.)
Groenewold, John (BSA Guns Ltd.; Webley & Scott Ltd.)
GSI, Inc. (Mauser Werke Oberndorf; Merkel Freres; Steyr-Mannlicher AG)
G.U., Inc. (New SKB Arms Co.; SKB Arms Co.)
Gun Shop, The (Ugartechea S.A., Ignacio)
Gunsite Custom Shop (Accuracy International Precision Rifles)
Gunsite Training Center (Accuracy International Precision Rifles)
Gunsmithing, Inc. (Anschutz GmbH)
Hammerli USA (Hammerli Ltd.)
Hanus Birdguns, Bill (Ugartechea S.A., Ignacio)
Heckler & Koch, Inc. (Benelli Armi S.p.A.; Heckler & Koch, GmbH)
IAR, Inc. (Uberti, Kimar, Armi San Marco, S.I.A.C.E.)
Imperial Magnum Corp. (Imperial Magnum Corp.)
Import Sports Inc. (Llama Gabilondo Y Cia)
Interarms (Helwan; Howa Machinery Ltd.; Interarms; Korth; Norinco; Rossi S.A., Amadeo; Star Bonifacio Echeverria S.A.; Walther GmbH, Carl)
Israel Arms International, Inc. (KSN Industries, Ltd.)
Ithaca Gun Co., LLC (Fabarm S.p.A.)
JägerSport, Ltd. (Voere-KGH m.b.H.)
J.R. Distributing (Wolf competition guns)
K.B.I., Inc. (FEG; Miroku, B.C./Daly, Charles)
Kemen American (Armas Kemen S.A.)
Keng's Firearms Specialty, Inc. (Lapua Ltd.; Ultralux)
Kongsberg America L.L.C. (Kongsberg)
Krieghoff International, Inc (Krieghoff Gun Co., H.)
K-Sports Imports, Inc.
Labanu, Inc. (Rutten; air rifles)
Lion Country Supply (Ugartechea S.A., Ignacio)
London Guns Ltd. (London Guns Ltd.)
Mac-1 Distributors
Magnum Research, Inc.
MagTech Recreational Products, Inc. (MagTech)
Mandall Shooting Supplies, Inc. (Arizaga; Atamec-Bretton; Cabanas; Crucelegui, Hermanos; Erma Werke GmbH; Firearms Co. Ltd./Alpine; Hammerli Ltd.; Korth; Krico Jagd-und Sportwaffen GmbH; Morini; SIG; Tanner; Zanoletti, Pietro; blackpowder arms)
Marx, Harry (FERLIB)
MCS, Inc. (Pardini)
MEC-Gar U.S.A., Inc. (MEC-Gar S.R.L.)
Moore & Co., Wm. Larkin (Garbi; Piotti; Rizzini, Battista; Rizzini F.lli)
Nationwide Sports Distributors, Inc. (Daewoo Precision Industries Ltd.)
Navy Arms Co. (Navy Arms Co. Pedersoli and Co., Davide; Pietta; Uberti, Aldo; blackpowder and cartridge arms)

GUNS, FOREIGN—MANUFACTURERS (Importers)

Accuracy International Precision Rifles (Gunsite Custom Shop; Gunsite Training Center)
Air Arms (World Class Airguns)
Anics (Anics Firm, Inc.)
Anschutz GmbH (Accuracy International; AcuSport Corporation; Champion Shooters' Supply; Champion's Choice; Gunsmithing, Inc.)
Arizaga (Mandall Shooting Supplies, Inc.)
Armas Kemen S.A. (Kemen America; USA Sporting Inc.)
Armi Perazzi S.p.A. (Perazzi USA, Inc.)
Armi San Marco (Taylor's & Co., Inc.; Cimarron Arms; IAR, Inc.)
Armi Sport (Cape Outfitters; Taylor's & Co., Inc.)
Arms Corporation of the Philippines
Arrieta, S.L. (Griffin & Howe, Inc. New England Arms Co.; The Orvis Co., Inc.; Quality Arms, Inc.; Wingshooting Adventures)
Astra Sport, S.A. (E.A.A. Corp.; P.S.M.G. Gun Co.)
Atamec-Bretton (Mandall Shooting Supplies, Inc.)
AYA (New England Custom Gun Service)
BEC Scopes (BEC, Inc.)
Beeman Precision Airguns, Inc. (Beeman Precision Airguns)
Benelli Armi S.p.A. (Heckler & Koch, Inc.; Whitestone Lumber Co.)
Beretta S.p.A., Pietro (Beretta U.S.A. Corp.)
Bernardelli S.p.A., Vincenzo (Armsport, Inc.)
Bersa S.A. (Eagle Imports, Inc.)
Bertuzzi (New England Arms Co.)
Blaser Jagdwaffen GmbH (Autumn Sales, Inc.)
Bondini Paolo (blackpowder arms)
Borovnik KG, Ludwig
Bosis (New England Arms Co.)
Brenneke KG, Wilhelm (Dynamit Nobel-RWS, Inc.)
BRNO (Bohemia Arms Co.)
Brocock Ltd.
Browning Arms Co. (Browning Arms Co.)
BSA Guns Ltd. (Groenewold, John; Precision Sales International, Inc.)
Cabanas (Mandall Shooting Supplies, Inc.)
CBC
Chapuis Armes (Champlin Firearms, Inc.; Chapuis USA)
Churchill (Ellett Bros.)
Cosmi Americo & Figlio s.n.c. (New England Arms Co.)
Crucelegui, Hermanos (Mandall Shooting Supplies, Inc.)
Daewoo Precision Industries Ltd. (Nationwide Sports Distributors, Inc.)
Daisy Mfg. Co. (Daisy Mfg. Co.)
Dakota (EMF Co. Inc.)
Diana (Dynamit Nobel-RWS, Inc.)
Dumoulin, Ernest (New England Arms Co.)
EAW (New England Custom Gun Service)
Effebi SNC-Dr. Franco Beretta (Nevada Cartridge Co.)
Erma Werke GmbH (Amtec 2000, Inc.; Mandall Shooting Supplies, Inc.)
Fabarm S.p.A. (Ithaca Gun Co., LLC)
F.A.I.R. Techni-Mec s.n.c.
FAS (Nygord Precision Products)
Fausti Cav. Stefano & Figlie snc (American Arms, Inc.)
FEG (Century International Arms, Inc.; K.B.I., Inc.)
FERLIB (Marx, Harry)
Fiocchi Munizioni S.p.A. (Fiocchi of America, Inc.)
Firearms Co. Ltd./Alpine (Mandall Shooting Supplies, Inc.)
FN Herstal
Franchi S.p.A (American Arms, Inc.)
FWB (Beeman Precision Airguns)
Gamba S.p.A.-Societa Armi Bresciane Srl., Renato (Gamba, USA)
Gamo (Arms United Corp.; Daisy Mfg. Co.; Dynamit Nobel-RWS, Inc.; Gamo USA, Inc.)
Garbi, Armas Urki (Moore & Co., Wm. Larkin)
Gaucher Armes S.A.
Glock GmbH (Glock, Inc.)
Grulla Armes (American Arms, Inc.)
Hammerli Ltd. (Hammerli USA; Mandall Shooting Supplies, Inc.; Sigarms, Inc.)
Hartford (EMF Co., Inc.)
Hartmann & Weiss GmbH
Heckler & Koch, GmbH (Heckler & Koch, Inc.)
Hege Jagd-u. Sporthandels, GmbH
Helwan (Interarms)
Holland & Holland Ltd.
Howa Machinery Ltd. (Interarms)
I.A.B. (Taylor's & Co., Inc.)
IGA (Stoeger Industries)
IMI
Imperial Magnum Corp. (Imperial Magnum Corp.)
Interarms (Interarms; P.S.M.G. Gun Co.)
JSL Ltd. (Specialty Shooters Supply, Inc.)
Kimar (IAR, Inc.)
Kongsberg (Kongsberg America L.L.C.)
Korth (Interarms; Mandall Shooting Supplies, Inc.)
Krico Jagd-und Sportwaffen GmbH (Mandall Shooting Supplies, Inc.)
Krieghoff Gun Co., H. (Krieghoff International, Inc.)
KSN Industries, Ltd. (Israel Arms International, Inc.)
Lakefield Arms Ltd. (Savage Arms, Inc.)
Lanber Armas, S.A.
Lapua Ltd. (Keng's Firearms Specialty, Inc.)
Laurona Armas Eibar S.A.D. (Galaxy Imports Ltd., Inc.)
Lebeau-Courally (New England Arms Co.)
Llama Gabilondo Y Cia (Import Sports Inc.)
London Guns Ltd. (London Guns Ltd.)
MagTech (MagTech Recreational Products, Inc.)
Marocchi F.lli S.p.A. (Precision Sales International, Inc.)
Mauser Werke Oberndorf (GSI, Inc.)
MEC-Gar S.R.L. (MEC-Gar M.S.A., Inc.)
Merkel Freres (GSI, Inc.)
Miroku, B.C./Daly, Charles (Bell's Legendary Country Wear; K.B.I., Inc.)
Morini (Mandall Shooting Supplies; Nygord Precision Products)
M.Thys (Champlin Firearms, Inc.)
Navy Arms Co. (Forgett Jr., Valmore J.; Navy Arms Co.)
New SKB Arms Co. (G.U., Inc.)
Norica, Avnda Otaola
Norinco (Century International Arms, Inc.; Interarms)
Norma Precision AB (Dynamit Nobel-RWS Inc.; The Paul Co., Inc.)
Para-Ordnance Mfg., Inc. (Para-Ordnance, Inc.)
Pardini Armi Srl. (Nygord Precision Products; MCS, Inc.)
Pedersoli and Co., Davide (Beauchamp & Son, Inc.; Cabela's; Cape Outfitters; Cimarron Arms; Dixie Gun Works, Inc.; EMF Co., Inc.; Navy Arms Co.; Track of the Wolf, Inc.)
Perugini-Visini & Co. s.r.l.
Peters Stahl GmbH (Franzen International, Inc.)
Pietta (Navy Arms Co.; Taylor's & Co., Inc.)
Piotti (Moore & Co., Wm. Larkin)
Powell & Son Ltd., William (Powell Agency, The, William)
Rigby & Co., John
Rizzini, Battista (Moore & Co., Wm. Larkin; New England Arms Co.)
Rizzini F.lli (Moore & Co., Wm. Larkin; New England Arms Co.)
Rossi S.A., Amadeo (Interarms)
Rutten (Labanu, Inc.)
RWS (Dynamit Nobel-RWS, Inc.)
Sabatti S.R.L. (E.A.A. Corp.)
Sako Ltd. (Stoeger Industries)
San Marco (Cape Outfitters; EMF Co., Inc.)
S.A.R.L. G. Granger
Sauer (Paul Co., The; Sigarms, Inc.)
Savage Arms (Canada), Inc. (Savage Arms, Inc.)
S.I.A.C.E. (IAR, Inc.)
SIG (Mandall Shooting Supplies, Inc.)
SIG-Sauer (Sigarms, Inc.)
Societa Armi Bresciane Srl. (Cape Outfitters; Gamba, USA)
Sphinx Engineering SA (Sphinx USA Inc.)
Springfield, Inc. (Springfield, Inc.)
Star Bonifacio Echeverria S.A. (E.A.A. Corp.; Interarms; P.S.M.G. Gun Co.)
Steyr-Mannlicher AG (GSI, Inc.; Nygord Precision Products)
Tanfoglio Fratelli S.r.l. (E.A.A. Corp.)
Tanner (Mandall Shooting Supplies, Inc.)
Taurus International Firearms (Taurus Firearms, Inc.)
Taurus S.A., Forjas
Techno Arms (Auto-Ordnance Corp.)
T.F.C. S.p.A.
Tikka (Stoeger Industries)
TOZ (Nygord Precision Products)
Turkish Firearms Corp. (Turkish Firearms Corp.)
Uberti, Aldo (American Arms, Inc.; Cabela's; Cimarron Arms; Dixie Gun Works, Inc.; EMF Co., Inc.; Forgett Jr., Valmore J.; IAR, Inc.; Navy Arms Co.; Taylor's & Co., Inc.; Uberti USA, Inc.)

Nevada Cartridge Co. (Effebi SNC-Dr. Franco Beretta)
New England Arms Co. (Arrieta, S.L.; Bertuzzi; Bosis; Cosmi Americo & Figlio s.n.c.; Dumoulin, Ernest; Lebeau-Courally; Rizzini, Battista; Rizzini F.lli)
New England Custom Gun Service (AYA; EAW)
Nygord Precision Products (FAS; Morini; Pardini Armi Srl; Steyr-Mannlicher AG; TOZ; Unique/M.A.P.F.)
OK Weber, Inc. (target rifles)
Orvis Co., Inc., The (Arrieta, S.L.)
Pachmayr Ltd.
Para-Ordnance, Inc. (Para-Ordnance Mfg., Inc.)
Paul Co., The (Norma Precision AB; Sauer)
Pelaire Products (Whiscombe air rifle)
Perazzi USA, Inc. (Armi Perazzi S.p.A.)
Powell Agency, William, The (William Powell & Son [Gunmakers] Ltd.)
Precision Sales International, Inc. (BSA Guns Ltd.; Marocchi o/u shotguns)
P.S.M.G. Gun Co. (Astra Sport, S.A.; Interarms; Star Bonifacio Echeverria S.A.; Walther GmbH, Carl)
Quality Arms, Inc. (Arrieta, S.L.)
Sarco, Inc.
Savage Arms, Inc. (Lakefield Arms Ltd.; Savage Arms [Canada], Inc.)
Schuetzen Pistol Works (Peters Stahl GmbH)
Sigarms, Inc. (Hammerli Ltd.; Sauer rifles; SIG-Sauer)
SKB Shotguns (SKB Arms Co.)
Specialty Shooters Supply, Inc. (JSL Ltd.)
Sphinx USA Inc. (Sphinx Engineering SA)
Springfield, Inc. (Springfield, Inc.)
Stoeger Industries (IGA; Sako Ltd.; Tikka; target pistols)
Stone Enterprises Ltd. (airguns)
Swarovski Optik North America Ltd.
Taurus Firearms, Inc. (Taurus International Firearms)
Taylor's & Co., Inc. (Armi San Marco; Armi Sport; I.A.B.; Pedersoli and Co., Davide; Pietta; Uberti, Aldo)
Track of the Wolf, Inc. (Pedersoli and Co., Davide)
Tradewinds, Inc. (blackpowder arms)
Tristar Sporting Arms, Ltd. (Turkish, German, Italian and Spanish made firearms)
Trooper Walsh
Turkish Firearms Corp. (Turkish Firearms Corp.)
Uberti USA, Inc. (Uberti, Aldo; blackpowder arms)
USA Sporting Inc. (Armas Kemen S.A.)
Vintage Arms, Inc.
Weatherby, Inc. (Weatherby, Inc.)
Westley Richards Agency USA (Westley Richards)
Whitestone Lumber Corp. (Heckler & Koch; Bennelli Armi S.p.A.)
Wingshooting Adventures (Arrieta, S.L.)
World Class Airguns (Air Arms)

30th EDITION, 1998 **299**

DIRECTORY OF THE ARMS TRADE

Ugartechea S.A., Ignacio (Aspen Outfitting Co.; Gun Shop, The; Hanus Birdguns, Bill; Lion Country Supply)
Ultralux (Keng's Firearms Specialty, Inc.)
Unique/M.A.P.F. (Nygord Precision Products)
Voere-KGH m.b.H. (JägerSport, Ltd.)
Walther GmbH, Carl (Champion's Choice; Interarms; P.S.M.G. Gun Co.)
Weatherby, Inc. (Weatherby, Inc.)
Webley & Scott Ltd. (Beeman Precision Airguns; Groenewold, John)
Weihrauch KG, Hermann (Beeman Precision Airguns; E.A.A. Corp.)
Westley Richards & Co. (Westley Richards Agency USA)
Whiscombe (Pelaire Products)
Wolf (J.R. Distributing)
Zabala, Hermanos S.A. (American Arms, Inc.)
Zanoletti, Pietro (Mandall Shooting Supplies, Inc.)
Zoli, Antonio
Wesson Firearms, Dan
Wildey, Inc.
Wilkinson Arms
Z-M Weapons

GUNS, U.S.-MADE

A.A. Arms, Inc.
Accu-Tek
Airrow
Allred Bullet Co.
American Arms, Inc.
American Derringer Corp.
American Frontier Firearms Co.
A.M.T.
ArmaLite, Inc.
A-Square Co., Inc.
Auto-Ordnance Corp.
Baer Custom, Inc., Les
Barrett Firearms Mfg., Inc.
Bar-Sto Precision Machine
Beretta S.p.A., Pietro
Beretta U.S.A. Corp.
Big Bear Arms & Sporting Goods, Inc.
Bond Arms, Inc.
Braverman Corp., R.J.
Brolin Arms
Brown Co., E. Arthur
Brown Products, Inc., Ed
Browning Arms Co. (Parts & Service)
Bushmaster Firearms
Calhoun Varmint Bullets, James
Calico Light Weapon Systems
Cape Outfitters
Casull Arms Corp.
Century Gun Dist., Inc.
Champlin Firearms, Inc.
Colt's Mfg. Co., Inc.
Competitor Corp., Inc.
Connecticut Shotgun Mfg. Co.
Connecticut Valley Classics
Coonan Arms
Cooper Arms
Cumberland Arms
Cumberland Mountain Arms
CVA
CVC
Dakota Arms, Inc.
Dangler, Homer L.
Davis Industries
Dayton Traister
Dixie Gun Works, Inc.
Downsizer Corp.
Eagle Arms, Inc.
Emerging Technologies, Inc.
Essex Arms
FN Herstal
Forgett Jr., Valmore J.
Fort Worth Firearms
Frank Custom Classic Arms, Ron
Freedom Arms, Inc.
Fullmer, Geo. M.
Galazan
Genecco Gun Works, K.
Gibbs Rifle Co., Inc.
Gilbert Equipment Co., Inc.
Gonic Arms, Inc.
Gunsite Custom Shop
Gunsite Gunsmithy
H&R 1871, Inc.
Harris Gunworks
Harrington & Richardson
Hawken Shop, The
Heritage Firearms
Heritage Manufacturing, Inc.
Hesco-Meprolight
High Standard Mfg. Co., Inc.
Hi-Point Firearms
HJS Arms, Inc.
Holston Ent. Inc.
H-S Precision, Inc.
IAR, Inc.
Imperial Russian Armory
Intratec
Ithaca Gun Co., LLC
Jones, J.D.
J.P. Enterprises, Inc.
JS Worldwide DBA
Kahr Arms
Kelbly, Inc.
Kel-Tec CNC Industries, Inc.
Kimber of America, Inc.
K.K. Arms Co.
Knight's Mfg. Co.
LaFrance Specialties
Lakefield Arms Ltd.
L.A.R. Mfg., Inc.
Laseraim, Inc.
Lever Arms Service Ltd.
Ljutic Industries, Inc.
Lorcin Engineering Co., Inc.
Mag-Na-Port International, Inc.
Magnum Research, Inc.
Marlin Firearms Co.
Maverick Arms, Inc.
McBros Rifle Co.
Miller Arms, Inc.
MKS Supply, Inc.
M.O.A. Corp.
Montana Armory, Inc.
Mossberg & Sons, Inc., O.F.
Mountain Rifles Inc.
New England Firearms
NCP Products, Inc.
North American Arms, Inc.
North Star West
Nowlin Mfg. Co.
Olympic Arms, Inc.
Oregon Arms, Inc.
Pacific Rifle Co.
Phillips & Rogers, Inc.
Phoenix Arms
Precision Small Arms
Professional Ordnance, Inc.
Quality Parts Co.
Raptor Arms Co., Inc.
Recoilless Technologies, Inc.
Remington Arms Co., Inc.
Republic Arms, Inc.
Rifle Works & Armory
Rocky Mountain Arms, Inc.
Rogue Rifle Co., Inc.
Rogue River Rifleworks
RPM
Ruger
Savage Arms (Canada), Inc.
Scattergun Technologies, Inc.
Seecamp Co., Inc., L.W.
Sharps Arms Co., Inc., C.
Shepherd & Turpin Dist. Company
Shiloh Rifle Mfg.
Small Arms Specialties
Smith & Wesson
Sporting Arms Mfg., Inc.
Springfield, Inc.
SSK Industries
STI International
Stoeger Industries
Sturm, Ruger & Co., Inc.
Sundance Industries, Inc.
Sunny Hill Enterprizes, Inc.
Survival Arms, Inc.
Swivel Machine Works, Inc.
Tar-Hunt Custom Rifles, Inc.
Texas Armory
Taurus Firearms, Inc.
Taylor & Robbins
Texas Longhorn Arms, Inc.
Thompson/Center Arms
Time Precision, Inc.
Tristar Sporting Arms, Ltd.
Ultra Light Arms, Inc.
UFA, Inc.
U.S. Repeating Arms Co.
Weatherby, Inc.
Wells, Fred F.
Wescombe, Bill
Wesson Firearms Co., Inc.

GUNS AND GUN PARTS, REPLICA AND ANTIQUE

Armi San Paolo
Auto-Ordnance Corp.
Bear Mountain Gun & Tool
Beauchamp & Son, Inc.
Billings Gunsmiths, Inc.
Bob's Gun Shop
British Antiques
Buckskin Machine Works
Buffalo Arms Co.
Burgess & Son Gunsmiths, R.W.
Cache La Poudre Rifleworks
Cape Outfitters
Chambers Flintlocks Ltd., Jim
Chicasaw Gun Works
Cochran, Oliver
Cogar's Gunsmithing
Cole's Gun Works
Colonial Arms, Inc.
Colonial Repair
Custom Riflestocks, Inc.
Dangler, Homer L.
Day & Sons, Inc., Leonard
Delhi Gun House
Delta Arms Ltd.
Dilliott Gunsmithing, Inc.
Dixon Muzzleloading Shop, Inc.
Dyson & Son., Ltd. Peter
Ed's Gun House
Flintlocks, Etc.
Forgett, Valmore J., Jr.
Getz Barrel Co.
Golden Age Arms Co.
Goodwin, Fred
Groenewold, John
Gun Parts Corp., The
Gun Works, The
Guns
Gun-Tec
Hastings Barrels
Hunkeler, A.
IAR, Inc.
Kokolus, Michael
Liberty Antique Gunworks
List Precision Engineering
L&R Lock Co.
Lucas, Edw. E.
Mandall Shooting Supplies, Inc.
Martin's Gun Shop
McKee Publications
McKinney, R.P.
Mountain Bear Rifle Works, Inc.
Mountain State Muzzleloading Supplies, Inc.
Munsch Gunsmithing, Tommy
Museum of Historical Arms, Inc.
Navy Arms Co.
Neumann GmbH
North Star West
October Country
Pasadena Gun Center
Pecatonica River Longrifle
PEM's Mfg. Co.
Pony Express Sport Shop, Inc.
Precise Metalsmithing Enterprises
Quality Firearms of Idaho, Inc.
Ranch Products
Randco UK
Ravell Ltd.
Retting, Inc., Martin B.
R.G.-G., Inc.
S&S Firearms
Sarco, Inc.
Shootin' Shack, Inc.
Silver Ridge Gun Shop
Simmons Gun Repair, Inc.
Southern Ammunition Co., Inc.
Starnes Gunmaker, Ken
Stott's Creek Armory, Inc.
Taylor's & Co., Inc.
Tennessee Valley Mfg.
Triple-K Mfg. Co., Inc.
Uberti USA, Inc.
Vintage Industries, Inc.
Vortek Products, Inc.
Walker Arms Co., Inc.
Weisz Parts
Wescombe, Bill
Winchester Sutler, Inc., The

GUNS, SURPLUS—PARTS AND AMMUNITION

Ad Hominem
Alpha 1 Drop Zone
Armscorp USA, Inc.
Arundel Arms & Ammunition, Inc., A.
Ballistica Maximus North
Bohemia Arms Co.
Bondini Paolo
Century International Arms, Inc.
Chuck's Gun Shop
Cole's Gun Works
Combat Military Ordnance Ltd.
Delta Arms Ltd.
Ed's Gun House
First, Inc., Jack
Flaig's
Fleming Firearms
Forgett, Valmore J., Jr.
Forrest, Inc., Tom
Frankonia Jagd
Fulton Armory
Garcia National Gun Traders, Inc.
Goodwin, Fred
Gun City
Gun Parts Corp., The
Hart & Son, Inc., Robert W.
Hege Jagd-u. Sporthandels, GmbH
Hofmann & Co.
Interarms
Jackalope Gun Shop
LaRocca Gun Works, Inc.
Lever Arms Service Ltd.
Lomont Precision Bullets
Mandall Shooting Supplies, Inc.
Navy Arms Co.
Nevada Pistol Academy Inc.
Oil Rod and Gun Shop
Paragon Sales & Services, Inc.
Parts & Surplus
Pasadena Gun Center
Perazone, Brian
Quality Firearms of Idaho, Inc.
Raptor Arms. Inc., Co.
Ravell Ltd.
Retting, Inc., Martin B.
Samco Global Arms, Inc.
San Francisco Gun Exchange
Sanders Custom Gun Service
Sarco, Inc.
Shootin' Shack, Inc.
Silver Ridge Gun Shop
Simmons Gun Repair, Inc.
Sportsmen's Exchange & Western Gun Traders, Inc.
Springfield Sporters, Inc.
Starnes Gunmaker, Ken
Tarnhelm Supply Co., Inc.
T.F.C. S.p.A.
Thurston Sports, Inc.
Westfield Engineering
Williams Shootin' Iron Service
Whitestone Lumber Corp.
GUNSMITHS, CUSTOM (see Custom Gunsmiths)
GUNSMITHS, HANDGUN (see Pistolsmiths)

GUNSMITH SCHOOLS

Bull Mountain Rifle Co.
Colorado Gunsmithing Academy
Colorado School of Trades
Cylinder & Slide, Inc.
Lassen Community College, Gunsmithing Dept.
Laughridge, William R.
Modern Gun Repair School
Montgomery Community College
Murray State College
North American Correspondence Schools
Nowlin Mfg. Co.
NRI Gunsmith School

PRODUCT DIRECTORY

Pennsylvania Gunsmith School
Piedmont Community College
Pine Technical College
Professional Gunsmiths of America
Southeastern Community College
Smith & Wesson

Spencer's Custom Guns
Trinidad State Junior College
   Gunsmithing Dept.
Wright's Hardwood Gunstock
   Blanks
Yavapai College

GUNSMITH SUPPLIES, TOOLS, SERVICES

Ace Custom 45's, Inc.
Actions by "T"
Aldis Gunsmithing & Shooting Supply
Allred Bullet Co.
American Frontier Firearms Co.
Baer Custom, Inc., Les
Bar-Sto Precision Machine
Bauska Barrels
Bear Mountain Gun & Tool
Belt MTN Arms
Bengtson Arms Co., L.
Biesen, Al
Biesen, Roger
Bill's Gun Repair
Blue Ridge Machinery & Tools, Inc.
Bowen Classic Arms Corp.
Break-Free, Inc.
Briganti, A.J.
Briley Mfg., Inc.
Brownells, Inc.
B-Square Co., Inc.
Bull Mountain Rifle Co.
Burgess & Son Gunsmiths, R.W.
Carbide Checkering Tools
Chapman Manufacturing Co.
Chem-Pak, Inc.
Choate Machine & Tool Co., Inc.
Chopie Mfg., Inc.
Chuck's Gun Shop
Clark Custom Guns, Inc.
Colonial Arms, Inc.
Colorado School of Trades
Conetrol Scope Mounts
Craig Custom Ltd.
CRR, Inc./Marble's Inc.
Cumberland Arms
Cumberland Mountain Arms
Custom Checkering Service
Custom Gun Products
D&J Bullet Co. & Custom Gun
   Shop, Inc.
Dakota Arms, Inc.
Decker Shooting Products
Dem-Bart Checkering Tools, Inc.
Dever Co., Jack
Dewey Mfg. Co., Inc., J.
Dremel Mfg. Co.
Du-Lite Corp.
Danuser Machine Co.
Dutchman's Firearms, Inc., The
Dyson & Son, Ltd., Peter
Echols & Co., D'Arcy
EGW Evolution Gun Works
Faith Associates, Inc.
FERLIB
Fisher, Jerry A.
Forgreens Tool Mfg., Inc.
Forkin, Ben
Forster, Kathy
Frazier Brothers Enterprises
Gentry Custom Gunmaker, David
Gilkes, Anthony W.
Grace Metal Products, Inc.
Greider Precision
Gr,-Tan Rifles
Gunline Tools
Gun-Tec
Half Moon Rifle Shop
Halstead, Rick
Hammon Custom Guns Ltd.
Hastings Barrels
Henriksen Tool Co., Inc.
High Performance International
Hoelscher, Virgil
Holland's
Huey Gun Cases
Ivanoff, Thomas G.
J&R Engineering
J&S Heat Treat
Jacobson, Teddy
Jantz Supply
JGS Precision Tool Mfg.
Kasenit Co., Inc.
KenPatable Ent., Inc.
Kimball, Gary
Kleinendorst, K.W.
Kmount

Kopp Professional Gunsmithing,
   Terry K.
Korzinek Riflesmith, J.
Kwik Mount Corp.
LaBounty Precision Reboring
Lawson Co., Harry
Lea Mfg. Co.
Lee's Red Ramps
Lee Supplies, Mark
List Precision Engineering
London Guns Ltd.
Mahovsky's Metalife
Marble Arms
Marsh, Mike
McKillen & Heyer, Inc.
Meier Works
Menck, Thomas W.
Metalife Industries
Metaloy Inc.
Michael's Antiques
MMC
Morrow, Bud
Mo's Competitor Supplies
Mowrey's Guns & Gunsmithing
N&J Sales
New England Custom Gun Service
Nowlin Custom Mfg.
Nu-Line Guns, Inc.
Ole Frontier Gunsmith Shop
Parker Gun Finishes
PEM's Mfg. Co.
Perazone, Brian
P.M. Enterprises, Inc.
Power Custom, Inc.
Practical Tools, Inc.
Precision Metal Finishing
Precision Specialties
Professional Gunsmiths of America
Prolix® Lubricants
Ransom International Corp.
Reardon Products
Rice, Keith
Robar Co.'s, Inc., The
Romain's Custom Guns, Inc.
Roto Carve
Royal Arms Gunstocks
Rusteprufe Laboratories
Scott, McDougall & Associates
Shooter's Choice
Simmons Gun Repair, Inc.
Smith Abrasives, Inc.
Spradlin's
Starrett Co., L.S.
Stiles Custom Guns
Sullivan, David S.
Texas Platers Supply
Theis, Terry
Time Precision, Inc.
Tom's Gun Repair
Track of the Wolf, Inc.
Trinidad State Junior College
   Gunsmithing Dept.
Trulock Tool
Turnbull Restoration, Doug
Van Gorden & Son, Inc., C.S.
Venco Industries, Inc.
Volquartsen Custom Ltd.
Warne Manufacturing Co.
Washita Mountain Whetstone Co.
Weaver Arms Corp. Gun Shop
Weigand Combat Handguns, Inc.
Welsh, Bud
Wessinger Custom Guns
   & Engraving
Westfield Engineering
Westrom, John
Westwind Rifles, Inc.
White Rock Tool & Die
Wilcox All-Pro Tools & Supply
Will-Burt Co.
Williams Gun Sight Co.
Williams Shootin' Iron Service
Willow Bend
Wilson's Gun Shop
Wise Guns, Dale
Wright's Hardwood Gunstock Blanks
Yavapai College

HANDGUN ACCESSORIES

A.A. Arms, Inc.
Ace Custom 45's, Inc.
Action Direct, Inc.
ADCO Sales, Inc.
Adventurer's Outpost
African Import Co.
Aimpoint U.S.A.
Aimtech Mount Systems
Ajax Custom Grips, Inc.
Alpha Gunsmith Division
American Derringer Corp.
American Frontier Firearms Co.
Arms Corporation of the Philippines
Aro-Tek, Ltd.
Astra Sport, S.A.
Baer Custom, Inc., Les
Bar-Sto Precision Machine
BEC, Inc.
Behlert Precision, Inc.
Blue and Gray Products, Inc.
Bond Custom Firearms
Bowen Classic Arms Corp.
Broken Gun Ranch
Brown Products, Inc., Ed
Brownells, Inc.
Bucheimer, J.M.
Bushmaster Firearms
Bushmaster Hunting & Fishing
Butler Creek Corp.
C3 Systems
Centaur Systems, Inc.
Central Specialties Ltd.
Clark Custom Guns, Inc.
Conetrol Scope Mounts
Craig Custom Ltd.
CRR, Inc./Marble's Inc.
D&L Industries
Dade Screw Machine Products
Dayson Arms Ltd.
Delhi Gun House
D.J. Marketing
Doskocil Mfg. Co., Inc
E&L Mfg., Inc.
E.A.A. Corp.
Eagle International Sporting Goods, Inc.
European American Armory Corp.
Faith Associates, Inc.
Feminine Protection, Inc.
Fisher Custom Firearms
Flashette Co.
Fleming Firearms
Flores Publications, Inc., J.
Frielich Police Equipment
FWB
Gage Manufacturing
Galati International
GALCO International Ltd.
G.G. & G.
Glock, Inc.
Greider Precision
Gremmel Enterprises
Gun Parts Corp., The
Gun-Alert
Gun-Ho Sports Cases
Hebard Guns, Gil
Heinie Specialty Products
Henigson & Associates
Hill Speed Leather, Ernie
H.K.S. Products
Hoppe's Div.
Hunter Co., Inc.
Impact Case Co.
Jarvis, Inc.

JB Custom
Jeffredo Gunsight
Jones, J.D.
J.P. Enterprises, Inc.
Jumbo Sports Products
KeeCo Impressions
KK Air International
Keller Co., The
King's Gun Works
K.K. Arms Co.
L&S Technologies, Inc.
Lee's Red Ramps
Lem Sports, Inc.
Loch Leven Industries
Lohman Mfg. Co., Inc.
Mac's .45 Shop
Mag-Na-Port International, Inc.
Magnolia Sports, Inc.
Marble Arms
Markell, Inc.
Maxi-Mount
MCA Sports
McCormick Corp., Chip
MEC-Gar S.R.L.
Merkuria Ltd.
Mid-America Guns and Ammo
Middlebrooks Custom Shop
Millett Sights
MTM Molded Products Co., Inc.
MCA Sports
Noble Co., Jim
No-Sho Mfg. Co.
Omega Sales
Ox-Yoke Originals, Inc.
PAST Sporting Goods, Inc.
Pearce Grip, Inc.
Penguin Industries, Inc.
Phoenix Arms
Power Custom, Inc.
Practical Tools, Inc.
Protector Mfg. Co., Inc., The
Quality Parts Co.
Ram-Line Blount, Inc.
Ranch Products
Ransom International Corp.
Recoilless Technologies, Inc.
Redfield, Inc.
Robar Co.'s, Inc., The
Round Edge, Inc.
RPM
Simmons Gun Repair, Inc.
Slings 'N Things, Inc.
Southwind Sanctions
SSK Industries
STI International
TacStar Industries, Inc.
TacTell, Inc.
Tanfoglio Fratelli S.r.l.
T.F.C. S.p.A.
Thompson/Center Arms
Trigger Lock Division
Trijicon, Inc.
Triple-K Mfg. Co., Inc.
Tyler Manufacturing & Distributing
Valor Corp.
Volquartsen Custom Ltd.
Waller & Son, Inc., W.
Weigand Combat Handguns, Inc.
Wessinger Custom Guns & Engraving
Western Design
Wichita Arms, Inc.
Wilson Gun Shop

HANDGUN GRIPS

A.A. Arms, Inc.
Ahrends, Kim
Ajax Custom Grips, Inc.
Altamont Co.
American Derringer Corp.
American Frontier Firearms Co.
American Gripcraft
Arms Corporation of the Philippines
Art Jewel Enterprises Ltd.
Baelder, Harry
Baer Custom, Inc., Les
Barami Corp.
Bear Hug Grips, Inc.
Big Bear Arms & Sporting Goods, Inc.
Boone's Custom Ivory Grips, Inc.
Boyds' Gunstock Industries, Inc.
Brooks Tactical Systems
Brown Products, Inc., Ed
CAM Enterprises

Cole-Grip
Colonial Repair
Custom Firearms
Dayson Arms Ltd.
E.A.A. Corp.
EMF Co., Inc.
Essex Arms
European American Armory Corp.
Eyears Insurance
Fisher Custom Firearms
Fitz Pistol Grip Co.
Forrest, Inc., Tom
FWB
Harrison-Hurtz Enterprises, Inc.
Herrett's Stocks, Inc.
Hogue Grips
Huebner, Corey O.
KeeCo Impressions
Knight's Mfg. Co.

30th EDITION, 1998 **301**

DIRECTORY OF THE ARMS TRADE

Korth
Lee's Red Ramps
Lett Custom Grips
Linebaugh Custom Sixgun
   & Rifle Works
Mac's .45 Shop
Masen Co., Inc., John
Michaels of Oregon Co.
Mid-America Guns and Ammo
Millett Sights
N.C. Ordnance Co.
Newell, Robert H.
Nickels, Paul R.
Pacific Rifle Co.
Pardini Armi Srl
Phoenix Arms

Pilgrim Pewter, Inc.
Radical Concepts
Recoilless Technologies, Inc.
Rosenberg & Sons, Jack A.
Roy's Custom Grips
Sile Distributors, Inc.
Smith & Wesson
Speedfeed, Inc.
Spegel, Craig
Stoeger Industries
Taurus Firearms, Inc.
Tyler Manufacturing & Distributing
Uncle Mike's
Vintage Industries, Inc.
Volquartsen Custom Ltd.
Western Gunstock Mfg. Co.

HEARING PROTECTORS

Aero Peltor
Ajax Custom Grips, Inc.
Autauga Arms, Inc.
Brown Co., E. Arthur
Brown Products, Inc., Ed
Browning Arms Co.
Clark Co., Inc., David
E-A-R, Inc.
Electronic Shooters Protection, Inc.
Faith Associates, Inc.
Flents Products Co., Inc.

Gentex Corp.
Hoppe's Div.
Kesselring Gun Shop
North Specialty Products
Paterson Gunsmithing
Peltor, Inc.
Penguin Industries, Inc.
R.E.T. Enterprises
Rucker Dist. Inc.
Silencio/Safety Direct
Willson Safety Prods. Div.

HOLSTERS AND LEATHER GOODS

A&B Industries, Inc.
Action Direct, Inc.
Action Products, Inc.
Aker Leather Products
Alessi Holsters, Inc.
American Sales & Kirkpatrick
Arratoonian, Andy
Bagmaster Mfg., Inc.
Baker's Leather Goods, Roy
Bandcor Industries
Bang-Bang Boutique
Barami Corp.
Bear Hug Grips, Inc.
Beretta S.p.A., Pietro
Bianchi International, Inc.
Bill's Custom Cases
Blocker Holsters, Inc., Ted
Brauer Bros. Mfg. Co.
Brown, H.R.
Browning Arms Co.
Bucheimer, J.M.
Bull-X, Inc.
Bushwacker Backpack & Supply Co.
Carvajal Belts & Holsters
Cathey Enterprises, Inc.
Chace Leather Products
Churchill Glove Co., James
Cimarron Arms
Clements' Custom Leathercraft, Chas
Cobra Sport
Colonial Repair
Counter Assault
Creedmoor Sports, Inc.
Davis Leather Co., G. Wm.
Delhi Gun House
DeSantis Holster & Leather
   Goods, Inc.
Dixie Gun Works, Inc.
D-Max, Inc.
Easy Pull Outlaw Products
Ekol Leather Care
El Dorado Leather (c/o Dill)
El Paso Saddlery Co.
EMF Co., Inc.
Eutaw Co., Inc., The
F&A Inc.
Faust, Inc., T.G.
Feminine Protection, Inc.
Ferdinand, Inc.
Flores Publications, Inc., J.
Fobus International Ltd.
Forgett Jr., Valmore J.
Frankonia Jagd
Gage Manufacturing
GALCO International Ltd.
GML Products, Inc.
Gould & Goodrich
Gun Leather Limited
Gunfitters, The
Gun Works, The
Gusty Winds Corp.
Hafner Creations, Inc.
HandiCrafts Unltd.
Hank's Gun Shop

Hebard Guns, Gil
Heinie Specialty Products
Hellweg Ltd.
Henigson & Associates, Steve
Hill Speed Leather, Ernie
Hofmann & Co.
Holster Shop, The
Horseshoe Leather Products
Hoyt Holster Co., Inc.
Hume, Don
Hunter Co., Inc.
John's Custom Leather
Jumbo Sports Products
Kane Products, Inc.
Keller Co., The
Kirkpatrick Leather Co.
Kolpin Mfg., Inc.
Korth
Kramer Handgun Leather, Inc.
L.A.R. Mfg., Inc.
Law Concealment Systems, Inc.
Lawrence Leather Co.
Leather Arsenal
Lone Star Gunleather
Magnolia Sports, Inc.
Markell, Inc.
Michaels of Oregon Co.
Minute Man High Tech Industries
Mixson Corp.
Noble Co., Jim
No-Sho Mfg. Co.
Null Holsters Ltd., K.L.
October Country
Ojala Holsters, Arvo
Oklahoma Leather Products, Inc.
Old West Reproductions, Inc.
Pathfinder Sports Leather
PWL Gunleather
Recoilless Technologies, Inc.
Renegade
Ringler Custom Leather Co.
Rybka Custom Leather Equipment, Thad
Safariland Ltd., Inc.
Safety Speed Holster, Inc.
Schulz Industries
Second Chance Body Armor
Shoemaker & Sons, Inc., Tex
ShurKatch Corporation
Silhouette Leathers
Smith Saddlery, Jesse W.
Southwind Sanctions
Sparks, Milt
Stalker, Inc.
Starr Trading Co., Jedediah
Strong Holster Co.
Stuart, V. Pat
Tabler Marketing
Texas Longhorn Arms, Inc.
Top-Line USA Inc.
Torel, Inc.
Triple-K Mfg. Co., Inc.
Tristar Sporting Arms, Ltd.
Tyler
Tyler Manufacturing & Distributing

Uncle Mike's
Valor Corp.
Venus Industries
Viking Leathercraft, Inc.
Walt's Custom Leather

HUNTING AND CAMP GEAR, CLOTHING, ETC.

A&M Waterfowl, Inc.
Ace Sportswear, Inc.
Action Direct, Inc.
Action Products, Inc.
Adventure Game Calls
Adventure 16, Inc.
Allen Co., Bob
Allen Sportswear, Bob
Armor
Atlanta Cutlery Corp.
Atsko/Sno-Seal, Inc.
Baekgaard Ltd.
Bagmaster Mfg., Inc.
Barbour, Inc.
Bauer, Eddie
Bear Archery
Beaver Park Products, Inc
Beretta S.p.A., Pietro
Better Concepts Co.
Big Beam Emergency Systems, Inc.
Boss Manufacturing Co.
Brown, H.R.
Brown Manufacturing
Browning Arms Co.
Buck Stop Lure Co., Inc.
Bushmaster Hunting & Fishing
Camofare Co.
Carhartt, Inc.
Catoctin Cutlery
Chippewa Shoe Co.
Churchill Glove Co., James
Clarkfield Enterprises, Inc.
Coghlan's Ltd.
Coleman Co., Inc.
Coulston Products, Inc.
Creedmoor Sports, Inc.
D&H Prods. Co., Inc.
Dakota Corp.
Danner Shoe Mfg. Co.
DeckSlider of Florida
Deer Me Products
Dr. O's Products Ltd.
Dunham Co.
Duofold, Inc.
Dynalite Products, Inc.
E-A-R, Inc.
Ekol Leather Care
Erickson's Mfg., Inc., C.W.
Eutaw Co., Inc., The
F&A Inc.
Flores Publications, Inc., J.
Flow-Rite of Tennessee, Inc.
Forrest Tool Co.
Fox River Mills, Inc.
G&H Decoys, Inc.
Game Winner, Inc.
Gander Mountain, Inc.
Gerber Legendary Blades
Glacier Glove
H&B Forge Co.
Hafner Creations, Inc.
Hinman Outfitters, Bob
Hodgman, Inc.
Houtz & Barwick
Hunter's Specialties, Inc.
Just Brass, Inc.
K&M Industries, Inc.

Westley Richards & Co.
Whinnery, Walt
Wild Bill's Originals
Wilson Gun Shop

Kamik Outdoor Footwear
Kolpin Mfg., Inc.
LaCrosse Footwear, Inc.
Langenberg Hat Co.
Lectro Science, Inc.
Liberty Trouser Co.
L.L. Bean, Inc.
MAG Instrument, Inc.
Mag-Na-Port International, Inc.
Marathon Rubber Prods. Co., Inc.
McCann's Machine & Gun Shop
Melton Shirt Co., Inc.
Molin Industries
Mountain Hollow Game Calls
Nelson/Weather-Rite, Inc.
North Specialty Products
Northlake Outdoor Footwear
Original Mink Oil, Inc.
Orvis Co., The
Outdoor Connection, Inc., The
Palsa Outdoor Products
Partridge Sales Ltd., John
Pointing Dog Journal
Porta Blind, Inc.
Powell & Son (Gunmakers) Ltd.,
   William
Pro-Mark
Pyromid, Inc.
Randolph Engineering, Inc.
Ranger Mfg. Co., Inc.
Ranging, Inc.
Rattlers Brand
Red Ball
Refrigiwear, Inc.
Re-Heater, Inc.
Rocky, Shoes & Boots
Ringler Custom Leather Co.
Scansport, Inc.
Sceery Game Calls
Servus Footwear Co.
ShurKatch Corporation
Silhouette Leathers
Simmons Outdoor Corp.
Slings 'N Things, Inc.
Sno-Seal
Streamlight, Inc.
Swanndri New Zealand
10-X Products Group
Thompson, Norm
T.H.U. Enterprises, Inc.
Tink's Safariland Hunting Corp.
Thompson/Center Arms
Torel, Inc.
TrailTimer Co.
Triple-K Mfg. Co., Inc.
Venus Industries
Wakina by Pic
Walker Co., B.B.
Walls Industries
Wilderness Sound Products Ltd.
Willson Safety Prods. Div.
Winchester Sutler, Inc., The
Wolverine Boots & Outdoor Footwear
   Division
Woolrich Woolen Mills
Wyoming Knife Corp.
Yellowstone Wilderness Supply

KNIVES AND KNIFEMAKER'S SUPPLIES FACTORY AND MAIL ORDER

Action Direct, Inc.
Adventure 16, Inc.
Aitor-Cuchilleria Del Norte, S.A.
All Rite Products, Inc.
American Target Knives
Aristocrat Knives
Art Jewel Enterprises Ltd.
Atlanta Cutlery Corp.
B&D Trading Co., Inc.
Barteaux Machetes, Inc.
Benchmark Knives
Beretta S.p.A., Pietro
Beretta U.S.A. Corp.
Big Bear Arms & Sporting Goods, Inc.
Bill's Custom Cases
Blackjack Knives, Ltd.
Boker USA, Inc.
Bowen Knife Co. Inc.

Brown, H.R.
Browning Arms Co.
Buck Knives, Inc.
Buster's Custom Knives
CAM Enterprises
Camillus Cutlery Co.
Case & Sons Cutlery Co., W.R.
Catoctin Cutlery
Chicago Cutlery Co.
Christopher Firearms Co., Inc., E.
Clements' Custom Leathercraft, Chas
Cold Steel, Inc.
Coleman Co., Inc.
Colonial Knife Co., Inc.
Compass Industries, Inc.
Creative Craftsman, Inc., The
Crosman Blades
CRR, Inc./Marble's Inc.

PRODUCT DIRECTORY

Cutco Cutlery
Cutlery Shoppe
DAMASCUS-U.S.A.
Dan's Whetstone Co., Inc.
Degen Inc.
Delhi Gun House
DeSantis Holster & Leather Goods, Inc.
Diamontd Machining Technology, Inc.
EdgeCraft Corp./P.B. Tuminello
EK Knife Co.
Empire Cutlery Corp.
Eze-Lap Diamond Prods.
Flitz International Ltd.
Flores Publications, Inc., J.
Forrest Tool Co.
Forthofer's Gunsmithing & Knifemaking
Fortune Products, Inc.
Frank Knives
Frost Cutlery Co.
Gerber Legendary Blades
Gibbs Rifle Co., Inc.
Glock, Inc.
Golden Age Arms Co.
Gun Room, The
H&B Forge Co.
HandiCrafts Unltd.
Harrington Cutlery, Inc., Russell
Harris Publications
Henckels Zwillingswerk, Inc., J.A.
High North Products, Inc.
Hoppe's Div.
Hubertus Schneidwarenfabrik
Hunter Co., Inc.
Hunting Classics
Ibberson (Sheffield) Ltd., George
Imperial Schrade Corp.
Iron Mountain Knife Co.
J.A. Blades, Inc.
Jackalope Gun Shop
Jantz Supply
Jenco Sales, Inc.
Johnson Wood Products
KA-BAR Knives
Kasenit Co., Inc.
Kershaw Knives
Knife Importers, Inc.
Koval Knives
Lamson & Goodnow Mfg. Co.
Leatherman Tool Group, Inc.
Linder Solingen Knives
Marble Arms
Mar Knives, Inc., Al
Matthews Cutlery
McCann's Machine & Gun Shop
Molin Industries
Mountain State Muzzleloading Supplies, Inc.
Murphy Co., Inc., R.
Normark Corp.
October Country
Outdoor Edge Cutlery Corp.
Penguin Industries, Inc.
Pilgrim Pewter, Inc.
Plaza Cutlery, Inc.
Precise International
Queen Cutlery Co.
R&C Knives & Such
Randall-Made Knives
Rodgers & Sons Ltd., Joseph
Russell Knives, Inc., A.G.
Scansport, Inc.
Schiffman, Mike
Schrimsher's Custom Knifemaker's Supply, Bob
Sheffield Knifemakers Supply, Inc.
Shell Shack
Silhouette Leathers
Smith Saddlery, Jesse W.
Soque River Knives
Spyderco, Inc.
Starr Trading Co., Jedediah
Swiss Army Knives, Inc.
T.F.C. S.p.A.
Theis, Terry
Traditions, Inc.
Tru-Balance Knife Co.
United Cutlery Corp.
Utica Cutlery Co.
Venus Industries
Walt's Custom Leather
Washita Mountain Whetstone Co.
Weber Jr., Rudolf
Wells Creek Knife & Gun Works
Western Cutlery Co.
Whinnery, Walt
Wostenholm
Wyoming Knife Corp.

LABELS, BOXES, CARTRIDGE HOLDERS

Ballistic Products, Inc.
Berry's Mfg., Inc.
Brown Co., E. Arthur
Cabinet Mountain Outfitters Scents & Lures
Cape Outfitters
Crane & Crane Ltd.
Del Rey Products
DeSantis Holster & Leather Goods, Inc.
Fitz Pistol Grip Co.
Flambeau Products Corp.
J&J Products Co.
Kolpin Mfg., Inc.
Liberty Shooting Supplies
Midway Arms, Inc.
MTM Molded Products Co., Inc.
Pendleton Royal

LOAD TESTING AND PRODUCT TESTING, (Chronographing, Ballistic Studies)

Ballistic Research
Bartlett, Don
Briese Bullet Co., Inc.
Buck Stix—SOS Products Co.
Clearview Products
Clerke Co., J.A.
D&H Precision Tooling
Defense Training International, Inc.
DGR Custom Rifles
Duane's Gun Repair
Henigson & Associates, Steve
Hensler, Jerry
Hoelscher, Virgil
Jackalope Gun Shop
Jensen Bullets
Lomont Precision Bullets
Maionchi-L.M.I.
MAST Technology
McMurdo, Lynn
Middlebrooks Custom Shop
Multiplex International
Oil Rod and Gun Shop
Rupert's Gun Shop
SOS Products Co.
Spencer's Custom Guns
Vancini, Carl
Vulpes Ventures, Inc.
Wells Custom Gunsmith, R.A.
White Laboratory, Inc., H.P.
X-Spand Target Systems

MISCELLANEOUS

Actions, Rifle
 Hall Manufacturing
Accurizing, Rifle
 Richards, John
 Stoney Baroque Shooters Supply
Adapters, Cartridge
 Alex, Inc.
Adapters, Shotshell
 PC Co.
Airgun Accessories
 BSA Guns Ltd.
 Groenewold, John
Airgun Repair
 Airgun Repair Centre
 Groenewold, John
 Nationwide Airgun Repairs
Assault Rifle Accessories
 Ram-Line Blount, Inc.
Barrel Stress Relieving
 300° Below Services
 Cryo-Accurizing
Bi-Pods
 B.M.F. Activator, Inc.
Body Armor
 A&B Industries, Inc.
 Faust, Inc., T.G.
 Second Chance Body Armor
 Top-Line USA Inc.
Bore Illuminator
 Flashette Co.
Bore Lights
 MDS, Inc.
Brass Catcher
 Gage Manufacturing
 M.A.M. Products, Inc.
Bullets, Rubber
 CIDCO
Bullets, Moly Coat
 NECO
 Starke Bullet Company
Calendar, Gun Shows
 Stott's Creek Printers
Cannons, Miniature Replicas
 Furr Arms
Dehumidifiers
 Buenger Enterprises
 Hydrosorbent Products
Dryers
 Peet Shoe Dryer, Inc.
E-Z Loader
 Del Rey Products
Field Carts
 Ziegel Engineering
Firearm Refinishers
 Armoloy Co. of Ft. Worth
Firearm Restoration
 Adair Custom Shop, Bill
 Burgess & Son Gunsmiths, R.W.
 Johns Master Engraver, Bill
 Mazur Restoration, Pete
 Moeller, Steve
 Nicholson Custom
FFL Record Keeping
 Basics Information Systems, Inc.
 PFRB Co.
 R.E.T. Enterprises
Hunting Trips
 J/B Adventures & Safaris, Inc.
 Professional Hunter Specialties
 Safaris Plus
 Wild West Guns
Hypodermic Rifles/Pistols
 Multipropulseurs
Industrial Dessicants
 WAMCO—New Mexico
Insert Barrels
 MCA Sports
Lettering Restoration System
 Pranger, Ed G.
Locks, Gun
 Brown Manufacturing
 Central Specialties Ltd.
 L&R Lock Co.
 Trigger Lock Division
 Voere-KGH m.b.H.
 Master Lock Co.
Magazines
 Mag-Pack Corp.
Mats
 Brigade Quartermasters
Military Equipment/Accessories
 Amherst Arms
 Ruvel & Co., Inc.
Photographers, Gun
 Bilal, Mustafa
 Hanusin, John
 Macbean, Stan
 Payne Photography, Robert
 Radack Photography, Lauren
 Smith, Michael
 Weyer International
 White Pine Photographic Services
Pistol Barrel Maker
 Bar-Sto Precision Machine
Power Tools, Rotary Flexible Shaft
 Foredom Electric Co.
RF Barrel Vibration Reducer
 Hoehn Sales, Inc.
Saddle Rings, Studs
 Silver Ridge Gun Shop
Safety Devices
 P&M Sales and Service
Safeties
 P.M. Enterprises, Inc.
Scents and Lures
 Buck Stop Lure Co., Inc.
 Cabinet Mountain Outfitters Scents & Lures
 Dr. O's Products Ltd.
 Flow-Rite of Tennessee, Inc.
 Mountain Hollow Game Calls
 Russ Trading Post
 Tink's Safariland Hunting Corp.
 Tinks & Ben Lee Hunting Products
 Wellington Outdoors
 Wildlife Research Center, Inc.
 Wyant's Outdoor Products, Inc.
Scrimshaw
 Boone's Custom Ivory Grips, Inc.
 Dolbare, Elizabeth
 Hoover, Harvey
 Lovestrand, Erik
 Reno, Wayne
Shooting Range Equipment
 Caswell International Corp.
Shotgun Barrel Maker
 Baker, Stan
 Eyster Heritage Gunsmiths, Inc., Ken
Shotgun Conversion Tubes
 Dina Arms Corporation
Silencers
 AWC Systems Technology
 Ciener, Jonathan Arthur
 DLO Mfg.
 Fleming Firearms
 S&H Arms Mfg. Co.
 S.C.R.C.
 Sound Technology
 Ward Machine
Silver Sportsmen's Art
Bell, Sid
Heritage Wildlife Carvings
Slings and Swivels
 DTM International, Inc.
 High North Products, Inc.
 Leather Arsenal
 Pathfinder Sports Leather
 Schulz Industries
 Torel, Inc.
Stock Finish
 Richards, John
Treestands and Steps
 A&J Products
 Dr. O's Products Ltd.
 Russ Trading Post
 Silent Hunter
 Summit Specialties, Inc.
 Trax America, Inc.
 Treemaster
 Warren & Sweat Mfg. Co.
Trophies
 Blackinton & Co., Inc., V.H.
Ventilated Rib
 Simmons Gun Repair, Inc.
Ventilation
 ScanCo Environmental Systems
Video Tapes
 American Pioneer Video
 Calibre Press, Inc.
 Cedar Hill Game Calls, Inc.
 Clements' Custom Leathercraft, Chas
 Eastman Products, R.T.
 Foothills Video Productions, Inc.
 HandiCraft Unltd.
 Lethal Force Institute
 New Historians Productions, The
 Police Bookkshelf
 Primos, Inc.
 Trail Visions
 Wells Creek Knife & Gun Works
 Wilderness Sound Products Ltd.
Wind Flags
 Time Precision, Inc.
Xythos-Miniature Revolver
 Andres & Dworsky

MUZZLE-LOADING GUNS, BARRELS AND EQUIPMENT

Accuracy Unlimited (Littleton, CO)
Adkins, Luther
Aimtech Mount Systems
Allen Manufacturing
Anderson Manufacturing Co., Inc.
Armi San Paolo
Armoury, Inc., The
Bauska Barrels
Beauchamp & Son, Inc.
Beaver Lodge
Bentley, John
Big Bore Bullets
Birdsong & Associates, W.E.
Blackhawk West
Black Powder Products
Blue and Gray Products, Inc.

DIRECTORY OF THE ARMS TRADE

Bridgers Best
Buckskin Bullet Co.
Buckskin Machine Works
Burgess & Son Gunsmiths, R.W.
Butler Creek Corp.
Cache La Poudre Rifleworks
California Sights
Cash Manufacturing Co., Inc.
CenterMark
Chambers Flintlocks, Ltd., Jim
Chopie Mfg., Inc.
Cimarron Arms
Cogar's Gunsmithing
Colonial Repair
Colt Blackpowder Arms Co.
Conetrol Scope Mounts
Cousin Bob's Mountain Products
Cumberland Arms
Cumberland Mountain Arms
Cumberland Knife & Gun Works
Curly Maple Stock Blanks
CVA
Dangler, Homer L.
Day & Sons, Inc., Leonard
Dayton Traister
deHaas Barrels
Delhi Gun House
Dixie Gun Works, Inc.
Dyson & Son Ltd., Peter
EMF Co., Inc.
Euroarms of America, Inc.
Eutaw Co., Inc., The
Fautheree, Andy
Feken, Dennis
Fellowes, Ted
Fire'n Five
Flintlocks, Etc.
Forgett Jr., Valmore J.
Fort Hill Gunstocks
Fowler, Bob
Frankonia Jagd
Frontier
Gain Twist Barrel Co.
Getz Barrel Co.
Golden Age Arms Co.
Gonic Arms, Inc.
Green Mountain Rifle Barrel Co., Inc.
Gun Works, The
Hastings Barrels
Hawken Shop, The
Hege Jagd-u. Sporthandels, GmbH
Hofmann & Co.
Hodgdon Powder Co., Inc.
Hoppe's Div.
Hornady Mfg. Co.
House of Muskets, Inc., The
Hunkeler, A.
Impact Case Co.
Jamison's Forge Works
Jones Co., Dale
K&M Industries, Inc.
Kennedy Firearms
Knight Rifles
Knight's Mfg. Co.
Kwik-Site Co.
L&R Lock Co.
L&S Technologies, Inc.
Legend Products Corp.
Lestrom Laboratories, Inc.
Log Cabin Sport Shop
Lone Star Rifle Company
Lothar Walther Precision Tool, Inc.
Lyons Gunworks, Larry

Lyman
Marlin Firearms Co.
Mathews & Son, Inc., George E.
McCann's Muzzle-Gun Works
Michaels of Oregon Co.
MMP
Modern MuzzleLoading, Inc.
Montana Precision Swaging
Mountain State Muzzleloading
   Supplies, Inc.
Mowrey Gun Works
MSC Industrial Supply Co.
Mt. Alto Outdoor Products
Mushroom Express Bullet Co.
Muzzleloaders Etcetera, Inc.
Naval Ordnance Works
North Star West
October Country
Oklahoma Leather Products, Inc.
Olson, Myron
Orion Rifle Barrel Co.
Ox-Yoke Originals, Inc.
Pacific Rifle Co.
Parker Gun Finishes
Pecatonica River Longrifle
Pedersoli and Co., Davide
Penguin Industries, Inc.
Pioneer Arms Co.
Prairie River Arms
R.E. Davis
Rusty Duck Premium Gun
   Care Products
R.V.I.
S&B Industries
S&S Firearms
Selsi Co., Inc.
Shiloh Creek
Shooter's Choice
Sile Distributors
Simmons Gun Repair, Inc.
Sklany's Machine Shop
Slings 'N Things, Inc.
Smokey Valley Rifles
South Bend Replicas, Inc.
Southern Bloomer Mfg. Co.
Starr Trading Co., Jedediah
Stone Mountain Arms
Taylor's & Co., Inc.
Tennessee Valley Mfg.
Thompson Bullet Lube Co.
Thompson/Center Arms
Thunder Mountain Arms
Tiger-Hunt
Track of the Wolf, Inc.
Traditions, Inc.
Treso, Inc.
UFA, Inc.
Uberti, Aldo
Uncle Mike's
Upper Missouri Trading Co.
Venco Industries, Inc.
Voere-KGH m.b.H.
Walters, John
Warren Muzzleloading Co., Inc.
Wescombe, Bill
White Owl Enterprises
White Muzzleloading Systems
White Shooting Systems, Inc.
Williams Gun Sight Co.
Woodworker's Supply
Wright's Hardwood Gunstock Blanks
Young Country Arms

Dilliott Gunsmithing, Inc.
EGW Evolution Gun Works
Ellicott Arms, Inc./Woods
   Pistolsmithing
Ferris Firearms
Fisher Custom Firearms
Forkin, Ben
Francesca, Inc.
Frielich Police Equipment
Garthwaite, Pistolsmith, Inc., Jim
G.G. & G.
Gonzalez Guns, Ramon B.
Greider Precision
Gun Room Press, The
Guncraft Sports, Inc.
Gunsite Custom Shop
Gunsite Gunsmithy
Gunsite Training Center
Hamilton, Alex B.
Hamilton, Keith
Hammond Custom Guns Ltd.
Hank's Gun Shop
Hanson's Gun Center, Dick
Harwood, Jack O.
Harris Gunworks
Hawken Shop, The
Hebard Guns, Gil
Heinie Specialty Products
High Bridge Arms, Inc.
Highline Machine Co.
Hoag, James W.
Irwin, Campbell H.
Island Pond Gun Shop
Ivanoff, Thomas G.
J&S Heat Treat
Jacobson, Teddy
Jarvis, Inc.
Jensen's Custom Ammunition
Johnston, James
Jones, J.D.
Jungkind, Reeves C.
K-D, Inc.
Kaswer Custom, Inc.
Ken's Gun Specialties
Kilham & Co.
Kimball, Gary
Kopp, Terry K.
La Clinique du .45
LaFrance Specialties
LaRocca Gun Works, Inc.
Lathrop's, Inc.
Lawson, John G.
Lee's Red Ramps
Leckie Professional Gunsmithing
Liberty Antique Gunworks
Linebaugh Custom Sixguns & Rifle
   Works
List Precision Engineering
Long, George F.
Mac's .45 Shop
Mag-Na-Port International, Inc.
Mahony, Philip Bruce
Mandall Shooting Supplies, Inc.
Marent, Rudolf

Marvel, Alan
Maxi-Mount
McCann's Machine & Gun Shop
MCS, Inc.
Middlebrooks Custom Shop
Miller Custom
Mitchell's Accuracy Shop
MJK Gunsmithing, Inc.
Mo's Competitor Supplies
Mountain Bear Rifle Works, Inc.
Mowrey's Guns & Gunsmithing
Mullis Guncraft
Nastoff's 45 Shop, Inc., Steve
NCP Products, Inc.
Novak's Inc.
Nowlin Custom Mfg.
Nygord Precision Products
Oglesby & Oglesby Gunmakers, Inc.
Paris, Frank J.
Pace Marketing, Inc.
Pasadena Gun Center
Peacemaker Specialists
PEM's Mfg. Co.
Performance Specialists
Pierce Pistols
Plaxco, J. Michael
Precision Specialties
Randco UK
Ries, Chuck
Rim Pac Sports, Inc.
Robar Co.'s, Inc., The
RPM
Score High Gunsmithing
Scott, McDougall & Associates
Seecamp Co., Inc., L.W.
Shooter Shop, The
Shooters Supply
Shootin' Shack, Inc.
Sight Shop, The
Singletary, Kent
Sipes Gun Shop
Spokhandguns, Inc.
Springfield, Inc.
SSK Industries
Starnes Gunmaker, Ken
Steger, James R.
Swenson's 45 Shop, A.D.
Swift River Gunworks
Ten-Ring Precision, Inc.
Thompson, Randall
300 Gunsmith Service, Inc.
Thurston Sports, Inc.
Tom's Gun Repair
Vic's Gun Refinishing
Volquartsen Custom Ltd.
Walker Arms Co., Inc.
Walters Industries
Wardell Precision Handguns Ltd.
Weigand Combat Handguns, Inc.
Wessinger Custom Guns & Engraving
Williams Gun Sight Co.
Williamson Precision Gunsmithing
Wilson Gun Shop
Wichita Arms, Inc.

PISTOLSMITHS

Acadian Ballistic Specialties
Accuracy Gun Shop
Accuracy Unlimited (Glendale, AZ)
Actions by "T"
Adair Custom Shop, Bill
Ahlman Guns
Ahrends, Kim
Aldis Gunsmithing & Shooting Supply
Alpha Precision, Inc.
Alpine's Precision Gunsmithing &
   Indoor Shooting Range
Armament Gunsmithing Co., Inc.
Arundel Arms & Ammunition, Inc., A.
Baer Custom, Inc., Les
Bain & Davis, Inc.
Baity's Custom Gunworks
Banks, Ed
Bear Arms
Behlert Precision, Inc.
Bellm Contenders
Belt MTN Arms

Bengtson Arms Co., L.
Bowen Classic Arms Corp.
Broken Gun Ranch
Campbell, Dick
Cannon's, Andy Cannon
Caraville Manufacturing
Carter's Gun Shop
Chicasaw Gun Works
Clark Custom Guns, Inc.
Cochran, Oliver
Colonial Repair
Colorado School of Trades
Corkys Gun Clinic
Craig Custom Ltd.
Curtis Custom Shop
Custom Firearms
Custom Gunsmiths
D&D Gunsmiths, Ltd.
D&L Sports
Davis Service Center, Bill
Dayton Traister

REBORING AND RERIFLING

A.M.T.
Arundel Arms & Ammunition, Inc., A.
BlackStar AccuMax Barrels
BlackStar Barrel Accurizing
Chicasaw Gun Works
Cochran, Oliver
Ed's Gun House
Flaig's
Gun Works, The
IAI
H&S Liner Service
Ivanoff, Thomas G.
Jackalope Gun Shop
K-D, Inc.
Kopp, Terry K.
LaBounty Precision Reboring
Matco, Inc.
NCP Products, Inc.

Pence Precision Barrels
Pro-Port Ltd.
Ranch Products
Redman's Rifling & Reboring
Rice, Keith
Ridgetop Sporting Goods
Shaw, Inc., E.R.
Siegrist Gun Shop
Simmons Gun Repair, Inc.
Stratco, Inc.
300 Gunsmith Service, Inc.
Time Precision, Inc.
Tom's Gun Repair
Van Patten, J.W.
West, Robert G.
White Rock Tool & Die
Zufall, Joseph F.

RELOADING TOOLS AND ACCESSORIES

Action Bullets, Inc.
Advance Car Mover Co., Rowell Div.
Alaska Bullet Works, Inc.
American Products Inc.
Ames Metal Products
Ammo Load, Inc.
Anderson Manufacturing Co., Inc.
Armite Laboratories
Arms Corporation of the Philippines

Atlantic Rose, Inc.
Atsko/Sno-Seal, Inc.
Bald Eagle Precision Machine Co.
Ballistic Products, Inc.
Ballisti-Cast, Inc.
Belltown, Ltd.
Ben's Machines
Berger Bullets, Ltd.
Berry's Mfg., Inc.

PRODUCT DIRECTORY

Birchwood Casey
Blount, Inc., Sporting Equipment Div.
Blue Ridge Machinery & Tools, Inc.
Bonanza
Break-Free, Inc.
Brobst, Jim
Brown Co., E. Arthur
Bruno Shooters Supply
Brynin, Milton
B-Square Co., Inc.
Buck Stix—SOS Products Co.
Bull Mountain Rifle Co.
Bullet Swaging Supply, Inc.
Bullseye Bullets
C&D Special Products
Camdex, Inc.
Camp-Cap Products
Canyon Cartridge Corp.
Carbide Die & Mfg. Co., Inc.
Case Sorting System
CFVentures
C-H Tool & Die Corp.
Chem-Pak, Inc.
CheVron Case Master
Claybuster Wads & Harvester Bullets
Cleanzoil Corp.
Clearview Products
Clymer Manufacturing Co., Inc.
Coats, Mrs. Lester
Colorado Shooter's Supply
CONKKO
Cook Engineering Service
Cooper-Woodward
Crouse's Country Cover
Cumberland Arms
Custom Products, Neil A. Jones
CVA
Davis, Don
Davis Products, Mike
D.C.C. Enterprises
Denver Bullets, Inc.
Denver Instrument Co.
Dever Co., Jack
Dewey Mfg. Co., Inc., J.
Dillon Precision Products, Inc.
Dropkick
Dutchman's Firearms, Inc., The
E&L Mfg., Inc.
Eagan, Donald V.
Eezox, Inc.
Eichelberger Bullets, Wm.
Elkhorn Bullets
Engineered Accessories
Enguix Import-Export
Estate Cartridge, Inc.
E-Z-Way Systems
F&A Inc.
Federal Cartridge Co.
Federated-Fry
Feken, Dennis
Ferguson, Bill
First, Inc., Jack
Fisher Custom Firearms
Fitz Pistol Grip Co.
Flambeau Products Corp.
Flitz International Ltd.
Forgett Jr., Valmore J.
Forgreens Tool Mfg., Inc.
Forster Products
4-D Custom Die Co.
4W Ammunition
Fremont Tool Works
Fry Metals
Fusilier Bullets
G&C Bullet Co., Inc.
GAR
Gehmann, Walter
Goddard, Allen
Gozon Corp., U.S.A.
Graf & Sons
"Gramps" Antique Cartridges
Graphics Direct
Graves Co.
Green, Arthur S.
Greenwood Precision
Gun Works, The
Hanned Line, The
Hanned Precision
Harrell's Precision
Harris Enterprises
Harrison Bullets
Haselbauer Products, Jerry
Haydon Shooters' Supply, Russ
Heidenstrom Bullets
Hensley & Gibbs
Hirtenberger Aktiengesellschaft

Hobson Precision Mfg. Co.
Hoch Custom Bullet Moulds
Hodgdon Powder Co., Inc.
Hoehn Sales, Inc.
Hoelscher, Virgil
Holland's Gunsmithing
Hollywood Engineering
Hondo Industries
Hornady Mfg. Co.
Howell Machine
Hunters Supply
Huntington Die Specialties
Image Ind. Inc.
IMI Services USA, Inc.
Imperial
Imperial Magnum Corp.
INTEC International, Inc.
Iosso Products
Javelina Lube Products
JGS Precision Tool Mfg.
J&L Superior Bullets
JLK Bullets
Jonad Corp.
Jones Custom Products, Neil A.
Jones Moulds, Paul
K&M Services
K&S Mfg. Inc.
Kapro Mfg. Co., Inc.
King & Co.
Kleen-Bore, Inc.
Korzinek Riflesmith, J.
Lane Bullets, Inc.
Lapua Ltd.
LBT
Le Clear Industries
Lee Precision, Inc.
Legend Products Corp.
Liberty Metals
Liberty Shooting Supplies
Lightning Performance Innovations, Inc.
Lithi Bee Bullet Lube
Littleton, J.F.
Lortone, Inc.
Loweth Firearms, Richard H.R.
Luch Metal Merchants, Barbara
Lyman Instant Targets, Inc.
Lyman Products Corp.
M&D Munitions Ltd.
MA Systems
Magma Engineering Co.
MarMik, Inc.
Marquart Precision Co.
MAST Technology
Match Prep—Doyle Gracey
Mayville Engineering Co.
McKillen & Heyer, Inc.
MCRW Associates Shooting Supplies
MCS, Inc.
MEC, Inc.
Midway Arms, Inc.
Miller Engineering
MI-TE Bullets
MMP
Mo's Competitor Supplies
Montana Armory, Inc.
Mountain State Muzzleloading
  Supplies, Inc.
Mt. Baldy Bullet Co.
MTM Molded Products Co., Inc.
Multi-Scale Charge Ltd.
MWG Company
Necromancer Industries, Inc.
NEI Handtools, Inc.
Niemi Engineering, W.B.
North Devon Firearms Services
Northern Precision Custom
  Swaged Bullets
October Country
Old West Bullet Moulds
Omark Industries
Original Box, Inc.
Paco's
Pattern Control
Pease Accuracy, Bob
Pedersoli and Co., Davide
Peerless Alloy, Inc.
Pend Oreille Sport Shop
Pinetree Bullets
Plum City Ballistic Range
Pomeroy, Robert
Ponsness/Warren
Prairie River Arms
Precision Castings & Equipment, Inc.
Precision Reloading, Inc.
Prime Reloading
Professional Hunter Supplies

Prolix® Lubricants
Pro-Shot Products, Inc.
Protector Mfg. Co., Inc., The
Quinetics Corp.
R&D Engineering & Manufacturing
Rapine Bullet Mould Mfg. Co.
Raytech
RCBS
R.D.P. Tool Co., Inc.
Redding Reloading Equipment
R.E.I.
Reloading Specialties, Inc.
Rice, Keith
Riebe Co., W.J.
RIG Products
R.I.S. Co., Inc.
Roberts Products
Rochester Lead Works, Inc.
Rolston, Inc., Fred. W.
Rooster Laboratories
Rorschach Precision Products
Rosenthal, Brad and Sallie
SAECO
Sandia Die & Cartridge Co.
Saunders Gun & Machine Shop
Saville Iron Co.
Scharch Mfg., Inc.
Scot Powder Co. of Ohio, Inc.
Scott, Dwight
Seebeck Assoc., R.E.
Sharp Shooter Supply
Sharps Arms Co. Inc., C.
Shiloh Creek
Shiloh Rifle Mfg.
Shooter's Choice
ShurKatch Corporation
Sierra Specialty Prod. Co.
Silhouette, The
Silver Eagle Machining
Simmons, Jerry
Sinclair International, Inc.
Skip's Machine
S.L.A.P. Industries
Small Custom Mould & Bullet Co.
Sno-Seal
SOS Products Co.
Spence, George W.
Spencer's Custom Guns
SPG, Inc.
Sport Flite Manufacturing Co.

Sportsman Supply Co.
Stalwart Corp.
Star Custom Bullets
Star Machine Works
Starr Trading Co., Jedediah
Stillwell, Robert
Stoney Point Products, Inc.
Stratco, Inc.
Taracorp Industries
TCCI
TCSR
TDP Industries, Inc.
Tetra Gun Lubricants
Thompson Bullet Lube Co.
Thompson/Center Arms
Timber Heirloom Products
Time Precision, Inc.
TMI Products
TR Metals Corp.
Trammco, Inc.
Tru-Square Metal Prods., Inc.
TTM
Varner's Service
Vega Tool Co.
Venco Industries, Inc.
VibraShine, Inc.
Vibra-Tek Co.
Vihtavuori Oy/Kaltron-Pettibone
Vitt/Boos
Von Minden Gunsmithing Services
Walters, John
Webster Scale Mfg. Co.
WD-40 Co.
Welsh, Bud
Wells Custom Gunsmith, R.A.
Werner, Carl
Westfield Engineering
White Rock Tool & Die
Whitetail Design & Engineering Ltd.
Widener's Reloading & Shooting
  Supply
William's Gun Shop, Ben
Wilson, Inc., L.E.
Wise Guns, Dale
Wolf's Western Traders
Woodleigh
WTA Manufacturing, Bill Wood
Yesteryear Armory & Supply
Young Country Arms

RESTS—BENCH, PORTABLE—AND ACCESSORIES

Accuright
Adventure 16, Inc.
Armor Metal Products
Bald Eagle Precision Machine Co.
Bartlett Engineering
Borden's Accuracy
Browning Arms Co.
B-Square Co., Inc.
Bull Mountain Rifle Co.
Canons Delcour
Chem-Pak, Inc.
Clift Mfg., L.R.
Clift Welding Supply
Cravener's Gun Shop
Decker Shooting Products
Desert Mountain Mfg.
Erickson's Mfg., Inc., C.W.
F&A Inc.
Greenwood Precision
Harris Engineering, Inc.
Hidalgo, Tony
Hoehn Sales, Inc.

Hoelscher, Virgil
Hoppe's Div.
Kolpin Mfg., Inc.
Kramer Designs
Midway Arms, Inc.
Millett Sights
MJM Manufacturing
Outdoor Connection, Inc., The
PAST Sporting Goods, Inc.
Penguin Industries, Inc.
Portus, Robert
Protektor Model
Ransom International Corp
Saville Iron Co.
ShurKatch Corporation
Stoney Point Products, Inc.
Thompson Target Technology
T.H.U. Enterprises, Inc.
Tonoloway Tack Drivers
Varner's Service
Wichita Arms, Inc.
Zanotti Armor, Inc.

RIFLE BARREL MAKERS (See also Muzzle-Loading Guns, Barrels and Equipment)

Airrow
Arundel Arms & Ammunition, Inc., A.
A.M.T.
BlackStar AccuMax Barrels
BlackStar Barrel Accurizing
Border Barrels Ltd.
Broad Creek Rifle Works
Broughton Rifle Barrels
Brown Co., E. Arthur
Bullberry Barrel Works, Ltd.
Canons Delcour
Carter's Gun Shop
Chicasaw Gun Works
Christensen Arms
Cincinnati Swaging
Clerke Co., J.A.
Competition Limited

D&J Bullet Co. & Custom Gun
  Shop, Inc.
deHaas Barrels
Dilliott Gunsmithing, Inc.
DKT, Inc.
Donnelly, C.P.
Douglas Barrels, Inc.
Gaillard Barrels
Gain Twist Barrel Co.
Getz Barrel Co.
Green Mountain Rifle Barrel Co., Inc.
Gun Works, The
Half Moon Rifle Shop
Harold's Custom Gun Shop, Inc.
Harris Gunworks
Hart Rifle Barrels, Inc.
Hastings Barrels

30th EDITION, 1998 **305**

DIRECTORY OF THE ARMS TRADE

Hoelscher, Virgil
H-S Precision, Inc.
IAI
Jackalope Gun Shop
K-D, Inc.
KOGOT
Kopp, Terry K.
Krieger Barrels, Inc.
LaBounty Precision Reboring
Lilja Precision Rifle Barrels
Lothar Walther Precision Tool, Inc.
Mac's .45 Shop
Matco, Inc.
McGowen Rifle Barrels
McMillan Rifle Barrels
Mid-America Recreation, Inc.
Nowlin Custom Mfg.
Obermeyer Rifled Barrels
Olympic Arms, Inc.
Pac-Nor Barreling
Pell, John T.
Pence Precision Barrels
Perazone, Brian
Raptor Arms Co., Inc.
Rocky Mountain Rifle Works Ltd.
Rosenthal, Brad and Sallie
Sabatti S.R.L.
Schneider Rifle Barrels, Inc., Gary
Shaw, Inc., E.R.
Shilen, Inc.
Siskiyou Gun Works
Small Arms Mfg. Co.
Sonora Rifle Barrel Co.
Specialty Shooters Supply, Inc.
Strutz Rifle Barrels, Inc., W.C.
Swift River Gunworks
Swivel Machine Works, Inc.
Unmussig Bullets, D.L.
Verney-Carron
Wells, Fred F.
Wilson Arms Co., The
Wiseman and Co., Bill

SCOPES, MOUNTS, ACCESSORIES, OPTICAL EQUIPMENT

Accuracy Innovations, Inc.
Ackerman, Bill
ADCO Sales, Inc.
Adventurer's Outpost
Aimpoint U.S.A.
Aimtech Mount Systems
Air Venture
Alley Supply Co.
Anderson Manufacturing Co., Inc.
Apel GmbH, Ernst
A.R.M.S., Inc.
Armscorp USA, Inc.
Autauga Arms, Inc.
Baer Custom, Inc., Les
Barrett Firearms Mfg., Inc.
Bausch & Lomb Sports Optics Div.
Beaver Park Products, Inc.
BEC, Inc.
Blount, Inc., Sporting Equipment Div.
Bohemia Arms Co.
Boonie Packer Products
Borden's Accuracy
Brown Co., E. Arthur
Brownells, Inc.
Brunton U.S.A.
B-Square Co., Inc.
Bull Mountain Rifle Co.
Burris
Bushnell Sports Optics Worldwide
Butler Creek Corp.
California Grip
Celestron International
Center Lock Scope Rings
Clearview Mfg. Co., Inc.
Combat Military Ordnance Ltd.
Compass Industries, Inc.
Concept Development Corp.
Conetrol Scope Mounts
CRDC Laser Systems Group
Creedmoor Sports, Inc.
Crimson Trace
Custom Quality Products, Inc.
D&H Prods. Co., Inc.
D.C.C. Enterprises
Del-Sports, Inc.
DHB Products
Doctor Optic Technologies, Inc.
Eagle International Sporting Goods, Inc.
Edmund Scientific Co.
Ednar, Inc.
Eggleston, Jere D.
EGW Evolution Gun Works
Eclectic Technologies, Inc.
Emerging Technologies, Inc.
Europtik Ltd.
Excalibur Enterprises
Farr Studio, Inc.
Forgett Jr., Valmore J.
Fotar Optics
Frankonia Jagd
Fujinon, Inc.
G.G. & G.
Gentry Custom Gunmaker, David
Glaser Safety Slug, Inc.
Great Lakes Airguns
GSI, Inc.
Gun Accessories
Gun South, Inc.
Guns, (Div. of D.C. Engineering, Inc.)
Gunsmithing, Inc.
Hakko Co. Ltd.
Hammerli USA
Harris Gunworks
Harvey, Frank
Hermann Leather Co., H.J.
Hertel & Reuss
Hiptmayer, Armurier
Hiptmayer, Klaus
Hofmann & Co.
Holland's
Ironsighter Co.
Jaeger, Inc., Paul/Dunn's
JägerSport, Ltd.
Jeffredo Gunsight
Jewell Triggers, Inc.
Kahles, A Swarovski Company
KDF, Inc.
Kelbly, Inc.
KenPatable Ent., Inc.
Kesselring Gun Shop
Kimber of America, Inc.
Kmount
Knight's Mfg. Co.
Kowa Optimed, Inc.
Kris Mounts
KVH Industries, Inc.
Kwik Mount Corp.
Kwik-Site Co.
L&S Technologies, Inc.
L.A.R. Mfg., Inc.
Laser Devices, Inc.
Laseraim
LaserMax, Inc.
Leapers, Inc.
Lectro Science, Inc.
Lee Co., T.K.
Leica USA, Inc.
Leupold & Stevens, Inc.
Lightforce U.S.A. Inc.
List Precision Engineering
Lohman Mfg. Co., Inc.
London Guns Ltd.
Lyte Optronics
Mac's .45 Shop
Mag-Na-Port International, Inc.
Masen Co., Inc., John
Maxi-Mount
McBros Rifle Co.
McCann's Machine & Gun Shop
McMillan Optical Gunsight Co.
MCS, Inc.
MDS
Meier Works
Merit Corp.
Michaels of Oregon Co.
Military Armament Corp.
Millett Sights
Mirador Optical Corp.
Mitchell Optics Inc.
Mo's Competitor Supplies
Mountain Rifles Inc.
MWG Co.
New England Custom Gun Service
Nic Max, Inc.
Nightforce
Nikon, Inc.
Norincoptics
Oakshore Electronic Sights, Inc.
Olympic Optical Co.
Optical Services Co.
Orchard Park Enterprise
Oregon Arms, Inc.
Outdoor Connection, Inc., The
Parsons Optical Mfg. Co.
PECAR Herbert Schwarz, GmbH
PEM's Mfg. Co.
Pentax Corp.
Perazone, Brian
P.M. Enterprises, Inc.
Precise Metalsmithing Enterprises
Precision Sport Optics
Premier Reticles
Quarton USA, Ltd. Co.
Ram-Line Blount, Inc.
Ranch Products
Randolph Engineering, Inc.
Ranging, Inc.
Redfield, Inc.
Rice, Keith
Rocky Mountain High Sports Glasses
Rogue Rifle Co., Inc.
RPM
S&K Mfg. Co.
Saunders Gun & Machine Shop
Schmidt & Bender, Inc.
Scope Control Inc.
ScopLevel
Score High Gunsmithing
Seattle Binocular & Scope Repair Co.
Segway Industries
Selsi Co., Inc.
Shepherd Scope Ltd.
Sightron, Inc.
Simmons Enterprises, Ernie
Simmons Outdoor Corp.
Six Enterprises
SKAN A.R.
SKB Shotguns
Slug Group, Inc.
Sportsmatch U.K. Ltd.
Springfield, Inc.
STI International
Stoeger Industries
SwaroSports, Inc.
Swarovski Optik North America Ltd.
Swift Instruments, Inc.
TacStar Industires, Inc.
Talley, Dave
Tasco Sales, Inc.
Tele-Optics
Thompson/Center Arms
Trijicon, Inc.
Ultra Dot Distribution
Uncle Mike's
Unertl Optical Co., Inc., John
United Binocular Co.
United States Optics Tech., Inc.
Valor Corp.
Voere-KGH m.b.H.
Warne Manufacturing Co.
Warren Muzzleloading Co., Inc.
WASP Shooting Systems
Weatherby, Inc.
Weaver Products
Weaver Scope Repair Service
Weigand Combat Handguns, Inc.
Westfield Engineering
Westley Richards & Co.
White Muzzleloading Systems
White Shooting Systems, Inc.
White Rock Tool & Die
Wideview Scope Mount Corp.
Williams Gun Sight Co.
York M-1 Conversions
Zanotti Armor, Inc.
Zeiss Optical, Carl

SHOOTING/TRAINING SCHOOLS

Accuracy Gun Shop
Alpine Precision Gunsmithing &
 Indoor Shooting Range
American Small Arms Academy
Auto Arms
Barsotti, Bruce
Bob's Tactical Indoor Shooting Range
 & Gun Shop
Cannon's, Andy Cannon
Chapman Academy of Practical
 Shooting
Chelsea Gun Club of New York
 City, Inc.
CQB Training
Daisy Mfg. Co.
Defense Training International, Inc.
Dowtin Gunworks
Executive Protection Institute
Feminine Protection, Inc.
Ferris Firearms
Firearm Training Center, The
Firearms Academy of Seattle
G.H. Enterprises Ltd.
Front Sight Firearms Training Institute
Gunsite Training Center
Guncraft Sports, Inc.
Henigson & Associates, Steve
International Shootists, Inc.
Jensen's Custom Ammunition
Jensen's Firearms Acadamy
L.L. Bean, Inc.
McMurdo, Lynn
Mendez, John A.
Montgomery Community College
NCP Products, Inc.
Nevada Pistol Academy Inc.
North American Shooting Systems
North Mountain Pine Training Center
Pacific Pistolcraft
Performance Specialists
Quigley's Personal Protection
 Strategies, Paxton
River Road Sporting Clays
SAFE
Shooter's World
Shooting Gallery, The
Shotgun Shop, The
Smith & Wesson
Specialty Gunsmithing
Starlight Training Center, Inc.
Steger, James R.
Tactical Defense Institute
300 Gunsmith Service, Inc.
Western Missouri Shooters Alliance
Yankee Gunsmith
Yavapai Firearms Academy Ltd.

SIGHTS, METALLIC

Accura-Site
All's, The Jim J. Tembelis Co., Inc.
Alpec Team, Inc.
Andela Tool & Machine, Inc.
Anschutz GmbH
Armsport, Inc.
Aro-Tek, Ltd.
Baer Custom, Inc., Les
BEC, Inc.
Bo-Mar Tool & Mfg. Co.
Bond Custom Firearms
Bowen Classic Arms Corp.
Bradley Gunsight Co.
Brown Co., E. Arthur
Brown Products, Inc., Ed
California Sights
Cape Outfitters
Carter's Gun Shop
Center Lock Scope Rings
C-More Systems
Colonial Repair
CRR, Inc./Marble's Inc.
DHB Products
Eagle International Sporting Goods, Inc.
Engineered Accessories
Evans, Andrew
Evans Gunsmithing
Farr Studio, Inc.
Fautheree, Andy
Forgett Jr., Valmore J.
G.G. & G.
Gun Doctor, The
Gun Works, The
Hank's Gun Shop
Heinie Specialty Products
Hesco-Meprolight
Hiptmayer, Armurier
Hiptmayer, Klaus
Innovative Weaponry, Inc.
Innovision Enterprises
Jaeger, Inc., Paul/Dunn's
J.P. Enterprises, Inc.
Kris Mounts
Lee's Red Ramps
List Precision Engineering
London Guns Ltd.
L.P.A. Snc
Lyman Instant Targets, Inc.
Lyman Products Corp.
Mac's .45 Shop
Marble Arms
MCS, Inc.
MEC-Gar S.R.L.
Meier Works

PRODUCT DIRECTORY

Meprolight
Merit Corp.
Mid-America Recreation, Inc.
Middlebrooks Custom Shop
Millett Sights
MMC
Mo's Competitor Supplies
Montana Armory, Inc.
Montana Vintage Arms
New England Custom Gun Service
Novak's Inc.
Oakshore Electronic Sights, Inc.
OK Weber, Inc.
Pachmayr Ltd.
PEM's Mfg. Co.
P.M. Enterprises, Inc.
Quarton USA, Ltd. Co.
Redfield, Inc.
RPM
Sharps Arms Co. Inc., C.
Slug Site
STI International
Stiles Custom Guns
Talley, Dave
T.F.C. S.p.A.
Thompson/Center Arms
Trijicon, Inc.
United States Optics Technologies, Inc.
WASP Shooting Systems
Wichita Arms, Inc.
Williams Gun Sight Co.
Wilson Gun Shop

STOCKS (Commercial and Custom)

A&W Repair
Accuracy Unlimited (Glendale, AZ)
Ackerman & Co.
Acra-Bond Laminates
Ahlman Guns
Amrine's Gun Shop
Arms Ingenuity Co.
Artistry In Wood
Arundel Arms & Ammunition, Inc., A.
Baelder, Harry
Bain & Davis, Inc.
Balickie, Joe
Bansner's Gunsmithing Specialties
Barnes Bullets, Inc.
Bartlett, Don
Beitzinger, George
Belding's Custom Gun Shop
Bell & Carlson, Inc.
Benchmark Guns
Biesen, Al
Biesen, Roger
Billeb, Stephen L.
Billings Gunsmiths, Inc.
Blount, Inc., Sporting Equipment Div.
Boltin, John M.
Borden's Accuracy
Bowerly, Kent
Boyds' Gunstock Industries, Inc.
Burgess & Son Gunsmiths, R.W.
Brace, Larry D.
Brgoch, Frank
Briganti, A.J.
Brown Co., E. Arthur
Brown Precision, Inc.
Buckhorn Gun Works
Bull Mountain Rifle Co.
Bullberry Barrel Works, Ltd.
Burkhart Gunsmithing, Don
Burres, Jack
Butler Creek Corp.
Cali'co Hardwoods, Inc.
Cape Outfitters
Camilli, Lou
Campbell, Dick
Caywood, Shane J.
Chambers Flintlocks Ltd., Jim
Chicasaw Gun Works
Christman Jr., Gunmaker, David
Churchill, Winston
Clark Custom Guns, Inc.
Claro Walnut Gunstock Co.
Clifton Arms, Inc.
Cloward's Gun Shop
Cochran, Oliver
Coffin, Charles H.
Coffin, Jim
Colonial Repair
Colorado Gunsmithing Academy
Colorado School of Trades
Conrad, C.A.
Crane Sales Co., George S.
Creedmoor Sports, Inc.
Curly Maple Stock Blanks
Custom Checkering Service
Custom Gun Products
Custom Gun Stocks
Custom Riflestocks, Inc.
D&D Gunsmiths, Ltd.
D&G Presicion Duplicators
D&J Bullet Co. & Custom Gun Shop, Inc.
Dangler, Homer L.
D.D. Custom Stocks
de Treville & Co., Stan
Dever Co., Jack
Devereaux, R.H. "Dick"
DGR Custom Rifles
DGS, Inc.
Dilliott Gunsmithing, Inc.
Dillon, Ed
Dowtin Gunworks
Dressel Jr., Paul G.
Duane Custom Stocks, Randy
Duane's Gun Repair
Duncan's Gunworks, Inc.
Echols & Co., D'Arcy
Eggleston, Jere D.
Erhardt, Dennis
Eversull Co., Inc., K.
Fajen, Inc., Reinhart
Farmer-Dressel, Sharon
Fibron Products, Inc.
Fisher, Jerry A.
Flaig's
Folks, Donald E.
Forster, Kathy
Forster, Larry L.
Forthofer's Gunsmithing & Knifemaking
Francotte & Cie S.A., Auguste
Frank Custom Classic Arms, Ron
Game Haven Gunstocks
Gene's Custom Guns
Gervais, Mike
Gillmann, Edwin
Gilman-Mayfield, Inc.
Giron, Robert E.
Goens, Dale W.
Golden Age Arms Co.
Gordie's Gun Shop
Goudy Classic Stocks, Gary
Grace, Charles E.
Great American Gun Co.
Green, Roger M.
Greene, M.L.
Greene Precision Duplicators
Greenwood Precision
Griffin & Howe, Inc.
Gun Shop, The
Guns
Gunsmithing Ltd.
Hallberg Gunsmith, Fritz
Halstead, Rick
Hamilton, Jim
Hanson's Gun Center, Dick
Harper's Custom Stocks
Harris Gunworks
Hart & Son, Inc., Robert W.
Harwood, Jack O.
Hastings Barrels
Hecht, Hubert J.
Heilmann, Stephen
Hensley, Darwin
Heppler, Keith M.
Heydenberk, Warren R.
High Tech Specialties, Inc.
Hillmer Custom Gunstocks, Paul D.
Hiptmayer, Armurier
Hiptmayer, Klaus
Hoelscher, Virgil
Hoenig & Rodman
H-S Precision, Inc.
Huebner, Corey O.
Ide, Kenneth G.
Island Pond Gun Shop
Ivanoff, Thomas G.
Jackalope Gun Shop
Jaeger, Inc., Paul/Dunn's
Jamison's Forge Works
Jarrett Rifles, Inc.
Johnson Wood Products
J.P. Gunstocks, Inc.
KDF, Inc.
Keith's Custom Gunstocks
Ken's Rifle Blanks
Kilham & Co.
Klein Custom Guns, Don
Klingler Woodcarving
Knippel, Richard
Knight's Mfg. Co.
Kokolus, Michael M.
Lawson Co., Harry
Lind Custom Guns, Al
Ljutic Industries, Inc.
Lock's Philadelphia Gun Exchange
Lynn's Custom Gunstocks
Lyons Gunworks, Larry
Mac's .45 Shop
Marple & Associates, Dick
Masen Co., Inc., John
McBros Rifle Co.
McCann's Muzzle-Gun Works
McCament, Jay
McCullough, Ken
McDonald, Dennis
McFarland, Stan
McGowen Rifle Barrels
McGuire, Bill
McKinney, R.P.
McMillan Fiberglass Stocks, Inc.
Mercer Custom Stocks, R.M.
Mid-America Recreation, Inc.
Miller Arms, Inc.
Morrison Custom Rifles, J.W.
MPI Fiberglass Stocks
MWG Co.
NCP Products, Inc.
Nelson, Stephen
Nettestad Gun Works
New England Arms Co.
New England Custom Gun Service
Newman Gunshop
Nickels, Paul R.
Norman Custom Gunstocks, Jim
Oakland Custom Arms, Inc.
Oil Rod and Gun Shop
OK Weber, Inc.
Old World Gunsmithing
One Of A Kind
Or-šn
Orvis Co., The
Ottmar, Maurice
P&S Gun Service
Pacific Research Laboratories, Inc.
Pagel Gun Works, Inc.
Paulsen Gunstocks
Pecatonica River Longrifle
PEM's Mfg. Co.
Pentheny de Pentheny
Perazone, Brian
Perazzi USA, Inc.
Pohl, Henry A.
Powell & Son (Gunmakers) Ltd., William
R&J Gun Shop
Ram-Line Blount, Inc.
Rampart International
Raptor Arms Co., Inc.
Reagent Chemical and Research, Inc.
Reiswig, Wallace E.
Richards Micro-Fit Stocks
Rimrock Rifle Stocks
RMS Custom Gunsmithing
Robinson, Don
Robinson Firearms Mfg. Ltd.
Rogers Gunsmithing, Bob
Roto Carve
Royal Arms Gunstocks
Ryan, Chad C.
Sanders Custom Gun Service
Saville Iron Co.
Schiffman, Curt
Schiffman, Mike
Schumakers Gun Shop
Schwartz Custom Guns, David W.
Schwartz Custom Guns, Wayne E.
Score High Gunsmithing
Shell Shack
Sile Distributors, Inc.
Simmons Gun Repair, Inc.
Six Enterprises
Skeoch, Brian R.
Smith, Sharmon
Snider Stocks, Walter S.
Speedfeed, Inc.
Speiser, Fred D.
Stiles Custom Guns
Storey, Dale A.
Stott's Creek Armory, Inc.
Strawbridge, Victor W.
Sturgeon Valley Sporters
Swan, D.J.
Swift River Gunworks
Szweda, Robert
Talmage, William G.
Taylor & Robbins
Tecnolegno S.p.A.
T.F.C. S.p.A.
Thompson/Center Arms
Tiger-Hunt
Tirelli
Tom's Gun Repair
Track of the Wolf, Inc.
Trevallion Gunstocks
Tucker, James C.
Turkish Firearms Corp.
Tuttle, Dale
Vest, John
Vic's Gun Refinishing
Vintage Industries, Inc.
Volquartsen Custom Ltd.
Von Minden Gunsmithing Services
Walker Arms Co., Inc.
Walnut Factory, The
Weber & Markin Custom Gunsmiths
Weems, Cecil
Wells Custom Gunsmith, R.A.
Wells, Fred F.
Wenig Custom Gunstocks, Inc.
Werth, T.W.
Wessinger Custom Guns & Engraving
West, Robert G.
Western Gunstock Mfg. Co.
Williams Gun Sight Co.
Williamson Precision Gunsmithing
Windish, Jim
Winter, Robert M.
Working Guns
Wright's Hardwood Gunstock Blanks
Yee, Mike
Zeeryp, Russ

TARGETS, BULLET AND CLAYBIRD TRAPS

Action Target, Inc.
American Target
American Whitetail Target Systems
A-Tech Corp.
Autauga Arms, Inc.
Beomat of America Inc.
Birchwood Casey
Blount, Inc., Sporting Equipment Div.
Blue and Gray Products, Inc.
Brown Manufacturing
Bull-X, Inc.
Camp-Cap Products
Caswell International Corp.
Champion Target Co.
Cunningham Co., Eaton
Dapkus Co., Inc., J.G.
Datumtech Corp.
Dayson Arms Ltd.
D.C.C. Enterprises
Detroit-Armor Corp.
Diamond Mfg. Co.
Estate Cartridge, Inc.
Federal Champion Target Co.
Freeman Animal Targets
G.H. Enterprises Ltd.
Gun Parts Corp., The
Hiti-Schuch, Atelier Wilma
H-S Precision, Inc.
Hunterjohn
Innovision Enterprises
JWH: Software
Kennebec Journal
Kleen-Bore, Inc.
Lakefield Arms Ltd.
Littler Sales Co.
Lyman Instant Targets, Inc.
Lyman Products Corp.
M&D Munitions Ltd.
Mendez, John A.
MSR Targets
National Target Co.
N.B.B., Inc.
North American Shooting Systems
Nu-Teck
Outers Laboratories, Div. of Blount
Ox-Yoke Originals, Inc.
Passive Bullet Traps, Inc.
PlumFire Press, Inc.
Quack Decoy & Sporting Clays
Redfield, Inc.
Remington Arms Co., Inc.
Rockwood Corp., Speedwell Div.
Rocky Mountain Target Co.
Savage Arms (Canada), Inc.

DIRECTORY OF THE ARMS TRADE

Savage Range Systems, Inc.
Schaefer Shooting Sports
Seligman Shooting Products
Shooters Supply
Shoot-N-C Targets
Shotgun Shop, The
Thompson Target Technology

Trius Products, Inc.
White Flyer Targets
World of Targets
X-Spand Target Systems
Z's Metal Targets & Frames
Zriny's Metal Targets

Noble Co., Jim
Palsa Outdoor Products
PAST Sporting Goods, Inc.
Penguin Industries, Inc.
Perazzi USA, Inc.
Pro-Port Ltd.
Protektor Model
Quack Decoy & Sporting Clays
Remington Arms Co., Inc.

Rhodeside, Inc.
Shootin' Accessories, Ltd.
Shooting Specialties
Shotgun Shop, The
ShurKatch Corporation
Titus, Daniel
Trius Products, Inc.
X-Spand Target Systems

TAXIDERMY

African Import Co.
Jonas Appraisers—Taxidermy Animals, Jack
Kulis Freeze Dry Taxidermy

Montgomery Community College
Parker, Mark D.
World Trek, Inc.

TRIGGERS, RELATED EQUIPMENT

A.M.T.
B&D Trading Co., Inc.
Baer Custom, Inc., Les
Behlert Precision, Inc.
Bond Custom Firearms
Boyds' Gunstock Industries, Inc.
Bull Mountain Rifle Co.
Canjar Co., M.H.
Cycle Dynamics, Inc.
Dayton Traister
Electronic Trigger Systems, Inc.
Eversull Co., Inc., K.
FWB
Galati International
Gentry Custom Gunmaker, David
Hastings Barrels
Hawken Shop, The
Hoelscher, Virgil
Holland's
IAI
Impact Case Co.
Jacobson, Teddy
Jaeger, Inc., Paul/Dunn's

Jewell Triggers, Inc.
J.P. Enterprises, Inc.
KK Air International
L&R Lock Co.
List Precision Engineering
Mahony, Philip Bruce
Masen Co., Inc., John
Master Lock Co.
Miller Single Trigger Mfg. Co.
NCP Products, Inc.
OK Weber, Inc.
PEM's Mfg. Co.
Penrod Precision
Perazone, Brian
Perazzi USA, Inc.
Raptor Arms Co., Inc.
S&B Industries
Shilen, Inc.
Simmons Gun Repair, Inc.
Slug Group, Inc.
STI International
Timney Mfg., Inc.

TRAP AND SKEET SHOOTER'S EQUIPMENT

Allen Co., Bob
Allen Sportswear, Bob
Bagmaster Mfg., Inc.
Baker, Stan
Beomat of America Inc.
Beretta S.p.A., Pietro
Cape Outfitters
Clearview Products
Clymer Manufacturing Co., Inc.
Crane & Crane Ltd.
Danuser Machine Co.
Dayson Arms Ltd.
Decker Shooting Products
Estate Cartridge, Inc.
F&A Inc.
Fiocchi of America, Inc.

Gander Mountain, Inc.
G.H. Enterprises Ltd.
Hastings Barrels
Hillmer Custom Gunstocks, Paul D.
Hoppe's Div.
Hunter Co., Inc.
K&T Co.
Lakewood Products, LLC
Ljutic Industries, Inc.
Lynn's Custom Gunstocks
Mag-Na-Port International, Inc.
Maionchi-L.M.I.
Meadow Industries
Moneymaker Guncraft Corp.
MTM Molded Products Co., Inc.
NCP Products, Inc.

MANUFACTURERS' DIRECTORY

A

A&B Industries, Inc. (See Top-Line USA, Inc.)
A&J Products, Inc., 5791 Hall Rd., Muskegon, MI 49442-1964
A&M Waterfowl, Inc., P.O. Box 102, Ripley, TN 38063/901-635-4003; FAX: 901-635-2320
A&W Repair, 2930 Schneider Dr., Arnold, MO 63010/314-287-3725
A.A. Arms, Inc., 4811 Persimmont Ct., Monroe, NC 28110/704-289-5356, 800-935-1119; FAX: 704-289-5859
Abel Safe & File, Inc., 124 West Locust St., Fairbury, IL 61739/800-346-9280, 815-692-2131; FAX: 815-692-3350
A.B.S. III, 9238 St. Morritz Dr., Fern Creek, KY 40291
AC Dyna-tite Corp., 155 Kelly St., P.O. Box 0984, Elk Grove Village, IL 60007/847-593-5566; FAX: 847-593-1304
Acadian Ballistic Specialties, P.O. Box 61, Covington, LA 70434
Acculube II, Inc., 4366 Shackleford Rd., Norcross, GA 30093-2912
Accupro Gun Care, 15512-109 Ave., Surrey, BC U3R 7E8, CANADA/604-583-7807
Accuracy Den, The, 25 Bitterbrush Rd., Reno, NV 89523/702-345-0225
Accuracy Gun Shop, 7818 Wilkerson Ct., San Diego, CA 92111/619-282-8500
Accuracy Innovations, Inc., P.O. Box 376, New Paris, PA 15554/814-839-4517; FAX: 814-839-2601
Accuracy International, 9115 Trooper Trail, P.O. Box 2019, Bozeman, MT 59715/406-587-7922; FAX: 406-585-9434
Accuracy International Precision Rifles (See U.S. importer—Gunsite Custom Shop; Gunsite Training Center)
Accuracy Unlimited, 7479 S. DePew St., Littleton, CO 80123
Accuracy Unlimited, 16036 N. 49 Ave., Glendale, AZ 85306/602-978-9089; FAX: 602-978-9089
Accura-Site (See All's, The Jim Tembelis Co., Inc.)
Accurate Arms Co., Inc., 5891 Hwy. 230 West, McEwen, TN 37101/615-729-4207, 800-416-3006; FAX 615-729-4211
Accurate Bullet Co., 159 Creek Road, Glen Mills, PA 19342/610-399-6584
Accuright, RR 2 Box 397, Sebeka, MN 56477/218-472-3383
Accu-Tek, 4510 Carter Ct., Chino, CA 91710/909-627-2404; FAX: 909-627-7817
Ace Custom 45's, Inc., 1880 1/2 Upper Turtle Creek Rd., Kerrville, TX 78028/210-257-4290; FAX: 210-257-5724
Ace Sportswear, Inc., 700 Quality Rd., Fayetteville, NC 28306/919-323-1223; FAX: 919-323-5392
Ackerman & Co., 16 Cortez St., Westfield, MA 01085/413-568-8008
Ackerman, Bill (See Optical Services Co.)
Acra-Bond Laminates (See Artistry in Wood)
Action Bullets, Inc., RR 1, P.O. Box 189, Quinter, KS 67752/913-754-3609; FAX: 913-754-3629
Action Direct, Inc., P.O. Box 830760, Miami, FL 33283/305-559-4652; FAX: 305-559-4652
Action Products, Inc., 22 N. Mulberry St., Hagerstown, MD 21740/301-797-1414; FAX: 301-733-2073
Action Target, Inc., P.O. Box 636, Provo, UT 84603/801-377-8033; FAX: 801-377-8096
Actions by "T," Teddy Jacobson, 16315 Redwood Forest Ct., Sugar Land, TX 77478/281-277-4008
ACTIV Industries, Inc., 1000 Zigor Rd., P.O. Box 339, Kearneysville, WV 25430/304-725-0451; FAX: 304-725-2080
AcuSport Corporation, 1 Hunter Place, Bellefontaine, OH 43311-3001/513-593-7010; FAX: 513-592-5625
Ad Hominem, 3130 Gun Club Lane, RR Orillia, Ont. L3V 6H3, CANADA/705-689-5303; FAX: 705-689-5303
Adair Custom Shop, Bill, 2886 Westridge, Carrollton, TX 75006
Adams & Son Engravers, John J., 87 Acorn Rd., Dennis, MA 02638/508-385-7971
Adams Jr., John J., 87 Acorn Rd., Dennis, MA 02638/508-385-7971
ADC, Inc., 33470 Chinook Plaza, Scappoose, OR 97056/503-543-5088
ADCO Sales Inc., 10 Cedar St., Unit 17, Woburn, MA 01801/617-935-1799; FAX: 617-935-1011
Adkins, Luther, 1292 E. McKay Rd., Shelbyville, IN 46176-9353/317-392-3795
Advance Car Mover Co., Rowell Div., P.O. Box 1, 240 N. Depot St., Juneau, WI 53039/414-386-4464; FAX: 414-386-4416
Adventure 16, Inc., 4620 Alvarado Canyon Rd., San Diego, CA 92120/619-283-6314
Adventure Game Calls, R.D. 1, Leonard Rd., Spencer, NY 14883/607-589-4611
Adventurer's Outpost, P.O. Box 70, Cottonwood, AZ 86326/800-762-7471; FAX: 602-634-8781
Aero Peltor, 90 Mechanic St., Southbridge, MA 01550/508-764-5500; FAX: 508-764-0188
African Import Co., 20 Braunecker Rd., Plymouth, MA 02360/508-746-8552
AFSCO Ammunition, 731 W. Third St., P.O. Box L, Owen, WI 54460/715-229-2516
Ahlman Guns, 9525 W. 230th St., Morristown, MN 55052/507-685-4243; FAX: 507-685-4280
Ahrends, Kim, Custom Firearms, Inc., Box 203, Clarion, IA 50525/515-532-3449; FAX: 515-532-3926
Aimpoint U.S.A, 420 W. Main St., Geneseo, IL 61254/309-944-1702
Aimtech Mount Systems, P.O. Box 223, 101 Inwood Acres, Thomasville, GA 31799/912-226-4313; FAX: 912-227-0222
Air Arms, Hailsham Industrial Park, Diplocks Way, Hailsham, E. Sussex, BN27 3JF ENGLAND/011-0323-845853 (U.S. importers—World Class Airguns)
Air Rifle Specialists, P.O. Box 138, 130 Holden Rd., Pine City, NY 14871-0138/607-734-7340; FAX: 607-733-3261
Air Venture, 9752 E. Flower St., Bellflower, CA 90706/310-867-6355
Airgun Repair Centre, 3227 Garden Meadows, Lawrenceburg, IN 47025/812-637-1463; FAX: 812-637-1463
Airrow (See Swivel Machine Works, Inc.)
Aitor-Cuchilleria Del Norte, S.A., Izelaieta, 17, 48260 Ermua (Vizcaya), SPAIN/43-17-08-50; FAX: 43-17-00-01
Ajax Custom Grips, Inc., 9130 Viscount Row, Dallas, TX 75247/214-630-8893; FAX: 214-630-4942; WEB: http://www.ajaxgrips.com
Aker International, Inc., 2248 Main St., Suite 6, Chula Vista, CA 91911/619-423-5182; FAX: 619-423-1363
Alaska Bullet Works, Inc., 9978 Crazy Horse Drive, Juneau, AK 99801/907-783-3834; FAX: 907-789-3433
Alcas Cutlery Corp. (See Cutco Cutlery)
Alco Carrying Cases, 601 W. 26th St., New York, NY 10001/212-675-5820; FAX: 212-691-5935
Aldis Gunsmithing & Shooting Supply, 502 S. Montezuma St., Prescott, AZ 86303/602-445-6723; FAX: 602-445-6763
Alessi Holsters, Inc., 2465 Niagara Falls Blvd., Amherst, NY 14228-3527/716-691-5615
Alex, Inc., Box 3034, Bozeman, MT 59772/406-282-7396; FAX: 406-282-7396
Alfano, Sam, 36180 Henry Gaines Rd., Pearl River, LA 70452/504-863-3364; FAX: 504-863-7715
All American Lead Shot Corp., P.O. Box 224566, Dallas, TX 75062
All Rite Products, Inc., 5752 N. Silverstone Circle, Mountain Green, UT 84050/801-876-3330; 801-876-2216
All's, The Jim J. Tembelis Co., Inc., P.O. Box 108, Winnebago, WI 54985-0108/414-725-5251; FAX: 414-725-5251
Allard, Gary, Creek Side Metal & Woodcrafters, Fishers Hill, VA 22626/703-465-3903
Allen Co., Bob, 214 SW Jackson, P.O. Box 477, Des Moines, IA 50315/515-283-2191; 800-685-7020; FAX: 515-283-0779
Allen Co., Inc., 525 Burbank St., Broomfield, CO 80020/303-469-1857, 800-876-8600; FAX: 303-466-7437
Allen Firearm Engraving, 339 Grove Ave., Prescott, AZ 86301/520-778-1237
Allen Mfg., 6449 Hodgson Rd., Circle Pines, MN 55014/612-429-8231
Allen Sportswear, Bob (See Allen Co., Bob)
Alley Supply Co., P.O. Box 848, Gardnerville, NV 89410/702-782-3800
Alliant Techsystems, Smokeless Powder Group, 200 Valley Rd., Suite 305, Mt. Arlington, NJ 07856/800-276-9337; FAX: 201-770-2528
Allred Bullet Co., 932 Evergreen Drive, Logan, UT 84321/801-752-6983; FAX: 801-752-6983
Alpec Team, Inc., 201 Ricken Backer Cir., Livermore, CA 94550/510-606-8245; FAX: 510-606-4279
Alpha 1 Drop Zone, 2121 N. Tyler, Wichita, KS 67212/316-729-0800
Alpha Gunsmith Division, 1629 Via Monserate, Fallbrook, CA 92028/619-723-9279, 619-728-2663
Alpha LaFranck Enterprises, P.O. Box 81072, Lincoln, NE 68501/402-466-3193
Alpha Precision, Inc., 2765-B Preston Rd. NE, Good Hope, GA 30641/770-267-6163
Alpine's Precision Gunsmithing & Indoor Shooting Range, 2401 Government Way, Coeur d'Alene, ID 83814/208-765-3559; FAX: 208-765-3559
Altamont Co., 901 N. Church St., P.O. Box 309, Thomasboro, IL 61878/217-643-3125, 800-626-5774; FAX: 217-643-7973
Alumna Sport by Dee Zee, 1572 NE 58th Ave., P.O. Box 3090, Des Moines, IA 50316/800-798-9899
AmBr Software Group Ltd., P.O. Box 301, Reisterstown, MD 21136-0301/800-888-1917; FAX: 410-526-7212
American Ammunition, 3545 NW 71st St., Miami, FL 33147/305-835-7400; FAX: 305-694-0037
American Arms, Inc., 715 Armour Rd., N. Kansas City, MO 64116/816-474-3161; FAX: 816-474-1225
American Derringer Corp., 127 N. Lacy Dr., Waco, TX 76705/800-642-7817, 817-799-9111; FAX: 817-799-7935
American Display Co., 55 Cromwell St., Providence, RI 02907/401-331-2464; FAX: 401-421-1264

30th EDITION, 1998 **309**

DIRECTORY OF THE ARMS TRADE

American Frontier Firearms Mfg. Inc., P.O. 744, Aguanga, CA 92536/909-763-0014; FAX: 909-763-0014
American Gas & Chemical Co., Ltd., 220 Pegasus Ave., Northvale, NJ 07647/201-767-7300
American Gripcraft, 3230 S. Dodge 2, Tucson, AZ 85713/602-790-1222
American Handgunner Magazine, 591 Camino de la Reina, Suite 200, San Diego, CA 92108/619-297-5350; FAX: 619-297-5353
American Pioneer Video, P.O. Box 50049, Bowling Green, KY 42102-2649/800-743-4675
American Products Inc., 14729 Spring Valley Road, Morrison, IL 61270/815-772-3336; FAX: 815-772-8046
American Safe Arms, Inc., 1240 Riverview Dr., Garland, UT 84312/801-257-7472; FAX: 801-785-8156
American Sales & Kirkpatrick, P.O. Box 677, Laredo, TX 78042/210-723-6893; FAX: 210-725-0672
American Security Products Co., 11925 Pacific Ave., Fontana, CA 92337/909-685-9680, 800-421-6142; FAX: 909-685-9685
American Small Arms Academy, P.O. Box 12111, Prescott, AZ 86304/602-778-5623
American Target, 1328 S. Jason St., Denver, CO 80223/303-733-0433; FAX: 303-777-0311
American Target Knives, 1030 Brownwood NW, Grand Rapids, MI 49504/616-453-1998
American Whitetail Target Systems, P.O. Box 41, 106 S. Church St., Tennyson, IN 47637/812-567-4527
Americase, P.O. Box 271, 1610 E. Main, Waxahachie, TX 75165/800-880-3629; FAX: 214-937-8373
Ames Metal Products, 4323 S. Western Blvd., Chicago, IL 60609/773-523-3230; FAX: 773-523-3854
Amherst Arms, P.O. Box 1457, Englewood, FL 34295/941-475-2020; FAX: 941-473-1212
Ammo Load, Inc., 1560 E. Edinger, Suite G, Santa Ana, CA 92705/714-558-8858; FAX: 714-569-0319
Amrine's Gun Shop, 937 La Luna, Ojai, CA 93023/805-646-2376
Amsec, 11925 Pacific Ave., Fontana, CA 92337
A.M.T., 6226 Santos Diaz St., Irwindale, CA 91702/818-334-6629; FAX: 818-969-5247
Amtec 2000, Inc., 84 Industrial Rowe, Gardner, MA 01440/508-632-9608; FAX: 508-632-2300
Analog Devices, Box 9106, Norwood, MA 02062
Andela Tool & Machine, Inc., RD3, Box 246, Richfield Springs, NY 13439
Anderson Manufacturing Co., Inc., 22602 53rd Ave. SE, Bothell, WA 98021/206-481-1858; FAX: 206-481-7839
Andres & Dworsky, Bergstrasse 18, A-3822 Karlstein, Thaya, Austria, EUROPE, 0 28 44-285
Angelo & Little Custom Gun Stock Blanks, P.O. Box 240046, Dell, MT 59724-0046
Anics Firm, Inc., 3 Commerce Park Square, 23200 Chagrin Blvd., Suite 240, Beechwood, OH 44122/216-292-4363, 800-556-1582; FAX: 216-292-2588
Anschutz GmbH, Postfach 1128, D-89001 Ulm, Donau, GERMANY (U.S. importers—Accuracy International; AcuSport Corporation; Champion Shooters' Supply; Champion's Choice; Gunsmithing, Inc.)
Ansen Enterprises, Inc., 1506 W. 228th St., Torrance, CA 90501-5105/310-534-1837; FAX: 310-534-3162
Answer Products Co., 1519 Westbury Drive, Davison, MI 48423/810-653-2911
Anthony and George Ltd., Rt. 1, P.O. Box 45, Evington, VA 24550/804-821-8117
Antique American Firearms (See Carlson, Douglas R.)
Antique Arms Co., 1110 Cleveland Ave., Monett, MO 65708/417-235-6501
AO Safety Products, Div. of American Optical Corp. (See E-A-R, Inc., Div. of Cabot Safety Corp.
Apel GmbH, Ernst, Am Kirschberg 3, D-97218 Gerbrunn, GERMANY/0 (931) 707192
Aplan Antiques & Art, James O., HC 80, Box 793-25, Piedmont, SD 57769/605-347-5016
Arcadia Machine & Tool, Inc. (See AMT)
Arco Powder, HC-Rt. 1, P.O. Box 102, County Rd. 357, Mayo, FL 32066/904-294-3882; FAX: 904-294-1498
Aristocrat Knives, 1701 W. Wernsing Ave., Effingham, IL 62401/800-953-3436; FAX: 217-347-3083
Arizaga (See U.S. importer—Mandall Shooting Supplies, Inc.)
Arizona Ammunition, Inc., 21421 No. 14th Ave., Suite E, Phoenix, AZ 85727/602-516-9004; FAX: 602-516-9012
Arizona Custom Case, 1015 S. 23rd St., Phoenix, AZ 85034/602-273-0220
Arkansas Mallard Duck Calls, Rt. Box 182, England, AR 72046/501-842-3597
Arkfeld Mfg. & Dist. Co., Inc., 1230 Monroe Ave., Norfolk, NE 68702-0054/402-371-9430; 800-533-0676
ArmaLite, Inc., P.O. Box 299, Geneseo, IL 61254/309-944-6939; FAX: 309-944-6949
Armament Gunsmithing Co., Inc., 525 Rt. 22, Hillside, NJ 07205/908-686-0960
Armas Kemen S.A. (See U.S. importers—Kemen America; USA Sporting)
Armfield Custom Bullets, 4775 Caroline Drive, San Diego, CA 92115/619-582-7188; FAX: 619-287-3238
Armi Perazzi S.p.A., Via Fontanelle 1/3, 1-25080 Botticino Mattina, ITALY/030-2692591; FAX: 030 2692594 (U.S. importer—Perazzi USA, Inc.)
Armi San Marco (See U.S. importers—Taylor's & Co., Inc.; Cimarron Arms; IAR, Inc.)
Armi San Paolo, via Europa 172-A, I-25062 Concesio, 030-2751725 (BS) ITALY
Armi Sport (See U.S. importers—Cape Outfitters; Taylor's & Co., Inc.)

Armite Laboratories, 1845 Randolph St., Los Angeles, CA 90001/213-587-7768; FAX: 213-587-5075
Armoloy Co. of Ft. Worth, 204 E. Daggett St., Fort Worth, TX 76104/817-332-5604; FAX: 817-335-6517
Armor (See Buck Stop Lure Co., Inc.)
Armor Metal Products, P.O. Box 4609, Helena, MT 59604/406-442-5560; FAX: 406-442-5650
Armory Publications, Inc., 2615 N. 4th St., No. 620, Coeur d'Alene, ID 83814-3781/208-664-5061; FAX: 208-664-9906
Armoury, Inc., The, Rt. 202, Box 2340, New Preston, CT 06777/860-868-0001; FAX: 860-868-2919
A.R.M.S., Inc., 230 W. Center St., West Bridgewater, MA 02379-1620/508-584-7816; FAX: 508-588-8045
Arms & Armour Press, Wellington House, 125 Strand, London WC2R 0BB ENGLAND/0171-420-5555; FAX: 0171-240-7265
Arms Corporation of the Philippines, Bo. Parang Marikina, Metro Manila, PHILIPPINES/632-941-6243, 632-941-6244; FAX: 632-942-0682
Arms Craft Gunsmithing, 1106 Linda Dr., Arroyo Grande, CA 93420/805-481-2830
Arms Ingenuity Co., P.O. Box 1, 51 Canal St., Weatogue, CT 06089/203-658-5624
Arms, Programming Solutions (See Arms Software)
Arms Software, P.O. Box 1526, Lake Oswego, OR 97035/800-366-5559, 503-697-0533; FAX: 503-697-3337
Arms United Corp., 1018 Cedar St., Niles, MI 49120/616-683-6837
Armscorp USA, Inc., 4424 John Ave., Baltimore, MD 21227/410-247-6200; FAX: 410-247-6205
Armsport, Inc., 3950 NW 49th St., Miami, FL 33142/305-635-7850; FAX: 305-633-2877
Arnold Arms Co., Inc., P.O. Box 1011, Arlington, WA 98223/800-371-1011, 360-435-1011; FAX: 360-435-7304
Aro-Tek, Ltd., 206 Frontage Rd. North, Suite C, Pacific, WA 98047/206-351-2984; FAX: 206-833-4483
Arratoonian, Andy (See Horseshoe Leather Products)
Arrieta, S.L., Morkaiko, 5, 20870 Elgoibar, SPAIN/34-43-743150; FAX: 34-43-743154 (U.S. importers—Griffin & Howe; Jansma, Jack J.; New England Arms Co.; The Orvis Co., Inc.; Quality Arms, Inc.; Wingshooting Adventures)
Art Jewel Enterprises Ltd., Eagle Business Ctr., 460 Randy Rd., Carol Stream, IL 60188/708-260-0400
Art's Gun & Sport Shop, Inc., 6008 Hwy. Y, Hillsboro, MO 63050
Artistry in Leather (See Stuart, V. Pat)
Artistry in Wood, 134 Zimmerman Rd., Kalispell, MT 59901/406-257-9003
Arundel Arms & Ammunition, Inc., A., 24A Defense St., Annapolis, MD 21401/410-224-8683
Ashby Turkey Calls, P.O. Box 1466, Ava, MO 65608-1466/417-967-3787
Aspen Outfitting Co., 520 East Cooper Ave., Aspen, CO 81611
A-Square Co., Inc., One Industrial Park, Bedford, KY 40006-9667/502-255-7456; FAX: 502-255-7657
Astra Sport, S.A., Apartado 3, 48300 Guernica, Espagne, SPAIN/34-4-6250100; FAX: 34-4-6255186 (U.S. importer—E.A.A. Corp.; P.S.M.G. Gun Co.)
Atamec-Bretton, 19, rue Victor Grignard, F-42026 St.-Etienne (Cedex 1) FRANCE/77-93-54-69; FAX: 33-77-93-57-98 (U.S. importer—Mandall Shooting Supplies, Inc.)
A-Tech Corp., P.O. Box 1281, Cottage Grove, OR 97424
Atlanta Cutlery Corp., 2143 Gees Mill Rd., Box 839 CIS, Conyers, GA 30207/800-883-0300; FAX: 404-388-0246
Atlantic Mills, Inc., 1295 Towbin Ave., Lakewood, NJ 08701-5934/800-242-7374
Atlantic Research Marketing Systems (See A.R.M.S., Inc.)
Atlantic Rose, Inc., P.O. Box 1305, Union, NJ 07083
Atsko/Sno-Seal, Inc., 2530 Russell SE, Orangeburg, SC 29115/803-531-1820; FAX: 803-531-2139
Audette, Creighton, 19 Highland Circle, Springfield, VT 05156/802-885-2331
Austin's Calls, Bill, Box 284, Kaycee, WY 82639/307-738-2552
Autauga Arms, Inc., Pratt Plaza Mall No. 13, Prattville, AL 36067/800-262-9563; FAX: 334-361-2961
Auto Arms, 738 Clearview, San Antonio, TX 78228/512-434-5450
Automatic Equipment Sales, 627 E. Railroad Ave., Salesburg, MD 21801
Auto-Ordnance Corp., Williams Lane, West Hurley, NY 12491/914-679-4190
Autumn Sales, Inc. (Blaser) 1320 Lake St., Fort Worth, TX 76102/817-335-1634; FAX: 817-338-0119
AWC Systems Technology, P.O. Box 41938, Phoenix, AZ 85080-1938/602-780-1050
AYA (See U.S. importer—New England Custom Gun Service)
A Zone Bullets, 2039 Walter Rd., Billings, MT 59105/800-252-3111; 406-248-1961
Aztec International Ltd., P.O. Box 2616, Clarkesville, GA 30523/706-754-8282; FAX: 706-754-6889

B

B&D Trading Co., Inc., 3935 Fair Hill Rd., Fair Oaks, CA 95628/800-334-3790, 916-967-9366; FAX: 916-967-4873
B&G Bullets (See Northside Gun Shop)
Badger Shooters Supply, Inc., P.O. Box 397, Owen, WI 54460/800-424-9069; FAX: 715-229-2332
Baekgaard Ltd., 1855 Janke Dr., Northbrook, IL 60062/708-498-3040; FAX: 708-493-3106
Baelder, Harry, Alte Goennebeker Strasse 5, 24635 Rickling, GERMANY/04328-722732; FAX: 04328-722733

MANUFACTURERS' DIRECTORY

Baer Custom, Les, Inc., 29601 34th Ave., Hillsdale, IL 61257/309-658-2716; FAX: 309-658-2610
Baer's Hollows, P.O. Box 284, Eads, CO 81036/719-438-5718
Bagmaster Mfg., Inc., 2731 Sutton Ave., St. Louis, MO 63143/314-781-8002; FAX: 314-781-3363; WEB: http://www.bagmaster.com
Bain & Davis, Inc., 307 E. Valley Blvd., San Gabriel, CA 91776-3522/818-573-4241, 213-283-7449
Baker, Stan, 10,000 Lake City Way, Seattle, WA 98125/206-522-4575
Baker's Leather Goods, Roy, P.O. Box 893, Magnolia, AR 71753/501-234-0344
Balaance Co., 340-39 Ave. S.E. Box 505, Calgary, AB, T2G 1X6 CANADA
Bald Eagle Precision Machine Co., 101-A Allison St., Lock Haven, PA 17745/717-748-6772; FAX: 717-748-4443
Balickie, Joe, 408 Trelawney Lane, Apex, NC 27502/919-362-5185
Ballard Built, P.O. Box 1443, Kingsville, TX 78364/512-592-0853
Ballard Industries, 10271 Lockwood Dr., Suite B, Cupertino, CA 95014/408-996-0957; FAX: 408-257-6828
Ballistic Engineering & Software, Inc., 185 N. Park Blvd., Suite 330, Lake Orion, MI 48362/313-391-1074
Ballistic Products, Inc., 20015 75th Ave. North, Corcoran, MN 55340-9456/612-494-9237; FAX: 612-494-9236
Ballistic Program Co., Inc., The, 2417 N. Patterson St., Thomasville, GA 31792/912-228-5739, 800-368-0835
Ballistic Research, 1108 W. May Ave., McHenry, IL 60050/815-385-0037
Ballistica Maximus North, 107 College Park Plaza, Johnstown, PA 15904/814-266-8380
Ballisti-Cast, Inc., Box 383, Parshall, ND 58770/701-862-3324; FAX: 701-862-3331
Bandcor Industries, Div. of Man-Sew Corp., 6108 Sherwin Dr., Port Richey, FL 34668/813-848-0432
Bang-Bang Boutique (See Holster Shop, The)
Banks, Ed, 2762 Hwy. 41 N., Ft. Valley, GA 31030/912-987-4665
Bansner's Gunsmithing Specialties, 261 East Main St. Box VH, Adamstown, PA 19501/800-368-2379; FAX: 717-484-0523
Barami Corp., 6689 Orchard Lake Rd. No. 148, West Bloomfield, MI 48322/810-738-0462; FAX: 810-855-4084
Barbour, Inc., 55 Meadowbrook Dr., Milford, NH 03055/603-673-1313; FAX: 603-673-6510
Barnes Bullets, Inc., P.O. Box 215, American Fork, UT 84003/801-756-4222, 800-574-9200; FAX: 801-756-2465; WEB: http://www.itsnet.com/home/bbullets
Baron Technology, 62 Spring Hill Rd., Trumbull, CT 06611/203-452-0515; FAX: 203-452-0663
Barraclough, John K., 55 Merit Park Dr., Gardena, CA 90247/310-324-2574
Barramundi Corp., P.O. Drawer 4259, Homosassa Springs, FL 32687/904-628-0200
Barrett Firearms Manufacturer, Inc., P.O. Box 1077, Murfreesboro, TN 37133/615-896-2938; FAX: 615-896-7313
Barsotti, Bruce (See River Road Sporting Clays)
Bar-Sto Precision Machine, 73377 Sullivan Rd., P.O. Box 1838, Twentynine Palms, CA 92277/619-367-2747; FAX: 619-367-2407
Barta's Gunsmithing, 10231 US Hwy. 10, Cato, WI 54206/414-732-4472
Barteaux Machete, 1916 SE 50th Ave., Portland, OR 97215-3238/503-233-5880
Bartlett, Don, P.O. Box 55, Colbert, WA 99005/509-467-5009
Bartlett Engineering, 40 South 200 East, Smithfield, UT 84335-1645/801-563-5910
Basics Information Systems, Inc., 1141 Georgia Ave., Suite 515, Wheaton, MD 20902/301-949-1070; FAX: 301-949-5326
Bates Engraving, Billy, 2302 Winthrop Dr., Decatur, AL 35603/205-355-3690
Bauer, Eddie, 15010 NE 36th St., Redmond, WA 98052
Baumgartner Bullets, 3011 S. Alane St., W. Valley City, UT 84120
Bausch & Lomb Sports Optics Div. (See Bushnell Sports Optics Worldwide)
Bauska Barrels, 105 9th Ave. W., Kalispell, MT 59901/406-752-7706
Bear Archery, RR 4, 4600 Southwest 41st Blvd., Gainesville, FL 32601/904-376-2327
Bear Arms, 121 Rhodes St., Jackson, SC 29831/803-471-9859
Bear Hug Grips, Inc., P.O. Box 16649, Colorado Springs, CO 80935-6649/800-232-7710
Bear Mountain Gun & Tool, 120 N. Plymouth, New Plymouth, ID 83655/208-278-5221; FAX: 208-278-5221
Beartooth Bullets, P.O. Box 491, Dept. HLD, Dover, ID 83825-0491/208-448-1865
Beauchamp & Son, Inc., 160 Rossiter Rd., P.O. Box 181, Richmond, MA 01254/413-698-3822; FAX: 413-698-3866
Beaver Lodge (See Fellowes, Ted)
Beaver Park Products, Inc., 840 J St., Penrose, CO 81240/719-372-6744
BEC, Inc., 1227 W. Valley Blvd., Suite 204, Alhambra, CA 91803/818-281-5751; FAX:818-293-7073
Beeline Custom Bullets Limited, P.O. Box 85, Yarmouth, Nova Scotia CANADA B5A 4B1/902-648-3494; FAX: 902-648-0253
Beeman Precision Airguns, 5454 Argosy Dr., Huntington Beach, CA 92649/714-890-4800; FAX: 714-890-4808
Behlert Precision, Inc., P.O. Box 288, 7067 Easton Rd., Pipersville, PA 18947/215-766-8681, 215-766-7301; FAX: 215-766-8681
Beitzinger, George, 116-20 Atlantic Ave., Richmond Hill, NY 11419/718-847-7661
Belding's Custom Gun Shop, 10691 Sayers Rd., Munith, MI 49259/517-596-2388
Bell & Carlson, Inc., Dodge City Industrial Park/101 Allen Rd., Dodge City, KS 67801/800-634-8586, 316-225-6688; FAX: 316-225-9095
Bell Alaskan Silversmith, Sid (See Heritage Wildlife Carvings)
Bell Reloading, Inc., 1725 Harlin Lane Rd., Villa Rica, GA 30180
Bell's Gun & Sport Shop, 3309-19 Mannheim Rd, Franklin Park, IL 60131
Bell's Legendary Country Wear, 22 Circle Dr., Bellmore, NY 11710/516-679-1158
Bellm Contenders, P.O. Box 459, Cleveland, UT 84518/801-653-2530
Belltown, Ltd., 11 Camps Rd., Kent, CT 06757/860-354-5750
Belt MTN Arms, 107 10th Ave. SW, White Sulphur Springs, MT 59645/406-586-4495
Ben's Machines, 1151 S. Cedar Ridge, Duncanville, TX 75137/214-780-1807; FAX: 214-780-0316
Benchmark Guns, 12593 S. Ave. 5 East, Yuma, AZ 85365
Benchmark Knives (See Gerber Legendary Blades)
Benelli Armi, S.p.A., Via della Stazione, 61029 Urbino, ITALY/39-722-307-1; FAX: 39-722-327427 (U.S. importers—Heckler & Koch, Inc.; Whitestone Lumber Co.)
Bengtson Arms Co., L., 6345-B E. Akron St., Mesa, AZ 85205/602-981-6375
Benjamin/Sheridan Co., Crossman, Rts. 5 and 20, E. Bloomfield, NY 14443/716-657-6161; FAX: 716-657-5405
Bentley, John, 128-D Watson Dr., Turtle Creek, PA 15145
Beomat of America Inc., 300 Railway Ave., Campbell, CA 95008/408-379-4829
Beretta S.p.A., Pietro, Via Beretta, 18-25063 Gardone V.T. (BS) ITALY/XX39/30-8341.1; FAX: XX39/30-8341.421 (U.S. importer—Beretta U.S.A. Corp.)
Beretta U.S.A. Corp., 17601 Beretta Drive, Accokeek, MD 20607/301-283-2191; FAX: 301-283-0435
Berger Bullets, Ltd., 5342 W. Camelback Rd., Suite 200, Glendale, AZ 85301/602-842-4001; FAX: 602-934-9083
Bergman & Williams, 2450 Losee Rd., Suite F, Las Vegas, NV 89030/702-642-1901; FAX: 702-642-1540
Bernardelli S.p.A., Vincenzo, 125 Via Matteotti, P.O. Box 74, Gardone V.T., Brescia ITALY, 25063/39-30-8912851-2-3; FAX: 39-30-8910249 (U.S. importer—Armsport, Inc.)
Berry's Bullets (See Berry's Mfg., Inc.)
Berry's Mfg., Inc., 401 North 3050 East St., St. George, UT 84770/801-634-1682; FAX: 801-634-1683
Bersa S.A., Gonzales Castillo 312, 1704 Ramos Mejia, ARGENTINA/541-656-2377; FAX: 541-656-2093 (U.S. importer—Eagle Imports, Inc.)
Bertram Bullet Co., P.O. Box 313, Seymour, Victoria 3660, AUSTRALIA/61-57-922912; FAX: 61-57-991650
Bertuzzi (See U.S. importer—New England Arms Co.)
Better Concepts Co., 663 New Castle Rd., Butler, PA 16001/412-285-9000
Beverly, Mary, 3201 Horseshoe Trail, Tallahassee, FL 32312
Bianchi International, Inc., 100 Calle Cortez, Temecula, CA 92590/909-676-5621; FAX: 909-676-6777
Biesen, Al, 5021 Rosewood, Spokane, WA 99208/509-328-9340
Biesen, Roger, 5021 W. Rosewood, Spokane, WA 99208/509-328-9340
Big Beam Emergency Systems, Inc., 290 E. Prairie St., Crystal Lake, IL 60039
Big Bear Arms & Sporting Goods, Inc., 1112 Milam Way, Carrollton, TX 75006/972-416-8051, 800-400-BEAR; FAX: 972-416-0771
Big Bore Bullets of Alaska, P.O. Box 872785, Wasilla, AK 99687/907-373-2673; FAX: 907-373-2673
Big Bore Express, 7154 W. State St., Boise, ID 83703/800-376-4010; FAX:208-376-4020
Big Sky Racks, Inc., P.O. Box 729, Bozeman, MT 59771-0729/406-586-9393; FAX: 406-585-7378
Big Spring Enterprises "Bore Stores", P.O. Box 1115, Big Spring, Rd./Yellville, AR 72687
501-449-5297; FAX: 501-449-4446; E-MAIL: BIGSPRNG@mtnhome.com
Bilal, Mustafa, 908 NW 50th St., Seattle, WA 98107-3634/206-782-4164
Bill's Custom Cases, P.O. Box 2, Dunsmuir, CA 96025/916-235-0177; FAX: 916-235-4959
Bill's Gun Repair, 1007 Burlington St., Mendota, IL 61342/815-539-5786
Billeb, Stephen L., 1101 N. 7th St., Burlington, IA 52601/319-753-2110
Billings Gunsmiths, Inc., 1841 Grand Ave., Billings, MT 59102/406-256-8390
Billingsley & Brownell, P.O. Box 25, Dayton, WY 82836/307-655-9344
Birchwood Casey, 7900 Fuller Rd., Eden Prairie, MN 55344/800-328-6156, 612-937-7933; FAX: 612-937-7979
Birdsong & Assoc., W.E., 1435 Monterey Rd., Florence, MS 39073-9748/601-366-8270
Bismuth Cartridge Co., 3500 Maple Ave., Suite 1650, Dallas, TX 75219/800-759-3333, 214-521-5880; FAX: 214-521-9035
Bison Studios, 1409 South Commerce St., Las Vegas, NV 89102/702-388-2891; FAX: 702-383-9967
Bitterroot Bullet Co., Box 412, Lewiston, ID 83501-0412/208-743-5635
Black Belt Bullets (See Big Bore Express)
Black Hills Ammunition, Inc., P.O. Box 3090, Rapid City, SD 57709-3090/605-348-5150; FAX: 605-348-9827
Black Hills Shooters Supply, P.O. Box 4220, Rapid City, SD 57709/800-289-2506
Black Powder Products, 67 Township Rd. 1411, Chesapeake, OH 45619/614-867-8047
Black Sheep Brand, 3220 W. Gentry Parkway, Tyler, TX 75702/903-592-3853; FAX: 903-592-0527
Blackhawk East, Box 2274, Loves Park, IL 61131
Blackhawk West, Box 285, Hiawatha, KS 66434
Blackinton & Co., Inc., V.H., 221 John L. Dietsch, Attleboro Falls, MA 02763-0300/508-699-4436; FAX: 508-695-5349
Blackjack Knives, Ltd., 1307 W. Wabash, Effingham, IL 62401/217-347-7700; FAX: 217-347-7737

DIRECTORY OF THE ARMS TRADE

Blacksmith Corp., 830 N. Road No. 1 E., P.O. Box 1752, Chino Valley, AZ 86323/520-636-4456; FAX: 520-636-4457
BlackStar AccuMax Barrels, 11501 Brittmoore Park Drive, Houston, TX 77041/281-721-6040; FAX: 281-721-6041
BlackStar Barrel Accurizing (See BlackStar AccuMax Barrels)
Blacktail Mountain Books, 42 First Ave. W., Kalispell, MT 59901/406-257-5573
Blair Engraving, J.R., P.O. Box 64, Glenrock, WY 82637/307-436-8115
Blammo Ammo, P.O. Box 1677, Seneca, SC 29679/803-882-1768
Blaser Jagdwaffen GmbH, D-88316 Isny Im Allgau, GERMANY (U.S. importer—Autumn Sales, Inc.)
Bleile, C. Roger, 5040 Ralph Ave., Cincinnati, OH 45238/513-251-0249
Blocker Holsters, Inc., Ted., Clackamas Business Park Bld. A, 14787 S.E. 82nd, Dr./Clackamas, OR 97015 503-557-7757; FAX: 503-557-3771
Blount, Inc., Sporting Equipment Div., 2299 Snake River Ave., P.O. Box 856, Lewiston, ID 83501/800-627-3640, 208-746-2351; FAX: 208-799-3904
Blue and Gray Products, Inc. (See Ox-Yoke Originals, Inc.)
Blue Book Publications, Inc., One Appletree Square, 8009 34th Ave. S., Suite, 175/Minneapolis, MN 55425 800-877-4867, 612-854-5229; FAX: 612-853-1486
Blue Mountain Bullets, HCR 77, P.O. Box 231, John Day, OR 97845/541-820-4594
Blue Ridge Machinery & Tools, Inc., P.O. Box 536-GD, Hurricane, WV 25526/800-872-6500; FAX: 304-562-5311
BMC Supply, Inc., 26051 - 179th Ave. S.E., Kent, WA 98042
B.M.F. Activator, Inc., 803 Mill Creek Run, Plantersville, TX 77363/409-894-2005, 800-527-2881
Bob's Gun Shop, P.O. Box 200, Royal, AR 71968/501-767-1970
Bob's Tactical Indoor Shooting Range & Gun Shop, 122 Lafayette Rd., Salisbury, MA 01952/508-465-5561
Boessler, Erich, Am Vogeltal 3, 97702 Munnerstadt, GERMANY/9733-9443
Boggs, Wm., 1816 Riverside Dr. C, Columbus, OH 43212/614-486-6965
Bohemia Arms Co., 17101 Los Modelos, Fountain Valley, CA 92708/619-442-7005; FAX: 619-442-7005
Boker USA, Inc., 1550 Balsam Street, Lakewood, CO 80215/303-462-0662; FAX: 303-462-0668
Boltin, John M., P.O. Box 644, Estill, SC 29918/803-625-2185
Bo-Mar Tool & Mfg. Co., Rt. 8, Box 405, Longview, TX 75604/903-759-4784; FAX: 903-759-9141
Bonanza (See Forster Products)
Bond Arms, Inc., P.O. Box 1296, Granbury, TX 76048/817-573-5733; FAX: 817-573-5636
Bond Custom Firearms, 8954 N. Lewis Ln., Bloomington, IN 47408/812-332-4519
Bondini Paolo, Via Sorrento, 345, San Carlo di Cesena, ITALY I-47020/0547 663 240; FAX: 0547 663 780 (U.S. importer—blackpowder arms)
Bone Engraving, Ralph, 718 N. Atlanta, Owasso, OK 74055/918-272-9745
Boone Trading Co., Inc., P.O. Box BB, Brinnan, WA 98320
Boone's Custom Ivory Grips, Inc., 562 Coyote Rd., Brinnon, WA 98320/206-796-4330
Boonie Packer Products, P.O. Box 12204, Salem, OR 97309/800-477-3244, 503-581-3244; FAX: 503-581-3191
Borden's Accuracy, RD 1, Box 250BC, Springville, PA 18844/717-965-2505; FAX: 717-965-2328
Border Barrels Ltd., Riccarton Farm, Newcastleton SCOTLAND U.K. TD9 0SN
Borovnik KG, Ludwig, 9170 Ferlach, Bahnhofstrasse 7, AUSTRIA/042 27 24 42; FAX: 042 26 43 49
Bosis (See U.S. importer—New England Arms Co.)
Boss Manufacturing Co., 221 W. First St., Kewanee, IL 61443/309-852-2131, 800-447-4581; FAX: 309-852-0848
Bostick Wildlife Calls, Inc., P.O. Box 728, Estill, SC 29918/803-625-2210, 803-625-4512
Bowen Classic Arms Corp., P.O. Box 67, Louisville, TN 37777/423-984-3583
Bowen Knife Co., Inc., P.O. Box 590, Blackshear, GA 31516/912-449-4794
Bowerly, Kent, 26247 Metolius Meadows Dr., Camp Sherman, OR 97730/541-595-6028
Boyds' Gunstock Industries, Inc., 3rd & Main, P.O. Box 305, Geddes, SD 57342/605-337-2125; FAX: 605-337-3363
Boyt, 509 Hamilton, P.O. Drawer 668, Iowa Falls, IA 50126/515-648-4626; FAX: 515-648-2385
Brace, Larry D., 771 Blackfoot Ave., Eugene, OR 97404/503-688-1278
Bradley Gunsight Co., P.O. Box 340, Plymouth, VT 05056/860-589-0531; FAX: 860-582-6294
Brass and Bullet Alloys, P.O. Box 1238, Sierra Vista, AZ 85636/602-458-5321; FAX: 602-458-9125
Brass Eagle, Inc., 7050A Bramalea Rd., Unit 19, Mississauga, Ont. L4Z 1C7, CANADA/416-848-4844
Bratcher, Dan, 311 Belle Air Pl., Carthage, MO 64836/417-358-1518
Brauer Bros. Mfg. Co., 2020 Delman Blvd., St. Louis, MO 63103/314-231-2864; FAX: 314-249-4952
Braverman Corp., R.J., 88 Parade Rd., Meridith, NH 03293/800-736-4867
Break-Free, Inc., P.O. Box 25020, Santa Ana, CA 92799/714-953-1900; FAX: 714-953-0402
Brenneke KG, Wilhelm, Ilmenauweg 2, 30851 Langenhagen, GERMANY/ 0511/97262-0; FAX: 0511/97262-62 (U.S. importer—Dynamit Nobel-RWS, Inc.)
Bretton (See Atamec-Bretton)
Bridgers Best, P.O. Box 1410, Berthoud, CO 80513

Briese Bullet Co., Inc., RR1, Box 108, Tappen, ND 58487/701-327-4578; FAX: 701-327-4579
Brigade Quartermasters, 1025 Cobb International Blvd., Dept. VH, Kennesaw, GA 30144-4300/404-428-1248, 800-241-3125; FAX: 404-426-7726
Briganti, A.J., 512 Rt. 32, Highland Mills, NY 10930/914-928-9573
Briley Mfg., Inc., 1230 Lumpkin, Houston, TX 77043/800-331-5718, 713-932-6995; FAX: 713-932-1043
British Antiques, P.O. Box 7, Latham, NY 12110/518-783-0773
British Sporting Arms, RR1, Box 130, Millbrook, NY 12545/914-677-8303
BRNO (See U.S. importers—Bohemia Arms Co.)
Broad Creek Rifle Works, 120 Horsey Ave., Laurel, DE 19956/302-875-5446
Brobst, Jim, 299 Poplar St., Hamburg, PA 19526/215-562-2103
Brockman's Custom Gunsmithing, P.O. Box 357, Gooding, ID 83330/208-934-5050
Brocock Ltd., 43 River Street, Digbeth, Birmingham, B5 5SA ENGLAND/011-021-773-1200
Broken Gun Ranch, 10739 126 Rd., Spearville, KS 67876/316-385-2587; FAX: 316-385-2597
Brolin Arms, 2755 Thompson Creek Rd., Pomona, CA 91767/909-392-2350; FAX: 909-392-2354
Brooker, Dennis, Rt. 1, Box 12A, Derby, IA 50068/515-533-2103
Brooks Tactical Systems, 279-A Shorewood Ct., Fox Island, WA 98333/800-410-4747; FAX: 206-572-6797
Brown Co., E. Arthur, 3404 Pawnee Dr., Alexandria, MN 56308/320-762-8847
Brown, H.R. (See Silhouette Leathers)
Brown Dog Ent., 2200 Calle Camelia, 1000 Oaks, CA 91360/805-497-2318; FAX: 805-497-1618
Brown Manufacturing, P.O. Box 9219, Akron, OH 44305/800-837-GUNS
Brown Precision, Inc., 7786 Molinos Ave., Los Molinos, CA 96055/916-384-2506; FAX: 916-384-1638
Brown Products, Inc., Ed, Rt. 2, Box 492, Perry, MO 63462/573-565-3261; FAX: 573-565-2791
Brownell Checkering Tools, W.E., 9390 Twin Mountain Circle, San Diego, CA 92126/619-695-2479; FAX: 619-695-2479
Brownells, Inc., 200 S. Front St., Montezuma, IA 50171/515-623-5401; FAX: 515-623-3896
Browning Arms Co., One Browning Place, Morgan, UT 84050/801-876-2711; FAX: 801-876-3331
Browning Arms Co. (Parts & Service), 3005 Arnold Tenbrook Rd., Arnold, MO 63010-9406/314-287-6800; FAX: 314-287-9751
BRP, Inc., High Performance Cast Bullets, 1210 Alexander Rd., Colorado Springs, CO 80909/719-633-0658
Bruno Shooters Supply, 111 N. Wyoming St., Hazleton, PA 18201/717-455-2281; FAX: 717-455-2211
Brunton U.S.A., 620 E. Monroe Ave., Riverton, WY 82501/307-856-6559; FAX: 307-856-1840
Brynin, Milton, P.O. Box 383, Yonkers, NY 10710/914-779-4333
BSA Guns Ltd., Armoury Rd. Small Heath, Birmingham, ENGLAND B11 2PX/011-021-772-8543; FAX: 011-021-773-0845 (U.S. importers—Groenewold, John; Precision Sales International, Inc.)
B-Square Company, Inc., P.O. Box 11281, 2708 St. Louis Ave., Ft. Worth, TX 76110/817-923-0964, 800-433-2909; FAX: 817-926-7012
Bucheimer, J.M., Jumbo Sports Products, 721 N. 20th St., St. Louis, MO 63103/314-241-1020
Buck Knives, Inc., 1900 Weld Blvd., P.O. Box 1267, El Cajon, CA 92020/619-449-1100, 800-326-2825; FAX: 619-562-5774, 800-729-2825
Buck Stix—SOS Products Co., Box 3, Neenah, WI 54956
Buck Stop Lure Co., Inc., 3600 Grow Rd. NW, P.O. Box 636, Stanton, MI 48888/517-762-5091; FAX: 517-762-5124
Buckeye Custom Bullets, 6490 Stewart Rd., Elida, OH 45807/419-641-4463
Buckhorn Gun Works, 8109 Woodland Dr., Black Hawk, SD 57718/605-787-6472
Buckskin Bullet Co., P.O. Box 1893, Cedar City, UT 84721/801-586-3286
Buckskin Machine Works, A. Hunkeler, 3235 S. 358th St., Auburn, WA 98001/206-927-5412
Budin, Dave, Main St., Margaretville, NY 12455/914-568-4103; FAX: 914-586-4105
Buenger Enterprises/Goldenrod Dehumidifier, 3600 S. Harbor Blvd., Oxnard, CA 93035/800-451-6797, 805-985-5828; FAX: 805-985-1534
Buffalo Arms Co., 3355 Upper Gold Creek Rd., Samuels, ID 83864/208-263-6953; FAX: 208-265-2096
Buffalo Bullet Co., Inc., 12645 Los Nietos Rd., Unit A, Santa Fe Springs, CA 90670/310-944-0322; FAX: 310-944-5054
Buffalo Rock Shooters Supply, R.R. 1, Ottawa, IL 61350/815-433-2471
Buffer Technologies, P.O. Box 104930, Jefferson City, MO 65110/573-634-8529; FAX: 573-634-8522
Bull Mountain Rifle Co., 6327 Golden West Terrace, Billings, MT 59106/406-656-0778
Bullberry Barrel Works, Ltd., 2430 W. Bullberry Ln. 67-5, Hurricane, UT 84737/801-635-9866; FAX: 801-635-0348
Bullet, Inc., 3745 Hiram Alworth Rd., Dallas, GA 30132
Bullet'n Press, 19 Key St., Eastport, Maine 04631/207-853-4116
Bullet Swaging Supply, Inc., P.O. Box 1056, 303 McMillan Rd, West Monroe, LA 71291/318-387-7257; FAX: 318-387-7779
BulletMakers Workshop, The, RFD 1 Box 1755, Brooks, ME 04921
Bullseye Bullets, 1610 State Road 60, No. 12, Valrico, FL 33594/813-654-6563
Bull-X, Inc., 520 N. Main, Farmer City, IL 61842/309-928-2574, 800-248-3845 orders only; FAX: 309-928-2130

MANUFACTURERS' DIRECTORY

Burgess, Byron, P.O. Box 6853, Los Osos, CA 93412/805-528-1005
Burgess & Son Gunsmiths, R.W., P.O. Box 3364, Warner Robins, GA 31099/912-328-7487
Burkhart Gunsmithing, Don, P.O. Box 852, Rawlins, WY 82301/307-324-6007
Burnham Bros., P.O. Box 1148, Menard, TX 78659/915-396-4572; FAX: 915-396-4574
Burres, Jack, 10333 San Fernando Rd., Pacoima, CA 91331/818-899-8000
Burris Co., Inc., P.O. Box 1747, 331 E. 8th St., Greeley, CO 80631/970-356-1670; FAX: 970-356-8702
Bushmann Hunters & Safaris, P.O. Box 293088, Lewisville, TX 75029/214-317-0768
Bushmaster Firearms (See Quality Parts Co./Bushmaster Firearms)
Bushmaster Hunting & Fishing, 451 Alliance Ave., Toronto, Ont. M6N 2J1 CANADA/416-763-4040; FAX: 416-763-0623
Bushnell Sports Optics Worldwide, 9200 Cody, Overland Park, KS 66214/913-752-3400, 800-423-3537; FAX: 913-752-3550
Bushwacker Backpack & Supply Co. (See Counter Assault)
Bustani, Leo, P.O. Box 8125, W. Palm Beach, FL 33407/305-622-2710
Buster's Custom Knives, P.O. Box 214, Richfield, UT 84701/801-896-5319
Butler Creek Corp., 290 Arden Dr., Belgrade, MT 59714/800-423-8327, 406-388-1356; FAX: 406-388-7204
Butler Enterprises, 834 Oberting Rd., Lawrenceburg, IN 47025/812-537-3584
Butterfield & Butterfield, 220 San Bruno Ave., San Francisco, CA 94103/415-861-7500
Buzztail Brass (See Grayback Wildcats)
B-West Imports, Inc., 2425 N. Huachuca Dr., Tucson, AZ 85745-1201/602-628-1990; FAX: 602-628-3602

C

C3 Systems, 678 Killingly St., Johnston, RI 02919
C&D Special Products (See Claybuster Wads & Harvester Bullets)
C&H Research, 115 Sunnyside Dr., Box 351, Lewis, KS 67552/316-324-5445
C&J Enterprises, Inc., 7101 Jurupa Ave., No. 12, Riverside, CA 92504/909-689-7758
C&T Corp. TA Johnson Brothers, 1023 Wappoo Road, Charleston, SC 29407-5960
Cabanas (See U.S. importer—Mandall Shooting Supplies, Inc.)
Cabela's, 812-13th Ave., Sidney, NE 69160/308-254-6644, 800-237-4444; FAX: 308-254-6745
Cabinet Mtn. Outfitters Scents & Lures, P.O. Box 766, Plains, MT 59859/406-826-3970
Cache La Poudre Rifleworks, 140 N. College, Ft. Collins, CO 80524/303-482-6913
Cadre Supply (See Parts & Surplus)
Calhoon Varmint Bullets, James, Shambo Rt., 304, Havre, MT 59501/406-395-4079
Calibre Press, Inc., 666 Dundee Rd., Suite 1607, Northbrook, IL 60062-2760/800-323-0037; FAX: 708-498-6869
Cali'co Hardwoods, Inc., 3580 Westwind Blvd., Santa Rosa, CA 95403/707-546-4045; FAX: 707-546-4027
Calico Light Weapon Systems, 405 E. 19th St., Bakersfield, CA 93305/805-323-1327; FAX: 805-323-7844
California Magnum, 20746 Dearborn St., Chatsworth, CA 91313/818-341-7302; FAX: 818-341-7304
California Sights (See Fautheree, Andy)
Camdex, Inc., 2330 Alger, Troy, MI 48083/810-528-2300; FAX: 810-528-0989
Cameron's, 16690 W. 11th Ave., Golden, CO 80401/303-279-7365; FAX: 303-628-5413
Camilli, Lou, 600 Sandtree Dr., Suite 212, Lake Park, FL 33403-1538
Camillus Cutlery Co., 54 Main St., Camillus, NY 13031/315-672-8111; FAX: 315-672-8832
Campbell, Dick, 20,000 Silver Ranch Rd., Conifer, CO 80433/303-697-0150
Camp-Cap Products, P.O. Box 173, Chesterfield, MO 63006/314-532-4340; FAX: 314-532-4340
Canjar Co., M.H., 500 E. 45th Ave., Denver, CO 80216/303-295-2638; FAX: 303-295-2638
Cannon's, Andy Cannon, Box 1026, 320 Main St., Polson, MT 59860/406-887-2048
Cannon Safe, Inc., 9358 Stephens St., Pico Rivera, CA 90660/310-692-0636, 800-242-1055; FAX: 310-692-7252
Canons Delcour, Rue J.B. Cools, B-4040 Herstal, BELGIUM 32.(0)42.40.61.40; FAX: 32(0)42.40.22.88
Canyon Cartridge Corp., P.O. Box 152, Albertson, NY 11507/FAX: 516-294-8946
Cape Outfitters, 599 County Rd. 206, Cape Girardeau, MO 63701/314-335-4103; FAX: 314-335-1555
Caraville Manufacturing, P.O. Box 4545, Thousand Oaks, CA 91359/805-499-1234
Carbide Checkering Tools (See J&R Engineering)
Carbide Die & Mfg. Co., Inc., 15615 E. Arrow Hwy., Irwindale, CA 91706/818-337-2518
Carhartt, Inc., P.O. Box 600, 3 Parklane Blvd., Dearborn, MI 48121/800-358-3825, 313-271-8460; FAX: 313-271-3455
Custom Gun Engraving, D-97422 Schweinfurt, Siedlerweg 17, GERMANY/01149-9721-41446; FAX: 01149-9721-44413
Carlson, Douglas R., Antique American Firearms, P.O. Box 71035, Dept. GD, Des Moines, IA 50325/515-224-6552
Carnahan Bullets, 17645 110th Ave. SE, Renton, WA 98055
Carolina Precision Rifles, 1200 Old Jackson Hwy., Jackson, SC 29831/803-827-2069

Carrell's Precision Firearms, 643 Clark Ave., Billings, MT 59101-1614/406-962-3593
Carroll Bullets (See Precision Reloading, Inc.)
Carry-Lite, Inc., 5203 W. Clinton Ave., Milwaukee, WI 53223/414-355-3520; FAX: 414-355-4775
Carter's Gun Shop, 225 G St., Penrose, CO 81240/719-372-6240
Carter's Wildlife Calls, Inc., Garth, P.O. Box 821, Cedar City, UT 84720/801-586-7639
Cartridge Transfer Group, Pete de Coux, 235 Oak St., Butler, PA 16001/412-282-3426
Carvajal Belts & Holsters, 422 Chestnut, San Antonio, TX 78202/210-222-1634
Cascade Bullet Co., Inc., 2355 South 6th St., Klamath Falls, OR 97601/503-884-9316
Cascade Shooters, 2155 N.W. 12th St., Redwood, OR 97756
Case & Sons Cutlery Co., W.R., Owens Way, Bradford, PA 16701/814-368-4123, 800-523-6350; FAX: 814-768-5369
Case Sorting System, 12695 Cobblestone Creek Rd., Poway, CA 92064/619-486-9340
Cash Mfg. Co., Inc., P.O. Box 130, 201 S. Klein Dr., Waunakee, WI 53597-0130/608-849-5664; FAX: 608-849-5664
Caspian Arms Ltd., 14 North Main St., Hardwick, VT 05843/802-472-6454; FAX: 802-472-6709
Cast Performance Bullet Company, 12441 U.S. Hwy. 26, Riverton, WY 82501/307-856-4347
Casull Arms Corp., P.O. Box 1629, Afton, WY 83110/307-886-0200
Caswell International Corp., 1221 Marshall St. NE, Minneapolis, MN 55413-1055/612-379-2000; FAX: 612-379-2367
Catco-Ambush, Inc., P.O.Box 300, Corte Madera, CA 94926
Cathey Enterprises, Inc., P.O. Box 2202, Brownwood, TX 76804/915-643-2553; FAX: 915-643-3653
Cation, 2341 Alger St., Troy, MI 48083/810-689-0658; FAX: 810-689-7558
Catoctin Cutlery, P.O. Box 188, 17 S. Main St., Smithsburg, MD 21783/301-824-7416; FAX: 301-824-6138
Caywood, Shane J., P.O. Box 321, Minocqua, WI 54548/715-277-3866 evenings
CBC, Avenida Humberto de Campos, 3220, 09400-000 Ribeirao Pires-SP-BRAZIL/55-11-742-7500; FAX: 55-11-459-7385
CCG Enterprises, 5217 E. Belknap St., Halton City, TX 76117/800-819-7464
CCI, Div. of Blount, Inc., Sporting Equipment Div., 2299 Snake River Ave.,, P.O. Box 856/Lewiston, ID 83501
800-627-3640, 208-746-2351; FAX: 208-746-2915
Cedar Hill Game Calls, Inc., Rt. 2 Box 236, Downsville, LA 71234/318-982-5632; FAX: 318-368-2245
Celestron International, P.O. Box 3578, 2835 Columbia St., Torrance, CA 90503/310-328-9560; FAX: 310-212-5835
Centaur Systems, Inc., 1602 Foothill Rd., Kalispell, MT 59901/406-755-8609; FAX: 406-755-8609
Center Lock Scope Rings, 9901 France Ct., Lakeville, MN 55044/612-461-2114
CenterMark, P.O. Box 4066, Parnassus Station, New Kensington, PA 15068/412-335-1319
Central Specialties Ltd. (See Trigger Lock Division/Central Specialties Ltd.)
Century Gun Dist., Inc., 1467 Jason Rd., Greenfield, IN 46140/317-462-4524
Century International Arms, Inc., P.O. Box 714, St. Albans, VT 05478-0714/802-527-1252, 800-527-1252; FAX: 802-527-0470; WEB: tp://www.centuryarms.com
CFVentures, 509 Harvey Dr., Bloomington, IN 47403-1715
C-H Tool & Die Corp. (See 4-D Custom Die Co.)
CHAA, Ltd., P.O. Box 565, Howell, MI 48844/800-677-8737; FAX: 313-894-6930
Chace Leather Products, 507 Alden St., Fall River, MA 02722/508-678-7556; FAX: 508-675-9666
Chadick's Ltd., P.O. Box 100, Terrell, TX 75160/214-563-7577
Chambers Flintlocks Ltd., Jim, Rt. 1, Box 513-A, Candler, NC 28715/704-667-8361
Champion Shooters' Supply, P.O. Box 303, New Albany, OH 43054/614-855-1603; FAX: 614-855-1209
Champion Target Co., 232 Industrial Parkway, Richmond, IN 47374/800-441-4971
Champion's Choice, Inc., 201 International Blvd., LaVergne, TN 37086/615-793-4066; FAX: 615-793-4070
Champlin Firearms, Inc., P.O. Box 3191, Woodring Airport, Enid, OK 73701/405-237-7388; FAX: 405-242-6922
Chapman Academy of Practical Shooting, 4350 Academy Rd., Hallsville, MO 65255/573-696-5544, 573-696-2266
Chapman Manufacturing Co., 471 New Haven Rd., P.O. Box 250, Durham, CT 06422/203-349-9228; FAX: 203-349-0084
Chapuis Armes, 21 La Gravoux, BP15, 42380 St. Bonnet-le-Chateau, FRANCE/(33)77.50.06.96 (U.S. importer—Champlin Firearms, Inc.; Chapuis USA)
Chapuis USA, 416 Business Park, Bedford, KY 40006
Checkmate Refinishing, 370 Champion Dr., Brooksville, FL 34601/904-799-5774
Cheddite France, S.A., 99, Route de Lyon, F-26501 Bourg-les-Valence, FRANCE/33-75-56-4545; FAX: 33-75-56-3587
Chelsea Gun Club of New York City, Inc., 237 Ovington Ave., Apt. D53, Brooklyn, NY 11209/718-836-9422, 718-833-2704
Chem-Pak, Inc., 11 Oates Ave., P.O. Box 1685, Winchester, VA 22604/800-336-9828, 703-667-1341; FAX: 703-722-3993
Cherry's Fine Guns, P.O. Box 5307, Greensboro, NC 27435-0307/919-854-4182
Chesapeake Importing & Distributing Co. (See CIDCO)
CheVron Bullets, RR1, Ottawa, IL 61350/815-433-2471
CheVron Case Master (See CheVron Bullets)
Chicago Cutlery Co., 1536 Beech St., Terre Haute, IN 47804/800-457-2665
Chicasaw Gun Works, 4 Mi. Mkr., Pluto Rd., Box 868, Shady Spring, WV 25918-0868/304-763-2848; FAX: 304-763-2848

DIRECTORY OF THE ARMS TRADE

Chipmunk (See Oregon Arms, Inc.)
Chippewa Shoe Co., P.O. Box 2521, Ft. Worth, TX 76113/817-332-4385
Choate Machine & Tool Co., Inc., P.O. Box 218, 116 Lovers Ln., Bald Knob, AR 72010/501-724-6193, 800-972-6390; FAX: 501-724-5873
Chopie Mfg., Inc., 700 Copeland Ave., LaCrosse, WI 54603/608-784-0926
Christensen Arms, 192 East 100 North, Fayette, UT 84630/801-528-7999; FAX: 801-528-7494
Christie's East, 219 E. 67th St., New York, NY 10021/212-606-0400
Christman Jr., David, Gunmaker, 937 Lee Hedrick Rd., Colville, WA 99114/509-684-1438
Christopher Firearms Co., Inc., E., Route 128 & Ferry St., Miamitown, OH 45041/513-353-1321
Chu Tani Ind., Inc., P.O. Box 2064, Cody, WY 82414-2064
Chuck's Gun Shop, P.O. Box 597, Waldo, FL 32694/904-468-2264
Churchill (See U.S. importer—Ellett Bros.)
Churchill, Winston, Twenty Mile Stream Rd., RFD P.O. Box 29B, Proctorsville, VT 05153/802-226-7772
Churchill Glove Co., James, P.O. Box 298, Centralia, WA 98531
CIDCO, 21480 Pacific Blvd., Sterling, VA 22170/703-444-5353
Ciener, Inc., Jonathan Arthur, 8700 Commerce St., Cape Canaveral, FL 32920/407-868-2200; FAX: 407-868-2201
Cimarron Arms, P.O. Box 906, Fredericksburg, TX 78624-0906/210-997-9090; FAX: 210-997-0802
Cincinnati Swaging, 2605 Marlington Ave., Cincinnati, OH 45208
Citadel Mfg., Inc., 5220 Gabbert Rd., Moorpark, CA 93021/805-529-7294; FAX: 805-529-7297
C.J. Ballistics, Inc., P.O. Box 132, Acme, WA 98220/206-595-5001
Clark Co., Inc., David, P.O. Box 15054, Worcester, MA 01615-0054/508-756-6216; FAX: 508-753-5827
Clark Custom Guns, Inc., 336 Shootout Lane, Princeton, LA 71067/318-949-9884; FAX: 318-949-9829
Clark Firearms Engraving, P.O. Box 80746, San Marino, CA 91118/818-287-1652
Clarkfield Enterprises, Inc., 1032 10th Ave., Clarkfield, MN 56223/612-669-7140
Claro Walnut Gunstock Co., 1235 Stanley Ave., Chico, CA 95928/916-342-5188
Classic Arms Corp., P.O. Box 106, Dunsmuir, CA 96025-0106/916-235-2000
Classic Guns, Inc., Frank S. Wood, 3230 Medlock Bridge Rd., Suite 110, Norcross, GA 30092/404-242-7944
Claybuster Wads & Harvester Bullets, 309 Sequoya Dr., Hopkinsville, KY 42240/800-922-6287, 800-284-1746, 502-885-8088; FAX: 502-885-1951
Clearview Mfg. Co., Inc., 413 S. Oakley St., Fordyce, AR 71742/501-352-8557; FAX: 501-352-7120
Clearview Products, 3021 N. Portland, Oklahoma City, OK 73107
Cleland's Gun Shop, Inc., 10306 Airport Hwy., Swanton, OH 43558/419-865-4713
Clements' Custom Leathercraft, Chas, 1741 Dallas St., Aurora, CO 80010-2018/303-364-0403; FAX:303-739-9824
Clenzoil Corp., P.O. Box 80226, Sta. C, Canton, OH 44708-0226/330-833-9758; FAX: 330-833-4724
Clerke Co., J.A., P.O. Box 627, Pearblossom, CA 93553-0627/805-945-0713
Clift Mfg., L.R., 3821 Hammonton Rd., Marysville, CA 95901/916-755-3390; FAX: 916-755-3393
Clift Welding Supply & Cases, 1332-A Colusa Hwy., Yuba City, CA 95993/916-755-3390; FAX: 916-755-3393
Cloward's Gun Shop, 4023 Aurora Ave. N, Seattle, WA 98103/206-632-2072
Clymer Manufacturing Co., Inc., 1645 W. Hamlin Rd., Rochester Hills, MI 48309-1530/810-853-5555, 810-853-5627; FAX: 810-853-1530
C-More Systems, P.O. Box 1750, 7553 Gary Rd., Manassas, VA 22110/703-361-2663; FAX: 703-361-5881
Coats, Mrs. Lester, 300 Luman Rd., Space 125, Phoenix, OR 97535/503-535-1611
Cobalt Mfg., Inc., 1020 Shady Oak Dr., Denton, TX 76205/817-382-8986; FAX: 817-383-4281
Cobra Sport S.r.l., Via Caduti Nei Lager No. 1, 56020 San Romano, Montopoli v/Arno (Pi), ITALY/0039-571-450490; FAX: 0039-571-450492
Coffin, Charles H., 3719 Scarlet Ave., Odessa, TX 79762/915-366-4729
Coffin, Jim (See Working Guns)
Cogar's Gunsmithing, P.O. Box 755, Houghton Lake, MI 48629/517-422-4591
Coghlan's Ltd., 121 Irene St., Winnipeg, Man., CANADA R3T 4C7/204-284-9550; FAX: 204-475-4127
Cold Steel, Inc., 2128-D Knoll Dr., Ventura, CA 93003/800-255-4716, 800-624-2363 (in CA); FAX: 805-642-9727
Cole's Gun Works, Old Bank Building, Rt. 4, Box 250, Moyock, NC 27958/919-435-2345
Cole-Grip, 16135 Cohasset St., Van Nuys, CA 91406/818-782-4424
Coleman Co., Inc., 250 N. St. Francis, Wichita, KS 67201
Coleman's Custom Repair, 4035 N. 20th Rd., Arlington, VA 22207/703-528-4486
Collings, Ronald, 1006 Cielta Linda, Vista, CA 92083
Colonial Arms, Inc., P.O. Box 636, Selma, AL 36702-0636/334-872-9455; FAX: 334-872-9540
Colonial Knife Co., Inc., P.O. Box 3327, Providence, RI 02909/401-421-1600; FAX: 401-421-2047
Colonial Repair, P.O. Box 372, Hyde Park, MA 02136-9998/617-469-4951
Colorado Gunsmithing Academy, 27533 Highway 287 South, Lamar, CO 81052/719-336-4099, 800-754-2046; FAX: 719-336-9642
Colorado School of Trades, 1575 Hoyt St., Lakewood, CO 80215/800-234-4594; FAX: 303-233-4723

Colorado Shooter's Supply, 1163 W. Paradise Way, Fruita, CO 81521/303-858-9191
Colorado Sutlers Arsenal (See Cumberland States Arsenal)
Colt Blackpowder Arms Co., 110 8th Street, Brooklyn, NY 11215/212-925-2159; FAX: 212-966-4986
Colt's Mfg. Co., Inc., P.O. Box 1868, Hartford, CT 06144-1868/800-962-COLT, 203-236-6311; FAX: 203-244-1449
Combat Military Ordnance Ltd., 3900 Hopkins St., Savannah, GA 31405/912-238-1900; FAX: 912-236-7570
Companhia Brasileira de Cartuchos (See CBC)
Compass Industries, Inc., 104 East 25th St., New York, NY 10010/212-473-2614, 800-221-9904; FAX: 212-353-0826
Compasseco, Ltd., 151 Atkinson Hill Ave., Bardtown, KY 40004/502-349-0910
Competition Electronics, Inc., 3469 Precision Dr., Rockford, IL 61109/815-874-8001; FAX: 815-874-8181
Competitive Pistol Shop, The, 5233 Palmer Dr., Ft. Worth, TX 76117-2433/817-834-8479
Competitor Corp., Inc., Appleton Business Center, 30 Tricnit Road, Unit 16, New Ipswich, NH 03071-0508/603-878-3891; FAX: 603-878-3950
Component Concepts, Inc., 10240 SW Nimbus Ave., Suite L-8, Portland, OR 97223/503-684-9262; FAX: 503-620-4285
Concept Development Corp., 14715 N. 78th Way, Suite 300, Scottsdale, AZ 85260/800-472-4405; FAX: 602-948-7560
Condon, Inc., David, 109 E. Washington St., Middleburg, VA 22117/703-687-5642
Conetrol Scope Mounts, 10225 Hwy. 123 S., Seguin, TX 78155/210-379-3030, 800-CONETROL; FAX: 210-379-3030
CONKKO, P.O. Box 40, Broomall, PA 19008/215-356-0711
Connecticut Shotgun Mfg. Co., P.O. Box 1692, 35 Woodland St., New Britain, CT 06051-1692/860-225-6581; FAX: 860-832-8707
Connecticut Valley Classics (See CVC)
Conrad, C.A., 3964 Ebert St., Winston-Salem, NC 27127/919-788-5469
Continental Kite & Key (See CONKKO)
Cook Engineering Service, 891 Highbury Rd., Vermont VICT 3133 AUSTRALIA
Coonan Arms (JS Worldwide DBA), 1745 Hwy. 36 E., Maplewood, MN 55109/612-777-3156; FAX: 612-777-3683
Cooper Arms, P.O. Box 114, Stevensville, MT 59870/406-777-5534; FAX: 406-777-5228
Cooper-Woodward, 3800 Pelican Rd., Helena, MT 59602/406-458-3800
Copperhead Bullets, Inc., P.O. Box 662, Butte, MT 59703/406-723-6300
Corbin Mfg. & Supply, Inc., 600 Industrial Circle, P.O. Box 2659, White City, OR 97503/541-826-5211; FAX: 541-826-8669
Cor-Bon Bullet & Ammo Co., 1311 Industry Rd., Sturgis, SD 57785/800-626-7266; FAX: 800-923-2666
Corkys Gun Clinic, 4401 Hot Springs Dr., Greeley, CO 80634-9226/970-330-0516
Corry, John, 861 Princeton Ct., Neshanic Station, NJ 08853/908-369-8019
Cosmi Americo & Figlio s.n.c., Via Flaminia 307, Ancona, ITALY I-60020/071-888208; FAX: 39-071-887008 (U.S. importer—New England Arms Co.)
Costa, David (See Island Pond Gun Shop)
Coulston Products, Inc., P.O. Box 30, 201 Ferry St., Suite 212, Easton, PA 18044-0030/215-253-0167, 800-445-9927; FAX: 215-252-1511
Counter Assault, Box 4721, Missoula, MT 59806/406-728-6241; FAX: 406-728-8800
Country Armourer, The, P.O. Box .308, Ashby, MA 01431-0308/508-827-6797; FAX: 508-827-4845
Cousin Bob's Mountain Products, 7119 Ohio River Blvd., Ben Avon, PA 15202/412-766-5114; FAX: 412-766-5114
Cox, Ed C., RD 2, Box 192, Prosperity, PA 15329/412-228-4984
CP Bullets, 340-1 Constance Dr., Warminster, PA 18974
CQB Training, P.O. Box 1739, Manchester, MO 63011
Craftguard, 3624 Logan Ave., Waterloo, IA 50703/319-232-2959; FAX: 319-234-0804
Craig Custom Ltd., Research & Development, 629 E. 10th, Hutchinson, KS 67501/316-669-0601
Crandall Tool & Machine Co., 19163 21 Mile Rd., Tustin, MI 49688/616-829-4430
Crane & Crane Ltd., 105 N. Edison Way 6, Reno, NV 89502-2355/702-856-1516; FAX: 702-856-1616
Crane Sales Co., George S., P.O. Box 385, Van Nuys, CA 91408/818-505-8337
CRDC Laser Systems Group, 3972 Barranca Parkway, Ste. J-484, Irvine, CA 92714/714-586-1295; FAX: 714-831-4823
Creative Craftsman, Inc., The, 95 Highway 29 North, P.O. Box 331, Lawrenceville, GA 30246/404-963-2112; FAX: 404-513-9488
Creedmoor Sports, Inc., P.O. Box 1040, Oceanside, CA 92051/619-757-5529
Creek Side Metal & Woodcrafters (See Allard, Gary)
Creekside Gun Shop, Inc., Main St., Holcomb, NY 14469/716-657-6338; FAX: 716-657-7900
Crimson Trace, 1433 N.W. Quimby, Portland, OR 97209/503-295-2406; 503-295-0005
Crit'R Call (See Rocky Mountain Wildlife Products)
Crosman Airguns, Rts. 5 and 20, E. Bloomfield, NY 14443/716-657-6161; FAX: 716-657-5405
Crosman Blades (See Coleman Co., Inc.)
Crosman Products of Canada Ltd., 1173 N. Service Rd. West, Oakville, Ontario, L6M 2V9 CANADA/905-827-1822
Crouse's Country Cover, P.O. Box 160, Storrs, CT 06268/860-423-8736
CRR, Inc./Marble's Inc., 420 Industrial Park, P.O. Box 111, Gladstone, MI 49837/906-428-3710; FAX: 906-428-3711
Crucelegui, Hermanos (See U.S. importer—Mandall Shooting Supplies, Inc.)

MANUFACTURERS' DIRECTORY

Cryo-Accurizing, 2101 East Olive, Decatur, IL 62526/217-423-3070; FAX: 217-423-3075
Cubic Shot Shell Co., Inc., 98 Fatima Dr., Campbell, OH 44405/216-755-0349; FAX: 216-755-0349
Cullity Restoration, Daniel, 209 Old County Rd., East Sandwich, MA 02537/508-888-1147
Cumberland Arms, 514 Shafer Road, Manchester, TN 37355/800-797-8414
Cumberland Knife & Gun Works, 5661 Bragg Blvd., Fayetteville, NC 28303/919-867-0009
Cumberland Mountain Arms, P.O. Box 710, Winchester, TN 37398/615-967-8414; FAX: 615-967-9199
Cumberland States Arsenal, 1124 Palmyra Road, Clarksville, TN 37040
Cummings Bullets, 1417 Esperanza Way, Escondido, CA 92027
Cunningham Co., Eaton, 607 Superior St., Kansas City, MO 64106/816-842-2600
Cupp, Alana, Custom Engraver, P.O. Box 207, Annabella, UT 84711/801-896-4834
Curly Maple Stock Blanks (See Tiger-Hunt)
Curtis Custom Shop, RR1, Box 193A, Wallingford, KY 41093/703-659-4265
Curtis Cast Bullets, 119 W. College, Bozeman, MT 59715/406-587-4934
Curtis Gun Shop (See Curtis Cast Bullets)
Custom Barreling & Stocks, 937 Lee Hedrick Rd., Colville, WA 99114/509-684-5686 (days), 509-684-3314 (evenings)
Custom Bullets by Hoffman, 2604 Peconic Ave., Seaford, NY 11783
Custom Calls, 607 N. 5th St., Burlington, IA 52601/319-752-4465
Custom Checkering Service, Kathy Forster, 2124 SE Yamhill St., Portland, OR 97214/503-236-5874
Custom Chronograph, Inc., 5305 Reese Hill Rd., Sumas, WA 98295/360-988-7801
Custom Firearms (See Ahrends, Kim)
Custom Gun Products, 5021 W. Rosewood, Spokane, WA 99208/509-328-9340
Custom Gun Stocks, Rt. 6, P.O. Box 177, McMinnville, TN 37110/615-668-3912
Custom Gunsmiths, 4303 Friar Lane, Colorado Springs, CO 80907/719-599-3366
Custom Hunting Ammo & Arms (See CHAA, Ltd.)
Custom Products (See Jones Custom Products, Neil A.)
Custom Quality Products, Inc., 345 W. Girard Ave., P.O. Box 71129, Madison Heights, MI 48071/810-585-1616; FAX: 810-585-0644
Custom Riflestocks, Inc., Michael M. Kokolus, 7005 Herber Rd., New Tripoli, PA 18066/610-298-3013
Custom Shop, The, 890 Cochrane Crescent, Peterborough, Ont. K9H 5N3 CANADA/705-742-6693
Custom Tackle and Ammo, P.O. Box 1886, Farmington, NM 87499/505-632-3539
Cutco Cutlery, P.O. Box 810, Olean, NY 14760/716-372-3111
Cutlery Shoppe, 5461 Kendall St., Boise, ID 83706-1248/800-231-1272
Cutsinger Bench Rest Bullets, RR 8, Box 161-A, Shelbyville, IN 46176/317-729-5360
CVA, 5988 Peachtree Corners East, Norcross, GA 30071/800-251-9412; FAX: 404-242-8546
CVC, 48 Commercial Street, Holyoke, MA 01040/413-552-3184; FAX: 413-552-3276
Cylinder & Slide, Inc., William R. Laughridge, 245 E. 4th St., Fremont, NE 68025/402-721-4277; FAX: 402-721-0263
CZ USA, 40356 Oak Park Way, Suite W, Oakhurst, CA 93664

D

D&D Gunsmiths, Ltd., 363 E. Elmwood, Troy, MI 48083/810-583-1512; FAX: 810-583-1524
D&G Precision Duplicators (See Greene Precision Duplicators)
D&H Precision Tooling, 7522 Barnard Mill Rd., Ringwood, IL 60072/815-653-4011
D&H Prods. Co., Inc., 465 Denny Rd., Valencia, PA 16059/412-898-2840, 800-776-0281; FAX: 412-898-2013
D&J Bullet Co. & Custom Gun Shop, Inc., 426 Ferry St., Russell, KY 41169/606-836-2663; FAX: 606-836-2663
D&L Industries (See D.J. Marketing)
D&L Sports, P.O. Box 651, Gillette, WY 82717/307-686-4008
D&R Distributing, 308 S.E. Valley St., Myrtle Creek, OR 97457/503-863-6850
Dade Screw Machine Products, 2319 NW 7th Ave., Miami, FL 33127/305-573-5050
Daewoo Precision Industries Ltd., 34-3 Yeoeuido-Dong, Yeongdeungoo-GU, 15th, Fl./Seoul, KOREA (U.S. importer—Nationwide Sports Distributors, Inc.)
Daisy Mfg. Co., P.O. Box 220, Rogers, AR 72757/501-636-1200; FAX: 501-636-1601
Dakota (See U.S. importer—EMF Co., Inc.)
Dakota Arms, Inc., HC 55, Box 326, Sturgis, SD 57785/605-347-4686; FAX: 605-347-4459
Dakota Corp., 77 Wales St., P.O. Box 543, Rutland, VT 05701/802-775-6062, 800-451-4167; FAX: 802-773-3919
Daly, Charles (See B.C. Miroku/Charles Daly)
DAMASCUS-U.S.A., 149 Deans Farm Rd., Tyner, NC 27980/919-221-2010; FAX: 919-221-2009
Dan's Whetstone Co., Inc., 130 Timbs Place, Hot Springs, AR 71913/501-767-1616; FAX: 501-767-9598
Dangler, Homer L., Box 254, Addison, MI 49220/517-547-6745
Danner Shoe Mfg. Co., 12722 NE Airport Way, Portland, OR 97230/503-251-1100, 800-345-0430; FAX: 503-251-1119
Danuser Machine Co., 550 E. Third St., P.O. Box 368, Fulton, MO 65251/573-642-2246; FAX: 573-642-2340
Dapkus Co., Inc., J.G., Commerce Circle, P.O. Box 293, Durham, CT 06422
Dara-Nes, Inc. (See Nesci Enterprises, Inc.)

Darlington Gun Works, Inc., P.O. Box 698, 516 S. 52 Bypass, Darlington, SC 29532/803-393-3931
Data Tech Software Systems, 19312 East Eldorado Drive, Aurora, CO 80013
Datumtech Corp., 2275 Wehrle Dr., Buffalo, NY 14221
Dave's Gun Shop, 555 Wood Street, Powell, Wyoming 82435/307-754-9724
Davidson, Jere, Rt. 1, Box 132, Rustburg, VA 24588/804-821-3637
Davis, Don, 1619 Heights, Katy, TX 77493/713-391-3090
Davis Co., R.E., 3450 Pleasantville NE, Pleasantville, OH 43148/614-654-9990
Davis Industries, 15150 Sierra Bonita Ln., Chino, CA 91710/909-597-4726; FAX: 909-393-9771
Davis Leather Co., Gordon Wm., P.O. Box 2270, Walnut, CA 91788/909-598-5620
Davis Products, Mike, 643 Loop Dr., Moses Lake, WA 98837/509-765-6178, 509-766-7281 orders only
Davis Service Center, Bill, 7221 Florin Mall Dr., Sacramento, CA 95823/916-393-4867
Day & Sons, Inc., Leonard, P.O. Box 122, Flagg Hill Rd., Heath, MA 01346/413-337-8369
Dayson Arms Ltd., P.O. Box 532, Vincennes, IN 47591/812-882-8680; FAX: 812-882-8680
Daystate Ltd., Birch House Lanee, Cotes Heath, Staffs, ST15.022 ENGLAND/01782-791755; FAX: 01782-791617
Dayton Traister, 4778 N. Monkey Hill Rd., P.O. Box 593, Oak Harbor, WA 98277/360-679-4657; FAX:360-675-1114
DBASE Consultants (See Arms, Peripheral Data Systems)
DBI Books, Division of Krause Publications, (Editorial office) 935 Lakeview Parkway, Suite 101, Vernon Hills, IL 60061/847-573-8530; FAX: 847-573-8534; For consumer orders, see Krause Publications
D-Boone Ent., Inc., 5900 Colwyn Dr., Harrisburg, PA 17109
D.C.C. Enterprises, 259 Wynburn Ave., Athens, GA 30601
D.D. Custom Stocks, R.H. "Dick" Devereaux, 5240 Mule Deer Dr., Colorado Springs, CO 80919/719-548-8468
de Coux, Pete (See Cartridge Transfer Group)
de Treville & Co., Stan, 4129 Normal St., San Diego, CA 92103/619-298-3393
Dead Eye's Sport Center, RD 1, Box 147B, Shickshinny, PA 18655/717-256-7432
Decker Shooting Products, 1729 Laguna Ave., Schofield, WI 54476/715-359-5873
DeckSlider of Florida, 27641-2 Reahard Ct., Bonita Springs, FL 33923/800-782-1474
Deepeeka Exports Pvt. Ltd., D-78, Saket, Meerut-250-006, INDIA/011-91-121-512889, 011-91-121-545363; FAX: 011-91-121-542988, 011-91-121-511599
Deer Me Products Co., Box 34, 1208 Park St., Anoka, MN 55303/612-421-8971; FAX: 612-422-0526
Defense Training International, Inc., 749 S. Lemay, Ste. A3-337, Ft. Collins, CO 80524/303-482-2520; FAX: 303-482-0548
Degen Inc. (See Aristocrat Knives)
deHaas Barrels, RR 3, Box 77, Ridgeway, MO 64481/816-872-6308
Del Rey Products, P.O. Box 91561, Los Angeles, CA 90009/213-823-0494
Delhi Gun House, 1374 Kashmere Gate, Delhi, INDIA 110 006/(011)2940974 2940814; FAX: 91-11-2917344
Delorge, Ed, 2231 Hwy. 308, Thibodaux, LA 70301/504-447-1633
Del-Sports, Inc., Box 685, Main St., Margaretville, NY 12455/914-586-4103; FAX: 914-586-4105
Delta Arms Ltd., P.O. Box 1000, Delta, VT 84624-1000
Delta Enterprises, 284 Hagemann Drive, Livermore, CA 94550
Delta Frangible Ammunition, LLC, P.O. Box 2350, Stafford, VA 22555-2350/540-720-5778, 800-339-1933; FAX: 540-720-5667
Dem-Bart Checkering Tools, Inc., 6807 Bickford Ave., Old Hwy. 2, Snohomish, WA 98290/360-568-7356; FAX: 360-568-1798
Denver Bullets, Inc., 1811 W. 13th Ave., Denver, CO 80204/303-893-3146; FAX: 303-893-9161
Denver Instrument Co., 6542 Fig St., Arvada, CO 80004/800-321-1135, 303-431-7255; FAX: 303-423-4831
DeSantis Holster & Leather Goods, Inc., P.O. Box 2039, 149 Denton Ave., New Hyde Park, NY 11040-0701/516-354-8000; FAX: 516-354-7501
Desert Mountain Mfg., P.O. Box 2767, Columbia Falls, MT 59912/800-477-0762, 406-892-7772
Detroit-Armor Corp., 720 Industrial Dr. No. 112, Cary, IL 60013/708-639-7666; FAX: 708-639-7694
Dever Co., Jack, 8590 NW 90, Oklahoma City, OK 73132/405-721-6393
Devereaux, R.H. "Dick" (See D.D. Custom Stocks)
Dewey Mfg. Co., Inc., J., P.O. Box 2014, Southbury, CT 06488/203-264-3064; FAX: 203-262-6907
DGR Custom Rifles, RR1, Box 8A, Tappen, ND 58487/701-327-8135
DGS, Inc., Dale A. Storey, 1117 E. 12th, Casper, WY 82601/307-237-2414
DHB Products, P.O. Box 3092, Alexandria, VA 22302/703-836-2648
Diamond Machining Techonology (See DMT—Diamond Machining Technology)
Diamond Mfg. Co., P.O. Box 174, Wyoming, PA 18644/800-233-9601
Diana (See U.S. importer—Dynamit Nobel-RWS, Inc.)
Dibble, Derek A., 555 John Downey Dr., New Britain, CT 06051/203-224-2630
Dietz Gun Shop & Range, Inc., 421 Range Rd., New Braunfels, TX 78132/210-885-4662
Dilliott Gunsmithing, Inc., 657 Scarlett Rd., Dandridge, TN 37725/423-397-9204
Dillon, Ed, 1035 War Eagle Dr. N., Colorado Springs, CO 80919/719-598-4929; FAX: 719-598-4929
Dillon Precision Products, Inc., 8009 East Dillon's Way, Scottsdale, AZ 85260/602-948-8009, 800-762-3845; FAX: 602-998-2786

DIRECTORY OF THE ARMS TRADE

Dina Arms Corporation, P.O. Box 46, Royersford, PA 19468/610-287-0266; FAX: 610-287-0266
Division Lead Co., 7742 W. 61st Pl., Summit, IL 60502
Dixie Gun Works, Inc., Hwy. 51 South, Union City, TN 38261/901-885-0561, order 800-238-6785; FAX: 901-885-0440
Dixon Muzzleloading Shop, Inc., RD 1, Box 175, Kempton, PA 19529/610-756-6271
D.J. Marketing, 10602 Horton Ave., Downey, CA 90241/310-806-0891; FAX: 310-806-6231
DKT, Inc., 14623 Vera Drive, Union, MI 49130-9744/616-641-7120, 800-741-7083 orders only; FAX: 616-641-2015
DLO Mfg., 10807 SE Foster Ave., Arcadia, FL 33821-7304
D-Max, Inc., RR1, Box 473, Bagley, MN 56621/218-785-2278
DMT—Diamond Machining Technology, Inc., 85 Hayes Memorial Dr., Marlborough, MA 01752/508-481-5944; FAX: 508-485-3924
Doctor Optic Technologies, Inc., 4685 Boulder Highway, Suite A, Las Vegas, NV 89121/800-290-3634, 702-898-7161; FAX: 702-898-3737
Dogtown Varmint Supplies, 1048 Irvine Ave. No. 333, Newport Beach, CA 92660/714-642-3997
Dohring Bullets, 100 W. 8 Mile Rd., Ferndale, MI 48220
Dolbare, Elizabeth, P.O. Box 222, Sunburst, MT 59482-0222
Donnelly, C.P., 405 Kubli Rd., Grants Pass, OR 97527/541-846-6604
Doskocil Mfg. Co., Inc., P.O. Box 1246, 4209 Barnett, Arlington, TX 76017/817-467-5116; FAX: 817-472-9810
Double A Ltd., P.O. Box 11306, Minneapolis, MN 55411/612-522-0306
Douglas Barrels, Inc., 5504 Big Tyler Rd., Charleston, WV 25313-1398/304-776-1341; FAX: 304-776-8560
Downsizer Corp., P.O. Box 710316, Santee, CA 92072-0316/619/448-5510; FAX: 619-448-5780
Dowtin Gunworks, Rt. 4, Box 930A, Flagstaff, AZ 86001/602-779-1898
Dr. O's Products Ltd., P.O. Box 111, Niverville, NY 12130/518-784-3333; FAX: 518-784-2800
Drain, Mark, SE 3211 Kamilche Point Rd., Shelton, WA 98584/206-426-5452
Dremel Mfg. Co., 4915-21st St., Racine, WI 53406
Dressel Jr., Paul G., 209 N. 92nd Ave., Yakima, WA 98908/509-966-9233; FAX: 509-966-3365
Dri-Slide, Inc., 411 N. Darling, Fremont, MI 49412/616-924-3950
Dropkick, 1460 Washington Blvd., Williamsport, PA 17701/717-326-6561; FAX: 717-326-4950
DTM International, Inc., 40 Joslyn Rd., P.O. Box 5, Lake Orion, MI 48362/313-693-6670
Duane Custom Stocks, Randy, 110 W. North Ave., Winchester, VA 22601/703-667-9461; FAX: 703-722-3993
Duane's Gun Repair (See DGR Custom Rifles)
Dubber, Michael W., P.O. Box 312, Evansville, IN 47702/812-424-9000; FAX: 812-424-6551
Duck Call Specialists, P.O. Box 124, Jerseyville, IL 62052/618-498-9855
Duffy (See Guns Antique & Modern DBA/Charles E. Duffy)
Du-Lite Corp., Charles E., 171 River Rd., Middletown, CT 06457/203-347-2505; FAX: 203-347-9404
Dumoulin, Ernest, Rue Florent Boclinville 8-10, 13-4041 Votten, BELGIUM/41 27 78 92 (U.S. importer—New England Arms Co.)
Duncan's Gun Works, Inc., 1619 Grand Ave., San Marcos, CA 92069/619-727-0515
Dunham Co., P.O. Box 813, Brattleboro, VT 05301/802-254-2316
Dunphy, Ted, W. 5100 Winch Rd., Rathdrum, ID 83858/208-687-1399; FAX: 208-687-1399
Duofold, Inc., RD 3 Rt. 309, Valley Square Mall, Tamaqua, PA 18252/717-386-2666; FAX: 717-386-3652
DuPont (See IMR Powder Co.)
Dutchman's Firearms, Inc., The, 4143 Taylor Blvd., Louisville, KY 40215/502-366-0555
Dybala Gun Shop, P.O. Box 1024, FM 3156, Bay City, TX 77414/409-245-0866
Dykstra, Doug, 411 N. Darling, Fremont, MI 49412/616-924-3950
Dynalite Products, Inc., 215 S. Washington St., Greenfield, OH 45123/513-981-2124
Dynamit Nobel-RWS, Inc., 81 Ruckman Rd., Closter, NJ 07624/201-767-7971; FAX: 201-767-1589
Dyson & Son Ltd., Peter, 3 Cuckoo Lane, Honley, Huddersfield, Yorkshire HD7 2BR, ENGLAND/44-1484-661062; FAX: 44-1484-663709

E

E&L Mfg., Inc., 4177 Riddle by Pass Rd., Riddle, OR 97469/541-874-2137; FAX: 541-874-3107
E.A.A. Corp., P.O. Box 1299, Sharpes, FL 32959/407-639-4842, 800-536-4442; FAX: 407-639-7006
Eagan, Donald V., P.O. Box 196, Benton, PA 17814/717-925-6134
Eagle Arms (See ArmaLite, Inc.)
Eagle Grips, Eagle Business Center, 460 Randy Rd., Carol Stream, IL 60188/800-323-6144, 708-260-0400; FAX: 708-260-0486
Eagle Imports, Inc., 1750 Brielle Ave., Unit B1, Wanamassa, NJ 07712/908-493-0333; FAX: 908-493-0301
Eagle International Sporting Goods, Inc., P.O. Box 67, Zap, ND 58580/888-932-4536; FAX: 701-948-2282
E-A-R, Inc., Div. of Cabot Safety Corp., 5457 W. 79th St., Indianapolis, IN 46268/800-327-3431; FAX: 800-488-8007

Eastman Products, R.T., P.O. Box 1531, Jackson, WY 83001/307-733-3217, 800-624-4311
Easy Pull Outlaw Products, 316 1st St. East, Polson, MT 59860/406-883-6822
EAW (See U.S. importer—New England Custom Gun Service)
Echols & Co., D'Arcy, 164 W. 580 S., Providence, UT 84332/801-753-2367
Eclectic Technologies, Inc., 45 Grandview Dr., Suite A, Farmington, CT 06034
Eckelman Gunsmithing, 3125 133rd St. SW, Fort Ripley, MN 56449/218-829-3176
Ed's Gun House, P.O. Box 62, Minnesota City, MN 55959/507-689-2925
Edenpine, Inc. c/o Six Enterprises, Inc., 320 D Turtle Creek Ct., San Jose, CA 95125/408-999-0201; FAX: 408-999-0216
EdgeCraft Corp./P.B. Tuminello, 825 Southwood Road, Avondale, PA 19311/610-268-0500, 800-342-3255; FAX: 610-268-3545
Edmisten Co., P.O. Box 1293, Boone, NC 28607
Edmund Scientific Co., 101 E. Gloucester Pike, Barrington, NJ 08033/609-543-6250
Ednar, Inc., 2-4-8 Kayabacho, Nihonbashi, Chuo-ku, Tokyo, JAPAN 103/81(Japan)-3-3667-1651; FAX: 81-3-3661-8113
Eezox, Inc., P.O. Box 772, Waterford, CT 06385-0772/860-447-8282, 800-462-3331; FAX: 860-447-3484
Effebi SNC-Dr. Franco Beretta, via Rossa, 4, 25062 Concesio, Italy/030-2751955; FAX: 030-2180414 (U.S. importer—Nevada Cartridge Co.
Eggleston, Jere D., 400 Saluda Ave., Columbia, SC 29205/803-799-3402
EGW Evolution Gun Works, 4050 B-8 Skyron Dr., Doylestown, PA 18901/215-348-9892; FAX: 215-348-1056
Eichelberger Bullets, Wm., 158 Crossfield Rd., King of Prussia, PA 19406
EK Knife Co., c/o Blackjack Knives, Ltd., 1307 Wabash Ave., Effingham, IL 62401
Ekol Leather Care, P.O. Box 2652, West Lafayette, IN 47906/317-463-2250; FAX: 317-463-7004
El Dorado Leather (c/o Dill), P.O. Box 566, Benson, AZ 85602/520-586-4791; FAX: 520-586-4791
El Paso Saddlery Co., P.O. Box 27194, El Paso, TX 79926/915-544-2233; FAX: 915-544-2535
Eldorado Cartridge Corp. (See PMC/Eldorado Cartridge Corp.)
Electro Prismatic Collimators, Inc., 1441 Manatt St., Lincoln, NE 68521
Electronic Shooters Protection, Inc., 11997 West 85th Place, Arvada, CO 80005/303-456-8964; 800-797-7791; FAX: 303-456-7179
Electronic Trigger Systems, Inc., P.O. Box 13, 230 Main St. S., Hector, MN 55342/320-848-2760; FAX: 320-848-2760
Eley Ltd., P.O. Box 705, Witton, Birmingham, B6 7UT, ENGLAND/021-356-8899; FAX: 021-331-4173
Elite Ammunition, P.O. Box 3251, Oakbrook, IL 60522/708-366-9006
Elk River, Inc., 1225 Paonia St., Colorado Springs, CO 80915/719-574-4407
Elkhorn Bullets, P.O. Box 5293, Central Point, OR 97502/541-826-7440
Ellett Bros., 267 Columbia Ave., P.O. Box 128, Chapin, SC 29036/803-345-3751, 800-845-3711; FAX: 803-345-1820
Ellicott Arms, Inc./Woods Pistolsmithing, 3840 Dahlgren Ct., Ellicott City, MD 21042/410-465-7979
Elliott Inc., G.W., 514 Burnside Ave., East Hartford, CT 06108/203-289-5741; FAX: 203-289-3137
Elsen, Inc., Pete, 1529 S. 113th St., West Allis, WI 53214
Emerging Technologies, Inc. (See Laseraim Technologies, Inc.)
EMF Co., Inc., 1900 E. Warner Ave. Suite 1-D, Santa Ana, CA 92705/714-261-6611; FAX: 714-756-0133
Empire Cutlery Corp., 12 Kruger Ct., Clifton, NJ 07013/201-472-5155; FAX: 201-779-0759
Engineered Accessories, 1307 W. Wabash Ave., Effingham, IL 62401/217-347-7700; FAX: 217-347-7737
English, Inc., A.G., 708 S. 12th St., Broken Arrow, OK 74012/918-251-3399
Englishtown Sporting Goods Co., Inc., David J. Maxham, 38 Main St., Englishtown, NJ 07726/201-446-7717
Engraving Artistry, 36 Alto Rd., RFD 2, Burlington, CT 06013/203-673-6837
Enguix Import-Export, Alpujarras 58, Alzira, Valencia, SPAIN 46600/(96) 241 43 95; FAX: (96) (241 43 95) 240 21 53
Enhanced Presentations, Inc., 5929 Market St., Wilmington, NC 28405/910-799-1622; FAX: 910-799-5004
Enlow, Charles, 895 Box, Beaver, OK 73932/405-625-4487
Ensign-Bickford Co., The, 660 Hopmeadow St., Simsbury, CT 06070
EPC, 1441 Manatt St., Lincoln, NE 68521/402-476-3946
Epps, Ellwood (See "Gramps" Antique Cartridges)
Erhardt, Dennis, 3280 Green Meadow Dr., Helena, MT 59601/406-442-4533
Erickson's Mfg., Inc., C.W., 530 Garrison Ave. N.E., P.O. Box 522, Buffalo, MN 55313/612-682-3665; FAX: 612-682-4328
Erma Werke GmbH, Johan Ziegler St., 13/15/FeldiglSt., D-8060 Dachau, GERMANY (U.S. importers—Amtec 2000, Inc.; Mandall Shooting Supplies, Inc.)
Eskridge Rifles, Steven Eskridge, 218 N. Emerson, Mart, TX 76664/817-876-3544
Essex Arms, P.O. Box 345, Island Pond, VT 05846/802-723-4313
Essex Metals, 1000 Brighton St., Union, NJ 07083/800-282-8369
Estate Cartridge, Inc., 12161 FM 830, Willis, TX 77378/409-856-7277; FAX: 409-856-5486
Euber Bullets, No. Orwell Rd., Orwell, VT 05760/802-948-2621
Euroarms of America, Inc., P.O. Box 3277, Winchester, VA 22604/540-662-1863; FAX: 540-662-4464
European American Armory Corp. (See E.A.A. Corp.)

MANUFACTURERS' DIRECTORY

Eutaw Co., Inc., The, P.O. Box 608, U.S. Hwy. 176 West, Holly Hill, SC 29059/803-496-3341
Evans, Andrew, 2325 NW Squire St., Albany, OR 97321/541-928-3190; FAX: 541-928-4128
Evans Engraving, Robert, 332 Vine St., Oregon City, OR 97045/503-656-5693
Evans Gunsmithing (See Evans, Andrew)
Eversull Co., Inc., K., 1 Tracemont, Boyce, LA 71409/318-793-8728; FAX: 318-793-5483
Excalibur Enterprises, P.O. Box 400, Fogelsville, PA 18051-0400/610-391-9105; FAX: 610-391-9220
Exe, Inc., 18830 Partridge Circle, Eden Prairie, MN 55346/612-944-7662
Executive Protection Institute, Rt. 2, Box 3645, Berryville, VA 22611/540-955-1128
C. Eyears, Roland, 576 Binns Blvd., Columbus, OH 43204-2441
Eyster Heritage Gunsmiths, Inc., Ken, 6441 Bishop Rd., Centerburg, OH 43011/614-625-6131
Eze-Lap Diamond Prods., P.O. Box 2229, 15164 Weststate St., Westminster, CA 92683/714-847-1555; FAX: 714-897-0280
E-Z-Way Systems, P.O. Box 4310, Newark, OH 43058-4310/614-345-6645, 800-848-2072; FAX: 614-345-6600

F

F&A Inc. (See ShurKatch Corporation)
Fabarm S.p.A., Via Averolda 31, 25039 Travagliato, Brescia, ITALY/030-6863629; FAX: 030-6863684 (U.S. importer—Ithaca Gun Co., LLC)
Fagan & Co., William, 22952 15 Mile Rd., Clinton Township, MI 48035/810-465-4637; FAX: 810-792-6996
Fair Game International, P.O. Box 77234-34053, Houston, TX 77234/713-941-6269
F.A.I.R. Techni-Mec s.n.c. di Isidoro Rizzini & C., Via Gitti, 41 Zona I, dustri-ale/25060 Marcheno (Brescia), ITALY 030/861162-8610344; FAX: 030/8610179
Faith Associates, Inc., 1139 S. Greenville Hwy., Hendersonville, NC 28792/704-692-1916; FAX: 704-697-6827
Fajen, Inc., Reinhart, Route 1, P.O. Box 214-A, Lincoln, MO 65338/816-547-3030; FAX: 816-547-2215
Famas (See U.S. importer—Century International Arms, Inc.)
Fanzoj GmbH, Griesgasse 1, 9170 Ferlach, AUSTRIA 9170/(43) 04227-2283; FAX: (43) 04227-2867
Far North Outfitters, Box 1252, Bethel, AK 99559
Farm Form Decoys, Inc., 1602 Biovu, P.O. Box 748, Galveston, TX 77553/409-744-0762, 409-765-6361; FAX: 409-765-8513
Farmer-Dressel, Sharon, 209 N. 92nd Ave., Yakima, WA 98908/509-966-9233; FAX: 509-966-3365
Farr Studio, Inc., 1231 Robinhood Rd., Greeneville, TN 37743/615-638-8825
Farrar Tool Co., Inc., 12150 Bloomfield Ave., Suite E, Santa Fe Springs, CA 90670/310-863-4367; FAX: 310-863-5123
FAS, Via E. Fermi, 8, 20019 Settimo Milanese, Milano, ITALY/02-3285844; FAX: 02-33500196 (U.S. importer—Nygord Precision Products)
Faulhaber Wildlocker, Dipl.-Ing. Norbert Wittasek, Seilergasse 2, A-1010 Wien, AUSTRIA/OM-43-1-5137001; FAX: OM-43-1-5137001
Faulk's Game Call Co., Inc., 616 18th St., Lake Charles, LA 70601/318-436-9726
Faust, Inc., T.G., 544 Minor St., Reading, PA 19602/610-375-8549; FAX: 610-375-4488
Fausti Cav. Stefano & Figlie snc, Via Martiri Dell Indipendenza, 70, Marcheno, ITALY 25060 (U.S. importer—American Arms, Inc.)
Fautheree, Andy, P.O. Box 4607, Pagosa Springs, CO 81157/303-731-5003
Feather, Flex Decoys, 1655 Swan Lake Rd., Bossier City, LA 71111/318-746-8596; FAX: 318-742-4815
Federal Cartridge Co., 900 Ehlen Dr., Anoka, MN 55303/612-323-2300; FAX: 612-323-2506
Federal Champion Target Co., 232 Industrial Parkway, Richmond, IN 47374/800-441-4971; FAX: 317-966-7747
Federated-Fry (See Fry Metals)
FEG, Budapest, Soroksariut 158, H-1095 HUNGARY (U.S. importers—Century International Arms, Inc.; K.B.I., Inc.)
Feinwerkbau Westinger & Altenburger GmbH (See FWB)
Feken, Dennis, Rt. 2 Box 124, Perry, OK 73077/405-336-5611
Fellowes, Ted, Beaver Lodge, 9245 16th Ave. SW, Seattle, WA 98106/206-763-1698
Feminine Protection, Inc., 10514 Shady Trail, Dallas, TX 75220/214-351-4500; FAX: 214-352-4686
Ferdinand, Inc., P.O. Box 5, 201 Main St., Harrison, ID 83833/208-689-3012, 800-522-6010 (U.S.A.), 800-258-5266 (Canada); FAX: 208-689-3142
Ferguson, Bill, P.O. Box 1238, Sierra Vista, AZ 85636/520-458-5321; FAX: 520-458-9125
FERLIB, Via Costa 46, 25063 Gardone V.T. (Brescia) ITALY/30-89-12-586; FAX: 30-89-12-586 (U.S. importer—Harry Marx)
Ferris Firearms, 7110 F.M. 1863, Bulverde, TX 78163/210-980-4424
Fibron Products, Inc., P.O. Box 430, Buffalo, NY 14209-0430/716-886-2378; FAX: 716-886-2394
Finch Custom Bullets, 40204 La Rochelle, Prairieville, LA 70769
Fiocchi Munizioni S.p.A. (See U.S. importer—Fiocchi of America, Inc.)
Fiocchi of America, Inc., 5030 Fremont Rd., Ozark, MO 65721/417-725-4118, 800-721-2666; FAX: 417-725-1039

Firearm Training Center, The, 9555 Blandville Rd., West Paducah, KY 42086/502-554-5886
Firearms Academy of Seattle, P.O. Box 2814, Kirkland, WA 98083/206-820-4853
Firearms Co. Ltd./Alpine (See U.S. importer—Mandall Shooting Supplies, Inc.)
Firearms Engraver's Guild of America, 332 Vine St., Oregon City, OR 97045/503-656-5693
Fire'n Five, P.O. Box 11 Granite Rt., Sumpter, OR 97877
First, Inc., Jack, 1201 Turbine Dr., Rapid City, SD 57701/605-343-9544; FAX: 605-343-9420
Fish Mfg. Gunsmith Sptg. Co., Marshall F., Rd. Box 2439, Rt. 22 North, Westport, NY 12993/518-962-4897
Fisher, Jerry A., 553 Crane Mt. Rd., Big Fork, MT 59911/406-837-2722
Fisher Custom Firearms, 2199 S. Kittredge Way, Aurora, CO 80013/303-755-3710
Fisher Enterprises, Inc., 1071 4th Ave. S., Suite 303, Edmonds, WA 98020-4143/206-771-5382
Fisher, R. Kermit (See Fisher Enterprises, Inc.)
Fitz Pistol Grip Co., P.O. Box 610, Douglas City, CA 96024/916-778-0240
Flaig's, 2200 Evergreen Rd., Millvale, PA 15209/412-821-1717
Flambeau Products Corp., 15981 Valplast Rd., Middlefield, OH 44062/216-632-1631; FAX: 216-632-1581
Flannery Engraving Co., Jeff W., 11034 Riddles Run Rd., Union, KY 41091/606-384-3127
Flashette Co., 4725 S. Kolin Ave., Chicago, IL 60632/773-927-1302; FAX: 773-927-3083
Flayderman & Co., Inc., N., P.O. Box 2446, Ft. Lauderdale, FL 33303/305-761-8855
Fleming Firearms, 7720 E 126th St. N, Collinsville, OK 74021-7016/918-665-3624
Flents Products Co., Inc., P.O. Box 2109, Norwalk, CT 06852/203-866-2581; FAX: 203-854-9322
Flintlocks, Etc. (See Beauchamp & Son, Inc.)
Flitz International Ltd., 821 Mohr Ave., Waterford, WI 53185/414-534-5898; FAX: 414-534-2991
Flores Publications, Inc., J. (See Action Direct, Inc.)
Flow-Rite of Tennessee, Inc., 107 Allen St., P.O. Box 196, Bruceton, TN 38317/901-586-2271; FAX: 901-586-2300
Fluoramics, Inc., 18 Industrial Ave., Mahwah, NJ 07430/800-922-0075, 201-825-7035
Flynn's Custom Guns, P.O. Box 7461, Alexandria, LA 71306/318-455-7130
FN Herstal, Voie de Liege 33, Herstal 4040, BELGIUM/(32)41.40.82.83; FAX: (32)41.40.86.79
Fobus International Ltd., P.O. Box 64, Kfar Hess, ISRAEL/40692 972-9-7964170; FAX: 972-9-7964169
Folks, Donald E., 205 W. Lincoln St., Pontiac, IL 61764/815-844-7901
Foothills Video Productions, Inc., P.O. Box 651, Spartanburg, SC 29304/803-573-7023, 800-782-5358
Foredom Electric Co., Rt. 6, 16 Stony Hill Rd., Bethel, CT 06801/203-792-8622
Forgett Jr., Valmore J., 689 Bergen Blvd., Ridgefield, NJ 07657/201-945-2500; FAX: 201-945-6859; E-MAIL: ValForgett@msn.com
Forgreens Tool Mfg., Inc., P.O. Box 990, 723 Austin St., Robert Lee, TX 76945/915-453-2800; FAX: 915-453-2460
Forkin, Ben (See Belt MTN Arms)
Forrest, Inc., Tom, P.O. Box 326, Lakeside, CA 92040/619-561-5800; FAX: 619-561-0227
Forrest Tool Co., P.O. Box 768, 44380 Gordon Lane, Mendocino, CA 95460/707-937-2141; FAX: 717-937-1817
Forster, Kathy (See Custom Checkering Service)
Forster, Larry L., P.O. Box 212, 220 First St. NE, Gwinner, ND 58040-0212/701-678-2475
Forster Products, 82 E. Lanark Ave., Lanark, IL 61046/815-493-6360; FAX: 815-493-2371
Fort Hill Gunstocks, 12807 Fort Hill Rd., Hillsboro, OH 45133/513-466-2763
Fort Knox Security Products, 1051 N. Industrial Park Rd., Orem, UT 84057/801-224-7233, 800-821-5216; FAX: 801-226-5493
Fort Worth Firearms, 2006-B Martin Luther King Fwy., Ft. Worth, TX 76104-6303/817-536-0718; FAX: 817-535-0290
Forthofer's Gunsmithing & Knifemaking, 5535 U.S. Hwy 93S, Whitefish, MT 59937-8411/406-862-2674
Fortune Products, Inc., HC04, Box 303, Marble Falls, TX 78654/210-693-6111; FAX: 210-693-6394
Forty Five Ranch Enterprises, Box 1080, Miami, OK 74355-1080/918-542-5875
Fotar Optics, 1756 E. Colorado Blvd., Pasadena, CA 91106/818-579-3919; FAX: 818-579-7209
Fouling Shot, The, 6465 Parfet St., Arvada, CO 80004
Fountain Products, 492 Prospect Ave., West Springfield, MA 01089/413-781-4651; FAX: 413-733-8217
4-D Custom Die Co., 711 N. Sandusky St., P.O. Box 889, Mt. Vernon, OH 43050-0889/614-397-7214; FAX: 614-397-6600
Fowler, Bob (see Black Powder Products)
4W Ammunition (See Hunters Supply)
Fowler Bullets, 806 Dogwood Dr., Gastonia, NC 28054/704-867-3259
Fox River Mills, Inc., P.O. Box 298, 227 Poplar St., Osage, IA 50461/515-732-3798; FAX: 515-732-5128
Foy Custom Bullets, 104 Wells Ave., Daleville, AL 36322
Francesca, Inc., 3115 Old Ranch Rd., San Antonio, TX 78217/512-826-2584; FAX: 512-826-8211

DIRECTORY OF THE ARMS TRADE

Franchi S.p.A., Via del Serpente, 12, 25131 Brescia, ITALY/030-3581833; FAX: 030-3581554 (U.S. importer—American Arms, Inc.)
Francotte & Cie S.A., Auguste, rue du Trois Juin 109, 4400 Herstal-Liege, BELGIUM/32-4-248-13-18; FAX: 32-4-948-11-79
Frank Custom Classic Arms, Ron, 7131 Richland Rd., Ft. Worth, TX 76118/817-284-9300; FAX: 817-284-9300
Frank Knives, 13868 NW Keleka Pl., Seal Rock, OR 97376/541-563-3041; FAX: 541-563-3041
Frankonia Jagd, Hofmann & Co., D-97064 Wurzburg, GERMANY/09302-200; FAX: 09302-20200
Franzen International, Inc. (U.S. importer for Peters Stahl GmbH)
Frazier Brothers Enterprises, 1118 N. Main St., Franklin, IN 46131/317-736-4000; FAX: 317-736-4000
Freedom Arms, Inc., P.O. Box 1776, Freedom, WY 83120/307-883-2468, 800-833-4432 (orders only); FAX: 307-883-2005
Freeman Animal Targets, 5519 East County Road, 100 South, Plainsfield, IN 46168/317-272-2663; FAX: 317-272-2674; E-MAIL: Signs@indy.net; WEB: http://www.freemansighs.com
Fremont Tool Works, 1214 Prairie, Ford, KS 67842/316-369-2327
French, J.R., Artistic Engraving, 1712 Creek Ridge Ct., Irving, TX 75060/214-254-2654
Frielich Police Equipment, 211 East 21st St., New York, NY 10010/212-254-3045
Europtik Ltd., P.O. Box 319,, Dunmore, PA 18512/717-347-6049; FAX: 717-969-4330
Front Sight Firearms Training Institute, P.O. Box 2619, Aptos, CA 95001/800-987-7719; FAX: 408-684-2137
Frontier, 2910 San Bernardo, Laredo, TX 78040/210-723-5409; FAX: 210-723-1774
Frontier Arms Co., Inc., 401 W. Rio Santa Cruz, Green Valley, AZ 85614-3932
Frontier Products Co., 164 E. Longview Ave., Columbus, OH 43202/614-262-9357
Frontier Safe Co., 3201 S. Clinton St., Fort Wayne, IN 46806/219-744-7233; FAX: 219-744-6678
Frost Cutlery Co., P.O. Box 22636, Chattanooga, TN 37422/615-894-6079; FAX: 615-894-9576
Fry Metals, 4100 6th Ave., Altoona, PA 16602/814-946-1611
FTI, Inc., 72 Eagle Rock Ave., Box 366, East Hanover, NJ 07936-3104
Fujinon, Inc., 10 High Point Dr., Wayne, NJ 07470/201-633-5600; FAX: 201-633-5216
Fullmer, Geo. M., 2499 Mavis St., Oakland, CA 94601/510-533-4193
Fulmer's Antique Firearms, Chet, P.O. Box 792, Rt. 2 Buffalo Lake, Detroit Lakes, MN 56501/218-847-7712
Fulton Armory, 8725 Bollman Place No. 1, Savage, MD 20763/301-490-9485; FAX: 301-490-9547
Furr Arms, 91 N. 970 W., Orem, UT 84057/801-226-3877; FAX: 801-226-3877
Fusilier Bullets, 10010 N. 6000 W., Highland, UT 84003/801-756-6813
FWB, Neckarstrasse 43, 78727 Oberndorf a. N., GERMANY/07423-814-0; FAX: 07423-814-89 (U.S. importer—Beeman Precision Airguns)

G

G96 Products Co., Inc., River St. Station, P.O. Box 1684, Paterson, NJ 07544/201-684-4050; FAX: 201-684-3848
G&C Bullet Co., Inc., 8835 Thornton Rd., Stockton, CA 95209/209-477-6479; FAX: 209-477-2813
G&H Decoys, Inc., P.O. Box 1208, Hwy. 75 North, Henryetta, OK 74437/918-652-3314; FAX: 918-652-3400
Gage Manufacturing, 663 W. 7th St., A, San Pedro, CA 90731/310-832-3546
Gaillard Barrels, P.O. Box 21, Pathlow, Sask., S0K 3B0 CANADA/306-752-3769; FAX: 306-752-5969
Gain Twist Barrel Co., Rifle Works and Armory, 707 12th Street, Cody, WY 82414/307-587-4914; FAX: 307-527-6097
Galati International, P.O. Box 326, Catawissa, MO 63015/314-257-4837; FAX: 314-257-2268
Galaxy Imports Ltd., Inc., P.O. Box 3361, Victoria, TX 77903/512-573-4867; FAX: 512-576-9622
GALCO International Ltd., 2019 W. Quail Ave., Phoenix, AZ 85027/602-258-8295, 800-874-2526; FAX: 602-582-6854
Gamba S.p.A.-Societa Armi Bresciane Srl., Renato, Via Artigiani, 93, 25063 Gardone Val Trompia (BS), ITALY/30-8911640; FAX: 30-8911648 (U.S. importer—Gamba, USA)
Gamba, USA, P.O. Box 60452, Colorado Springs, CO 80960/719-578-1145; FAX: 719-444-0731
Gamco, 1316 67th Street, Emeryville, CA 94608/510-527-5578
Game Haven Gunstocks, 13750 Shire Rd., Wolverine, MI 49799/616-525-8257
Game Winner, Inc., 2625 Cumberland Parkway, Suite 220, Atlanta, GA 30339/770-434-9210; FAX: 770-434-9215
Gamo (See U.S. importers—Arms United Corp.; Daisy Mfg. Co.; Dynamit Nobel-RWS, Inc.; Gamo USA, Inc.)
Gamo USA, Inc., 3911 SW 47th Ave., Suite 914, Ft. Lauderdale, FL 33314/343-640-7248; FAX: 343-654-0300
Gander Mountain, Inc., P.O. Box 128, Hwy. W,, Wilmot, WI 53192/414-862-2331,Ext. 6423
GAR, 590 McBride Avenue, West Paterson, NJ 07424/201-754-1114; FAX: 201-754-1114
Garbi, Armas Urki, 12-14, 20.600 Eibar (Guipuzcoa) SPAIN/43-11 38 73 (U.S. importer—Moore & Co., Wm. Larkin)

Garcia National Gun Traders, Inc., 225 SW 22nd Ave., Miami, FL 33135/305-642-2355
Garrett Cartridges, Inc., P.O. Box 178, Chehalis, WA 98532/360-736-0702
Garthwaite, Pistolsmith, Inc., Jim, Rt. 2, Box 310, Watsontown, PA 17777/717-538-1566; FAX: 717-538-2965
Gator Guns & Repair, 6255 Spur Hwy., Kenai, AK 99611/907-283-7947
Gaucher Armes, S.A., 46, rue Desjoyaux, 42000 Saint-Etienne, FRANCE/04-77-33-38-92; FAX: 04-77-61-95-72
G.C.C.T., 4455 Torrance Blvd., Ste. 453, Torrance, CA 90509-2806
GDL Enterprises, 409 Le Gardeur, Slidell, LA 70460/504-649-0693
Gehmann, Walter (See Huntington Die Specialties)
Genco, P.O. Box 5704, Asheville, NC 28803
Genecco Gun Works, K., 10512 Lower Sacramento Rd., Stockton, CA 95210/209-951-0706
General Lead, Inc., 1022 Grand Ave., Phoenix, AZ 85007
Gene's Custom Guns, P.O. Box 10534, White Bear Lake, MN 55110/612-429-5105
Gentex Corp., 5 Tinkham Ave., Derry, NH 03038/603-434-0311; FAX: 603-434-3002
Gentner Bullets, 109 Woodlawn Ave., Upper Darby, PA 19082/610-352-9396
Gentry Custom Gunmaker, David, 314 N. Hoffman, Belgrade, MT 59714/406-388-GUNS
George & Roy's, 2950 NW 29th, Portland, OR 97210/503-228-5424, 800-553-3022; FAX: 503-225-9409
George, Tim, Rt. 1, P.O. Box 45, Evington, VA 24550/804-821-8117
Gerber Legendary Blades, 14200 SW 72nd Ave., Portland, OR 97223/503-639-6161, 800-950-6161; FAX: 503-684-7008
Gervais, Mike, 3804 S. Cruise Dr., Salt Lake City, UT 84109/801-277-7729
Getz Barrel Co., P.O. Box 88, Beavertown, PA 17813/717-658-7263
G.G. & G., 3602 E. 42nd Stravenue, Tucson, AZ 85713/520-748-7167; FAX: 520-748-7583
G.H. Enterprises Ltd., Bag 10, Okotoks, Alberta T0L 1T0 CANADA/403-938-6070
Giacomo Sporting USA, 6234 Stokes Lee Center Rd., Lee Center, NY 13363
Gibbs Rifle Co., Inc., Cannon Hill Industrial Park, Rt. 2, Box 214 Hoffman, Rd./Martinsburg, WV 25401
304-274-0458; FAX: 304-274-0078
Gilbert Equipment Co., Inc., 960 Downtowner Rd., Mobile, AL 36609/205-344-3322
Gilkes, Anthony W., 5950 Sheridan Blvd., Arvada, CO 80003/303-657-1873; FAX: 303-657-1885
Gillmann, Edwin, 33 Valley View Dr., Hanover, PA 17331/717-632-1662
Gilman-Mayfield, Inc., 3279 E. Shields, Fresno, CA 93703/209-221-9415; FAX: 209-221-9419
Gilmore Sports Concepts, 5949 S. Garnett, Tulsa, OK 74146/918-250-4867; FAX: 918-250-3845
Giron, Robert E., 1328 Pocono St., Pittsburgh, PA 15218/412-731-6041
Glacier Glove, 4890 Aircenter Circle, Suite 210, Reno, NV 89502/702-825-8225; FAX: 702-825-6544
Glaser Safety Slug, Inc., P.O. Box 8223, Foster City, CA 94404/800-221-3489, 510-785-7754; FAX: 510-785-6685
Glass, Herb, P.O. Box 25, Bullville, NY 10915/914-361-3021
Glimm, Jerome C., 19 S. Maryland, Conrad, MT 59425/406-278-3574
Glock GmbH, P.O. Box 50, A-2232 Deutsch Wagram, AUSTRIA (U.S. importer—Glock, Inc.)
Glock, Inc., P.O. Box 369, Smyrna, GA 30081/770-432-1202; FAX: 770-433-8719
GML Products, Inc., 394 Laredo Dr., Birmingham, AL 35226/205-979-4867
Gner's Hard Cast Bullets, 1107 11th St., LaGrande, OR 97850/503-963-8796
Goddard, Allen, 716 Medford Ave., Hayward, CA 94541/510-276-6830
Goens, Dale W., P.O. Box 224, Cedar Crest, NM 87008/505-281-5419
Goergen's Gun Shop, Inc., Rt. 2, Box 182BB, Austin, MN 55912/507-433-9280
GOEX, Inc., 1002 Springbrook Ave., Moosic, PA 18507/717-457-6724; FAX: 717-457-1130
Goldcoast Reloaders, Inc., 4260 NE 12th Terrace, Pompano Beach, FL 33064/954-783-4849; FAX: 954-942-3452
Golden Age Arms Co., 115 E. High St., Ashley, OH 43003/614-747-2488
Golden Bear Bullets, 3065 Fairfax Ave., San Jose, CA 95148/408-238-9515
Gonic Arms, Inc., 134 Flagg Rd., Gonic, NH 03839/603-332-8456, 603-332-8457
Gonic Bullet Works, P.O. Box 7365, Gonic, NH 03839
Gonzalez Guns, Ramon B., P.O. Box 370, 93 St. Joseph's Hill Road, Monticello, NY 12701/914-794-4515
Goodling's Gunsmithing, R.D. 1, Box 1097, Spring Grove, PA 17362/717-225-3350
Goodwin, Fred, Silver Ridge Gun Shop, Sherman Mills, ME 04776/207-365-4451
Gordie's Gun Shop, 1401 Fulton St., Streator, IL 61364/815-672-7202
Gotz Bullets, 7313 Rogers St., Rockford, IL 61111
Goudy Classic Stocks, Gary, 263 Hedge Rd., Menlo Park, CA 94025-1711/415-322-1338
Gould & Goodrich, P.O. Box 1479, Lillington, NC 27546/910-893-2071; FAX: 910-893-4742
Gournet, Geoffroy, 820 Paxinosa Ave., Easton, PA 18042/610-559-0710
Gozon Corp., U.S.A., P.O. Box 6278, Folson, CA 95763/916-983-2026; FAX: 916-983-9500
Grace, Charles E., 1305 Arizona Ave., Trinidad, CO 81082/719-846-9435
Grace Metal Products, Inc., P.O. Box 67, Elk Rapids, MI 49629/616-264-8133
Graf & Sons, Route 3 Highway 54 So., Mexico, MO 65265/573-581-2266; FAX: 573-581-2875
"Gramps" Antique Cartridges, Box 341, Washago, Ont. L0K 2B0 CANADA/705-689-5348

MANUFACTURERS' DIRECTORY

Grand Falls Bullets, Inc., P.O. Box 720, 803 Arnold Wallen Way, Stockton, MO 65785/816-229-0112
Granite Custom Bullets, Box 190, Philipsburg, MT 59858/406-859-3245
Grant, Howard V., Hiawatha 15, Woodruff, WI 54568/715-356-7146
Graphics Direct, P.O. Box 372421, Reseda, CA 91337-2421/818-344-9002
Graves Co., 1800 Andrews Ave., Pompano Beach, FL 33069/800-327-9103; FAX: 305-960-0301
Grayback Wildcats, 5306 Bryant Ave., Klamath Falls, OR 97603/541-884-1072
Graybill's Gun Shop, 1035 Ironville Pike, Columbia, PA 17512/717-684-2739
Great American Gunstock Co., 3420 Industrial Drive, Yuba City, CA 95993/916-671-4570; FAX: 916-671-3906
Great Lakes Airguns, 6175 S. Park Ave., Hamburg, NY 14075/716-648-6666; FAX: 716-648-5279
Green, Arthur S., 485 S. Robertson Blvd., Beverly Hills, CA 90211/310-274-1283
Green Genie, Box 114, Cusseta, GA 31805
Green Head Game Call Co., RR 1, Box 33, Lacon, IL 61540/309-246-2155
Green Mountain Rifle Barrel Co., Inc., P.O. Box 2670, 153 West Main St., Conway, NH 03818/603-447-1095; FAX: 603-447-1099
Green, Roger M., P.O. Box 984, 435 E. Birch, Glenrock, WY 82637/307-436-9804
Greene Precision Duplicators, M.L. Greene Engineering Services, P.O. Box, 1150, Glendo, CO 80402-1150/303-279-2383
Greenwald, Leon E. "Bud", 2553 S. Quitman St., Denver, CO 80219/303-935-3850
Greenwood Precision, P.O. Box 468, Nixa, MO 65714-0468/417-725-2330
Greg Gunsmithing Repair, 3732 26th Ave. North, Robbinsdale, MN 55422/612-529-8103
Greg's Superior Products, P.O. Box 46219, Seattle, WA 98146
Greider Precision, 431 Santa Marina Ct., Escondido, CA 92029/619-480-8892; FAX: 619-480-9800; E-MAIL: Greider@msn.com
Gremmel Enterprises, 2111 Carriage Drive, Eugene, OR 97408-7537/541-302-3000
Gré-Tan Rifles, 29742 W.C.R. 50, Kersey, CO 80644/970-353-6176; FAX: 970-356-9133
Grier's Hard Cast Bullets, 1107 11th St., LaGrande, OR 97850/503-963-8796
Griffin & Howe, Inc., 33 Claremont Rd., Bernardsville, NJ 07924/908-766-2287; FAX: 908-766-1068
Griffin & Howe, Inc., 36 W. 44th St., Suite 1011, New York, NY 10036/212-921-0980
Grifon, Inc., 58 Guinam St., Waltham, MS 02154
Groenewold, John, P.O. Box 830, Mundelein, IL 60060/847-566-2365
Group Tight Bullets, 482 Comerwood Court, San Francisco, CA 94080/415-583-1550
GRS Corp., Glendo, P.O. Box 1153, 900 Overlander St., Emporia, KS 66801/316-343-1084, 800-835-3519
Grulla Armes, Apartado 453, Avda Otaloa, 12, Eiber, SPAIN (U.S. importer—American Arms, Inc.)
GSI, Inc., 108 Morrow Ave., P.O. Box 129, Trussville, AL 35173/205-655-8299; FAX: 205-655-7078
G.U., Inc. (U.S. importer for New SKB Arms Co.; SKB Arms Co.)
Guardsman Products, 411 N. Darling, Fremont, MI 49412/616-924-3950
Gun Accessories (See Glaser Safety Slug, Inc.)
Gun-Alert, 1010 N. Maclay Ave., San Fernando, CA 91340/818-365-0864; FAX: 818-365-1308
Gun City, 212 W. Main Ave., Bismarck, ND 58501/701-223-2304
Gun Doctor, The, 435 East Maple, Roselle, IL 60172/708-894-0668
Gun Doctor, The, P.O. Box 39242, Downey, CA 90242/310-862-3158
Gun-Ho Sports Cases, 110 E. 10th St., St. Paul, MN 55101/612-224-9491
Gun Hunter Books (See Gun Hunter Trading Co.)
Gun Hunter Trading Co., 5075 Heisig St., Beaumont, TX 77705/409-835-3006
Gun Leather Limited, 116 Lipscomb, Ft. Worth, TX 76104/817-334-0225; 800-247-0609
Gun List (See Krause Publications, Inc.)
Gun Locker, Div. of Airmold, W.R. Grace & Co.-Conn., Becker Farms Ind. Park,, P.O. Box 610/Roanoke Rapids, NC 27870
800-344-5716; FAX: 919-536-2201
Gun Parts Corp., The, 226 Williams Lane, West Hurley, NY 12491/914-679-2417; FAX: 914-679-5849
Gun Room, The, 1121 Burlington, Muncie, IN 47302/317-282-9073; FAX: 317-282-5270
Gun Room Press, The, 127 Raritan Ave., Highland Park, NJ 08904/908-545-4344; FAX: 908-545-6686
Gun Shop, The, 5550 S. 900 East, Salt Lake City, UT 84117/801-263-3633
Gun Shop, The, 62778 Spring Creek Rd., Montrose, CO 81401
Gun Shop, The, 716-A South Rogers Road, Olathe, KS 66062
Gun South, Inc. (See GSI, Inc.)
Gun-Tec, P.O. Box 8125, W. Palm Beach, FL 33407
Gun Works, The, 247 S. 2nd, Springfield, OR 97477/541-741-4118; FAX: 541-988-1097
Guncraft Books (See Guncraft Sports, Inc.)
Guncraft Sports, Inc., 10737 Dutchtown Rd., Knoxville, TN 37932/423-966-4545; FAX: 423-966-4500
Gunfitters, The, P.O. 426, Cambridge, WI 53523-0426/608-764-8128
Gunline Tools, 2950 Saturn St., "O", Brea, CA 92821/714-993-5100; FAX: 714-572-4128
Gunnerman Books, P.O. Box 217, Owosso, MI 48867/517-729-7018; FAX: 517-725-9391
Guns, 81 E. Streetsboro St., Hudson, OH 44236/216-650-4563
Guns Antique & Modern DBA/Charles E. Duffy, Williams Lane, West Hurley, NY 12491/914-679-2997
Guns, Div. of D.C. Engineering, Inc., 8633 Southfield Fwy., Detroit, MI 48228/313-271-7111, 800-886-7623 (orders only); FAX: 313-271-7112
GUNS Magazine, 591 Camino de la Reina, Suite 200, San Diego, CA 92108/619-297-5350; FAX: 619-297-5353
Gunsight, The, 1712 North Placentia Ave., Fullerton, CA 92631
Gunsite Custom Shop, P.O. Box 451, Paulden, AZ 86334/520-636-4104; FAX: 520-636-1236
Gunsite Gunsmithy (See Gunsite Custom Shop)
Gunsite Training Center, P.O. Box 700, Paulden, AZ 86334/520-636-4565; FAX: 520-636-1236
Gunsmith in Elk River, The, 14021 Victoria Lane, Elk River, MN 55330/612-441-7761
Gunsmithing, Inc., 208 West Buchanan St., Colorado Springs, CO 80907/719-632-3795; FAX: 719-632-3493
Gunsmithing Ltd., 57 Unquowa Rd., Fairfield, CT 06430/203-254-0436; FAX: 203-254-1535
Gurney, F.R., Box 13, Sooke, BC V0S 1N0 CANADA/604-642-5282; FAX: 604-642-7859
Gusdorf Corp., 11440 Lackland Rd., St. Louis, MO 63146/314-567-5249
Gusty Winds Corp., 2950 Bear St., Suite 120, Costa Mesa, CA 92626/714-536-3587
Gwinnell, Bryson J., P.O. Box 248C, Maple Hill Rd., Rochester, VT 05767/802-767-3664

H

H&B Forge Co., Rt. 2 Geisinger Rd., Shiloh, OH 44878/419-895-1856
H&P Publishing, 7174 Hoffman Rd., San Angelo, TX 76905/915-655-5953
H&R 1871, Inc., 60 Industrial Rowe, Gardner, MA 01440/508-632-9393; FAX: 508-632-2300
H&S Liner Service, 515 E. 8th, Odessa, TX 79761/915-332-1021
Hafner Creations, Inc., P.O. Box 1987, Lake City, FL 32055/904-755-6481; FAX: 904-755-6595
Hagn Rifles & Actions, Martin, P.O. Box 444, Cranbrook, B.C. VIC 4H9, CANADA/604-489-4861
Hakko Co. Ltd., Daini-Tsunemi Bldg., 1-13-12, Narimasu, Itabashiku Tokyo 175, JAPAN/03-5997-7870/2; FAX: 81-3-5997-7840
Hale, Engraver, Peter, 800 E. Canyon Rd., Spanish Fork, UT 84660/801-798-8215
Half Moon Rifle Shop, 490 Halfmoon Rd., Columbia Falls, MT 59912/406-892-4409
Hall Manufacturing, 142 CR 406, Clanton, AL 35045/205-755-4094
Hall Plastics, Inc., John, P.O. Box 1526, Alvin, TX 77512/713-489-8709
Hallberg Gunsmith, Fritz, P.O. Box 339, 160 N. Oregon St., Ontario, OR 97914/541-889-3135; FAX: 541-889-2633
Hallowell & Co., 340 W. Putnam Ave., Greenwich, CT 06830/203-869-2190; FAX: 203-869-0692
Hally Caller, 443 Wells Rd., Doylestown, PA 18901/215-345-6354
Halstead, Rick, RR4, Box 272, Miami, OK 74354/918-540-0933
Hamilton, Alex B. (See Ten-Ring Precision, Inc.)
Hamilton, Jim, Rte. 5, Box 278, Guthrie, OK 73044/405-282-3634
Hamilton, Keith, P.O. Box 871, Gridley, CA 95948/916-846-2316
Hammans, Charles E., P.O. Box 788, 2022 McCracken, Stuttgart, AR 72106/501-673-1388
Hammerli USA, 19296 Oak Grove Circle, Groveland, CA 95321/209-962-5311; FAX: 209-962-5931
Hammerli Ltd., Seonerstrasse 37, CH-5600 Lenzburg, SWITZERLAND/064-50 11 44; FAX: 064-51 38 27 (U.S. importer—Hammerli USA; Mandall Shooting Supplies, Inc.; Sigarms, Inc.)
Hammets VLD Bullets, P.O. Box 479, Rayville, LA 71269/318-728-2019
Hammond Custom Guns Ltd., 619 S. Pandora, Gilbert, AZ 85234/602-892-3437
Hammonds Rifles, RD 4, Box 504, Red Lion, PA 17356/717-244-7879
Handgun Press, P.O. Box 406, Glenview, IL 60025/847-657-6500; FAX: 847-724-8831
HandiCrafts Unltd. (See Clements' Custom Leathercraft, Chas)
Hands Engraving, Barry Lee, 26192 E. Shore Route, Bigfork, MT 59911/406-837-0035
Hank's Gun Shop, Box 370, 50 West 100 South, Monroe, UT 84754/801-527-4456
Hanned Line, The, P.O. Box 2387, Cupertino, CA 95015-2387
Hanned Precision (See Hanned Line, The)
Hansen & Co. (See Hansen Cartridge Co.)
Hansen Cartridge Co., 244-246 Old Post Rd., Southport, CT 06490/203-259-6222, 203-259-7337; FAX: 203-254-3832
Hanson's Gun Center, Dick, 233 Everett Dr., Colorado Springs, CO 80911
Hanus Birdguns, Bill, P.O. Box 533, Newport, OR 97365/541-265-7433; FAX: 541-265-7400
Hanusin, John, 3306 Commercial, Northbrook, IL 60062/708-564-2706
Hardin Specialty Dist., P.O. Box 338, Radcliff, KY 40159-0338/502-351-6649
Harold's Custom Gun Shop, Inc., Broughton Rifle Barrels, Rt. 1, Box 447, Big Spring, TX 79720/915-394-4430
Harper's Custom Stocks, 928 Lombrano St., San Antonio, TX 78207/210-732-5780
Harrell's Precision, 5756 Hickory Dr., Salem, VA 24153/703-380-2683
Harrington & Richardson (See H&R 1871, Inc.)
Harrington Cutlery, Inc., Russell, Subs. of Hyde Mfg. Co., 44 River St., Southbridge, MA 01550/617-765-0201
Harris Engineering, Inc., 999 Broadway, Barlow, KY 42024/502-334-3633; FAX: 502-334-3000
Harris Enterprises, P.O. Box 105, Bly, OR 97622/503-353-2625

30th EDITION, 1998 **319**

DIRECTORY OF THE ARMS TRADE

Harris Hand Engraving, Paul A., 113 Rusty Lane, Boerne, TX 78006-5746/512-391-5121
Harris Gunworks, 3840 N. 28th Ave., Phoenix, AZ 85017-4733/602-230-1414; FAX: 602-230-1422
Harris Publications, 1115 Broadway, New York, NY 10010/212-807-7100; FAX: 212-627-4678
Harrison Bullets, 6437 E. Hobart St., Mesa, AZ 85205
Harrison-Hurtz Enterprises, Inc., P.O. Box 268, RR1, Wymore, NE 68466/402-645-3378; FAX: 402-645-3606
Hart & Son, Inc., Robert W., 401 Montgomery St., Nescopeck, PA 18635/717-752-3655, 800-368-3656; FAX: 717-752-1088
Hart Rifle Barrels, Inc., P.O. Box 182, 1690 Apulia Rd., Lafayette, NY 13084/315-677-9817; FAX: 315-677-9610
Hartford (See U.S. importer— EMF Co., Inc.)
Hartmann & Weiss GmbH, Rahlstedter Bahnhofstr. 47, 22143 Hamburg, GERMANY/(40) 677 55 85; FAX: (40) 677 55 92
Harvey, Frank, 218 Nightfall, Terrace, NV 89015/702-558-6998
Harwood, Jack O., 1191 S. Pendlebury Lane, Blackfoot, ID 83221/208-785-5368
Haselbauer Products, Jerry, P.O. Box 27629, Tucson, AZ 85726/602-792-1075
Hastings Barrels, 320 Court St., Clay Center, KS 67432/913-632-3169; FAX: 913-632-6554
Hawk, Inc., 849 Hawks Bridge Rd., Salem, NJ 08079/609-299-2700; FAX: 609-299-2800
Hawk Laboratories, Inc. (See Hawk, Inc.)
Hawken Shop, The (See Dayton Traister)
Haydel's Game Calls, Inc., 5018 Hazel Jones Rd., Bossier City, LA 71111/318-746-3586, 800-HAYDELS; FAX: 318-746-3711
Haydon Shooters' Supply, Russ, 15018 Goodrich Dr. NW, Gig Harbor, WA 98329/253-857-7557; FAX: 253-857-7884
Heatbath Corp., P.O. Box 2978, Springfield, MA 01101/413-543-3381
Hebard Guns, Gil, 125-129 Public Square, Knoxville, IL 61448
HEBB Resources, P.O. Box 999, Mead, WA 99021-09996/509-466-1292
Hecht, Hubert J., Waffen-Hecht, P.O. Box 2635, Fair Oaks, CA 95628/916-966-1020
Heckler & Koch GmbH, P.O. Box 1329, 78722 Oberndorf, Neckar, GERMANY/49-7423179-0; FAX: 49-7423179-2406 (U.S. importer—Heckler & Koch, Inc.)
Heckler & Koch, Inc., 21480 Pacific Blvd., Sterling, VA 20166-8903/703-450-1900; FAX: 703-450-8160
Hege Jagd-u. Sporthandels, GmbH, P.O. Box 101461, W-7770 Ueberlingen a. Bodensee, GERMANY
Heidenstrom Bullets, Urds GT 1 Heroya, 3900 Porsgrunn, NORWAY
Heilmann, Stephen, P.O. Box 657, Grass Valley, CA 95945/916-272-8758
Heinie Specialty Products, 301 Oak St., Quincy, IL 62301-2500/309-543-4535; FAX: 309-543-2521
Heintz, David, 800 N. Hwy. 17, Moffat, CO 81143/719-256-4194
Hellweg Ltd., 40356 Oak Park Way, Suite H, Oakhurst, CA 93644/209-683-3030; FAX: 209-683-3422
Helwan (See U.S. importer—Interarms)
Henckels Zwillingswerk, J.A., Inc., 9 Skyline Dr., Hawthorne, NY 10532/914-592-7370
Hendricks, Frank E., Inc., Master Engravers, HC03, Box 434, Dripping Springs, TX 78620/512-858-7828
Hendricks Gun Works, 1162 Gillionville Rd., Albany, GA 31707/912-439-2003
Henigson & Associates, Steve, P.O. Box 2726, Culver City, CA 90231/310-305-8288; FAX: 310-305-1905
Henriksen Tool Co., Inc., 8515 Wagner Creek Rd., Talent, OR 97540/541-535-2309
Henry Repeating Arms Co., 110 8th St., Brooklyn, NY 11215/718-499-5600
Hensler, Jerry, 6614 Country Field, San Antonio, TX 78240/210-690-7491
Hensley & Gibbs, Box 10, Murphy, OR 97533/541-862-2341
Hensley, Gunmaker, Darwin, P.O. Box 329, Brightwood, OR 97011/503-622-5411
Heppler, Keith M., Keith's Custom Gunstocks, 540 Banyan Circle, Walnut Creek, CA 94598/510-934-3509; FAX: 510-934-3143
Heppler's Machining, 2240 Calle Del Mundo, Santa Clara, CA 95054/408-748-9166; FAX: 408-988-7711
Hercules, Inc. (See Alliant Techsystems, Smokeless Powder Group)
Heritage Firearms (See Heritage Manufacturing, Inc.)
Heritage Manufacturing, Inc., 4600 NW 135th St., Opa Locka, FL 33054/305-685-5966; FAX: 305-687-6721
Heritage/VSP Gun Books, P.O. Box 887, McCall, ID 83638/208-634-4104; FAX: 208-634-3101
Heritage Wildlife Carvings, 2145 Wagner Hollow Rd., Fort Plain, NY 13339/518-993-3983
Hermann Leather Co., H.J., Rt. 1, P.O. Box 525, Skiatook, OK 74070/918-396-1226
Herrett's Stocks, Inc., P.O. Box 741, Twin Falls, ID 83303/208-733-1498
Hertel & Reuss, Werk für Optik und Feinmechanik GmbH, Quellhofstrasse, 67/34 127 Kassel, GERMANY 0561-83006; FAX: 0561-893308
Herter's Manufacturing, Inc., 111 E. Burnett St., P.O. Box 518, Beaver Dam, WI 53916/414-887-1765; FAX: 414-887-8444
Hesco-Meprolight, 2139 Greenville Rd., LaGrange, GA 30241/706-884-7967; FAX: 706-882-4683
Heydenberk, Warren R., 1059 W. Sawmill Rd., Quakertown, PA 18951/215-538-2682
Hickman, Jaclyn, Box 1900, Glenrock, WY 82637
Hidalgo, Tony, 12701 SW 9th Pl., Davie, FL 33325/954-476-7645
High Bridge Arms, Inc., 3185 Mission St., San Francisco, CA 94110/415-282-8358

High North Products, Inc., P.O. Box 2, Antigo, WI 54409/715-627-2331
High Performance International, 5734 W. Florist Ave., Milwaukee, WI 53218/414-466-9040
High Standard Mfg. Co., Inc., 4601 S. Pinemont, Suite 144, Houston, TX 77041/713-462-4200; FAX: 713-462-6437
High Tech Specialties, Inc., P.O. Box 387R, Adamstown, PA 19501/215-484-0405, 800-231-9385
Highline Machine Co., 654 Lela Place, Grand Junction, CO 81504/970-434-4971
Hill, Loring F., 304 Cedar Rd., Elkins Park, PA 19117
Hill Speed Leather, Ernie, 4507 N. 195th Ave., Litchfield Park, AZ 85340/602-853-9222; FAX: 602-853-9235
Hillmer Custom Gunstocks, Paul D., 7251 Hudson Heights, Hudson, IA 50643/319-988-3941
Hinman Outfitters, Bob, 1217 W. Glen, Peoria, IL 61614/309-691-8132
Hi-Grade Imports, 8655 Monterey Rd., Gilroy, CA 95021/408-842-9301; FAX: 408-842-2374
Hi-Point Firearms, 5990 Philadelphia Dr., Dayton, OH 45415/513-275-4991; FAX: 513-522-8330
Hi-Performance Ammunition Company, 484 State Route 366, Apollo, PA 15613/412-327-8100
Hiptmayer, Armurier, RR 112 750, P.O. Box 136, Eastman, Quebec J0E 1P0, CANADA/514-297-2492
Hiptmayer, Heidemarie, RR 112 750, P.O. Box 136, Eastman, Quebec J0E 1P0, CANADA/514-297-2492
Hiptmayer, Klaus, RR 112 750, P.O. Box 136, Eastman, Quebec J0E 1P0, CANADA/514-297-2492
Hirtenberger Aktiengesellschaft, Leobersdorferstrasse 31, A-2552 Hirtenberg, AUSTRIA/43(0)2256 81184; FAX: 43(0)2256 81807
HiTek International, 484 El Camino Real, Redwood City, CA 94063/415-363-1404, 800-54-NIGHT; FAX: 415-363-1408
Hiti-Schuch, Atelier Wilma, A-8863 Predlitz, Pirming Y1 AUSTRIA/0353418278
HJS Arms, Inc., P.O. Box 3711, Brownsville, TX 78523-3711/800-453-2767, 210-542-2767
H.K.S. Products, 7841 Founion Dr., Florence, KY 41042/606-342-7841, 800-354-9814; FAX: 606-342-5865
Hoag, James W., 8523 Canoga Ave., Suite C, Canoga Park, CA 91304/818-998-1510
Hobbie Gunsmithing, Duane A., 2412 Pattie Ave., Wichita, KS 67216/316-264-8266
Hobson Precision Mfg. Co., Rt. 1, Box 220-C, Brent, AL 35034/205-926-4662
Hoch Custom Bullet Moulds (See Colorado Shooter's Supply)
Hodgdon Powder Co., 6231 Robinson, Shawnee Mission, KS 66202/913-362-9455; FAX: 913-362-1307; WEB: http://www.hodgdon.com
Hodgman, Inc., 1750 Orchard Rd., Montgomery, IL 60538/708-897-7555; FAX: 708-897-7558
Hodgson, Richard, 9081 Tahoe Lane, Boulder, CO 80301
Hoehn Sales, Inc., 2045 Kohn Road, Wright City, MO 63390/314-745-8144; FAX: 314-745-8144
Hoelscher, Virgil, 11047 Pope Ave., Lynwood, CA 90262/310-631-8545
Hoenig & Rodman, 6521 Morton Dr., Boise, ID 83704/208-375-1116
Hofer Jagdwaffen, P., Buchsenmachermeister, Kirchgasse 24, A-9170 Ferlach, AUSTRIA/04227-3683
Hoffman New Ideas, 821 Northmoor Rd., Lake Forest, IL 60045/312-234-4075
Hogue Grips, P.O. Box 1138, Paso Robles, CA 93447/800-438-4747, 805-239-1440; FAX: 805-239-2553
Holland & Holland Ltd., 33 Bruton St., London, ENGLAND 1W1/44-171-499-4411; FAX: 44-171-408-7962
Holland, Dick, 422 NE 6th St., Newport, OR 97365/503-265-7556
Holland's Gunsmithing, P.O. Box 69, Powers, OR 97466/541-439-5155; FAX: 541-439-5155
Hollis Gun Shop, 917 Rex St., Carlsbad, NM 88220/505-885-3782
Hollywood Engineering, 10642 Arminta St., Sun Valley, CA 91352/818-842-8376
Holster Shop, The, 720 N. Flagler Dr., Ft. Lauderdale, FL 33304/305-463-7910; FAX: 305-761-1483
Homak, 5151 W. 73rd St., Chicago, IL 60638-6613/312-523-3100, FAX: 312-523-9455
Home Shop Machinist, The, Village Press Publications, P.O. Box 1810, Traverse City, MI 49685/800-447-7367; FAX: 616-946-3289
Hondo Ind., 510 S. 52nd St.,l04, Tempe, AZ 85281
Hoover, Harvey, 5750 Pearl Dr., Paradise, CA 95969-4829
Hoppe's Div., Penguin Industries, Inc., Airport Industrial Mall, Coatesville, PA 19320/610-384-6000
Horizons Unlimited, P.O. Box 426, Warm Springs, GA 31830/706-655-3603; FAX: 706-655-3603
Hornady Mfg. Co., P.O. Box 1848, Grand Island, NE 68802/800-338-3220, 308-382-1390; FAX: 308-382-5761
Horseshoe Leather Products, Andy Arratoonian, The Cottage Sharow, Ripon HG4 5BP ENGLAND/44-1765-605858
Horst, Alan K., 3221 2nd Ave. N., Great Falls, MT 59401/406-454-1831
Horton Dist. Co., Inc., Lew, 15 Walkup Dr., Westboro, MA 01581/508-366-7400; FAX: 508-366-5332
House of Muskets, Inc., The, P.O. Box 4640, Pagosa Springs, CO 81157/970-731-2295
Houtz & Barwick, P.O. Box 435, W. Church St., Elizabeth City, NC 27909/800-775-0337, 919-335-4191; FAX: 919-335-1152
Howa Machinery, Ltd., Sukaguchi, Shinkawa-cho, Nishikasugai-gun, Aichi 452, JAPAN (U.S. importer—Interarms)

MANUFACTURERS' DIRECTORY

Howell Machine, 815½ D St., Lewiston, ID 83501/208-743-7418
Hoyt Holster Co., Inc., P.O. Box 69, Coupeville, WA 98239-0069/360-678-6640; FAX: 360-678-6549
H-S Precision, Inc., 1301 Turbine Dr., Rapid City, SD 57701/605-341-3006; FAX: 605-342-8964
HT Bullets, 244 Belleville Rd., New Bedford, MA 02745/508-999-3338
Hubertus Schneidwarenfabrik, P.O. Box 180 106, D-42626 Solingen, GERMANY/01149-212-59-19-94; FAX: 01149-212-59-19-92
Huebner, Corey O., P.O. Box 2074, Missoula, MT 59806-2074/406-721-7168
Huey Gun Cases, P.O. Box 22456, Kansas City, MO 64113/816-444-1637; FAX: 816-444-1637
Hugger Hooks Co., 3900 Easley Way, Golden, CO 80403/303-279-0600
Hughes, Steven Dodd, P.O. Box 545, Livingston, MT 59047/406-222-9377
Hume, Don, P.O. Box 351, Miami, OK 74355/918-542-6604; FAX: 918-542-4340
Hungry Horse Books, 4605 Hwy. 93 South, Whitefish, MT 59937/406-862-7997
Hunkeler, A. (See Buckskin Machine Works)
Hunter Co., Inc., 3300 W. 71st Ave., Westminster, CO 80030/303-427-4626; FAX: 303-428-3980
Hunters Supply, Rt. 1, P.O. Box 313, Tioga, TX 76271/800-868-6612; FAX: 817-437-2228
Hunter's Specialties, Inc., 6000 Huntington Ct. NE, Cedar Rapids, IA 52402-1268/319-395-0321; FAX: 319-395-0326
Hunterjohn, P.O. Box 477, St. Louis, MO 63166/314-531-7250
Hunting Classics Ltd., P.O. Box 2089, Gastonia, NC 28053/704-867-1307; FAX: 704-867-0491
Huntington Die Specialties, 601 Oro Dam Blvd., Oroville, CA 95965/916-534-1210; FAX: 916-534-1212
Hutton Rifle Ranch, P.O. Box 45236, Boise, ID 83711/208-345-8781
Hydrosorbent Products, P.O. Box 437, Ashley Falls, MA 01222/413-229-2967; FAX: 413-229-8743
Hyper-Single, Inc., 520 E. Beaver, Jenks, OK 74037/918-299-2391

I

I.A.B. (See U.S. importer—Taylor's & Co., Inc.)
IAI (See A.M.T.)
IAR, Inc., 33171 Camino Capistrano, San Juan Capistrano, CA 92675/714-443-3642; FAX: 714-443-3647
Ibberson (Sheffield) Ltd., George, 25-31 Allen St., Sheffield, S3 7AW ENGLAND/0114-2766123; FAX: 0114-2738465
ICI-America, P.O. Box 751, Wilmington, DE 19897/302-575-3000
I.D.S.A. Books, 1324 Stratford Drive, Piqua, OH 45356/937-773-4203; FAX: 937-778-1922.
IGA (See U.S. importer—Stoeger Industries)
Illinois Lead Shop, 7742 W. 61st Place, Summit, IL 60501
Image Ind. Inc., 864 Lively, Wood Dale, IL 60191/630-616-1340; FAX: 630-616-1341
Image Ind. Inc., 382 Balm Court, Wood Dale, IL 60191/630-766-2402; FAX: 630-766-7373
IMI, P.O. Box 1044, Ramat Hasharon 47100, ISRAEL/972-3-5485617;FAX: 972-3-5406908
IMI Services USA, Inc., 2 Wisconsin Circle, Suite 420, Chevy Chase, MD 20815/301-215-4800; FAX: 301-657-1446
Impact Case Co., P.O. Box 9912, Spokane, WA 99209-0912/800-262-3322, 509-467-3303; FAX: 509-326-5436
Imperial (See E-Z-Way Systems)
Imperial Magnum Corp., P.O. Box 249, Oroville, WA 98844/604-495-3131; FAX: 604-495-2816
Imperial Russian Armory, 10547 S. Post Oak, Houston, TX 77035/1-800-MINIATURE
Imperial Schrade Corp., 7 Schrade Ct., Box 7000, Ellenville, NY 12428/914-647-7601; FAX: 914-647-8701
Import Sports Inc., 1750 Brielle Ave., Unit B1, Wanamassa, NJ 07712/908-493-0302; FAX: 908-493-0301
IMR Powder Co., 1080 Military Turnpike, Suite 2, Plattsburgh, NY 12901/518-563-2253; FAX: 518-563-6916
I.N.C., Inc. (See Kick Eez)
Independent Machine & Gun Shop, 1416 N. Hayes, Pocatello, ID 83201
Info-Arm, P.O. Box 1262, Champlain, NY 12919/514-955-0355; FAX: 514-955-0357
Ingle, Engraver, Ralph W., 112 Manchester Ct., Centerville, GA 31028/912-953-5824
Innovative Weaponry, Inc., 337 Eubank NE, Albuquerque, NM 87123/800-334-3573, 505-296-4645; FAX: 505-271-2633
Innovision Enterprises, 728 Skinner Dr., Kalamazoo, MI 49001/616-382-1681; FAX: 616-382-1830
INTEC International, Inc., P.O. Box 5708, Scottsdale, AZ 85261/602-483-1708
Interarms, 10 Prince St., Alexandria, VA 22314/703-548-1400; FAX: 703-549-7826
Intercontinental Munitions Distributors, Ltd., P.O. Box 815, Beulah, ND 58523/701-948-2260; FAX: 701-948-2282
International Shooters Service (See I.S.S.)
Intratec, 12405 SW 130th St., Miami, FL 33186-6224/305-232-1821; FAX: 305-253-7207
Iosso Products, 1485 Lively Blvd., Elk Grove Village, IL 60007/847-437-8400; FAX: 847-437-8478
Iron Bench, 12619 Bailey Rd., Redding, CA 96003/916-241-4623
Iron Mountain Knife Co., P.O. Box 2146, Sparks, NV 89432-2146/702-356-3632; FAX: 702-356-3640
Ironside International Publishers, Inc., P.O. Box 55, 800 Slaters Lane, Alexandria, VA 22313/703-684-6111; FAX: 703-683-5486
Ironsighter Co., P.O. Box 85070, Westland, MI 48185/313-326-8731; FAX: 313-326-3378
Irwin, Campbell H., 140 Hartland Blvd., East Hartland, CT 06027/203-653-3901
Island Pond Gun Shop, P.O. Box 428, Cross St., Island Pond, VT 05846/802-723-4546
Israel Arms International, Inc., 5709 Hartsdale, Houston, TX 77036/713-789-0745; FAX: 713-789-7513
Israel Military Industries Ltd. (See IMI)
I.S.S., P.O. Box 185234, Ft. Worth, TX 76181/817-595-2090
I.S.W., 106 E. Cairo Dr., Tempe, AZ 85282
Ithaca Gun Co., LLC, 891 Route 34-B, King Ferry, NY 13081/315-364-7171, 888-9ITHACA; FAX: 315-364-5134
Ivanoff, Thomas G. (See Tom's Gun Repair)

J

J-4, Inc., 1700 Via Burton, Anaheim, CA 92806/714-254-8315; FAX: 714-956-4421
J&D Components, 75 East 350 North, Orem, UT 84057-4719/801-225-7007
J&J Products, Inc., 9240 Whitmore, El Monte, CA 91731/818-571-5228, 800-927-8361; FAX: 818-571-8704
J&J Sales, 1501 21st Ave. S., Great Falls, MT 59405/406-453-7549
J&L Superior Bullets (See Huntington Die Specialties)
J&R Engineering, P.O. Box 77, 200 Lyons Hill Rd., Athol, MA 01331/508-249-9241
J&R Enterprises, 4550 Scotts Valley Rd., Lakeport, CA 95453
J&S Heat Treat, 803 S. 16th St., Blue Springs, MO 64015/816-229-2149; FAX: 816-228-1135
J.A. Blades, Inc. (See Christopher Firearms Co., Inc., E.)
Jackalope Gun Shop, 1048 S. 5th St., Douglas, WY 82633/307-358-3441
Jaeger, Inc./Dunn's, Paul, P.O. Box 449, 1 Madison Ave., Grand Junction, TN 38039/901-764-6909; FAX: 901-764-6503
JägerSport, Ltd., One Wholesale Way, Cranston, RI 02920/800-962-4867, 401-944-9682; FAX: 401-946-2587
Jamison's Forge Works, 4527 Rd. 6.5 NE, Moses Lake, WA 98837/509-762-2659
Jantz Supply, P.O. Box 584-GD, Davis, OK 73030-0584/405-369-2316; FAX: 405-369-3082; WEB: http//www.jantzsupply.com; E-MAIL: jantz@brightok.net
Jarrett Rifles, Inc., 383 Brown Rd., Jackson, SC 29831/803-471-3616
Jarvis, Inc., 1123 Cherry Orchard Lane, Hamilton, MT 59840/406-961-4392
JAS, Inc., P.O. Box 0, Rosemount, MN 55068/612-890-7631
Javelina Lube Products, P.O. Box 337, San Bernardino, CA 92402/714-882-5847; FAX: 714-434-6937
J/B Adventures & Safaris, Inc., 2275 E. Arapahoe Rd. Ste. 109, Littleton, CO 80122-1521/303-771-0977
JB Custom, P.O. Box 6912, Leawood, KS 66206/913-381-2329
Jeffredo Gunsight, P.O. Box 669, San Marcos, CA 92079/619-728-2695
Jenco Sales, Inc., P.O. Box 1000, Manchaca, TX 78652/800-531-5301; FAX: 800-266-2373
Jenkins Recoil Pads, Inc., 5438 E. Frontage Ln., Olney, IL 62450/618-395-3416
Jensen Bullets, 86 North, 400 West, Blackfoot, ID 83221/208-785-5590
Jensen's Custom Ammunition, 5146 E. Pima, Tucson, AZ 85712/602-325-3346; FAX: 602-322-5704
Jensen's Firearms Academy, 1280 W. Prince, Tucson, AZ 85705/602-293-8516
Jericho Tool & Die Co. Inc., RD 3 Box 70, Route 7, Bainbridge, NY 13733-9494/607-563-8222; FAX: 607-563-8560
Jester Bullets, Rt. 1 Box 27, Orienta, OK 73737
Jewell Triggers, Inc., 3620 Hwy. 123, San Marcos, TX 78666/512-353-2999
J-Gar Co., 183 Turnpike Rd., Dept. 3, Petersham, MA 01366-9604
JGS Precision Tool Mfg., 1141 S. Summer Rd., Coos Bay, OR 97420/541-267-4331; FAX:541-267-5996
Jim's Gun Shop (See Spradlin's)
Jim's Precision, Jim Ketchum, 1725 Moclips Dr., Petaluma, CA 94952/707-762-3014
J.I.T., Ltd., P.O. Box 230, Freedom, WY 83120/708-494-0937
JLK Bullets, 414 Turner Rd., Dover, AR 72837/501-331-4194
Johanssons Vapentillbehor, Bert, S-430 20 Veddige, SWEDEN
John's Custom Leather, 523 S. Liberty St., Blairsville, PA 15717/412-459-6802
Johns Master Engraver, Bill, 7927 Ranch Roach 965, Fredericksburg, TX 78624-9545/210-997-6795
Johnson's Gunsmithing, Inc., Neal, 208 W. Buchanan St., Suite B, Colorado Springs, CO 80907/800-284-8671 (orders), 719-632-3795; FAX: 719-632-3493
Johnson Wood Products, 34968 Crystal Road, Strawberry Point, IA 52076/319-933-4930
Johnston Bros. (See C&T Corp. TA Johnson Brothers)
Johnston, James (See North Fork Custom Gunsmithing)
Jonad Corp., 2091 Lakeland Ave., Lakewood, OH 44107/216-226-3161
Jonas Appraisals & Taxidermy, Jack, 1675 S. Birch, Suite 506, Denver, CO 80222/303-757-7347; FAX: 303-639-9655
Jones Co., Dale, 680 Hoffman Draw, Kila, MT 59920/406-755-4684
Jones Custom Products, Neil A., 17217 Brookhouser Road, Saegertown, PA 16433/814-763-2769; FAX: 814-763-4228
Jones Moulds, Paul, 4901 Telegraph Rd., Los Angeles, CA 90022/213-262-1510
Jones, J.D. (See SSK Industries)
J.P. Enterprises, Inc., P.O. Box 26324, Shoreview, MN 55126/612-486-9064; FAX: 612-482-0970
J.P. Gunstocks, Inc., 4508 San Miguel Ave., North Las Vegas, NV 89030/702-645-0718

30th EDITION, 1998

DIRECTORY OF THE ARMS TRADE

JP Sales, Box 307, Anderson, TX 77830
J.R. Distributing, 2976 E. Los Angeles Ave., Simi Valley, CA 93065/805-527-1090; FAX: 805-529-2368
JRP Custom Bullets, RR2 2233 Carlton Rd., Whitehall, NY 12887/518-282-0084 (a.m.), 802-438-5548 (p.m.)
JRW, 2425 Taffy Ct., Nampa, ID 83687
JS Worldwide DBA (See Coonan Arms)
JSL Ltd. (See U.S. importer—Specialty Shooters Supply, Inc.)
Juenke, Vern, 25 Bitterbush Rd., Reno, NV 89523/702-345-0225
Jumbo Sports Products (See Bucheimer, J.M.)
Jungkind, Reeves C., 5001 Buckskin Pass, Austin, TX 78745-2841/512-442-1094
Jurras, L.E., P.O. Box 680, Washington, IN 47501/812-254-7698
JWH: Software, 6947 Haggerty Rd., Hillsboro, OH 45133/513-393-2402

K

K&M Industries, Inc., Box 66, 510 S. Main, Troy, ID 83871/208-835-2281; FAX: 208-835-5211
K&M Services, 5430 Salmon Run Rd., Dover, PA 17315/717-292-3175
K&P Gun Co., 1024 Central Ave., New Rockford, ND 58356/701-947-2248
K&S Mfg., 2611 Hwy. 40 East, Inglis, FL 34449/904-447-3571
K&T Co., Div. of T&S Industries, Inc., 1027 Skyview Dr., W. Carrollton, OH 45449/513-859-8414
KA-BAR Knives, 1116 E. State St., Olean, NY 14760/800-282-0130; FAX: 716-373-6245
Ka Pu Kapili, P.O. Box 745, Honokaa, HI 96727/808-776-1644; FAX: 808-776-1731
Kahles, A Swarovski Company, 1 Wholesale Way, Cranston, RI 02920-5540/800-426-3089; FAX: 401-946-2587
Kahr Arms, P.O. Box 220, 630 Route 303, Blauvelt, NY 10913/914-353-5996; FAX: 914-353-7833
Kalispel Case Line, P.O. Box 267, Cusick, WA 99119/509-445-1121
Kamik Outdoor Footwear, 554 Montee de Liesse, Montreal, Quebec, H4T 1P1 CANADA/514-341-3950; FAX: 514-341-1861
Kamyk Engraving Co., Steve, 9 Grandview Dr., Westfield, MA 01085-1810/413-568-0457
Kandel, P.O. Box 4529, Portland, OR 97208
Kane, Edward, P.O. Box 385, Ukiah, CA 95482/707-462-2937
Kane Products, Inc., 5572 Brecksville Rd., Cleveland, OH 44131/216-524-9962
Kapro Mfg. Co., Inc. (See R.E.I.)
Kasenit Co., Inc., 13 Park Ave., Highland Mills, NY 10930/914-928-9595; FAX: 914-928-7292
Kasmarsik Bullets, 4016 7th Ave. SW, Puyallup, WA 98373
Kaswer Custom, Inc., 13 Surrey Drive, Brookfield, CT 06804/203-775-0564; FAX: 203-775-6872
K.B.I., Inc., P.O. Box 6625, Harrisburg, PA 17112/717-540-8518; FAX: 717-540-8567
K-D, Inc., Box 459, 585 N. Hwy. 155, Cleveland, UT 84518/801-653-2530
KDF, Inc., 2485 Hwy. 46 N., Seguin, TX 78155/210-379-8141; FAX: 210-379-5420
KeeCo Impressions, Inc., 346 Wood Ave., North Brunswick, NJ 08902/800-468-0546
Keeler, R.H., 817 "N" St., Port Angeles, WA 98362/206-457-4702
Kehr, Roger, 2131 Agate Ct. SE, Lacy, WA 98503/360-456-0831
Keith's Bullets, 942 Twisted Oak, Algonquin, IL 60102/708-658-3520
Keith's Custom Gunstocks (See Heppler, Keith M.)
Kelbly, Inc., 7222 Dalton Fox Lake Rd., North Lawrence, OH 44666/216-683-4674; FAX: 216-683-7349
Keller Co., The, 4215 McEwen Rd., Dallas, TX 75244/214-770-8585
Kelley's, P.O. Box 125, Woburn, MA 01801/617-935-3389
Kellogg's Professional Products, 325 Pearl St., Sandusky, OH 44870/419-625-6551; FAX: 419-625-6167
Kelly, Lance, 1723 Willow Oak Dr., Edgewater, FL 32132/904-423-4933
Kel-Tec CNC Industries, Inc., P.O. Box 3427, Cocoa, FL 32924/407-631-0068; FAX: 407-631-1169
Kemen America, 2550 Hwy. 23, Wrenshall, MN 55797
Ken's Kustom Kartridges, 331 Jacobs Rd., Hubbard, OH 44425/216-534-4595
Ken's Gun Specialties, Rt. 1, Box 147, Lakeview, AR 72642/501-431-5606
Ken's Rifle Blanks, Ken McCullough, Rt. 2, P.O. Box 85B, Weston, OR 97886/503-566-3879
Keng's Firearms Specialty, Inc., 875 Wharton Dr., P.O. Box 44405, Atlanta, GA 30336-1405/404-691-7611; FAX: 404-505-8445
Kennebec Journal, 274 Western Ave., Augusta, ME 04330/207-622-6288
Kennedy Firearms, 10 N. Market St., Muncy, PA 17756/717-546-6695
KenPatable Ent., Inc., P.O. Box 19422, Louisville, KY 40259/502-239-5447
Kent Cartridge Mfg. Co. Ltd., Unit 16, Branbridges Industrial Estate, East, Peckham/Tonbridge, Kent, TN12 5HF ENGLAND 622-872255; FAX: 622-872645
Keowee Game Calls, 608 Hwy. 25 North, Travelers Rest, SC 29690/864-834-7204; FAX: 864-834-7831
Kershaw Knives, 25300 SW Parkway Ave., Wilsonville, OR 97070/503-682-1966, 800-325-2891; FAX: 503-682-7168
Kesselring Gun Shop, 400 Hwy. 99 North, Burlington, WA 98233/206-724-3113; FAX: 206-724-7003
Ketchum, Jim (See Jim's Precision)
Kick Eez, P.O. Box 12767, Wichita, KS 67277/316-721-9570; FAX: 316-721-5260
Kilham & Co., Main St., P.O. Box 37, Lyme, NH 03768/603-795-4112
Kimar (See U.S. importer—IAR, Inc.)
Kimball, Gary, 1526 N. Circle Dr., Colorado Springs, CO 80909/719-634-1274

Kimber of America, Inc., 1 Lawton St., Yonkers, NY 10705/800-880-2418
King & Co., P.O. Box 1242, Bloomington, IL 61702/309-473-3964
King's Gun Works, 1837 W. Glenoaks Blvd., Glendale, CA 91201/818-956-6010; FAX: 818-548-8606
Kingyon, Paul L. (See Custom Calls)
Kirk Game Calls, Inc., Dennis, RD1, Box 184, Laurens, NY 13796/607-433-2710; FAX: 607-433-2711
Kirkpatrick Leather Co., 1910 San Bernardo, Laredo, TX 78040/210-723-6631; FAX: 210-725-0672
KJM Fabritek, Inc., P.O. Box 162, Marietta, GA 30061/770-426-8251; FAX: 770-426-8252
KK Air International (See Impact Case Co.)
K.K. Arms Co., Star Route Box 671, Kerrville, TX 78028/210-257-4718; FAX: 210-257-4891
KLA Enterprises, P.O. Box 2028, Eaton Park, FL 33840/941-682-2829; FAX: 941-682-2829
Kleen-Bore, Inc., 16 Industrial Pkwy., Easthampton, MA 01027/413-527-0300; FAX: 413-527-2522
Klein Custom Guns, Don, 433 Murray Park Dr., Ripon, WI 54971/414-748-2931
Kleinendorst, K.W., RR 1, Box 1500, Hop Bottom, PA 18824/717-289-4687
Klingler Woodcarving, P.O. Box 141, Thistle Hill, Cabot, VT 05647/802-426-3811
Kmount, P.O. Box 19422, Louisville, KY 40259/502-239-5447
Kneiper, James, P.O. Box 1516, Basalt, CO 81621-1516/303-963-9880
Knife Importers, Inc., P.O. Box 1000, Manchaca, TX 78652/512-282-6860
Knight & Hale Game Calls, Box 468 Industrial Park, Cadiz, KY 42211/502-924-1755; FAX: 502-924-1763
Knight Rifles (See Modern MuzzleLoading, Inc.)
Knight's Mfg. Co., 7750 9th St. SW, Vero Beach, FL 32968/561-562-5697; FAX: 561-569-2955
Knippel, Richard, 1455 Jubal Ct., Oakdale, CA 95361-9669/209-869-1469
Knock on Wood Antiques, 355 Post Rd., Darien, CT 06820/203-655-9031
Knoell, Doug, 9737 McCardle Way, Santee, CA 92071
Kodiak Custom Bullets, 8261 Henry Circle, Anchorage, AK 99507/907-349-2282
Koevenig's Engraving Service, Box 55 Rabbit Gulch, Hill City, SD 57745
KOGOT, 410 College, Trinidad, CO 81082/719-846-9406
Kokolus, Michael M. (See Custom Riflestocks, Inc.)
Kolpin Mfg., Inc., P.O. Box 107, 205 Depot St., Fox Lake, WI 53933/414-928-3118; FAX: 414-928-3687
Kongsberg America L.L.C., P.O. Box 252, Fairfield, CT 06430/203-259-0938; FAX: 203-259-2566
Kopec Enterprises, John (See Peacemaker Specialists)
Kopp Professional Gunsmithing, Terry K., Route 1, Box 224F, Lexington, MO 64067/816-259-2636
Korth, Robert-Bosch-Str. 4, P.O. Box 1320, 23909 Ratzeburg, GERMANY/451-4991497; FAX: 451-4993230 (U.S. importer—Interarms; Mandall Shooting Supplies, Inc.)
Korzinek Riflesmith, J., RD 2, Box 73D, Canton, PA 17724/717-673-8512
Koval Knives, 5819 Zarley St., Suite A, New Albany, OH 43054/614-855-0777; FAX: 614-855-0945
Kowa Optimed, Inc., 20001 S. Vermont Ave., Torrance, CA 90502/310-327-1913; FAX: 310-327-4177
Kramer Designs, P.O. Box 129, Clancy, MT 59634/406-933-8658; FAX: 406-933-8658
Kramer Handgun Leather, P.O. Box 112154, Tacoma, WA 98411/206-564-6652; FAX: 206-564-1214
Krause Publications, Inc., 700 E. State St., Iola, WI 54990/715-445-2214; FAX: 715-445-4087; Consumer orders only 800-258-0929
Krico Jagd-und Sportwaffen GmbH, Nurnbergerstrasse 6, D-90602 Pyrbaum GERMANY/09180-2780; FAX: 09180-2661 (U.S. importer—Mandall Shooting Supplies, Inc.)
Krieger Barrels, Inc., N114 W18697 Clinton Dr., Germantown, WI 53022/414-255-9593; FAX: 414-255-9586
Krieghoff Gun Co., H., Boschstrasse 22, D-89079 Ulm, GERMANY/731-401820; FAX: 731-4018270 (U.S. importer—Krieghoff International, Inc.)
Krieghoff International, Inc., 7528 Easton Rd., Ottsville, PA 18942/610-847-5173; FAX: 610-847-8691
Kris Mounts, 108 Lehigh St., Johnstown, PA 15905/814-539-9751
KSN Industries, Ltd. (See U.S. importer—Israel Arms International, Inc.)
K-Sports Imports, Inc., 2755 Thompson Creek Rd., Pomona, CA 91767/909-392-2345; FAX: 909-392-2354
Kudlas, John M., 622 14th St. SE, Rochester, MN 55904/507-288-5579
Kulis Freeze Dry Taxidermy, 725 Broadway Ave., Bedford, OH 44146/216-232-8352; FAX: 216-232-7305; WEB: http://www.kastaway.com; E-Mail: jkulis@kastaway.com
KVH Industries, Inc., 110 Enterprise Center, Middletown, RI 02842/401-847-3327; FAX: 401-849-0045
Kwik Mount Corp., P.O. Box 19422, Louisville, KY 40259/502-239-5447
Kwik-Site Co., 5555 Treadwell, Wayne, MI 48184/313-326-1500; FAX: 313-326-4120

L

L&R Lock Co., 1137 Pocalla Rd., Sumter, SC 29150/803-775-6127; FAX: 803-775-5171
L&S Technologies, Inc. (See Aimtech Mount Systems)

MANUFACTURERS' DIRECTORY

La Clinique du .45, 1432 Rougemont, Chambly, Quebec, J3L 2L8 CANADA/514-658-1144
Labanu, Inc., 2201-F Fifth Ave., Ronkonkoma, NY 11779/516-467-6197; FAX: 516-981-4112
LaBounty Precision Reboring, P.O. Box 186, 7968 Silver Lk. Rd., Maple Falls, WA 98266/360-599-2047
LaCrosse Footwear, Inc., P.O. Box 1328, La Crosse, WI 54602/608-782-3020, 800-323-2668; FAX: 800-658-9444
Lady Clays, P.O. Box 457, Shawnee Mission, KS 66201/913-268-8006
LaFrance Specialties, P.O. Box 178211, San Diego, CA 92177-8211/619-293-3373
Lage Uniwad, P.O. Box 2302, Davenport, IA 52809/319-388-LAGE; FAX: 319-388-LAGE
Lair, Sam, 520 E. Beaver, Jenks, OK 74037/918-299-2391
Lake Center, P.O. Box 38, St. Charles, MO 63302/314-946-7500
Lakefield Arms Ltd. (See Savage Arms, Inc.)
Lakewood Products, LLC, 275 June St., Berlin, WI 54923/800-US-BUILT; FAX: 414-361-7719
Lampert, Ron, Rt. 1, Box 177, Guthrie, MN 56461/218-854-7345
Lamson & Goodnow Mfg. Co., 45 Conway St., Shelburne Falls, MA 03170/413-625-6331; FAX: 413-625-9816
Lanber Armas, S.A., Zubiaurre 5, Zaldibar, SPAIN 48250/34-4-6827702; FAX: 34-4-6827999
Lane Bullets, Inc., 1011 S. 10th St., Kansas City, KS 66105/913-621-6113, 800-444-7468
Lane Publishing, P.O. Box 459, Lake Hamilton, AR 71951/501-525-7514; FAX: 501-525-7519
Langenberg Hat Co., P.O. Box 1860, Washington, MO 63090/800-428-1860; FAX: 314-239-3151
Lanphert, Paul, P.O. Box 1985, Wenatchee, WA 98807
Lapua Ltd., P.O. Box 5, Lapua, FINLAND SF-62101/6-310111; FAX: 6-4388991 (U.S. importer—Keng's Firearms Specialty, Inc.
L.A.R. Mfg., Inc., 4133 W. Farm Rd., West Jordan, UT 84088/801-280-3505; FAX: 801-280-1972
LaRocca Gun Works, Inc., 51 Union Place, Worcester, MA 01608/508-754-2887; FAX: 508-754-2887
Laseraim Technologies, Inc., P.O. Box 3548, Little Rock, AR 72203/501-375-2227; FAX: 501-372-1445
Laser Devices, Inc., 2 Harris Ct. A-4, Monterey, CA 93940/408-373-0701; FAX: 408-373-0903
LaserMax, Inc., 3495 Winton Place, Bldg. B, Rochester, NY 14623-2807/716-272-5420; FAX: 716-272-5427
Lassen Community College, Gunsmithing Dept., P.O. Box 3000, Hwy. 139, Susanville, CA 96130/916-251-8800; FAX: 916-251-8838
Lathrop's, Inc., 5146 E. Pima, Tucson, AZ 85712/520-881-0266, 800-875-4867; FAX: 520-322-5704
Laughridge, William R. (See Cylinder & Slide, Inc.)
Laurel Mountain Forge, P.O. Box 224C, Romeo, MI 48065/810-749-5742
Laurona Armas Eibar, S.A.L., Avenida de Otaola 25, P.O. Box 260, 20600 Eibar, SPAIN/34-43-700600; FAX: 34-43-700616 (U.S. importer—Galaxy Imports Ltd., Inc.)
Law Concealment Systems, Inc., P.O. Box 3952, Wilmington, NC 28406/919-791-6656, 800-373-0116 orders; FAX: 910-791-8388
Lawrence Brand Shot (See Precision Reloading, Inc.)
Lawrence Leather Co., P.O. Box 1479, Lillington, NC 27546/910-893-2071; FAX: 910-893-4742
Lawson Co., Harry, 3328 N. Richey Blvd., Tucson, AZ 85716/520-326-1117
Lawson, John G. (See Sight Shop, The)
Lazzeroni Arms Co., 1415 S. Cherry Ave., Tucson, AZ 85726/520-577-7500; FAX: 520-624-4250
LBT, HCR 62, Box 145, Moyie Springs, ID 83845/208-267-3588
Le Clear Industries (See E-Z-Way Systems)
Lea Mfg. Co., 237 E. Aurora St., Waterbury, CT 06720/203-753-5116
Lead Bullets Technology (See LBT)
Leapers, Inc., 7675 Five Mile Rd., Northville, MI 48167/810-486-1231; FAX: 810-486-1430
Leather Arsenal, 27549 Middleton Rd., Middleton, ID 83644/208-585-6212
Leatherman Tool Group, Inc., 12106 NE Ainsworth Cir., P.O. Box 20595, Portland, OR 97294/503-253-7826; FAX: 503-253-7830
Lebeau-Courally, Rue St. Gilles, 386, 4000 Liege, BELGIUM/042-52-48-43; FAX: 32-042-52-20-08 (U.S. importer—New England Arms Co.)
Leckie Professional Gunsmithing, 546 Quarry Rd., Ottsville, PA 18942/215-847-8594
Lectro Science, Inc., 6410 W. Ridge Rd., Erie, PA 16506/814-833-6487; FAX: 814-833-0447
Ledbetter Airguns, Riley, 1804 E. Sprague St., Winston Salem, NC 27107-3521/919-784-0676
Lee Precision, Inc., 4275 Hwy. U, Hartford, WI 53027/414-673-3075; FAX: 414-673-9273
Lee Supplies, Mark, 9901 France Ct., Lakeville, MN 55044/612-461-2114
Lee's Red Ramps, 4 Kristine Ln., Silver City, NM 88061/505-538-8529
Lee Co., T.K., One Independence Plaza, Suite 520, Birmingham, AL 35209/205-913-5222
LeFever Arms Co., Inc., 6234 Stokes, Lee Center Rd., Lee Center, NY 13363/315-337-6722; FAX: 315-337-1543
Legend Products Corp., 21218 Saint Andrews Blvd., Boca Raton, FL 33433-2435
Leibowitz, Leonard, 1205 Murrayhill Ave., Pittsburgh, PA 15217/412-361-5455
Leica USA, Inc., 156 Ludlow Ave., Northvale, NJ 07647/201-767-7500; FAX: 201-767-8666
LEM Gun Specialties, Inc., The Lewis Lead Remover, P.O. Box 2855, Peachtree City, GA 30269-2024
Lem Sports, Inc., P.O. Box 2107, Aurora, IL 60506/815-286-7421, 800-688-8801 (orders only)
Lenahan Family Enterprise, P.O. Box 46, Manitou Springs, CO 80829
Lestrom Laboratories, Inc., P.O. Box 628, Mexico, NY 13114-0628/315-343-3076; FAX: 315-592-3370
Lethal Force Institute (See Police Bookshelf)
Lett Custom Grips, 672 Currier Rd., Hopkinton, NH 03229-2652
Leupold & Stevens, Inc., P.O. Box 688, Beaverton, OR 97075/503-646-9171; FAX: 503-526-1455
Lever Arms Service Ltd., 2131 Burrard St., Vancouver, B.C. V6J 3H7 CANADA/604-736-0004; FAX: 604-738-3503
Lewis Lead Remover, The (See LEM Gun Specialties, Inc.)
Liberty Antique Gunworks, 19 Key St., P.O. Box 183, Eastport, ME 04631/207-853-4116
Liberty Metals, 2233 East 16th St., Los Angeles, CA 90021/213-581-9171; FAX: 213-581-9351
Liberty Safe, 1060 N. Spring Creek Pl., Springville, UT 84663/800-247-5625; FAX: 801-489-6409
Liberty Shooting Supplies, P.O. Box 357, Hillsboro, OR 97123/503-640-5518
Liberty Trouser Co., 3500 6 Ave S., Birmingham, AL 35222-2406/205-251-9143
Light Optronics (See TacStar Industries, Inc.)
Lightfield Ammunition Corp. (See Slug Group, Inc.)
Lightforce U.S.A. Inc., 19226 66th Ave. So., L-103, Kent, WA 98032/206-656-1577; FAX:206-656-1578
Lightning Performance Innovations, Inc., RD1 Box 555, Mohawk, NY 13407/315-866-8819, 800-242-5873; FAX: 315-866-8819
Lilja Precision Rifle Barrels, P.O. Box 372, Plains, MT 59859/406-826-3084; FAX: 406-826-3083
Lincoln, Dean, Box 1886, Farmington, NM 87401
Lind Custom Guns, Al, 7821 76th Ave. SW, Tacoma, WA 98498/206-584-6361
Linder Solingen Knives, 4401 Sentry Dr., Tucker, GA 30084/770-939-6915; FAX: 770-939-6738
Lindsay, Steve, RR 2 Cedar Hills, Kearney, NE 68847/308-236-7885
Lindsley Arms Cartridge Co., P.O. Box 757, 20 College Hill Rd., Henniker, NH 03242/603-428-3127
Linebaugh Custom Sixguns, Route 2, Box 100, Maryville, MO 64468/816-562-3031
Lion Country Supply, P.O. Box 480, Port Matilda, PA 16870
List Precision Engineering, Unit 1, Ingley Works, 13 River Road, Barking, Essex 1G11 0HE ENGLAND/011-081-594-1686
Lithi Bee Bullet Lube, 1728 Carr Rd., Muskegon, MI 49442/616-788-4479
"Little John's" Antique Arms, 1740 W. Laveta, Orange, CA 92668
Little Trees Ramble (See Scott Pilkington, Little Trees Ramble)
Littler Sales Co., 20815 W. Chicago, Detroit, MI 48228/313-273-6889; FAX: 313-273-1099
Littleton, J.F., 275 Pinedale Ave., Oroville, CA 95966/916-533-6084
Ljutic Industries, 732 N. 16th Ave., Suite 22, Yakima, WA 98902/509-248-0476; FAX: 509-576-8233
Llama Gabilondo Y Cia, Apartado 290, E-01080, Victoria, SPAIN (U.S. importer—Import Sports, Inc.)
L.L. Bean, Inc., Freeport, ME 04032, 207-865-4761; FAX: 207-552-2802
Load From A Disk, 9826 Sagedale, Houston, TX 77089/713-484-0935; FAX: 281-484-0935
Loch Leven Industries, P.O. Box 2751, Santa Rosa, CA 95405/707-573-8735; FAX: 707-573-0369
Lock's Philadelphia Gun Exchange, 6700 Rowland Ave., Philadelphia, PA 19149/215-332-6225; FAX: 215-332-4800
Lodewick, Walter H., 2816 NE Halsey St., Portland, OR 97232/503-284-2554
Log Cabin Sport Shop, 8010 Lafayette Rd., Lodi, OH 44254/216-948-1082
Logan, Harry M., Box 745, Honokaa, HI 96727/808-776-1644
Lohman Mfg. Co., Inc., 4500 Doniphan Dr., P.O. Box 220, Neosho, MO 64850/417-451-4438; FAX: 417-451-2576
Lomont Precision Bullets, RR 1, Box 34, Salmon, ID 83467/208-756-6819; FAX: 208-756-6824
London Guns Ltd., Box 3750, Santa Barbara, CA 93130/805-683-4141; FAX: 805-683-1712
Lone Star Gunleather, 1301 Brushy Bend Dr., Round Rock, TX 78681/512-255-1805
Lone Star Rifle Company, 11231 Rose Road, Conroe, Texas 77303/409-856-3363
Long, George F., 1500 Rogue River Hwy., Ste. F, Grants Pass, OR 97527/541-476-7552
Lorcin Engineering Co., Inc., 10427 San Sevaine Way, Ste. A, Mira Loma, CA 91752
Lortone, Inc., 2856 NW Market St., Seattle, WA 98107/206-789-3100
Lothar Walther Precision Tool, Inc., 2190 Coffee Rd., Lithonia, GA 30058/770-482-4253; Fax: 770-482-9344
Lovestrand, Erik, 206 Bent Oak Circle, Harvest, AL 35749-9334
Loweth (Firearms), Richard H.R., 29 Hedgerow Lane, Kirby Muxloe, Leics. LE9 2BN ENGLAND/(0)116 238 6295
L.P.A. Snc, Via Alfieri 26, Gardone V.T., Brescia, ITALY 25063/30-891-14-81; FAX: 30-891-09-51
LPS Laboratories, Inc., 4647 Hugh Howell Rd., P.O. Box 3050, Tucker, GA 30084/404-934-7800

DIRECTORY OF THE ARMS TRADE

Lucas, Edward E., 32 Garfield Ave., East Brunswick, NJ 08816/201-251-5526
Lucas, Mike, 1631 Jessamine Rd., Lexington, SC 29073/803-356-0282
Luch Metal Merchants, Barbara, 48861 West Rd., Wixon, MI 48393/800-876-5337
Lutz Engraving, Ron, E. 1998 Smokey Valley Rd., Scandinavia, WI 54977/715-467-2674
Lyman Instant Targets, Inc. (See Lyman Products Corp.)
Lyman Products Corp., 475 Smith Street, Middletown, CT 06457-1541/860-632-2020, 800-22-LYMAN; FAX: 860-632-1699
Lynn's Custom Gunstocks, RR 1, Brandon, IA 52210/319-474-2453
Lyons Gunworks, Larry, 110 Hamilton St., Dowagiac, MI 49047/616-782-9478
Lyte Optronics (See TracStar Industries, Inc.,)

M

M&D Munitions Ltd., 127 Verdi St., Farmingdale, NY 11735/800-878-2788, 516-752-1038; FAX: 516-752-1905
M&M Engineering (See Hollywood Engineering)
M&N Bullet Lube, P.O. Box 495, 151 NE Jefferson St., Madras, OR 97741/503-255-3750
MA Systems, P.O. Box 1143, Chouteau, OK 74337/918-479-6378
Mac-1 Distributors, 13974 Van Ness Ave., Gardena, CA 90249/310-327-3582
Mac's .45 Shop, P.O. Box 2028, Seal Beach, CA 90740/310-438-5046
Macbean, Stan, 754 North 1200 West, Orem, UT 84057/801-224-6446
Madis Books, 2453 West Five Mile Pkwy., Dallas, TX 75233/214-330-7168
Madis, George, P.O. Box 545, Brownsboro, TX 75756/903-852-6480
MAG Instrument, Inc., 1635 S. Sacramento Ave., Ontario, CA 91761/909-947-1006; FAX: 909-947-3116
Mag-Na-Port International, Inc., 41302 Executive Dr., Harrison Twp., MI 48045-1306/810-469-6727; FAX: 810-469-0425
Mag-Pack Corp., P.O. Box 846, Chesterland, OH 44026
Magma Engineering Co., P.O. Box 161, 20955 E. Ocotillo Rd., Queen Creek, AZ 85242/602-987-9008; FAX: 602-987-0148
Magnolia Sports, Inc., 211 W. Main, Magnolia, AR 71753/501-234-8410, 800-530-7816; FAX: 501-234-8117
Magnum Grips, Box 801G, Payson, AZ 85547
Magnum Power Products, Inc., P.O. Box 17768, Fountain Hills, AZ 85268
Magnum Research, Inc., 7110 University Ave. NE, Minneapolis, MN 55432/800-772-6168, 612-574-1868; FAX: 612-574-0109; WEB:http://www.magnumresearch.com
Magnus Bullets, P.O. Box 239, Toney, AL 35773/205-420-8359; FAX: 205-420-8360
MagSafe Ammo Co., 2725 Friendly Grove Rd NE, Olympia, WA 98506/360-357-6383; FAX: 360-705-4715
MagTech Recreational Products, Inc., 5030 Paradise Rd., Suite A104, Las Vegas, NV 89119/702-736-2043; FAX: 702-736-2140
Mahony, Philip Bruce, 67 White Hollow Rd., Lime Rock, CT 06039-2418/203-435-9341
Mahovsky's Metalife, R.D. 1, Box 149a Eureka Road, Grand Valley, PA 16420/814-436-7747
Maine Custom Bullets, RFD 1, Box 1755, Brooks, ME 04921
Maionchi-L.M.I., Via Di Coselli-Zona Industriale Di Guamo, Lucca, ITALY 55060/011 39-583 94291
Makinson, Nicholas, RR 3, Komoka, Ont. N0L 1R0 CANADA/519-471-5462
Malcolm Enterprises, 1023 E. Prien Lake Rd., Lake Charles, LA 70601
Mallardtone Game Calls, 2901 16th St., Moline, IL 61265/309-762-8089
M.A.M. Products, Inc., 153 B Cross Slope Court, Englishtown, NJ 07726/908-536-3604; FAX:908-972-1004
Mandall Shooting Supplies, Inc., 3616 N. Scottsdale Rd., Scottsdale, AZ 85252/602-945-2553; FAX: 602-949-0734
Manufacture D'Armes Des Pyrenees Francaises (See Unique/M.A.P.F.)
Mar Knives, Inc., Al, 5755 SW Jean Rd., Suite 101, Lake Oswego, OR 97035/503-635-9229; FAX: 503-223-0467
Marathon Rubber Prods. Co., Inc., 510 Sherman St., Wausau, WI 54401/715-845-6255
Marble Arms (See CRR, Inc./Marble's Inc.)
Marchmon Bullets, 8191 Woodland Shore Dr., Brighton, MI 48116
Marent, Rudolf, 9711 Tiltree St., Houston, TX 77075/713-946-7028
Markell, Inc., 422 Larkfield Center 235, Santa Rosa, CA 95403/707-573-0792; FAX: 707-573-9867
Markesbery Muzzle Loaders, Inc., 7785 Foundation Dr., Ste. 6, Florence, KY 41042/800-875-0121; 606-342-2380
Marksman Products, 5482 Argosy Dr., Huntington Beach, CA 92649/714-898-7535, 800-822-8005; FAX: 714-891-0782
Marlin Firearms Co., 100 Kenna Dr., North Haven, CT 06473/203-239-5621; FAX: 203-234-7991
MarMik, Inc., 2116 S. Woodland Ave., Michigan City, IN 46360/219-872-7231; FAX: 219-872-7231
Marocchi F.lli S.p.A, Via Galileo Galilei 8, I-25068 Zanano di Sarezzo, ITALY/ (U.S. importers—Precision Sales International, Inc.)
Marple & Associates, Dick, 21 Dartmouth St., Hooksett, NH 03106/603-627-1837; FAX: 603-627-1837
Marquart Precision Co., P.O. Box 1740, Prescott, AZ 86302/520-445-5646
Marsh, Johnny, 1007 Drummond Dr., Nashville, TN 37211/615-833-3259
Marsh, Mike, Croft Cottage, Main St., Elton, Derbyshire DE4 2BY, ENGLAND/01629 650 669
Marshall Enterprises, 792 Canyon Rd., Redwood City, CA 94062

Martin Bookseller, J., P.O. Drawer AP, Beckley, WV 25802/304-255-4073; FAX 304-255-4077
Martin's Gun Shop, 937 S. Sheridan Blvd., Lakewood, CO 80226/303-922-2184
Martz, John V., 8060 Lakeview Lane, Lincoln, CA 95648/916-645-2250
Marvel, Alan, 3922 Madonna Rd., Jarretsville, MD 21084/301-557-6545
Marx, Harry (U.S. importer for FERLIB)
Maryland Paintball Supply, 8507 Harford Rd., Parkville, MD 21234/410-882-5607
Masen Co., Inc., John, 1305 Jelmak, Grand Prairie, TX 75050/817-430-8732; FAX: 817-430-1715
MAST Technology, 4350 S. Arville, Suite 3, Las Vegas, NV 89103/702-362-5043; FAX: 702-362-9554
Master Engravers, Inc. (See Hendricks, Frank E.)
Master Lock Co., 2600 N. 32nd St., Milwaukee, WI 53245/414-444-2800
Master Products, Inc. (See Gun-Alert/Master Products, Inc.)
Match Prep—Doyle Gracey, P.O. Box 155, Tehachapi, CA 93581/805-822-5383
Matco, Inc., 1003-2nd St., N. Manchester, IN 46962/219-982-8282
Mathews & Son, George E., Inc., 10224 S. Paramount Blvd., Downey, CA 90241/562-862-6719; FAX: 562-862-6719
Matthews Cutlery, 4401 Sentry Dr., Tucker, GA 30084/770-939-6915
Mauser Werke Oberndorf Waffensysteme GmbH, Postfach 1349, 78722 Oberndorf/N. GERMANY/ (U.S. importer—GSI, Inc.)
Maverick Arms, Inc., 7 Grasso Ave., P.O. Box 497, North Haven, CT 06473/203-230-5300; FAX: 203-230-5420
Maxi-Mount, P.O. Box 291, Willoughby Hills, OH 44094-0291/216-944-9456; FAX: 216-944-9456
Maximum Security Corp., 32841 Calle Perfecto, San Juan Capistrano, CA 92675/714-493-3684; FAX: 714-496-7733
Mayville Engineering Co. (See MEC, Inc.)
Mazur Restoration, Pete, 13083 Drummer Way, Grass Valley, CA 95949/916-268-2412
MCA Sports, P.O. Box 8868, Palm Springs, CA 92263/619-770-2005
McBros Rifle Co., P.O. Box 86549, Phoenix, AZ 85080/602-582-3713; FAX: 602-581-3825
McCament, Jay, 1730-134th St. Ct. S., Tacoma, WA 98444/206-531-8832
McCann's Machine & Gun Shop, P.O. Box 641, Spanaway, WA 98387/206-537-6919; FAX: 206-537-6993
McCann's Muzzle-Gun Works, 14 Walton Dr., New Hope, PA 18938/215-862-2728
McCluskey Precision Rifles, 10502 14th Ave. NW, Seattle, WA 98177/206-781-2776
McCombs, Leo, 1862 White Cemetery Rd., Patriot, OH 45658/614-256-1714
McCormick Corp., Chip, 1825 Fortview Rd., Ste. 115, Austin, TX 78704/800-328-CHIP, 512-462-0004; FAX: 512-462-0009
McCullough, Ken (See Ken's Rifle Blanks)
McDonald, Dennis, 8359 Brady St., Peosta, IA 52068/319-556-7940
McFarland, Stan, 2221 Idella Ct., Grand Junction, CO 81505/970-243-4704
McGowen Rifle Barrels, 5961 Spruce Lane, St. Anne, IL 60964/815-937-9816; FAX: 815-937-4024
McGuire, Bill, 1600 N. Eastmont Ave., East Wenatchee, WA 98802/509-884-6021
McKee Publications, 121 Eatons Neck Rd., Northport, NY 11768/516-575-8850
McKonzio, Lynton, 6940 N. Alvernon Way, Tucson, AZ 85718/520-299-5090
McKillen & Heyer, Inc., 35535 Euclid Ave. Suite 11, Willoughby, OH 44094/216-942-2044
McKinney, R.P. (See Schuetzen Gun Co.)
McMillan Fiberglass Stocks, Inc., 21421 N. 14th Ave., Suite B, Phoenix, AZ 85027/602-582-9635; FAX: 602-581-3825
McMillan Optical Gunsight Co., 28638 N. 42nd St., Cave Creek, AZ 85331/602-585-7868; FAX: 602-585-7872
McMillan Rifle Barrels, P.O. Box 3427, Bryan, TX 77805/409-690-3456; FAX: 409-690-0156
McMurdo, Lynn (See Specialty Gunsmithing)
MCRW Associates Shooting Supplies, R.R. 1 Box 1425, Sweet Valley, PA 18656/717-864-3967; FAX: 717-864-2669
MCS, Inc., 34 Delmar Dr., Brookfield, CT 06804/203-775-1013; FAX: 203-775-9462
McWelco Products, 6730 Santa Fe Ave., Hesperia, CA 92345/619-244-8876; FAX: 619-244-9398
McWhorter Custom Rifles, 4460 SW 35th Terrace, Suite 310, Gainesville, FL 32608/352-373-9057; FAX: 352-377-3816
MDS, P.O. Box 1441, Brandon, FL 33509-1441/813-653-1180; FAX: 813-684-5953
Meadow Industries, 24 Club Lane, Palmyra, VA 22963/804-589-7672; FAX: 804-589-7672
Measurement Group, Inc., Box 27777, Raleigh, NC 27611
MEC, Inc., 715 South St., Mayville, WI 53050/414-387-4500; FAX: 414-387-5802
MEC-Gar S.r.l., Via Madonnina 64, Gardone V.T., Brescia, ITALY 25063/39-30-8912687; FAX: 39-30-8910065 (U.S. importer—MEC-Gar U.S.A., Inc.)
MEC-Gar U.S.A., Inc., Box 112, 500B Monroe Turnpike, Monroe, CT 06468/203-635-8662; FAX: 203-635-8662
Meier Works, P.O. Box 423, Tijeras, NM 87059/505-281-3783
Meister Bullets (See Gander Mountain)
Mele, Frank, 201 S. Wellow Ave., Cookeville, TN 38501/615-526-4860
Melton Shirt Co., Inc., 56 Harvester Ave., Batavia, NY 14020/716-343-8750; FAX: 716-343-6887
Men-Metallwerk Elisenhuette, GmbH, P.O. Box 1263, D-56372 Nassau/Lahn, GERMANY/2604-7819
Menck, Gunsmith Inc., T.W., 5703 S. 77th St., Ralston, NE 68127
Mendez, John A., P.O. Box 620984, Orlando, FL 32862/407-344-2791
Meprolight (See Hesco-Meprolight)

MANUFACTURERS' DIRECTORY

Mercer Custom Stocks, R.M., 216 S. Whitewater Ave., Jefferson, WI 53549/414-674-5130
Merit Corp., Box 9044, Schenectady, NY 12309/518-346-1420
Merkel Freres, Strasse 7 October, 10, Suhl, GERMANY/ (U.S. importer—GSI, Inc.)
Merkuria Ltd., Argentinska 38, 17005 Praha 7, CZECH REPUBLIC/422-875117; FAX: 422-809152
Metal Merchants, 48861 West Rd., Wixom, MI 48393
Metal Products Co. (See MPC)
Metalife Industries (See Mahovsky's Metalife)
Metaloy Inc., Rt. 5, Box 595, Berryville, AR 72616/501-545-3611
Metals Hand Engraver/European Hand Engraving, Ste. 216, 12 South First St., San Jose, CA 95113/408-293-6559
Michael's Antiques, Box 591, Waldoboro, ME 04572
Michaels of Oregon Co., P.O. Box 13010, Portland, OR 97213/503-255-6890; FAX: 503-255-0746
Micro Sight Co., 242 Harbor Blvd., Belmont, CA 94002/415-591-0769; FAX: 415-591-7531
Microfusion Alfa S.A., Paseo San Andres N8, P.O. Box 271, Eibar, SPAIN 20600/34-43-11-89-16; FAX: 34-43-11-40-38
Mid-America Guns and Ammo, 1205 W. Jefferson, Suite E, Effingham, IL 62401/800-820-5177
Mid-America Recreation, Inc., 1328 5th Ave., Moline, IL 61265/309-764-5089; FAX: 309-764-2722
Middlebrooks Custom Shop, 7366 Colonial Trail East, Surry, VA 23883/757-357-0881; FAX: 757-365-0442
Midway Arms, Inc., 5875 W. Van Horn Tavern Rd., Columbia, MO 65203/800-243-3220, 573-445-6363; FAX: 573-446-1018
Midwest Gun Sport, 1108 Herbert Dr., Zebulon, NC 27597/919-269-5570
Midwest Sport Distributors, Box 129, Fayette, MO 65248
Military Armament Corp., P.O. Box 120, Mt. Zion Rd., Lingleville, TX 76461/817-965-3253
Miller Arms, Inc., P.O. Box 260 Purl St., St. Onge, SD 57779/605-642-5160; FAX: 605-642-5160
Miller Custom, 210 E. Julia, Clinton, IL 61727/217-935-9362
Miller Co., David, 3131 E. Greenlee Rd., Tucson, AZ 85716-1267/520-326-3117
Miller Enterprises, Inc., R.P., 1557 E. Main St., P.O. Box 234, Brownsburg, IN 46112/317-852-8187
Miller Single Trigger Mfg. Co., Rt. 209 Box 1275, Millersburg, PA 17061/717-692-3704
Millett Sights, 7275 Murdy Circle, Adm. Office, Huntington Beach, CA 92647/714-842-5575, 800-645-5388; FAX: 714-843-5707
Mills Jr., Hugh B., 3615 Canterbury Rd., New Bern, NC 28560/919-637-4631
Milstor Corp., 80-975 E. Valley Pkwy. C-7, Indio, CA 92201/619-775-9998; FAX: 619-772-4990
Miniature Machine Co. (See MMC)
Minute Man High Tech Industries, 10611 Canyon Rd. E., Suite 151, Puyallup, WA 98373/800-233-2734
Mirador Optical Corp., P.O. Box 11614, Marina Del Rey, CA 90295-7614/310-821-5587; FAX: 310-305-0386
Miroku, B.C./Daly, Charles (See U.S. importer—Bell's Legendary Country Wear; K.B.I., Inc.; U.S. distributor—Outdoor Sports Headquarters, Inc.)
Mitchell Bullets, R.F., 430 Walnut St., Westernport, MD 21562
Mitchell Optics Inc., 2072 CR 1100 N, Sidney, IL 61877/217-688-2219, 217-621-3018; FAX: 217-688-2505
Mitchell's Accuracy Shop, 68 Greenridge Dr., Stafford, VA 22554/703-659-0165
MI-TE Bullets, R.R. 1 Box 230, Ellsworth, KS 67439/913-472-4575
Mittermeier, Inc., Frank, P.O. Box 2G, 3577 E. Tremont Ave., Bronx, NY 10465/718-828-3843
Mixson Corp., 7435 W. 19th Ct., Hialeah, FL 33014/305-821-5190, 800-327-0078; FAX: 305-558-9318
MJK Gunsmithing, Inc., 417 N. Huber Ct., E. Wenatchee, WA 98802/509-884-7683
MJM Mfg., 3283 Rocky Water Ln. Suite B, San Jose, CA 95148/408-270-4207
MKS Supply, Inc. (See Hi-Point Firearms)
MMC, 2513 East Loop 820 North, Ft. Worth, TX 76118/817-595-0404; FAX: 817-595-3074
MMP, Rt. 6, Box 384, Harrison, AR 72601/501-741-5019; FAX: 501-741-3104
M.O.A. Corp., 2451 Old Camden Pike, Eaton, OH 45320/513-456-3669
Modern Gun Repair School, P.O. Box 92577, Southlake, TX 76092/800-493-4114; FAX: 800-556-5112
Modern MuzzleLoading, Inc., 234 Airport Rd., P.O. Box 130, Centerville, IA 52544/515-856-2626; FAX: 515-856-2628
Moeller, Steve, 1213 4th St., Fulton, IL 61252/815-589-2300
Molin Industries, Tru-Nord Division, P.O. Box 365, 204 North 9th St., Brainerd, MN 56401/218-829-2870
Mo's Competitor Supplies (See MCS, Inc.)
MoLoc Bullets, P.O. Box 2810, Turlock, CA 95381-2810/209-632-1644
Monell Custom Guns, 228 Red Mills Rd., Pine Bush, NY 12566/914-744-3021
Moneymaker Guncraft Corp., 1420 Military Ave., Omaha, NE 68131/402-556-0226
Montana Armory, Inc. (See C. Sharps Arms Co. Inc.)
Montana Outfitters, Lewis E. Yearout, 308 Riverview Dr. E., Great Falls, MT 59404/406-761-0859
Montana Precision Swaging, P.O. Box 4746, Butte, MT 59702/406-782-7502
Montana Vintage Arms, 2354 Bear Canyon Rd., Bozeman, MT 59715
Montgomery Community College, P.O. Box 787-GD, Troy, NC 27371/910-576-6222, 800-839-6222; FAX: 910-576-2176

Moore & Co., Wm. Larkin, 8727 E. Via de Commencio, Suite A, Scottsdale, AZ 85258/602-951-8913; FAX: 602-951-8913
Morini (See U.S. importers—Mandall Shooting Supplies, Inc.; Nygord Precision Products)
Morrison Custom Rifles, J.W., 4015 W. Sharon, Phoenix, AZ 85029/602-978-3754
Morrow, Bud, 11 Hillside Lane, Sheridan, WY 82801-9729/307-674-8360
Morton Booth Co., P.O. Box 123, Joplin, MO 64802/417-673-1962; FAX: 417-673-3642
Moschetti, Mitchell R., P.O. Box 27065, Denver, CO 80227
Moss Double Tone, Inc., P.O. Box 1112, 2101 S. Kentucky, Sedalia, MO 65301/816-827-0827
Mossberg & Sons, Inc., O.F, 7 Grasso Ave., North Haven, CT 06473/203-230-5300; FAX: 203-230-5420
Mountain Bear Rifle Works, Inc., 100 B Ruritan Rd., Sterling, VA 20164/703-430-0420; FAX: 703-430-7068
Mountain Hollow Game Calls, Box 121, Cascade, MD 21719/301-241-3282
Mountain Plains, Inc., 244 Glass Hollow Rd., Alton, VA 22920/800-687-3000
Mountain Rifles Inc., P.O. Box 2789, Palmer, AK 99645/907-373-4194; FAX: 907-373-4195
Mountain South, P.O. Box 381, Barnwell, SC 29812/FAX: 803-259-3227
Mountain State Muzzleloading Supplies, Inc., Box 154-1, Rt. 2, Williamstown, WV 26187/304-375-7842; FAX: 304-375-3737
Mountain States Engraving, Kenneth W. Warren, P.O. Box 2842, Wenatchee, WA 98802/509-663-6123
Mountain View Sports, Inc., Box 188, Troy, NH 03465/603-357-9690; FAX: 603-357-9691
Mowrey Gun Works, P.O. Box 246, Waldron, IN 46182/317-525-6181; FAX: 317-525-9595
Mowrey's Guns & Gunsmithing, 119 Fredericks St., Canajoharie, NY 13317/518-673-3483
MPC, P.O. Box 450, McMinnville, TN 37110-0450/615-473-5513; FAX: 615-473-5516
MPI Fiberglass Stocks, 5655 NW St. Helens Rd., Portland, OR 97210/503-226-1215; FAX: 503-226-2661
MSC Industrial Supply Co., 151 Sunnyside Blvd., Plainview, NY 11803-9915/516-349-0330
MSR Targets, P.O. Box 1042, West Covina, CA 91793/818-331-7840
Mt. Alto Outdoor Products, Rt. 735, Howardsville, VA 24562
Mt. Baldy Bullet Co., 12981 Old Hill City Rd., Keystone, SD 57751-6623/605-666-4725
M.Thys (See U.S. importer—Champlin Firearms, Inc.)
MTM Molded Products Co., Inc., 3370 Obco Ct., Dayton, OH 45414/513-890-7461; FAX: 513-890-1747
Mulhern, Rick, Rt. 5, Box 152, Rayville, LA 71269/318-728-2688
Mullins Ammunition, Rt. 2, Box 304K, Clintwood, VA 24228/540-926-6772; FAX: 540-926-6772
Mullis Guncraft, 3523 Lawyers Road E., Monroe, NC 28110/704-283-6683
Multi-Caliber Adapters (See MCA Sports)
Multiplex International, 26 S. Main St., Concord, NH 03301/FAX: 603-796-2223
Multipropulseurs, La Bertrandiere, 42580 L'Etrat, FRANCE/77 74 01 30; FAX: 77 93 19 34
Multi-Scale Charge Ltd., 3269 Niagara Falls Blvd., N. Tonawanda, NY 14120/905-566-1255; FAX: 905-276-6295
Mundy, Thomas A., 69 Robbins Road, Somerville, NJ 08876/201-722-2199
Munsch Gunsmithing, Tommy, Rt. 2, P.O. Box 248, Little Falls, MN 56345/612-632-6695
Murmur Corp., 2823 N. Westmoreland Ave., Dallas, TX 75222/214-630-5400
Murphy Co., Inc., R., 13 Groton-Harvard Rd., P.O. Box 376, Ayer, MA 01432/617-772-3481
Murray State College, 100 Faculty Dr., Tishomingo, OK 73460/405-371-2371 ext. 238
Muscle Products Corp., 112 Fennell Dr., Butler, PA 16001/800-227-7049, 412-283-0567; FAX: 412-283-8310
Museum of Historical Arms Inc., 2750 Coral Way, Suite 204, Miami, FL 33145/305-444-9199
Mushroom Express Bullet Co., 601 W. 6th St., Greenfield, IN 46140-1728/317-462-6332
Muzzleload Magnum Products (See MMP)
Muzzleloaders Etcetera, Inc., 9901 Lyndale Ave. S., Bloomington, MN 55420/612-884-1161
MWG Co., P.O. Box 971202, Miami, FL 33197/800-428-9394, 305-253-8393; FAX: 305-232-1247

N

N&J Sales, Lime Kiln Rd., Northford, CT 06472/203-484-0247
Nagel's Custom Bullets, 100 Scott St., Baytown, TX 77520-2849
Nalpak, 1937-C Friendship Drive, El Cajon, CA 92020/619-258-1200
Napoleon Bonaparte, Inc. (See Metals Hand Engraver)
Nastoff's 45 Shop, Inc., Steve, 12288 Mahoning Ave., P.O. Box 446, North Jackson, OH 44451/330-538-2977
National Bullet Co., 1585 E. 361 St., Eastlake, OH 44095/216-951-1854; FAX: 216-951-7761
National Security Safe Co., Inc., P.O. Box 39, 620 S. 380 E., American Fork, UT 84003/801-756-7706, 800-544-3829; FAX: 801-756-8043
National Target Co., 4690 Wyaconda Rd., Rockville, MD 20852/800-827-7060, 301-770-7060; FAX: 301-770-7892
Nationwide Airgun Repairs (See Airgun Repair Centre)

DIRECTORY OF THE ARMS TRADE

Nationwide Sports Distributors, Inc., 70 James Way, Southampton, PA 18966/215-322-2050, 800-355-3006; FAX: 702-358-2093
Naval Ordnance Works, Rt. 2, Box 919, Sheperdstown, WV 25443/304-876-0998
Navy Arms Co., 689 Bergen Blvd., Ridgefield, NJ 07657/201-945-2500; FAX: 201-945-6859
N.B.B., Inc., 24 Elliot Rd., Sterling, MA 01564/508-422-7538, 800-942-9444
N.C. Ordnance Co., P.O. Box 3254, Wilson, NC 27895/919-237-2440; FAX: 919-243-9845
NCP Products, Inc., 3500 12th St. N.W., Canton, OH 44708/330-456-5130: FAX: 330-456-5234
Necessary Concepts, Inc., P.O. Box 571, Deer Park, NY 11729/516-667-8509; 800-671-8881
NECO, 1316-67th St., Emeryville, CA 94608/510-450-0420
Necromancer Industries, Inc., 14 Communications Way, West Newton, PA 15089/412-872-8722
NEI Handtools, Inc., 51583 Columbia River Hwy., Scappoose, OR 97056/503-543-6776; FAX: 503-543-6799; E-MAIL: neiht@mcimail.com
Nelson, Gary K., 975 Terrace Dr., Oakdale, CA 95361/209-847-4590
Nelson, Stephen, 7365 NW Spring Creek Dr., Corvallis, OR 97330/541-745-5232
Nelson/Weather-Rite, Inc., 14760 Santa Fe Trail Dr., Lenexa, KS 66215/913-492-3200; FAX: 913-492-8749
Nesci Enterprises, Inc., P.O. Box 119, Summit St., East Hampton, CT 06424/203-267-2588
Nesika Bay Precision, 22239 Big Valley Rd., Poulsbo, WA 98370/206-697-3830
Nettestad Gun Works, RR 1, Box 160, Pelican Rapids, MN 56572/218-863-4301
Neumann GmbH, Am Galgenberg 6, 90575 Langenzenn, GERMANY/09101/8258; FAX: 09101/6356
Nevada Cartridge Co., 44 Montgomery St., Suite 500, San Francisco, CA 94104/415-925-9394; FAX: 415-925-9396
Nevada Pistol Academy Inc., 4610 Blue Diamond Rd., Las Vegas, NV 89139/702-897-1100
New England Ammunition Co., 1771 Post Rd. East, Suite 223, Westport, CT 06880/203-254-8048
New England Arms Co., Box 278, Lawrence Lane, Kittery Point, ME 03905/207-439-0593; FAX: 207-439-6726
New England Custom Gun Service, 438 Willow Brook Rd., RR2, Box 122W, W. Lebanon, NH 03784/603-469-3450; FAX: 603-469-3471
New England Firearms, 60 Industrial Rowe, Gardner, MA 01440/508-632-9393; FAX: 508-632-2300
New Historians Productions, The, 131 Oak St., Royal Oak, MI 48067/313-544-7544
New Orleans Jewelers Supply Co., 206 Charters St., New Orleans, LA 70130/504-523-3839; FAX: 504-523-3836
New SKB Arms Co., C.P.O. Box 1401, Tokyo, JAPAN/81-3-3943-9550; FAX: 81-3-3943-0695
New Win Publishing, Inc., 186 Center St., Clinton, NJ 08809/908-735-9701; FAX: 908-735-9703
Newark Electronics, 4801 N. Ravenswood Ave., Chicago, IL 60640
Newell, Robert H., 55 Coyote, Los Alamos, NM 87544/505-662-7135
Newman Gunshop, 119 Miller Rd., Agency, IA 52530/515-937-5775
NgraveR Co., The, 67 Wawecus Hill Rd., Bozrah, CT 06334/860-823-1533
Nicholson Custom, 17285 Thornlay Road, Hughesville, MO 65334/816-826-8746
Nickels, Paul R., 4789 Summerhill Rd., Las Vegas, NV 89121/702-435-5318
Nicklas, Ted, 5504 Hegel Rd., Goodrich, MI 48438/810-797-4493
Nic Max, Inc., 535 Midland Ave., Garfield, NJ 07026/201-546-7191; FAX: 201-546-7419
Niemi Engineering, W.B., Box 126 Center Road, Greensboro, VT 05841/802-533-7180 days, 802-533-7141 evenings
Nightforce (See Lightforce U.S.A. Inc.)
Nikon, Inc., 1300 Walt Whitman Rd., Melville, NY 11747/516-547-8623; FAX: 516-547-0309
Nitex, Inc., P.O. Box 1706, Uvalde, TX 78801/210-278-8843
Noble Co., Jim, 1305 Columbia St., Vancouver, WA 98660/360-695-1309; FAX: 360-695-6835
Noreen, Peter H., 5075 Buena Vista Dr., Belgrade, MT 59714/406-586-7383
Norica, Avnda Otaola, 16, Apartado 68, 20600 Eibar, SPAIN
Norin, Dave, Schrank's Smoke & Gun, 2010 Washington St., Waukegan, IL 60085/708-662-4034
Norinco, 7A, Yun Tan N Beijing, CHINA/ (U.S. importers—Century International Arms, Inc.; Interarms)
Norincoptics (See BEC, Inc.)
Norma Precision AB (See U.S. importers—Dynamit Nobel-RWS Inc.; Paul Co. Inc., The)
Norman Custom Gunstocks, Jim, 14281 Cane Rd., Valley Center, CA 92082/619-749-6252
Normark Corp., 10395 Yellow Circle Dr., Minnetonka, MN 55343-9101/612-933-7060; FAX: 612-933-0046
Norrell Arms, John, 2608 Grist Mill Rd., Little Rock, AR 72207/501-225-7864
North American Arms, Inc., 2150 South 950 East, Provo, UT 84606-6285/800-821-5783, 801-374-9990; FAX: 801-374-9998
North American Correspondence Schools, The Gun Pro School, Oak & Pawney St., Scranton, PA 18515/717-342-7701
North American Munitions, P.O. Box 815, Beulah, ND 58523/701-948-2260; FAX: 701-948-2282
North American Shooting Systems, P.O. Box 306, Osoyoos, B.C. V0H 1V0 CANADA/604-495-3131; FAX: 604-495-2816

North Devon Firearms Services, 3 North St., Braunton, EX33 1AJ ENGLAND/01271 813624; FAX: 01271 813624
North Fork Custom Gunsmithing, James Johnston, 428 Del Rio Rd., Roseburg, OR 97470/503-673-4467
North Mountain Pine Training Center (See Executive Protection Institute)
North Specialty Products, 2664-B Saturn St., Brea, CA 92621/714-524-1665
North Star West, P.O. Box 488, Glencoe, CA 95232/209-293-7010
North Wind Decoy Co., 1005 N. Tower Rd., Fergus Falls, MN 56537/218-736-4378; FAX: 218-736-7060
Northern Precision Custom Swaged Bullets, 329 S. James St., Carthage, NY 13619/315-493-1711
Northlake Outdoor Footwear, P.O. Box 10, Franklin, TN 37065-0010/615-794-1556; FAX: 615-790-8005
Northside Gun Shop, 2725 NW 109th, Oklahoma City, OK 73120/405-840-2353
No-Sho Mfg. Co., 10727 Glenfield Ct., Houston, TX 77096/713-723-5332
Nosler, Inc., P.O. Box 671, Bend, OR 97709/800-285-3701, 541-382-3921; FAX: 541-388-4667
Novak's, Inc., 1206½ 30th St., P.O. Box 4045, Parkersburg, WV 26101/304-485-9295; FAX: 304-428-6722
Now Products, Inc., 1045 South Edward Drive, Tempe, AZ 85281/602-966-6100; FAX: 602-966-0890
Nowlin Mfg. Co., Rt. 1, Box 308, Claremore, OK 74017/918-342-0689; FAX: 918-342-0624
NRI Gunsmith School, 4401 Connecticut Ave. NW, Washington, D.C. 20008
Nu-Line Guns, Inc., 1053 Caulks Hill Rd., Harvester, MO 63304/314-441-4500, 314-447-4501; FAX: 314-447-5018
Null Holsters Ltd., K.L., 161 School St. NW, Hill City Station, Resaca, GA 30735/706-625-5643; FAX: 706-625-9392
Numrich Arms Corp., 203 Broadway, W. Hurley, NY 12491
Nu-Teck, 30 Industrial Park Rd., Box 37, Centerbrook, CT 06409/203-767-3573; FAX: 203-767-9137
NW Sinker and Tackle, 380 Valley Dr., Myrtle Creek, OR 97457-9717
Nygord Precision Products, P.O. Box 12578, Prescott, AZ 86304/520-717-2315; FAX: 520-717-2198

O

Oakland Custom Arms, Inc., 4690 W. Walton Blvd., Waterford, MI 48329/810-674-8261
Oakman Turkey Calls, RD 1, Box 825, Harrisonville, PA 17228/717-485-4620
Oakshore Electronic Sights, Inc., P.O. Box 4470, Ocala, FL 32678-4470/904-629-7112; FAX: 904-629-1433
Obermeyer Rifled Barrels, 23122 60th St., Bristol, WI 53104/414-843-3537; FAX: 414-843-2129
October Country, P.O. Box 969, Dept. GD, Hayden, ID 83835/208-772-2068; FAX: 208-772-9230
Oehler Research, Inc., P.O. Box 9135, Austin, TX 78766/512-327-6900, 800-531-5125; FAX: 512-327-6903
Oglesby & Oglesby Gunmakers, Inc., RR 5, Springfield, IL 62707/217-487-7100
Oil Rod and Gun Shop, 69 Oak St., East Douglas, MA 01516/508-476-3687
Ojala Holsters, Arvo, P.O. Box 98, N. Hollywood, CA 91603/503-669-1404
Oker's Engraving, 365 Bell Rd., P.O. Box 126, Shawnee, CO 80475/303-838-6042
Oklahoma Ammunition Co., 3701A S. Harvard Ave., No. 367, Tulsa, OK 74135-2265/918-396-3187; FAX: 918-396-4270
Oklahoma Leather Products, Inc., 500 26th NW, Miami, OK 74354/918-542-6651; FAX: 918-542-6653
OK Weber, Inc., P.O. Box 7485, Eugene, OR 97401/541-747-0458; FAX: 541-747-5927
Old Wagon Bullets, 32 Old Wagon Rd., Wilton, CT 06897
Old West Bullet Moulds, P.O. Box 519, Flora Vista, NM 87415/505-334-6970
Old West Reproductions, Inc., R.M. Bachman, 446 Florence S. Loop, Florence, MT 59833/406-273-2615; FAX: 406-273-2615
Old Western Scrounger, Inc., 12924 Hwy. A-l2, Montague, CA 96064/916-459-5445; FAX: 916-459-3944
Old World Gunsmithing, 2901 SE 122nd St., Portland, OR 97236/503-760-7681
Old World Oil Products, 3827 Queen Ave. N., Minneapolis, MN 55412/612-522-5037
Ole Frontier Gunsmith Shop, 2617 Hwy. 29 S., Cantonment, FL 32533/904-477-8074
Olson, Myron, 989 W. Kemp, Watertown, SD 57201/605-886-9787
Olson, Vic, 5002 Countryside Dr., Imperial, MO 63052/314-296-8086
Olt Co., Philip S., P.O. Box 550, 12662 Fifth St., Pekin, IL 61554/309-348-3633; FAX: 309-348-3300
Olympic Optical Co., P.O. Box 752377, Memphis, TN 38175-2377/901-794-3890, 800-238-7120; FAX: 901-794-0676, 800-748-1669
Omark Industries, Div. of Blount, Inc., 2299 Snake River Ave., P.O. Box 856, Lewiston, ID 83501/800-627-3640, 208-746-2351
Omega Sales, P.O. Box 1066, Mt. Clemens, MI 48043/810-469-7323; FAX: 810-469-0425
One Of A Kind, 15610 Purple Sage, San Antonio, TX 78255/512-695-3364
Op-Tec, P.O. Box L632, Langhorn, PA 19047/215-757-5037
Optical Services Co., P.O. Box 1174, Santa Teresa, NM 88008-1174/505-589-3833
Orchard Park Enterprise, P.O. Box 563, Orchard Park, NY 14227/616-656-0356
Ordnance Works, The, 2969 Pidgeon Point Road, Eureka, CA 95501/707-443-3252
Oregon Arms, Inc. (See Rogue Rifle Co., Inc.)
Oregon Trail Bullet Company, P.O. Box 529, Dept. P, Baker City, OR 97814/800-811-0548; FAX: 514-523-1803

MANUFACTURERS' DIRECTORY

Original Box, Inc., 700 Linden Ave., York, PA 17404/717-854-2897; FAX: 717-845-4276
Original Mink Oil, Inc., 10652 NE Holman, Portland, OR 97220/503-255-2814, 800-547-5895; FAX: 503-255-2487
Orion Rifle Barrel Co., RR2, 137 Cobler Village, Kalispell, MT 59901/406-257-5649
Or-Un, Tahtakale Menekse Han 18, Istanbul, TURKEY 34460/90212-522-5912; FAX: 90212-522-7973
Orvis Co., The, Rt. 7, Manchester, VT 05254/802-362-3622 ext. 283; FAX: 802-362-3525
Ottmar, Maurice, Box 657, 113 E. Fir, Coulee City, WA 99115/509-632-5717
Outa-Site Gun Carriers, 219 Market St., Laredo, TX 78040/210-722-4678, 800-880-9715; FAX: 210-726-4858
Outdoor Connection, Inc., The, 201 Cotton Dr., P.O. Box 7751, Waco, TX 76714-7751/800-533-6076; 817-772-5575; FAX: 817-776-3553
Outdoor Edge Cutlery Corp., 2888 Bluff St., Suite 130, Boulder, CO 80301/303-652-8212; FAX: 303-652-8238
Outdoor Enthusiast, 3784 W. Woodland, Springfield, MO 65807/417-883-9841
Outdoor Sports Headquarters, Inc., 967 Watertower Ln., West Carrollton, OH 45449/513-865-5855; FAX: 513-865-5962
Outdoorsman's Bookstore, The, Llangorse, Brecon, Powys LD3 7UE, U.K./44-1874-658-660; FAX: 44-1874-658-650
Outers Laboratories, Div. of Blount, Inc., Sporting Equipment Div., Route 2,, P.O. Box 39/Onalaska, WI 54650
608-781-5800; FAX: 608-781-0368
Ox-Yoke Originals, Inc., 34 Main St., Milo, ME 04463/800-231-8313, 207-943-7351; FAX: 207-943-2416
Ozark Gun Works, 11830 Cemetery Rd., Rogers, AR 72756/501-631-6944; FAX: 501-631-6944

P

P&M Sales and Service, 5724 Gainsborough Pl., Oak Forest, IL 60452/708-687-7149
P&S Gun Service, 2138 Old Shepardsville Rd., Louisville, KY 40218/502-456-9346
Pac-Nor Barreling, 99299 Overlook Rd., P.O. Box 6188, Brookings, OR 97415/503-469-7330; FAX: 503-469-7331
Pace Marketing, Inc., P.O. Box 2039, Stuart, FL 34995/561-871-9682; FAX: 561-871-6552
Pachmayr, Ltd., 1875 S. Mountain Ave., Monrovia, CA 91016/818-357-7771, 800-423-9704; FAX: 818-358-7251
Pacific Cartridge, Inc., 2425 Salashan Loop Road, Ferndale, WA 98248/360-366-4444; FAX: 360-366-4445
Pacific Pistolcraft, 1810 E. Columbia Ave., Tacoma, WA 98404/206-474-5465
Pacific Precision, 755 Antelope Rd., P.O. Box 2549, White City, OR 97503/503-826-5808; FAX: 503-826-5304
Rimrock Rifle Stocks, P.O. Box 589, Vashon Island, WA 98070/206-463-5551; FAX: 206-463-2526
Pacific Research Laboratories, Inc. (See Rimrock Rifle Stocks)
Pacific Rifle Co., 1040-D Industrial Parkway, Newberg, OR 97132/503-538-7437
Paco's (See Small Custom Mould & Bullet Co.)
P.A.C.T., Inc., P.O. Box 531525, Grand Prairie, TX 75053/214-641-0049
Page Custom Bullets, P.O. Box 25, Port Moresby Papua, NEW GUINEA
Pagel Gun Works, Inc., 1407 4th St. NW, Grand Rapids, MN 55744/218-326-3003
Paintball Games International Magazine (Aceville Publications), Castle House, 97 High St./Colchester, Essex, CO1 1TH ENGLAND
011-44-206-564840
Paintball Sports Magazine, 540 Main St., Mt. Kisco, NY 10549/914-241-7400
Palmer Manufacturing Co., Inc., C., P.O. Box 220, West Newton, PA 15089/412-872-8200; FAX: 412-872-8302
Palmer Security Products, 2930 N. Campbell Ave., Chicago, IL 60618/800-788-7725; FAX: 773-267-8080
Palsa Outdoor Products, P.O. Box 81336, Lincoln, NE 68501/402-488-5288, 800-456-9281; FAX: 402-488-2321
PanaVise Products, Inc., 7540 Colbert Drive, Sparks, NV 89431/702-850-2900; FAX: 702-850-2929
Para-Ordnance Mfg., Inc., 980 Tapscott Rd., Scarborough, Ont. M1X 1E7, CANADA/416-297-7855; FAX: 416-297-1289 (U.S. importer—Para-Ordnance, Inc.)
Para-Ordnance, Inc., 1919 NE 45th St., Ft. Lauderdale, FL 33308
Paragon Sales & Services, Inc., P.O. Box 2022, Joliet, IL 60434/815-725-9212; FAX: 815-725-8974
Pardini Armi Srl, Via Italica 154, 55043 Lido Di Camaiore Lu, ITALY/584-90121; FAX: 584-90122 (U.S. importers—Nygord Precision Products;MCS, Inc.)
Paris, Frank J., 17417 Pershing St., Livonia, MI 48152-3822
Park Rifle Co., Ltd., The, Unit 6a, Dartford Trade Park, Power Mill Lane, Dartford, Kent, ENGLAND DA7 7NX/011-0322-222512
Parker Div. Reageant Chemical (See Parker Reproductions)
Parker Gun Finishes, 9337 Smokey Row Rd., Strawberry Plains, TN 37871/423-933-3286
Parker Reproductions, 124 River Rd., Middlesex, NJ 08846/908-469-0100; FAX: 908-469-9692
Parker, Mark D., 1240 Florida Ave. 7, Longmont, CO 80501/303-772-0214
Parsons Optical Mfg. Co., P.O. Box 192, Ross, OH 45061/513-867-0820; FAX: 513-867-8380
Parts & Surplus, P.O. Box 22074, Memphis, TN 38122/901-683-4007
Partridge Sales Ltd., John, Trent Meadows, Rugeley, Staffordshire, WS15 2HS ENGLAND/0889-584438

Pasadena Gun Center, 206 E. Shaw, Pasadena, TX 77506/713-472-0417; FAX: 713-472-1322
Passive Bullet Traps, Inc. (See Savage Range Systems, Inc.)
PAST Sporting Goods, Inc., P.O. Box 1035, Columbia, MO 65205/314-445-9200; FAX: 314-446-6606
Paterson Gunsmithing, 438 Main St., Paterson, NJ 07502/201-345-4100
Pathfinder Sports Leather, 2920 E. Chambers St., Phoenix, AZ 85040/602-276-0016
Patrick Bullets, P.O. Box 172, Warwick QSLD 4370 AUSTRALIA
Pattern Control, 114 N. Third St., P.O. Box 462105, Garland, TX 75046/214-494-3551; FAX: 214-272-8447
Paul Co., The, 27385 Pressonville Rd., Wellsville, KS 66092/913-883-4444; FAX: 913-883-2525
Paulsen Gunstocks, Rt. 71, Box 11, Chinook, MT 59523/406-357-3403
Payne Photography, Robert, P.O. Box 141471, Austin, TX 78714/512-272-4554
PC Bullet/ADC, Inc., 52700 NE First, Scappoose, OR 97056-3212/503-543-5088; FAX: 503-543-5990
PC Co., 5942 Secor Rd., Toledo, OH 43623/419-472-6222
Peacemaker Specialists, P.O. Box 157, Whitmore, CA 96096/916-472-3438
Pearce Grip, Inc., P.O. Box 187, Bothell, WA 98041-0187/206-485-5488; FAX:206-488-9497
Pease Accuracy, Bob, P.O. Box 310787, New Braunfels, TX 78131/210-625-1342
PECAR Herbert Schwarz, GmbH, Kreuzbergstrasse 6, 10965 Berlin, GERMANY/004930-785-7383; FAX: 004930-785-1934
Pecatonica River Longrifle, 5205 Nottingham Dr., Rockford, IL 61111/815-968-1995; FAX: 815-968-1996
Pedersen, C.R., 2717 S. Pere Marquette Hwy., Ludington, MI 49431/616-843-2061
Pedersen, Rex C., 2717 S. Pere Marquette Hwy., Ludington, MI 49431/616-843-2061
Pedersoli and Co., Davide, Via Artigiani 57, Gardone V.T., Brescia, ITALY 25063/030-8912402; FAX: 030-8911019 (U.S. importers—Beauchamp & Son, Inc.; Cabela's; Cape Outfitters; Cimarron Arms; Dixie Gun Works; EMF Co., Inc.; Navy Arms Co.; Track of the Wolf, Inc.)
Peerless Alloy, Inc., 1445 Osage St., Denver, CO 80204-2439/303-825-6394, 800-253-1278
Peet Shoe Dryer, Inc., 130 S. 5th St., P.O. Box 618, St. Maries, ID 83861/208-245-2095, 800-222-PEET; FAX: 208-245-5441
Peifer Rifle Co., P.O. Box 192, Nokomis, IL 62075-0192/217-563-7050; FAX: 217-563-7060
Pejsa Ballistics, 2120 Kenwood Pkwy., Minneapolis, MN 55405/612-374-3337; FAX: 612-374-3337
Pelaire Products, 5346 Bonky Ct., W. Palm Beach, FL 33415/561-439-0691; FAX: 561-967-0052
Pell, John T. (See KOGOT)
Peltor, Inc. (See Aero Peltor)
PEM's Mfg. Co., 5063 Waterloo Rd., Atwater, OH 44201/216-947-3721
Pence Precision Barrels, 7567 E. 900 S., S. Whitley, IN 46787/219-839-4745
Pend Oreille Sport Shop, 3100 Hwy. 200 East, Sandpoint, ID 83864/208-263-2412
Pendleton Royal, c/o Swingler Buckland Ltd., 4/7 Highgate St., Birmingham, ENGLAND B12 0XS/44 121 440 3060, 44 121 446 5898; FAX: 44 121 446 4165
Pendleton Woolen Mills, P.O. Box 3030, 220 N.W. Broadway, Portland, OR 97208/503-226-4801
Penguin Industries, Inc., Airport Industrial Mall, Coatesville, PA 19320/610-384-6000; FAX: 610-857-5980
Penn Bullets, P.O. Box 756, Indianola, PA 15051
Penn's Woods Products, Inc., 19 W. Pittsburgh St., Delmont, PA 15626/412-468-8311; FAX: 412-468-8975
Pennsylvania Gun Parts, 1701 Mud Run Rd., York Springs, PA 17372/717-259-8010; FAX: 717-259-0057
Pennsylvania Gunsmith School, 812 Ohio River Blvd., Avalon, Pittsburgh, PA 15202/412-766-1812
Penrod Precision, 312 College Ave., P.O. Box 307, N. Manchester, IN 46962/219-982-8385
Pentax Corp., 35 Inverness Dr. E., Englewood, CO 80112/800-709-2020; FAX: 303-643-0393
Pentheny de Pentheny, 2352 Baggett Ct., Santa Rosa, CA 95401/707-573-1390; FAX: 707-573-1390
Perazone-Gunsmith, Brian, Cold Spring Rd., Roxbury, NY 12474/607-326-4088; FAX: 607-326-3140
Perazzi m.a.p. S.p.A. (See Armi Perazzi S.p.A.)
Perazzi USA, Inc., 1207 S. Shamrock Ave., Monrovia, CA 91016/818-303-0068; FAX: 818-303-2081
Peregrine Sporting Arms, Inc., 14155 Brighton Rd., Brighton, CO 80601/303-654-0850
Performance Specialists, 308 Eanes School Rd., Austin, TX 78746/512-327-0119
Peripheral Data Systems (See Arms Software)
Personal Protection Systems, RD 5, Box 5027-A, Moscow, PA 18444/717-842-1766
Perugini Visini & Co. S.r.l., Via Camprelle, 126, 25080 Nuvolera (Bs.), ITALY
Peters Stahl GmbH, Stettiner Strasse 42, D-33106 Paderborn, GERMANY/05251-750025; FAX: 05251-75611 (U.S. importer—Franzen International, Inc.)
Petersen Publishing Co., 6420 Wilshire Blvd., Los Angeles, CA 90048/213-782-2000; FAX: 213-782-2867
Peterson Gun Shop, Inc., A.W., 4255 W. Old U.S. 441, Mt. Dora, FL 32757-3299/352-383-4258; FAX: 352-735-1001

DIRECTORY OF THE ARMS TRADE

Petro-Explo, Inc., 7650 U.S. Hwy. 287, Suite 100, Arlington, TX 76017/817-478-8888
Pettinger Books, Gerald, Rt. 2, Box 125, Russell, IA 50238/515-535-2239
Pflumm Mfg. Co., 10662 Widmer Rd., Lenexa, KS 66215/800-888-4867; FAX: 913-451-7857
PFRB Co., P.O. Box 1242, Bloomington, IL 61702/309-473-3964; FAX: 309-473-2161
Phil-Chem, Inc. (See George & Roy's)
Phillippi Custom Bullets, Justin, P.O. Box 773, Ligonier, PA 15658/412-238-9671
Phillips, Jerry, P.O. Box L632, Langhorne, PA 19047/215-757-5037
Phillips & Rodgers, Inc., 100 Hilbig, Suite C, Conroe, TX 77301/409-756-1001, 800-682-2247; FAX: 409-756-0976
Phoenix Arms, 1420 S. Archibald Ave., Ontario, CA 91761/909-947-4843; FAX: 909-947-6798
Photronic Systems Engineering Company, 6731 Via De La Reina, Bonsall, CA 92003/619-758-8000
Piedmont Community College, P.O. Box 1197, Roxboro, NC 27573/910-599-1181
Pierce Pistols, 55 Sorrellwood Lane, Sharpsburg, GA 30277-9523/404-253-8192
Pietta (See U.S. importers—Navy Arms Co.; Taylor's & Co., Inc.)
Pilgrim Pewter, Inc. (See Bell Originals Inc., Sid)
Pilkington, Scott, Little Trees Ramble, P.O. Box 97, Monteagle, TN 37356/615-924-3475; FAX: 615-924-3489
Pine Technical College, 1100 4th St., Pine City, MN 55063/800-521-7463; FAX: 612-629-6766
Pinetree Bullets, 133 Skeena St., Kitimat BC, CANADA V8C 1Z1/604-632-3768; FAX: 604-632-3768
Pioneer Arms Co., 355 Lawrence Rd., Broomall, PA 19008/215-356-5203
Pioneer Guns, 5228 Montgomery Rd., Norwood, OH 45212/513-631-4871
Pioneer Research, Inc., 216 Haddon Ave., Suite 102, Westmont, NJ 08108/800-257-7742; FAX: 609-858-8695
Piotti (See U.S. importer—Moore & Co., Wm. Larkin)
Piquette, Paul R., 80 Bradford Dr., Feeding Hills, MA 01030/413-781-8300, Ext. 682
Plaxco, J. Michael, Rt. 1, P.O. Box 203, Roland, AR 72135/501-868-9787
Plaza Cutlery, Inc., 3333 Bristol, 161, South Coast Plaza, Costa Mesa, CA 92626/714-549-3932
Plum City Ballistic Range, N2162 80th St., Plum City, WI 54761-8622/715-647-2539
PlumFire Press, Inc., 30-A Grove Ave., Patchogue, NY 11772-4112/800-695-7246; FAX:516-758-4071
PMC/Eldorado Cartridge Corp., P.O. Box 62508, 12801 U.S. Hwy. 95 S., Boulder City, NV 89005/702-294-0025; FAX: 702-294-0121
P.M. Enterprises, Inc., 146 Curtis Hill Rd., Chehalis, WA 98532/360-748-3743; FAX: 360-748-1802
Poburka, Philip (See Bison Studios)
Pohl, Henry A. (See Great American Gun Co.)
Pointing Dog Journal, Village Press Publications, P.O. Box 968, Dept. PGD, Traverse City, MI 49685/800-272-3246; FAX: 616-946-3289
Police Bookshelf, P.O. Box 122, Concord, NH 03301/603-224-6814; FAX: 603-226-3554
Polywad, Inc., P.O. Box 7916, Macon, GA 31209/912-477-0669
Pomeroy, Robert, RR1, Box 50, E. Corinth, ME 04427/207-285-7721
Ponsness/Warren, P.O. Box 8, Rathdrum, ID 83858/208-687-2231; FAX: 208-687-2233
Pony Express Reloaders, 608 E. Co. Rd. D, Suite 3, St. Paul, MN 55117/612-483-9406; FAX: 612-483-9884
Pony Express Sport Shop, Inc., 16606 Schoenborn St., North Hills, CA 91343/818-895-1231
Porta Blind, Inc., 2700 Speedway, Wichita Falls, TX 76308/817-723-6620
Portus, Robert, 130 Ferry Rd., Grants Pass, OR 97526/503-476-4919
Potts, Wayne E., 912 Poplar St., Denver, CO 80220/303-355-5462
Powder Horn Antiques, P.O. Box 4196, Ft. Lauderdale, FL 33338/305-565-6060
Powder Horn, Inc., The, P.O. Box 114 Patty Drive, Cusseta, GA 31805/404-989-3257
Powell & Son (Gunmakers) Ltd., William, 35-37 Carrs Lane, Birmingham B4 7SX ENGLAND/121-643-0689; FAX: 121-631-3504 (U.S. importer—The William Powell Agency)
Powell Agency, William, The, 22 Circle Dr., Bellmore, NY 11710/516-679-1158
Power Custom, Inc., RR 2, P.O. Box 756AB, Gravois Mills, MO 65037/314-372-5684
Powley Computer (See Hutton Rifle Ranch)
Practical Tools, Inc., Div. Behlert Precision, 7067 Easton Rd., P.O. Box 133, Pipersville, PA 18947/215-766-7301; FAX: 215-766-8681
Pragotrade, 307 Humberline Dr., Rexdale, Ontario, CANADA M9W 5V1/416-675-1322
Prairie River Arms, 1220 N. Sixth St., Princeton, IL 61356/815-875-1616, 800-445-1541; FAX: 815-875-1402
Pranger, Ed G., 1414 7th St., Anacortes, WA 98221/206-293-3488
Precise International, 15 Corporate Dr., Orangeburg, NY 10962/914-365-3500; FAX: 914-425-4700
Precise Metalsmithing Enterprises, 146 Curtis Hill Rd., Chehalis, WA 98532/206-748-3743; FAX: 206-748-8102
Precision Airgun Sales, Inc., 5139 Warrensville Center Rd., Maple Hts., OH 44137-1906/216-587-5005
Precision Cartridge, 176 Eastside Rd., Deer Lodge, MT 59722/800-397-3901, 406-846-3900
Precision Cast Bullets, 101 Mud Creek Lane, Ronan, MT 59864/406-676-5135
Precision Castings & Equipment, Inc., P.O. Box 326, Jasper, IN 47547-0135/812-634-9167
Precision Components, 3177 Sunrise Lake, Milford, PA 18337/717-686-4414
Precision Components and Guns, Rt. 55, P.O. Box 337, Pawling, NY 12564/914-855-3040
Precision Delta Corp., P.O. Box 128, Ruleville, MS 38771/601-756-2810; FAX: 601-756-2590
Precision Metal Finishing, John Westrom, P.O. Box 3186, Des Moines, IA 50316/515-288-8680; FAX: 515-244-3925
Precision Munitions, Inc., P.O. Box 326, Jasper, IN 47547
Precision Reloading, Inc., P.O. Box 122, Stafford Springs, CT 06076/860-684-7979; FAX: 860-684-6788
Precision Sales International, Inc., P.O. Box 1776, Westfield, MA 01086/413-562-5055; FAX: 413-562-5056
Precision Shooting, Inc., 222 McKee St., Manchester, CT 06040/860-645-8776; FAX: 860-643-8215
Precision Small Arms, 9777 Wilshire Blvd., Suite 1005, Beverly Hills, CA 90212/310-859-4867; FAX: 310-859-2868
Precision Specialties, 131 Hendom Dr., Feeding Hills, MA 01030/413-786-3365; FAX: 413-786-3365
Precision Sport Optics, 15571 Producer Lane, Unit G, Huntington Beach, CA 92649/714-891-1309; FAX: 714-892-6920
Premier Reticles, 920 Breckinridge Lane, Winchester, VA 22601-6707/540-722-0601; FAX: 540-722-3522
Prescott Projectile Co., 1808 Meadowbrook Road, Prescott, AZ 86303
Preslik's Gunstocks, 4245 Keith Ln., Chico, CA 95926/916-891-8236
Pre-Winchester 92-90-62 Parts Co., P.O. Box 8125, W. Palm Beach, FL 33407
Price Bullets, Patrick W., 16520 Worthley Drive, San Lorenzo, CA 94580/510-278-1547
Prime Reloading, 30 Chiswick End, Meldreth, Royston SG8 6LZ UK/0763-260636
Primos, Inc., P.O. Box 12785, Jackson, MS 39236-2785/601-366-1288; FAX: 601-362-3274
PRL Bullets, c/o Blackburn Enterprises, 114 Stuart Rd., Ste. 110, Cleveland, TN 37312/423-559-0340
Pro Load Ammunition, Inc., 5180 E. Seltice Way, Post Falls, ID 83854/208-773-9444; FAX: 208-773-9441
Pro-Mark, Div. of Wells Lamont, 6640 W. Touhy, Chicago, IL 60648/312-647-8200
Pro-Port Ltd., 41302 Executive Dr., Harrison Twp., MI 48045-1306/810-469-7323; FAX: 810-469-0425
Pro-Shot Products, Inc., P.O. Box 763, Taylorville, IL 62568/217-824-9133; FAX: 217-824-8861
Professional Firearms Record Book Co. (See PFRB Co.)
Professional Gunsmiths of America, Inc., Route 1, Box 224F, Lexington, MO 64067/816-259-2636
Professional Hunter Supplies (See Star Custom Bullets)
Professional Ordnance, Inc., 1215 E. Airport Dr., Box 182, Ontario, CA 91761/909-923-5559; FAX: 909-923-0899
Prolix® Lubricants, P.O. Box 1348, Victorville, CA 92393/800-248-LUBE, 619-243-3129; FAX: 619-241-0148
Protecto Plastics, Div. of Penguin Ind., Airport Industrial Mall, Coatesville, PA 19320/215-384-6000
Protector Mfg. Co., Inc., The, 443 Ashwood Place, Boca Raton, FL 33431/407-394-6011
Protektor Model, 1-11 Bridge St., Galeton, PA 16922/814-435-2442
Prototech Industries, Inc., Rt. 1, Box 81, Delia, KS 66418/913-771-3571; FAX: 913-771-2531
ProWare,Inc., 15847 NE Hancock St., Portland, OR 97230/503-239-0159
P.S.M.G. Gun Co., 10 Park Ave., Arlington, MA 02174/617-646-8845; FAX: 617-646-2133
PWL Gunleather, P.O. Box 450432, Atlanta, GA 31145/770-822-1640; FAX: 770-822-1704
Pyromid, Inc., 3292 S. Highway 97, Redmond, OR 97756/503-548-1041; FAX: 503-923-1004

Q

Quack Decoy & Sporting Clays, 4 Ann & Hope Way, P.O. Box 98, Cumberland, RI 02864/401-723-8202; FAX: 401-722-5910
Quaker Boy, Inc., 5455 Webster Rd., Orchard Parks, NY 14127/716-662-3979; FAX: 716-662-9426
Quality Arms, Inc., Box 19477, Dept. GD, Houston, TX 77224/713-870-8377; FAX: 713-870-8524
Quality Firearms of Idaho, Inc., 659 Harmon Way, Middleton, ID 83644-3065/208-466-1631
Quality Parts Co./Bushmaster Firearms, 999 Roosevelt Trail, Bldg. 3, Windham, ME 04062/800-998-7928, 207-892-2005; FAX: 207-892-8068
Quarton USA, Ltd. Co., 7042 Alamo Downs Pkwy., Suite 370, San Antonio, TX 78238-4518/800-520-8435, 210-520-8430; FAX: 210-520-8433
Que Industries, Inc., P.O. Box 2471, Everett, WA 98203/800-769-6930, 206-347-9843; FAX: 206-514-3266

MANUFACTURERS' DIRECTORY

Queen Cutlery Co., P.O. Box 500, Franklinville, NY 14737/800-222-5233; FAX: 716-676-5535
Quigley's Personal Protection Strategies, Paxton, 9903 Santa Monica Blvd.,, 300/Beverly Hills, CA 90212/310-281-1762

R

R&C Knives & Such, P.O. Box 1047, Manteca, CA 95336/209-239-3722; FAX: 209-825-6947
R&J Gun Shop, 133 W. Main St., John Day, OR 97845/503-575-2130
R&S Industries Corp., 8255 Brentwood Industrial Dr., St. Louis, MO 63144/314-781-5400
Rabeno, Martin, 92 Spook Hole Rd., Ellenville, NY 12428/914-647-4567
Radack Photography, Lauren, 21140 Jib Court L-12, Aventura, FL 33180/305-931-3110
Radiator Specialty Co., 1900 Wilkinson Blvd., P.O. Box 34689, Charlotte, NC 28234/800-438-6947; FAX: 800-421-9525
Radical Concepts, P.O. Box 1473, Lake Grove, OR 97035/503-538-7437
Rainier Ballistics Corp., 4500 15th St. East, Tacoma, WA 98424/800-638-8722, 206-922-7589; FAX: 206-922-7854
Ram-Line Blount, Inc., P.O. Box 39, Onalaska, WI 54650-0039
Rampart International, 2781 W. MacArthur Blvd., #B-283, Santa Ana, CA 92704/800-976-7240, 714-557-6405
Ranch Products, P.O. Box 145, Malinta, OH 43535/313-277-3118; FAX: 313-565-8536
Randall-Made Knives, P.O. Box 1988, Orlando, FL 32802/407-855-8075
Randco UK, 286 Gipsy Rd., Welling, Kent DA16 1JJ, ENGLAND/44 81 303 4118
Randolph Engineering, Inc., 26 Thomas Patten Dr., Randolph, MA 02368/800-541-1405; FAX: 800-875-4200
Range Brass Products Company, P.O. Box 218, Rockport, TX 78381
Ranger Mfg. Co., Inc., 1536 Crescent Dr., P.O. Box 14069, Augusta, GA 30919-0069/706-738-2023; FAX: 404-738-3608
Ranger Products, 2623 Grand Blvd., Suite 209, Holiday, FL 34609/813-942-4652, 800-407-7007; FAX: 813-942-6221
Ranger Shooting Glasses, 26 Thomas Patten Dr., Randolph, MA 02368/800-541-1405; FAX: 617-986-0337
Ranging, Inc., Routes 5 & 20, East Bloomfield, NY 14443/716-657-6161; FAX: 716-657-5405
Ransom International Corp., P.O. Box 3845, 1040-A Sandretto Dr., Prescott, AZ 86302/520-778-7899; FAX: 520-778-7993; E-MAIL: ransom@primenet.com; WEB: http://www.primenet.com/˜ransom
Rapine Bullet Mould Mfg. Co., 9503 Landis Lane, East Greenville, PA 18041/215-679-5413; FAX: 215-679-9795
Raptor Arms Co., Inc., 115 S. Union St., Suite 308, Alexandria, VA 22314/703-683-0018; FAX: 703-683-5592
Rattlers Brand, P.O. Box 311, 115 E. Main St., Thomaston, GA 30286/706-647-7131, 800-825-7131; FAX: 706-646-5090
Ravell Ltd., 289 Diputacion St., 08009, Barcelona SPAIN/34(3) 4874486; FAX: 34(3) 4881394
Ray's Gunsmith Shop, 3199 Elm Ave., Grand Junction, CO 81504/970-434-6162; FAX: 970-434-6162
Raytech, Div. of Lyman Products Corp., 475 Smith Street, Middletown, CT 06457-1541/860-632-2020; FAX: 860-632-1699
RCBS, Div. of Blount, Inc., Sporting Equipment Div., 605 Oro Dam Blvd., Oroville, CA 95965/800-533-5000, 916-533-5191; FAX: 916-533-1647
Reagent Chemical & Research, Inc. (See Calico Hardwoods, Inc.)
Reardon Products, P.O. Box 126, Morrison, IL 61270/815-772-3155
Recoilless Technologies, Inc., 3432 W. Wilshire Dr., Suite 11, Phoenix, AZ 85009/602-278-8903; FAX: 602-272-5946
Red Ball, 100 Factory St., Nashua, NH 03060/603-881-4420
Red Cedar Precision Mfg., W. 485 Spruce Dr., Brodhead, WI 53520/608-897-8416
Red Diamond Dist. Co., 1304 Snowdon Dr., Knoxville, TN 37912
Redding Reloading Equipment, 1089 Starr Rd., Cortland, NY 13045/607-753-3331; FAX: 607-756-8445
Redfield, Inc., 5800 E. Jewell Ave., Denver, CO 80224/303-757-6411; FAX: 303-756-2338
Redman's Rifling & Reboring, 189 Nichols Rd., Omak, WA 98841/509-826-5512
Redwood Bullet Works, 3559 Bay Rd., Redwood City, CA 94063/415-367-6741
Reed, Dave, Rt. 1, Box 374, Minnesota City, MN 55959/507-689-2944
Refrigiwear, Inc., 71 Inip Dr., Inwood, Long Island, NY 11696
R.E.I., P.O. Box 88, Tallevast, FL 34270/813-755-0085
Reiswig, Wallace E. (See Claro Walnut Gunstock Co.)
Reloaders Equipment Co., 4680 High St., Ecorse, MI 48229
Reloading Specialties, Inc., Box 1130, Pine Island, MN 55463/507-356-8500; FAX: 507-356-8800
Remington Arms Co., Inc., 870 Remington Drive, P.O. Box 700, Madison, NC 27025-0700/800-243-9700; 910-548-8700
Remington Double Shotguns, 7885 Cyd Dr., Denver, CO 80221/303-429-6947
Renegade, P.O. Box 31546, Phoenix, AZ 85046/602-482-6777; FAX: 602-482-1952
Renfrew Guns & Supplies, R.R. 4, Renfrew, Ontario K7V 3Z7 CANADA/613-432-7080
Reno, Wayne, 2808 Stagestop Rd., Jefferson, CO 80456/719-836-3452
Republic Arms, Inc., 15167 Sierra Bonita Lane, Chino, CA 91710/909-597-3873; FAX:909-393-9771
R.E.T. Enterprises, 2608 S. Chestnut, Broken Arrow, OK 74012/918-251-GUNS; FAX: 918-251-0587

Retting, Inc., Martin B., 11029 Washington, Culver City, CA 90232/213-837-2412
R.F.D. Rifles, 8230 Wilson Dr., Ralston, NE 68127/402-331-9529
R.G.-G., Inc., P.O. Box 1261, Conifer, CO 80433-1261/303-697-4154; FAX: 303-697-4154
Rhino, P.O. Box 787, Locust, NC 28097/704-753-2198
Rhodeside, Inc., 1704 Commerce Dr., Piqua, OH 45356/513-773-5781
Rice, Keith (See White Rock Tool & Die)
Richards, John, Richards Classic Oil Finish, Rt. 2, Box 325, Bedford, KY 40006/502-255-7222
Richards Micro-Fit Stocks, 8331 N. San Fernando Ave., Sun Valley, CA 91352/818-767-6097; FAX: 818-767-7121
Rickard, Inc., Pete, RD 1, Box 292, Cobleskill, NY 12043/800-282-5663; FAX: 518-234-2454
Ridgetop Sporting Goods, P.O. Box 306, 42907 Hilligoss Ln. East, Eatonville, WA 98328/360-832-6422; FAX: 360-832-6422
Riebe Co., W.J., 3434 Tucker Rd., Boise, ID 83703
Ries, Chuck, 415 Ridgecrest Dr., Grants Pass, OR 97527/503-476-5623
Rifle Works & Armory, 707 12th St., Cody, WY 82414/307-587-4919
Rifles Inc., 873 W. 5400 N., Cedar City, UT 84720/801-586-5996; FAX: 801-586-5996
RIG Products, 87 Coney Island Dr., Sparks, NV 89431-6334/702-331-5666; FAX: 702-331-5669
Rigby & Co., John, 66 Great Suffolk St., London SE1 0BU, ENGLAND/0171-620-0690; FAX: 0171-928-9205
Riggs, Jim, 206 Azalea, Boerne, TX 78006/210-249-8567
Riling Arms Books Co., Ray, 6844 Gorsten St., P.O. Box 18925, Philadelphia, PA 19119/215-438-2456; FAX: 215-438-5395
Rim Pac Sports, Inc., 1034 N. Soldano Ave., Azusa, CA 91702-2135
Ringler Custom Leather Co., 31 Shining Mtn. Rd., Powell, WY 82435/307-645-3255
Ripley Rifles, 42 Fletcher Street, Ripley, Derbyshire, DE5 3LP ENGLAND/011-0773-748353
R.I.S. Co., Inc., 718 Timberlake Circle, Richardson, TX 75080/214-235-0933
River Road Sporting Clays, Bruce Barsotti, P.O. Box 3016, Gonzales, CA 93926/408-675-2473
Rizzini, Battista, Via 2 Giugno, 7/7Bis-25060 Marcheno (Brescia), ITALY/ (U.S. importers—Wm. Larkin Moore & Co.; New England Arms Co.)
Rizzini F.lli (See U.S. importers—Moore & Co., Wm. Larkin; New England Arms Co.)
RLCM Enterprises, 110 Hill Crest Drive, Burleson, TX 76028
R.M. Precision, Inc., Attn. Greg F. Smith Marketing, P.O. Box 210, LaVerkin, UT 84745/801-635-4656; FAX: 801-635-4430
RMS Custom Gunsmithing, 4120 N. Bitterwell, Prescott Valley, AZ 86314/520-772-7626
Robar Co.'s, Inc., The, 21438 N. 7th Ave., Suite B, Phoenix, AZ 85027/602-581-2648; FAX: 602-582-0059
Roberts/Engraver, J.J., 7808 Lake Dr., Manassas, VA 22111/703-330-0448
Roberts Products, 25328 SE Iss. Beaver Lk. Rd., Issaquah, WA 98029/206-392-8172
Robinett, R.G., P.O. Box 72, Madrid, IA 50156/515-795-2906
Robinson, Don, Pennsylvania Hse., 36 Fairfax Crescent, Southowram, Halifax, W. Yorkshire HX3 9SQ, ENGLAND/0422-364458
Robinson Firearms Mfg. Ltd., 1699 Blondeaux Crescent, Kelowna, B.C. CANADA V1Y 4J8/604-868-9596
Robinson H.V. Bullets, 3145 Church St., Zachary, LA 70791/504-654-4029
Rochester Lead Works, 76 Anderson Ave., Rochester, NY 14607/716-442-8500; FAX: 716-442-4712
Rockwood Corp., Speedwell Division, 136 Lincoln Blvd., Middlesex, NJ 08846/908-560-7171, 800-243-8274; FAX: 980-560-7475
Rocky Fork Enterprises, P.O. Box 427, 878 Battle Rd., Nolensville, TN 37135/615-941-1307
Rocky Mountain Arms, Inc., 600 S. Sunset, Unit C, Longmont, CO 80501/303-768-8522; FAX: 303-678-8766
Rocky Mountain High Sports Glasses, 8121 N. Central Park Ave., Skokie, IL 60076/847-679-1012, 800-323-1418; FAX: 847-679-0184
Rocky Mountain Rifle Works Ltd., 1707 14th St., Boulder, CO 80302/303-443-9189
Rocky Mountain Target Co., 3 Aloe Way, Leesburg, FL 34788/352-365-9598
Rocky Mountain Wildlife Products, P.O. Box 999, La Porte, CO 80535/970-484-2768; FAX: 970-484-0807
Rocky Shoes & Boots, 294 Harper St., Nelsonville, OH 45764/800-848-9452, 614-753-1951; FAX: 614-753-4024
Rod Guide Co., Box 1149, Forsyth, MO 65653/800-952-2774
Rodgers & Sons Ltd., Joseph (See George Ibberson (Sheffield) Ltd.)
Rogers Gunsmithing, Bob, P.O. Box 305, 344 S. Walnut St., Franklin Grove, IL 61031/815-456-2685; FAX: 815-288-7142
Rogue Rifle Co., Inc., P.O. Box 20, Prospect, OR 97536/541-560-4040; FAX: 541-560-4041
Rogue River Rifleworks, 1317 Spring St., Paso Robles, CA 93446/805-227-4706; FAX: 805-227-4723
Rohner, Hans, 1148 Twin Sisters Ranch Rd., Nederland, CO 80466-9600
Rohner, John, 710 Sunshine Canyon, Boulder, CO 80302/303-444-3841
Rolston, Fred W., Inc., 210 E. Cummins St., Tecumseh, MI 49286/517-423-6002, 800-314-9061 (orders only); FAX: 517-423-6002
Romain's Custom Guns, Inc., RD 1, Whetstone Rd., Brockport, PA 15823/814-265-1948

DIRECTORY OF THE ARMS TRADE

Rooster Laboratories, P.O. Box 412514, Kansas City, MO 64141/816-474-1622; FAX: 816-474-1307
Rorschach Precision Products, P.O. Box 151613, Irving, TX 75015/214-790-3487
Rosenberg & Sons, Jack A., 12229 Cox Ln., Dallas, TX 75234/214-241-6302
Rosenthal, Brad and Sallie, 19303 Ossenfort Ct., St. Louis, MO 63038/314-273-5159; FAX: 314-273-5149
Ross & Webb (See Ross, Don)
Ross, Don, 12813 West 83 Terrace, Lenexa, KS 66215/913-492-6982
Rosser, Bob, 1824 29th Ave., Suite 214, Birmingham, AL 35209/205-870-4422; FAX: 205-870-4421
Rossi S.A., Amadeo, Rua: Amadeo Rossi, 143, Sao Leopoldo, RS, BRAZIL 93030-220/051-592-5566 (U.S. importer—Interarms)
Roto Carve, 2754 Garden Ave., Janesville, IA 50647
Round Edge, Inc., P.O. Box 723, Lansdale, PA 19446/215-361-0859
Rowe Engineering, Inc. (See R.E.I.)
Royal Arms Gunstocks, 919 8th Ave. NW, Great Falls, MT 59404/406-453-1149
Roy's Custom Grips, Rt. 3, Box 174-E, Lynchburg, VA 24504/804-993-3470
RPM, 15481 N. Twin Lakes Dr., Tucson, AZ 85739/520-825-1233; FAX: 520-825-3333
Rubright Bullets, 1008 S. Quince Rd., Walnutport, PA 18088/215-767-1339
Rucker Dist. Inc., P.O. Box 479, Terrell, TX 75160/214-563-2094
Rudnicky, Susan, 9 Water St., Arcade, NY 14009/716-492-2450
Ruger (See Sturm, Ruger & Co., Inc.)
Rundell's Gun Shop, 6198 Frances Rd., Clio, MI 48420/313-687-0559
Robert P. Runge, 94 Grove St., Ilion, NY 13357/315-894-3036
Rupert's Gun Shop, 2202 Dick Rd., Suite B, Fenwick, MI 48834/517-248-3252
Russ Trading Post, 23 William St., Addison, NY 14801-1326/607-359-3896
Russell Knives, Inc., A.G., 1705 Hwy. 71B North, Springdale, AR 72764/501-751-7341
Rusteprufe Laboratories, 1319 Jefferson Ave., Sparta, WI 54656/608-269-4144
Rusty Duck Premium Gun Care Products, 7785 Foundation Dr., Suite 6, Florence, KY 41042/606-342-5553; FAX: 606-342-5556
Rutgers Book Center, 127 Raritan Ave., Highland Park, NJ 08904/908-545-4344; FAX: 908-545-6686
Rutten (See U.S. importer—Labanu, Inc.)
Ruvel & Co., Inc., 4128-30 W. Belmont Ave., Chicago, IL 60641/773-286-9494; FAX: 773-286-9323
R.V.I. (See Fire'n Five)
RWS (See U.S. importer—Dynamit Nobel-RWS, Inc.)
Ryan, Chad L., RR 3, Box 72, Cresco, IA 52136/319-547-4384
Rybka Custom Leather Equipment, Thad, 134 Havilah Hill, Odenville, AL 35120

S

S&B Industries, 11238 McKinley Rd., Montrose, MI 48457/810-639-5491
S&K Mfg. Co., P.O. Box 247, Pittsfield, PA 16340/814-563-7808; FAX: 814-563-4067
S&S Firearms, 74-11 Myrtle Ave., Glendale, NY 11385/718-497-1100; FAX: 718-497-1105
Sabatti S.r.l., via Alessandro Volta 90, 25063 Gardone V.T., Brescia, ITALY/030-8912207-831312; FAX: 030-8912059 (U.S. importer—E.A.A. Corp.)
SAECO (See Redding Reloading Equipment)
Saf-T-Lok, 5713 Corporate Way, Suite 100, W. Palm Beach, FL 33407
Safari Outfitters Ltd., 71 Ethan Allan Hwy., Ridgefield, CT 06877/203-544-9505
Safari Press, Inc., 15621 Chemical Lane B, Huntington Beach, CA 92649/714-894-9080; FAX: 714-894-4949
Safariland Ltd., Inc., 3120 E. Mission Blvd., P.O. Box 51478, Ontario, CA 91761/909-923-7300; FAX: 909-923-7400
SAFE, P.O. Box 864, Post Falls, ID 83854/208-773-3624
Safety Speed Holster, Inc., 910 S. Vail Ave., Montebello, CA 90640/213-723-4140; FAX: 213-726-6973
Sako Ltd. (See U.S. importer—Stoeger Industries)
Salter Calls, Inc., Eddie, Hwy. 31 South-Brewton Industrial Park, Brewton, AL 36426/205-867-2584; FAX: 206-867-9005
Samco Global Arms, Inc., 6995 NW 43rd St., Miami, FL 33166/305-593-9782
Sampson, Roger, 430 N. Grove, Mora, MN 55051/320-679-4868
San Francisco Gun Exchange, 124 Second St., San Francisco, CA 94105/415-982-6097
San Marco (See U.S. importers—Cape Outfitters; EMF Co., Inc.)
Sanders Custom Gun Service, 2358 Tyler Lane, Louisville, KY 40205/502-454-3338
Sanders Gun and Machine Shop, 145 Delhi Road, Manchester, IA 52057
Sandia Die & Cartridge Co., 37 Atancacio Rd. NE, Albuquerque, NM 87123/505-298-5729
Sarco, Inc., 323 Union St., Stirling, NJ, Stirling, NJ 07980/908-647-3800; FAX: 908-647-9413
S.A.R.L. G. Granger, 66 cours Fauriel, 42100 Saint Etienne, FRANCE/04 77 25 14 73; FAX: 04 77 38 66 99
Sauer (See U.S. importers—Paul Co., The; Sigarms, Inc.)
Saunders Gun & Machine Shop, R.R. 2, Delhi Road, Manchester, IA 52057
Savage Arms, Inc., 100 Springdale Rd., Westfield, MA 01085/413-568-7001; FAX: 413-562-7764
Savage Arms (Canada), Inc., 248 Water St., P.O. Box 1240, Lakefield, Ont. K0L 2H0, CANADA/705-652-8000; FAX: 705-652-8431

Savage Range Systems, Inc., 100 Springdale RD., Westfield, MA 01085/413-568-7001; FAX: 413-562-1152
Saville Iron Co. (See Greenwood Precision)
Savino, Barbara J., P.O. Box 1104, Hardwick, VT 05843-1104
Scanco Environmental Systems, 5000 Highlands Parkway, Suite 180, Atlanta, GA 30082/770-431-0025; FAX: 770-431-0028
Scansport, Inc., P.O. Box 700, Enfield, NH 03748/603-632-7654
Scattergun Technologies, Inc., 620 8th Ave. S., Nashville, TN 37203/615-254-1441; FAX: 615-254-1449
Sceery Game Calls, P.O. Box 6520, Sante Fe, NM 87502/505-471-9110; FAX: 505-471-3476
Schaefer Shooting Sports, P.O. Box 1515, Melville, NY 11747-0515/516-379-4900; FAX: 516-379-6701
Scharch Mfg., Inc., 10325 CR 120, Salida, CO 81201/719-539-7242, 800-836-4683; FAX: 719-539-3021
Scherer, Box 250, Ewing, VA 24240/615-733-2615; FAX: 615-733-2073
Schiffman, Curt, 3017 Kevin Cr., Idaho Falls, ID 83402/208-524-4684
Schiffman, Mike, 8233 S. Crystal Springs, McCammon, ID 83250/208-254-9114
Schiffman, Norman, 3017 Kevin Cr., Idaho Falls, ID 83402/208-524-4684
Schmidtke Group, 17050 W. Salentine Dr., New Berlin, WI 53151-7349
Schmidt & Bender, Inc., Brook Rd., P.O. Box 134, Meriden, NH 03770/603-469-3565, 800-468-3450; FAX: 603-469-3471
Schmidtman Custom Ammunition, 6 Gilbert Court, Cotati, CA 94931
Schneider Bullets, 3655 West 214th St., Fairview Park, OH 44126
Schneider Rifle Barrels, Inc., Gary, 12202 N. 62nd Pl., Scottsdale, AZ 85254/602-948-2525
School of Gunsmithing, The, 6065 Roswell Rd., Atlanta, GA 30328/800-223-4542
Schrimsher's Custom Knifemaker's Supply, Bob, P.O. Box 308, Emory, TX 75440/903-473-3330; FAX: 903-473-2235
Schroeder Bullets, 1421 Thermal Ave., San Diego, CA 92154/619-423-3523; FAX: 619-423-8124
Schuetzen Pistol Works, 620-626 Old Pacific Hwy. SE, Olympia, WA 98513/360-459-3471; FAX: 360-491-3447
Schulz Industries, 16247 Minnesota Ave., Paramount, CA 90723/213-439-5903
Schumakers Gun Shop, 512 Prouty Corner Lp. A, Colville, WA 99114/509-684-4848
Schwartz Custom Guns, David W., 2505 Waller St., Eau Claire, WI 54703/715-832-1735
Schwartz Custom Guns, Wayne E., 970 E. Britton Rd., Morrice, MI 48857/517-625-4079
Scobey Duck & Goose Calls, Glynn, Rt. 3, Box 37, Newbern, TN 38059/901-643-6241
Scope Control, Inc., 5775 Co. Rd. 23 SE, Alexandria, MN 56308/612-762-7295
ScopLevel, 151 Lindbergh Ave., Suite C, Livermore, CA 94550/510-449-5052; FAX: 510-373-0861
Score High Gunsmithing, 9812-A, Cochiti SE, Albuquerque, NM 87123/800-326-5632, 505-292-5532; FAX: 505-292-2592; E-MAIL: scorehi@rt66.com; WEB: http://www.rt66.com/~scorehi/home.htm
Scot Powder, Rt.1 Box 167, McEwen, TN 37101/800-416-3006; FAX: 615-729-4211
Scot Powder Co. of Ohio, Inc., Box GD96, Only, TN 37140/615-729-4207, 800-416-3006; FAX: 615-729-4217
Scott Fine Guns, Inc., Thad, P.O. Box 412, Indianola, MS 38751/601-887-5929
Scott, McDougall & Associates, 7950 Redwood Dr., Cotati, CA 94931/707-546-2264; FAX: 707-795-1911
Scott, Dwight, 23089 Englehardt St., Clair Shores, MI 48080/313-779-4735
S.C.R.C., P.O. Box 660, Katy, TX 77492-0660/FAX: 713-578-2124
Scruggs' Game Calls, Stanley, Rt. 1, Hwy. 661, Cullen, VA 23934/804-542-4241, 800-323-4828
Seattle Binocular & Scope Repair Co., P.O. Box 46094, Seattle, WA 98146/206-932-3733
Second Chance Body Armor, P.O. Box 578, Central Lake, MI 49622/616-544-5721; FAX: 616-544-9824
Security Awareness & Firearms Education (See SAFE)
Seebeck Assoc., R.E., P.O. Box 59752, Dallas, TX 75229
Seecamp Co., Inc., L.W., P.O. Box 255, New Haven, CT 06502/203-877-3429
Segway Industries, P.O. Box 783, Suffern, NY 10901-0783/914-357-5510
Seligman Shooting Products, Box 133, Seligman, AZ 86337/602-422-3607
Selsi Co., Inc., P.O. Box 10, Midland Park, NJ 07432-0010/201-935-0388; FAX: 201-935-5851
Semmer, Charles (See Remington Double Shotguns)
Sentinel Arms, P.O. Box 57, Detroit, MI 48231/313-331-1951; FAX: 313-331-1456
Serva Arms Co., Inc., RD 1, Box 483A, Greene, NY 13778/607-656-4764
Service Armament, 689 Bergen Blvd., Ridgefield, NJ 07657
Servus Footwear Co., 1136 2nd St., Rock Island, IL 61204-3610/309-786-7741; FAX: 309-786-9808
S.G.S. Sporting Guns Srl., Via Della Resistenza, 37, 20090 Buccinasco (MI) ITALY/2-45702446; FAX: 2-45702464
Shanghai Airguns, Ltd. (U.S. importer—Sportsman Airguns, Inc.)
Shappy Bullets, 76 Milldale Ave., Plantsville, CT 06479/203-621-3704
Shaw, Inc., E.R. (See Small Arms Mfg. Co.)
Sharp Shooter Supply, 4970 Lehman Road, Delphos, OH 45833/419-695-3179
C. Sharps Arms Co. Inc., 100 Centennial, Box 885, Big Timber, MT 59011/406-932-4353
Shay's Gunsmithing, 931 Marvin Ave., Lebanon, PA 17042
Sheffield Knifemakers Supply, Inc., P.O. Box 741107, Orange City, FL 32774-1107/904-775-6453; FAX: 904-774-5754

MANUFACTURERS' DIRECTORY

Shell Shack, 113 E. Main, Laurel, MT 59044/406-628-8986
Shepherd & Turpin Distributing Co., P.O. Box 40, Washington, UT 84780/801-635-2001
Shepherd Scope Ltd., Box 189, Waterloo, NE 68069/402-779-2424; FAX: 402-779-4010
Sheridan USA, Inc., Austin, P.O. Box 577, 36 Haddam Quarter Rd., Durham, CT 06422/203-349-1772; FAX: 203-349-1771
Sherwood, George, 46 N. River Dr., Roseburg, OR 97470/541-672-3159
Shilen, Inc., 205 Metro Park Blvd., Ennis, TX 75119/972-875-5318; FAX: 972-875-5402
Shiloh Creek, Box 357, Cottleville, MO 63338/314-925-1842; FAX: 314-925-1842
Shiloh Rifle Mfg., 201 Centennial Dr., Big Timber, MT 59011/406-932-4454; FAX: 406-932-5627
Shockley, Harold H., 204 E. Farmington Rd., Hanna City, IL 61536/309-565-4524
Shoemaker & Sons, Inc., Tex, 714 W. Cienega Ave., San Dimas, CA 91773/909-592-2071; FAX: 909-592-2378
Shooten' Haus, The, 102 W. 13th, Kearney, NE 68847/308-236-7929
Shooter Shop, The, 221 N. Main, Butte, MT 59701/406-723-3842
Shooter's Choice, 16770 Hilltop Park Place, Chagrin Falls, OH 44023/216-543-8808; FAX: 216-543-8811
Shooter's Edge, Inc., P.O.Box 769, Trinidad, CO 81082
Shooter's World, 3828 N. 28th Ave., Phoenix, AZ 85017/602-266-0170
Shooters Supply, 1120 Tieton Dr., Yakima, WA 98902/509-452-1181
Shootin' Accessories, Ltd., P.O. Box 6810, Auburn, CA 95604/916-889-2220
Shootin' Shack, Inc., 1065 Silver Beach Rd., Riviera Beach, FL 33403/561-842-0990
Shooting Chrony, Inc., 3269 Niagara Falls Blvd., N. Tonawanda, NY 14120/905-276-6292; FAX: 416-276-6295
Shooting Components Marketing, P.O. Box 1069, Englewood, CO 80150/303-987-2543; FAX: 303-989-3508
Shooting Gallery, The, 8070 Southern Blvd., Boardman, OH 44512/216-726-7788
Shooting Specialties (See Titus, Daniel)
Shooting Star, 1825 Fortview Rd., Ste. 115, Austin, TX 78747/512-462-0009
Shoot-N-C Targets (See Birchwood Casey)
Shotgun Shop, The, 14145 Proctor Ave., Suite 3, Industry, CA 91746/818-855-2737; FAX: 818-855-2735
Shotguns Unlimited, 2307 Fon Du Lac Rd., Richmond, VA 23229/804-752-7115
ShurKatch Corporation, 50 Elm St., Richfield Springs, NY 13439/315-858-1470; FAX: 315-858-2969
S.I.A.C.E. (See U.S. importer—IAR, Inc.)
Siegrist Gun Shop, 8754 Turtle Road, Whittemore, MI 48770
Sierra Bullets, 1400 W. Henry St., Sedalia, MO 65301/816-827-6300; FAX: 816-827-6300; WEB: http://www.sierrabullets.com
Sierra Specialty Prod. Co., 1344 Oakhurst Ave., Los Altos, CA 94024/FAX: 415-965-1536
SIG, CH-8212 Neuhausen, SWITZERLAND/ (U.S. importer—Mandall Shooting Supplies, Inc.)
Sigarms, Inc., Corporate Park, Exeter, NH 03833/603-772-2302; FAX: 603-772-9082
SIG-Sauer (See U.S. importer—Sigarms, Inc.)
Sight Shop, The, John G. Lawson, 1802 E. Columbia Ave., Tacoma, WA 98404/206-474-5465
Sightron, Inc., 1672B Hwy. 96, Franklinton, NC 27525/919-528-8783; FAX: 919-528-0995
Signet Metal Corp., 551 Stewart Ave., Brooklyn, NY 11222/718-384-5400; FAX: 718-388-7488
Sile Distributors, Inc., 7 Centre Market Pl., New York, NY 10013/212-925-4111; FAX: 212-925-3149
Silencio/Safety Direct, 56 Coney Island Dr., Sparks, NV 89431/800-648-1812, 702-354-4451; FAX: 702-359-1074
Silent Hunter, 1100 Newton Ave., W. Collingswood, NJ 08107/609-854-3276
Silhouette Leathers, P.O. Box 1161, Gunnison, CO 81230/303-641-6639
Silhouette, The, P.O. Box 1509, Idaho Falls, ID 83403
Silver Eagle Machining, 18007 N. 69th Ave., Glendale, AZ 85308
Silver Ridge Gun Shop (See Goodwin, Fred)
Silver-Tip Corp., RR2, Box 184, Gloster, MS 39638-9520
Simmons, Jerry, 715 Middlebury St., Goshen, IN 46526/219-533-8546
Simmons Enterprises, Ernie, 709 East Elizabethtown Rd., Manheim, PA 17545/717-664-4040
Simmons Gun Repair, Inc., 700 S. Rogers Rd., Olathe, KS 66062/913-782-3131; FAX: 913-782-4189
Simmons Outdoor Corp., 201 Plantation Oak Parkway, Thomasville, GA 31792/912-227-9053; FAX: 912-227-9054
Sinclair International, Inc., 2330 Wayne Haven St., Fort Wayne, IN 46803/219-493-1858; FAX: 219-493-2530
Singletary, Kent, 2915 W. Ross, Phoenix, AZ 85027/602-582-4900
Sipes Gun Shop, 7415 Asher Ave., Little Rock, AR 72204/501-565-8480
Siskiyou Gun Works (See Donnelly, C.P.)
Six Enterprises, 320-D Turtle Creek Ct., San Jose, CA 95125/408-999-0201; FAX: 408-999-0216
SKAN A.R., 4 St. Catherines Road, Long Melford, Suffolk, CO10 9JU ENGLAND/011-0787-312942
SKB Arms Co. (See New SKB Arms Co.)
SKB Shotguns, 4325 S. 120th St., P.O. Box 37669, Omaha, NE 68137/800-752-2767; FAX: 402-330-8029
Skeoch, Brian R., P.O. Box 279, Glenrock, WY 82637/307-436-9655; FAX: 307-436-9034
Skip's Machine, 364 29 Road, Grand Junction, CO 81501/303-245-5417
Sklany's Machine Shop, 566 Birch Grove Dr., Kalispell, MT 59901/406-755-4257
SKR Industries, POB 1382, San Angelo, TX 76902/915-658-3133
S.L.A.P. Industries, P.O. Box 1121, Parklands 2121, SOUTH AFRICA/27-11-788-0030; FAX: 27-11-788-0030
Slezak, Jerome F., 1290 Marlowe, Lakewood (Cleveland), OH 44107/216-221-1668
Slings 'N Things, Inc., 8909 Bedford Circle, Suite 11, Omaha, NE 68134/402-571-6954; FAX: 402-571-7082
Slug Group, Inc., P.O. Box 376, New Paris, PA 15554/814-839-4517; FAX: 814-839-2601
Slug Site, Ozark Wilds, Rt. 2, Box 158, Versailles, MO 65084/573-378-6430
Small Arms Mfg. Co., 5312 Thoms Run Rd., Bridgeville, PA 15017/412-221-4343; FAX: 412-221-4303
Small Arms Specialties, 29 Bernice Ave., Leominster, MA 01453/800-635-9290
Small Custom Mould & Bullet Co., Box 17211, Tucson, AZ 85731
Smart Parts, 1203 Spring St., Latrobe, PA 15650/412-539-2660; FAX: 412-539-2298
Smires, C.L., 5222 Windmill Lane, Columbia, MD 21044-1328
Smith & Wesson, 2100 Roosevelt Ave., Springfield, MA 01102/413-781-8300; FAX: 413-731-8980
Smith, Art, 230 Main St. S., Hector, MN 55342/320-848-2760; FAX: 320-848-2760
Smith, Mark A., P.O. Box 182, Sinclair, WY 82334/307-324-7929
Smith, Michael, 620 Nye Circle, Chattanooga, TN 37405/615-267-8341
Smith, Ron, 5869 Straley, Ft. Worth, TX 76114/817-732-6768
Smith, Sharmon, 4545 Speas Rd., Fruitland, ID 83619/208-452-6329
Smith Abrasives, Inc., 1700 Sleepy Valley Rd., P.O. Box 5095, Hot Springs, AR 71902-5095/501-321-2244; FAX: 501-321-9232
Smith Saddlery, Jesse W., 16909 E. Jackson Road, Elk, WA 99009-9600/509-325-0622
Smokey Valley Rifles (See Lutz Engraving, Ron E.)
Snapp's Gunshop, 6911 E. Washington Rd., Clare, MI 48617/517-386-9226
Snider Stocks, Walter S., Rt. 2 P.O. Box 147, Denton, NC 27239
Sno-Seal (See Atsko/Sno-Seal)
Societa Armi Bresciane Srl. (See U.S. importer—Cape Outfitters; Gamba, USA)
Sonora Rifle Barrel Co., 14396 D. Tuolumne Rd., Sonora, CA 95370/209-532-4139
Soque River Knives, P.O. Box 880, Clarkesville, GA 30523/706-754-8500; FAX: 706-754-7263
SOS Products Co. (See Buck Stix—SOS Products Co.)
Sotheby's, 1334 York Ave. at 72nd St., New York, NY 10021/212-606-7260
Sound Technology, Box 391, Pelham, AL 35124/205-664-5860; Summer phone: 907-486-2825
South Bend Replicas, Inc., 61650 Oak Rd., South Bend, IN 46614/219-289-4500
Southeastern Community College, 1015 S. Gear Ave., West Burlington, IA 52655/319-752-2731
Southern Ammunition Co., Inc., 4232 Meadow St., Loris, SC 29569-3124/803-756-3262; FAX: 803-756-3583
Southern Armory, The, 25 Millstone Road, Woodlawn, VA 24381/703-238-1343; FAX: 703-238-1453
Southern Bloomer Mfg. Co., P.O. Box 1621, Bristol, TN 37620/615-878-6660; FAX: 615-878-8761
Southern Security, 1700 Oak Hills Dr., Kingston, TN 37763/423-376-6297; 800-251-9992
Southwind Sanctions, P.O. Box 445, Aledo, TX 76008/817-441-8917
Sparks, Milt, 605 E. 44th St. No. 2, Boise, ID 83714-4800
Spartan-Realtree Products, Inc., 1390 Box Circle, Columbus, GA 31907/706-569-9101; FAX: 706-569-0042
Specialty Gunsmithing, Lynn McMurdo, P.O. Box 404, Afton, WY 83110/307-886-5535
Specialty Shooters Supply, Inc., 3325 Griffin Rd., Suite 9mm, Fort Lauderdale, FL 33317
Speedfeed, Inc., 3820 Industrial Way, Suite N, Benicia, CA 94510/707-746-1221; FAX: 707-746-1888
Speer Products, Div. of Blount, Inc., Sporting Equipment Div., P.O. Box 856, Lewiston, ID 83501/208-746-2351; FAX: 208-746-2915
Spegel, Craig, P.O. Box 3108, Bay City, OR 97107/503-377-2697
Speiser, Fred D., 2229 Dearborn, Missoula, MT 59801/406-549-8133
Spence, George W., 115 Locust St., Steele, MO 63877/314-695-4926
Spencer Reblue Service, 1820 Tupelo Trail, Holt, MI 48842/517-694-7474
Spencer's Custom Guns, Rt. 1, Box 546, Scottsville, VA 24590/804-293-6836
Spezial Waffen (See U.S. importer—American Bullets)
SPG, Inc., P.O. Box 761, Livingston, MT 59047/406-222-8416; FAX: 406-222-8416
Sphinx Engineering SA, Ch. des Grandes-Vies 2, CH-2900 Porrentruy, SWITZERLAND/41 66 66 73 81; FAX: 41 66 66 30 90 (U.S. importer—Sphinx USA Inc.)
Sphinx USA Inc., 998 N. Colony Rd., Meriden, CT 06450/203-238-1399; FAX: 203-238-1375
Spokhandguns, Inc., 1206 Fig St., Benton City, WA 99320/509-588-5255

DIRECTORY OF THE ARMS TRADE

Sport Flite Manufacturing Co., P.O. Box 1082, Bloomfield Hills, MI 48303/810-647-3747
Sporting Arms Mfg., Inc., 801 Hall Ave., Littlefield, TX 79339/806-385-5665; FAX: 806-385-3394
Sports Innovations, Inc., P.O. Box 5181, 8505 Jacksboro Hwy., Wichita Falls, TX 76307/817-723-6015
Sportsman Safe Mfg. Co., 6309-6311 Paramount Blvd., Long Beach, CA 90805/800-266-7150, 310-984-5445
Sportsman Supply Co., 714 East Eastwood, P.O. Box 650, Marshall, MO 65340/816-886-9393
Sportsman's Communicators, 588 Radcliffe Ave., Pacific Palisades, CA 90272/800-538-3752
Sportsmatch U.K. Ltd., 16 Summer St., Leighton Buzzard, Bedfordshire, LU7 8HT ENGLAND/01525-381638; FAX: 01525-851236
Sportsmen's Exchange & Western Gun Traders, Inc., 560 S. "C" St., Oxnard, CA 93030/805-483-1917
Spradlin's, 113 Arthur St., Pueblo, CO 81004/719-543-9462; FAX: 719-543-9465
Springfield, Inc., 420 W. Main St., Geneseo, IL 61254/309-944-5631; FAX: 309-944-3676
Springfield Sporters, Inc., RD 1, Penn Run, PA 15765/412-254-2626; FAX: 412-254-9173
Spyderco, Inc., 4565 N. Hwy. 93, P.O. Box 800, Golden, CO 80403/303-279-8383, 800-525-7770; FAX: 303-278-2229
SSK Industries, 721 Woodvue Lane, Wintersville, OH 43952/614-264-0176; FAX: 614-264-2257
Stackpole Books, 5067 Ritter Rd., Mechanicsburg, PA 17055-6921/717-796-0411; FAX: 717-796-0412
Stalker, Inc., P.O. Box 21, Fishermans Wharf Rd., Malakoff, TX 75148/903-489-1010
Stalwart Corporation, 76 Imperial, Unit A, Evanston, WY 82930/307-789-7687; FAX: 307-789-7688
Stanley Bullets, 2085 Heatheridge Ln., Reno, NV 89509
Star Ammunition, Inc., 5520 Rock Hampton Ct., Indianapolis, IN 46268/317-872-5840, 800-221-5927; FAX: 317-872-5847
Star Bonifacio Echeverria S.A., Torrekva 3, Eibar, SPAIN 20600/43-107340; FAX: 43-101524 (U.S. importer—E.A.A. Corp.; Interarms; P.S.M.G. Gun Co.)
Star Custom Bullets, P.O. Box 608, 468 Main St., Ferndale, CA 95536/707-786-9140; FAX: 707-786-9117
Star Machine Works, 418 10th Ave., San Diego, CA 92101/619-232-3216
Star Master-Match Bullets (See Star Ammunition, Inc.)
Star Reloading Co., Inc. (See Star Ammunition, Inc.)
Starke Bullet Company, P.O. Box 400, 605 6th St. NW, Cooperstown, ND 58425/888-797-3431
Starkey Labs, 6700 Washington Ave. S., Eden Prairie, MN 55344
Starkey's Gun Shop, 9430 McCombs, El Paso, TX 79924/915-751-3030
Stark's Bullet Mfg., 2580 Monroe St., Eugene, OR 97405
Starline, 1300 W. Henry St., Sedalia, MO 65301/816-827-6640; FAX: 816-827-6650
Starlight Training Center, Inc., Rt. 1, P.O. Box 88, Bronaugh, MO 64728/417-843-3555
Starnes Gunmaker, Ken, 32900 SW Laurelview Rd., Hillsboro, OR 97123/503-628-0705; FAX: 503-628-6005
Starr Trading Co., Jedediah, P.O. Box 2007, Farmington Hills, MI 48333/810-683-4343; FAX: 810-683-3282
Starrett Co., L.S., 121 Crescent St., Athol, MA 01331/617-249-3551
State Arms Gun Co., 815 S. Division St., Waunakee, WI 53597/608-849-5800
Steelman's Gun Shop, 10465 Beers Rd., Swartz Creek, MI 48473/810-735-4884
Steiner (See Pioneer Research, Inc.)
Steffens, Ron, 18396 Mariposa Creek Rd., Willits, CA 95490/707-485-0873
Stegall, James B., 26 Forest Rd., Wallkill, NY 12589
Steger, James R., 1131 Dorsey Pl., Plainfield, NJ 07062
Steves House of Guns, Rt. 1, Minnesota City, MN 55959/507-689-2573
Stewart Game Calls, Inc., Johnny, P.O. Box 7954, 5100 Fort Ave., Waco, TX 76714/817-772-3261; FAX: 817-772-3670
Stewart's Gunsmithing, P.O. Box 5854, Pietersburg North 0750, Transvaal, SOUTH AFRICA/01521-89401
Steyr Mannlicher AG & CO KG, Mannlicherstrasse 1, A-4400 Steyr, AUSTRIA /0043-7252-78621; FAX: 0043-7252-68621 (U.S. importer—GSI, Inc.; Nygord Precision Products)
STI International, 114 Halmar Cove, Georgetown, TX 78628/800-959-8201; FAX: 512-819-0465
Stiles Custom Guns, RD3, Box 1605, Homer City, PA 15748/412-479-9945, 412-479-8666
Stillwell, Robert, 421 Judith Ann Dr., Schertz, TX 78154
Stoeger Industries, 5 Mansard Ct., Wayne, NJ 07470/201-872-9500, 800-631-0722; FAX: 201-872-2230
Stoeger Publishing Co. (See Stoeger Industries)
Stone Enterprises Ltd., Rt. 609, P.O. Box 335, Wicomico Church, VA 22579/804-580-5114; FAX: 804-580-8421
Stone Mountain Arms, 5988 Peachtree Corners E., Norcross, GA 30071/800-251-9412
Stoney Baroque Shooters Supply, John Richards, Rt. 2, Box 325, Bedford, KY 40006/502-255-7222
Stoney Point Products, Inc., P.O. Box 234, 1815 North Spring Street, New Ulm, MN 56073-0234/507-354-3360; FAX: 507-354-7236
Storage Tech, 1254 Morris Ave., N. Huntingdon, PA 15642/800-437-9393
Storey, Dale A. (See DGS, Inc.)
Storm, Gary, P.O. Box 5211, Richardson, TX 75083/214-385-0862
Stott's Creek Armory, Inc., 2526 S. 475W, Morgantown, IN 46160/317-878-5489
Stott's Creek Printers, 2526 S. 475W, Morgantown, IN 46160/317-878-5489
Stratco, Inc., P.O. Box 2270, Kalispell, MT 59901/406-755-1221; FAX: 406-755-1226
Strawbridge, Victor W., 6 Pineview Dr., Dover, NH 03820/603-742-0013
Streamlight, Inc., 1030 W. Germantown Pike, Norristown, PA 19403/215-631-0600; FAX: 610-631-0712
Strong Holster Co., 39 Grove St., Gloucester, MA 01930/508-281-3300; FAX: 508-281-6321
Strutz Rifle Barrels, Inc., W.C., P.O. Box 611, Eagle River, WI 54521/715-479-4766
Stuart, V. Pat, Rt.1, Box 447-S, Greenville, VA 24440/804-556-3845
Sturgeon Valley Sporters, K. Ide, P.O. Box 283, Vanderbilt, MI 49795/517-983-4338
Sturm, Ruger & Co., Inc., 200 Ruger Rd., Prescott, AZ 86301/520-541-8820; FAX: 520-541-8850
"Su-Press-On," Inc., P.O. Box 09161, Detroit, MI 48209/313-842-4222 7:30-11p.m. Mon-Thurs.
Sullivan, David S. (See Westwind Rifles, Inc.)
Summit Specialties, Inc., P.O. Box 786, Decatur, AL 35602/205-353-0634; FAX: 205-353-9818
Sundance Industries, Inc., 25163 W. Avenue Stanford, Valencia, CA 91355/805-257-4807
Sunny Hill Enterprizes, Inc., W1790 Cty. HHH, Malone, WI 53049/414-795-4822
Sun Welding Safe Co., 290 Easy St. No.3, Simi Valley, CA 93065/805-584-6678, 800-729-SAFE; FAX: 805-584-6169
Surecase Co., The, 233 Wilshire Blvd., Ste. 900, Santa Monica, CA 90401/800-92ARMLOC
Sure-Shot Game Calls, Inc., P.O. Box 816, 6835 Capitol, Groves, TX 77619/409-962-1636; FAX: 409-962-5465
Survival Arms, Inc., P.O. Box 965, Orange, CT 06477/203-924-6533; FAX: 203-924-2581
Svon Corp., 280 Eliot St., Ashland, MA 01721/508-881-8852
Swampfire Shop, The (See Peterson Gun Shop, Inc., A.W.)
Swann, D.J., 5 Orsova Close, Eltham North, Vic. 3095, AUSTRALIA/03-431-0323
Swanndri New Zealand, 152 Elm Ave., Burlingame, CA 94010/415-347-6158
SwaroSports, Inc. (See JägerSport, Ltd.)
Swarovski Optik North America Ltd., One Wholesale Way, Cranston, RI 02920/401-946-2220, 800-426-3089; FAX: 401-946-2587
Sweet Home, Inc., P.O. Box 900, Orrville, OH 44667-0900
Swenson's 45 Shop, A.D., P.O. Box 606, Fallbrook, CA 92028
Swift Bullet Co., P.O. Box 27, 201 Main St., Quinter, KS 67752/913-754-3959; FAX: 913-754-2359
Swift Instruments, Inc., 952 Dorchester Ave., Boston, MA 02125/617-436-2960; FAX: 617-436-3232
Swift River Gunworks, 450 State St., Belchertown, MA 01007/413-323-4052
Swiss Army Knives, Inc., 151 Long Hill Crossroads, 37 Canal St., Shelton, CT 06484/800-243-4032
Swivel Machine Works, Inc., 11 Monitor Hill Rd., Newtown, CT 06470/203-270-6343
Szweda, Robert (See RMS Custom Gunsmithing)

T

Tabler Marketing, 2554 Lincoln Blvd., Suite 555, Marina Del Rey, CA 90291/818-755-4565; FAX: 818-755-0972
TacStar Industries, Inc., 218 Justin Drive, P.O. Box 70, Cottonwood, AZ 86326/602-639-0072; FAX: 602-634-8781
TacTell, Inc., P.O. Box 5654, Maryville, TN 37802/615-982-7855; FAX: 615-558-8294
Tactical Defense Institute, 574 Miami Bluff Ct., Loveland, OH 45140/513-677-8229
Talbot QD Mounts, 2210 E. Grand Blanc Rd., Grand Blanc, MI 48439-8113/810-695-2497
Talley, Dave, P.O. Box 821, Glenrock, WY 82637/307-436-8724, 307-436-9315
Talmage, William G., 10208 N. County Rd. 425 W., Brazil, IN 47834/812-442-0804
Talon Mfg. Co., Inc., 621 W. King St., Martinsburg, WV 25401/304-264-9714; FAX: 304-264-9725
Tamarack Products, Inc., P.O. Box 625, Wauconda, IL 60084/708-526-9333; FAX: 708-526-9353
Tanfoglio Fratelli S.r.l., via Valtrompia 39, 41, 25068 Gardone V.T., Brescia, ITALY/30-8910361; FAX: 30-8910183 (U.S. importer—E.A.A. Corp.)
Tanglefree Industries, 1261 Heavenly Dr., Martinez, CA 94553/800-982-4868; FAX: 510-825-3874
Tank's Rifle Shop, P.O. Box 474, Fremont, NE 68026-0474/402-727-1317; FAX: 402-721-2573
Tanner (See U.S. importer—Mandall Shooting Supplies, Inc.)
Taracorp Industries, Inc., 1200 Sixteenth St., Granite City, IL 62040/618-451-4400

MANUFACTURERS' DIRECTORY

Tar-Hunt Custom Rifles, Inc., RR3, P.O. Box 572, Bloomsburg, PA 17815-9351/717-784-6368; FAX: 717-784-6368
Tarnhelm Supply Co., Inc., 431 High St., Boscawen, NH 03303/603-796-2551; FAX: 603-796-2918
Tasco Sales, Inc., 7600 NW 26th St., Miami, FL 33122-1494/305-591-3670; FAX: 305-592-5895
Taurus Firearms, Inc., 16175 NW 49th Ave., Miami, FL 33014/305-624-1115; FAX: 305-623-7506
Taurus International Firearms (See U.S. importer—Taurus Firearms, Inc.)
Taurus S.A., Forjas, Avenida Do Forte 511, Porto Alegre, RS BRAZIL 91360/55-51-347-4050; FAX: 55-51-347-3065
Taylor & Robbins, P.O. Box 164, Rixford, PA 16745/814-966-3233
Taylor's & Co., Inc., 304 Lenoir Dr., Winchester, VA 22603/540-722-2017; FAX: 540-722-2018
TCCI, P.O. Box 302, Phoenix, AZ 85001/602-237-3823; FAX: 602-237-3858
TCSR, 3998 Hoffman Rd., White Bear Lake, MN 55110-4626/800-328-5323; FAX: 612-429-0526
TDP Industries, Inc., 606 Airport Blvd., Doylestown, PA 18901/215-345-8687; FAX: 215-345-6057
Techni-Mec (See F.A.I.R. Techni-Mec s.n.c. di Isidoro Rizzini & C.)
Techno Arms (See U.S. importer—Auto-Ordnance Corp.)
Tecnolegno S.p.A., Via A. Locatelli, 6, 10, 24019 Zogno, ITALY/0345-55111; FAX: 0345-55155
Tele-Optics, 5514 W. Lawrence Ave., Chicago, IL 60630/773-283-7757; FAX: 773-283-7757
Ten-Ring Precision, Inc., Alex B. Hamilton, 1449 Blue Crest Lane, San Antonio, TX 78232/210-494-3063; FAX: 210-494-3066
10-X Products Group, 2915 Lyndon B. Johnson Freeway, Suite 133, Dallas, TX 75234/972-243-4016, 800-433-2225; FAX: 972-243-4112
Tennessee Valley Mfg., P.O. Box 1175, Corinth, MS 38834/601-286-5014
Tepeco, P.O. Box 342, Friendswood, TX 77546/713-482-2702
Testing Systems, Inc., 220 Pegasus Ave., Northvale, NJ 07647
Teton Arms, Inc., P.O. Box 411, Wilson, WY 83014/307-733-3395
Tetra Gun Lubricants (See FTI, Inc.)
Texas Armory (See Bond Arms, Inc.)
Texas Longhorn Arms, Inc., 5959 W. Loop South, Suite 424, Bellaire, TX 77401/713-660-6323; FAX: 713-660-0493
Texas Platers Supply Co., 2453 W. Five Mile Parkway, Dallas, TX 75233/214-330-7168
T.F.C. S.p.A., Via G. Marconi 118, B, Villa Carcina, Brescia 25069, ITALY/030-881271; FAX: 030-881826
Theis, Terry, P.O. Box 535, Fredericksburg, TX 78624/210-997-6778
Theoben Engineering, Stephenson Road, St. Ives, Huntingdon, Cambs., PE17 4WJ ENGLAND/011-0480-461718
Thiewes, George W., 14329 W. Parada Dr., Sun City West, AZ 85375
Things Unlimited, 235 N. Kimbau, Casper, WY 82601/307-234-5277
Thirion Gun Engraving, Denise, P.O. Box 408, Graton, CA 95444/707-829-1876
Thomas, Charles C., 2600 S. First St., Springfield, IL 62794/217-789-8980; FAX: 217-789-9130
Thompson, Norm, 18905 NW Thurman St., Portland, OR 97209
Thompson, Randall (See Highline Machine Co.)
Thompson Bullet Lube Co., P.O. Box 472343, Garland, TX 75047-2343/972-271-8063; FAX: 972-840-6743
Thompson/Center Arms, P.O. Box 5002, Rochester, NH 03866/603-332-2394; FAX: 603-332-5133
Thompson Precision, 110 Mary St., P.O. Box 251, Warren, IL 61087/815-745-3625
Thompson Target Technology, 618 Roslyn Ave., SW, Canton, OH 44710/216-453-7707; FAX: 216-478-4723
Thompson Tool Mount (See TTM)
3-D Ammunition & Bullets, 112 W. Plum St., P.O. Box J, Doniphan, NE 68832/402-845-2285, 800-255-6712; FAX: 402-845-6546
300° Below Services (See Cryo-Accurizing)
300 Gunsmith Service, Inc., at Cherry Creek State Park Shooting Center, 12500 E. Belleview Ave./Englewood, CO 80111 303-690-3300
3-Ten Corp., P.O. Box 269, Feeding Hills, MA 01030/413-789-2086; FAX: 413-789-1549
T.H.U. Enterprises, Inc., P.O. Box 418, Lederach, PA 19450/215-256-1665; FAX: 215-256-9718
Thunder Mountain Arms, P.O. Box 593, Oak Harbor, WA 98277/206-679-4657; FAX: 206-675-1114
Thunderbird Cartridge Co., Inc. (See TCCI)
Thurston Sports, Inc., RD 3 Donovan Rd., Auburn, NY 13021/315-253-0966
Tiger-Hunt, Box 379, Beaverdale, PA 15921/814-472-5161
Tikka (See U.S. importer—Stoeger Industries)
Timber Heirloom Products, 618 Roslyn Ave. SW, Canton, OH 44710/216-453-7707; FAX: 216-478-4723
Time Precision, Inc., 640 Federal Rd., Brookfield, CT 06804/203-775-8343
Timney Mfg., Inc., 3065 W. Fairmont Ave., Phoenix, AZ 85017/602-274-2999; FAX: 602-241-0361
Tink's Safariland Hunting Corp., P.O. Box 244, 1140 Monticello Rd., Madison, GA 30650/706-342-4915; FAX: 706-342-7568
Tinks & Ben Lee Hunting Products (See Wellington Outdoors)
Tioga Engineering Co., Inc., P.O. Box 913, 13 Cone St., Wellsboro, PA 16901/717-724-3533, 717-662-3347

Tippman Pneumatics, Inc., 3518 Adams Center Rd., Fort Wayne, IN 46806/219-749-6022; FAX: 219-749-6619
Tirelli, Snc Di Tirelli Primo E.C., Via Matteotti No. 359, Gardone V.T., Brescia, ITALY 25063/030-8912819; FAX: 030-832240
Titus, Daniel, Shooting Specialties, 119 Morlyn Ave., Bryn Mawr, PA 19010-3737/215-525-8829
TMI Products (See Haselbauer Products, Jerry)
TM Stockworks, 6355 Maplecrest Rd., Fort Wayne, IN 46835/219-485-5389
Tom's Gun Repair, Thomas G. Ivanoff, 76-6 Rt. Southfork Rd., Cody, WY 82414/307-587-6949
Tom's Gunshop, 3601 Central Ave., Hot Springs, AR 71913/501-624-3856
Tomboy, Inc., P.O. Box 846, Dallas, OR 97338/503-623-8405
Tombstone Smoke`n'Deals, 3218 East Bell Road, Phoenix, AZ 85032/602-905-7013; Fax: 602-443-1998
Tonoloway Tack Drives, HCR 81, Box 100, Needmore, PA 17238
Tooley Custom Rifles, 516 Creek Meadow Dr., Gastonia, NC 28054/704-864-7525
Top-Line USA, Inc., 7920-28 Hamilton Ave., Cincinnati, OH 45231/513-522-2992, 800-346-6699; FAX: 513-522-0916
Torel, Inc., 1708 N. South St., P.O. Box 592, Yoakum, TX 77995/512-293-2341; FAX: 512-293-3413
Totally Dependable Products (See TDP Industries, Inc.)
TOZ (See U.S. importer—Nygord Precision Products)
TR Metals Corp., 1 Pavilion Ave., Riverside, NJ 08075/609-461-9000; FAX: 609-764-6340
Track of the Wolf, Inc., P.O. Box 6, Osseo, MN 55369-0006/612-424-2500; FAX: 612-424-9860
TracStar Industries, Inc., 218 Justin Dr., Cottonwood, AZ 86326/520-639-0072; FAX: 520-634-8781
Tradewinds, Inc., P.O. Box 1191, 2339-41 Tacoma Ave. S., Tacoma, WA 98401/206-272-4887
Traditions, Inc., P.O. Box 776, 1375 Boston Post Rd., Old Saybrook, CT 06475/860-388-4656; FAX: 860-388-4657
Trafalgar Square, P.O. Box 257, N. Pomfret, VT 05053/802-457-1911
Traft Gunshop, P.O. Box 1078, Buena Vista, CO 81211
TrailTimer Co., 1992-A Suburban Ave., P.O. Box 19722, St. Paul, MN 55119/612-738-0925
Trail Visions, 5800 N. Ames Terrace, Glendale, WI 53209/414-228-1328
Trammco, 839 Gold Run Rd., Boulder, CO 80302
Trappers Trading, P.O. Box 26946, Austin, TX 78755/800-788-9334
Trax America, Inc., P.O. Box 898, 1150 Eldridge, Forrest City, AR 72335/501-633-0410, 800-232-2327; FAX: 501-633-4788
Treadlok Gun Safe, Inc., 1764 Granby St. NE, Roanoke, VA 24012/800-729-8732, 703-982-6881; FAX: 703-982-1059
Treemaster, P.O. Box 247, Guntersville, AL 35976/205-878-3597
Treso, Inc., P.O. Box 4640, Pagosa Springs, CO 81157/303-731-2295
Trevallion Gunstocks, 9 Old Mountain Rd., Cape Neddick, ME 03902/207-361-1130
de Treville & Co., Stan, 4129 Normal St., San Diego, CA 92103/619-298-3393
Trico Plastics, 590 S. Vincent Ave., Azusa, CA 91702
Trigger Lock Division/Central Specialties Ltd., 1122 Silver Lake Road, Cary, IL 60013/847-639-3900; FAX: 847-639-3972
Trijicon, Inc., 49385 Shafer Ave., P.O. Box 930059, Wixom, MI 48393-0059/810-960-7700; FAX: 810-960-7725
Trilux Inc., P.O. Box 24608, Winston-Salem, NC 27114/910-659-9438; FAX: 910-768-7720
Trinidad State Junior College, Gunsmithing Dept., 600 Prospect St., Trinidad, CO 81082/719-846-5631; FAX: 719-846-5667
Triple-K Mfg. Co., Inc., 2222 Commercial St., San Diego, CA 92113/619-232-2066; FAX: 619-232-7675
Tristar Sporting Arms, Ltd., 1814-16 Linn St., P.O. Box 7496, N. Kansas City, MO 64116/816-421-1400; FAX: 816-421-4182
Trius Products, Inc., P.O. Box 25, 221 S. Miami Ave., Cleves, OH 45002/513-941-5682; FAX: 513-941-7970
Trooper Walsh, 2393 N. Edgewood St., Arlington, VA 22207
Trophy Bonded Bullets, Inc., 900 S. Loop W., Suite 190, Houston, TX 77054/713-645-4499, 888-308-3006; FAX: 713-741-6393
Trotman, Ken, 135 Ditton Walk, Unit 11, Cambridge CB5 8PY, ENGLAND/01223-211030; FAX: 01223-212317
Tru-Balance Knife Co., P.O. Box 140555, Grand Rapids, MI 49514/616-453-3679
Tru-Square Metal Prods., Inc., 640 First St. SW, P.O. Box 585, Auburn, WA 98071/206-833-2310; FAX: 206-833-2349
True Flight Bullet Co., 5581 Roosevelt St., Whitehall, PA 18052/610-262-7630; FAX: 610-262-7806
Trulock Tool, Broad St., Whigham, GA 31797/912-762-4678
TTM, 1550 Solomon Rd., Santa Maria, CA 93455/805-934-1281
Tucker, James C., P.O. Box 575, Raymond, NH 03077
Tucson Mold, Inc., 930 S. Plumer Ave., Tucson, AZ 85719/520-792-1075; FAX: 520-792-1075
Turkish Firearms Corp., 522 W. Maple St., Allentown, PA 18101/610-821-8660; FAX: 610-821-9049
Turnbull Restoration, Doug, 6426 County Rd. 30, P.O. Box 471, Bloomfield, NY 14469/716-657-6338; WEB: http://gunshop.com/dougt.htm
Tuttle, Dale, 4046 Russell Rd., Muskegon, MI 49445/616-766-2250

DIRECTORY OF THE ARMS TRADE

Twin Pine Armory, P.O. Box 58, Hwy. 6, Adna, WA 98522/360-748-4590; FAX: 360-748-1802
Tyler Manufacturing & Distributing, 3804 S. Eastern, Oklahoma City, OK 73129/405-677-1487, 800-654-8415

U

Uberti USA, Inc., P.O. Box 469, Lakeville, CT 06039/860-435-8068; FAX: 860-435-8146
Uberti, Aldo, Casella Postale 43, I-25063 Gardone V.T., ITALY/ (U.S. importers—American Arms, Inc.; Cabela's; Cimarron Arms; Dixie Gun Works; EMF Co., Inc.; Forgett Jr., Valmore J.; IAR, Inc.; Navy Arms Co; Taylor's & Co., Inc.; Uberti USA, Inc.)
UFA, Inc., 6927 E. Grandview Dr., Scottsdale, AZ 85254/800-616-2776
Ugartechea S.A., Ignacio, Chonta 26, Eibar, SPAIN 20600/43-121257; FAX: 43-121669 (U.S. importer–Aspen Outfitting Co.; The Gun Shop; Bill Hanus Birdguns; Lion Country Supply)
Ultimate Accuracy, 121 John Shelton Rd., Jacksonville, AR 72076/501-985-2530
Ultra Dot Distribution, 2316 N.E. 8th Rd., Ocala, FL 34470
Ultra Light Arms, Inc., P.O. Box 1270, 214 Price St., Granville, WV 26505/304-599-5687; FAX: 304-599-5687
Ultralux (See U.S. importer—Keng's Firearms Specialty, Inc.)
UltraSport Arms, Inc., 1955 Norwood Ct., Racine, WI 53403/414-554-3237; FAX: 414-554-9731
Uncle Bud's, HCR 81, Box 100, Needmore, PA 17238/717-294-6000; FAX: 717-294-6005
Uncle Mike's (See Michaels of Oregon Co.)
Unertl Optical Co., Inc., John, 308 Clay Ave., P.O. Box 818, Mars, PA 16046-0818/412-625-3810
Unique/M.A.P.F., 10, Les Allees, 64700 Hendaye, FRANCE 64700/33-59 20 71 93 (U.S. importer—Nygord Precision Products)
UniTec, 1250 Bedford SW, Canton, OH 44710/216-452-4017
United Binocular Co., 9043 S. Western Ave., Chicago, IL 60620
United Cutlery Corp., 1425 United Blvd., Sevierville, TN 37876/615-428-2532, 800-548-0835; FAX: 615-428-2267
United States Ammunition Co. (See USAC)
United States Optics Technologies, Inc., 5900 Dale St., Buena Park, CA 90621/714-994-4901; FAX: 714-994-4904
United States Products Co., 518 Melwood Ave., Pittsburgh, PA 15213/412-621-2130
Unmussig Bullets, D.L., 7862 Brentford Drive, Richmond, VA 23225/804-320-1165
Upper Missouri Trading Co., 304 Harold St., Crofton, NE 68730/402-388-4844
USAC, 4500-15th St. East, Tacoma, WA 98424/206-922-7589
U.S.A. Magazines, Inc., P.O. Box 39115, Downey, CA 90241/800-872-2577
USA Sporting Inc., 1330 N. Glassell, Unit M, Orange, CA 92667/714-538-3109, 800-538-3109; FAX: 714-538-1334
U.S. Patent Fire Arms, No. 25-55 Van Dyke Ave., Hartford, CT 06106/800-877-2832; FAX: 800-644-7265
U.S. Repeating Arms Co., Inc., 275 Winchester Ave., Morgan, UT 84050-9333/801-876-3440; FAX: 801-876-3737
Utica Cutlery Co., 820 Noyes St., Utica, NY 13503/315-733-4663; FAX: 315-733-6602
Uvalde Machine & Tool, P.O. Box 1604, Uvalde, TX 78802

V

Valade Engraving, Robert, 931 3rd Ave., Seaside, OR 97138/503-738-7672
Valmet (See Tikka/U.S. importer—Stoeger Industries)
Valor Corp., 5555 NW 36th Ave., Miami, FL 33142/305-633-0127; FAX: 305-634-4536
Van Epps, Milton (See Van's Gunsmith Service)
Van's Gunsmith Service, 224 Route 69-A, Parish, NY 13131/315-625-7251
Van Gorden & Son, Inc., C.S., 1815 Main St., Bloomer, WI 54724/715-568-2612
Van Horn, Gil, P.O. Box 207, Llano, CA 93544
Van Patten, J.W., P.O. Box 145, Foster Hill, Milford, PA 18337/717-296-7069
Vancini, Carl (See Bestload, Inc.)
Vann Custom Bullets, 330 Grandview Ave., Novato, CA 94947
Varner's Service, 102 Shaffer Rd., Antwerp, OH 45813/419-258-8631
Vega Tool Co., c/o T.R. Ross, 4865 Tanglewood Ct., Boulder, CO 80301/303-530-0174
Venco Industries, Inc. (See Shooter's Choice)
Venus Industries, P.O. Box 246, Sialkot-1, PAKISTAN/FAX: 92 432 85579
Verney-Carron, B.P. 72, 54 Boulevard Thiers, 42002 St. Etienne Cedex 1, FRANCE/33-477791500; FAX: 33-477790702; E-MAIL: Verney-Carron@mail.com
Versa-Pod (See Keng's Firearms Specialty, Inc.
Vest, John, P.O. Box 1552, Susanville, CA 96130/916-257-7228
VibraShine, Inc., P.O. Box 577, Taylorsville, MS 39168/601-785-9854; FAX: 601-785-9874
Vibra-Tek Co., 1844 Arroya Rd., Colorado Springs, CO 80906/719-634-8611; FAX: 719-634-6886
Vic's Gun Refinishing, 6 Pineview Dr., Dover, NH 03820-6422/603-742-0013
Victory USA, P.O. Box 1021, Pine Bush, NY 12566/914-744-2060; FAX: 914-744-5181
Vihtavuori Oy, FIN-41330 Vihtavuori, FINLAND/358-41-3779211; FAX: 358-41-3771643
Vihtavuori Oy/Kaltron-Pettibone, 1241 Ellis St., Bensenville, IL 60106/708-350-1116; FAX: 708-350-1606
Viking Leathercraft, Inc., 1579A Jayken Way, Chula Vista, CA 91911/800-262-6666; FAX: 619-429-8268
Viking Video Productions, P.O. Box 251, Roseburg, OR 97470
Vincent's Shop, 210 Antoinette, Fairbanks, AK 99701
Vintage Arms, Inc., 6003 Saddle Horse, Fairfax, VA 22030/703-968-0779; FAX: 703-968-0780
Vintage Industries, Inc., 781 Big Tree Dr., Longwood, FL 32750/407-831-8949; FAX: 407-831-5346
Viper Bullet and Brass Works, 11 Brock St., Box 582, Norwich, Ontario, CANADA N0J 1P0
Viramontez, Ray, 601 Springfield Dr., Albany, GA 31707/912-432-9683
Visible Impact Targets, Rts. 5 & 20, E. Bloomfield, NY 14443/716-657-6161; FAX: 716-657-5405
Vitt/Boos, 2178 Nichols Ave., Stratford, CT 06497/203-375-6859
Voere-KGH m.b.H., P.O. Box 416, A-6333 Kufstein, Tirol, AUSTRIA/0043-5372-62547; FAX: 0043-5372-65752 (U.S. importers—JäagerSport, Ltd.)
Volquartsen Custom Ltd., 24276 240th Street, P.O. Box 271, Carroll, IA 51401/712-792-4238; FAX: 712-792-2542
Vom Hoffe (See Old Western Scrounger, Inc., The)
Von Minden Gunsmithing Services, 2403 SW 39 Terrace, Cape Coral, FL 33914/813-542-8946
Vorhes, David, 3042 Beecham St., Napa, CA 94558/707-226-9116
Vortek Products, Inc., P.O. Box 871181, Canton, MI 48187-6181/313-397-5656; FAX:313-397-5656
VSP Publishers (See Heritage/VSP Gun Books)
Vulpes Ventures, Inc., Fox Cartridge Division, P.O. Box 1363, Bolingbrook, IL 60440-7363/708-759-1229

W

Wagoner, Vernon G., 2325 E. Encanto, Mesa, AZ 85213/602-835-1307
Wakina by Pic, 24813 Alderbrook Dr., Santa Clarita, CA 91321/800-295-8194
Waldron, Herman, Box 475, 80 N. 17th St., Pomeroy, WA 99347/509-843-1404
Walker Arms Co., Inc., 499 County Rd. 820, Selma, AL 36701/334-872-6231; FAX: 334-872-6262
Walker Mfg., Inc., 8296 S. Channel, Harsen's Island, MI 48028
Walker Co., B.B., P.O. Box 1167, 414 E. Dixie Dr., Asheboro, NC 27203/910-625-1380; FAX: 910-625-8125
Wallace, Terry, 385 San Marino, Vallejo, CA 94589/707-642-7041
Waller & Son, Inc., W., 2221 Stoney Brook Road, Grantham, NH 03753-7706/603-863-4177
Walls Industries, Inc., P.O. Box 98, 1905 N. Main, Cleburne, TX 76031/817-645-4366; FAX: 817-645-7946
Walnut Factory, The, 235 West Rd. No. 1, Portsmouth, NH 03801/603-436-2225; FAX: 603-433-7003
Walt's Custom Leather, Walt Whinnery, 1947 Meadow Creek Dr., Louisville, KY 40218/502-458-4361
Walters Industries, 6226 Park Lane, Dallas, TX 75225/214-691-6973
Walters, John, 500 N. Avery Dr., Moore, OK 73160/405-799-0376
Walther GmbH, Carl, B.P. 4325, D-89033 Ulm, GERMANY/ (U.S. importer—Champion's Choice; Interarms; P.S.M.G. Gun Co.)
WAMCO, Inc., Mingo Loop, P.O. Box 337, Oquossoc, ME 04964-0337/207-864-3344
WAMCO—New Mexico, P.O. Box 205, Peralta, NM 87042-0205/505-869-0826
Ward & Van Valkenburg, 114 32nd Ave. N., Fargo, ND 58102/701-232-2351
Ward Machine, 5620 Lexington Rd., Corpus Christi, TX 78412/512-992-1221
Wardell Precision Handguns Ltd., 48851 N. Fig Springs Rd., New River, AZ 85027-8513/602-465-7995
Warenski, Julie, 590 E. 500 N., Richfield, UT 84701/801-896-5319; FAX: 801-896-5319
Warne Manufacturing Co., 9039 SE Jannsen Rd., Clackamas, OR 97015/503-657-5590, 800-683-5590; FAX: 503-657-5695
Warren & Sweat Mfg. Co., P.O. Box 350440, Grand Island, FL 32784/904-669-3166; FAX: 904-669-7272
Warren Muzzleloading Co., Inc., Hwy. 21 North, P.O. Box 100, Ozone, AR 72854/501-292-3268
Warren, Kenneth W. (See Mountain States Engraving)
Washita Mountain Whetstone Co., P.O. Box 378, Lake Hamilton, AR 71951/501-525-3914
Wasmundt, Jim, P.O. Box 511, Fossil, OR 97830
WASP Shooting Systems, Rt. 1, Box 147, Lakeview, AR 72642/501-431-5606
Waterfield Sports, Inc., 13611 Country Lane, Burnsville, MN 55337/612-435-8339
Watson Bros., 39 Redcross Way, London Bridge, London, United Kingdom, SE1 1HG/FAX: 44-171-403-3367
Watson Trophy Match Bullets, 2404 Wade Hampton Blvd., Greenville, SC 29615/864-244-7948; 941-635-7948 (Florida)
Watsontown Machine & Tool Co., 309 Dickson Ave., Watsontown, PA 17777/717-538-3533
Wayne Firearms for Collectors and Investors, James, 2608 N. Laurent, Victoria, TX 77901/512-578-1258; FAX: 512-578-3559
Wayne Specialty Services, 260 Waterford Drive, Florissant, MO 63033/413-831-7083

334 GUNS ILLUSTRATED

MANUFACTURERS' DIRECTORY

WD-40 Co., 1061 Cudahy Pl., San Diego, CA 92110/619-275-1400; FAX: 619-275-5823
Weatherby, Inc., 3100 El Camino Real, Atascadero, CA 93422/805-466-1767, 800-227-2016, 800-334-4423 (Calif.); FAX: 805-466-2527
Weaver Arms Corp. Gun Shop, RR 3, P.O. Box 266, Bloomfield, MO 63825-9528
Weaver Products, P.O. Box 39, Onalaska, WI 54650/800-648-9624, 608-781-5800; FAX: 608-781-0368
Weaver Scope Repair Service, 1121 Larry Mahan Dr., Suite B, El Paso, TX 79925/915-593-1005
Webb, Bill, 6504 North Bellefontaine, Kansas City, MO 64119/816-453-7431
Weber & Markin Custom Gunsmiths, 4-1691 Powick Rd., Kelowna, B.C. CANADA V1X 4L1/250-762-7575; FAX: 250-861-3655
Weber Jr., Rudolf, P.O. Box 160106, D-5650 Solingen, GERMANY/0212-592136
Webley and Scott Ltd., Frankley Industrial Park, Tay Rd., Rubery, Rednal, Birmingham B45 0PA, ENGLAND/011-021-453-1864; FAX: 021-457-7846 (U.S. importer—Beeman Precision Airguns; Groenewold, John)
Webster Scale Mfg. Co., P.O. Box 188, Sebring, FL 33870/813-385-6362
Weems, Cecil, P.O. Box 657, Mineral Wells, TX 76067/817-325-1462
Weigand Combat Handguns, Inc., 685 South Main Rd., Mountain Top, PA 18707/717-868-8358; FAX: 717-868-5218
Weihrauch KG, Hermann, Industriestrasse 11, 8744 Mellrichstadt, GERMANY/09776-497-498 (U.S. importers—Beeman Precision Airguns; E.A.A. Corp.)
Weisz Parts, P.O. Box 20038, Columbus, OH 43220-0038/614-45-70-500; FAX: 614-846-8585
Welch, Sam, CVSR 2110, Moab, UT 84532/801-259-8131
Wellington Outdoors, P.O. Box 244, 1140 Monticello Rd., Madison, GA 30650/706-342-4915; FAX: 706-342-7568
Wells Creek Knife & Gun Works, 32956 State Hwy. 38, Scottsburg, OR 97473/541-587-4202; FAX: 541-587-4223
Wells Custom Gunsmith, R.A., 3452 1st Ave., Racine, WI 53402/414-639-5223
Wells, Fred F., Wells Sport Store, 110 N. Summit St., Prescott, AZ 86301/520-445-3655
Wells, Rachel, 110 N. Summit St., Prescott, AZ 86301/520-445-3655
Welsh, Bud, 80 New Road, E. Amherst, NY 14051/716-688-6344
Wenig Custom Gunstocks, Inc., 103 N. Market St., P.O. Box 249, Lincoln, MO 65338/816-547-3334; FAX: 816-547-2881
Werner, Carl, P.O. Box 492, Littleton, CO 80160
Werth, T.W., 1203 Woodlawn Rd., Lincoln, IL 62656/217-732-1300
Wescombe, Bill (See North Star West)
Wessinger Custom Guns & Engraving, 268 Limestone Rd., Chapin, SC 29036/803-345-5677
Wesson Firearms, Dan, 119 Kemper Lane, Norwich, NY 13815/607-336-1174; FAX: 607-336-2730
West, Jack L., 1220 W. Fifth, P.O. Box 427, Arlington, OR 97812
West, Robert G., 3973 Pam St., Eugene, OR 97402/541-344-3700
Western Cutlery (See Camillus Cutlery Co.)
Western Design (See Alpha Gunsmith Division)
Western Gunstock Mfg. Co., 550 Valencia School Rd., Aptos, CA 95003/408-688-5884
Western Missouri Shooters Alliance, P.O. Box 11144, Kansas City, MO 64119/816-597-3950; FAX: 816-229-7350
Western Munitions (See North American Munitions)
Western Nevada West Coast Bullets, 2307 W. Washington St., Carson City, NV 89703/702-246-3941; FAX: 702-246-0836
Westfield Engineering, 6823 Watcher St., Commerce, CA 90040/FAX: 213-928-8270
Westley Richards Agency USA (U.S. importer for Westley Richards & Co.)
Westley Richards & Co., 40 Grange Rd., Birmingham, ENGLAND B29 6AR/010-214722953 (U.S. importer—Westley Richards Agency USA)
Westrom, John (See Precision Metal Finishing)
Westwind Rifles, Inc., David S. Sullivan, P.O. Box 261, 640 Briggs St., Erie, CO 80516/303-828-3823
Weyer International, 2740 Nebraska Ave., Toledo, OH 43607/419-534-2020; FAX: 419-534-2697
Whildin & Sons Ltd., E.H., RR2, Box 119, Tamaqua, PA 18252/717-668-6743; FAX: 717-668-6745
Whinnery, Walt (See Walt's Custom Leather)
Whiscombe (See U.S. importer—Pelaire Products)
White Flyer Targets, 124 River Road, Middlesex, NJ 08846/908-469-0100, 602-972-7528 (Export); FAX: 908-469-9692, 602-530-3360 (Export)
White Laboratory, Inc., H.P., 3114 Scarboro Rd., Street, MD 21154/410-838-6550; FAX: 410-838-2802
White Owl Enterprises, 2583 Flag Rd., Abilene, KS 67410/913-263-2613; FAX: 913-263-2613
White Pine Photographic Services, Hwy. 60, General Delivery, Wilno, Ontario K0J 2N0 CANADA/613-756-3452
White Rock Tool & Die, 6400 N. Brighton Ave., Kansas City, MO 64119/816-454-0478
White Muzzleloading Systems, 25 E. Hwy. 40, Suite 330-12, Roosevelt, UT 84066/801-722-5996; FAX: 801-722-5909
White Shooting Systems (See White Muzzleloading Systems)
Whitehead, James D., 204 Cappucino Way, Sacramento, CA 95838
Whitestone Lumber Corp., 148-02 14th Ave., Whitestone, NY 11357/718-746-4400; FAX: 718-767-1748
Whitetail Design & Engineering Ltd., 9421 E. Mannsiding Rd., Clare, MI 48617/517-386-3932

Whits Shooting Stuff, Box 1340, Cody, WY 82414
Wichita Arms, Inc., 923 E. Gilbert, P.O. Box 11371, Wichita, KS 67211/316-265-0661; FAX: 316-265-0760
Wick, David E., 1504 Michigan Ave., Columbus, IN 47201/812-376-6960
Widener's Reloading & Shooting Supply, Inc., P.O. Box 3009 CRS, Johnson City, TN 37602/615-282-6786; FAX: 615-282-6651
Wideview Scope Mount Corp., 13535 S. Hwy. 16, Rapid City, SD 57701/605-341-3220; FAX: 605-341-9142
Wiebe, Duane, 33604 Palm Dr., Burlington, WI 53105-9260
Wiest, M.C., 10737 Dutchtown Rd., Knoxville, TN 37932/423-966-4545
Wilcox All-Pro Tools & Supply, 4880 147th St., Montezuma, IA 50171/515-623-3138; FAX: 515-623-3104
Wild Bill's Originals, P.O. Box 13037, Burton, WA 98013/206-463-5738
Wild West Guns, 7521 Old Seward Hwy, Unit A, Anchorage, AK 99518/800-992-4570, 907-344-4500; FAX: 907-344-4005
Wilderness Sound Products Ltd., 4015 Main St. A, Springfield, OR 97478/503-741-0263, 800-437-0006; FAX: 503-741-7648
Wildey, Inc., P.O. Box 475, Brookfield, CT 06804/203-355-9000; FAX: 203-354-7759
Wildlife Research Center, Inc., 1050 McKinley St., Anoka, MN 55303/612-427-3350, 800-USE-LURE; FAX: 612-427-8354
Wilkinson Arms, 26884 Pearl Rd., Parma, ID 83660/208-722-6771; FAX: 208-722-5197
Will-Burt Co., 169 S. Main, Orrville, OH 44667
William's Gun Shop, Ben, 1151 S. Cedar Ridge, Duncanville, TX 75137/214-780-1807
Williams Bullet Co., J.R., 2008 Tucker Rd., Perry, GA 31069/912-987-0274
Williams Gun Sight Co., 7389 Lapeer Rd., Box 329, Davison, MI 48423/810-653-2131, 800-530-9028; FAX: 810-658-2140
Williams Mfg. of Oregon, 110 East B St., Drain, OR 97435/503-836-7461; FAX: 503-836-7245
Williams Shootin' Iron Service, The Lynx-Line, 8857 Bennett Hill Rd., Central Lake, MI 49622/616-544-6615
Williamson Precision Gunsmithing, 117 W. Pipeline, Hurst, TX 76053/817-285-0064
Willow Bend, P.O. Box 203, Chelmsford, MA 01824/508-256-8508; FAX: 508-256-8508
Willson Safety Prods. Div., P.O. Box 622, Reading, PA 19603-0622/610-376-6161; FAX: 610-371-7725
Wilson Arms Co., The, 63 Leetes Island Rd., Branford, CT 06405/203-488-7297; FAX: 203-488-0135
Wilson Case, Inc., P.O. Box 1106, Hastings, NE 68902-1106/800-322-5493; FAX: 402-463-5276
Wilson, Inc., L.E., Box 324, 404 Pioneer Ave., Cashmere, WA 98815/509-782-1328
Wilson Gun Shop, Box 578, Rt. 3, Berryville, AR 72616/870-545-3618; FAX: 870-545-3310
Winchester (See U.S. Repeating Arms Co., Inc.)
Winchester Div., Olin Corp., 427 N. Shamrock, E. Alton, IL 62024/618-258-3566; FAX: 618-258-3599
Winchester Press (See New Win Publishing, Inc.)
Winchester Sutler, Inc., The, 270 Shadow Brook Lane, Winchester, VA 22603/540-888-3595; FAX: 540-888-4632
Windish, Jim, 2510 Dawn Dr., Alexandria, VA 22306/703-765-1994
Windjammer Tournament Wads, Inc., 750 W. Hampden Ave. Suite 170, Englewood, CO 80110/303-781-6329
Wingshooting Adventures, 0-1845 W. Leonard, Grand Rapids, MI 49544/616-677-1980; FAX: 616-677-1986
Winkle Bullets, R.R. 1 Box 316, Heyworth, IL 61745
Winter, Robert M., P.O. Box 484, 42975-287th St., Menno, SD 57045/605-387-5322
Wise Guns, Dale, 333 W. Olmos Dr., San Antonio, TX 78212/210-828-3388
Wiseman and Co., Bill, P.O. Box 3427, Bryan, TX 77805/409-690-3456; FAX: 409-690-0156
Wolf's Western Traders, 40 E. Works, No. 3F, Sheridan, WY 82801/307-674-5352
Wolfe Publishing Co., 6471 Airpark Dr., Prescott, AZ 86301/520-445-7810, 800-899-7810; FAX: 520-778-5124
W.C. Wolff Co., P.O. Box 458, Newtown Square, PA 19073/610-359-9600, 800-545-0077
Wolverine Footwear Group, 9341 Courtland Dr. NE, Rockford, MI 49351/616-866-5500; FAX: 616-866-5658
Wood, Frank (See Classic Guns, Inc.)
Wood, Mel, P.O. Box 1255, Sierra Vista, AZ 85636/602-455-5541
Woodleigh (See Huntington Die Specialties)
Woods Wise Products, P.O. Box 681552, 2200 Bowman Rd., Franklin, TN 37068/800-735-8182; FAX: 615-726-2637
Woodstream, P.O. Box 327, Lititz, PA 17543/717-626-2125; FAX: 717-626-1912
Woodworker's Supply, 1108 North Glenn Rd., Casper, WY 82601/307-237-5354
Woolrich Inc., Mill St., Woolrich, PA 17701/800-995-1299; FAX: 717-769-6234/6259
Working Guns, 250 Country Club Lane, Albany, OR 97321/541-928-4391
World of Targets (See Birchwood Casey)
World Class Airguns, 2736 Morningstar Dr., Indianapolis, IN 46229/317-897-5548
World Trek, Inc., 7170 Turkey Creek Rd., Pueblo, CO 81007-1046/719-546-2121; FAX: 719-543-6886
Worthy Products, Inc., RR 1, P.O. Box 213, Martville, NY 13111/315-324-5298
Wosenitz VHP, Inc., Box 741, Dania, FL 33004/305-923-3748; FAX: 305-925-2217

DIRECTORY OF THE ARMS TRADE

Wostenholm (See Ibberson [Sheffield] Ltd., George)
Wright's Hardwood Gunstock Blanks, 8540 SE Kane Rd., Gresham, OR 97080/503-666-1705
W. Square Enterprises (See Load From A Disk)
WTA Manufacturing, Bill Wood, P.O. Box 164, Kit Carson, CO 80825/800-700-3054, 719-962-3570
Wyant Bullets, Gen. Del., Swan Lake, MT 59911
Wyant's Outdoor Products, Inc., P.O. Box B, Broadway, VA 22815
Wyoming Bonded Bullets, Box 91, Sheridan, WY 82801/307-674-8091
Wyoming Custom Bullets, 1626 21st St., Cody, WY 82414
Wyoming Knife Corp., 101 Commerce Dr., Ft. Collins, CO 80524/303-224-3454

X, Y

X-Spand Target Systems, 26-10th St. SE, Medicine Hat, AB T1A 1P7 CANADA/403-526-7997; FAX: 403-528-2362
Yankee Gunsmith, 2901 Deer Flat Dr., Copperas Cove, TX 76522/817-547-8433
Yavapai College, 1100 E. Sheldon St., Prescott, AZ 86301/602-776-2359; FAX: 602-776-2193
Yavapai Firearms Academy Ltd., P.O. Box 27290, Prescott Valley, AZ 86312/520-772-8262
Yearout, Lewis E. (See Montana Outfitters)
Yee, Mike, 29927 56 Pl. S., Auburn, WA 98001/206-839-3991
Yellowstone Wilderness Supply, P.O. Box 129, W. Yellowstone, MT 59758/406-646-7613
Yesteryear Armory & Supply, P.O. Box 408, Carthage, TN 37030
York M-1 Conversions, 803 Mill Creek Run, Plantersville, TX 77363/800-527-2881, 713-477-8442
Young, Paul A., RR 1 Box 694, Blowing Rock, NC 28605-9746
Young Country Arms, P.O. Box 3615, Simi Valley, CA 93093
Yukon Arms Classic Ammunition, 1916 Brooks, P.O. Box 223, Missoula, MT 59801/406-543-9614

Z

Z's Metal Targets & Frames, P.O. Box 78, South Newbury, NH 03255/603-938-2826
Zabala Hermanos S.A., P.O. Box 97, Eibar, SPAIN 20600/43-768085, 43-768076; FAX: 34-43-768201 (U.S. importer—American Arms, Inc.)
Zander's Sporting Goods, 7525 Hwy 154 West, Baldwin, IL 62217-9706/800-851-4373 ext. 200; FAX: 618-785-2320
Zanoletti, Pietro, Via Monte Gugielpo, 4, I-25063 Gardone V.T., ITALY/ (U.S. importer—Mandall Shooting Supplies, Inc.)
Zanotti Armor, Inc., 123 W. Lone Tree Rd., Cedar Falls, IA 50613/319-232-9650
Z-Coat Industrial Coatings, Inc., 3375 U.S. Hwy. 98 S. No. A, Lakeland, FL 33803-8365/813-665-1734
ZDF Import Export Inc., 2975 South 300 West, Salt Lake City, UT 84115/801-485-1012; FAX: 801-484-4363
Zeeryp, Russ, 1601 Foard Dr., Lynn Ross Manor, Morristown, TN 37814/615-586-2357
Zeiss Optical, Carl, 1015 Commerce St., Petersburg, VA 23803/804-861-0033, 800-388-2984; FAX: 804-733-4024
Zero Ammunition Co., Inc., 1601 22nd St. SE, P.O. Box 1188, Cullman, AL 35056-1188/800-545-9376; FAX: 205-739-4683
Ziegel Engineering, 2108 Lomina Ave., Long Beach, CA 90815/310-596-9481; FAX: 310-598-4734
Zim's Inc., 4370 S. 3rd West, Salt Lake City, UT 84107/801-268-2505
Z-M Weapons, 203 South St., Bernardston, MA 01337
Zoli, Antonio, Via Zanardelli 39, Casier Postal 21, I-25063 Gardone V.T., ITALY
Zonie Bullets, 790 N. Lake Havasu Ave., Suite 26, Lake Havasu City, AZ 86403/520-680-6303; FAX: 520-680-6201
Zriny's Metal Targets (See Z's Metal Targets & Frames)
Zufall, Joseph F., P.O. Box 304, Golden, CO 80402-0304